Insanity and Genius

Insanity and Genius: Masks of Madness and the Mapping of Meaning and Value (Second Edition)

By

Harry Eiss

CAMBRIDGE
SCHOLARS
PUBLISHING

Insanity and Genius:
Masks of Madness and the Mapping of Meaning and Value
(Second Edition),
by Harry Eiss

This book first published 2014

Cambridge Scholars Publishing

12 Back Chapman Street, Newcastle upon Tyne, NE6 2XX, UK

British Library Cataloguing in Publication Data
A catalogue record for this book is available from the British Library

ISBN (10): 1-4438-5885-4, ISBN (13): 978-1-4438-5885-4

Dedicated to my brother John with respect and affection

At first
Senseless as beasts I gave men sense, possessed them of
mind . . .
In the beginning, seeing, they saw amiss, and hearing,
heard not,
but like phantoms huddled
In dreams, the perplexed story of their days
Confounded.

—Aeschylus, *Prometheus Bound*

To sleep—perchance to dream: ay, there's the rub . . .

—William Shakespeare, *Hamlet*

. . . just as one has to turn the entire body for the eyes to see the light instead of the darkness, it is necessary to turn the entire soul away from the changing world of the senses for the eyes to see beyond them to the reality, the ultimate splendor of the higher truth. Thus, there may well be an art whose purpose it is to accomplish this very thing, the conversion of the soul's eye, not to give it that higher sight, which it already has, but to connect that sight with the world beyond it.

—Plato, *The Republic*

TABLE OF CONTENTS

PREFACE

A crisp, clean November afternoon gave the campus a still, silent quality. Except for the shoveled sidewalks surrounding and crisscrossing the open ground between the buildings, a thin sheet of crusted snow covered the ground, as if, at least for a moment, the world had paused to reflect upon itself. Even the occasional student walking rapidly from building to building could not disturb the feeling of a scene briefly frozen in time.

The seminar room had a full wall of windows jutting out from the side of the building, offering a panoramic view of nature's meditation, and the silence surrounding us gave our conversation a forbidden quality. It was just the two of us, Dr. Norton Kinghorn, who was the chair of the English department, and me. The conversation turned to the one great work of literature to come out of Spain, the work often said to have fathered the modern novel.

Norton smiled. He had a wonderful smile, and a certain Don Quixote childlike gleam that often filled his eyes. "I once had a professor who said Don Quixote was cracked in the head," he began and stopped in mid thought. The pregnant pause, the endearing smile, the mischievous eyes. Norton should have been an entertainer, for he knew how to capture the moment.

Again the silence filled the room, and the high ceiling gave it a certain spiritual quality. I waited for the punch line.

"But the crack let out a beautiful light."

****

Pablo Picasso said "We all know that Art is not truth. Art is a lie that makes us realize truth, at least the truth that is given us to understand." John Keats expressed the same in the climatic couplet of *Ode on a Grecian Urn*, "Beauty is truth, truth beauty,—that is all / Ye know on earth and all ye need to know." On September 8, 1888, Vincent Van Gogh, referring to his painting *The Night Café*, wrote to his brother Theo, "I have tried to express the terrible passions of humanity by means of red and green."

This is what I have struggled with, this higher truth, and its messengers: drama, dance, sculpture, painting—all of the arts, and such other disciplines as philosophy, psychology and neurology. It is what led me, innocent of all the implications and reasons for it, to first submerse myself in literature and music in my desperate search for meaning as a child following my father's death.

In his book about the discovery of the structure of DNA, James Watson wrote, "So we had lunch, telling ourselves that a structure this pretty just had to exist." Indeed, the question most often asked by scientists about a scientific theory is "Is it beautiful?" Yes, truth does equal beauty. Scientists know; mathematicians know.

But the beauties, the truths of math and science were not the truths I needed as a child, and I intuitively knew it, intuitively knew that the truths I needed come from a different way of knowing, a way of knowing not of the world of logic and reason and explanation (though they help lead us to them), but rather a way of knowing that is of the world of expression.

Neurologists, led by such scientists as Roger W. Sperry, winner of the 1981 Nobel Prize in Physiology or Medicine for his pioneering work, are right now mapping it out in the human brain. I'm sure you've all heard of at least the simplified left/right brain theories based on such research, if not the details, support for, and extensions of them. Well, simply speaking, neurologists have, in fact, shown that humans think in two distinct ways, one way in the left hemisphere of the brain, the other in the right. The left hemisphere gives us our literal, non-artistic forms of thinking. It applies scientific theory and deduction to its mapping out of the world. It is denotative and deals in the truths of the physical world.

The right hemisphere is where we go beyond this to the world of meaning and value. Good, bad, right, wrong—these are not literal objects, not literal truths. They are judgments made by humans, and they exist in the human mind, in the right hemisphere, in the world of the arts (both ethical and beyond ethical to the realities of the world of faith, of the sublime, of the numinous). And these are what place humans beyond all else in existence. Only humans deal with meaning and value; it is our gift to the world. Anthropologists, evolutionists and the like say true humans appear on the scene not when crude tools are found, or weapons. No, it is when cave wall paintings are found. Why is this? It's actually rather simple and straight-forward. Paintings are a form of symbolic thought, as are all of the arts, and only humans do this. Other creatures don't attempt to influence the hunt or appease the gods with paintings or rituals or myths. Other creatures do not have this invisible world of the mind that gives a value to an otherwise meaningless universe. Only humans have eaten of the apple of the knowledge of good and evil. Only humans consciously give up physical existence for some higher or other non-physical form of existence, an existence of the spirit or soul.

Loren Eiseley in *The Immense Journey* writes:

> Symbolic communication had begun. Man had escaped out of the eternal present of the animal world into a knowledge of past and future. The unseen gods, the powers behind the world of phenomenal appearance, began to stalk through his dreams.
>
> Nature, one might say, through the powers of this mind, grossly superstitious though it might be in its naïve examination of wind and water, was beginning to reach out into the dark behind itself. Nature was beginning to evade its own limitations in the shape of this strange, dreaming and observant brain. It was a weird multi-headed universe, going on, unseen and immaterial save as its thoughts smoldered in the eyes of hunters huddled by night fires, or were translated into pictures upon cave walls, or were expressed in the trappings of myth or ritual. The Eden of the eternal present that the animal world had known for ages was shattered at last. Through the human mind, time and darkness, good and evil, would enter and possess the world.

Rollo May gives us the pinnacle of this in the following from *Man's Search for Himself*:

> Man's consciousness of himself is the source of his highest qualities. It underlies his ability to distinguish between I and the world. It gives him the capacity to keep time, which is simply the ability to stand outside the present and to imagine oneself back in yesterday or ahead in the day after tomorrow. Thus human beings can learn from the past and plan for the future. And thus man is the historical mammal in that he can stand outside and look at his history; and thereby he can influence his own development as a person, and to a minor extent he can influence the march of history in his nation and society as a whole. The capacity for consciousness of self also underlies man's ability to use symbols, which is a way of disengaging something from what it is, such as the two sounds which make up the word "table," and agreeing that these sounds will stand for a whole class of things. Thus man can think in abstractions like "beauty," "reason," and "goodness."
>
> This capacity for consciousness of ourselves gives us the ability to see ourselves as others see us and to have empathy with others. It underlies our remarkable capacity to transport ourselves into someone else's parlor where we will be in reality next week, and then in imagination to think and plan how we will act. And it enables us to imagine ourselves in someone else's place, and to ask how we would feel and what we would do if we were this other person.
>
> No matter how poorly we use or fail to use or even abuse these capacities, they are the rudiments of our ability to begin to love our neighbor, to have ethical sensitivity, to see truth, to create beauty, to devote ourselves to ideals, and to die for them if need be.

> To fulfill these potentialities is to be a person.

Isn't empathy, after all, the basis for all of the arts? We the readers empathize with the characters in a work of fiction, we actually laugh and cry for characters who are not even real, who we know do not literally exist; but they do exist, don't they, in the world of meaning and value, the same world where empathy exists, the only world that really matters.

Abelard's perspective on the Crucifixion of Christ is that it is in his suffering that he brings salvation, because his suffering causes compassion, empathy in those who are exposed to it, both in humans and in God, and it is through compassion, through humans and God coming together in passion that salvation takes place.

It is in the knowing beyond explanation, the knowing beyond logic and reason, the knowing that explanation must struggle to support, not to deny, that I immerse my students, and I watch as their eyes light up and suddenly the classroom becomes not a jail cell but a key to unlock the mental jail cells they had not thought possible to open, likely were unaware even existed. And I am humbled, for I am but the messenger, and as they connect with the message, it is clear to me that something wonderful has taken place, and I have had the privilege of being a part of it, of witnessing the transformation. Sounds a bit like a fantasy, like an overly sentimental movie, I know, yet I've witnessed it again and again.

This book takes us there, not to offer answers to the mysteries of life, but to make us aware of the beauty of them.

A Few Notes

Today, *insanity* is mainly a legal term, a defense or excuse for actions deemed illegal, actions that the culture in the form of established laws has defined as wrong and deserving of some form of punishment. The disciplines of neurology and psychology no longer use the term, as they feel it is degrading, and have substituted other, less dramatic terms, *euphemisms*. While the current legal definitions and use of the term are interesting, they are not how I am using it. I am using it in its more emotional, historic sense. Furthermore, while current clinical terms being substituted for it are more precise within their specific disciplines, they are ironically more dehumanizing in their stubborn determination to eliminate the powerful connotations that give language much of its value. I have no interest in destroying the power of language or science by eliminating the value of meaning beyond explanation in an attempt to be *politically correct*, and the only other term I considered using instead was *madness*. As a reading of the text will prove, there is nothing at all *degrading* about being *insane*, at least not from my perspective. In fact, those deemed *insane* are perhaps the only ones who fit the category of *genius*, the artists, shamans, perhaps even saviors, condemned, ridiculed, some confined to institutions, at least one crucified. They are the humans of the highest level, the ones who have connected to the mysteries of existence beyond the meaningless physical world of the body, the ones giving us the only maps that really matter, the maps of meaning and value.

I have purposely applied as many different disciplines of human thinking to the discussions as possible, believing that interdisciplinary approaches to meaning and value are superior and that the fragmentation of our current world is represented and promoted in the contemporary embracement of discipline specific cages in academia.

My references to Vincent van Gogh simply as Vincent throughout my discussions of him are because that's what he expressly stated he wanted to be called, and that's the single name he used to sign his paintings.

The sections titled "Broken Windmills" and "Don Quixote" are revised works previously published, the first in *Children's Literature and Culture* as "Broken Windmills" the second in *Metaesthetics* as a portion of "Children of a Greater God." Such revising is a necessary part of my work. As William Butler Yeats expresses so simply and eloquently in his preliminary poem in *The Collected Works in Verse and Prose of William Butler Yeats*:

> "The friends that have it I do wrong / Whenever I remake a song / Should know what issue is at stake / It is myself that I remake." There can be no better reason for writing a book.

CHAPTER ONE

THE FAIRY-FELLER'S MASTER STROKE

It is to their madness that we owe the many benefits that the Pythia of Delphi and priestesses of Dodona were able to bestow upon Greece, both privately and in public life, for when they were in their right minds their achievements amounted to little or nothing.

The Fairy-Feller's Master Stroke, Richard Dadd

Part I

The spring-knife safely hidden beneath his black waistcoat, Richard Dadd sits at the Ship Inn, waiting for his father to return. Osiris has been with him much of late, and now he is certain of what must be done.

The dinner talk has skirted the reason for it. Instead, discussions of the world of Titania and Oberon, of Puck, of the whole fairy world of Shakespeare's *A Midsummer Night's Dream* have filled the uneasy conversation. Robert Dadd has praised his son's four illustrations of Robin Goodfellow in *The Book of British Ballads*, highlighting the excellent use of dramatic lighting, a result of a better understanding of the potential of wood engraving than the other artists, carefully ignoring the darker content of a laughing Puck emerging from a shadowy world of half-born elves and frog-like images imprisoned in giant dew-drops, the knife-sharp lettering impaling one elf, the eerie, bulging eyes of the goblins.

It is a hot, humid August day, with little wind. Richard has not consumed much, three boiled eggs rolled in salt, but a pint of ale. Yet his stomach twists uncomfortably. Perhaps, he thinks, a taste of grog, sure medicine for a nervous digestive system.

He is about to order one when his father returns. "Well, are we ready then?"

Richard looks up at his father, a man who has stood staunchly and affectionately by him through the recent afflictions, insisting that they are nothing more than the effects of sunstroke and will pass with rest and quiet. But then, on the urging of others, has recently taken him to see a specialist on disorders of the mind, a cautionary measure, not to be misconstrued as more than a father's love for his son.

Osiris thunders in Richard's brain—the driving commands, the pounding headache.

He knows sunstroke is but a convenient dismissal, trivializing the tremendous transformation that he has experienced. Something took hold of him during his recent trip through the Middle East with his patron Sir Thomas Phillips. The stop in Cairo, where he experienced the exotic bazaar, a seemingly endless market-place where strange looking people wearing turbans and dark flowing robes dyed Persian red and India black mixed with wealthy Europeans in white suits and broad brimmed hats; the short jaunt to Giza, where he rode ornery camels and felt small and insignificant beneath the other-worldly Sphinx and three huge pyramids; the languid journey up the Nile through the wheat fields and white desert sands that lead the eyes into a liquid horizon; the slow, leisurely boat ride to Thebes, during which time he sketched crocodiles sliding in and out of the mysterious waters; the endless visits to the huge Egyptian palaces and temples; the night he sat beneath a pale moon on the deck of the boat and heard a strange, low chant, a chant that aroused his curiosity and led him to peer intently over the roped rail of the boat until he could see the shadowy forms of the Egyptian and Nubian crew holding hands, forming a circle on the desert sand, twisting and writhing in an ever growing frenzy, accompanying their movements by intoned passages from the Koran, finally falling senseless in mythic submission to the ritual—all of these exotic experiences came together within him, and something took hold, something that has informed him ever since, and he knows now, now knows for certain, this man, his father, is not the kind, gentle man he appears. He pushes back his chair and stands. "Let us go."

"To Lord Darnley's," his father replies and turns toward the heavy oak door. Recently he was persuaded to call upon Dr. Alexander Sutherland, the famed alienist at St. Luke's Hospital, who did not tell him what he wanted to hear. Perhaps this visit will help. After all, his son has promised to unburden his mind. A walk through Cobham Park where Richard often sketched as a child—perhaps that is just the thing to snap him out of his recent malady.

Robert has an uneasy stomach, probably from the two bangers, the bitter ale. A trip to the wash room has not produced the results he'd hoped for. Some people, he reasons, simply have nervous digestive systems, and he has been cursed with one.

One last task before the anticipated walk.The plan is to spend the night. So, first it is necessary to stop at a nearby boarding house and reserve a room. Richard's disconnected comments during the short passage about not needing beds because they will be spending the night consorting with the witches of Macbeth are disconcerting, but Robert is getting used to such strange asides and has taught himself to discount them.

Richard, however, is becoming more agitated, speaking louder: "When shall we three meet again? In thunder, lightning, or in rain? What's that you say? When the hurly-burly's done. When the battle's lost and won. That will be ere the set of sun." (2)

"Come, Richard, come away."

"I come, Graymalkin. Paddock calls. Anon!"

Robert takes hold of his son's shoulder and steers him toward the inn. "Come, come, let us get the arrangements made. Come before the setting sun takes away our walk."

Richard smiles, then laughs. "Fair is foul, and foul is fair. Hover through the fog and filthy air."

The clerk is an old woman, her gray hair pulled back into a carelessly pinned bun, her dress faded but still exhibiting the Scottish squares-of-red-and-green common to the current style of the region. As expected, a room is available. The woman's thick Liverpool accent reveals her origin, and combined with her insistence on mumbling through the entire exchange makes her nearly undecipherable. At first Robert thinks her comments are aimed at him and asks for clarification, but then realizes she is in her own world, her blurred commentary constituting a personal conversation that does not include him or even recognize he stands in front of her. Furthermore, to his surprise, her seemingly undirected body is able to handle the necessary basics of the transaction without interrupting this conversation in the least. And Richard is doing much the same thing, only in a louder, more disjointed fashion. Momentarily Robert again makes the mistake of thinking the comments are aimed at him, but quickly grasps that it is but an internal affair, and Richard is oblivious to the reality of the situation.

"A witch indeed, and where hast thou been, sister? Killing swine. Killing swine sayest thou? Killing swine!"

Robert takes a deep breath. This visit, which got off to a promising beginning during the meal, is now turning decidedly negative.

However, once outside, beneath the blue skies, he feels better. At least for the moment, Richard has returned to a more normal countenance. Robert forces his thoughts to be positive, strong. First, he thinks, the day itself, the weather. Admittedly, it is humid, but then that is to be expected on an August day. Yes, he concludes, nature is telling me it *is* a good day for a walk.

A good *day* for a walk, a good day for a *walk,* a *good* day for a walk. He takes hold of this phrase, playing with the emphasis, refusing to let it slip away, forcing it to deny the psychic shadows that threaten to flood in upon him, demanding his thoughts remain upbeat. And with each repetition of the phrase, his thoughts become ever more forceful. He begins to elaborate, to expand the mantra. We've had *enough* pleasantries over a meal, *enough* making lodging arrangements. It is time for what *will* be more than just a leisurely stroll, though it *will* be that. It is time for what *will* be more than just a pleasant visit, though it *will* be that as well. It is time to find out what it is that has so taken Richard, what *demon* it is that has gotten inside of his *wonderful, loving,* and *talented* soul. Indeed, Robert's hopes are gaining strength, perhaps the strength of desperation because he fears the obligation being forced upon him to take away his son's freedom.

This, then, is his state of mind as the walk begins.

The road is dry and dust becomes momentary shallow clouds of light brown-and-gray particles as Richard occasionally kicks pebbles and fine gravel into the clumps of green and yellow grass that mixes with pink, red and purple fuchsia growing along the road edge. They walk for a time in silence. The voices, Richard calls them *secret admonitions*, keep at him. Osiris wants him to fulfill his assigned duties. He is the exterminator, and he needs to rid the world of those possessed by demons, to purify the world of evil.

He remembers when it all became clear, that night in Egypt, with the old Arab men smoking a "hubbly-bubby." He joined them, and if memory serves spent five straight days and nights smoking their strange herbs. No one spoke, but he became convinced the sound of the pipe was a form of language, a means for Osiris, the Egyptian god killed by his brother, to speak to him. That was the beginning, the breakthrough. It all made sense, life out of death, resurrection, salvation. Hadn't Osiris floated down the Nile just as he had floated down the Nile? The journey, yes, the same journey. Down the Nile, into another world, not just an exotic world, but a mythic world, a world where mere physical existence merges with the spiritual, the world of the gods, a place where the rules differ, where logic and reason must bow to higher truths, truths that cannot be explained, but truths that can be known, if only one has found a way through the fabric of illusion. Hadn't Osiris, *after death*, given birth to Horus?

The days following his revelation were filled with constant headaches. Periods of depression alternated with periods of great energy and excitement when he would walk agitatedly about Alexandria or lay down at night with his "imagination so full of wild vagaries that at times" he wrote in his journal "I have really and truly doubted of my own sanity." (3) Thomas had suggested to him that his behavior had changed, had grown strange, that perhaps he was suffering from the hot, overly bright sun. But Richard had known it was more, had been in no mood for such condescending remarks, and had begun to wish he could end the journey and the entire arrangement with this patron he had once thought so promising, but now knew had not been *chosen* as he had, had not realized that the journey down the Nile was more than just a superficial experiencing of exotic sights and customs.

But whatever manifestations of divinity it might have taken on for Richard, the trip had been carefully arranged and there were still more visits to make. Rome had been promising. A chance to view the Pope. Thomas had chatted on and on about how exciting it was to be. But Richard had seen through the silly, babbling face, and when the viewing took place had been immediately possessed to want to attack this Pope, this agent of evil. The men in Egypt, the dancing, the rituals, the silent words. The wonderful hubbly-bubbly. Osiris had known. And Osiris had wanted revenge!

But, at least for the moment, Osiris would be denied. The Pope had been well protected.

By late spring, 1843, the once friends, now disillusioned traveling companions had reached Paris, nearing the end of their lengthy journey together. By now Richard had come to see Thomas as an agent for the Devil, one of an ever growing number of people it was his mission to murder. Unaware of how seriously his life was in danger, Thomas had experienced enough of Richard's strange behavior to know it could no longer be simply attributed to the brilliance of an Egyptian sun. It was time for Richard to go home.

Back in England, Richard could reunite with his family. He had been looking forward to it. They also had anticipated a positive return. But the reunion wasn't the happy reconnection expected. Richard had changed and was prone to violent outbursts. Attempts to cover up, to ignore the erratic behavior had proven impossible. Distressed and reluctant, the family had admitted the possibility of madness, and he been brought to St. Luke's Hospital, where he had been diagnosed "non compos mentis," not sound of mind. But Richard wasn't about to be shut away. He had convinced his dad all he needed was some rest.

And now the father-son visit, an evening meal, a walk, a final chance to end this troubling behavior. How pedestrian, Richard thinks, how unenlightened. His father has not floated the Nile, can never know the higher truths.

"So," says the senior Dadd, "what is it you want to unburden? You promised a talk, an explanation for your illness. You know I wish only to help, to see you through this terrible disease. Come. It is time. Let us talk."

But Richard has grown morose, silent. Robert can see his mind is elsewhere. The shadows threaten. A *good* day for a walk. A good *day* for a walk. A good day for a *walk.*

"Come, come. Let us talk. The sun is shining. It is a *good* day. A father and his son on a walk through the friendly landscape of our past. What better way to spend a summer evening?"

Richard stops, turns his eyes upon his father's face, opens and then shuts his mouth.

Robert returns the look, trying to give his eyes a friendly, inviting welcome. It seems to him Richard is ready, finally ready to open up, to explain and reconnect.

A dark scowl crosses Richard's continence, but quickly dissipates.

His father stands patient, waiting. He has seen such changing masks much of late, knows his son is struggling. It makes no sense to him, but he is determined to see his son, his *favorite* son through it. In truth it seems his whole family is falling apart. Not only is Richard struggling, but the younger son George is exhibiting similar hallucinations, similar dangerous tendencies. Perhaps it is just something in the air, something his sons need a medicine to put aside. (4)

The silence is all he hears. Though robins and sparrows chatter away in peace, he hears none of it, just the silence that surrounds his son. He waits, patiently, he waits.

But no words come.

Robert nearly breaks, nearly gives in to the shadows. It *is* a good day. It *is* a good day. The mantra has shifted slightly, but he does not notice. He is too focused on the repetition, on the needed blocking mechanism. He refuses to enter the nightmare that threatens him.

They resume the walk, at first slow, but then gaining rapidity as Richard grows more anxious, rushed, determined, focused. Now the gravel spreads from his shoes not because he casually kicks at it, but because he pushes so forcefully against the ground to gain momentum.

Robert walks faster to keep up with his son. Such a promising artist. Brilliant. Already proclaimed one of the best of the young artists in England. Already having created *Come unto these Yellow Sands, Titania Sleeping*, and the highly acclaimed panels for Lord Foley. (5) Now such a mystery. This disease. What is one to make of it?

Stumbling, almost tripping and falling from the abruptness of it, Richard turns off Cobham road, and pounds straight through the oak trees and wheat grass, through the pole fence, down toward the Paddock Hole chalk pit.

Robert follows, not sure what his son is up to, but hoping against hope something will snap here, and he can bring his son back, back from wherever this illness has taken him. It must be the heat, sunstroke—the sun *is* too brilliant in Egypt, too much to take. Just a bit more time, he keeps telling himself, just a bit more time, soon he will recover.

Then, right above Paddock Hole, Richard comes to a sudden stop, finds the razor in his picket, the words of Osiris all he can hear. Now! Now is the moment! Now the deed must be done! Quick! Quick!"

He turns, nearly knocking Robert over, tries to bring the razor across the exposed throat, but an arm gets in the way. A clean cut isn't possible. Osiris! The spring-knife! His waistcoat! Thrusting! Thrusting! Thrusting!

There is no cry out. It doesn't matter.

From that day on, Richard no longer remembers his father.

Osiris has largely taken over. But that doesn't mean he is completely helpless. He has a passport to France ready. It is nothing to hire a boat at Dover for 10 pounds. Such actions suggest a carefully planned murder. However, the sane, reasoned aspects make his behavior more bizarre when mixed with the obvious insanity, and one can only guess at his reasons or lack thereof for making his escape in the same bloodied clothes he wore to kill his father. He doesn't bother to change or wash up until he has gotten all the way to Calais. Once there, he proceeds on his next assignment to kill the emperor of Austria.

The London police find the dismembered body, think at first that Richard Dadd, rather than being the criminal, is also a likely victim of the same person who has killed his father, and begin searching about Cobham for his body or artifacts or any clue that might lead to the real killer. However, upon notifying his brother, the police are alerted to the recent odd behavior. They search his apartment in London and find the world of a deranged man, over three hundred eggs, the carpet covered with egg shells, bottles upon bottles of ale, and a sketchbook filled with portraits of family and friends, all depicted with throats slashed. Now the search for Richard Dadd is no longer to save him but to save others *from* him. But he has already left the country.

On his way, traveling in a diligence through a forest of Valance near Fontainebleau, he becomes entranced with the cravat and collar of a fellow passenger, who tolerantly lets him play with it for a time. However, when this man asks for it back, Richard produces a razor and attempts to cut his throat. Once again the razor attack does not work. This time he is subdued and turned over to the French authorities, where apparently Osiris leaves him for a time, as he promptly confesses to his activities, including the murder of his father, and even turns over what money he has as recompense for his latest victim's welfare.

The police find in his possession a list of people "who must die" with his father at the top.

Later, Richard himself will write: "These and the like, coupled with the idea of a descent from the Egyptian god Osiris, induced me to put a period to the existence of him whom I had always regarded as a parent, but whom the secret admonishings I had counseled me was the author of the ruin of my race. I inveigled him, by false pretences, into Cobham Park, and slew him with a knife, with which I stabbed him, after having vainly endeavoured to cut his throat." (6)

Under French law there is no need for a trial. Instead Richard Dadd is immediately committed and passed through various insane asylums in Melun, Brie, St. Denis and Luzarches to Clarmont.

On May 1, 1844, the *Art Union,* reports:

> He took no notice whatever of pencils, colours, and canvas they [his family] had sent him; retaining, it would appear, no sort of memory of his former pursuits, and never giving the slightest indication of a desire to produce a picture. His employment all day is to stand in the courtyard, with up-turned eyes, gazing at the sun, which he calls his father. (7)

The following July he is extradited to England, where two preliminary hearings are held at Rochester Magistrates' Court, and on August 22, 1844, he is transferred to the State Criminal Lunatic Asylum, Bethlem Hospital, famous as a cruel institution for the insane, conditions and treatment so horrible that it gives the English language the origin of the word Bedlam.

This institution where Richard is to spend the next twenty years originated as Bethlehem, located on Bishopsgate Street (current site of Liverpool Street Station) in 1247, founded by Simon Fitz Mary (or Fitzmary; sources are contradictory here), an alderman and sheriff of the city, as a priory for the Order of St. Mary of Bethlehem, "originally intended for the poor suffering from any ailment and for such as might have no other lodging, hence its name *Bethlehem*, in Hebrew, "the house of bread." (8) In 1330 it was granted a license as a hospital, and by the 1400s there are records that people deemed insane were being lodged there. Because of his separation from the Catholic Church, Henry VIII issued The First Act of the Dissolution of the Monasteries in 1536, and in 1547 put Bethlem under the jurisdiction of the City of London, with the governors of the Bridewell House of Correction in charge. For the next three centuries, its own uneven records and other writings indicate that it became ever more an asylum for the lodging or imprisonment of the insane, and by the 1600s a place for the public to view the insane under the most horrific confinement and display. After the Great Fire of London in 1666, Robert Hooke designed a new hospital, Bethlem in Moorfields, modeled on Tuileries in France, and this new facility opened in 1676. *Mania* and *Melancholy*, statues by Caius Gabriel Cibber depicting the inmates chained and in states of insanity, were meant to draw visitors into the hospital, which now became a popular and featured tourist site, a human zoo.

Ebenezer Haskell paraphrases and then quotes Edward Wakefield's speech to the Committee of the House of Commons about his two visits to Bethlem Hospital April 25 and May 2, 1814, providing a clear and disturbing picture of how insanity was understood at this time.

First, paraphrasing Edward Wakefield, Haskell writes:

> In the women's galleries, one of the side rooms contained about ten patients, each chained by one arm or leg to the wall, the chain allowing them merely to stand up by the bench or form fixed to the wall, or to sit down on it. The nakedness of each patient was covered by a blanket, made into something like a dressing-gown, but with nothing to fasten it in front. This was the whole covering, the feet being naked. In another part he [Edward Wakefield] found many of the unfortunate women locked up in their cells, naked, and chained on straw, with only one blanket for a covering. In the men's wing, in the side room, six patients were chained close to the wall, five hand cuffed, and one locked to the wall by the right arm, as well as by the right leg; he was very noisy; all were naked except as to the blanket-gown or small rug on the shoulders, and without shoes—their nakedness and their mode of confinement gave this room the complete appearance of a dog-kennel.

Then directly quoting Wakefield, he continues:

> "In one of the cells of the lower gallery we saw William Norris. He stated himself to be fifty-five years of age and that he had been confined about fourteen years; that in the consequence of attempting to defend himself from what he conceived the improper treatment of his keeper, he was fastened by a long chain, which, passing through a partition, enabled the keeper, by going into the next cell, to draw him close to the wall at pleasure; that to prevent this, Norris muffled the chain with straw, so as to hinder it passing through the wall; that he afterwards was confined in the manner we saw him, namely, a stout iron ring was riveted round his neck, from which a short chain passed to a ring made to slide upwards and downwards, on an upright massive iron bar, more than six feet high, inserted into the wall; round his body a strong iron bar about two inches wide, was riveted; on each side of the bar was a circular projection, which, being fastened to and enclosing each of his arms, pinioned them close to his sides. This waist-bar was secured by two similar bars, which passing over his shoulders, were riveted to the waist-bar, both before and behind. The iron ring round his neck was connected to the bars on his shoulders by a double link; from each of these bars another short chain passed to the ring on the upright bar. We were informed he was enabled to raise himself so as to stand against the wall on the pillow of his bed in the trough bed in which he lay; but it is impossible for him to advance from the wall in which the iron bar is soldered, on account of the shortness of the chains, which were only twelve inches long. It was, I conceive, equally out of his power to repose in any other position than on his back, the projections on each side of the waist-bar enclosed his arms, rendering it impossible for him to lie on his side, even if the length of the chain from his neck and shoulders would permit it. His right leg was chained to the trough, in

> which he had remained thus encaged and chained for more than twelve years."
>
> A marvelous example, truly, of what the human constitution may accommodate itself to. Poor Norris was not released from this confinement until about three weeks or a month before his death, which, notwithstanding the opinion of Dr. Monro, the physician then at the head of the medical department, may well be thought to have been hastened by the treatment to which he was subjected. That gentleman being asked whether he did not think that the pulmonary complaint of which Norris died might have been produced by the great quantity of iron he wore for many years, replied, "I think not."
>
> When asked—"Do you think a person could have had about him a weight of iron say six or eight-and-twenty points; that he could have been confined to his bed without being allowed to turn around for nine years, or without being able to get out and sit on the edge of his bed, being chained by the head by a chain only twelve inches from the iron stanchion, and that would have no effect upon his general health?"
>
> "It did not appear to have any general effect upon his health—he was in very good health till within a very short period of his death."
>
> The apothecary coincided in his colleague's opinion, and even became enthusiastic about the excellence of restraint by means of irons. In reply to a question, he asserted that to secure the patient with irons was a thousand times less objectionable than the strait-waistcoat, and that, footlocked and manacled, he was rendered an "innoxious animal." (9)

A year later, in 1815, partially the result of Wakefield's public condemnations of the current state of the institution, Bethlem was moved to Southgate, and the next year was officially established as a place for insane criminals, and remained as such until 1864.

By the time Richard Dadd is situated there in 1844, conditions have improved from Edward Wakefield's reported experience. Though Richard is transported to his new residency in a strait-jacket, once admitted he is never again subject to any kind of physical restraint, and does not experience the horrible abuse of previous patients.

That does not mean conditions are either good or defensible. At the time of his interment, the institution has been divided into two blocks, one for women and one for men. No other separations are made, and thus Richard is crammed together with the most violent and degraded of English criminals.

Indeed, Bethlem retains many of the same characteristics of the original structure. The gallery is some 100 feet long, lit by but one small, heavily barred window at each end. The inmates' sleeping rooms open off each side of the gallery. Dining tables sit in the middle of the basement gallery, and there are two small sink and water-closets on either side of the main passage. The three floors above are much the same, except also holding a bath room. Several portions of the galleries are divided by iron rods of wires into cages to allow separation of the patients without locking them in their rooms.

The physicians to the institution are not permanent residents but, rather, simply visit the hospital on occasion. Records are almost nonexistent. Thus, beyond knowing the obviously unpleasant conditions and lack of much being done to understand and help any of the inmates, including Richard Dadd, it is hard to know just what his physical and mental conditions are. However, an article from the *Art Union* suggests a surprising change has taken place in just one year:

> He is in good health; and we have lately seen some drawings recently executed by him which exhibit all the power, fancy, and judgement [sic] for which his works were eminent previous to his insanity. They are absolutely wonderful in delicate finish. They consist principally of landscapes—memories of Eastern scenes, or wrought from a small sketch book in his possession. One is, however, of an avenue of close box-tree, terminated by the tall gate of a mansion. It is a marvelous production, such as scarcely any of our living painters could surpass. This drawing was, we believe, produced within the last few weeks. (10)

The reasons for both his improved mental condition and his renewed interest in his art must remain speculative. However, it is at least likely that two of the attending physicians at the time, Dr. Edward Thomas Monro and Dr. Alexander Morison have some responsibility for it.

Dr. Monro is a fourth generation physician at the hospital and holds a highly respected position as an expert on insanity or lunacy. Court records at the time indicate that he testifies at 400 trials involving madness, and only twice do his findings not match the verdict (and in both of those cases, the original verdicts were later suspended). More important in terms of Dadd's renewed interest in his art is that Monro's family is famous for its support of artists. Monro's grandfather, John Monro seriously studied engraving, and his father Thomas Monro is considered the most important patron of the English school of watercolor painting. (11) In 1976, an exhibition of the work of Dr. Edward Thomas Monro and the artists of his circle is held at Victoria and Albert Museum. A catalogue is published, titled, *Dr. Monro and the Monro Academy*. (12) His brother is considered a brilliant draughtsman. Indeed, he grew up surrounded by the artists and art works of his day—Turner, Linnel, Hunt, Cotman, De Wint, and the Hoppners. Certainly Dr. Monro knows what is going on in the art world surrounding him, and since Dadd is highly acclaimed, there can be no doubt Monro knows he has a famous/infamous painter in his asylum. So it is likely he supplies Dadd with materials and encourages him to continue with his art.

However, though Monro is both an art connoisseur and an acclaimed expert on insanity, his mismanagement of Bedlam, negligence and mistreatment of the patients, and nearly nonexistent records result in a scandal, revealing that, if he does encourage Dadd's artistic activities, it is not for therapeutic reasons, but rather for the celebrity Monro might gain from it.

The other attending physician, Dr. Alexander Morison is not at the center of the English art scene. But he is a highly respected specialist in mental diseases, whose publications include *Outlines of Lectures in Mental Diseases* (1826), *Cases of Mental Diseases* (1828) and *The Physiognomy of Mental Diseases* (1840). While he does advocate a bit less cruel treatment of patients then what is in general use, he still believes in the use of strait-jackets, and as with Monro, he does not give his patients at Bedlam much time or attention, preferring instead to focus on his lectures on insanity in London and Edinburgh, his other patients at Hanwell, and mainly his aristocratic clientele in private practice.

Morison becomes especially intriguing in trying to understand Dadd's relations with his physicians because of a portrait Dadd completes of him in 1852.

Morison believes in the now largely discounted doctrine of physiognomy, the theory that a patient's facial expression reveals his underlying mental state. In the portrait, Dadd puts Morison's own physiognomy to the test. Using a sketch by Morison's daughter Ann of the grounds of Anchorfield, Alexander Morison's childhood home on the shores of the Firth of Forth, Dadd, never having seen the original landscape, creates a strange scene including two sailing ships on the Fife coastline and two Newhaven fishwives. It would be interesting to know what Morison thinks of the physiognomy presented. With only the painting itself to judge by, it might be claimed Morison is austere, though a hint of kindness might be found in the mouth. He is 73 at the time of the painting, has recently witnessed the death of his wife of nearly 50 years, and has just been forced to retire from Bethlem the result of the horrendous conditions there. Nevertheless, though presenting a man separate from the rest of the scene, and perhaps lonely, the painting does not seem an overly negative portraiture or meant to carry any serious distain. In fact, the simple fact Dadd completes it suggests a friendly relationship.

Whatever encouragement and friendship these two highly respected mental physicians possibly offer Dadd, and whatever the reasons might be for their actions or lack thereof, they are not giving Bethlam the attention it needs.

In 1852 The Report of the Commissioners in Lunacy to the Secretary of State on Bethlem Hospital removes both from their positions in a highly public scandal, and a resident physician superintendent position is established to take the place of the visiting physicians. (13) Dr. William Charles Hood is appointed to this position and is "one of the most outstanding in Bethlem's history." (14) He immediately begins making life less harsh for the patients, providing every ward with an aviary, enlarging the windows to bring in more light, increasing the living space, even adding pictures, plants and statues.

Nevertheless, improvements are relative. Even as enlightened a physician superintendent as Hood finds his attempts at improvement slow going, and in 1855 still describes Bethlam as having "dark, gloomy, and inefficiently warmed" wards where "the windows are small and unnecessarily laden with iron bars, the staircases with iron gates, . . . [and where] the absence of all workshops for occupation, and the scanty means of amusement at our disposal render the daily life of all the patients irksome." (15)

In 1857 *The Quarter Review,* sounding much the same as reports from previous centuries, describes the wards as "more like those which enclose the fiercer carnivore at the Zoological Gardens than anything we have elsewhere seen employed of the detention of afflicted humanity." (16)

These galleries are where Dadd spends most of his first thirteen years at Bethlem. Occasionally patients are allowed into the exercise yard, little more than an open, featureless yard surrounded by walls. Occasionally friends or family visit, though it seems that for Dadd such visits are rare.

Especially in his earlier years, Dadd is as likely to be as belligerent as any of the other patients, but as time goes on he becomes less so. Certainly, his psyche is multi-dimensional, and in 1854, Hood writes of him as being "a very sensible and agreeable companion" and showing himself in his conversations to have "a mind once well educated and thoroughly informed in all the particulars of his profession in which he still shines." (17) However, this does not mean he is free of his delusions, and though he might have become less prone to being disagreeable, it is clear he is still capable of violent emotions. Osiris remains a strong part of his mental life to the day he dies.

In 1857 Hood is able to convert a ward in the main hospital and move Dadd along with about forty of the other "better class" patients into more friendly accommodations which include access to a billiard table and a library. In an article on March 31, 1860, the *Illustrated London News* praises Hood for the "admirable and highly useful improvements that have taken place." (18) Upon visiting the hospital in 1863, W. M. Rossetti reports finding Dadd in a "large airy room." (19)

It is also clear that Hood and Dadd have a friendly personal relationship. While there is little to go on here it is obvious that Hood not only encourages Dadd to continue with his art but collects many of Dadd's paintings and drawings, possessing a total of thirty-three upon his death, and in addition Dadd gives Hood perhaps his second greatest work, *Contradiction. Oberon and Titania*, a painting he has worked on specifically for Hood for four years.

Within a year of Rossetti's visit, Dadd is moved into even better conditions. The new criminal asylum in Berkshire, Broadmoor Hospital is completed, and all criminal patients are transferred. One can only imagine Dadd's emotions on the train seeing the English countryside for the first time in twenty years. And the new quarters have to delight him. No longer is life to be restricted to the small bedroom, the crowded gallery and the dank prison yard. Broadmoor purposely opens up the environment to the outside world as much as possible. Terraces in the back cover up the brick walls and allow what must seem once-upon-a-time views of the wooded fields that stretch for miles. Rhododendrons and laurels intertwine the iron railings in Victorian elegance; perhaps enough to allow for the illusion of his family's cultured life when he was a child.

During the years he spends here, his general mental state appears to remain about the same. His major artwork at Broadmoor consists of a number of theatre murals and other works for the stage, most of which no longer survive. He also reads a good deal of classical literature, history, and poetry; and is considered a skilled violinist.

However, while his outbursts become less frequent and he is able at times to separate himself from his feelings of being controlled by Osiris and other spirits, and to talk about this part of himself in a detached manner, as if resigned to being misunderstood, he is never able to completely free himself from his torments. An article in *The World* (20) writes of him being still held by "thick-coming horrors and portentous visions—meditative, gloomy, abstracted . . . a recluse doing the honours of his modest unpretending abode; a pleasant-visaged old man with a long and flowing snow-white beard, with mild blue eyes that beam benignly through spectacles when in conversation, or turn up when in reverie till their pupils are nearly lost to sight. He is dressed with extreme simplicity in gray 'dittos'; his manner is unassuming, but impressive and perfectly courteous; his utterance slow, not as though ideas were lacking, but as if he wished to weigh carefully his words before he spoke them." His hair is snow-white, and though he comes from a family that is robust into old age, he is not so.

He will live still nine more years, becoming ill with consumption in 1885 and dying January 8, 1886. By this time his family has dispersed and the only person informed of his death is Elizabeth Langley, a childhood friend whose sister has married his brother Robert, and who has previously written the hospital to inquire about him on her sister's behalf. He is buried in a peaceful, tiny cemetery at Broadmoor. His sister, Mary Ann, living at the time in America, writes in her epitaph for him: "I am truly thankful to know him at rest. It is less grief to me, than it was to think of him in the changed condition in which he has lived for many years past, his life has been to me a living death." (21)

The Getty Museum online site succinctly states the dry facts:

> Richard Dadd was born August 1, 1817 in Chatham, Kent, England. The fourth of eleven children of a chemist and druggist, Richard Dadd began his artistic training at the age of thirteen. He studied miniature painting, portraiture, and landscapes and was accepted to the prestigious Royal Academy for further study after turning twenty. Academy professors remarked on his gentleness, cheerful good nature, and great promise as an artist.
>
> Dadd's Neoclassical paintings of ancient subjects, theatrical in concept, followed in the tradition of Lawrence Alma Tadema. He painted scaled-down human figures in small, luminous landscapes of carefully rendered plants and flowers.
>
> In 1842 Dadd left England on a one-year drawing trip throughout Europe and the Middle East. Shortly after his return, he lost his sanity, as had three of his siblings. Dadd murdered his father, attempted to kill a stranger, and spoke of killing the pope and the emperor of Austria, insisting that the Egyptian god Osiris requested these acts. In 1843 the courts and his family committed him to an insane asylum, where he remained for the next forty-three years, continuing to paint. (22)

That Dadd kills his father and is confined in Bethlem and later Broadmoor might make for an interesting, though depressing biography, especially if an imaginative writer can embellish it or if a historian can use it to help reveal the conditions and views of the insane in 19th century England. However, in and of itself, it does not make for an exceptional life.

Part II

The Fairy Feller's Master Stroke hangs in the Tate Gallery. It is oil-on-canvas, about 21 ¼ by 15 ½ inches, and there can be no doubt but that it was commissioned by George Henry Haydon, Dr. Hood's assistant at Bethlem, as the complete title states it: *Fairy-Feller's Master Stroke. Painted for G. H. Hayden, Esqr by Rd. Dadd / quasi 1855 – 1864.*

This commission and other materials indicate that Dadd and George Henry Haydon had a friendship based on a mutual embracement of artistic expression. Though Haydon never considered himself an artist, he drew constantly, mostly humorous sketches exhibiting notable skills, illustrations for books, and drawings for *Punch.* As a member of the Langham Sketching Club he became friends with such important illustrators as George Cruikshank, Charles Keene, and John Leech. (23) According to his obituary in *Under the Dome: The Quarterly Magazine of the Bethlem Royal Hospital* he was a very friendly man who "made numerous friends and no enemies during his term of office." (24)

The title of the work, by including the word "*quasi*" also confirms that Dadd did not consider the painting finished, though he had worked on it for nine years, and a close study of it reveals that some of the nuts, a piece of grass, and the woodman's axe are only sketched in, in contrast with the microscopic detail of the rest of it. (25) What is likely is that he was interrupted from finishing it when he was moved to Broadmoor in 1864.

However, though the actual painting was put aside, he did not put aside his connections to it. He repainted it in watercolor, the details extraordinarily similar, especially since he did it from memory. (26) Perhaps, however, as Patricia Allderidge suggests, some of the changes "such as the fact that the grasses are now in flower, and the great proliferation of calligraphic swirls across the surface, as well as the completely different colour and tonality, suggest that he did not intend to make a literal copy but rather a translation into another style and mood." (27)

He also wrote *Elimination of a Picture & Its Subject—Called the Feller's Master Stroke* the following January, a "long, rambling and sometimes incoherent poem" where he "explains the action in the painting . . . and digresses on a number of subjects, some tenuously related to it, and some which seem to have slipped in while no one was looking." (28)

Two copies of the poem have been found and, according to John MacGregor, are identical. Having personally viewed the autographed copy in possession of Haydon's relatives, MacGregor makes a good case for it having been written especially intended to accompany the painting Dadd had done for Haydon, perhaps simply to help clarify the painting, probably also as a chance to communicate thoughts to a friend. (29) In it, Dadd "explains how Haydon, the Steward at Bethlem, liked the picture of *Contradiction* on which he had been working for a year, and wanted something like it for himself" based on some verses about fairies one of his friends had written. (30) The second copy might have been Dadd's own, perhaps what served as a somewhat strange journal or something he saw as a work in progress or simply his own record of it. He was a very detail oriented man, and it can easily be suggested he wanted this written record. However, as we will see, perhaps there was much more to this than clarification or a record.

The poem/explanation of the painting, an oxymoron of form and content in itself, both reveals and mystifies the meanings. Patricia Allderidge comments on this in terms of the title:

> The use of the word "elimination" in the title is open to speculation, and it could be suggested that Dadd intended something significant by it, perhaps trying to convey that by explaining, he was somehow eliminating the picture from his mind: but in view of his fondness for punning and word play shown elsewhere, for example in one of his early letters to Firth, a likely explanation is that it is a play on some word such as "illumination," or "elucidation." It could even be a simple mistake." (31)

On the one hand there are some excellent objective descriptions of the characters in the painting. On the other hand the juxtaposition of realities, the obscure references, and often undecipherable passages leave everything in a state of temptation, of sudden truths sparkling out of the mysterious shadows of the depths of the psyche.

Just above the center of the painting is the white, bearded face of the Patriarch, hand raised, ready to give the command. According to Erick Fromm "Patriarchies are distinguished by a respect for man-made laws, the favouring of works of art and craft, and obedience to the hierarchy." (32) This, then, is a father figure, a giver of laws, a man to be obeyed, and interestingly a man who promotes artistic expression. Since Dadd specifically named this character the Patriarch, we know he intended the meanings, the associations of *patriarch* to be attached to that male entity, and all of Erick Fromm's characteristics can easily be applied. If so, a problem for Dadd had to be that he ended up having two *patriarchs*, his own father (a literal, physical patriarch) and Osiris (a spiritual, mythical patriarch), one demanding the death of the other.

According to Carl Jung, "In dreams, it is always the father-figure from whom the decisive convictions, prohibitions, and wise counsels emanate." (33) This corresponds to what Erick Fromm claimed, and it fits precisely with both what Dadd struggled with in real life and what he presented in the painting, as it is clear the entire scene is poised, waiting for the Patriarch to give the command to begin the action. Jung's discussion continues, "The invisibility of this source is frequently emphasized by the fact that it consists simply of an authoritative voice which passes final judgments." Again, it is easy to see how Dadd was caught here by the authoritative voice of his spiritual father Osiris and the opposing real world voice of his biological father. "Mostly," Carl Jung continues, "it is the figure of a 'wise old man' who symbolizes the spiritual factor. Sometimes the part is played by a 'real' spirit, namely the ghost of one dead, or, more rarely, by grotesque gnomelike figures . . . ." (34) Certainly, the Patriarch in the painting is a gnome-like figure, and his "wise old man" qualities are emphasized by his long, flowing hair and beard.

Jung's discussion continues:

> It can never be established with one-hundred-per-cent certainty whether the spirit-figures in dreams are morally good. Very often they show all the signs of duplicity, if not of outright malice. I must emphasize, however, that the grand plan on which the unconscious life of the psyche is constructed is so inaccessible to our understanding that we can never know what evil may not be necessary in order to produce good by enantiodromia, and what good my very possibly lead to evil. Sometimes the *probate spiritus* recommended by John cannot, with the best will in the world, be anything other than a cautious and patient waiting to see how things will finally turn out. (35)

In the case of Dadd and the painting, this applies first to his carrying out of Osiris' command to kill his own father. While even he had to admit that on the surface it appeared an act of evil, still, if Osiris was correct, in the end, it might be that the initial evil in fact would result in a more important good. For Dadd himself, and for those of us judging him, whether or not his obedience to the command of his *Patriarch* Osiris is evil or good remains to be seen. This is expressed in the painting by having it hold us at that moment just before the command is given and obeyed, a state of expectation, not resolution. Dadd stated in his commentary that this *Patriarch* "casts a glance adown dim vistas of the pregnant coming bustle to note if there is aught to stay or hustle." (36)

Jung's discussion further connects all this:

> The figure of the wise old man can appear so plastically, not only in dreams but also in visionary meditation (or what we call 'active imagination'), that, as is sometimes apparently the case in India, it takes over the role of a guru. The wise old man appears in dreams in the guise of a magician, doctor, priest, teacher, professor, grandfather, or any other person possessing authority. The archetype of spirit in the shape of a man, hobgoblin, or animal always appears in a situation where insight, understanding, good advice, determination, planning, etc., are needed but cannot be mustered on one's own resources. The archetype compensates this state of spiritual deficiency by contents designed to fill the gap. (37)

This connects precisely with what Dadd can be seen to be doing in this painting he worked at so carefully for nine years and then when it was taken from him continued to work in by repainting it and by writing his lengthy poem about it. It is his need to give plastic or visual (and later poetic) form to his spiritual struggles, specifically in terms of the Patriarch to his inability to sort out the demands of his own demons.

Here the work of art and the dream come together, the invisible and visible expressions of the same archetypal demands.

Jung clarifies it:

> An excellent example of this is the dream about the white and black magicians, which tried to compensate the spiritual difficulties of a young theological student. I did not know the dreamer myself, so the question of my personal influence is ruled out. He dreamed *he was standing in the presence of a sublime hieratic figure called the "white magician," who was nevertheless clothed in a long black robe. This magician had just ended a lengthy discourse with the words "And for that we require the help of the black magician." Then the door suddenly opened and another old man came in, the "black magician," who however was dressed in a white robe. He too looked noble and sublime. The black magician evidently wanted to speak with the white, but hesitated to do so in the presence of the dreamer. At that the white magician, pointing to the dreamer, said, "Speak, he is an innocent." So the black magician began to relate a strange story of how he had found the lost keys of Paradise and did not know how to use them. He had, he said, come to the white magician for an explanation of the secret of the keys. He told hem that the kind of the country in which he lived was seeking a suitable tomb for himself. His subjects had chanced to dig up an old sarcophagus containing the mortal remains of a virgin. The king opened the sarcophagus, threw away the bones, and had the empty sarcophagus buried again for later use. But no sooner had the bones seen the light of day than the being to whom they once had belonged—the virgin—changed into a black horse that galloped off into the desert. The black magician pursued it across the sandy wastes and beyond, and there after many vicissitudes and difficulties he found the*

lost keys of Paradise. That was the end of his story, and also, unfortunately, of the dream.

Here the compensation certainly did not fall out as the dreamer would wish, by handing him a solution on a plate; rather it confronted him with a problem to which I have already alluded, and one which life is always bringing us up against: namely, the uncertainty of all moral valuation, the bewildering interplay of good and evil, and the remorseless concatenation of guilt, suffering, and redemption. This path to the primordial religious experience is the right one, but how many can recognize it? It is like a still small voice, and it sounds from afar. It is ambiguous, questionable, dark, presaging danger and hazardous adventure; a razor-edged path, to be trodden for God's sake only, without assurance and without sanction. (38)

Part III

It is easy to imagine Dadd caught within this interplay of good and evil, guilt, suffering and redemption, struggling between *ethical* and *spiritual* demands. The story of Abraham and Isaac comes immediately to mind:

1. And it came to pass . . . that God did tempt Abraham, and said unto him, Abraham: and he said, Behold, here I am.

2. And he said, Take now thy son, thine only son Isaac, whom thou lovest, and get thee into the land of Mo-ri-ah; and offer him there for a burnt offering upon one of the mountains which I will tell thee of. (39)

In obeying this command from God, Abraham became the ultimate expression of faith. In the end, once God was certain Abraham would indeed kill his own son in obedience, God showed his mercy and rescinded the command. But the key is the knowledge that there is no doubt but that Abraham was going to carry it out. The passage continues:

3. And Abraham rose up early in the morning, and saddled his ass, and took two of his young men with him, and Isaac his son, and clave the wood fro the burnt offering, and rose up and went unto the place of which God had told him.

4. Then on the third day Abraham lifted up his eyes, and saw the place afar off.

5. And Abraham said unto his young men, Abide ye here with the ass; and I and the lad will go yonder and worship, and come again to you.

6. And Abraham took the wood of the burnt offering, and laid it upon Isaac his son; and he took the fire in his hand, and a knife; and they went both of them together.

> 7. And Isaac spake unto Abraham his father, and said, My father; and he said, Here *am I*, my son. And he said, Behold the fire and the wood; but where *is* the lamb for a burnt offering?
>
> 8. And Abraham said, My son, God will provide himself a lamb for a burnt offering: so they went both of them together.
>
> 9. And they came to the place which God had told him of; and Abraham built an altar there, and laid the wood in order, and bound Isaac his son, and laid him on the altar upon the wood.
>
> 10. And Abraham stretched forth his hand, and took the knife to slay his son. (40)

This story became the central one for Soren Kierkegaard's similar struggles with what he called the *teleological suspension of the ethical*. In other words, faith demands an acceptance of the *absurd*, beyond even *ethos*. For Soren Kierkegaard, a leap of faith is based on a willingness to sacrifice all reason and logic--all seemingly correct behavior. He wrote in his journal at the time he was working on his famous expression of these views in the book *Fear and Trembling*:

> Faith . . . hopes also for this life, but, be it noted by virtue of the absurd, not by virtue of human understanding, otherwise it is only practical wisdom, not faith. Faith is therefore what the Greeks called the divine madness." (41)

As the initial quote for this chapter indicates, Plato expressed this "divine madness" through Socrates:

> It is to their madness that we owe the many benefits that the Pythia of Delphi and priestesses of Dodona were able to bestow upon Greece, both privately and in public life, for when they were in their right minds their achievements amounted to little or nothing.

The passage surrounding this quote gives three contexts for madness: prophecy, divine origin, and inspiration for artistic expression.

> Socrates: "I told a lie when I said that the beloved ought to accept the non-lover when he might have the lover, because the one is sane, and the other mad. It might be so if madness were simply an evil; but there is also a madness which is a divine gift, and the source of the greatest blessings granted to men. For prophecy is a madness, and the prophetess at Delphi and the priestesses at Dodona when out of their senses have conferred great benefits on Hellas, both in public and private life, but when in their senses few or none. And I might also tell you how the Sibyl and other inspired persons have given to many an intimation of the future which has saved

them from falling. But it would be tedious to speak of what every one knows.

There will be more reason in appealing to the ancient inventors of names, who would never have connected prophecy (mantike) which foretells the future and is the noblest of arts, with madness (manike), or called them both by the same name, if they had deemed madness to be a disgrace or dishonour;-they must have thought that there was an inspired madness which was a noble thing; for the two words, mantike and manike, are really the same, and the letter t is only a modern and tasteless insertion. And this is confirmed by the name which was given by them to the rational investigation of futurity, whether made by the help of birds or of other signs, for as much as it is an art which supplies from the reasoning faculty mind (nous) and information (istoria) to human thought (oiesis) they originally termed oionoistike, but the word has been lately altered and made sonorous by the modern introduction of the letter Omega (oionoistike and oionistike), and in proportion prophecy (mantike) is more perfect and august than augury, both in name and fact, in the same proportion, as the ancients testify, is madness superior to a sane mind (sophrosune) for the one is only of human, but the other of divine origin. Again, where plagues and mightiest woes have bred in certain families, owing to some ancient blood-guiltiness, there madness has entered with holy prayers and rites, and by inspired utterances found a way of deliverance for those who are in need; and he who has part in this gift, and is truly possessed and duly out of his mind, is by the use of purifications and mysteries made whole and except from evil, future as well as present, and has a release from the calamity which was afflicting him. The third kind is the madness of those who are possessed by the Muses; which taking hold of a delicate and virgin soul, and there inspiring frenzy, awakens lyrical and all other numbers; with these adorning the myriad actions of ancient heroes for the instruction of posterity. But he who, having no touch of the Muses' madness in his soul, comes to the door and thinks that he will get into the temple by the help of art, he, I say, and his poetry are not admitted; the sane man disappears and is nowhere when he enters into rivalry with the madman.

I might tell of many other noble deeds which have sprung from inspired madness. And therefore, let no one frighten or flutter us by saying that the temperate friend is to be chosen rather than the inspired, but let him further show that love is not sent by the gods for any good to lover or beloved; if he can do so we will allow him to carry off the palm. And we, on our part, will prove in answer to him that the madness of love is the greatest of heaven's blessings and the proof shall be one which the wise will receive, and the witling disbelieve. (1)

The more we dig into the oracle at Delphi, the more obvious it becomes that the people of Classical Greece put this "divine madness" at the center of all of their beliefs.

Delphi was located on what was considered to be the *axis mundi*, the sacred center of the world, originally a sacred site of the pre-Greek inhabitants ruled by the god Gee or Gaia, who interestingly enough was protected by a large snake, a *pythia*. According to Hesiod's *Theogony*, the Python was a female dragon, daughter of Gaia herself. This snake/dragon carries the standard transformation qualities discussed elsewhere in this essay (including the Garden of Eden expression of the gaining of the knowledge of ethical existence). Thus, the snake here serves a standard mythological function as the guardian and messenger of knowledge beyond mere physical existence (i.e., prophecy).

Another story of the source of the site claims that a shepherd came upon a place where the earth was cracked and a mist came out of the crack that, when inhaled, allowed people to see into the future. While not as rich a story, this still stresses the prophetic associations, and contains the more earth-centered view of human origins, because here the "source" of the "divine madness" is mother earth herself. Delphi literally translates as "womb," and, thus, this site is the connection between the mysterious spiritual world of mother earth (Gaia), the womb of the earth, and human existence.

When the snake takes on dragon qualities (dragons have wings), the sky as cosmic ocean also gets brought into the equation. When the Greeks took over this site, they brought their whole mythology of the sky gods with them, including the sacred world of the gods on Mount Olympus. Thus, there is a conflict in mythologies. Are humans born out of the earth or the sky? The fact that Zeus (most important of the sky gods), at times, gives birth without any female involved represents this conflict. According to the Homeric *Hymn to Python Apollo*, it all ignited when Zeus birthed Athena out of his head, ignoring Hera (granddaughter of Gaia) and denying her as female the exclusive right to bear divine children. After a less-then-ideal attempt to match his single god birth in giving birth to a malformed Hephaestus, Hera bore Typhon, a monstrous serpent whose strength rivaled Zeus himself, and she put Typhon in the care of Python, the very dragon/snake that guarded the female womb of Mother Earth.

Undeterred, Zeus impregnated Leto, the daughter of Titans Coeus and Phoebe. In revenge, Hera terrorized the earth, hoping to prevent Leto from finding a safe place to birth what would become the twins Apollo and Artemis. Leto however found refuge on Ortygia, a barren, floating island (later to become Delos, an important religious center). Hera responded by not allowing Eileithyia, goddess of birth, to leave Mount Olympus, causing Leto to remain in labor for nine days. However, in the often less-than-noble and less-than-intelligent fashion of the Greek gods and goddesses, Hera was persuaded to rescind her prohibition when Athena bribed her with a huge gold-and-amber necklace.

But the battle of the sexes was not yet over. Python immediately began terrorizing the land, slaughtering and eating both animals and humans. Apollo's worth and the victory of the new order of the sky gods got immediately emphasized when, at the age of three days, he shot an arrow through Python.

Next came the need to somehow justify this ethically messy dispute. So Apollo took responsibility for his murder, exiled himself from Mount Olympus, and worked as a shepherd for King Admentus to purge himself for whatever guilt he took on in killing the Python. Furthermore, he commemorated the Python, putting his oracles or sibyls or *pythia* in place as the voice of his prophecies.

The purging aspect of all of this is important, and the site took on a strong cleansing-of-sins quality, giving the Classical Greek world an ethical quality.

The initial *pythia* were attractive young virgins who were chosen to spend their lives performing various spiritual rituals, and once a year, on February 7, considered to be the birthday of Apollo, provided answers to those selected from the many travelers who came seeking divine prophecy. They were said to bathe in, perhaps even drink, the sacred water of a nearby stream in order to prepare for the state of ecstasy necessary. The site was also associated with laurel leaves, and apparently the sibyls ate these, possibly being affected by them. Their responses were always enigmatic, not to be understood literally but in the same sense as a dream.

As time went on, the attractive young virgins were replaced by older, less attractive women from the lower classes, perhaps because there were practical problems with keeping young attractive virgins completely focused on their spiritual duties, perhaps because the wealthy rulers were manipulating to get their own relatives in place and influence the prophecies, perhaps to suggest that the prophecies came, not from the *pythia* themselves, but only through them. Certainly, at least later on, the *pythia's* prophecies would consist of incoherent babbling in a trance, a state of ecstasy that was incomprehensible to outsiders, but that an attending male priest would interpret.

As this oracle became more and more important to Greece and the surrounding countries, it took on ever more complex elaboration and symbolism. The *omphalos* (Greek for navel) stone was located here, in the *adyton* (restricted area; the word means "do not enter" or "inaccessible"; it is the sacred place where only the *pythia* were allowed entrance). According to the story, Zeus sent out two eagles to fly over the world and meet at its center, the *axis mundi* or navel of the world, and this stone designated where the meeting took place. Though this *axis mundi*, this *omphalos* appeared at other places, it was most commonly thought to be at Delphi, and the actual stone, or most likely a copy of it, had a knotted net over it, a hollow center, and a widening at the base.

It later came to be designated as the stone which Rhea wrapped in swaddling clothes, pretending it was Zeus to deceive Cronus, his father, who swallowed his children so they could not grow up and replace him as he had replaced his own father Uranus.

Such stones were said to allow direct communication with the gods. It has been suggested that the stone was hollow to allow vapors to pass through, which the *pythia* breathed to reach a state of ecstasy. This, of course, connects up to the story of Apollo conquering the previous Gee or Gaia spirit, suggesting that the original temple represented an earth spirit, and the designation of it by the term *delphi*, which translates as hollow or womb, suggesting an early form of Gaia worship took place there.

Statues of sphinx were also erected. While the famous Sphinx of Egypt was a male monster, in Greece it became female, and added the tail of a snake, the daughter of Echidna (half woman, half snake), and Typhon (a male monster with wings, a body of snakes, and fists that were snake's heads). Thus, there is the obvious connection of the sphinx to the snake and the Pythia to the snake. These statues at Delphi were votive columns given by wealthy donors who had sworn an oath or vow to Apollo in repayment for a positive oracle. While the sphinx were seen in a positive light as protectors, they were also ferocious beasts, not to be tangled with, members of the family of winged demons which included the Furies, the Keres, the Harpies and the Sirens, belonging to a class of vampire-like demons, spirits of the dead who ravage and carry off the living, thus having the same dualities as the snake and its relatives.

The use of a goat to determine who gets to ask for a prophecy became a common practice. What would happen is that the supplicant would bring a goat with him, a cold pale of water would be splashed over the goat, and if the goat shook from the cold water, then the prophecy requested would be granted, but if the goat did not shake the privilege would be denied. This connection of goats went all the way back to the story of the shepherd's discovery of Delphi as a sacred place. In fact, the goats were the first to encounter the crack in the earth, breath the vapor and begin behaving strangely--emitting sounds that goats would not make and erratically jumping about. Goats, then, could be seen as an intermediary, and as such were a common sacrifice.

The eternal fire also burned there, and each city-state would come to the oracle to light, and then carry this fire to preside over their city, much the way the current Olympic flame is carried to each new Olympic contest. Delphi, in fact, had its own version of the Olympics, second only to them in popularity, and expanding the athletic contests to artistic contests as well.

It is important to note that the site was not Apollo's alone. Dionysus was also worshipped there, Apollo in charge during the summer months, and Dionysus during the winter months. Apollo was the god of the sun, of *reason*, of music, and of prophecy. Dionysus was the god of nature, of ecstasy and of the vine--of that state of *madness* that takes place beyond reason. Each winter, maenads (from the word mania), women followers of Dionysus, would follow a pipe-playing boy (obvious associations with Pan) into the wild woods in the mountains surrounding Delphi, where they would participate in a frenzied dance to celebrate fertility and sexuality, striving to reach a state of madness, where they would kill and eat wild animals and move beyond reason into a higher psychic level. As an aside, it is worth mentioning that these gatherings were the beginnings of Theatre, and Dionysus is considered to be the god of Theatre. This all fits, as Theatre is of the same world as all of the arts, which is of the same world as that of the dream and knowing beyond explanation and ethos.

It is possible, and perhaps even likely that the combination of Apollo and Dionysus here is a result of the same commingling of earth and sky worship the myths expressed. Certainly, Walter F. Otto's account of the origins of the cult and subsequent myths of Dionysus support a view that it was in place before the epics of Homer, and it can be suggested that the female earth rituals of Dionysus were in place first, were the original spiritual connections to Delphi that the followers of Apollo supplanted, or perhaps augmented, since there is a blending rather than a clean replacement (similar to the sometimes strange incorporation of pre-Christian mythos and rituals into the Christian mythology; i.e., a rabbit hiding eggs at Easter and the intertwining of Halloween and All Saints' Day).

In reference to this, it is worth noting that the resurrection of Christ has been said to have developed out of or at least been prefigured in the mythos of Dionysus. In fact, believers in Jung's theories of archetypes can have a field day with the multiple correspondences between the rebirths of Christ and Dionysius. In one of the two most accepted versions, Dionysus is the son of Zeus (think supreme God here) and Semele, a mortal woman (think Virgin Mary here), the result not of a physical sexual act, but a spiritual union, in this case a bolt of lightning (this carrying the transforming symbolism of both fire and water, as lightning is nearly always accompanied by rain, and of a phallic shape, and of the realm of the sky). Expressing the same symbolism of the birth of Apollo, the woman is immediately replaced by Zeus, as Semele dies from the lightning bolt and Zeus takes Dionysus into his body (his thigh, with its sensual symbolism--not the intellect, not of the world of reason and logic).

In both cases, it must be made clear that they are the sons of both a human and a god. If they had only gods for parents, then they would simply be gods and could not offer humans the chance at death/rebirth or resurrection into the spiritual because the human world of physical existence would not be involved. If their parents were both human then the spiritual connection would be missing. It is the coming together of the two worlds that is the key. Walter Otto writes:

> At his conception the earthly was touched by the splendor of divine heaven. But in this union of the heavenly with the earthly, which is expressed in the myth of the double birth, man's tear-filled lot was not dissolved but preserved, rather, in sharp contrast to superhuman majesty. He who was born in this way is not only the exultant god, the god who brings man joy. He is the suffering and dying god, the god of tragic contrast. And the inner force of this dual reality is so great that he appears among men like a storm, he staggers them, and he tames their opposition with the whip of madness. All tradition, all order must be shattered. Life becomes suddenly an ecstasy—an ecstasy of blessedness, but an ecstasy, no less, of terror. (42)

Joseph Campbell explains this idea to Bill Moyers:

> The virgin birth comes into Christianity by way of the Greek tradition. When you read the four gospels, for example, the only one in which the virgin birth appears is the Gospel According to Luke, and Luke was a Greek.

Moyers responds:

> And in the Greek tradition there were images, legends, mythos of virgin births?

Campbell:

> Oh yes . . .

Moyers:

This was not a new idea, then, in Bethlehem. But what is the meaning of the virgin birth?

Campbell:

I think the best way to answer that is to talk about a system they have in India that describes stages of spiritual development. In India, there is a system of seven psychological centers up the spine. They represent psychological planes of concern and consciousness and action. The first is at the rectum, representing alimentation, the basic, life-sustaining function. The serpent well represents this compulsion—as a kind of traveling esophagus going along just eating, eating, eating. None of us would be here if we weren't forever eating. What you eat is always something that just a moment before was alive. This is the sacramental mystery of food and eating, which doesn't often come to our minds when we sit ourselves down to eat. If we say grace before meals, we thank this figure out of the Bible for our food. But in earlier mythologies, when people would sit down to eat, they would thank the animal they were about to consume for having given of itself as a willing sacrifice. . . .

Now, the second psychological center is symbolized in the Indian order of spiritual development by the sex organs, which is to say the urge to procreation. A third center is at the level of the navel, and here is the center of the will to power, to mastery and achievement, or, in its negative aspect, to the conquering, mastering, smashing, and trashing of others. This is the third, or aggressive, function. And as we are given to recognize in the symbolism of the Indian psychological system, the first function, alimentation, is of a animal instinct; the second, procreation, is of an animal instinct; and the third, mastery and conquest, is also of an animal instinct—and these three centers are located symbolically in the pelvic basin.

The next, or fourth, center is at the level of the heart; and this is of the opening to compassion. Here you move out of the field of animal action into a field that is properly human and spiritual.

And for each of these four centers there is envisioned a symbolic form. At the base, for example, the first one, the symbol is the lingam and yoni, the male and female organs in conjunction. And at the heart center, there is *again* the lingam and yoni, that is to say, male and female organs in conjunction, but here they are represented in gold as symbolic of the virgin birth, that is to say, it is the birth of spiritual man out of the animal man. . . .

It happens when you awaken at the level of the heart to compassion, com-passion, shared suffering: experienced participation in the suffering of another person. That's the beginning of humanity. And the meditations of religion properly are on that level, the heart level.

Moyers:

You say that's the beginning of humanity. But in these stories, that's the moment when gods are born. The virgin birth—it's a god who emerges.

Campbell:

And do you know who that god is? It's *you*. All of these symbols in mythology refer to you. You can get stuck out *there*, and think it's all out *there*. So you're thinking about Jesus with all the sentiments relevant to how he suffered—out there. But that suffering is what ought to be going on in you. Have you been spiritually reborn? Have you died to your animal nature and come to life as a human incarnation of compassion?

Moyers:

Why is it significant that this is of a virgin?

Campbell:

The begetter is of the spirit. This is a spiritual birth. The virgin conceived of the word through the ear.

Moyers:

The word came like a shaft of light.

Campbell:

Yes. And the Buddha, with the same meaning, is said to have been born from his mother's side from the level of the heart chakra. . . . That's a symbolic birth. He wasn't physically born from his mother's side, but symbolically.

Moyers:

But the Christ came the way you and I did.

Campbell:

Yes, but of a virgin. And then, according to Roman Catholic doctrine, her virginity was restored. So nothing happened physically, you might say. What is symbolically referred to is not Jesus' physical birth but his spiritual significance. That's what the virgin birth represents. Heroes and

> demigods are born that way as being motivated by compassion and not mastery, sexuality, or self-preservation.
>
> This is the sense of the second birth, when you begin to live out of the heart center. The lower three centers are not to be refuted but transcended, when they become subject to and servant to the heart. (43)

Thus, in both Dionysus and Christ, the virgin birth represents the birth of spirit out of matter, the coming together in humans of the two worlds, physical and spiritual. It is this union that defines human existence. Walter Otto writes:

> The myth of his [Dionysus'] birth . . . is the most sublime expression of his Being. Just as the amazing image of Athena's ascent from the head of her father can be conceived only in the spirit of the genuine revelation of her Being, so beneath the lightning flashes of Dionysus grew the certainty that the enigmatic god, the spirit of a dual nature and of paradox, had a human mother and, therefore, was already by his birth a native of two realms. (44)

If the virgin birth connects both the physical and spiritual, brings the spiritual into the physical, the other half of this duality is the exit. Thus, the human or physical part of Dionysus must include a physical death as well, a leaving of the physical world, but most importantly allows for the possibility of a rebirth into the spiritual. In other words, the spiritual world, the invisible world beyond the physical world does exist within humans. Furthermore, if the world of meaning and value, the world of the spirit can come into the physical world of the human body, then that world exists. A web of questions having to do with the relationship of these two worlds immediately results. If the spiritual and physical can come together in humans, as seems to be the case, what happens when the human body dies? Does the spirit live on? If so, is it an individualized spirit? In other words, does the individual human have self-awareness after death? Does the world of the spirit exist without the human mind? Does the human mind exist without the human brain? Does this human coming together of the physical world of the brain and the spiritual world of the mind only exist within each individual human while the human is physically alive? Is there such a thing as an individual existence or its opposite, a universal humanity?

Here the whole death and rebirth of Dionysus takes front stage. If he is both of the physical world, then his death to it and rebirth into it offers a desirable answer to these important questions, allowing both for meaning and value in human existence and for a continuance of self-awareness after the death of the body. Thus, it is the death/rebirth of Dionysus (prefiguring Christ) that gets stressed. Walter Otto writes:

> Other gods, like Apollo, also go off into the distance and return. But only Dionysus disappears in an incomprehensible manner from the circle of his followers or is swallowed up in the deep. As surprising as is his coming, so is his going away. In the Agrionia festival in Chaeronea the women searched for him and returned finally with the tidings that he had fled to the Muses and lay concealed among them. According to the belief of the Argives, he had plunged into the lake of Lerna. That signified, at the same time, his plunge into the underworld; for a sacrifice was thrown down to the "guardian of the door" of the underworld at the festival of Dionysus' return from below. The story is also told that Perseus vanquished the god and hurled him into the lake. An Orphic hymn says that he rested for two years in the sacred house of Persephone after his departure.
>
> And now the one who had disappeared was supposed to reappear suddenly with his tipsy look and his dazed smile, or he was supposed to burst forth out of the darkness in the form of a savage bull. (45)

It is not surprising that the reappearance of Dionysus is filled with images that go against reason and logic, confront ethics, and are filled with horror, terror, and debauchery, for he represents life escaping the boundaries of human comprehension, certainly the bounds of human *explanation*. Only in denying the seeming truths of existence, the logic of human existence can there be something more, something beyond mere meaningless physical existence. And Dionysus takes us to the edge between what can be known and what flirts with human comprehension, the source of life itself, the cutting edge between order and chaos, between meaning and that which is beyond meaning.

Walter Otto describes the scene:

> They were waiting for him—the choruses of women, true images of those higher beings, who followed Dionysus everywhere. In Elis it was the dancing chorus of "the sixteen women," who invoked the god with the words: "Come, Lord Dionysus, attended by the Graces, into the holy temple of Elis, rushing into the temple with your bull's hoof, venerable bull, venerable bull!" They knew, in short, that the one who would appear would be a wild creature who would bring, through his demonic violence, a breathtaking excitement. (46)

This enthralling entrance is filled with images of *transformation*. The elemental forces of life shatter the routines, the beliefs, the *explanations* that have caged the world. Truth is broken or reinvented as the *inexplicable* becomes the norm to be embraced. "The earth flows with milk, flows with wine, flows with the nectar of bees. And there is a vapor in the air as of Syrian frankincense." (47) The women "gird themselves with snakes and give such to fawns and wolf cubs as if they were infants at the breast. (48) "Fierce bulls fall to the ground, victims to numberless, tearing female hands (49) Sturdy trees are torn up by the roots. (50) Chains break apart, walls fall into ruins, locked doors swing open, future and past mix with the present, for Dionysus is the *liberator.*

The paradox, the logic of illogic, the truth that the birth of birth also births death is emphasized. The maenads, the women followers of Dionysus not only go into the forests, the world of nature outside of civilization, join with nature and suckle animals, but they also slaughter and reveal in the primitive bloody eating of the raw flesh of animals:

> . . . With one voice they cried aloud:
> "*O Iacchus! Son of Zeus!*" "*O bromius!*" they cried
> until the beasts and all the mountain seemed
> wild with divinity. And when they ran,
> everything ran with them.
> It happened, however,
> that Agave ran near the ambush where I lay
> concealed. Leaping up, I tried to seize her,
> but she gave a cry: "Hounds who run with me,
> men are hunting us down! Follow, follow me!
> Use your wands for weapons."
> At this we fled
> and barely missed being torn to pieces by the women.
> Unarmed, they swooped down upon the herds of cattle
> grazing there on the green of the meadow. And then
> you could have seen a single woman with bare hands
> tear a fat calf, still bellowing with fright,
> in two, while others clawed the heifers to pieces.
> There were ribs and cloven hooves scattered everywhere,
> and scraps smeared with blood hung from the fir trees.
> And bulls, their raging fury gathered in their horns,
> lowered their heads to charge, then fell, stumbling
> to the earth, pulled down by hordes of women
> and stripped of flesh and skin more quickly, sire,
> than you could blink your royal eyes. Then,
> carried up by their own speed, they flew like birds
> across the spreading fields along Asopus' stream
> where most of all the ground is good for harvesting.
> Like invaders they swooped on Hysiae
> and on Erythrae in the foothills of Cithaeron.
> Everything in sight they pillaged and destroyed.

> They snatched the children from their homes. And when
> they piled their plunder on their backs, it stayed in place,
> untied. Nothing, neither bronze nor iron,
> fell to the dark earth. Flames flickered
> in their curls and did not burn them. Then the villagers,
> furious at what the women did, took to arms.
> And *there*, sire, was something terrible to see.
> For the men's spears were pointed and sharp, and yet
> drew no blood, whereas the wands the women threw
> inflicted wounds. And then the men *ran*,
> routed by women! Some god, I say, was with them.
> The Bacchae then returned where they had started,
> by the springs the god had made, and washed their hands
> while the snakes licked away the drops of blood
> that dabbled their cheeks. (51)

The horrific ecstasy unites these women with nature; they mate with, make love to animals, certainly experience the *Eros*, the sexual form of love or desire, and they kill and eat these same animals. The reason Dionysus' followers are women seems obvious enough for it to be somewhat patronizing to go into, but it's best to make sure everything is clear. In the world of the body, the physical world, women are the source of life, the ones who give birth, who represent transformation. And, as this whole discussion began in the conflict between the male Zeus and female Gaia about who has the power of giving birth, here Dionysus represents the earth-based symbolism of Gaia.

But then comes the other side of transformation, the birth out of physical existence—death. The paradox is that the birth of life is also the birth of death. And this gets emphasized in various myths of Dionysus and his followers. Walter Otto provides the following example:

> According to the myth, they [the daughters of Minyas] were the only ones who did not wish to listen to the summons of Dionysus but remained at home, modest and diligent women, awaiting the return of their husbands. There suddenly the spirit of Dionysus came upon them with marvels and terrors, and in the madness which seized them they cast lots for their little boys. The story goes that they had developed a violent lust for human flesh. The lot fell on Leucippe's little son, who was then torn into pieces by the three. The gruesome savagery of these women, who had been previously so virtuous and motherly, . . . was the wildest eruption of the destructive madness which belongs to the reality of Dionysus as much as do the ecstasy and abandon which accompany him. This same savagery also re-occurs . . . in the cult sacrifices made to Dionysus. And as the counterpart of the maternal instinct, it repeats itself even among the beasts

> of the wilderness, whose young maenads not only suckle at their breasts but tear apart and devour—a representation of those mothers who pounce upon their own infants with hideous desire. (52)

The madness of a mother killing and eating her own child goes beyond any possible logical, reasonable view of human existence. It is a horrific display of the paradox of birth and death, and that is the point of it.

Dionysus is the god of two worlds, both physical and spiritual.This duality is the key, and it permeates everything about him. He is the nurturer, the god of rapture, the god praised for giving wine, which removes us from sorrow and leads us to ecstasy, the god who brings about birth, the god who heals, the god who delights, the god who brings joy, the god of the dance and of love and desire. But he is also the god of horror and savagery, the god who eats raw flesh, the god who delights in violence and bloodshed and human sacrifice, the monstrous god of darkness and most importantly the god of death. He even gets equated with Hades, the Greek god of the afterlife. Heraclites stated it, "Hades and Dionysus, for whom they [the Maenads] go mad and rage, are one and the same." (53) *Orphic Hymn 46* states that he grew up in the house of Persephone, queen of Hades, and *Hymn 53* that he sleeps there between his appearances in the world of the living. (54) Erwin Rohde claimed Dionysus was thought to be the "Lord of the Souls," (55) and Walter Otto claimed that the great festival of the Anthesteria, "the most important festival of the dead in Athens and among the Ionians . . ." probably included "the entrance of Dionysus in a ship followed by the mixing of the sacred wine and the subsequent drinking contest and, perhaps too, the marriage of the god with the Basilinna . . . a time when the dead came to visit the living and remained with them until a ceremonial pronouncement was made which signified to them that their time was up, and they had to take their leave." (56)

Thus, we have here a myth and ritual combining birth and death, and the tie to the marriage of the physical world to the spiritual world. Furthermore, the very women who were the followers of Dionysus, the givers of life, and the takers of life, were also persecuted. Hesychius wrote that the Agrionia in Argos were days of the dead. (57) Erwin Rhode claimed the same for the Boeotian Agrionia. (58) Walter Otto writes:

> When it is said that the Argive Agrionia were celebrated in honor of a daughter of Proetus, we must associate with this the myth in which the daughters of Proetus, who were driven mad by Dionysus for the same reasons as the daughters of Minyas, were hunted down, and the oldest of them, Iphinoe, died as a result of this persecution. Women were hunted

> down in this way in the ritual of the Boeotian Agrionia, as we know, and this was not without its element of tragedy either, for the women were threatened with the sword, and when one of them could not save herself, she was killed. What happened here in the cult, the myth of Lycurgus tells us. He fell upon the "nurses" of Dionysus with a deadly weapon and forced the god himself to seek refuge by leaping into the sea. Is not this feature repeated in the ritual of the Agrionia when in Chaeronea the women searched long for the god, who had escaped, and finally brought back the news that he had fled to the Muses and had concealed himself there.
>
> Thus, the Argive festival which the sources call a festival of the dead was clearly the same as the Boeotian festival which was intended for Dionysus. (59)

Dionysus, then, becomes the god of duality and paradox, the inexplicable god of the Oneness of the mystery of human existence, both its ecstasy and its horror, the violent coming together of birth and death, of logic and illogic, of meaning and that which is beyond meaning, of knowing and that which is beyond knowing, and perhaps even more frightening the knowing that there is something beyond knowing.

Indeed, the first reference to him in existence, that of Homer in *The Iliad,* (60) gives him an epitaph translated as *rapturous* or *mad*. (61)

He MUST be a *mad* god, because only in madness can meaning and value exist. If reason, logic, rationality, common sense, math and science are his source, then life is but a somewhat complex playing out of computer programs, nothing more than an elaborate predestined machine without the possibility of individual responsibility, and thus without either praise or blame, without either condemnation or salvation, merely the automatic robotic living out of the programs of some form of logic.

But humans *sense* something more.

Part IV

Another very important connection with this and all of Classical Greek thinking is the saying on the wall of the Delphi temple: *know thyself. Self-knowledge* is the key to human existence. With this in mind, it is also interesting to note that the *pythia* declared Socrates, who stressed *know thyself*, to be the wisest of all men.

It must be emphasized that this form of self-knowledge is attached not just to the forms of knowing of Apollo, the knowing of the bright light of the sun, of reason and logic and ethos, but also to the forms of knowing of Dionysus, for that state of ecstasy reached in the worship of Dionysus is precisely where one finds the *sublime!*

This is why Soren Kierkegaard referred directly to the *pythic* experience in his notes, the *divine madness* as explained by Socrates, the man declared the wisest of all men by the oracle at Delphi, in an attempt to clarify the human relationship to the spiritual, to God, and to *that which is beyond explanation*.

Indeed, Soren Kierkegaard's theory takes us to the *sublime* in human psychological terms, the ultimate faith that must be if there is to be a God at all, *expressed* in the story of Abraham commanded by his God to kill his long awaited son Isaac for no reason other than the command of God. An extreme challenge, a moment to face the *sublime* horror and find a way to give it meaning and value.

Here Soren Kierkegaard's theory takes us to where beauty and horror collide. A European dialogue on the *sublime* had begun in the early 1700s. Anthony Ashley Cooper and John Dennis used the term *sublime* in an attempt to clarify their feelings of being overwhelmed, positively affected by a feeling of fear in certain grandiose landscapes. Joseph Addison, in turn, attempted to synthesize and put a frame over these ideas in *The Spectator* and *Pleasures of the Imagination.* All three had made trips over the Alps and had been impressed by the *aesthetic* beauty of them in conjunction with feelings of terror or fear, what Addison described as an "agreeable kind of horror." Joseph Addison's writings, along with those of Mark Akenside, his main work also titled *Pleasures of the Imagination*, and Edward Young, in a work titled *Night Thoughts*, are believed to have sparked the important work by Edmund Burke, *A Philosophical Inquiry into the Origin of Our Ideas of the Sublime and Beautiful*, considered the first to argue that the *sublime* and the *beautiful* are mutually exclusive, not just opposite, but rather opposed, just as light and darkness. (62) *Beauty* might be made more or less beautiful by the accents of light, but intense light or dark is *sublime* to the extent it completely obliterates the sight of the object. In this sense, the imagination is moved to awe and horror by what is "dark, uncertain, and confused." This breaks the idea that *beauty* and *ugly* are simple opposites; *beauty* resulting in an *aesthetic* experience based on a realization of positive form, *ugly* the absence of this experience. In this sense, intense negative feelings resulting from an experience of the *ugly* produce similar feelings of pleasure to those resulting from an experience of *beauty.* In either case, both produce strong emotional reactions, and the pleasure comes from the intensity of the emotional experience. Thus, the dichotomy comes more from the idea of an emotional experience and the lack of one, emotions meant to be understood as more than mere physical reactions, yet the result of physical reactions. In both an *aesthetic* experience and a *sublime* experience the mind is stimulated to be engaged in its most dynamic activity.

Seven years later Immanuel Kant published *Observations on the Feeling of the Beautiful and Sublime.* Distinguishing between the courser, thoughtless emotions, what might be called instinctual emotions, and the higher feelings of the intellect, those involving sensitivity, intellectual excellence, talent and virtue, he put forth that these higher emotions are of two kinds, the *beautiful* and the *sublime,* those of the *beautiful* "joyous and smiling," those of the *sublime* also causing "enjoyment but with horror." He further divided the *sublime* into three categories, the *terrifying sublime* that which is accompanied with dread or melancholy, the *noble sublime* that which consists of a quiet wonder, and the *splendid sublime* that which is filled with *beauty*. He claimed, furthermore, that the *sublime* and the *beautiful* can be totally separate or intertwined, and gave tragedy as an example of an expression that arouses feelings of the *sublime*, as opposed to comedy, which arouses feelings of the *beautiful.* (63)

Arthur Schopenhauer, in *The World as Will and Representation*, published in 1819 and later republished with additions in 1844, the same year as Soren Kierkegaard's *Fear and Trembling*, put forth that the *sublime* is the pleasure in experiencing an overpowering force, a tremendous space or vast malignant object, one that could destroy the observer. (64)

Thus, at the time of Soren Kierkegaard, this philosophical discourse on the *sublime* was in full force, and his use of the term *sublime* was meant to be understood in context of it. For him, the *sublime* is beyond all *ethical*, all *aesthetic,* all *sociological* judgment and comprehension. And in depicting a leap of faith as an embracement of the absurd he was presenting his theories in the context of an experience of the *sublime*.

Under the pseudonym of Johannes de Silentio, Soren Kierkegaard gave four variations on this story of faith from the Bible, four attempts to understand what went through Abraham's mind, a poetic format for exploring the conflict between ethos and faith, between logic and that which is beyond logic, between physical existence and spiritual existence. And a key here is to focus on, not the rejection of reason and logic and ethos, but the struggle between two different forms of knowing, both valid both necessary. Human existence takes shape, not just in the physical world of logic or just in the spiritual world of faith, but in the "paradoxical" union of the two, the combination of the brain and mind

In a startling coincidence, Soren Kierkegaard was writing this famous formation of what was to become the beginning of Existentialism at the same time Richard Dadd was living it out.

Furthermore, it is clear Dadd both believed in his connections to Osiris (both heard the voice of his god just as Abraham heard the voice of his god) and yet also realized his actions were ethically wrong. In fact, he even commented on the seeming necessity of being mad for creative genius, stating in penciled marginalia to *Lectures on Painting and Design*: "I am of the opinion that there is a great deal of secret in the matter that it is explained by one's own second self which is perhaps as obstinate and vicious a devil as we could desire to oppose or thwart one and that few can overcome it, . . . A strong genius is most likely antagonized by a strong beast or devil a secret worth knowing." (65)

Soren Kierkegaard felt the same way about his own struggles, stating in his *Journal*: "there never existed great genius without some madness: *Nullum exstitit magnum ingenium sine aliqua dementia*. This is the worldly expression for the religious affirmation that him whom God blesses He *eo ipso* curses in a worldly sense. Thus it must be. The first is due to the limitations of nature, the second to its duplicity." (66)

Part V

Now we enter the realm of the *shadow* and the connection of the *shadow* with the *trickster*.

The dove returns: it found no resting place;
It was in flight all night above the shaken seas;
Beneath dark eaves
The dove shall magnify the tiger's bed;
Give the dove peace.
The split-tailed swallow leaves the sill at dawn;
At dusk, blue swallows shall return.
On the third day the crow shall fly,
The crow, the crow, the spider-colored crow,
The crow shall find new mud to walk upon.

—Robert Bly (67)

Taking a Babylonian version of the story of Noah and the Great Flood, of the sending out of a bird to find the newly reborn land, the new world, a version where three different birds are sent out on successive days, Bly clearly expresses the value of the *shadow* in the image of a black bird, an intelligent black bird, a bird of the night, connected to both the spider and mud.

The dove is the bird of innocence--pure, spiritually clean and untouched by darkness, the bird of peace and love, of the world of Adam and Eve before the fall. It is not the bird of transformation and energy, not the bird of dark drives and desires, not the bird of creativity and imagination.

The swallow is not as pure as the dove, but neither is it a bird of the night. It is blue, and blue is the color of spirituality and of the intellect, of the sky, but not of the *blood* and the *libido*.

The crow is the bird that will find the new mud to reshape the world, the wet clay necessary for the creation and evolution of life.

The raven is a large black bird of the crow family, the one chosen by Edgar Allan Poe to represent the dark, mental sufferings of a lover for his lost love, the exact image of the *shadow* of the human psyche working its mysterious, uncompromising mantra on a man unable to resist its temptations, even more a man desiring its call to self-destruction. (68)

Poe begins by setting the scene, one of a man caught in that twilight state between two worlds, the conscious and the unconscious, the world of the dream about to descend. It is no accident that the connection takes place at midnight, the moment when supernatural entities emerge to replace the natural spirits of the sun:

> Once upon a midnight dreary, while I pondered, weak and weary,
> Over many a quaint and curious volume of forgotten lore—
> While I nodded, nearly napping, suddenly there came a tapping,
> As if some one gently rapping, rapping at my chamber door.
> "Tis some visitor," I muttered, "tapping at my chamber door—
> Only this and nothing more."

This is not just a physical door, not just a literal chamber. It is all a wonderfully symbolic entrance to the psyche, a "weak and weary" psyche, one filled with "many a quaint and curious volume of forgotten lore."

The narration continues:

> Ah, distinctly I remember it was in the bleak December,
> And each separate dying ember wrought its ghost upon the floor.
> Eagerly I wished the morrow;—vainly I had sought to borrow
> From my books surcease of sorrow—sorrow for the lost Lenore—
> For the rare and radiant maiden whom the angels name Lenore—
> Nameless here forevermore.

Here the dark images and symbols continue, "bleak December," "dying ember," "ghost," "sorrow," 'lost." Of course, it takes place in "December," the last month of the year, the month when the nights become the longest, just *before* the winter solstice (the rebirth). Poe uses the name Lenore to represent a lost love in several poems, including one titled *Lenore*, and she does not refer to any specific person, simply to a *lost* love that the narrator here both wants to forget and yet to remember, the combination of agony and ecstasy, a form of amore, in this case one that cannot be satisfied, but that can only torment, and this lover seems to ravel in his own self-perpetuating anguish and destruction. Joseph Campbell stresses the sense of loss (of time always moving forward, never to be reclaimed) in his discussion of the necessary *sorrow* of life (the first rule of Buddhism).

The narration continues:

> And the silken sad uncertain rustling of each purple curtain
> Thrilled me—filled me with fantastic terrors never felt before;
> So that now, to still the beating of my heart, I stood repeating:
> "Tis some visitor entreating entrance at my chamber door—
> Some late visitor entreating entrance at my chamber door;
> This it is and nothing more."

Notice how he is both "thrilled" and "filled . . . with fantastic terrors," heart "beating" with anticipation. Also, keep in mind that the physical space, the room he is in is his own psyche, and the enticing "silken" fabric (that precious, royal material) of each of his "purple" (that noble, redish color, neither compleley "red," as in the body's blood, nor blue, as in the world of the "spirit" or "soul) "curtains" (what cover the windows, the eyes, the entrances to a person's true self or soul) "rustle" in a "sad," "uncertain" manner. What, after all, is causing this movement? Something from outside of the psyche, from that unknown "other" world is touching him, is making contact. This is huge, without bothering to elaborate on it's assumptions. It suggests that, indeed, there exists some kind of life beyond the psyche.

This is worth some elaboration. Whatever artistic works we deal with, a common assumption they put in place is the existence of a world beyond the self, beyond mere physical existence, beyond the mind of the individual. Indeed, if we go into the entire discipline of philosophy, modern philosophy is based on René Descartes' famous *cogito ergo sum,* originally stated in French, *je pense, donc je suis (Discours de la Méthode,* 1637):

> *Ainsi, à cause que nos sens nous trompent quelquefois, je voulus supposer qu'il n'y avoit aucune chose qui fût telle qu'ils nous la font imaginer; et parce qu'il y a des hommes qui se méprennent en raisonnant, même touchant les plus simples matières de géométrie, et y font des paralogismes, jugeant que j'étois sujet à faillir autant qu'aucun autre, je rejetai comme fausses toutes les raisons que j'avois prises auparavant pour démonstrations; et enfin, considérant que toutes les mêmes pensées*

> *que nous avons étant éveillés nous peuvent aussi venir quand nous dormons, sans qu'il y en ait aucune pour lors qui soit vraie, je me résolus de feindre que toutes les choses qui m'étoient jamais entrées en l'esprit n'étoient non plus vraies que les illusions de mes songes. Mais aussitôt après je pris garde que, pendant que je voulois ainsi penser que tout étoit faux, il falloit nécessairement que moi qui le pensois fusse quelque chose: et remarquant que cette vérité, **je pense, donc je suis** [italics in original], étoit si ferme et si assurée, que toutes les plus extravagantes suppositions des sceptiques n'étoient pas capables de l'ébranler, je jugeai que je pouvois la recevoir sans scrupule pour le premier principe de la philosophie que je cherchois.*

Though not exactly correct, in popular useage, it translates into English as "I think therefore I am," and the French passage can be translated as follows:

> Accordingly, seeing that our senses sometimes deceive us, I was willing to suppose that there existed nothing really such as they presented to us; and because some men err in reasoning, and fall into paralogisms, even on the simplest matters of geometry, I, convinced that I was as open to error as any other, rejected as false all the reasonings I had hitherto taken for demonstrations; and finally, when I considered that the very same thoughts (presentations) which we experience when awake may also be experienced when we are asleep, while there is at that time not one of them true, I supposed that all the objects (presentations) that had ever entered into my mind when awake, had in them no more truth than the illusions of my dreams. But immediately upon this I observed that, whilst I thus wished to think that all was false, it was absolutely necessary that I, who thus thought, should be somewhat; and as I observed that this truth, ***I think, therefore I am***, was so certain and of such evidence that no ground of doubt, however extravagant, could be alleged by the sceptics capable of shaking it, I concluded that I might, without scruple, accept it as the first principle of the philosophy of which I was in search. (*Discourse on the Method of Rightly Conducting the Reason, and Seeking Truth in the Sciences*).

If we stop for a moment and consider the enormity of what is being expressed by Poe in this context, it is much easier to understand why the narrator of this poem is both "thrilled" and "filled . . . "with fantastic terrors never felt before." There are endless examples of this central aspect of human existence, everything from a recent song by Pink Floyd, titled appropriately "Is There Anybody Out There?," to T.S. Eliot's *The Love Song of J. Alfred Prufrock,* which begins with a quote from Dante's *Inferno* (XXVII, 61-66):

S'io credesse che mia risposta fosse
A persona che mai tornasse al mondo,
Questa fiamma staria sense piu scosse.
Ma perciocche giammai di questo fondo
Non torno vivo alcun, s'I'odo il vero,
Senza tema d'infamia ti rispondo.

[If I but thought that my response were made to one perhaps returning to the world, this tongue of flame would cease to flicker. But since, up from these depths, no one has yet returned alive, if what I hear is true, I answer without fear of being shamed.]

In *The Inferno*, the epigraph refers to a meeting between Dante and Guido da Montefeltro, condemned to the eighth circle of Hell for providing counsel to Pope Boniface VIII, who wished to use Guido's advice for evil purpose. It is possible to see this as an ironic perspective on Prufrock's intent. Like Guido, Prufrock never intended his story to be told, and so by quoting Guido, Eliot expresses his view of Prufrock's love song (a confession that must never be revealed to a living soul). But it can also be seen in less convoluted ways as simply the fear of being misunderstood, as a comment on the impossibility of tru human communication and the fear of rejection in the most important sphere, opening up one's self to a beloved. In other words, if one opens the "window" to the "self," especially in that deepest emotion, love, one is voluneralbe and can be badly damaged (die on various levels).

Pink Floyd's song comes from their album *The Wall*, 1979 (subsequent movie by the same name, 1982, and endless reinventions as live performances), one of the more interesting "concept" albums that began appearing in rock-and-roll music at this time, this one expressing numerous intertwining themes and realities but certainly centering on the idea of isolation, alienation and a self-imposed reclusion (yet another expression of the fear of that "other" world outside the psyche).

Poe continues:

Presently my soul grew stronger; hesitating then no longer,
"Sir," said I, "or Madam, truly your forgiveness I implore;
But the fact is I was napping, and so gently you came rapping,
And so faintly you came tapping, tapping at my chamber door,

That I scarce was sure I heard you"—here I opened wide the door;—
Darkness there and nothing more.

How to face the unknown, that dark world, not just in terms of sight, but of human experience and knowing, perhaps beyond human knowing. Should we just keep the "window" closed, the "curtains" drawn? Do our senses deceive us? Is there really something, some kind of energy, outside? Is that tapping real? Or is it the result of an overactive imagination? If so, does that mean there is a desire involved? Do humans *want* something outside of themselves? Perhaps they pretend it exists, find ways to fool themselves to satisfy a desire, a need.

And why the apology? How long has that "sir" or "madam" been trying to catch Poe's attention? Something trying to catch the attention of a human "soul," and the soul was napping (literally or symbolically?), because of some kind of defense mechanism or for some other reason--a need to develop, to learn, to grow, to evolve, to be able to wake up to some "other" reality beyond low level, instinctual, physical, animal existence?

It is "faint," a "gentle" tapping, perhaps that slow gaining of knowledge or of a necessary awakening into a world beyond mere conscious, daylight existence, that nagging thought that one can't quite bring to consciousness but that won't go away.

After a time of not noticing it (perhaps at first an automatic, psychological dismissal of it, an unconscious rejection, a "natural" defense system that filters out the "static," the unimportant messages so that the important ones get the attentention they deserve), he invites the *dream* or *nightmare* in, opens the door to his psyche, his inner chamber. And what does he find on the other side? "Darkness there and nothing more." Now what? Is "darkness," is "nothing" something? What of it? How does one understand the cosmos, the endless universe? If space ends, then still there must be something--nothing must be something beyond the end of something, beyond the limit, whether it take the form of some kind of wall (transparent or opaque), a cliff one might step off and "float" into without gravity or electric or magnetic forces to shape it, or a shoreline where one can watch the slowly dissipating particles or energy forces fading into nothingness, as if standing on the edge of an endless, black ocean.

What of time? Before the Big Bang that began the universe—what existed before the beginning? Time cannot simply not exist, can it?

Certainly there is a sense in which time is a human perception, a human construct, and yet *certainly* humans sense a continuum *in* time, a past and a future that have and will take place no matter what humans do (and lengthy discussions can and have expanded on how this knowledge is a uniquely human quality—other animals and living entities cannot mentally travel the maps of time beyond "short term memory," and though they can, indeed, know they must do certain things in certain "times" of the year or at certain times in life, it is also intuitively known (and has been proven to the extent anything can be proven) by humans that these are not of the same world of time that humans know (just as animals cannot use language or think in a symbolic way as humans do, and even though they might be taught a limited vocabulary or have some kind of pattern of communication, as bees do to signal finding honey, they cannot step outside of the pattern to develop it and use it for meanings beyond the literal—similar to computers "thinking" faster and with less errors than humans, perhaps composing a thousand symphonies in an hour, yet having no idea which ones are meaningful and which are not (though, today, as neurology and computer programming come together in various disciplines of bio-technology, the human brain/mind is changing traditional assumptions, changing the fundamental truths in ways most humans are not currently ready to accept, but, if they live many more decades will experience, and as the human condition changes, the human relation to time will also change—in one basic way, if humans learn how to live, to have self-awareness connected in some way to the physical world, noticeably longer, say, 200 years, time's insistent drive toward death will be less ominous, less demanding).

But here we have slide into talking about time in its chronological sense. There are other ways of understanding time (and space). There are those who use the terms *infinity* and *eternity* to mean something other than endless time and space, to mean all time and space at the same time and in the same space. Joseph Campbell discusses this, and, for the purposes of his discussion distinguishes *eternity* (all time at the same time) from *forever* (the linear path through time).

Such ever more complex attempts to understand the world are valuable, give humans a larger, richer existence; but nevertheless, to play word games doesn't deny the ideas, the concepts, the realizations that literal space and time both extend beyond human comprehension.

Back to *The Raven*. Intuitively, poetically, Poe has opened the door between the "I" and the "other," and yet all he can know about the other is "darkness and nothing more."

But there is this. There is the dream. The dream can and perhaps must fill the void. Is it true? What he says--has he really dreamed the dream no mortal "has ever dared to dream before"? Well, he believes he has. And, though it isn't much, he does get a clue, a response from the emptiness:

> Deep into that darkness peering, long I stood there wondering, fearing,
> Doubting, dreaming dreams no mortals ever dared to dream before;
> But the silence was unbroken, and the stillness gave no token,
> And the only word there spoken was the whispered word, "Lenore!"

This I whispered, and an echo murmured back the word, "Lenore!"—
Merely this and nothing more.

The echo response highlights that he is talking to himself. It is his own *projection*, in Jungian terms the placing upon another one's own psychic image or complexes. Is the world outside of each human psyche simply a projection of the human psyche? Is that ultimately all there is?

What about a God?

Has the door in this poem opened to the extensive theological and philosophical discussions of how humans are connected to or separate from their gods? And, just to make this opened door more honest, does it have to include the great mythological maps, the spiritual writings?

Is it possible to claim humans are not separate from God, but in one way or another connect to Him (or Her or Whatever)? The Christian Bible certainly offers this possibility, even more, states it. God does, after all, breathe life (His own essence, His own spirit) into the first man, then, depending on which version of the story, also into woman or creates woman out of man's rib and thus is already in her as well. Certainly Christ is meant to be the child of both a human and God, a comingling of the two, the connecting link of physical and spiritual, visible and invisible worlds. He is born of a virgin, the birth of spirit out of or into flesh.

Needless to say, interpretations of all this are wide-ranging, and though often interesting and valuable, beyond the current discussion. And then this expands to all of those other scriptural writings and what they are expressing, how they are attempting to grasp this interphase between the sensual or knowable "I" and that which is beyond it, the "other." Within that discussion is the possibility that God is a projection, a human attempt, conscious or unconscious, to understand the unknown.

The Mayan culture thought echoes were an attribute of the great Underworld god, the Jaguar, which fits here, as the only response by Lenore would come from the afterlife, the underworld, the otherworld, the unconscious, the *numinous*. In Greek mythology, Echo is a nymph scorned by Narcissus (whose name is derived from the Greek *narke*, numbness), who drowned in a pool while admiring his own beauty, thus representing vanity, self love, in Jungian terms an *ego* unable to connect with the *self* or with another, and water symbolizes the unconscious, so it is an *ego* drowning in its own *unconscious*, certainly appropriate to what's going on in this poem, and with the *shadow* in the form of a raven about to enter into this dream, it becomes the *ego* unable to resist or gain some kind of balance with its *shadow*.

Lenore, the flower the name comes from was used in initiation ceremonies associated with Demeter, Mother Earth, the Greek goddess of the seasons, the stages of life and death, and the relation of numbness to death resulted in narcissus flowers being planted on graves, not necessarily to symbolize death but to symbolize a death which is, perhaps, nothing more than a sleep, an entrance into the world of the dream. Demeter's daughter Persephone was drugged by the narcissus flower and carried off to the Underworld by Hades. All these connections include the possibility of rebirth. Winter does give way to spring. Persephone does return from Hades. It is possible to wake from a dream, though, as the poem will make clear, that will not happen here.

This is a man lost forever in his own nightmare.

The narration continues:

> Back into the chamber turning, all my soul within me burning,
> Soon again I heard a tapping, something louder than before.
> "Surely," said I, "surely that is something at my window lattice;
> Let me see, then, what thereat is, and this mystery explore—
> Let my heart be still a moment, and this mystery explore;—
> 'Tis the wind and nothing more."

"Burning" brings in the whole symbolism of "fire," that powerful force of change, of transformation, and its connection specifically to the "soul" emphasizes the context. According to Hesiod's *Theogony* (lines 507–616), fire is the gift Promethius steals from the gods and gives to humans (similar to Satan's "gift" of the forbidden fruit from the tree of the knowledge of good and evil, both representing a form of *enlightenment*, the gaining of a mind similar to that of the gods), and Zeus is furious when he finds out about it, punishing Promethius by chaining him to a rock on the Caucasus, Kazbek Mountain, where his liver (the soul or spirit) is eaten daily by an eagle, only to be regenerated by night (he is, after all, a god, immortal). As usual, the loss of the soul takes place during the bright light of the day (conscious), and the forces of night (unconscious) bring it back.

In the *Works and Days* (lines 42–105), Hesoid elaborates on the story, not only having Zeus withhold fire from humanity but "the means of life" as well, stating that, if it were not for Prometheus, "You would easily do work enough in a day to supply you for a full year even without working; soon would you put away your rudder over the smoke, and the fields worked by ox and sturdy mule would run to waste." (lines 44-47)

Same thing God does in "Genesus":

1 Now the serpent was more subtle than any beast of the field
which the Lord God had made. And he said unto the woman,
Yea, hath God said, Ye shall not eat of every tree of the
garden?
2 And the woman said unto the serpent, We may eat of the fruit
of the trees of the garden:
3 But of the fruit of the tree which is in the midst of the garden,
God hath said, Ye shall not eat of it, neither shall ye touch it,
lest ye die.
4 And the serpent said unto the woman, Ye shall not surely die:
5 For God doth know that in the day ye eat thereof, then your
eyes shall be opened, and ye shall be as gods, knowing good
and evil.
6 And when the woman saw that the tree was good for food,
and that it was pleasant to the eyes, and a tree to be desired to
make one wise, she took of the fruit thereof, and did eat, and
gave also unto her husband with her; and he did eat.
7 And the eyes of them both were opened, and they knew that
they were naked; and they sewed fig leaves together, and made
themselves aprons.
8 And they heard the voice of the Lord God walking in the
garden in the cool of the day: and Adam and his wife hid
themselves from the presence of the Lord God amongst the
trees of the garden.
9 And the Lord God called unto Adam, and said unto him,
Where art thou?
10 And he said, I heard thy voice in the garden, and I was
afraid, because I was naked; and I hid myself.
11 And he said, Who told thee that thou wast naked? Hast thou
eaten of the tree, whereof I commanded thee that thou
shouldest not eat?
12 And the man said, The woman whom thou gavest to be with
me, she gave me of the tree, and I did eat.
13 And the Lord God said unto the woman, What is this that
thou hast done? And the woman said, The serpent beguiled me,
and I did eat.
14 And the Lord God said unto the serpent, Because thou hast
done this, thou art cursed above all cattle, and above every
beast of the field; upon thy belly shalt thou go, and dust shalt
thou eat all the days of thy life:
15 And I will put enmity between thee and the woman, and
between thy seed and her seed; it shall bruise thy head, and
thou shalt bruise his heel.
16 Unto the woman he said, I will greatly multiply thy sorrow
and thy conception; in sorrow thou shalt bring forth children;

> and thy desire shall be to thy husband, and he shall rule over
> thee.
> 17 And unto Adam he said, Because thou hast hearkened unto
> the voice of thy wife, and hast eaten of the tree, of which I
> commanded thee, saying, Thou shalt not eat of it: cursed is the
> ground for thy sake; in sorrow shalt thou eat of it all the days
> of thy life;
> 18 Thorns also and thistles shall it bring forth to thee; and thou
> shalt eat the herb of the field;
> 19 In the sweat of thy face shalt thou eat bread, till thou return
> unto the ground; for out of it wast thou taken: for dust thou art,
> and unto dust shalt thou return.
> 20 And Adam called his wife's name Eve; because she was the
> mother of all living.
> 21 Unto Adam also and to his wife did the Lord God make
> coats of skins, and clothed them.
> 22 And the Lord God said, Behold, the man is become as one of
> us, to know good and evil: and now, lest he put forth his hand,
> and take also of the tree of life, and eat, and live for ever:
> 23 Therefore the Lord God sent him forth from the garden of
> Eden, to till the ground from whence he was taken.
> 24 So he drove out the man; and he placed at the east of the
> garden of Eden Cherubims, and a flaming sword which turned
> every way, to keep the way of the tree of life. (*King James
> Bible*)

Hesiod also expands upon *Theogony*'s story of the first woman, now explicitly called Pandora (*all gifts*). After Prometheus' theft of fire, Zeus sent Pandora in retaliation. Despite Prometheus' warning, Epimetheus accepted this "gift" from the gods. Pandora carried a jar with her, from which were released "evils, harsh pain and troublesome diseases which give men death." (lines 91-92) Pandora shut the lid of the jar too late to contain all the evil plights that escaped, but foresight remained in the jar, giving humanity hope. (In both versions, woman is depicted as largely evil, which is not surpriseing, since woman is the human entity that gives birth to the physical world, the world of sensual experience and death (she is a creator, a human creator, and as such takes on the role of the gods). Also, every creation must include destruction. Creation is change, and change means something that once was no longer is. With birth comes death; and, as the Garden of Eden story stresses, with knowledge comes death, comes explusion from innocence and that childlike world without end (without death).

A window is a place both of separation and of transformation, as it divides two separate worlds, but one can see through it, see from one world to the other world, and ultimately could pass through it from one into the other (opposite of a mirror, which reflects back; though, of course a mirror has many facits, for its reflection is not a true one. If I stand in front of a mirror with a book in one hand, then turn around to try and duplicate the mirror image, the book ends up being in the wrong hand. Furthermore, a mirror is also a kind of entrance to that "other," not quite the same or faithfully opposite world, but a world that is not of the physical world, an enchanted world where things don't work by the same rules as the world it reflects, and as such it offers a passage out of the "normal," literal world into one that works more in the manner of a dream, filled with images that both are and are not real at the same time, in some ways similar to a shadow, connected to a person and moving in conjunction to a person's movements and yet not quite identical to that person, not really a person.

Interestingly, the window in *The Raven* has lattice, suggesting prison bars, perhaps a feeling of being caged in or perhaps more likely a protection from what is on the other side, which as we will immediately see, the narrator cannot resist.

The narration continues:

> Open here I flung the shutter, when, with many a flirt and flutter,
> In there stepped a stately Raven of the saintly days of yore.
> Not the least obeisance made he; not a minute stopped or stayed he,
> But, with mien of lord or lady, perched above my chamber door—
> Perched upon a bust of Pallas just above my chamber door—
> Perched, and sat, and nothing more.

Now the shutter is flung open, the world of the dream allowed in, and the life force of the "other," the dream world, the Raven appears--a powerful, stately, saintly Raven, with the dignity of a lord or lady, perched upon the head of Athena, a Raven from the days of *yore*, the days of the great Classical Greek gods and goddesses, a Raven from the birth of European consciousness, the world, as stated earlier, the narrator has been seeking in volumes of forgotten lore. It is an excellent choice to represent that energy, that spark of life from the dark world, a *bird* (one of the creatures of the sky (the unconscious and spiritual realm), a *black* bird, black being the opposite of white (the opposite of innocence, of purity, of the emptiness of life before the "fall"). No, this is a bird that knows the darkness, knows the ways of the world, is well acquainted with death.

And it is a "stately" Raven, a dignified, majestic, ceremonious, courtly, imposing, impressive, solemn, awe-inspiring, regal, elegant, grand, glorious, splendid, magnificent, resplendent Raven. Yes, this is that messanger of knowledge and death that carries a certain aura, a chrisma, a sense of power, a confidence. This is a bird that one will follow, will trust, will serve.

And Poe wants to emphasize all this, so he connects his magnificent Raven to those "*saintly* days of yore." Of course, a "saint" is special, more than special, is exceptional—angelic, pious, upstanding, beatific, blameless, blessed, devout, divine, god-fearing, godly, holy, pure, religious, sinless, virtuous, worthy. The original Christian meaning referred to any believer who is "in Christ" and in whom Christ dwells, whether in heaven or in earth. In Orthodox and Catholic teachings, all Christians in Heaven are considered to be saints, but some are considered to be worthy of higher honor, emulation, or veneration--officially recognized through canonization or glorification, a complex process that includes determining if the person has performed at least two miracles (described as wonders performed by supernatural power as signs of some special mission or gift and explicitly ascribed to God—extraordinary happenings that cannot be explained by the laws of the natural world).

This Raven, then, is a creature, a living entity from the "saintly days of yore," a time before the world of logic, reason, math and science emerged to take away the sense of the miraculous. Oh yes!, this Raven is a powerful bird from a powerful "other" world, the mythical once-upon-a-time world where miracles could happen.

This grand Raven, then, this black bird of death, has the "mien of lord or lady," now an emphasis on its aristocratic visage, a quality of superiority (and anyone who has studied the storied history of European royalty knows how hard they worked to distinguish themselves as clearly superior (rememver, "the land and the kind are one, and if the king is suffering spiritually, the land becomes a spiritual and by extension literal wasteland" as well).

Furthermore, this noble Raven perches on the bust of "Pallas," almost certainly referring to Athena (though there are many characters with that name, especially in Greek mythology), and, if so, then all of her attributes become representative of the bird, and she is *impressive*, beginning with her beginning, being born out of Zeus' head—hard to be more blunt about her obvious intellectual abilities—born out of the head of the most powerful of all the gods; talk about bloodlines). In addition to her intelligence (often equated with wisdom), she is also the goddess of courage, inspiration, civilization, law and justice, "just" or honorable warfare (mainly referring to military strategy), mathematics, strength, the arts, crafts, and skill; and as such an impressive goddess, she plays major roles in the major Greek Epics, serving as a guide to Odysseus in the most complex, complete hero's adventure to come out of Greece; oh, yes, there is the *Iliad*, the earlier, companion Epic, where she is the main goddess supporting the Greeks in their war with the Torjans (by the way, on one level this war is a conflict between her and Posiden, both claiming to have created the horse—which is why we have the "Trogen Horse" used to trick them). It is not surprising that she is the patroness of Athens (who built the beautiful Parthenon on the equally beautiful Acropolis in her honor), the center of the entire Classical Greek World.

It's also interesting that she is a virgin and one born out of a male's head instead of the normal way. This, then, is the bust that serves as the Raven's perch (his foundation, his podium—again fodder for a lengthy discussion, especially in relation to the previous ideas about a woman's symbolism; in other words, all that the Raven represents calls to the male's psyche, at first faintly heard, but then louder and more insistend, until he opens his psyche to it and allows it in, and it immediately perches on what a powerful female, born out of a male's head, represents (in Jungian terms, the man's *anima*). Here again is the dance of opposites that are somehow intertwined, two aspects of the total, the dance of moonlight (*anima*) and *shadow* or/and *animus* on the rippling waters of the *eternal* (in both senses of that word) ocean within the male psyche).

The narration continues:

> Then this ebony bird beguiling my sad fancy into smiling,
> By the grave and stern decorum of the countenance it wore,
> "Though thy crest be shorn and shaven, thou," I said, "art sure no craven,
> Ghastly grim and ancient Raven wandering from the Nightly shore—
> Tell me what thy lordly name is on the Night's Plutonian shore!"
> Quoth the Raven, "Nevermore."

Perhaps the one word response from the Raven is a real sound, the sound a Raven does make; perhaps it is the result of the deranged man's determination to hear "Nevermore," to find a negative word in the wordless sound. If it really is a universal voice and judgment of the universal black bird, then it might well be a form of Jung's archetypal *shadow*. If it is but the man's own dark imaginings, then it is his personal *shadow*. It matters not, as either carries the other within it.

Perhaps "Nevermore" is a direct reference to suicide. If the Raven is death come to call, then its insistent demand might well be seen as the "beguiling" messenger of the Grim Reaper come to claim another life (remember, if it is related to the female, it is the woman who brings physical existence into being, and birth requires death). Notice the bird "beguils" him into "smiling," brings a relief to his "sad fancy," and it is no accident the next line has the word "grave" in it; the Raven wears a "grave and stern . . . "countenance," in Jung's terms, the *persona*, the mask or face of death. And again, the arrival of this noble black bird from the "saintly . . . days of yore," the once-upon-a-time otherworld pleases him, brings a "smile" to his "sad fancy."

It is no accident the man thinks the Raven is from "Night's Plutonian shore," since Pluto is the Roman god of the underworld, and thus, yet again, it is emphasized that the Raven is seen as a messenger of that world, a place of shadowy existence, of afterlife, not so much of death-as-an-end, but of death as transformation, of continued existence in a world of either suffering or, perhaps, of bliss. In Greek terms, Hades (both the realm and name of its ruler, whom the Romans simply renamed Pluto) is neither Hell nor Heaven, or, perhaps, is both, as it holds many terrors and punishments, but it also has Elysium (the Elysiun Fields) which are not hellish, but rather are a very pleasant place to go and are reserved for the exceptionally good people.

Just exactly how this "afterlife" world works changes through time and depending on location, but it is a place humans go after death, and certain writings about it are similar to Christian depictions, especially Catholic attempts to describe what they call Purgatory, which according to Catholic Church doctrine is an intermediary state after physical death in which those destined for it "undergo purification, so as to achieve the holiness necessary to enter the joy of heaven" Only those who die in the state of grace but have not in life reached a sufficient level of holiness can be in Purgatory, and therefore no one in Purgatory will remain forever in that state or go to Hell. By the time of Poe's writing, the meaning of Purgatory had slide more toward the negative, slide away from a state between Earthly existence and Heaven into a state between Earthly existence and Hell, along the lines of a state of suffering, but not as horrific as Hell and one that will end.

If the Raven is calling him to the more severe afterlife of a Christian Hell, and suicide would qualify one for that, then Poe can be envisioning the eternal (or forever) punishments in the Satanic fire (and *fire* comes full circle here, as now fire is not only the gift of Prometheus as a symbolic enlightenment, a gift of knowledge, so powerful a gift that even the gods fear humans reciving it, but it also becomes the literal oximornic punishment that accompanies the gift of knowledge. It is that ultimate paradox that a black bird, a raven represents—death and knowledge are inexorably intertwined).

John Boorman's *Excalibur*, a movie carefully following the deeper, symbolic meanings of the Arthurian Romance, has a wonderful scene in the "coils of the dragon" (a dragon is a form of serpent or snake—remember, the serpent in Eden entices humans to eat the forbidden fruit that brings opposites into human existence) where Merlin explains this to Morgana (notice the careful pairing of *opposites*):

> Merlin: Bat wing, snake skin, is this all you've learned, Morgana? To deal in potions and petty evil.
> Morgana: And where have your meddling arts brought the world? To the edge of ruin.
> Merlin: I'm worn thin and threadbare. I've tried to guide men, or meddled in their affairs . . . as you would have it, for far too long. The time has come for me to go.
> Morgana: Oh . . . Forgive me. Where are you going?
> Merlin: Whence I came.
> Morgana: Take me with you. The Charm of Making--you swore to pass it to me. [the spell of transformation]
> Merlin: You think you're ready? Are you worthy?
> Morgana: I am. . . . I am.
> Merlin: Then follow me.
> Morgana: What is this place?
> Merlin: Here, you enter the coils of the Dragon. Here, my power was born. Here, all things are possible, and all things meet their opposites.
> Morgana: The future?
> Merlin: And the past.
> Morgana: Desire?
> Merlin: And regret.
> Morgana: Knowledge?
> Merlin: And oblivion.
> Morgana: Love?
> Merlin: Oh, yes.

Knowledge and *oblivion*--this *is* the fall from innocence, the realization of *self*, *self*-knowledge, the separation of *self* from *other*. It is the exit from the Garden of Eden, the Garden of Eternal Bliss (and ignorance of self).

This is central to Christianity: It is what the serpent promises Eve, saying: "For God doth know that in the day ye eat thereof, then your eyes shall be opened, and ye shall be as gods, knowing god and evil." (*King James Version*, Genesis 3:5). Here is the entire Chapter:

Now the serpent was more subtil than any beast of the field which the
LORD God had made. And he said unto the woman, Yea, hath God said,
Ye shall not eat of every tree of the garden?
2 And the woman said unto the serpent, We may eat of the fruit of the trees
of the garden:
3 But of the fruit of the tree which *is* in the midst of the garden, God hath
said, Ye shall not eat of it, neither shall ye touch it, lest ye die.
4 And the serpent said unto the woman, Ye shall not surely die:
5 For God doth know that in the day ye eat thereof, then your eyes shall be
opened, and ye shall be as gods, knowing good and evil.
6 And when the woman saw that the tree *was* good for food, and that it *was*
pleasant to the eyes, and a tree to be desired to make *one* wise, she took of
the fruit thereof, and did eat, and gave also unto her husband with her; and
he did eat.
7 And the eyes of them both were opened, and they knew that they *were*
naked; and they sewed fig leaves together, and made themselves aprons.
8 And they heard the voice of the LORD God walking in the garden in the
cool of the day: and Adam and his wife hid themselves from the presence
of the LORD God amongst the trees of the garden.
9 And the LORD God called unto Adam, and said unto him, Where *art*
thou?
10 And he said, I heard thy voice in the garden, and I was afraid, because I
was naked; and I hid myself.
11 And he said, Who told thee that thou *wast* naked? Hast thou eaten of the
tree, whereof I commanded thee that thou shouldest not eat?
12 And the man said, The woman whom thou gavest *to be* with me, she
gave me of the tree, and I did eat.
13 And the LORD God said unto the woman, What *is* this *that* thou hast
done? And the woman said, The serpent beguiled me, and I did eat.
14 And the LORD God said unto the serpent, Because thou hast done this,
thou *art* cursed above all cattle, and above every beast of the field; upon
thy belly shalt thou go, and dust shalt thou eat all the days of thy life:
15 And I will put enmity between thee and the woman, and between thy
seed and her seed; it shall bruise thy head, and thou shalt bruise his heel.
16 Unto the woman he said, I will greatly multiply thy sorrow and thy
conception; in sorrow thou shalt bring forth children; and thy desire *shall*
be to thy husband, and he shall rule over thee.
17 And unto Adam he said, Because thou hast hearkened unto the voice of
thy wife, and hast eaten of the tree, of which I commanded thee, saying,
Thou shalt not eat of it: cursed *is* the ground for thy sake; in sorrow shalt
thou eat *of* it all the days of thy life;
18 Thorns also and thistles shall it bring forth to thee; and thou shalt eat the
herb of the field;
19 In the sweat of thy face shalt thou eat bread, till thou return unto the
ground; for out of it wast thou taken: for dust thou *art*, and unto dust shalt
thou return.

[20] And Adam called his wife's name Eve; because she was the mother of all living.
[21] Unto Adam also and to his wife did the LORD God make coats of skins, and clothed them.
[22] And the LORD God said, Behold, the man is become as one of us, to know good and evil: and now, lest he put forth his hand, and take also of the tree of life, and eat, and live for ever:
[23] Therefore the LORD God sent him forth from the Garden of Eden, to till the ground from whence he was taken.
[24] So he drove out the man; and he placed at the east of the garden of Eden Cherubims, and a flaming sword which turned every way, to keep the way of the tree of life. (*King James Version*)

Thus, by gaining knowledge (here *ethical* knowledge is stressed, though *wise* also gets used at one point, and there has been a lot of discussion about knowledge beyond Good and Evil, beyond ethos—ultimately, God and His world are not ethical, as Kierkegaard's discussion of Abraham stresses, see *Fear and Trembling*), humans (and it can be and has been argued specifically male humans) become equal with God, except for being able to live forever (gift of the tree of life, which God obviously does not want them to have; just as he did not want them to gain the knowledge of Good and Evil). Furthermore, ethical judgments are subjective, at least for humans. One's man's Good is another Man's evil; one Man's pain is another man's pleasure (as an obvious, though perhaps overly simplistic example, think of purposely punishing oneself or allowing oneself to be punished—masochism, for the pleasure of pain, or . . .).

But this is not just some aberrant psychic desire, or, if it is, it has been incorporated into Christianity (and other religions) in a formal manner. Pope John Paul II, sometimes called Blessed John Paul or John Paul the Great, was the head of the Catholic Church from October 16, 1978 to his death in 2005, the second longest-serving pope in history, and one of the most influential leaders of the 20th century--recognised for having a great deal of influence ending Communist rule in his native Poland and eventually all of Europe, for significantly improving the Catholic Church's relations with Judaism, Islam, the Eastern Orthodox Church, and the Anglican Communion. Furthermore, though controversial, indicative of his conservative values, he upheld the Church's traditional views against artificial contraception and the ordination of women, supported the Church's Second Vatican Council, and held firm orthodox Catholic stances. On the other hand, he is known for his implementation of several papal documents pertaining to the role of the Church in the modern world.

He was one of the most travelled world leaders in history, visiting 129 countries during his pontificate. Furthermoire, as part of his special emphasis on the universal call to holiness, he beatified 1,340 people and canonised 483 saints, more than the combined total of his predecessors during the preceding five centuries. By the time his reign ended, he had named most of the present College of Cardinals, consecrated or co-consecrated a large number of the world's past and current bishops, and ordained many priests. A key goal of his papacy was to transform and reposition the Catholic Church "at the heart of a new religious alliance that would bring together Jews, Muslims and Christians in a great [religious] armada."

On December 19, 2009, his successor, Pope Benedict XVI proclaimed him *venerable*, and he was *beatified* on May 1, 2011, after the Congregation for the Causes of Saints attributed one miracle to him, the healing of a French nun from Parkinson's disease, and a second miracle the result of a happening at his tomb when a nine-year-old Polish boy from Gdańsk, who was suffering from kidney cancer and was completely unable to walk, visited it and upon leaving St. Peter's Basilica said "I want to walk," and began walking normally. Pope John Paul II was *canonized* on April 27, 2014. In other words, the Catholic Church has recognized him as a *saint*.

Furthermore, this saint projected a warm, grandfatherly image to an adoring public who flocked in large numbers to hear his homilies.

Then, a close friend of John Paul II and highly respected member of the Church, Monsignor Slawomir Oder (with the help of writer Saverio Gaeta) published a well received, positive book about his friend, *Why He's a Saint: The Real John Paul II According to the Postulator of His Beatification Cause*.

It held a surprise. The Venerable, Beatified, St. John Paul II whipped himself with a belt and sometimes lay prostrate all night on the floor.

He obviously did not want his aides to know, going so far as to make his bed appear used by tossing around the sheets. But his friend Oder did know, and, in the book explained that the pope believed these acts of penance would affirm God's primacy and help him seek perfection.

While such self-flagellation or other forms of self-inflicted physical suffering are not common in the Catholic Church, other notables have performed them. Mother Teresa wore a cilice, a strap secured around the thigh that inflicts pain with inward-pointing spikes.

So here we are, embracing suffering, self-inflicted pain in the name of salvation.

How do Catholics explain this? Writing for the *National Catholic Register*, Jimmy Akin comes out swinging (couldn't resist the pun), turning the tables on the nay-sayers (on our "pleasure-obsessed culture"):

> Let's start with what we can say to fellow Christians (Catholic or not) who find themselves thinking this way: While not every person is called to the kind of self-mortification that John Paul II practiced, self-mortification is part of the Judeo-Christian tradition with roots going all the way back to the Bible, both Old and New Testaments.
>
> We read in the Old Testament, for example, of people fasting, wearing sackcloth (which abrades the skin; the Old Testament equivalent of a hairshirt), putting ashes on their heads, and lying tied-up in uncomfortable positions for long periods of time (Ezekiel 4:4-8).
>
> In the New Testament we also read of such practices, and of particular note are Jesus' own remarks about (and personal practice of) fasting. If Our Lord himself practiced fasting, then self-mortification could scarcely fail to find a place in Christian spirituality. Note also that in the Sermon on the Mount he doesn't say "if" you fast but "when" you fast—implying an expectation of his followers. . . .
>
> Once we have recognized this, the issue of self-mortification becomes one of degree and occasion, for the fundamental principle has been established. If a particular Christian's faith tradition (or personal view) hasn't made room for self-mortification then he needs to conduct an open-minded re-examination of the issue. . . .
>
> Self-mortification teaches humility by making us recognize that there are things more important than our own pleasure. It teaches compassion by giving us a window into the sufferings of others—who don't have a choice in whether they're suffering. And it strengthens self-control. As well as (here's the big one I've saved for last) encouraging us to follow the example of Our Lord, who made the central act of the Christian religion one of self-denial and (in his case) *literal* mortification to bring salvation to all mankind.

According to Oder, John Paul II believed suffering brought him closer to Christ, and referenced Colossians 1:24, "Now I rejoice in my sufferings for your sake, and in my flesh I am filling up what is lacking in Christ's afflictions for the sake of his body, that is, the church."

It might be pointed out that Paul's sufferings were not self-inflicted, and he even pleaded with God to relent (Cor 2 11-12). One response is that God allowed (or caused, depending on the argument one wants to make) the suffering in order to demonstrate to Paul that he had it within him to endure, and this can lead to the argument that it is through suffering and hardship that man grows and earns what he receives, that the trials are necessary to the revelations.

In the case of Poe's poem, it can be claimed that the suffering is self-induced. Once he realizes the Raven has only one response, he begins punishing himself by asking it questions to which he already knows it will respond negatively. The poem takes on an unrelenting psychic whipping, and the reader can hear the mental leather-strap snapping "nevermore" against the mind's weak flesh. This man is a masochist (or a saint, as the two spiril and spiril until they seem to merge into a beautiful, if horrific, Venetian waltz across an elegant, shadowy Vienna ballroom, as opposites do in that world beyond the knowledge of Good and Evil; and yet the beautiful is filled with the dangers hidden in the dark corners, the experience of the *sublime*, as this dance partakes of opposites, that of overwhelming beauty *and* horror, desire *and* repulsion, the self *and* the other).

But, in the quiet night of *The Raven*, has Poe found the balance, the dance of the *ego* and the *shadow*, or has the darkness overcome him. His *ego*, his *self* seems to have surrendered to his dark side.

The final four lines are so exactly representative of the negative side of the *shadow* they express it in the form of a poem long before Carl Jung and his followers identified and attempted to explain it:

> And his eyes have all the seeming of a demon's that is dreaming,
> And the lamp-light o'er him streaming throws his shadow on the floor;
> And my soul from out that shadow that lies floating on the floor
> Shall be lifted—nevermore!

Jung thinks of the *shadow* as the counterpart to the *ego*, sometimes depicting it as a primitive force, similar to the *id* or *libido*, sometimes giving it more of an ethical quality, the evil side to either balance or work against the good intentions of the *ego*. In 1912, he writes:

> It is a frightening thought that man also has a shadow side to him, consisting not just of little weaknesses and foibles, but of a positively demonic dynamism. The individual seldom knows anything of this; to him, as an individual, it is incredible that he should ever in any circumstances go beyond himself. But let these harmless creatures form a mass, and there emerges a raging monster; and each individual is only one tiny cell in the monster's body, so that for better or worse he must accompany it on its bloody rampages and even assist it to the utmost. Having a dark suspicion of these grim possibilities, man turns a blind eye to the shadow-side of human nature. Blindly he strives against the salutary dogma of original sin, which is yet so prodigiously true. Yes, he even hesitates to admit the conflict of which he is so painfully aware. (69)

And in another passage from the same year, he directly states that this conflict between the *ego* and *shadow* can result in a form of neuroses, in a dramatic mental collapse:

> We know that the wildest and most moving dramas are played not in the theatre but in the hearts of ordinary men and women who pass by without exciting attention, and who betray to the world nothing of the conflicts that rage within them except possibly by a nervous breakdown. What is so difficult for the layman to grasp is the fact that in most cases the patients themselves have no suspicion whatever of the internecine war raging in their unconscious. If we remember that there are many people who understand nothing at all about themselves, we shall be less surprised at the realization that there are also people who are utterly unaware of their actual conflicts. (70)

Certainly these ideas relate to those expressed by Socrates, Soren Kierkegaard, and Richard Dadd. Insight, creativity, transformation, the birth of meaning and value come not just from the conscious self of the *ego*, but from the conscious *ego* in conflict with the unconscious drives of the *shadow*, with madness, insanity, the loss of reason. Only by facing one's *shadow*, admitting it, empowering it, and then controlling it can one move beyond a shallow world of innocence and self-denial, can one in Jungian terms achieve Individuation. Jung states:

> The things that come to light brutally in insanity remain hidden in the background in neurosis, but they continue to influence consciousness nonetheless. When, therefore, the analysis penetrates the background of conscious phenomena, it discovers the same archetypal figures that activate the deliriums of psychotics. Finally, there is any amount of literary and historical evidence to prove that in the case of these archetypes we are dealing with normal types of fantasy that occur practically everywhere and not with the monstrous products of insanity. The pathological element does not lie in the existence of these ideas, but in the dissociation of consciousness that can no longer control the unconscious. In all cases of dissociation it is therefore necessary to integrate the unconscious into consciousness. This is a synthetic process which I have termed the "individuation process." (71)

This can be depicted in the symbol of the Ouroboros, a circular dragon or snake or serpent where the head is biting and perhaps eating its tail. In alchemy this is taken to represent a process of the *ego* (the serpent's head) of consciousness connecting to and consuming the *shadow* (the entire world of the unconscious). In other words, consciousness is assimilating its unconscious origins. In Depth Psychology, this is referred to as "integrating the *shadow*." In mythology, it is generally represented by a hero defeating a dragon, often one guarding gold (the riches of the unconscious, the energy of life) and/or a female (the *anima*, the agent of transformation). It is important that the hero partake of the dragon's flesh, taste the dragon's blood, somehow take the essence of the dragon into his being, thus connecting up with the natural impulses out of which consciousness is born. On one level, this is a uniting or marriage of the *ego* and the *anima* through an encounter with the dragon or serpent, the force that brings them together,

Jung's writing offers a worthwhile frame for understanding this better in a discussion about the *trickster*, a figure he puts forth as the precursor to the *shadow* entity.

He begins by comparing Adolf Bandelier's discussion of the *trickster* in Pueblo Indian mythologies of New Mexico (72) to the European carnival in the Medieval church, where the hierarchic order gets reversed, to the medieval depictions of the Devil as the ape of God (*simian dei*), to various folklore characters such as Stupid Hans and Tom Thumb, and noticeably to Mercurius, the alchemical entity known for "sly jokes and malicious pranks, his powers as a shape-shifter, his dual nature, half animal, half divine, his exposure to all kinds of tortures, and—last but not least—his approximation to the figure of a saviour." (73) In alchemical representation, this is the character that units opposites, life and death, sun and moon, man and woman.

Jung also connects the *trickster* to *shamans* and *medicine-men*, both known to play jokes on others, often only to fall victim to their own malicious tricks. This leads into their connection to suffering, where due to either or both the reversal of their *prankster* antics and the real taking on of the suffering of others in both physical and psychological realms they suffer, perhaps never to completely recover. Jung said the *trickster's* "'approximation to the saviour' is an obvious consequence of this, in confirmation of the mythological truth that the wounded wounder is the agent of healing, and that the sufferer takes away the suffering." (74)

After discussing examples of the *trickster* in various official ceremonies of the Medieval Church, the *tripudia* and *tripudium hypodiaconorum* (the most famous called the *festum stultorum*, Feast of Fools) where lesser clergy and lower classes held wild partying, including a mock election of a child to bishop, ultimately degenerating to grotesque masquerades filled with dancing, indecent songs, dice games, and profanity; and a particularly disrespectful celebration, the *festum asinorum*, in France, which began as a celebration of Mary's flight into Egypt, but kept getting more and more filled with wanton crudity, bringing donkeys right into the church, ultimately concluding the mass with the priest braying like a donkey three times; after discussing these and other examples, including pointing out that the Winnebago tribes still have similar customs, Jung comes to the conclusion that the *trickster*, in psychological terms, is the reflection of an earlier stage of consciousness.

Then he makes an important claim, a claim that these reflections still exist in the current psychic make-up as dissociations, as split off personalities that stand in direct complementary or compensatory relationship to the *ego-personality*, not as a personal complex but as something known to all, an aggregate character. In modern man, Jung claims, this figure appears in "counter tendencies," "accidents," that seem to be purposely thwarting him, seem to have malicious intent. Jung continues, "I have, I think, found a suitable designation for this character-component when I called it the *shadow*." (75) Thus, the *shadow* began as part of the collective unconscious, but as civilization pushed it aside, it became personalized, an object of personal responsibility. In other words, the *trickster* is of a pre-conscious world, and as humans gained the higher conscious levels of thinking, they were able to detach it and objectify it. Jung states, "He is both subhuman and superhuman, a bestial and divine being, whose chief and most alarming characteristic is his unconsciousness." (76) He is a creator god, the body out of which the world is made. At the same time he is simply stupid, less intelligent than even the lowest animals.

Why then, Jung continues, does modern man continue to retain this *shadow*. And this leads into a complex relationship between the *shadow* and the *ego-consciousness*. He states, "from the psychology of the individual, namely the appearance of an impressive shadow figure antagonistically confronting a personal consciousness: this figure does not appear merely because it still exists in the individual, but because it rests on a dynamism whose existence can only be explained in terms of his actual situation, for instance because the shadow is so disagreeable to his ego-consciousness that it has to be repressed into the unconscious." However, he continues, "This explanation does not quite meet the case here because the trickster obviously represents a vanishing level of consciousness which increasingly lacks the power to take, express and assert itself. Furthermore, repression would prevent it from vanishing, because repressed contents are the very ones that have the best chance of survival. . . ." Furthermore, he suggests that it is actually in the best interest of consciousness to keep the shadow as much in the conscious realm as possible, because that allows consciousness to critique it, to respond to it, to have some say in what it does.

Nevertheless, as consciousness becomes more and more powerful, the *shadow* does recede, not disappear, but retreating deeper into the unconscious, where it waits for a "favourable opportunity to reappear as a projection upon one's neighbour. If this trick is successful, there is immediately created between them that world of primordial darkness where everything that is characteristic of the trickster can happen—even on the highest plane of civilization." (77)

Jung continues: "The so-called civilized man has forgotten the trickster. He remembers him only figuratively and metaphorically, when, irritated by his own ineptitude, he speaks of fate playing tricks on him or of things being bewitched. He never suspects that his own hidden and apparently harmless shadow has qualities whose dangerousness exceeds his wildest dreams. As soon as people get together in masses and submerge the individual, the shadow is mobilized, and, as history shows, may even be personified and incarnated." (78)

This view, as Jung points out, comes from the idea that humans are born *tabula rasa*, empty vessels to be molded by the environment, and it leads to the loss of self-responsibility. It is not, in this view, the individual's fault for any of the problems in the world, including his own psychic being. The various environments, natural, political, economic are in control.

Before continuing with Jung's views here, I must jump on this, as it is the key to meaning and value in human existence. Without self-responsibility life is meaningless. If a human is not responsible for his own choices then he can neither be blamed nor praised for anything, nor can he achieve any form of salvation. He is simply the "mindless" respondent to his environment. (Life might consist of discovering one's self, learning what one is, but that is not the same as having self-responsibility for creating what one is.)

A note here: do not confuse ethos with meaning and value. Remember Soren Kierkegaard's dictum that there be a teleological suspension of the ethical. It is easy to fall prey to the idea that the *ego* is Good and the *shadow* is Bad. But it is Robert Bly's crow, not his dove that finds the new land for the next step forward.

At this point in his essay Jung has a foundation for why the *trickster/shadow* is preserved and developed. He writes, "It holds the earlier low intellectual and moral level before the eyes of the more highly developed individual, so that he shall not forget how things looked yesterday. We like," he continues, "to imagine that something which we do not understand does not help us in any way. But that is not always so. Seldom does a man understand with his head alone, least of all when he is a primitive. Because of its numinosity the myth has a direct effect on the unconscious, no matter whether it is understood or not. The fact that its repeated telling has not long since become obsolete can, I believe, be explained by its usefulness. The explanation is rather difficult because two contrary tendencies are at work: the desire on the one hand to get out of the earlier condition and on the other hand not to forget it." Jung quotes from Radin "Viewed psychologically, it might be contended that the history of civilization is largely the account of the attempts of man to forget his transformation from an animal into a human being. . . . So stubborn a refusal to forget is not an accident." (79) Jung continues, "And it is also no accident that we are forced to contradict ourselves as soon as we try to formulate man's paradoxical attitude to myth. Even the most enlightened of us will set up a Christmas-tree for his children without having the least idea what this custom means, and is invariably disposed to nip any attempt at interpretation in the bud. It is really astonishing to see how many so-called superstitions are rampant nowadays in town and country alike, but if one took hold of the individual and asked him, loudly and clearly, 'Do you believe in ghosts? in witches? In spells and magic?' he would deny it indignantly. It is a hundred to one he has never heard of such things and thinks it all rubbish. But in secret he is all for it, just like a jungle-dweller. The public knows very little of these things anyway, for everyone is convinced that in our enlightened society that kind of superstition has long since been eradicated, and it is part of the general convention to act as though one had never heard of such things, not to mention believing in them." (80)

[This easy suspension of logic and common sense is also tied to the human need for life to be more than logic and common sense, because, again, if there is nothing more, then life is meaningless, then humans do not have any control because everything is predetermined by the logic of math and science, and everything a human does has been put in place by this logic before he or she was ever born. Blame and praise is pointless. It would be the same thing as blaming or praising a computer, as offering an ethical or spiritual judgment on it.]

Jung continues, "But nothing is ever lost, not even the blood pact with the devil. Outwardly it is forgotten, but inwardly not at all." (80)

Furthermore, he states later, "The conflict between the two dimensions of consciousness is simply an expression of the polaristic structure of the psyche, which like any other energic system is dependent on the tension of opposites. . . . In other words the "figure works because secretly it participates in the observer's psyche and appears as its reflection, thought it is not recognized as such. It is split off from his consciousness and consequently behaves like an autonomous personality." (81)

He continues later, "If we take the trickster as a parallel of the individual shadow, then the question arises whether that trend towards meaning, which we saw in the trickster myth, can also be observed in the subjective and personal shadow. Since this shadow frequently appears in the phenomenology of dreams as a well-defined figure, we can answer this question positively: the shadow, although by definition a negative figure, sometimes has certain clearly discernible traits and associations which point to a quite different background. It is as though he were hiding meaningful contents under an unprepossessing exterior." (82)

Jung puts the *shadow* figure of the unconscious nearest to consciousness, the first one to be encountered and dealt with in the process of *individuation*, and insightfully relates it to the riddle of the Sphinx, which of course opens up the thick meanings of Oedipus Rex, who solved this riddle, and thus was able to begin his psychic journey to self-knowledge, to *individuation*.

Though Jung doesn't elaborate on the connection of *individuation* to *salvation*, he does bring his essay to a conclusion by referring again to the connection of the *trickster* myth with *salvation*, stating that this "comforting premonition or hope means that some calamity or other has happened and been consciously understood." Briefly elaborating, he states, "Only out of disaster can the longing for the saviour arise—in other words, the recognition and unavoidable integration of the shadow creates such a harrowing situation that nobody but a saviour can undo the tangled web of fate. . . . As in its collective, mythological form, so also the individual shadow contains within it the seed of an enantiodromia, of a conversion into its opposite." (83)

This discussion by Jung, while constantly suggesting the *shadow* has a relationship to such higher levels of meaning and value as *salvation*, nonetheless stresses the *shadow* as a primitive, lesser part of the psyche, to the point where its importance as the *counter-ego*, the dark side of the *ego* gets undermined, at least *suggesting* that the balancing is more of one of a higher form of consciousness opposing a lower form of psychic existence rather than an interplay of two equally developed levels.

In fact, as theories of the *shadow* have developed, its position as the counter-balance to the *cultural persona*, as a somewhat attractive and absolutely necessary psychic entity, perhaps even more the "I" within us than the *ego,* have gained prominence. Robert A. Johnson writes:

> We are born whole and, let us hope, will die whole. But somewhere early on our way, we eat one of the wonderful fruits of the tree of

> knowledge, things separate into good and evil, and we begin the shadow-making process; we divide our lives. In the cultural process we sort out our God-given characteristics into those that are acceptable to our society and those that have to be put away. This is wonderful and necessary, and there would be no civilized behavior without this sorting out of good and evil. But the refused and unacceptable characteristics do not go away; they only collect in the dark corners of our personality. When they have been hidden long enough, they take on a life of their own—the shadow life. The shadow is that which has not entered adequately into consciousness. It is the despised quarter of our being. It often has an energy potential nearly as great as that of our ego. If it accumulates more energy than our ego, it erupts as an overpowering rage or some indiscretion that slips past us; or we have a depression or an accident that seems to have its own purpose. The shadow gone autonomous is a terrible monster in our psychic house.
>
> The civilizing process, which is the brightest achievement of humankind, consists of culling out those characteristics that are dangerous to the smooth functioning of our ideals. Anyone who does not go through this process remains a "primitive" and can have no place in a cultivated society. We all are born whole but somehow the culture demands that we live out only part of our nature and refuse other parts of our inheritance. We divide the self into an ego and a shadow because our culture insists that we behave in a particular manner. This is our legacy from having eaten of the fruit of the tree of knowledge in the Garden of Eden. Culture takes away the simple human in us, but gives us more complex and sophisticated power. (84)

It is clear here that Johnson is attaching the *ego* to the cultural values and the *shadow* to the individual outside of the cultural values. He goes on to point out both the artificial quality of the cultural values (they vary in different cultures) and how they are both necessary and practical.

Then he writes: "It is also astonishing to find that some very good characteristics turn up in the shadow. Generally, the ordinary, mundane characteristics are the norm. Anything less than this goes into the shadow. But anything better also goes into the shadow! Some of the pure gold of our personality is relegated to the shadow because it can find no place in that great leveling process that is culture." (85)

In this view, the *ego* tends to lose its attraction, becoming a cultural pawn, at best boring, at worst a completely passive entity, meaningless. The *shadow*, on the other hand, becomes the individual thinking for itself that has been purposely hidden deep in the psyche.

Furthermore, Johnson writes: "Curiously, people resist the noble aspects of their shadow more strenuously than they hide the dark sides. To draw the skeletons out of the closet is relatively easy, but to own the gold in the shadow is terrifying. It is more disrupting to find that you have a profound nobility of character than to find out you are a bum." (86)

He completes his model of the ego and the shadow by claiming that they involve a process, the first step the learning of the culture's rules, the fall from innocence into a world of ethical values, where the *ego* is strongly related to the public *persona*, the cultural good person, and the *shadow* the cultural bad person, evil a necessary balance to good; the second step the restoring of the wholeness that got divided when the forbidden fruit was eaten, the new unity a bringing of what was unconscious into consciousness.

Robert Bly writes:

> When we were one or two years old we had what we might visualize as a 360-degree personality. Energy radiated out from all parts of our body and all parts of our psyche. A child running is a living globe of energy. We had a ball of energy, all right; but one day we noticed that our parents didn't like certain parts of that ball. They said things like: "Can't you be still?" Or "It isn't nice to try and kill your brother." Behind us we have an invisible bag, and the part of us our parents don't like, we, to keep our parents' love, put in the bag. By the time we go to school our bag is quite large. Then our teachers have their say: "Good children don't get angry over such little thing." So we take our anger and put it in the bag. By the time my brother and I were twelve in Madison, Minnesota we were known as "the nice Bly boys." Our bags were already a mile long.
>
> Then we do a lot of bag-stuffing in high school. This time it's no longer the evil grownups that pressure us, but people our own age. So the student's paranoia about grownups can be misplaced. I lied all through high school automatically to try to be more like the basketball players. Any part of myself that was a little slow went into the bag. My sons are going through the process now; I watched my daughters, who were older, experience it. I noticed with dismay how much they put into the bag, but there was nothing their mother or I could do about it. Often my daughters seemed to make their decision on the issue of fashion and collective ideas of beauty, and they suffered as much damage from other girls as they did from men. . . . (87)

After a few more examples, he writes: "We spend our life until we're twenty deciding what parts of our self to put into the bag, and we spend the rest of our lives trying to get them out again." (88)

Robert Bly's bag here is the same thing as the serpent's tail of the Ouroboros, the forbidden energy of the unconscious that must be brought into consciousness, incorporated into the adult psyche to be whole and complete.

The problem here, again, is the strong suggestion that the *shadow* is simply the ethical bad guy as defined by culture, and *we must go beyond ethics*. The *shadow* is not ethically good or bad. It exists beyond ethics, in the world of the *numinous*.

Part VI

In 1917, Rudolph Otto published *Das Heilige (The Idea of the Holy)*, and created the term *numinous* from the Latin word *numen*, which means god or the power of god or the will of god. According to Otto the *numinous* is a "category for the interpretation and evaluation" of nonrational manifestations of the sacred characterized by a "sense of one's creature state" (by which he means a being that has been created), mystical awe (*tremendum*), a presentiment of divine power (*majestas*), amazement in the face of the "completely other" (*mysterium*), demoniacal energy, and paradox. (89)

It is the mystery behind a religious experience. This experience is not simply an ordinary experience intensified; for example fear intensifying until it becomes dread, but an entirely different feeling, a dread aroused by intimations of or actual experience of the *numinous*.

This *numinous dread* or awe or awe-fullness, Otto called the *mysterium tremendum*, which has three components, tremendousness (*tremendum*), mysteriousness (*mysterium*), and fascination (*facinans*). The difference between the fear of physical danger and dread is that the experience of dread is that of the realization of the inexplicable. So, for example, the fear that comes from encountering a wild beast, a lion or tiger or bear that might harm one is not the same as that of encountering a ghost, which suggests an other realm of existence, not just as a subject for speculation, but as a reality.

This reality is absolutely other, unapproachable through any conscious form of understanding, overwhelmingly powerful, and of an urgency and energy that might be compared to the "wrath of God."

Since it is absolutely other, it results in a state of stupor, of blank amazement, of astonishment, where the soul reaches beyond the mind to a *trembling* shudder only known by the feelings, not by the intellect. The awareness of being a creature, of being but a creation of the creator is a sense, not of guilt, but of realization of being other than the spiritual, a feeling of nothingness in the presence of the other, the creator, the god. This has nothing to do with a moral judgment, but is in the realm of an awareness of the *numinous*. And it is this awareness of the *numinous* that frees the person experiencing it from being unfit, provides a shield, a form of grace.

Furthermore, according to Rudolph Otto a *mysterium fascinosum*, an attraction to the *numinous*, draws us to the experience, pulls us to an encounter with the *numen*, a mystical moment, when all of the oppositions come together in a strange harmony, what Otto called the *mysterium tremendum* and *fascinosum*.

Rudolph Otto believed humans have an a priori ability to experience the *numinous*, a potential for such an experience, some more so than others, but that does not necessarily mean each human will experience it. It must be evoked through some experience or series of experiences. It cannot be taught. Those who are especially receptive and who have the necessary experiences are the shamans and prophets who draw others to also experience it.

It is difficult but worthwhile trying to clarify the difference between the *numinous* and the *sublime*. Both are mysterious, inexplicable, and enticing; both involve the experience of human existence as small and weak and unimportant in the overpowering movements of nature and fate; yet in complete paradox include the joy of being a part of it all, even the necessity of human existence to the value of all existence, the strange truth inherent in the Biblical story that when God created man he considered man superior to the angels, considered man his greatest creation. While the lines are not clearly drawn, it might be said that the *sublime* is the experience of these truths in the physical world, while the *numinous* is the experience of the *other*, the world beyond this world. Thus, in an experiencing of one it is possible to be led to an experience of the other.

James Baldwin concludes "Sonny's Blues":

> All I know about music is that not many people ever really hear it. And even then, on the rare occasions when something opens within, and the music enters, what we mainly hear, or hear corroborated, are personal, private, vanishing evocations. But the man who creates the music is hearing something else, is dealing with the roar rising from the void and imposing order on it as it hits the air. What is evoked in him, then, is of another order, more terrible because it has no words, and triumphant, too, for that same reason. And his triumph, when he triumphs, is ours. I just watched Sonny's face. His face was troubled, he was working hard, but he wasn't with it. And I had the feeling that, in a way, everyone on the bandstand was waiting for him, both waiting for him and pushing him along. But as I began to watch Creole, I realized that it was Creole who held them all back. He had them on a short rein. Up there, keeping the beat with his whole body, wailing on the fiddle, with his eyes half closed, he was listening to everything, but he was listening to Sonny. He was having

a dialogue with Sonny. He wanted Sonny to leave the shoreline and strike out for the deep water. He was Sonny's witness that deep water and drowning were not the same thing—he had been there, and he knew. And he wanted Sonny to know. He was waiting for Sonny to do the things on the keys which would let Creole know that Sonny was in the water.

And, while Creole listened, Sonny moved, deep within, exactly like someone in torment. I had never before thought of how awful the relationship must be between the musician and his instrument. He has to fill it, this instrument, with the breath of life, his own. He has to make it do what he wants it to do. And a piano is just a piano. It's made out of so much wood and wires and little hammers and big ones, and ivory. While there's only so much you can do with it, the only way to find this out is to try; to try and make it do everything.

And Sonny hadn't been near a piano for over a year. And he wasn't on much better terms with his life, not the life that stretched before him now. He and the piano stammered, started one way, got scared, stopped; started another way, panicked, marked time, started again; then seemed to have found a direction, panicked again, got stuck. And the face I saw on Sonny I'd never seen before. Everything had been burned out of it, and, at the same time, things usually hidden were being burned in, by the fire and fury of the battle which was occurring in him up there.

Yet, watching Creole's face as they neared the end of the first set, I had the feeling that something had happened, something I hadn't heard. Then they finished, there was scattered applause, and then, without an instant's warning, Creole started into something else, it was almost sardonic, it was Am I Blue. And, as though he commanded, Sonny began to play. Something began to happen. And Creole let out the reins. The dry, low, black man said something awful on the drums, Creole answered, and the drums talked back. Then the horn insisted, sweet and high, slightly detached perhaps, and Creole listened, commenting now and then, dry, and driving, beautiful, calm and old. Then they all came together again, and Sonny was part of the family again. I could tell this from his face. He seemed to have found, right there, beneath his fingers, a damn brand-new piano. It seemed that he couldn't get over it. Then, for a while, just being happy with Sonny, they seemed to be agreeing with him that brand-new pianos certainly were a gas.

Then Creole stepped forward to remind them that what they were playing was the blues. He hit something in all of them, he hit something in me, myself, and the music tightened and deepened, apprehension began to beat the air. Creole began to tell us what the blues were all about. They were not about anything very new. He and his boys up there were keeping it new, at the risk of ruin, destruction, madness and death, in order to find new ways to make us listen. For, while the tale of how we suffer, and how we are delighted, and how we may triumph is never new, it must always be heard. There isn't any other tale to tell, it's the only light we've got in all this darkness.

> And this tale, according to that face, that body, those strong hands on those strings, has another aspect in every country, and a new depth in every generation. Listen, Creole seemed to be saying, listen. Now these are Sonny's blues. He made the little black man on the rums now it, and the bright, brown man on the horn. Creole wasn't trying any longer to get Sonny in the water. He was wishing him Godspeed. Then he stepped back, very slowly, filling the air with the immense suggestion that Sonny speak for himself.
>
> Then they all gathered around Sonny and Sonny played. Every now and again one of them seemed to say, amen. Sonny's fingers filled the air with life, his life. But that life contained so many others. And Sonny went all the way back, he really began with the spare, flat statement of the opening phrase of the song. Then he began to make it his. It was very beautiful because it wasn't hurried and it was no longer a lament. I seemed to hear with what burning he had made it his, with what burning we had yet to make ours, how we could cease lamenting. Freedom lurked around us and I understood at last how he could help us to be free if we would listen, that he would never be free until we did. Yet, there was no battle in his face now. I heard what he had gone through, and would continue to go through until he came to rest in earth. He had made it his: that long line, of which we knew only Mama and Daddy. And he was giving it back, as everything must be given back, so that, passing through death, it can live forever. I saw my mother's face again, and felt, for the first time, how the stones of the road she had walked on must have bruised her feet. I saw the moonlit road where my father's brother died. And it brought something else back to me, and carried me past it. I saw my little girl again and felt Isabel's tears again, and I felt my own tears begin to rise. And I was yet aware that this was only a moment, that the world waited outside, as hungry as a tiger, and that trouble stretched above us, longer than the sky.
>
> Then it was over. Creole and Sonny let out their breath, both soaking wet, and grinning. There was a lot of applause and some of it was real. In the dark, the girl came by and I asked her to take drinks to the bandstand. There was a long pause, while they talked up there in the indigo light and after awhile I saw the girl put a Scotch and milk on top of the piano for Sonny. He didn't seem to notice it, but just before they started playing again, he sipped from it and looked toward me, and nodded. Then he put it back on top of the piano. For me, then, as they began to play again, it glowed and shook above my brother's head like the very cup of trembling. (90)

Here it all is, the connection of music to one's heritage, one's traditions, one's community, to one's inner soul, one's psyche, to one's salvation, to touching the *numinous*, the knowledge of the *other,* the world of God. And the final sentence refers directly to Isaiah, 51, xvii-xxii:

> 17. Awake, awake, stand up, O Jerusalem, which hast drunk at the hand of the LORD the cup of his fury; thou hast drunken the dregs of the cup of trembling, *and* wrung *them* out.
>
> 18. *There is* none to guide her among all the sons *whom* she hath brought forth; neither is *there any* that taketh her by the hand of all the sons *that* she hath brought up.
>
> 19. These two *things* are come unto thee; who shall be sorry for thee? Desolation, and destruction, and the famine, and the sword: by whom shall I comfort thee?
>
> 20. Thy sons have fainted, they lie at the head of all the streets, as a wild bull in a net: they are full of the fury of the LORD, the rebuke of thy God.
>
> 21. Therefore hear now this, thou afflicted, and drunken, but not with wine:
>
> 22. Thus saith thy Lord the LORD, and thy God that pleadeth the cause of his people, Behold, I have taken out of thine hand the cup of trembling, *even* the dregs of the cup of my fury; thou shalt no more drink it again:
>
> 23. But I will put it into the hand of them that afflict thee; which have said to thy soul, Bow down, that we may go over: and thou hast laid thy body as the ground, and as the street, to them that went over. (91)

Here it is, God's wraith, "the dregs of the cup of trembling," the suffering he has unleashed on the Israelites, but now will turn on their enemies, now will put into the hands of their enemies.

But it is not nearly so simple, not nearly so shallow, superficial, and literal as this. The passage connects directly to the images throughout "Sonny's Blues," scenes of drunkenness and the dark corners of the streets of the ghetto, the struggles of addiction, whether it be heroin or alcohol, and it connects these types of suffering to those "afflicted, and drunken, but not with wine."

How does all this connect to the musician, the artist? Sonny here has just demonstrated that he can enter into, can experience, can express the fear, the pain, the sorrows and sadness of his people. And though it is clear that this *cup of trembling* contains the drug, the alcohol, the addiction that brings the suffering, it is offered to Sonny at a *moment of salvation*. It is a gift, a sign of acceptance. Certainly Sonny and his brother have now connected for the first time, have resolved or found a way beyond their conflicts (and there are strong intimations of the Cain and Able story throughout, so on one level this is a resolution to the conflict of brothers underlying all of Christianity). But why symbolize it with a *cup of trembling*, why an offering of alcohol and milk, why an acceptance of the addiction?

Perhaps it is an expression of the necessity of the artist to experience, to drink from the *cup of trembling*, the wraith of God. Perhaps it even goes further to say that the artist, in drinking from this cup is taking on the suffering of his community, of the entire human race, perhaps it connects directly to the last verse just quoted from Isaiah, and expresses the idea that the artist is the one who, in translating his own pain into music, "hast laid [his] body as the ground, and as the street, to them that went over." Perhaps the artist is taking on the suffering so that his listeners will not have to do so themselves, and the acceptance of it by the listeners, the audience, the community, the culture, the whole human race, is a realization of this.

It is not hard at all to take this to the conclusion that Sonny, representing the artist, performs a sacrifice, internalizes and then expresses back all of the anguish, fear, and joy of his listeners, absorbs the condemnation of God when he expels Adam and Eve from the Garden of Eden for disobedience and gives the means of salvation. Here the artist becomes Christ-like.

The various lines of thought converge. We have reached a point of *axis mundi*, a center. The *artist*, the *true artist*, and the *shaman* become one, and they are one with the *trickster*, the *shadow*, and the *savior*.

They exist beyond *ethos*, not subject to the tree of the knowledge of good and evil, but, rather, as mediums, as illuminated beings, and their expressions, their works of art are meant to take us there, to take us beyond mere physical existence to that *numinous* world we cannot enter until we pass beyond the body, but we can know, can sense, can feel, can realize exists, if we experience that *trembling* of the soul.

Jung, who has already used *numen* to designate the autonomy of psychic energy in relation to *mana*, spiritual power, (92) begins in the 1930s to simply incorporate *numinous* into his Analytical Psychology. For him, *numinous* refers to a psychic event, an experience when the person connects to that which is completely other and is beyond intellectual understanding, an experience that is indescribable and "puts the subject into a state of amazement or passive submission" (93). This is dangerous and can result in various forms of psychic neurosis. Consciousness is altered, possibly resulting in schizophrenia, egotism, possession, and fanaticism. (94).

Yet, in spite of the serious dangers, a *numinous* experience can also be positive, a necessary step toward psychic healing. It is a transforming experience, signaling the emergence of an archetype with a specific energy or emotional charge, and this can compensate for the unilateral attitude of consciousness. For example, the *numinosity* of the archetype of the *self* can prompt a man to realize the paradoxical totality of his being, conscious and unconscious, through the female the symbols of the quaternity that appear in his dreams. The *ego* is the determining factor—whether it is ready, in the right place for a positive transformation, or not. In this case, the attitude of the *ego* is the determining factor. Such an experience, Jung considered to be religious in the sense of *relegere* (sending forth), a reaching out, a connecting, "a careful and scrupulous observation of . . . the *numinosum.*" (95)

Does the artist, then, perform the same function as the shaman? The shadow? The trickster? The savior? Do these entities take on the suffering for the culture? Are they the wounded wounders who take on the wounds for all?

Can these ideas be applied to Richard Dadd? Do they apply to his masterpiece?

Part VII

If the Patriarch is the character in the painting of *The Fairy-Feller's Master Stroke* ready to initiate the action, the title character, the Fairy-feller standing in the lower third of the work, facing away from the viewer, axe raised, ready to split an acorn, is the character about to obey the command, to begin the needed vivisection.

This Fairy-feller easily connects to Dadd himself. He is, after all, the man at Bethlam known for having painted fairies.

Fairies and The Fairy School of Art give us yet another context, another world to consider in the intertwining realities of Dadd and his painting.

Several strands intertwined in Victorian England to spark an interest in fairies, driven at least partially by the English desire to regain their heritage, to rediscover and preserve their traditions and to embrace their unofficial, previously denigrated past. The cultural views of the eighteenth century, what has gained the designation the *Age of Reason*, had given way to the views of the *Romantic Movement*, all driven by the philosophies of such thinkers as John Locke and Jean Jacques Rousseau. Now, instead of reason and logic and form being designated the highest levels of existence, the world of creativity and the imagination gained the upper hand.

On the political level, the old aristocracies that had held sway for centuries, supporting their position with various religious and ethical systems that claimed certain people, certain blood lines were simply superior, had found the established aristocracy confronted, and then defeated by the new views of democracy. The first major defeat came at the hands of the leaders of the American Revolution, men putting forth the statement that "all men are created equal." This idea was not new, but this was the first time a major segment of the European civilization had used it as a basis for government. Of course, the fact that this young country across the ocean might separate from the control of an English king was not in itself a change in Europe. However, it was the spark, the ignition for what would change the whole European world, and what it sparked was the French revolution, the storming of Bastille by Napoleon.

With democracy on the political level came a desire to regain the cultural democracy outside of the official aristocratic cannons of history and culture, to regain the heritage and traditions of the people. In terms of the world of fairies, which by this time had gotten closely tied to those unimportant stories the uneducated people passed along, folktales, the first major thrust came from the Brothers Grimm in Germany. They were serious scholars, highly educated, members of the cultural elite, and now, since the former purveyors of culture had been condemned as corrupt, and a system was in place for embracing the formerly rejected culture of the outsiders, the poor, the peasants, they decided to do just that, to attempt to gather and preserve the heritage and traditions of the people. What was to become the most influential body of material of their scholarship was the world of the folktale, stories told by illiterate people to pass along their values and beliefs. In 1823 Edgar Taylor translated the major collections of the Brothers Grimm into English, illustrated by George Cruikshank, and England began its own quest to preserve its heritage, one filled with folktales, folktales that often included supernatural beings outside of and strongly condemned by the Christian churches of the time, such unchristian entities as leprechauns, elves, and of course fairies.

In terms of museum art, the stage had been set in the late 1700s by Henry Fuseli (1741-1845), who used Shakespeare's *A Midsummer Night's Dream* as the main source for such paintings as *Titania and Bottom* and *Titania's Awakening*, beginning a new genre, something of a poetic history form of painting, and establishing the initial format for it, a quotation from high art or literature, the addition of folkloric themes, and the establishment of a central narrative. (96)

All of these are evident in Dadd's *The Fairy Feller's Master Stroke*, which takes the most famous passage from Shakespeare about "dreams," the world of Queen Mab according to Mercutio from *Romeo and Juliet*. He and Romeo are having one of their clever word contests filled with Shakespeare's usual oxymorons, puns, and double entendres:

Romeo: I dreamt a dream tonight.
Mercutio. And so did I.
Romeo: Well, what was yours?
Mercutio: That dreamers often lie.

Romeo: In bed asleep, while they do dream things true.
Mercutio: O. then I see Queen Mab hath been with you.
She is the fairies' midwife, and she comes
In shape no bigger than an agate stone
On the forefinger of an alderman,
Drawn with a team of little atomies
Over men's noses as they lie asleep;
Her wagon spokes made of long spinners legs,
The cover, of the wings of grasshoppers;
Her traces, of the smallest spider web;
Her collars, of the moonshine's wat'ry beams;
Her whip, of cricket's bone; the lash, of film;
Her wagoner, a small gray-coated gnat,
Not half so big as a round little worm
Pricked from the lazy finger of a maid;
Her chariot is an empty hazelnut,
Made by the joiner squirrel or old grub,
Time out o' mind the fairies' coachmakers.
And in this state she gallops night by night
Through lovers' brains, and then they dream of love;
On courtiers' knees, that dream on curtsies straight;
O'er lawyers' fingers, who straight dream on fees;
O'er ladies' lips, who straight on kisses dream,
Which oft the angry Mab with blisters plagues,
Because their breath with sweetmeats tainted are,
Sometime she gallops o'er a courtier's nose,
And then dreams he of smelling out a suit;
And sometime comes she with a tithe pig's tail
Tickling a parson's nose as 'a lies asleep,
Then he dreams of another benefice.
Sometime she driveth o'er a solder's neck,
And then dreams he of cutting foreign throats,
Of breaches, ambuscades, Spanish blades,
Of healths five fathom deep; and then anon
Drums in his ear, at which he starts and wakes,
And being thus frighted, swears a prayer or two
And sleeps again. This is that very Mab
That bakes the elflocks in foul sluttish hairs,
Which once untangled much misfortune bodes.
This is the hag, when maids lie on their backs,
That presses them and learns them first to bear,
Making them women of good carriage.
This is she—
Romeo: Peace, peace, Mercutio, peace!
Thou talk'st of nothing.
Mercutio; True, I talk of dreams;

> Which are the children of an idle brain,
> Begot of nothing but vain fantasy;
> Which is as thin of substance as the air,
> And more inconstant than the wind, who woos
> Even now the frozen bosom of the North
> And, being angered, puffs away from thence,
> Turning his side to the dew-dropping South. (97)

In Dadd's painting, the Fairy-feller is about to crack in half the walnut that will serve as the carriage of Queen Mab, is, in other words, about to begin the dream. He just waits for the command of the Patriarch.

The world of fairy and its connection with the world of the dream, especially with Shakespeare's *A Midsummer Night's Dream*, also clearly captured the imagination of William Blake (1757-1827). Working in the media of engraving and watercolor, William Blake represented fairies as "rulers of the vegetable world." (98) Richard Schindler states: "In *Oberon, Titania, and Puck with Fairies Dancing* (c. 1785), the artist [Blake] conceives of fairies as nature worshippers, miniature druidic celebrants of the corporeal earth," differing from Fuseli's more grandiose historic genre by "concentrating solely on the diminutive participants (without comparison to normal-sized human beings) and giving fairies wings, which add to the airy feeling of the dance." (99) Blake also mixed literary works with popular beliefs about fairies in such works as *The Goblin* (c. 1816-20), a pen and watercolor illustration for Milton's poem "L'Allegro," where Milton's metaphors are visualized in conjunction with Robin Goodfellow, represented as a domestic fiend hurling himself into the morning sky after completing his nightly tasks, while other fairies punish a women for remaining in bed when she should be up and at her work, and another leads a foolish man astray. Even Queen Mab, mistress of the dream, is in attendance, eating her pudding and overseeing the general activity. (100)

Again, the influence on Richard Dadd is evident here. Patricia Allderidge points out that "For Dadd, fairy painting was a serious business, an act of the most intensely personal creation. It had nothing to do with the whimsy—despite his ability to charm and please by purely fanciful and playful touches—the voyeurism, sometimes the sadism with which his contemporaries often achieved popularity in this field. For fairy painting was also a way of exploring nature, as his landscape painting later became in Bethlem; but a way of exploring it by putting it under a microscope, discovering a world in a grain of sand." (101) This microscopic world, where the inhabitants are little larger than the size of the walnut to be cracked open and strands of flora that humans would simply walk across rise high above them is not just a cute fairy world, an escape from reality, but rather, just as for William Blake, a place to discover a higher reality, one filled with darkness and honesty and possibly salvation.

Prior to his ill fated journeys through the Middle East, Richard Dadd had been a popular, active, and highly respected student at the Royal Academy Schools, the Academy. In 1834, when he was seventeen, his father had moved the family to London, locating his house and business at 15 Suffolk Street, Pall Mall East, a highly respected neighborhood, establishing himself in the business of a carver, bronzist, and water-gilder, and making contacts with many people of importance in society and the world of art, including Clarkson Stanfield, who remains today a highly acclaimed marine painter, one of the founders of the Society of British Artists in 1823, its President in 1829, and an Associate of the Royal Academy and a Royal Academician in 1832 and 1835. Painting in both oil and watercolor, he specialized in shipping, coastal and river scenes, making regular visits to Italy, France and Holland and painting many Venetian views in the 1830s and Dutch scenes in the 1840s. He was commissioned to paint the opening of the New London Bridge and Portsmouth Harbor by King William IV, but probably his finest work was the *Battle of Trafalgar* in 1863, painted for the United Services Club in Pall Mall, London, where it is still exhibited. Other important works included *The Castle of Ischia* (1841), *The Day after the Wreck* (1844), *On the Dogger Bank* (1846), *The Battle of Roveredo* (1851), *Victory towed into Gibraltar* (1853), and *The Abandoned* (1856).

Before devoting himself to his marine art, Clarkson Stanfield worked with David Roberts painting the backdrops for theatrical productions. David Roberts would also move into a highly respected career in painting, known mostly for his paintings of exotic landscapes, especially interesting in relation to Richard Dadd because David Roberts' most famous exotic landscapes are the result of his travels through the Middle East and Egypt, where he also came down with a serious fever. Robert Dadd was a close enough friend of David Roberts that he even traveled with him to visit Scotland.

Since Richard Dadd's art has such obsessive miniaturist qualities, it's also interesting to note that Robert Dadd knew the famous miniaturist John Turmeau (1776-1846) who painted Robert Dadd's portrait in 1836.

In addition to a life populated by the most important artists in London, Robert Dadd's world was filled with such art collectors as Joseph Mayer and his agent William Clements, a wood engraver who also wrote articles on the arts, and print-seller Henri Josi, who became the Keeper of Prints and Drawings at the British Museum in 1836.

While it is impossible to know just how much Richard Dadd also interacted with these men, it is clear he was destined to become a painter, and it was Clarkson Stanfield himself who gave the necessary recommendation for Dadd to enter into the Academy in 1837.

That same year the Academy moved from Somerset house to the new National Gallery building in Trafalgar Square, just around the corner from Suffolk Street. The training there promoted a strict obedience to the established rules of art, mainly stressing a copying of antique masters borrowed from Dulwich College or those in the National Gallery. Visiting painters included Maclise, Mulready, Etty, Stanfield, Landseer and Turner. The main professor of painting was Henry Howard, who was apparently an indefatigable painter of fairy and poetic subjects, though little of his work has survived, and it is thought to have been not very original. (102) Nevertheless, he preached the representation of the very fairy and poetic subjects Dadd would later embrace, and thus must have had some influence. (103)

Richard Dadd's closest friend, William Powell Frith, entered the Academy the same year as he did, and was to leave the most complete records of Dadd at this time, giving a very positive picture of both Dadd's nature and talent: "Dadd was my superior in all respects; he drew infinitely better than I did . . . I can truly say, from a thorough knowledge of Dadd's character, that a nobler being, and one more free from the common failings of humanity, never breathed." (104)

Firth, a friend of Charles Dickens, whose literature served as the source for several of his earlier paintings, was to achieve a level of fame himself with such works as *Derby Day*, a painting so popular that it had to be protected by a specially installed rail when it was displayed at the Royal Academy of Arts, so his praise of Dadd is not to be taken lightly.

John Phillip, later to marry Dadd's youngest sister Maria Elizabeth, also entered the Academy at the same time and was perhaps only second to Firth in his friendship with Richard. Augustus Egg, Henry Nelson O'Neil, and Alfred Elmore, all students already situated at the Academy, also joined into what became a close knit group. Edward Matthew Ward, Thomas Jay, and William Bell Scott later joined in to various degrees, and a sketching club named The Clique was formed.

The Clique appears to have been more of a social group of young men who enjoyed one-another's company than anything else. They gathered in Dadd's rooms weekly. Subjects, generally from either Byron or Shakespeare, were chosen to be illustrated, and a competition was held to determine the best sketch each week. Dadd's sketches usually won, and he was considered by Firth and the group in general to be the most talented of them. Furthermore, as S. C. Hall noted "His personal appearance was in his favour. . . . He was somewhat tall, with good and expressive features, and gentlemanly demeanour." (105) A writer for the Art Union wrote "a person more invariably gentle, kind, considerate, and lavish in his praise did not exist. He was emphatically one who could not deliberately injure a fly . . . his full rich voice, as full of music as a joy-bell, and his sportive humour, exciting innocent mirth . . . he was satisfied with small praise for himself, but ready and lavish of his praise of others." And another was quoted saying he was "one of the kindest and the best, as well as the most gifted, of the children of genius it has ever been our lot to know." (106)

While Fuseli and Blake were well known to Dadd and the rest of the group of painters, an Irish artist Daniel Maclise (1806-1870), one of the visiting painters at the Academy, was idolized by them. (107) Daniel Maclise recognized early in his career the possibilities of fairy imagery, first publishing drawings in Thomas Crofton Croker's *Fairy Legends and Traditions of the South of Ireland* in 1826 (108), and by the beginning of the 1830s turning his attention to unique interpretations of historical genre painting, including fairy scenes. Certainly he was enamored with Shakespeare, and especially with *A Midsummer Night's Dream.* His painting *The Disenchantment of Bottom* in 1832 depicted an unhappily grimacing Bottom, awaking to his human self, surrounded by menacingly playful sprites, who remind the audience of his bestial alter ego through the suggestion of an ass's ears in the placement of two fairies on either side of his head. It is set at night, a night fitfully lit by a supernatural light, and filled with narrative details, including the reconciliation of Oberon and Titania in the upper right-hand corner and a fairy ring dancing around a toadstool in the lower left-hand corner, all in the manner of Henry Fuseli, yet filled with a dark imagination that takes it well beyond Fuseli. While such early paintings as this did not receive as much critical acclaim as such historical genre paintings as *Snap Apple Night*,1833, and *Merry Christmas in the Baron's Hall*, 1838, Dadd and his group certainly knew and understood the concepts behind them.

Maclise's techniques can be detected clearly in Dadd's early work. Patricia Allderidge writes:

> The emphasis on surface pattern, seen here at its most distinctive in the arabesques and ellipses of *Come unto These Yellow Sands*, the use of foliage to form a flat decorative border in *Puck* or the still more inventive proscenium arch of *Titania Sleeping*, the idea of setting part of the scene in a deep recess within the picture, used for the main group in Titania, and the creation of tension by dramatic lighting are all formal devices highly characteristic of Maclise. Equally characteristic are the meticulous

> draughtsmanship, love of fine detail and fertile imagination, none of which, however, could have been acquired by mere emulation had they not already been innate. (109)

However, she continues:

> For all their derivativeness there is no element of slavish copying in these pictures, and already Dadd shows a fluency and control in handling highly artificial designs which goes far beyond Maclise and is entirely his own; he especially never allows hiself to be encumbered by Maclise's craving for symmetry. Dadd uses intricate detail as an essential part of his total vision rather than mere decoration, a virtuoso embellishment which tends to dazzle and confuse when applied to the surface of large and crowded compositions. By working on a small scale he was able to keep a sense of proportion in a very literal sense, and his exploration of the highlights in a dewdrop or the texture of a flower petal was an important part of his way of looking at the minutely ordered universe which he created. (110)

While Fuseli, Blake, Maclise, and other lesser artists such as Henry Singleton (1766-1839), Henry Howard (1769-1847), Frank Howard (1805-66), and Joshua Cristall (1767-1847) provided a background for fairy painting, a good deal of fairy lore also came out of the writings of such folklorists as Sir Walter Scott (1771-1832), Nathan Drake (1766-1836), and Thomas Crofton Croker (1798-1854). Thomas Keightley (1789-1872) published a popular book on fairy mythology, and an English translation of Jakob and Wilhelm Grimm's *Kinder und Hausmarchen* by Taylor, illustrated by Cruickshank, who created an "ants eye view" similar to what Dadd would later do, came out in 1823.

Richard Dadd's work at this time was uneven, both in terms of the quality and in terms of the subject matter. At the Academy he won three silver metals, given for works finishing second and fourth in 1839, and in 1840 first for the best drawing in his class. (111) He also exhibited at the Society of British Artists and the British Institution in Pall Mall, and made his first sale with a scene from *Don Quixote.* This painting was to lead him into scenes from novels and ultimately into scenes from Shakespeare, including the above mentioned *Puck* and *Titania Sleeping* from *A Midsummer Night's Dream*, both purchased by Henry Farrer, and representing a clear step forward in vision and imagination. In 1841 he exhibited two fairy paintings, *Fairies Assembling at Sunset* and *Ever Let the Fancy Roam*, and in 1942, he sent *Come unto These Yellow Sands*, a fairy scene from *The Tempest*, to the Academy. Then, in the spring of 1842 came the commission to illustrate "The Ballad of Robin Goodfellow" for S. C. Hall's *Book of British Ballads*. At this same time, Dadd was one of the founders of the Printer's Etching Society. He had also been commissioned to do a series of panels for Lord Foley, the subjects to be taken from Tasso's *Jerusalem Delivered* and Byron's *Manfred*, a demon haunted poem that included the prophetic lines: "My solitude is solitude no more, / But peopled with the Furies. . . . / I have pray'd / For madness as a blessing." These panels received praise, though none remain and the content and quality can only be guessed at. One thing is certain; Dadd was beginning to gain recognition.

This earned him the attention of Sir Thomas Phillips, who was looking to establish himself as a patron of the arts and wanted a traveling companion to draw pictures of his trip through the Middle East and Egypt before setting down. David Roberts recommended Dadd, highlighting his "powers as a draughtsman, and his amiable qualities as a man," which would "render him as charming in companionship as he would be efficient as an artist." (112) This extended trip went from Ostend to Acona, along the Rhine to North Italy, the Bernese Alps, Venice, Bologna, Corfu. Patras, Delphi, Athens, Smyrna, Bodrum, Lycia, Rhodes, Cyprus, Tripoli, Beirut, Demascus, Jerusalem, Jordan, The Dead Sea, Jericho, Cairo and the pyramids of Giza, up the Nile to Thebes. It was hard traveling and by the time they reached Egypt Dadd was exhausted. He suffered sunstroke, and that might have been what sparked his insanity, though, perhaps it was the reverse, a mistaking of the insanity as sunstroke. Whatever the case, he began to act strange, even writing to Frith prophetically "my imagination [is] so full of wild vagaries that I have really and truly doubted of my own sanity." (113)

And now the biography circles back to the violent murder, or, rather, spirals above it, for the murder takes on larger meanings, contexts relevant to understanding multiplicities and paradoxes of human life. Dadd's actions become ever more erratic. He and Thomas continue the trip, spending three weeks near Naples, followed by a month in Rome, but by the end of May, 1843, Dadd has become so deranged Thomas seeks medical advice, finally leaving Dadd in Paris, from where Dadd heads directly home. His family and friends immediately see the change in him. Patricia Allderidge writes:

> It was soon obvious to his family and friends that Dadd was insane. His method of arrival, having posted home alone from Paris in such haste that all his money was used up, would have aroused some comment, and Sir Thomas soon followed with more details. So far as his companions could see, his insanity showed itself in a complete change of character: he was now watchful and suspicious, increasingly reserved and gloomy, and

> showed "a degree of shrewdness and sharpness about little things utterly foreign to his former character." [35] His actions became unpredictable and occasionally violent, and he seemed to believe that he was being watched: once he rushed away alone while a friend prepared to accompany him home; [36] another time he shut himself in his room and waved a knife under the door at visitors; [37] he is said to have cut a birthmark from his forehead, saying it was planted there by the devil. [33] He obviously saw and heard things which did not exist, and though he did not speak much of this, he let slip top a close friend that he was haunted by evil spirits, and was himself searching for the devil. (114)

Dadd's own writing reveals that he is both aware of his change and yet unable to fight it. Believing himself on a mission to rid the world of evil, both persecuted by the devil and prompted by voices of outside powers, and confusing it all with Egyptian mythology, he writes:

> On my return from travel, I was roused to a consideration of subjects which I had previously never dreamed of, or thought about, connected with self; and I had such ideas that, had I spoken of them openly, I must, if answered in the world's fashion, have been told I was unreasonable. I concealed, of course, these secret admonitions. I know not whence they came, although I could not question their propriety, nor could I separate myself from what appeared my fate. My religious opinions varied and do vary from the vulgar; I was inclined to fall in with the views of the ancients, and to regard the substitution of modern ideas thereon as not for the better. These and the like, coupled with the idea of a descent from the Egyptian god Osiris." (115)

It will be several years before Dadd returns to the genre of fairy paintings. Perhaps that can also be seen as a circling back, or maybe it is better also seen as a spiraling, either up or down, for as he will tell us there is no distinguishing between the two.

The Fairy-Feller's Master Stroke clearly reveals Dadd caught in several worlds. He is, of course, the painter of the work, and sees himself as being seen by others at Bethlem as the Fairy-Feller, the guy known for his paintings of fairies. At the same time, in the painting he has the Fairy-Feller about to make the master stroke to break open the walnut for Queen Mab, the queen of the world of both fairies and the dream, to use for her carriage, the nut, in other words, to carry the life force of the dream. Thus, he, Richard Dadd the man, is positioned, ready to open the world of dreams, which includes the spiritual world of Osiris, at the command of the Patriarch, already discussed. Art, theatre, dreams, the mythic, spiritual, and religious worlds, the whole realm of the psyche, and even the physical world, the real body existence of a real man all get intertwined here, and

Dadd is the at the center of them all, or, perhaps, more honestly, the relation of Dadd to the Patriarch is at the center, that invisible tension that exists between the Patriarch's eyes and those of the Fairy-Feller.

The rich intertwining becomes too much to reduce to explanation. The struggles with the suspension of the ethical in the name of faith, of some world of meaning and value beyond explanation, some absurd ultimate truth, some demand from an unethical god in order to gain salvation freeze the painting. Everyone in the painting is held motionless, waiting for the Fairy-Feller, who, in turn, waits for the command from this mysterious Patriarch, this demanding God of Abraham.

According to Graham Reynolds, the painting "has been interpreted as implying that all the figures are under a spell from which the Fairy Feller will release them as he cracks the nut." (116)

David Greysmith expresses it well:

> [The figures] live but without breathing. They wait, sometimes balefully, waiting for one's eyes to avert. There is no sense of time. The picture seems to be transmitting some subliminal message. Because we cannot wholly understand the world of the imagination which he inhabited his work is often unnerving: the accoutrements appear to be real but it is as if the focus were through alien eyes. But Richard [Dadd] explains each and every part of The Fairy-Feller and nothing is the result of automatism. Never for a moment does he lose control of his subject or paint anything unintentionally. (117)

What Greysmith is referring to in terms of Richard Dadd explaining it all is that Dadd details out the entire painting in the accompanying *Elimination*. Remember the title itself is a pun on elimination and illumination (an oxymoron presented in a single word to signal a careful reader to be alert to the endless multiplicities involved):

The duality continues. Dadd writes about it:

> . . . fay woodman holds aloft the axe
> Whose double edge virtue now they tax
> To do it singly & make single double
> Featly & neatly—equal without trouble.

Here he states it directly. He, in the role of the woodsman/Fairy-Feller, is going to split the seed of life into dualities, which later he says is "either good or bad / For not so the same terms are had." The dualities multiply. First, there is not only the reference to the worlds of painting and painter as woodman and seed of the world of Queen Mab, or the dream, but this splitting also resonates on the level of the Garden of Eden, where the biting of the fruit of the tree of the knowledge of Good and Evil brings duality into existence, man and woman, day and night, birth and death, and most important time and eternity.

Joseph Campbell offers some commentary in his interviews with Bill Moyers:

> Moyers: What is the myth of Adam and Eve trying to tell us about the pairs of opposites? What is the meaning?
>
> Campbell: It started with the sin, you see—in other words, moving out of the mythological dreamtime zone of the Garden of Paradise, where there is no time, and where mean and women don't even know that they are different from each other. The two are just creatures. God and man are practically the same. God walks in the cool of the evening in the garden where they are. And then they eat the apple, the knowledge of the opposites. And when they discover they are different, the man and woman cover their shame. You we, they had not thought of themselves as opposites. Male and female is one opposition. Another opposition is the human and God. Good and evil is a third opposition. The primary oppositions are the sexual and that between human beings and God. Then comes the idea of good and evil in the world. And so Adam and Eve have thrown themselves out of the Garden of Timeless Unity, you might say, just by that act of recognizing duality. To move out into the world, you have to act in terms of pairs of opposites.
>
> There's a Hindu image that shows a triangle, which is the Mother Goddess, and a dot in the center of the triangle, which is the energy of the transcendent entering the field of time. And then from this triangle there come pairs of triangles in all directions. Out of one comes two. All things in the field of time are pairs of opposites. So this is the shift of consciousness from the consciousness of identity to the consciousness of participation I duality. And then you are into the field of time.
>
> Moyers: Is the story trying to tell us that, prior to what happened in this Garden to destroy us, there was a unity of life?
>
> Campbell: It's a matter of planes of consciousness. It doesn't have to do with anything that happened. There is the plane of consciousness where you can identify yourself with that which transcends pairs of opposites.
>
> Moyers: Which is?
>
> Campbell: Unnamable. Unnamable. It is transcendent of all names.
>
> Moyers: God?
>
> Campbell: "God" is an ambiguous word in our language because it appears to refer to something that is known. But the transcendent is unknowable and unknown. God is transcendent, finally, of anything like the name "God." God is beyond names and forms. Meister Eckhart said that the ultimate and highest leave-taking is leaving God for God, leaving your notion of God for an experience of that which transcends all notions.

The mystery of life is beyond all human conception. Everything we know is within the terminology of the concepts of being and not being, many and single, true and untrue. We always think in terms of opposites. But God, the ultimate, is beyond the pairs of opposites, that is all there is to it.

Moyers: Why do we think in terms of opposites?

Campbell: Because we can't think otherwise.

Moyers: That's the nature of reality in our time.

Campbell: That's the nature of our experience of reality.

Moyers: Man-woman, life-death, good-evil—

Campbell: —I and you, this and that, true and untrue—every one of them has its opposite. But mythology suggests that behind that duality there is a singularity over which this plays like a shadow game. "Eternity is in love with the productions of time," says the poet Blake.

Moyers: What does that mean, "Eternity is in love with the productions of time"?

Campbell: The source of temporal life is eternity. Eternity pours itself into the world. It is a basic mythic idea of the god who becomes many in us. n India, the god who lies in me is called the "inhabitant" of the body. To identify with that divine immortal aspect of you is to identify yourself with divinity.

Now, eternity is beyond all categories of thought. This is an important point in all of the great Oriental religions. We want to think about God. God is a thought. God is a name. God is an idea. But its reference is to something that transcends all thinking. The ultimate mystery of being is beyond all categories of thought. As Kant said, the thing in itself is no thing. It transcends thingness, it goes past anything that could be thought. The best things can't be told because they transcend thought. The second best are misunderstood, because those are the thoughts that are supposed to refer to that which can't be thought about. The third best are what we talk about. And myth is that field of reference to what is absolutely transcendent. . . .

. . . transcendent means to "transcend" to go past duality. Everything in the field of time is dual. . . .

Moyers: Do you think there was such a place as the Garden of Eden?

Campbell: Of course not. The Garden of Eden is a metaphor for that innocence that is innocent of time, innocent of opposites, and that is the prime center out of which consciousness then becomes aware of the changes.

Moyers: But if there is in the idea of Eden this innocence, what happens to it? Isn't it shaken, dominated, and corrupted by fear?

Campbell: That's it. There is a wonderful story of the deity, of the Self that said, "I am." As soon as it said I am," it was afraid. (118)

Thus, in this sense, Richard Dadd is cracking open the hazelnut to all of existence, the deepest realization of all of life, the realization of *Self* or "*self*-realization," and the fear is in the realization of *Self*.

Moyers asks "Why?"

Joseph Campbell replies: "It was an entity now, in time. Then it thought, 'What should I be afraid of, I'm the only thing that is.' And as soon as it said that, it felt lonesome, and wished that there were another, and so it felt desire. It swelled, split in two, became male and female, and begot the world." (119)

This is what Sigmund Freud struggles with, that moment of *self*-realization, the moment when the self realizes its *Self*, and ends up deciding it takes place at the oedipal stage. Otto Rank breaks with Sigmund Freud and says it takes place at birth. He calls it "life fear," the fear of living autonomously, of being abandoned. Rollo May states, "This is the fear of self-actualization." (120) Joseph Campbell refers to psychiatrist Stanislav Grof as claiming it takes place in the womb just before birth, when the uterus begins the rhythms of labor, and there is the terror, "the horrific stage of getting born, the difficult passage through the birth canal, and then . . . light! . . . . That is the breaking into the world of light and the pairs of opposites." (121)

Furthermore, just as the Fairy-Feller/woodsman in the painting is representing the being about to experience self realization, he in turn is about to begin the process for the entire world, thus suggesting the creator, the artist is connecting to or is taking on the role of God.

James Joyce writes at the end of *A Portrait of the Artist as a Young Man*: "Welcome, O life! I go to encounter for the millionth time the reality of experience and to forge in the smithy of my soul the uncreated conscience of my race." (122)

Thus, conscience, the world of meaning and value, is not simply somewhere outside of the human psyche, but rather, as Rollo May states, "It is created, first of all, out of the inspiration derived from the artist's symbols and forms. Every authentic artist is engaged in this creating of the conscience of the race, even though he or she may be unaware of the fact." (123) Thus, creativity, Rollo May continues, involves "an active battle with the gods." (124)

By definition, creativity brings something new into existence, and also by definition this changes existence, re-creates existence. Archibald MacLeish quotes a Chinese poet: "We poets struggle with Non-being to force it to yield Being. We knock upon silence for an answering music." (125)

Rollo May presents exactly what we find in Richard Dadd when he states:

> A host of other riddles, which I can only cite without comment, are bound up with this major one. One is that genius and psychosis are so close to each other. Another is that creativity carries such an inexplicable guilt feeling. A third is that so many artists and poets commit suicide, and often at the very height of their achievement. (126)

The myths representing this are combining the birth of human consciousness, of *Self* awareness with the accompanying anger of the supernatural forces, of the gods. The biting of the forbidden fruit in Eden brings the wrath of God on Adam and Eve. They are condemned and lose their eternal *Self.* The same is true in the Greek myth of Prometheus. He stole fire from the gods and gave it to humans, thus representing giving humans both physical and psychological illumination, enlightenment, the ability to think and create. Zeus was outraged and condemned Prometheus to have his liver (thought at the time the myth was created to be the center of consciousness) eaten by a vulture (a black bird that lives on the flesh of death animals; black representing the opposite of innocence (white), a bird a creature of the unconscious (the sky), each day (the bright sun symbolizing conscious knowing) the liver growing back each night (night and the moon symbolizing the unconscious and regeneration), and the entire story involving the necessity of death once consciousness of *Self* as separate from unity with god is realized. Just to emphasize this, the story has a concluding possibility of release for Prometheus, where Zeus will free him from his chains if another immortal will give up his immortality (which is done by Chiron). Of course, in the Christian version this reunion with God and immortality is accomplished through the half-man-half-god Christ (both cases involving a willing sacrifice).

Whether or not Richard Dadd is consciously and logically planning or intuitively aware of his structuring, dualities serve to organize the entire painting formally as well. Just to the right of the Fairy feller, a dwarf monk's tonsure doubles the hazelnut. Just to the right of him and nearer to the viewer, an ostler from the fairy inn has his singular hair curled to match the Spanish curl of a female dwarf opposite. On the right of the painting two handsome gentlemen dressed in seventeenth century outfits, looking as if they belong in a Shakespeare play, balance off two lady's maids Dadd describes as "wenches rather smart," dressed in short, pleated skirts, with large, pointed breasts, full hips and calves, with "good legs and feet so small. Bavaria, Flanders, Germany and all Can shew no more fantastic limb." Furthermore, to emphasize the sexual quality of these voluptuous woman, he has a satyr looking up or out from beneath them at "what doth not lack an explanation," explaining and probably referring to himself as forgivable because he has been "shut out from natures game, banished from natures book of life," which he continues by claiming he has blanked out or removed from his memories and lives contented "in a paradise of fools," no doubt meaning multiple interpretations of this.

The pictorial dualities continue by having the gesture of the mandolin player matched by the woman who holds a mirror, a character he claims is "your fairy man upon the town that can clean out a swell or clown." Indeed, the entire painting is carefully balanced, the large hazelnuts in the lower right corner a counterpart to those in the upper left, the diagonal depth from lower right to upper left contrasted by the strands of grass crossing from the lower left to the upper right, the wandering path from lower left to upper right balanced by the wandering flowers from upper right to lower left, the vertical grass stems up the left side balanced by the horizontal ones across the bottom. And it is obvious that these dualities are meant to be three-dimensional. In other words, they are not just on a flat surface, but work into the depths of the painting. The central circle is meant to be seen as in the depth of the painting with characters focused on the central action in the painting, behind the surface, and their eyes all look toward a central point (the unseen woodsman's face) inside the work.

The "story" element common to this fairy painting genre, as suggested by some of the above named characters, gets both carefully detailed out and also left just as curiously mysterious, similar to a dream, capable of holding many stories and possible interpretations, but restricted from any final resolution or explanation. The microscopic detail becomes dual in itself, both illuminating and eliminating, and one of the ways Dadd does this is by having his careful designation of each of the characters in his accompanying explanation. He clarifies for us that next to the dwarf monk, just behind the ostler is a farmer's ploughman, above the axe, resting one foot across the other knee, is a satyr, complacent and fleshy that Dadd says is "proud . . . a foot and not a hoof to own. But he can put a hat upon his crown? His horns forbid." Just beneath the Patriarch, dressed in red cloak is a Politician, whom Dadd says is a "First Chop Englishman," in spite of the fact he looks very forbidding. Next to him, clad in green with fancy top hat is a fairy dandy flirting with a disgusting nymph, who, in turn are being watched by two childlike elves. At their feet is perhaps the character that most captures the eyes, a little old man with a large head, pointed ears, staring with squinting eyes passed the woodsman, directly out of the picture. Dadd calls him the Critic. He is apprehensive, judgmental. Dadd says "His business is to teach to do. Do it himself? Oh no!" Certainly,

Dadd is giving us an easy critique of a critic here. But, perhaps, he is meaning this also to mean those who stand by, who judge others in life but who are afraid to take self-responsibility themselves and "do" or "create" or simply "live" life. Those who critique and condemn others, but who don't have the talent or the guts or whatever it takes to put themselves on the line, to be alive and try to give life meaning and value. He must have thought about his own life and actions and how others had condemned him for doing what he thought he must. In this sense, it is extra interesting to note that the characters surrounding this critic are totally ignoring him.

Dadd explains that the man in the pointed hat on the left side of the painting is Lubin, the fairy tanner, the one who made the Fairy-feller's coat, cap, and leggings. The male dwarf beneath him is a conjuror and trickster, and he is taking bets on whether or not the nut will be split successfully, thus bringing in or at least highlighting another important aspect of it all. If the Patriarch does give the command, and the Fairy-feller does crack the nut, that doesn't mean it will be a successful opening up of all the worlds. What if it is flawed? A failure?

The spider hanging near him has wonderful symbolic qualities. It is a standard symbol of both creativity and of entrapment. Furthermore, the spider's gossamer webs are a common symbol of illusions. Here is one of nature's master artists weaving the webs of illusion, certainly capable of alluring, capturing, and even sucking the life blood out of its victims. Carl Jung points out that the spider sitting in its web symbolizes the center of the world, and hence is regarded in Maya as the eternal weaver of the web of illusion. (127) Marious Schneider claims that spiders in their endless weaving and killing, creating and destroying, symbolize the ceaseless alternation of the underlying forces of the universe, involving continuous sacrifice, which is a definition of man's continual transmutation or transformation throughout his life. (128) Even death here becomes but the thread used to spin out a new life, and thus is not an end but a transition from the past to the future. The moon is the ruler of the world of phenomenal forms (forms subject to growth and death) because the moon waxes and wanes though several phases (dies and is reborn), and thus the spider in being lunar is a symbol of transmutation or metamorphosis, weaving the treads of each man's destiny. On the psychic level, this means the spider is related to the imagination and creativity.

On the brim of the Patriarch's hat dance "motes" or "atomies." Queen Mab is shown on one side of the brim riding in her carriage. Some female centaurs, which Dadd says are somewhat fanciful, and two little pages playing at being Cupid and Psyche, and thus, this whole myth of the heart and mind can be incorporated into the intertwining strands. Above the brim stand Oberon and Titania, king and queen of this world of fairies, two characters Dadd has depicted throughout his career, and certainly ones that he relates to Shakespeare's *Midsummer Night's Dream*. Here they seem to be overwatching the event to take place with a bit of disdain.

Above them is a row of figures meant to represent the childhood game of counting fruit-stones or buttons in an attempt to predict adult occupations. Dadd writes, "The ragged soldier sure is mad, made so by wounds, debauch and glad but hard earned victory," points out that the sailor has a pleasure yacht which he sails for his own enjoyment, the tinker grinds shears for the tailor, who is showing a coat to the ploughboy, who, in turn, is having his pocket picked by a thief.

To the left, two boys identified as "a tatterdemalion" and a "junketeer" blow on horns, while a dragonfly hangs on grass nearby. From behind this grass, an elf peers out.

Is it possible to make the multiplicities any more complex? Well, interestingly, on the upper right corner of the painting Dadd depicts himself with pestle and mortar, perhaps mixing the necessary chemicals for the creation he is a part of, perhaps even consciously meaning to depict himself as an alchemist, the master of transformation, able to connect with the ultimate mysteries of life and bring the world of the spiritual into the world of the profane. Just how exactly he means such a reading is impossible to know. However, he gives enough clues to let us know he does have something of the sort in mind.

And in doing this Dadd becomes even more enigmatic because we are finally forced to admit his own knowledge of himself and his work, and we realize that he was able to exist in a twilight zone of overlapping, multiple realities. He indeed manages to create rich poetic spaces not just in his painting, nor just in the relationship of his commentary about his paintings and the expressions of his paintings, but in his own multiple personalities and ways of thinking. He is both insane and aware he is insane, both mad and yet more than simply sane because he is mad, his insanity the passage to higher or other realizations about existence. In Dadd, madness and genius coalesce, as if the two halves of a broken bone have grown together to form a more powerful bond than what was there before the break.

As we've seen, this realization of multiplicities gets expressed directly in his creative works. He concludes his frustrating/enticing poem/non-poem, analysis/drivel, real world/fantasy world commentary on the painting with one of the great quatrains in all of literature:

> But whether it be or be not so,
> You cannot afford to let this go,

For nought as nothing it explains
And nothing from nothing nothing gains.

Not only is the passage in and of itself a great expression of the paradoxical nature of artistic expression and its relationship to explanation, but it extends this into the workings of the mind itself. And Dadd is aware of this, stating earlier in the commentary "some dubious point to fairies only known to exist, or to the lonely thoughtful man recluse." He thus both draws us into his world or worlds and yet keeps them mysteriously beyond reduction to explanation.

The quatrain cannot help but suggest a similar passage from King Lear, especially since Dadd is enamored with Shakespeare and has done so many paintings based on his work:

"Can you make no use of nothing, nuncle?"
"Why no, boy; nothing can be made out of nothing."

Part VIII

Schizophrenia has been used in so many ways to describe so many different forms of mental illness that it becomes at times simply a catch-all term. The same is true of paranoia. However, in more strict uses, Paranoiac Schizophrenia is a firmly established category of mental illness, and it fits Richard Dadd.

The Mayo Clinic claims that the onset of paranoid schizophrenia occurs in men in their late teens or twenties. It involves delusions, such as being spied on by the FBI or perhaps being an agent for the FBI, in either case not being able to trust others because of some sense of them being the enemy. These delusions often become very complex stories that reinforce the person's non-realistic views. The person commonly feels a sense of grandeur, some sense of connecting up to superhuman powers or forces or spirits. Voices that no one else can hear are common, and at times they demand harmful actions to one-self or others.

Theories of its cause include the possibility that early brain development might involve too many or too few connections being made, specifically certain areas of the brain that are rich in the chemical dopamine. There are several different theories for why this might happen. One possibility is that there are risk-associated genes. Another is that it might be the result of certain viruses. A third is that stressful childhood environments might trigger increased amounts.

There are no blood or brain-imaging tests currently available for a diagnosis, and one form of schizophrenia has been known to mutate at times into another form. However, once schizophrenia develops, it does not go away. It is a chronic disease. (129)

Edgar Allan Poe first published "The Tell-Tale Heart" in James Russell Lowell's *The Pioneer*, January, 1843, the same year Dadd fell completely into his madness. This short story of a man both sane and insane is probably the best expression ever of such a condition. (130)

It begins in perfect schizophrenic voice:

> TRUE!—Nervous—VERY, very dreadfully nervous I had been and am; but why *will* you say that I am mad? The disease had sharpened my senses—not destroyed—not dulled them. Above all was the sense of hearing acute. I heard all things in the heaven and in the earth. I heard many things in hell. How, then, am I mad? Hearken! and observe how healthily—how calmly I can tell you the whole story.

Even in just this short initial passage, the dual insane and sane voices of the narrator are apparent, and it reads as if it could have been written by Richard Dadd himself.

As it continues, it seems almost as if it were Dadd discussing his relationship with his dad, almost as if Edgar Allan Poe had heard of Dadd and based the story on him:

> It is impossible to say how first the idea entered my brain; but once conceived, it haunted me day and night. Object there was one. Passion there was none. I loved the old man. He had never wronged me. He had never given me insult. For his gold I had no desire. I think it was he eye! yes, it was this! One of his eyes resembled that of a vulture—a pale blue eye, with a film over it. Whenever it fell upon me, my blood ran cold; and so by degree—very gradually—I made up my mind to take the life of the old man, and thus rid myself of the eye for ever.
>
> Now this is the point. You fancy me mad. Madmen know nothing. But you should have seen *me*. You should have seen how wisely I proceeded—with what caution—with what foresight—with what dissimulation I went to work! I was never kinder to the old man than during the whole week before I killed him. And every night, about midnight, I turned the latch of his door and opened it—oh so gently! And then, when I had made an opening sufficient for my head, I put in a dark lantern, all closed, closed, so that no light shone out, and then I thrust in my head. Oh, you would have laughed to see how cunningly I thrust it in! I moved it slowly—very, very slowly, so that I might not disturb the old man's sleep. It took me an hour to place my whole head within the opening so far that I could see him as he lay upon his bed. Ha!—would a madman have been so wise as this?

After describing how he crept in precisely at midnight for seven nights to gaze in on this father figure of a man, he describes how, on the eight night, once again with great caution, with great care and patience and cunning, he entered the old man's room, this time accidentally making a noise when his thumb slipped, but again displaying great patience in standing completely silent and still for a whole hour until the man again fell asleep. But then he describes how he began to hear the old man's heart beating, saying "It increased my fury, as the beating of a drum stimulates the soldier into courage." Finally, the beating drives him to kill the man. Then, he continues, "If still you think me mad, you will think so mo longer when I describe the wise precautions I took for the concealment of the body." And he proceeds to explain how he carefully cut up and hide the body under the floor. Then two officers arrive. A neighbor had heard a shriek during the night. But he is not concerned. Instead, he seems to enjoy it all. He says:

> I smiled,—for *what* had I to fear? I bade the gentlemen welcome. The shriek, I said, was my own in a dream. The old man, I mentioned, was absent in the country. I took my visitors all over the house. I bade them search—search *well*. I led them, at length, to his chamber. I showed them his treasures, secure, undisturbed. In the enthusiasm of my confidence, I brought chairs into the room, and desired them *here* to rest from their fatigues, while I myself, in the wild audacity of my perfect triumph, placed my own seat upon the very spot beneath which reposed the corpse of the victim.

He is in control. He is confident. He is enjoying his superiority over these lesser minds. But the insane part of him starts to intrude, the beating of the heart becomes louder and louder and drives him into a frenzy. He says:

> The officers were satisfied. My *manner* convinced them. I was singularly at ease. They sat, and while I answered cheerily, they chatted familiar things. But, ere long, I felt myself getting pale and wished them gone. My head ached, and I fancied a ringing in my ears: but still they sat and still chatted. The ringing became more distinct:—it continued and became more distinct: I talked more freely to get rid of the feeling: but it continued and gained definitiveness—until, at length, I found that the noise was *not* within my ears.
>
> No doubt I now grew *very* pale;—but I talked more fluently, and with a heightened voice. Yet the sound increased—and what could I do? It was *a low, dull, quick sound—much such a sound as a watch makes when enveloped in cotton.* I gasped for breath—an yet the officers heard it not. I talked more quickly—more vehemently; but the noise steadily increased. I

> arose and argued about the trifles, in a high key and with violent gesticulations, but the noise steadily increased. Why *would* they not be gone? I paced the floor to and fro with heavy strides, as if excited to fury by the observation of the men—but the noise steadily increased. Oh God! what *could* I do? I foamed—I raved—I swore! I swung the chair upon which I had been sitting, and grated it upon the boards, but the noise arose over all and continually increased. It grew louder—louder—*louder*! And still the men chatted pleasantly, and smiled. Was it possible they heard not? Almighty God!—no, no! They heard!—they suspected!—they *knew*!—they were making a mockery of my horror!—this I thought, and this I think. But any thing was better than this agony! Any thing was more tolerable than this derision! I could bear those hypocritical smiles no longer! I felt that I must scream or die!—and now—again!—hark! louder! louder! louder! *louder*!—
>
> "Villains!" I shrieked, "dissemble no more! I admit the deed!—tear up the planks!—here, here!—it is the beating of his hideous heart!"

Now consider this report about Richard Dadd by Dr. Hood:

> For some years after his admission he was considered a violent and dangerous patient, for he would jump up and strike a violent blow without any aggravation, and then beg pardon for the deed. This arose from some vague idea that filled his mind and still does to a certain extent that certain spirits have the power of possessing a man's body and compelling him to adopt a particular course whether he will or no. When he talks on this subject and on any other at all associated with the motives that influenced him to commit the crime for which he is confined here, he frequently becomes excited in his manner speaking, and soon rambles from the subject and becomes quite unintelligible. He is very eccentric and glories that he is not influenced by motives that other men pride themselves in possessing—thus he pays o sort of attention to decency in his acts or words, if he feels the least inclination to be otherwise; he is perfectly sensual being, a thorough animal, he will gorge himself with food till he actually vomits, and then return again to the meal. With all these disgusting points in his conduct he can be a very sensible and agreeable companion, and shew in conversation, a mind once well educated and thoroughly informed in all the particulars of his profession in which he still shines and would it is thought have pre-eminently excelled had circumstances not opposed." (131)

In "Schizophrenia—The Inward Journey," Joseph Campbell discusses paranoid schizophrenia as the psychological equivalent of a failed hero's adventure. (132) He has been invited to deliver a talk in conjunction with Dr. John Weir Perry, noted psychiatrist, obtains a paper of Weir on schizophrenia published in the *Annals of the New York Academy of*

Sciences (133), and realizes that the imagery of schizophrenia matches that of a mythological hero's adventure. His discussion begins:

> My own [theory] had been a work based on a comparative study of the mythologies of mankind, with only here and there passing references to the phenomenology of dream, hysteria, mystic visions, and the like. Mainly, it was an organization of themes and motifs common to all mythologies; and I had had no idea, in bringing these together, of the extent to which they would correspond to the fantasies of madness. According to my thinking, they were the universal, archetypal, psychologically based symbolic themes and motifs of all traditional mythologies; and now from this paper of Dr. Perry I was learning that the same symbolic figures of mind of modern individuals suffering from a complete schizophrenic breakdown: the condition of one who has lost touch with the life and thought of his community and is compulsively fantasizing out of his own completely cut-off base." (134)

After making this general comparison, Campbell continues the discussion with a more concrete correspondence:

> Very briefly: The usual pattern is, first, of a break away or departure from the local social order and context; next, a long, deep retreat inward and backward, as it were, in time, and inward, deep into the psyche; a chaotic series of encounters there, darkly terrifying experiences, and presently (if the victim is fortunate) encounters of a centering kind, fulfilling, harmonizing, giving new courage; and then finally, in such fortunate cases, a return journey of rebirth to life." (135)

According to Perry, the best way to handle such a schizophrenic breakdown might be to allow the patient to live out the fantasy, to let the psychic journey take its course in the belief it will, as does a hero's adventure, eventually work its way back to the world of light, back out of the nightmare to the world of day. This view is held by others.

Well known psychiatrist Rollo May writes:

> [The] autobiographical novel *I Never Promised You a Rose Garden*, tells the experience of a young schizophrenic woman, Deborah, in her actual treatment with a psychiatrist. The stirring events in the treatment of this girl read like a contemporary extraterrestrial film. In her therapy we see a constant and gripping interplay of myths. Deborah (as she is called) lived with the mythic figures of Idat, Yr. Anterrabae, Lactamaen, the Collect, all of whom inhabited the Kingdom of Yr. Since Deborah could communicate with no one else in the world, she desperately needed these mythic figures. She writes, "the gods of Yr had been companions—secret,

precisely *sharers of her loneliness*." She would flee to them when she was terrified or unbearably lonely in the so-called real world.

On the way to the sanatorium, as Deborah tells us, she and her parents stayed overnight in adjacent rooms in a motel.

> On the other side of the wall, Deborah stretched to sleep. The kingdom of Yr had a kind of neutral place which was called the Fourth Level. It was achieved only by accident and could not be reached by formula or an act of will. At the Fourth Level there was no emotion to endure, no past or future to grind against.
>
> Now, in bed, achieving the Fourth Level, a future was of no concern to her. The people in the next room were supposed to be her parents. Very well. But that was part of a shadowy world that was dissolving, and now she was being flung unencumbered into a new one in which she had not the slightest concern. In moving from the old world, she was also moving from the intricacies of Yr's kingdom, from the Collect of Others, the Censor, and the Yr gods. She rolled over and slept a deep, dreamless, and restful sleep.
>
> Next morning, she tells us, she felt the great reassurance and comfort the myths had given her. . . . it occurred to Deborah, as the car pulled away from the motel and out into the sunny day, that the trip might last forever and that the calm and marvelous freedom she felt might be a new gift from the usually too demanding gods and offices of Yr.

Not only are these gods in Deborah's scheme remarkable for their imaginative depth, but they are remarkable as well for their great similarity to what has been shown thirty years later in *E.T.*, *The Return of the Jedi*, *Close Encounters of the Third Kind*, and other extra-worldly films which attract millions of children and adults in our late twentieth century. Deborah was schizophrenic. But where one draws the line between schizophrenia and intensely creative imagination is a perpetual puzzle. Again Hannah Green (her pen name) writes:

> She began to fall, going with Anterrabe through this fire-framed darkness into Yr. This time the fall was far. There was utter darkness for a long time and then a graynes, seen only in bands across the eye. The place was familiar; it was the Pit. In this place gods and Collect moaned and shouted, but even they were unintelligible. Human sounds came, too, but they came without meaning. The world intruded, but it was a shattered world and unrecognizable.

The psychiatrist who served as therapist for Deborah at Chestnut Lodge, Frieda Fromm-Reichmann, wisely made clear to Deborah at the outset that she would not pull these gods away against Deborah's will. Dr. Frieda, as

> she is called the book, worded them into the treatment, suggesting sometimes to Deborah that she tell her gods such-and-such, or occasionally asking her what her gods say. What is most important is that Dr. Fromm-Reichmann respected Deborah's need for these mythic figures, and she sought to help Deborah to see that she, Deborah, had her part in creating them. In one session,
>
> > "Our time is over," the doctor said gently, "You have done well to tell me about the secret world. I want you to go back and tell those gods and Collect and Censor that I will not be cowed by them and that neither of us is going to stop working because of their power."
>
> But when Dr. Frieda had to go to Europe for a summer, Deborah was temporarily assigned to a younger psychiatrist who was imbued with the new rationalism. This psychiatrist marched in to destroy the "delusions" of Deborah with no understanding whatever of Deborah's need for her myths. The result was that Deborah, her whole system of gods and their extraterrestrial kingdom in shambles, deteriorated markedly. She regressed into a completely withdrawn world. She set fire to the sanatorium, burned and maimed herself, and behaved like a human being whose humanity is destroyed. For this is literally what had happened. Her soul—defined as the most intimate and fundamental function of her consciousness—was taken away, and she had literally nothing to hold on to. (136)

Dr. Perry introduced Joseph Campbell to "Shamans and Acute Schizophrenia," an article by Dr. Julian Silverman of the National Institute of Mental Health. (137) In addition to it describing a Shamanic experience and equating that to a psychological experience, a form of psychosis, the article distinguishes two types of schizophrenia, Essential Schizophrenia which involves a withdrawal from the experiences of the external world, and Paranoid Schizophrenia, where the person remains alert, sensitive to the external world, perhaps even highly social, but at the same time sees the external world as a projection of his own inner nightmare, his own fears, fantasies, and terrors. The external world, in other words, is a projection of his inner world.

In Essential Schizophrenia the person disassociates from the symbolism of the external world and must find psychic health in his own developed symbol systems. This fits the case of Deborah.

It also fits Richard Dadd. However, he does not withdraw from the external world and the symbolism there. In fact, he plunges into that very symbolism in complex details and multi-faceted ways, but he bends it to suit his own nightmarish world.

He is, in fact, a highly educated, cultivated man who knows his artistic and intellectual world well, an artisan of the first order, and if one is to attempt to understand his expressions of meaning and value, one must know often minute details of the artistic and intellectual works that make up European culture.

Obviously, however, his inner demons, his schizophrenic nightmare has taken over, and been projected onto this exterior world, people such as his own loving father seen as agents of evil to be coldly murdered in the name of some nightmarish god, in the service of the dream.

Joseph Campbell discusses the stages of the schizophrenic fall. First, he says, there is a sense of splitting. The world divides and the person sees himself in two roles. In one, he is but a clown, a ghost, an outsider, a fool, a person seen as inferior. In the other, he sees himself as the savior, the hero chosen for a high purpose. In an astonishing coincidence with Richard Dadd and his motionless *Fairy-Feller's Master Stroke*, Campbell quotes a man having such an experience saying "I have seen my father. . . . He is old now and has told me just to wait. I shall know when the time comes for me to take over." (138)

The second is of regression, falling back, becoming more one with the world outside of human existence, the exaltation of transcending personal bounds, what Freud referred to as the "oceanic feeling," the feeling of a new knowledge, of an ability to understand the deep mysteries of life. It is similar to the experience of a yogi, a saint, a shaman. In Richard Dadd's case, it can be argued that he suddenly, perhaps in starts and fits, feels he understands the mysterious exotic world of Egypt, the myths of Osiris, the dark rituals of the people who surrounded him and danced and smoked the sacred pipes.

Certainly Dadd also experiences Campbell's next stage, that of feeling an important task lay ahead, dangerous, but carrying with it the belief that invisible hands are there to help, to guide and protect, and that at the end of it, when all of the hardships, the tortures, the negative forces have been suffered, a great terrible rapture (something in the mode of a sublime experience) will result, an awakening, perhaps in the sense of the Buddha, the awakened one.

Psychiatrist R. D. Laing, in his book *The Politics of Experience*, discusses the case of Jesse Watkins, a former Royal Navy commodore, then a sculptor, who, at the culmination of his journey experiences a sense of sheer light, a terribly dangerous, over-powering light to be encountered and endured. (139)

Briefly, here is his experience. He is born the last day of the 19th century, December 31, 1899, goes to sea in 1916 during WWI on a tramp steamer, and ends WWII as a commander and Commodore of coastal convoys. During this time, he encounters ship wrecks, mutiny and murder.

He also paints throughout this life and has some short stories published about his adventures at sea.

After retiring from the service, he begins working long hours. During this time he grows emotionally, physically, and spiritually run down. Then he is bitten by a dog, and the wound will not heal. Finally he goes to the hospital, and is given an anesthetic. Upon his return from the hospital, he sits down in a chair and his son finds him acting strangely.

This when the "voyage" begins. He describes it for Laing in sessions taped twenty-seven years later:

> "Suddenly I looked at the clock and the wireless was on and then the music was playing . . . something like Revel's repetitive tune. . . . Suddenly I felt time going backwards.
>
> I even felt it so strongly I looked at the clock and in the same way I felt that the clock was reinforcing my own opinion of time going back . . . I felt alarmed because I suddenly felt as if I was moving somewhere on a kind of conveyor belt—and unable to do anything about it, as if I was slipping along and sliding down a chute. . . . I looked into the mirror at myself, and I looked in a strange way . . . as though I were looking at someone who—who was familiar but . . . very strange and different from myself.
>
> My wife became very . . . worried. . . . I was going back into sort of previous existences [My wife, the next door neighbor and the doctor who had been brought over] looked at me as if I were mad. . . And the next thing was that an ambulance came and I was taken off.
>
> I was put into bed [in and observation ward]. . . I had the feeling that I had died. And I felt that other people were in beds around me and I thought they were all other people that had died—and they were there-just waiting to pass on to the next department . . .

According to Dr. Laing, this experience is not representing a real death but an ego death, which is accompanied by a feeling of an enhanced significance of everything, of the entire world and the experiences one has, as Jesse's next statements indicate:

> I started going into . . . feelings of real regression in time. . . . experiencing everything . . . At one time I actually seemed to be wandering in a kind of landscape . . . as if I were an animal. . . . I felt as if I were a kind of rhinoceros or something like that and emitting sounds like a rhinoceros and being at the same time afraid and at the same time being aggressive and on guard. And then . . . going back to further periods of regression and even sort of when I was just struggling like something that had no brain . . . as if I were just struggling for my own existence against

> other things which were opposing me. . . . Then at time I felt as if I were like a baby . . . I could even hear myself cry like a baby.
>
> All these feeling were very acute and . . . real. . . . I was aware of these things happening to me. . . I was a sort of observer of myself but yet experiencing it. . . . I found that I had periods when I came right out of this state . . . comparatively lucid states. . . .
>
> I had a letter from my wife. . . . she said "The sun is shinning here" . . . and "It's a nice day." . . . And I felt that if this letter was from her, she was in a quite different world. She was in a world that I could never inhabit any more—and this gave me feelings of alarm and I felt somehow that I was—I'd gone off into a world that I could never move out of.

Here Jesse clearly sees the world as splitting into two worlds, the physical world he once inhabited and occasionally still enters and the other world, the psychological, spiritual, mythological world, precisely as we saw in Richard Dadd. Jesse is both unable to stop his hallucinations and yet able to step outside of them and see himself as an outsider would. This wandering back and forth between these two worlds has him frightened, has him lost, in a twilight zone, whereas if he had completely entered into the "other" world, the dream world, he would not have had these uneasy feelings, as he would have been unaware of the dual realities.

Jesse also thinks his enhanced powers can influence the real world, and discusses how he tells a nurse not to bother bandaging a cut on his finger because he can will it to heal itself, and how he feels he is able by mere thoughts to will other patients to lie down and stop making noises.

He continues:

> I felt that I had . . . tapped powers that I in some vague way . . . reincarnation. . . . [I felt] an enormous journey in front . . . , a fantastic journey, . . . and it seemed that I had got an understanding of things which I'd been trying to understand for a long time, problems of good and evil and so on. . . . I had come to the conclusion . . . that I was more–more then I had always imagined myself, not just existing now, but I had existed since the beginning— . . . from the lowest form of life to the present time. . . . and ahead of me was lying the most horrific journey . . . a journey to being aware of all—everything . . . and I felt this so strongly, it was such a horrifying experience to suddenly feel that, that I immediately shut myself off from it because I couldn't contemplate it. . . . I was unable to take it.

Dr. Laing discusses how Jesse feels there are three levels of existence, an antechamber level, a central level, and a higher world. He feels most people are in waiting in the antechamber to move into the next level, the one he has entered.

He continues:

> I had feelings of . . . gods, not only God but gods as it were, of beings which are far above us capable of . . . dealing with the situation that I was incapable of dealing with, that were in charge and were running things and . . . at the end of it, everybody had to take on the job at the top. . . . And at the same time I felt that . . . God himself was a madman . . . because he's got this enormous load of having to be aware and governing and running things . . . and that all of us had to come up and finally get to the point where we had to experience that ourselves. . . . and every single one of us has to go through it . . . the purpose of everything and the whole of existence is . . to equip you to take another step, and another step, and another step, and so on . . .

After several days, Jesse gets to the point where he decides it was time to return from his "voyage." The nurses have reached a decision to put him in a padded cell, apparently because he talks so loud all night he keeps them up. However, he asks a nurse to leave the door open, and she does. This physical door, then, becomes the psychological door he will walk through to return to sanity.

He says:

> I sat on the bed, and I though, well, somewhere or other I've got to sort of join up with my present . . . self.

The nurse offered him some medicine, but he declined, determined to no longer be mentally sick. He continued:

> I sat on the bed and I held my hands together, and as—I suppose in a clumsy way of linking myself up with my present self, I kept on saying my own name over and over again and all of a sudden, just like that—I suddenly realized that it was all over.

Dr. Laing concluded by emphasizing that "We can no longer assume that such a voyage is an illness that has to be treated," but rather that "this voyage is not what we need to be cured of, but that it is itself a natural way of healing our own appalling state of alienation called normality," and referred to Carl Jung as leading the way toward understanding this.

As mentioned, Campbell tied this psychological journey to the mythological journey of the Hero's Adventure, repeating Dr. Laing's conclusion by expanding it with the following claims:

> Something much the same was the view, also, of both Dr. Perry and Dr. Silverman . . . and, as I have most lately learned, the earliest documented proposal of this view was in a study published by C. G. Jung

> already in 1902, "On the Psychology and Pathology of So-called Occult Phenomena." (140)

"In sum," then, he continued,

> the inward journey of the mythological hero, the shaman, the mystic, and the schizophrenic are in principle the same; and when the return or remission occurs, it is experienced as a rebirth: the birth, that is to say, of a "twice born" ego, no longer bound in by its daylight-world horizon. It is now known to be but the reflex of a larger self, its proper function being to carry the energies of an archetypal instinct system into fruitful play in a contemporary space-time daylight situation. One is now no longer afraid of nature; nor of nature's child, society—which is monstrous too, and in fact cannot be otherwise; it would otherwise not survive. The new ego is in accord with all this, in harmony, at peace; and, as those who have returned from the journey tell, life is then richer, stronger, and more joyous.
>
> The whole problem, it would seem, is somehow to go through it, even time and again, without shipwreck: the answer being not that one should not be permitted to go crazy; but that one should have been taught something already of the scenery to be entered and the powers likely to be met, given a formula of some kind by which to recognize, subdue them, and incorporate their energies. (141)

Returning to Richard Dadd, we now have some interesting frames for trying to understand him, his psychological journey, his painting *The Fairy-Feller's Masterstroke*, and his accompanying poem. How well does he fit all of the discussion above? Well, it seems clear he did wander through a twilight zone of multiple realities. The world of fairies, the world of mythology, the world of the dream, the world of art, the world of psychology, and the world of simple physical existence certainly got intermingled for him, and all existed, all were real in some way. There can be no doubt that he saw himself at the center of all of these worlds. He was the Fairy-Feller in the painting waiting for the command from the Patriarch to cut open the hazelnut. As suggested earlier, this in itself opens multiple realities, for the Fairy-Feller in the painting ultimately exists in one world as merely a fictional character in a painting, the result of Richard Dadd's imagination, a character that populated the world of the fairy, which was also the world of the dream, and the world of mythology, and the world of the psyche. At the same time this character was Dadd himself. So, in depicting himself in this fictional world he was saying that he, that Dadd himself was the person capable of cracking open the hazelnut, and the hazelnut was the carriage of Queen Mab (of the main life force of the world of dreams). Thus, Dadd was in position to provide the vehicle for the dream to begin. But, he waited for the command from the Patriarch, which has been discussed earlier as that father figure, both real and yet spiritual, in Dadd's case clearly represented in his schizophrenic connection to Osiris.

Dadd was also capable of disassociating himself from his dual personalities, both the servant of Osiris, insane and condemned by the "real" world, seen as someone to be pitied, to be ostracized, to be locked up, and at the same time a person superior to this "real" world in having been chosen by a god as a special person with higher knowledge. And the other depiction of him in the painting, the creator of this world, the one who conjures it up, the alchemist, represented this.

Whether or not Dadd went through the schizophrenic journey or voyage described above can be debated. The various commentaries still in existence about him make it obvious he retained his beliefs in Osiris his entire life, and thus perhaps never did complete the journey, never did find his way back (which actually fits the current definitions of paranoid schizophrenia better than these stories of people who did make it back). He was seen by others consistently as a poor figure, someone to be pitied. At the same time, there are many references to his sane personality, especially as time went on in his imprisonment, and he made friends with those in charge of him, was even considered a teacher by Hayden in something of a role reversal. Also while there is no statement from Dadd or anyone else that he returned or recovered from his insanity, the *Fairy-Feller's Master Stroke* reveals a highly complex comprehension of what he had gone through and where he had emerged, much in line with shamanic experiences.

This becomes even more interesting when current views about the connections of artists with shamans and spirituality are explored.

Joseph Campbell states it again and again in his interviews with Bill Moyers:

> Moyers: Who interprets the divinity inherent in nature for us today? Who are our shamans? Who interprets unseen things for us!
>
> Campbell: It is the function of the artist to do this. The artist is the one who communicates myth for today. But he has to be an artist who understands mythology and humanity and isn't simply a sociologist with a program for you.
>
> . . . .
>
> Moyers: So shamans functioned in early societies as artists do now. They play a much more important role than simply being—
>
> Campbell: They played the role the priesthood traditionally plays in our society.
>
> Moyers: The shamans were priests?

Campbell: There's a major difference, as I see it, between a shaman and a priest. A priest is a functionary of a social sort. The society worships certain deities in a certain way, and the priest becomes ordained as a functionary to carry out that ritual. The deity to whom he is devoted is a deity that was there before he came along. But the shaman's powers are symbolized in his own familiars, deities of his own personal experience. His authority comes out of a psychological experience, not a social ordination.

Moyers: he shaman has been somewhere I haven't, . . .

Campbell: Also, . . . the shaman may translate some of his visions into ritual performances for his people. That's bringing the inner experience into the outer life of the people themselves.

Moyers: This is the beginning of religion?

Campbell: Personally, I think that's how religions began. But that's just a guess. We don't really know.

Moyers: Jesus goes into the wilderness, experiences a psychological transformation, comes back, and says to people, "Follow me." And this happens in these elementary cultures?

Campbell: That's the evidence we have. We find a shamanic aspect in practically all the hunting cultures.

Moyers: Why, particularly, in the hunting cultures?

Campbell: Because they're individual. The hunter is an individual in a way that no farmer will ever be. Toiling in the fields and waiting for nature to tell you when you're going to do it is one thing, but going off on a hunt—every hunt is a different hunt from the last one. And the hunters are trained in individual skills that require very special talents and abilities.

Moyers: So what happened to the shaman in human evolution?

Campbell: When this big emphasis came on the settled village life, the shaman lost power. In fact, there's a wonderful set of stories and myths of some of the Southwestern American Indians, the Navaho and Apache, who were originally hunting peoples who came down into an area where agriculture had been developed and took on an agricultural system of life. In their stories of the beginning, there is typically an amusing episode where the shamans are disgraced and the priests take over. The shamans say something that offends the sun, and the sun disappears, and then they say, "Oh, I can bring the sun back." Then they do all their tricks, and these are cynically, comically described. But their tricks don't bring back the sun back. The shamans are reduced, then, to a shaman society, a kind of clown society. They are magicians of a special power, but their power is now subordinate to a larger society. (142)

At another point, Campbell says:

Myth must be kept alive. The people who can keep it alive are artists of one kind or another. The function of the artist is the mythologization of the environment and the world.

Moyers: You mean artists are the mythmakers of our day?
Campbell: The mythmakers of earlier days were the counterparts of our artists.
Moyers: They do the painting on the walls, they perform the rituals.
Campbell: Yes. There's an old romantic idea in German, *das Volk dichtet,* which says that the ideas and poetry of the traditional cultures come out of the folk. They do not. They come out of an elite experience, the experience of people particularly gifted, whose ears are open to the song of the universe. These people speak to the folk, and there is an answer from the folk, which is then received as an interaction. But the first impulse in the shaping of a folk tradition comes from above, not from below.
Moyers: In these early elementary cultures, as you call them, who would have been the equivalent of the poets today?
Campbell: The shamans. The shaman is the person, male of female, who in his late childhood or early youth has an overwhelming experience that turns him totally inward. It's a kind of schizophrenic crack-up. The whole unconscious opens up, and the shaman falls into it. The shaman experience has been described many, many times. It occurs all the way from Siberia right through the Americas down to Tierra del Fuego. (143)

Later in the same set of interviews;

Moyers: What is illumination?
Campbell: The illumination is the recognition of the radiance of one eternity through all things, whether in the vision of time these things are judged as good or evil. To come to this, you must release yourself completely from desiring the goods of this world and fearing their loss. "Judge not that you be not judged," we read in the words of Jesus. "If the doors of perception were cleansed," wrote Blake, "man would see everything as it is, infinite."
Moyers: That's a tough trip.
Campbell: That's a heavenly trip.
Moyers: But is this really just for saints and monks?
Campbell: No, I think it's also for artists. The real artist is the one who has learned to recognize and to render what Joyce has called the "radiance" of all things, as an epiphany of showing forth of their truth. (144)

This is expressed most simply in John Keats' famous couplet from *Ode on a Grecian Urn*: "'Beauty is truth, truth beauty,—that is all / Ye know on earth, and all ye need to know.'" (145) Piet Mondrian calls it "pure reality," claiming that "the appearance of natural forms changes but reality remains constant." (146) Pablo Picasso states:

> We all know that Art is not truth. Art is a lie that makes us realize truth, at least the truth that is given us to understand. . . . And from the point of view of art there are no concrete or abstract forms, but only forms which are more or less convincing lies. That those lies are necessary to our mental selves is beyond any doubt, as it is through them that we form our aesthetic point of view of life. . . . To me there is no past or future in art. If a work of art cannot live always in the present it must not be considered at all. The art of the Greeks, of the Egyptians, of the great painters who lived in other times, is not an art of the past; perhaps it is more alive today than it ever was. (147)

Part IX

Some difficult concepts and the terms used to identify them are at play here. *Aesthetics* has been used in several senses through the years, all referring to some kind of beauty. It is being used by John Keats (representing all of the Romantic Movement poets) as equal to the highest truths, those beyond explanation, truths that we know, but cannot explain. These are the same truths to be found in the worlds of *faith*, the *sublime* and the *numinous, each term offering a slight variation or other category of knowing beyond explanation*. Carl Jung, in attempting to clarify what he means by *active imagination* (something he calls by many different terms, originally the *transcendent function*) uses *aesthetic knowing* as the beauty of form, stating "we could say that aesthetic formulation needs understanding of the meaning, and understanding needs aesthetic formulation. The two supplement each other to form the transcendent function." (148) Here he is trying to clarify that *aesthetics* alone does not lead to the higher truth, or to transformation, but that *aesthetics* applies to the *symbolic expressions of the unconscious*, which need *conscious interpretation* or *critical comprehension* to result in the final truth. In the language of neurology, the symbolic knowing of the right cerebral hemisphere (which works in the same manner or a very similar manner to the unconscious) needs to be combined with the literal, logical knowing of the left cerebral hemisphere (conscious knowing) to reach the final truth. Jung calls this left cerebral hemisphere form of thinking (conscious thinking) the *intellect* and the right cerebral hemisphere form of thinking (unconscious thinking) *aesthetics*. John Keats is considering the *final* truth an *aesthetic* truth. For him, meaning and logic, content and form have already been combined, this combination an expression of the highest truth. Ultimately, Carl Jung and John Keats are saying the same thing, but Jung uses the term aesthetics in a slightly different manner, tending to suggest they disagree, when they actually support each other.

I use the term *metaesthetics* to extend this, *metaesthetics* meaning to take the *explanations* of the left cerebral hemisphere or conscious or intellect or logical, literal forms of thinking as raw data for the right cerebral hemisphere or unconscious or symbolic forms of thinking. Since the highest truths are understood symbolically, they need to obtain the highest position, and thus, instead of using artistic *expressions* as raw data for *explanations*, it makes sense to reverse this. The process, however, is a continual back-and-forth exchange, a spiral to ever higher truths, the truths of meaning and value.

An expression of the highest truth results in an *epiphany*. Joseph Campbell explains:

> Joyce's formula for the aesthetic experience is that it does not move you to want to possess the object. A work of art that moves you to possess the object depicted, he calls pornography. Nor does the aesthetic experience move you to criticize and reject the object—such art he calls didactic, or social criticism in art. The aesthetic experience is a simple beholding of the object. Joyce says that you put a frame around it and see it first as one thing, and that, in seeing it as one thing, you then become aware of the relationship of part to part, each part to the whole, and the whole to each of its parts. This is the essential, aesthetic factor—rhythm, the harmonious rhythm of relationships. And when a fortunate rhythm has been struck by the artist, you experience a radiance. You are held in aesthetic arrest. That is the epiphany. And that is what might in religious terms be thought of as the all-informing Christ principle coming through. (149)

As Campbell immediately points out, and as should be obvious at this point, "The aesthetic experience transcends ethics and didactics." (150) It is in the realm of meaning and value that we already saw Soren Kierkegaard emphasize involves a *teleological suspension of the ethical.*

Oedipus solved the Riddle of the Sphinx, demonstrating his intelligence and releasing his city from its control. Thus Thebes saw Oedipus as a savior and made him its king.

However, a problem remained. Apollo had put a plague on Thebes, and it was thought that the plague would remain until the person responsible for murdering the former king, King Laius was found and brought to justice.

Oedipus vowed to once again save his city. He condemned this murderer as evil and pushed forward to find him. Curiously, others were not sure he should do this. His wife, the queen, once married to Laius, urged him not to continue with his search. The great blind seer Teriasies told him to let it go.

Why? It made no sense to Oedipus. Evil had been done, and the city remained under a plague. Certainly he as the king was obligated to save his people, to discover the truth, to right the wrong.

But then came the astonishing discovery. Oedipus himself was the man who had murdered Laius, and seemingly for not a very good reason, seemingly simply out of anger and pride. Now came the moment of truth. The moment in life when one faces the ultimate challenge, the ultimate horror inside oneself. The key was knowledge, not just any knowledge, but *self-knowledge*, the truth about who and what one is.

The gods had demanded that he do what he did. He had been *fated* to kill his father and marry his mother. That was not of his own making. Yet it was considered *ethically* evil to murder one's father and wed one's mother. He had said it, time and again; the man who murdered Laius was evil and must be punished. He had condemned himself to damnation.

But, now, now that he knew, now he had gained some power he did not have prior to this, the power that comes from *the knowledge of self*, now he could decide (be in charge of his own actions, his own ethos, now he was in a position to think for him self, to take *self-responsibility, the only way to give life meaning and value*. The choice was harsh, seemingly unfair. He could remain as king of Thebes, perhaps justify this action by saying the punishment was unfair. He had not known after all that it was his father he had killed, that he had married his own mother. Thus, he could deny his own darkness, his own Shadow desires (what Freud later named after him as the desire to kill one's father and marry one's mother; the Oedipus Complex, a standard stage in male development; Jung designated the female version of this the Electra Complex). The city would remain under a plague, condemned by Apollo, continuing as a wasteland, a place where frightening entities lived in the shadows of each person, and appeared literally in the form of punishment from the gods, but no one would be to blame for it. It was simply the human condition being lived out in the city of Thebes. It had been fated by the gods, a fate issued because of the sins of a previous generation.

Oedipus chose not to follow this line of reasoning. Instead, in what must be seen as an act of heroism, of self-sacrifice, he condemned himself, banned himself from Thebes, and in so doing took all of the darkness, all of the evil shadows of human existence into himself, freeing Thebes to be reborn in its original innocence.

Sigmund Freud thought this a perfect story to represent his views of early childhood desires to kill one's father and have sex with one's mother. For Freud, this became the initial emergence into consciousness, the initial and basic drive to life that gave birth to all of the other drives. He writes:

> If *Oedipus Rex* moves a modern audience no less than it did the contemporary Greek one, the explanation can only be that its effect does not lie in the contrast between destiny and human will, but is to be looked for in the particular nature of the material on which that contrast is exemplified. There must be something which makes a voice within us ready to recognize the compelling force of destiny in the Oedipus, while we can dismiss as merely arbitrary such dispositions as are laid down in *Die Ahnfrau* or other modern tragedies of destiny. And a factor of this kind is in fact involved in the story of King Oedipus. His destiny moves us only because it might have been ours—because the oracle laid the same curse upon us before our birth as upon him. It is the fate of all of us, perhaps, to direct our first sexual impulse towards our mother and our first hatred wish against our father. (151)

This psychological combination of self-consciousness, of gaining the knowledge of whom and what one is with the potential toward evil matches a standard interpretation of the Garden of Eden.

"In Adam's fall we sinned all" begins the famous Puritan alphabet used to teach children that humans are all born in sin as result of Adam and Eve's original fall from grace. (152)

In the minds of these God fearing Puritans that original sin meant that all mortal flesh was infected with an innate sinfulness! Ann Bradstreet, one of the acclaimed Puritan poets, expresses it in a poem titled "Childhood":

> Stained from birth with Adam's sinful fact,
> Thence I began to sin as soon as act:
> A perverse will, a love to what's forbid,
> A serpent's sting in pleasing face lay hid:
> A lying tongue as soon as it could speak
> And fifth commandment do daily break. (153)

Condemned by God himself because of that initial weakness, the Puritans saw life as but a continual battle against the forces of evil within, and only the most stringent discipline, the most masochistic existence could bring salvation. As Cotton Mather was known to preach, it was better for a child to be whipped than damned! (154)

After all, (and again we revisit those influential passages) the *King James Bible* makes it clear in Genesis:

6. And when the woman saw that the tree *was* good for food, and that it *was* pleasant to the eyes, and a tree to be desired to make *one* wise, she took of the fruit thereof, and did eat, and gave also unto her husband with her: he did eat. 7. And the eyes of them both were opened, and they knew that they *were* naked; and they sewed fig leaves together, and made themselves aprons.

8. And they heard the voice of the Lord God walking in the garden in the cool of the day: and Adam and his wife hid themselves from the presence of the Lord God amongst the trees of the garden.

9. And the Lord God called unto Adam, and said unto him, Where *art* thou?

10. And he said, I heard thy voice in the garden, and I was afraid, because I *was* naked; and I hid myself.

11. And he said, Who told thee that thou *wast* naked? Hast thou eaten of the tree, whereof I commanded thee that thou shouldest not eat?

12. And the man said, "The woman whom thou gavest *to be* with me, she gave me of the tree, and I did eat.

13. And the Lord God said unto the woman, What *is* this *that* thou hast done? And the woman said, "The serpent beguiled me, and I did eat.

14. And the Lord God said unto the serpent, Because thou hast done this, thou *art* cursed above all cattle, and above every beast of the field; upon they belly shalt thou go, and dust shalt thou eat all the days of they life:

15. And I will put enmity between thee and the woman, and between thy seed and her seed; it shall bruise thy head,

and thou shalt bruise thy his heel.

16. Unto woman he said, I will greatly multiply thy sorrow and thy conception; in sorrow thou shalt bring forth children, and thy desire *shall by* to thy husband, and he shall rule over thee.

17. And unto Adam he said, Because thou hast hearkened unto the voice of thy wife, and hast eaten of the tree,

of which I commanded thee, saying, Thou shalt not eat of it: cursed *is* the ground for thy sake; in sorrow shalt thou eat *of* it all the days of thy life; thorns also and thistles shall it bring forth to thee; and thou shalt eat the herb of the field: in the sweat of thy face shalt thou eat bread, till thou return unto the ground, for out of it wast thou taken: for dust thou *art*, and unto dust shalt thou return.

20. And Adam called his wife's name Eve; because she was the mother of all living.

21. Unto Adam also and to his wife did the Lord God make coats of skins, and clothed them.

22. And the Lord God said, Behold, the man is become as one of us, to know good and evil: and now, lest he put forth his hand, and take also of the tree of life, and eat, and live for ever:

23. Therefore the Lord God sent him forth from the Garden of Eden, to till the ground from whence he was taken.

24. So he drove out the man; and he placed at the east of the garden of Eden cherubim, and a flaming sword which turned every way, to keep the way of the tree of life. (155)

It was clear. Life was a curse! Children were a curse! Because Adam and Eve had eaten the forbidden fruit of the tree of the knowledge of good and evil, God had condemned them and all of their children to an eternal suffering, and, since all that now remained to separate humans and God was the fruit on the tree of eternal life, humans were forbidden to come near it.

The curse had been cast! Innocence had been lost, and once lost could never be regained! How unfortunate! What a horrible fall from grace!

But wait! How fortunate! A wondrous fall! The birth of meaning and value! The birth of human existence! The birth of mind! The human mind! In the human brain!

The intersection of two planes of existence!

The fall from Grace IS the miracle of human existence!

The *King James Bible* tells us that before this fall Adam and Eve "were both naked, the man and his wife, and were not ashamed." (156) That is, they were innocent to shame, were without a sense of good or evil. It was a state of animal existence. The human brain did not have a map for judging good and evil.

Thus, the first humans to be were unable to be either loyal and obedient or disloyal. They simply did what they were told to do. It did not matter who told them, as they could not distinguish good from evil. So, when snake told Eve to bite of the fruit because it was good, she could only see that snake was right and bite of it.

Genesis tells us:

1. Now the serpent was more subtle than any beast of the field which the Lord God had made. And he said unto the woman, Yea, hath God said, Ye shall not eat of every tree of the garden?

2. And the woman said unto the serpent, We may eat of the fruit of the trees of the garden:

3. But of the fruit of the tree which *is* in the midst of the garden, God hath said, Ye shall not eat of it, neither shall ye touch it, lest ye die.

4. And the serpent said unto the woman, Ye shall not surely die:

5. For God doth know that in the day ye eat thereof, then your eyes shall be opened, and ye shall be as gods, knowing good and evil.

6. And when the woman saw that the tree *was* good for food, and that it *was* pleasant to the eyes, and a tree to be desired to make *one* wise, she

> took of the fruit thereof, and did eat, and gave also unto her husband with her: and he did eat. (157)

It is not surprising that it is the woman who is first tempted to bring about the transformation, since it is obvious in nearly all of the natural world that the female is the giver of life, the one who literally brings new life into being, and, thus, she is also the one who births the new level of existence out of the old.

Bill Moyers questions Joseph Campbell: "It does seem that this story has done women a great disservice by casting Eve as responsible for the Fall. Why are women the ones held responsible for the downfall?"

Joseph Campbell replies simply: "They represent life. Man doesn't enter life except by women, and so it is woman who brings us into this world of pairs of opposites and suffering." (158)

And it is not surprising that it is a snake that tempts this transformation, as the snake is the creature who transforms, shedding its skin to be reborn. According to Ania Teillard, the snake is "an animal endowed with magnetic force. Because it sheds its skin, it symbolizes resurrection. Because of its sinuous movement '(and also because its coils are capable of strangling)' it signifies strength. Because of its viciousness, it represents the evil side of nature." (159) Heinrich Zimmer claims that the snake is the life force, determining both birth and rebirth. (160) Carl Jung claims that snakes are a well documented representation of transformation and renovation, a firm archetype, suggesting as an example the Egyptian Uraeus as the expression of the Kundalini (the Yoga concept of the snake as the coiled ring in at the bottom of the spinal cord that uncoils and climbs through all of the chakras up the spinal column to the third eye of Shiva, giving man the sense of the eternal; thus, representing the necessary sexual energy capable of spiritual awareness) on a higher plane. (161) J. E. Cirlot points out that Philo of Alexandria (representing a common view of ancient writers) "believed that when the snake shakes off its skin it likewise shakes off its old age, that it can both kill and cure and that it is therefore the symbol and attribute of the aggressive powers, positive and negative, which rule the world. (This is a Gnostic and Manichean idea of Persian provenance.) He decided finally that it is the 'most spiritual of animals'." (162)

The snake is also a creature of the swamp, the fertile birthing place of life, the place where earth mixes with water (water being the most powerful place of transformation, because it is the natural life-giving amniotic fluid of the womb, because it is the dominant element in living beings, and, in a tautology, because it is the place of the birth of life, as opposed to a desert, which is without life). J. E. Cirlot refers to Enel (163) in pointing out that the snake is "used, as are other reptiles, to refer to the primordial—the most primitive strata of life. In the *Book of the Dead* (XVII), the reptiles are the first to acclaim Ra when he appears above the surface of the waters of Nou (or Nu or Nun)." Marcia Eliade points out that, in India, snake cults or cults of the spirit of the snake are connected with the symbolism of the waters of the sea. Snakes are guardians of the springs of life and of immortality, and also of those superior riches of the spirit that are symbolized by hidden treasure. (164)

And the snake has obvious phallic qualities in its shape, its texture (slippery), its ability to be both soft and hard, its sinuous movements, its threatening tongue, its ability to spit out poison, its resemblance to a ligament, and its undulating body patterns.

Furthermore, the snake has close associations with women and seduction beyond Eve. According to Pail Diel the snake not only symbolizes personal sin, but the principle of evil inherent in all of life, as represented, for example, in the serpent of Midgard. (165) Eliade refers to Gresmann (166) who claimed that Eve was an archaic Phonetian goddess of the underworld who represented the snake, and Eliade supports this by referencing many Mediterranean deities who are shown carrying a snake (i.e., Artemis, Hecate, and Persephone), and further relating these to Cretan priestesses, and to Medusa and the Erinyes, mythic females with snakes for hair. (167)

Very interesting in context of all of this is a huge amount of myth or folklore surrounding Lilith, especially the specific versions involving the dual creation stories in Genesis. Genesis 1:27 reads, "So God created man in his own image, in the image of God he created him; male and female he created them." Here Adam and his wife are made from clay or dust, equal and adult.

The creation story in Genesis 2 has God create Adam, and then, showing some pity on Adam because Adam is lonely, create his wife Eve from his rib. Here man and woman certainly are not created at the same time or given equal status. While Biblical scholars and clergy have come up with numerous explanations for this discrepancy (and some others in the two stories), an important body of stories has evolved outside of the Bible that claim Adam had two wives. His first wife was Lilith.

Lilith, created equally and at the same time as Adam, considered herself his equal and refused to submit to him, most specifically refused to assume the standard missionary position beneath him for sex, thus violating the command to be fruitful and multiply. Instead, she left him. The various stories give various ways she left (one is that she simply spoke the Holy Name of God and disappeared).

But her disappearance didn't mean she was gone, and a whole body of stories developed around her as the negative female, one who consorted with the fallen angels and other demons, perhaps bore and killed children of Adam, and in terms of the fall from innocence in the Garden of Eden, either as the snake that tempted Eve or the one who prompted it to do so. (168)

J. E. Cirlot also points out connections of the snake to the tree, which, he says, "being unitary may be said to correspond to the masculine principle, in which case the ophidian would represent the feminine. Thus, the tree and the serpent are, in mythology, prefigurations of Adam and Eve (if the above story is incorporate, better stated as prefigurations of man and woman). Furthermore" Cirlot continues, "by analogy, we also have here a situation of symbolic Entanglement—the snake curled round the tree (or round the staff of Aesculapius)—and a symbolic image of moral dualism." (169)

The garden is also a place of growth and transformation, and this is God's garden, so the transformation would naturally be a growth into a spiritual or mental world. Also, since man was made from clay, the physical substance, but "God created man in his *own* image, in the image of God created he him," (Genesis 1, 27) man must have within him the spiritual, the mental level of existence.

The invisible world of meaning and value, the world of the human mind, joins with the visible world, the literal, physical world, in the brain. In other words, the mind is the reality beyond the limitations of matter, and the brain is where this mental world enters into and patterns the visible world of matter. It is, then, in the human mind/brain that the mapping out of meaning and value take place, and, thus, that is where the world of physical existence obtains an *ethical* and a *spiritual* quality. Physical existence (life) has meaning and value simply and solely because of this interaction in the human mind/brain. The mapping out of meaning and value is the purpose of human existence, and the fulfillment of it is the highest experience possible. Thus, denying mere meaningless physical existence for a higher mental or spiritual existence brings the highest *joy*, and *joy* here is meant as that experience of the wonder of the universe and the knowledge of being a central part of that wonder. *Joy is the realization of self, the affirmation of life.*

Part X

Referring to C. S. Lewis' famous phrase "surprised by joy," Abraham H. Maslow discusses this realization of self, this "self-actualization" as "experiencing fully, vividly, selflessly, with full concentration and total absorption . . . "without the self-consciousness of the adolescent" . . . , "a moment when the self is wholly and fully human." It is an ongoing process, a continuing sequence of choices (we create ourselves in our choices), and making these choices involves "listening to the impulse voices" within us rather than simply following the "voice of the Establishment, of the Elders, of authority, or of tradition." This requires being honest with ourselves and "taking self responsibility." If we do this throughout our lives, then we began to find our "destiny," what our "mission in life will be." And it must involve working to reach our highest potential, not settling for less than our best. When we do this, we have "peak experiences . . . , moments of ecstasy." (170)

Joseph Campbell discusses Abraham Maslow's theories of "peak experiences" as those moments in life when one operates at the highest level of one's potential, when everything comes together and a perfect unity is realized, and offers his own experiences when he was a member of his track and field relay team at Columbia: "During the second race, I knew I was going to win even though there was no reason for me to know this, because I was touched off as anchor in the relay with the leading runner thirty yards ahead of me. But I just knew, and it was my peak experience. Nobody could beat me that day. That's being in full form and really knowing it." (171)

This is not to suggest that a peak experience must be physical. Later in this same interview, Campbell relates it more directly to the spiritual experience: "But when you really realize the sound, "AUM," the sound of the mystery of the word everywhere, then you don't have to go out and die for anything because it's right there all around. Just sit still and see it and experience it and know it. That's a peak experience." (172) These are moments of *joy*, and the greatest of these is when a human enters the greatest moment of meaning and value. This occurs when mindless, physical existence, existence without ethos or that which is beyond ethos, is shattered by the tremendous power of the human mind to deny such random monotony; and the greater, the more dramatic the denial, the more meaningful the result. Thus, the endurance of the most *horrific suffering* brings the most *sublime salvation*, which is the transcending of terror in the realization that even the most horrifying experience cannot deny human consciousness of self, that it only exists in the consciousness of self, and thus, that it affirms consciousness of self, and, in turn, affirms both an ethical and a spiritual universe, a universe of meaning and value.

Abraham Maslow uses the phrase "self-actualization" as the highest "need" for human existence. It is self-fulfillment, the reaching of one's potential. And, since the highest potential for humans is the giving of meaning and value to existence, that is what all humans need to do to be human, and, thus, that is the journey of life (the meaning is in the journey), and the peak experiences are when it is being realized.

The answer to the *Absurd Hero* put forth most famously by Albert Camus in *The Myth of Sisyphus* (173) is that there is indeed meaning and value to existence because of the joining of mind/brain in the very humans whose purpose it is to give existence meaning and value. It is a tautology. As long as humans exist, life has meaning and value, because, by definition, humans are the givers of meaning and value. Furthermore, humans are not separate from existence, but a part of existence, and, thus, any existence that has humans in it must have meaning and value. Since humans are the givers of meaning and value, to deny this as the human place in existence is to deny human existence. And since denial is a decision, just as acceptance is, it is an ethical and a spiritual choice, and becomes a double negative, negating itself. In other words, humans are, in fact, giving existence meaning and value even in their attempt to deny giving it meaning and value. They cannot escape themselves no matter how hard they try. In fact, the harder they try to deny self and self-responsibility, the more they confirm it. The less they struggle with either acceptance or denial of self, the less human quality they possess, and the more they become simply mindless, and without meaning and value. Thus, ironically, humans, the more human they are, are fated to have free will!

Sisyphus rolls the rock up the mountain in Tartarus. It rolls back down. Sisyphus rolls the rock up the mountain. It rolls back down. It is an eternal punishment. Seems at first meaningless and boring and stupid. That is what the gods thought when they condemned him to it. After all, at least if we believe Homer, Sisyphus is considered the wisest of all humans. (174) The eternal, physical condemnation of such a meaningless task can only be a horrible punishment. But not for Sisyphus, because Sisyphus is human, which means he has a mind, and the human mind denies the meaningless physical world its meaninglessness, turning its limitations into a world of value and meaning. The landscape of the mind does not match that ethical and spiritual wasteland of the mindless body, either in time or space, and, in fact, given such a horrific physical situation, is also given the most ennobling challenge, a chance to succeed at something that's horrific, which, in fact, is the chance to demonstrate that the human mind does escape such physical limitations (unless they result in damaging or eliminating the mind; for example if the brain, the physical container of the mind is denied or damaged). In truth, the gods must fail in this punishment, must prove their own limitations, demonstrating that they do not comprehend the human mind, the human spirit, do not comprehend the human ability to rise above weakness and endure even the most harsh of physical punishments (because the mind both endures physical suffering for ethical and spiritual desires (for meaning and value), and escapes the cages of physical time and space).

Unless the assumption is that the gods wished Sisyphus to succeed, for the punishment is reversed, and Sisyphus can only fail if he refuses the demand. His refusal, especially since, as is demonstrated again and again in such stories, he is the best of humans (the wisest), would say that humans cannot rise above physical limitations. And, as with all humans operating at peak level of experience, because it is, ultimately, a chance at self-realization, Sisyphus does not refuse, but, rather, accepts this most horrifying punishment, and in so doing gives the universe meaning and value, at the same time demonstrating a quality only humans can achieve, nobility.

But the choice of a myth to express the human condition already predisposes that meaning and value exist, because myths are the very maps of meaning and value humans have always embraced. Literal, physical existence has already been superseded by the invisible existence of the mind in postulating gods who have the power to subdue the laws of nature. From a cold, scientific perspective, Sisyphus as mere physical being could not accomplish an eternal task, a task beyond his physical limitations of time in the physical world.

It is, in fact, in cave wall paintings in such places as Altamira, Spain, and Lascaux, France, that we first sense that distinguishing feature, that mythic world beyond the visible that only appears with the human world of the dream and imagination. Deep in dark caves within the earth, humans create paintings that can only have value as symbolic objects that cannot have any literal, practical value, unless they first have a mythic, spiritual value. Literally, the paintings accomplish nothing, cannot even be said to offer some kind of decorative value, as the caves were totally dark unless lit up with fire. But they can and must offer a symbolic value. Since they are of animals, and the hunt, they must have been meant to influence that hunt, to make the hunt successful, to protect the hunters, to justify the killing and eating of animals. They are the evidence from the deepest past that claims human existence. It is not the crude tools or weapons found from ancient history that demonstrate human existence. Animals will use a blunt object or a sharp stone to accomplish a task. But animals do not paint pictures, or, by extension, write poetry, compose music or perform complex rituals based on equally complex myths.

Loren Eiseley writes:

> Although there is still much that we do not understand, it is likely that the selective forces working upon the humanization of man lay essentially in nature of the socio-cultural world itself. Man, in other words, once he had "crossed over" into this new invisible environment, was being as rigorously selected for survival within it as the first fish that waddled up the shore on its fins. I have said that this new world was "invisible." I do so advisedly. It lay, not so much in his surroundings as in man's brain, in his way of looking at the world around him and at the social environment he was beginning to create in his tiny human groupings.
>
> He was becoming something the world had never seen before—a dream animal—living at least partially within a secret universe in his head with other, similar heads. Symbolic communication had begun. Man had escaped out of the eternal present of the animal world into a knowledge of past and future. The unseen gods, the powers behind the world of phenomenal appearance, began to stalk through his dreams.
>
> Nature, one might say, through the powers of this mind, grossly superstitious though it might be in its naive examination of wind and water, was beginning to reach out into the dark behind itself. Nature was beginning to evade its own limitations in the shape of this strange, dreaming and observant brain. It was a weird multiheaded universe, going on, unseen and immaterial save as its thoughts smoldered in the eyes of hunters huddled by night fires, or were translated into pictures upon cave walls, or were expressed in the trappings of myth or ritual. The Eden of the eternal present that the animal world had known for ages was shattered at last. Through the human mind, time and darkness, good and evil, would enter and possess the world. (175)

Rollo May extends this:

> Man's consciousness of himself is the source of his highest qualities. It underlies his ability to distinguish between "I" and the world. It gives him the capacity to keep time, which is simply the ability to stand outside the present and to imagine oneself back in yesterday or ahead in the day after tomorrow. Thus human beings can learn from the past and plan for the future. And thus man is the historical mammal in that he can stand outside and look at his history; and thereby he can influence his own development as a person, and to a minor extent he can influence the march of history in his nation and society as a whole. The capacity for consciousness of self also underlies man's ability to use symbols, which is a way of disengaging something from what it is, such as the two sounds which make up the word "table" and agreeing that these sounds will stand for a whole class of things. Thus man can think in abstractions like "beauty," "reason," and "goodness."
>
> This capacity for consciousness of ourselves gives us the ability to see ourselves as others see us and to have empathy with others. It underlies our remarkable capacity to transport ourselves into someone else's parlor

> where we will be in reality next week, and then in imagination to think and plan how we will act. And it enables us to imagine ourselves in someone else's place, and to ask how we would feel and what we would do if we were this other person. No matter how poorly we use or fail to use or even abuse these capacities, they are the rudiments of our ability to begin to love our neighbor, to have ethical sensitivity, to see truth, to create beauty, to devote ourselves to ideals, and to die for them if need be.
>
> To fulfill these potentialities is to be a person. (176)

Joseph Campbell presents the Christian expression of it:

> Abelard's idea was that Christ came to be crucified to evoke in man's heart the sentiment of compassion for the suffering of life, and so to remove man's mind from blind commitment to the goods of this world. It is in compassion with Christ that we turn to Christ, and the injured one becomes our Savior. . . It is the suffering that evokes the humanity of the human heart.
>
> Moyers: So you would agree with Abelard that mankind yearning for God and God yearning for mankind met in compassion at the cross?
>
> Campbell: Yes. (177)

Even Arthur Schopenhauer, a philosopher considered to offer the epitome of a pessimistic view of life, embraces a Christian morality (though not at all exclusively and strongly influenced by Eastern religions) that includes, in fact focuses on an embracement of *compassion* as the highest form of human existence. (178) His views here develop out of the belief that each human is but on aspect of the single act of *will* (*will* defined by him as an instinctual, irrational force that underlies all of existence) that is all of humanity, that the same humanity is within each person, that each person has all of the sufferings of all of existence within him, that because of this each person carries all of the sufferings, pains, and guilt of all of humanity. Thus, humans come-together in passion, i.e., com-passion is the unity of all humans. (179)

It should be emphasized, however, that this unity in compassion does not deny individuality or self-responsibility. In his conclusion to "Essay on the Freedom of the Will," he writes, that humans have "an unshakeable certainty that we are the doers of our deeds," resolving the contradiction between the claims that we could not have done other we have in life by claiming we are free in a higher sense, the result of an innate character that is self-determining and independent of experience. (180) This is similar to current *general wave* theory, which claims that, though we can know the general directions sub-atomic particles will take, we cannot know the paths of each individual particle. In Arthur Schopenhauer's philosophy, both the universal (as in the general wave direction) is in place, and the individual (as in the deviations of each particular particle) is in place. Thus, each individual's actions are the *spatial-temporal manifestations* of that individual person's *innate or intelligible character*.

Building in Immanuel Kant's views on the *intelligible character* (181), Arthur Schopenhauer claims that a person's *intelligible character* is a *timeless act of will*, which is the essence of a person, and can be thought of as the *subjective* part of the Platonic Idea, which *objectively* defines one's inner essence. (182) By joining this with Immanuel Kant's *empirical character* (the *intelligible character* as it is expressed in time and space), Arthur Schopenhauer finds resolution. Individual humans, thus, are both of a universal will (predetermined and without freedom) and of inexplicable individual character, thus escaping this universality.

If this is true, then we empower ourselves (manifest our intelligible character in our empirical character) through self-knowledge. As with any subject, the more we know about it, the more control we have over it, and the more potential we have to make something worthwhile through it. In other words, the more humans comprehend both their universal qualities and their individual qualities the more they can live artistic lives of value and "grace." (183) In terms of the underlying themes of Oedipus Rex, through self-knowledge, the knowledge of who and what we are, we empower ourselves with the possibility to achieve nobility, and, in the process, to gain lives of meaning and value.

But, but, but, Maurice Friedman points out:

> It is hard to escape the conclusion that Camus, in his insistence that one must imagine Sisyphus happy, has succumbed to romanticism. Certainly, there is no basis for happiness in the never-ending Sisyphic affirmation *despite* the absurd; for the absurd by Camus' definition is the absence of a meaningful relation to the world, and no amount of subjective rock pushing can change that fact. The Absurd Man of the early Camus, like the Absurd Man of Beckett, is imprisoned in the self." (184)

But *self* is not separate from physical existence. *Self* is a part of physical existence. If there were no part of the physical world in the *self* and no part of the *self* in the physical world, then there could be no relationship, no knowledge, no interaction, no existence, at least not as understood by humans. Joseph Campbell states:

> In that yin/yang figure from China, in the dark fish, or whatever you want to call it, there is the light spot. And in the light one, there's a dark spot. That's how they can relate. You couldn't relate at all to something in which you did not somehow participate. That's why the idea of God as the

> Absolute Other is a ridiculous idea. There could be no relationship to the Absolute Other. (185)

The human mind is connected to the human brain. Without mind, the brain is mere matter (has no thoughts, at least not in the sense we would consider thoughts (i.e., it might still have energy, in the sense that all matter might be considered to be a form of energy)), in essence has no connection to the world beyond the physical. Without brain, there is no mind, at least not interacting with the physical world.

Contrary to Maurice Friedman's position here, or his claims about Albert Camus' position, Albert Camus gives Sisyphus' relationship to the world a meaning! If Albert Camus elsewhere claims there is no meaningful relation to the world outside of the mind (the *absurd*) he must either mean it in some other context or be contradicting it here. Let us *not* leave Albert Camus as he leaves Sisyphus:

> I leave Sisyphus at the foot of the mountain! One always finds one's burden again. But Sisyphus teaches the higher fidelity that negates the gods and raises rocks. He too concludes that all is well. This universe henceforth without a master seems to him neither sterile nor futile. Each atom of that stone, each mineral flake of that night-filled mountain, in itself forms a world.
>
> The struggle itself toward the heights is enough to fill a man's heart. One must imagine Sisyphus happy. (186)

Sisyphus thinks about what he is doing, and he chooses to continue doing it, and he has a perspective on what he is doing, and in his mind, he can put whatever pattern he wishes upon what he does (depending on what the structure of the human brain allows), and he can travel beyond the physical existence he endures or embraces or suffers and enjoys, and he can choose to continue doing it, or not. But then the choice to not do it is to choose to not exist, at least to not exist as a human, the definition of which predisposes a world of meaning and value.

Perhaps, if Maurice Friedman chooses to deny Albert Camus' Sisyphus an authentic existence, it is because he wants to set the stage for his next claims about Albert Camus, which are that he moves beyond this "sentimental" "romantic" view in such works as *The Rebel* into a view where "the absurd is shaped into a positive image of man, one that loses none of the shock of the confrontation with the absurd yet finds a way toward authentic existence through that very confrontation," what Maurice Friedman calls a *Dialogue with the Absurd*. He writes:

> Camus' book *The Rebel* is a continuation of the absurdist reasoning of The *Myth of Sisyphus*, but one that sets forth an image of man in revolt as the latter book does not. The connecting link between the two books is the affirmation which leads the Absurd Man *not* to commit suicide despite the irrational silence that he encounters. The step beyond *The Myth of Sisyphus* is the recognition that his protest against the absurd is already the affirmation of values—of something worthwhile in the individual who rebels, in man, in the solidarity of all men. The Absurd Man recognizes that life is good since only through life can he continue his desperate encounter with the absurd. If it is good for him, it is good for all men. Therefore, the Absurd Man rejects suicide and murder alike. He accepts the contradiction inherent in the experience of the absurd—the exclusion of all values in favor of life and the recognition that life itself is a value judgment and that living means choice. If he believes in nothing else, the Absurd Man must at least believe in his protest, his rebellion. In so doing, he is demanding "order in the midst of chaos, and unity in the very heart of the ephemeral." The problem which Camus sets for himself is whether all rebellion must end in the justification of universal murder, as it did with Caligula, or whether, without laying claim to an impossible innocence, it can establish a genuine responsibility. (187)

Here, as Maurice Friedman sees it, instead of Sisyphus accepting the punishment of the gods and turning into his *self* to find a meaning in it, which Maurice Friedman sees as denying the external world, the *absurd*, the *rebel* confronts the *absurd*, and he does so simply by choosing life, for there is "something in man in the basis of which he rebels." (188) And here we come directly to man's role as *the giver of meaning and value*. Maurice Friedman quotes Albert Camus that "analysis of rebellion leads at least to the suspicion that, contrary to the postulates of contemporary thought, a human nature does exist, as the Greeks believed." (189) There is, in-other-words, something inherent in the structure of the human psyche that in-and-of-itself rebels against the *absurd,* the *absurd* here representing a world outside of the human psyche without meaning and value.

In *Resistance, Rebellion, and Death*, Albert Camus writes "I continue to believe that this world has no ultimate meaning. But I know that something in it has a meaning and that is man, because he is the only creature to insist on having one. This world has at least the truth of man, and our task is to provide its justifications against fate itself." (190) The struggle against fate (the demands of the gods, the absurd, the limitations of human existence) can hardly be said more directly, and Camus elaborates, connecting it clearly to the central realization of *Oedipus Rex*, "With all my being I shout to you that I mean not mutilating him and yet give a chance to the justice that man alone can conceive." (191) Man, here, in his insistence on rebelling against the absurdity of existence, becomes the giver of justice, of meaning and value. Albert Camus elaborates:

> Nothing is given to men, and the little they can conquer is paid for with unjust deaths. But man's greatness lies elsewhere. It lies in his decision to be stronger than his condition. And if his condition is unjust, he has only one way of overcoming it, which is to be just himself. (192)

This is how man gains meaning and value, and becomes noble. Only humans can achieve *nobility*, because only humans both live in the world of the *absurd* and *are aware of it*. And *nobility* can only be achieved through *self-sacrifice*, because meaning and value beyond *self-gratification*, in the physical world of *the absurd*, the world of the body, or brain emptied of its functions beyond instinctual demands, can only be gained by denying the *self* as the final goal. In other words, in order to have meaning and value, there must be some system of rules, some clarification of what justice is beyond the simple drive to exist, some "human nature" beyond the instincts to survive.

In this, then, there is a commonality in human existence, and the individual human willingly sacrifices his *self* for the common good, the creation and maintenance of an *ethos*.

About 12:45, January 2, 2007, Westley Autrey, a fifty-year-old construction worker waiting for the downtown local at 137th Street and Broadway in Manhattan with his two daughters, Syshe, age 4, and Shuqui, age 6, saw Cameron Hollopeter, age twenty, collapse in convulsions on the platform near him. Two women tried to help, but Cameron stumbled and fell onto the subway tracks. Without hesitation (there was no time for it, the train was rapidly approaching), Westley jumped onto the tracks and lay on top of the man, keeping him from inadvertently jerking high enough for the train to hit him. The train could not stop in time, and five cars passed over the two of them but some two inches above Westley's head. Once the train had stopped, and with the crowd anxiously anticipating the unlikely survival, Westley called up from under the tracks, "We're O.K. down here, but I've got two daughters up there. Let them know their father's O.K." The power to the train was cut, and the two men rescued. Cameron Hollopeter was taken to the hospital, and found not to have any serious injuries. Westley Autrey declined any treatment and, instead, simply went home and later to work on his night shift. Later he said, "I don't feel like I did something spectacular; I just saw someone who needed help. I did what I felt was right."

On August 16, 2007, three miners weren't as fortunate when they went into a dangerous mine shaft in an attempt to find and save fellow miners trapped below. The mine caved in, killing them and injuring six others. Utah Gov. Jon Huntsman Jr. said the three who died did so in a "remarkable act of selflessness," emphasizing this with "there is nothing more selfless than giving one's life while rescuing another."

September 9, 2007, three teenagers in Glenview, Illinois, a suburb of Chicago saw an elderly woman turn her car onto some train tracks and stop, a train rapidly approaching. Thomas Faust tells CNN on "American Morning":

> "Me and two of my friends, Tyler and Zach, we ran up to the car and started yelling and pounding on the window, and finally we got the door open, and we're like, 'Ma'am you need to get out of this car right now because there's a train coming toward it,' We unclipped her seat belt and basically just pulled her to safety and my friend, Tyler, ran and called the police, and ... we got maybe 10 feet away from the train, and it hit like six seconds after we got her out."

Joseph Campbell explains:

> There is a magnificent essay by Schopenhauer in which he asks, how is it that a human being can so participate in the peril or pain of another that without thought, spontaneously, he sacrifices his own life to the other? How can it happen that what we normally think of as the first law of nature and self-preservation is suddenly dissolved?
>
> In Hawaii some four or five years ago there was an extraordinary event that represents this problem. There is a place there called the Pali, where the trade winds from the north come rushing through a great ridge of mountains. People like to go up there to get their hair blown about or sometimes to commit suicide—you know, something like jumping off the Golden Gate Bridge.
>
> One day, two policemen were driving up the Pali road when they saw, just beyond the railing that keeps the cars from rolling over, a young man preparing to jump. The police car stopped, and the policeman on the right jumped out to grab the man but caught him just as he jumped, and he was himself being pulled over when the second cop arrived in time and pulled the two of them back.
>
> Do you realize what had suddenly happened to that policeman who had given himself to death with that unknown youth? Everything else in his life had dropped off—his duty to his family, his duty to his job, his duty to his own life—all of his wishes and hopes for his lifetime had just disappeared. He was about to die.
>
> Later, a newspaper reporter asked him, "Why didn't you let go? You would have been killed." And his reported answer was, "I couldn't let go.

> If I had let that young man go, I couldn't have lived another day of my life." How come?
>
> Schopenhauer's answer is that such a psychological crisis represents the breakthrough of a metaphysical realization, which is that you and that other are one, that you are two aspects of the one life, and that your apparent separateness is but an effect of the way we experience forms under the conditions of space and time. Our true reality is in our identity and unity with all life. This is a metaphysical truth which may become spontaneously realized under circumstances of crisis. For it is, according to Schopenhauer, the truth of your life. (193)

This compassion can be seen in the willingness to volunteer for military service, sacrificing everything, including one's life, for a set of values, whether they be those of a government, those of a religion, or those of a culture. Proof is everywhere displayed: the endless suicide bombings in Iraq, the helicopter rescue pilots in Vietnam, the Japanese *kamikaze* pilots in WWII. While some doubters might highlight political and religious propaganda, the physical benefits (monetary rewards and the like), and threats of punishment for not making the sacrifice, it is still impossible to deny the obvious belief in an ethical and/or religious system driving the self-sacrifices. Furthermore, these actions cover the whole range of human responses, from the immediate, instinctual response to the consciously thought-through actions that take place over, at times, an entire life. Humans do sacrifice their physical existence, their *selfs* for ethical and spiritual reasons, for meaning and value. *Universal* meaning and value takes the place of individual survival. These people are not risking death and dying because they want to die, but because meaning and value are more important than life itself.

This is the basis, then, for Albert Camus' statement "Man's solidarity is founded upon rebellion, and rebellion, in its turn, can only find its justification in that solidarity." (194) Solidarity here is that commonality of "human nature," which *must* give life meaning and value against the meaningless absurd outside of itself. Albert Camus continues:

> We have, then, the right to say that any rebellion which claims the right to deny or destroy this solidarity loses simultaneously its right to be called rebellion and becomes in reality an acquiescence in murder. In the same way, this solidarity, except in so far as religion is concerned, comes to life only on the level of rebellion. And so the real drama of revolutionary thought is announced. In order to exist, man must rebel, but rebellion must respect the limit it discovers in itself—a limit where minds meet and, in meeting, begin to exist. . . .

> In absurdist experience, suffering is individual. But from the moment when a movement of rebellion begins, suffering is seen as a collective experience. Therefore the first progressive step for a mind overwhelmed by the strangeness of things is to realize that this feeling of strangeness is shared with all mean and that human reality, in its entirety, suffers from the distance which separates it from the rest of the universe. The malady experienced by a single man becomes a mass plague. In our daily trials rebellion plays the same role as does the "cogito" in the realm of thought: it is the first pieces of evidence. But this evidence lures the individual from his solitude. It founds its first value on the whole human race. I rebel—therefore we exist. (195)

If Sisyphus is seen as simply accepting his fate, as passively continuing a seemingly meaningless and boring task given him by the gods, then he is not actively rebelling, is passive, and inauthentic, as Friedman claims. However, if Sisyphus is seen as choosing life, choosing to keep his psyche alive and thinking, even given the horrifically boring task of pushing a rock up a hill only to have it roll down and need to be pushed up again, then he is rebelling. He is, after all, choosing life over death, existence over non-existence. Furthermore, it is interesting to realize that humans hearing or reading this story connect to Sisyphus, have compassion and empathy for him, and knowing that the rules are that it is an eternal punishment, still hope for an escape for him, because that one gift given humans, *hope*, will not be denied. In a letter trying to clarify what his fellow Frenchmen have to withstand the seemingly inevitable defeat at the hands of the Nazis, Albert Camus writes that "hopeless hope is what sustains us." (196)

It important to keep in mind that the connections between people, this coming together in passion, this compassion, this realization of a unity of human existence behind each individual does not deny the individual and individual responsibility, just the opposite. Self-sacrifice can only be made by a self, and self must be somehow distinguishable from the community or it is incapable of choosing. This is key to Soren Kierkegaard's important distinction between two different kinds of knowing, two different kinds of truth. His discussions about obtaining an authentic life through a personal relationship with God (with the Absurd), stress the importance of the self, and the self's relation to the world as being grounded in self-reflection and introspection. Soren Kierkegaard claims that "subjectivity is truth" and "truth is subjectivity," clearly delineating what is objectively true and an individual's subjective relation to that truth. People who in *some sense* believe the same things may act differently in response to those beliefs. His primary concern is subjectivity with regard to religious and true faith, arguing that doubt is an element of faith and that it is impossible to gain any objective certainty about religious doctrines such as the existence of God or the life of Christ. It is possible to hope the conclusion that Christian doctrines are true, but to believe such doctrines only to the degree they seem likely to be true is not genuinely religious. Faith consists in a subjective relation of absolute commitment to these doctrines, which cannot be proven. (197)

Here the individual assumes responsibility, if he chooses to take it.

Furthermore, two levels of truth are taking place simultaneously, the literal, objective level, and the non-literal, subjective level. In other words, for humans, every happening in the physical world also is a happening in the invisible world of the mind. Every encounter, for the human mind, has both a literal and a symbolic meaning.

The founder of 18th-century Hasidism, Baal Shem Tov phrased it: "No encounter with a being or a thing in the course of our life lacks a hidden significance . . . The highest culture of soul remains basic and barren unless, day by day, waters of life pour forth into the soul from those little encounters to which we give our due (198) Martin Buber began with this statement and built on it to explain that it as our duty to try to grasp the hidden meaning of such exchanges—for beyond a doubt they have a meaning. Maintaining Soren Kierkegaard's personal relationship to a God and also incorporating a need for a community, a comradeship with others similar to Camus claims, Martin Buber said it is the individual's responsibility to interact, not only with the physical world, to take responsibility for turning what he calls an I-It (impersonal, meaningless interaction) into an I-Thou interaction, which requires an "I," thus self-responsibility, but also to an interaction with God. In his most important and famous book, translated as *I and Thou*, he enthusiastically (though perhaps not as clearly as he might) turns the relationship with God into a personal one, one not requiring a religion or social context, in fact, one where a religion might even get in the way. "The world is not comprehensible, but it is embraceable: through the embracing of one of its beings. Each thing and being has a twofold nature: passive, absorbable, usable, dissectible, comparable, combinable, rationalizable, and the other, the active, non-absorbable, unusable, undissectible, incomparable, noncombinable, and nonrationalizable. This is the confronting, the shaping, the bestowing of things. He who truly experiences a thing so that it springs up to meet him and embraces him of itself has in that thing known the world..." (199)

Buber's idealistic approach provoked bitter criticism; but to him his philosophy was no empty theory. It is essential to stand up for one's principles, he believed, and sounded much like Camus when he stated that even a "handful of just and honest men can prevent a society from becoming corrupt, if they speak out and say No!" Only thus, being fully himself and living his convictions, can man realize his potential as a human being. His actions were motivated by his striving toward humanism which, in his view, the Jewish people in particular should try to manifest. And if such an ideal was capable of affecting their inner life for the better, it had also to be translated into a change of behavior, individually and as a community. (200)

But now, in these views, we have reincorporated the "ethical," have found an "ethical" level of existence in the "absurd."

Nicolas Berdyaev, however, takes Soren Kierkegaard's absurd back into this process of what Jung calls "individuation," removing "ethics" from the universal as well, thus eliminating the distinction between sacrificing one's self to the cultural ethos and making a self-sacrifice to religious and the ethical the same thing, both going beyond the general cultural beliefs. He writes:

> Personal ethics signify just that going out from the "common" which Kierkegaard and Shestov consider a break with ethics, which they identify with standards of universal obligation. The personalistic transvaluation of values regards as immoral everything which is defined exclusively by its relation to the "common"—to society, the nation, the state, an abstract idea, abstract goodness, moral and logical law—and not to concrete man and his existence. (201)

He, in fact, considers the "society, the nation, the state," as basically evil, stating that "it is only as a spiritual being that man can know the good as such," as opposed to social man where "the feud between the Creator and the creature which shadows our whole existence" takes place. (202) Thus, for him, "Personality is recognized only as a subject, in the infinite subjectivity, in which is hidden the secret of existence." (203)

He is actually somewhat Freudian in seeing that the subconscious, probably better termed the unconscious, is where the dualities of spiritual man and natural man come into conflict. Man himself is the point of intersection; "in him there takes place the conflict between spirit and nature, freedom and necessity, independence and dependence." (204). This ultimately is a psychological version of existentialism, where the "I" is that inner part of the psyche, and the "Other" is the social and/or natural world the "I" must interact with. And, just as Freud did, he ends up stressing sex and the sexual drives, though he condemns them as a negative force binding humans to matter and necessity. For him, the desire is to separate the basic sexual drive from lust, purify it, thus sublimating and transforming the passions into a higher form of creativity, which includes a spiritual level, and, thus, in a sense, a human (for him, the concern is more a man) gains the highest love when the sexual desire is combined with a spiritual one, seeing, as it were, the vision of the face of the loved one in God. In this sense, he is establishing a basis for a concept put forth by others that eros, the drives of the body's organs, the lusts of sex, can be transformed through amore into agape. (205)

In more prosaic neurological terms, the brain is the world of nature, the physical world, and the mind is the spiritual world, the world beyond nature. It is in the coming together of the brain and mind in humans that the secret of existence takes place. Does the mind exist without the brain? Does the brain exist without the mind?

Can they be separated? Do the demands of nature, the logics of physiology deny self responsibility? Does the individual mind somehow escape the laws of science?

Paul Tillich takes the original insight of Descartes, the famous "*Cogito, ergo sum*," I think therefore I am" to a rephrasing of these questions. For Tillich, man cannot help but question his existence because "his very being is the question of his existence." (206) all that man can know is his own being, but he can only know this through a realization of its opposite, nonbeing. Existence only exists if there is also non-existence. Furthermore, man knows he will cease to exist, knows there was a time he did not exist, at least did not have self-awareness, and certainly expects that time will come again. This, then, produces anxiety, the fear of non-being.

Paul Tillich breaks down ontological anxiety into three categories: fate and death, emptiness and meaninglessness, and guilt and condemnation. These are all simply the result of the human condition and cannot be eliminated.

The anxiety of fate is a simple realization that things happen that seem to have no logic and are beyond human control. We do not, for example, have any control over whether or not we are born, or over the particular environment we are born into. Nor do we have a choice between life and death. Perhaps we can choose the time and manner of our death if we choose to commit suicide, but we cannot chose to not die at some time, and truthfully in a relatively short time span. Along with this realization is the potential that everything is simply meaningless. For Tillich, man "is human only by understanding and shaping reality, both his world and himself, according to meanings and values." Thus, only if man has some kind of free will beyond fate to participate in the creation of meaning and value can he exist in the sense of human existence. The anxiety of ultimate meaningless, then, is the result of a sense that, perhaps, in truth, there is no real meaning in existence. It is important here to stress that, for Tillich, this question is not one to be argued from a pragmatic, a metaphysical, or an intellectual perspective, but, rather, a simple realization that life without meaning results in human loss of motivation. In other words, there is something about human existence that necessitates meaning and value. Furthermore, man has no choice but to judge his own existence in terms of ethics. Within himself, and unable to know if anything exists outside himself, man must still find a means of judging himself. Here, it is not social or culture ethos determining the judgment, but a self imposed ethos.

However, even though self is all that man can affirm, since he can only affirm it in contrast to non-self or non-being, he must also exist in the surrounding worlds that he cannot affirm. In other words, self must interact with the environment, and the values of self must be confirmed or denied in these encounters. Here the being of the individual must believe in the non-being that exists outside of himself. This, if taken in a religious sense, means that a non-personal God, what Tillich calls a God beyond God. An all powerful or all omnipotent God denies individual existence. Thus, there is the need for that outside God, but at the same time the realization that the outside God does not exist except as it exists inside the individual.

In the end, Tillich can be said to be rephrasing the insight that humans are both imprisoned by and freed up by the realization that all that can be absolutely known is the existence of some *self,* which can only be realized if there is some *non-self*, some environment for the *self*, that this *self* needs to judge itself in terms of some ethical system which might or might not match that of the environment, and that however God is understood, God must be both all powerful and all knowing and yet not so, because an all powerful, all knowing God denies self determination, which, in turn denies any meaning in life, which humans cannot accept, because the very definition of human must include meaning and value.

Furthermore, a God, however defined, must escape reason and logic, must be represented as being beyond the rules, must be of the world of the *absurd*, for without the *absurd* there can be no dialogue, no chance for self-responsibility. There must be something beyond reason and logic (an *absurd*) or life is meaningless. If logic, if math and science can ultimately explain the *absurd*, then it is no longer *absurd*, but simply a form of inevitable processes, all governed by eternal rules, humans, existing in that physical world, lose self-responsibility, and life is meaningless.

Thus, there are two conflicting drives taking place in the psyche of humans, one the drive to *explain* (math and science), and the other the drive to *deny explanation* (spirituality and all of the arts, the world of *expression*). This becomes the ultimate paradox of human existence. It is the knowing of the unknowable, the sensing of something just beyond one's grasp, impossible to ever grasp, yet still existing. It does not exist, yet it does. And this saturates all of human thinking, both what falls under *explanation* and what falls under *expression*. There is in *explanation* the belief that all of existence must be able to be *explained*, even if the most sophisticated attempts seem to point to an ultimate paradox, and there is in expression the belief that one can intuitively grasp the ultimate knowing, even if it cannot be completely understand what is being grasped.

Here humans seem to have reached the edge of the abilities of the human brain, the place where it cannot hold a more knowing mind. How can eternity be comprehended? How can infinity be comprehended? Clever illusions can be created to give these ungraspable realizations metaphysical and even scientific masks, but even the most clever of us knows the masks are not that which is beyond the masks.

Albert Camus writes:

> Man stands face to face with the irrational. He feels within him his longing for happiness and for reason. The absurd is born of this confrontation between the human need and the unreasonable silence of the world. (207)

Rather than make Soren Kierkegaard's leap of faith to God from this realization, Albert Camus remains firmly in the realm of what humans can know, not that mysterious God beyond knowing, not an easy illusion to embrace without reason, but, rather, a "lucid reason noting its limits." It's the moment before the act, when self-awareness remains that Albert Camus wants to deal with. He writes:

> I don't know whether this world has a meaning that transcends it. But I know that I do not know that meaning and that it is impossible for me just now to know it. What can a meaning outside my condition mean to me? I can understand only in human terms. What I touch, what resists me—that is what I understand. And these two certainties—my appetite for the absolute and for unity and the impossibility of reducing this world to a rational and reasonable principle—I also know that I cannot reconcile them. What other truth can I admit without lying, without bringing in a hope I lack and which means nothing with the limits of my condition. (208)

The absurd here becomes, not a rejection of reason or self-awareness, but the opposite, lucid awareness of the limitations of human knowing (a clarity that denies the illusions used to hide the truths), and yet a conscious, determined affirmation of human existence even in awareness of the absurdity of that affirmation in the meaninglessness of the universe.

In an earlier quote from Maurice Friedman about Albert Camus' representation of Sisyphus, he groups Samuel Beckett, the central play-write from Theatre of the Absurd, with the early Albert Camus.

Friedman argues that Samuel Beckett works hard in his plays to strip away all possible maps of meaning and value in human existence. Nothingness. Meaninglessness. For Friedman, Samuel Beckett is determined to convince his audience and his readers that all else is merely illusion, attempts by humans to hide the truth. Friedman begins his discussion of Beckett's most famous play, *Waiting for Godot* by pointing out its similarities to Simone Weil's *Waiting for God,* strongly suggesting that *Waiting for Godot* is, intentionally or not, a parody of *Waiting for God*. If so, Beckett has taken one of the most highly praised spiritual writers of his time, a woman constantly praised as perhaps the one true *saint* in the modern world, and this work, though often difficult to work through, as perhaps the greatest expression of true faith in the contemporary world. Thus, if Beckett did indeed choose this work as the object of a satire, he purposely choose what he would have felt was one worthy of it.

Friedman writes:

> There is no reason to say that Mr. Godot is God, for there is no reason to suppose that in the world of the absurd God exists, even in Simone Weil's anguished sense. But there is an abundance of religious motifs which can be taken neither seriously nor ironically. In the total context, they appear as a parody of even such harsh religious consolation as Simone Weil offers. The parallels with Weil's *Waiting for God* are numerous and explicit: the discussion of the thief on the cross beside the Saviour; the repeated suggestion of the tree as the cross; the frequent mention of prayer, supplication, listening; the recognition that they have no rights any more; and the paradoxical waiting for someone whom they do not know, have not seen, and would not know if they saw him. There is the suggestion of man (Lucky) propitiating God (Pozzo) so that he will give up the idea of parting with him. There is a longing for death—"Will night never come?"—and the seeming acceptance of suffering from God for reasons that only God knows. There is the constant comparison with the crucified Christ, reminiscent of Weil's "I envy Christ his crucifixion" and even such explicit Weil-ian statements as "When you seek you hear. . . . That prevents you from finding." (209)

Such similar passages, however, though they suggest a parody, can be written off as the types of passages that are not unique to either Simone

Weil or Beckett, but simply more representative of what humans tend to say about such things. Friedman takes them another step, tying them more elaborately to his concerns:

> Above all, what is important is not that they [Lucky and Pozzo] know Godot exists, but that Godot should know that they exist. It is *their* existence that is in doubt, not as a fact but through lack of confirmation, lack of being needed, wanted, called. Waiting for Godot means, for them, waiting to pass away the time, or as Gogot puts it more exactly, "to give us the impression that we exist." (210)

Then, Friedman leads this through an undermining by Beckett of the wait for God (Godot) in a statement by Estragon, "We are not saints, but we have kept our appointment. How many people can boast as much?" To which Estragon responds, "Billions!"

This harsh comment on the wait for God, the meeting up with God, Friedman points out is the blunt final commentary on the previous overblown monologue from Vladimir:

> Let us do something, while we have the chance! It is not every day that we are needed. Not indeed that we personally are needed. Others would meet the case equally well, if not better. To all mankind they were addressed, those cries for help still ringing in our ears! But at this place, at this moment of time, all mankind is us, whether we like it or not. Let us make the most of it, before it is too late! Let us represent worthily for once the foul brood to which a cruel fate consigned us! . . . In this immense confusion one thing alone is clear. We are waiting for Godot to come. . . . Or for night to fall. (211)

Here, Friedman claims is the perspective of the absurd. Then he ties this to his claims of parody:

> Simone Weil says in *Waiting for God* that one thing is sure: when one asks God for bread, one may trust that HE will not give one a stone. But it is a stone—a burial stone—that the dog is given which, in the poem recited by Vladimir at the beginning of the second act of *Waiting for Godot*, comes to the kitchen to try to get some bread. And it is a stone that Vladimir and Estragon are given in *Waiting for Godot.* (212)

There can be little doubt that Friedman has given us some comparisons. He then goes on to suggest that some have found comparisons of Lucky's misguided thinking about his sufferings and his role as a much maligned and punished slave with Weil's suffering and personal God in *Waiting for God*, and suggests that, if so, then the interruption of such thoughts with the nonsensical "quaquaquaqua" babbling offers an obvious satire. Perhaps, we think, perhaps Beckett really did have Weil's book next to him as he wrote this play. Perhaps, however, we also think, Friedman has just been overly clever here, reaching for a particularized parody when in truth the parody is more of a general one. Ultimately, we might ask, does it matter?

Simply by suggesting a very conscious, careful parody, Friedman is undermining his larger claim, which is that Beckett is giving us a meaningless world. Friedman concludes his discussion of *Waiting for Godot*:

> All Beckett's works seem concerned with the long, slow process of dying. Beckett's characters endlessly and painfully drag themselves through a wasteland of nonpresence and nonexistence, the ultimate goal and relief of which is death. "Don't mind me. Don't take any notice of me. I do not exist. The fact is well known," says Mrs. Rooney in *All That Fall*, and Mrs. Fitt says: "I suppose the truth is I am not there, Mrs. Rooney, just not really there at all. I see, hear, smell, and so on, I go through the usual motions, but my heart is not in it." Such plot as there is in the play—the meeting of a train which is unexpectedly late because a child has fallen (or been thrown?) to its death from the train—is robbed of any possible dramatic tension by the general absurdity. "Did you ever wish to kill a child?" the blind Mr. Rooney asks his wife and adds, after a pause, "Nip some young doom in the bud." If human existence has neither meaning nor value, then the taking of a child's life does not mean a tragic foreshortening but sparing a doomed creature the misery of a lifelong dying. For the Buddha, too, existence was seen as suffering, but, however negatively stated, there was a positive deliverance from suffering in the cessation of existence in Nirvana. In Beckett's world there is not positive deliverance from suffering, only quick or slow death. (213)

If Friedman is correct, and the point of Beckett's *Waiting for Godot* is that life is meaningless then the writing of the play itself undermines this view, especially if it has been a carefully constructed parody. It is hard to get around the simple fact that anyone believing that life is meaningless would not work so hard to construct a play.

Furthermore, if this play is the ultimate expression of an existence where there is "no authentic existence which is possible, and hence no judgment or accountability . . . . no image of man in the sense of a direction of authentic existence . . . . only realism, honesty, and despair . . . anguish and boredom," why has it garnered so much attention, so much commentary, so many attempts at interpretation? Certainly in a meaningless world such a plethora of responses, even violent reactions both for and against it would not make sense. But the dialogue spirals on forever. One response might be that this is the point, that humans are driven to find meaning in a meaningless world, are driven to find meaning even in a play that purposely expresses meaninglessness, and in doing this make themselves out to be simply pathetic, simply the perfect example of the characters in the play waiting for a God that will never come. Perhaps the illusion of meaning is the final wall to acceptance of meaninglessness. But the spiral continues. If humans do find meaning even in their attempts to present a meaningless world doesn't that in itself support a belief that, at least for humans, meaning must exist. Humans, even with all of their mental powers, their logic and reason, their insatiable search for TRUTH, cannot escape meaning and value. The very search for the ultimate truth *must* involve a brain that supports meaning and value. Otherwise, there could be no realization of the whole concept of meaning and value. In other words, the mere fact that humans can realize such a thing as meaning and value means it must or at least can exist in the human brain/mind. All of the clever dialogues are the result of exactly what they are trying to deny. It might have been a convoluted, confusing, and frustrating journey, but hasn't it circled back to Descartes' simple claim, *Cogito, ergo sum*?

Part XI

It is worth noting the physical environment Beckett and Camus are struggling to clarify, the literal wasteland to match T. S. Eliot's poetic wasteland.

Certainly the possibility of a meaningless world was thrust upon the entire history of western thought in concrete terms over the first half of the nineteenth century in clear, dramatic terms with the advent of WWI followed by WWII, leaving the world in a state of shock, a dark foreboding landscape dominated by the very real potential to simply be blown-up into the end of human existence completely.

Such mass murder as took place in these wars, accented by the bombing of Hiroshima and Nagasaki, and the dramatic denial of the value of human life put forth very consciously, supported with elaborate logic and cruel precision in the Nazi concentration camps, had to have prompted concrete questions about human existence, questions infused with a raw, naked facing of life striped bare of all of the former systems of meaning and value.

But a few excerpts from those who were first on the scene give full force, or, rather, as much as we need stomach, to the nightmarish world that European civilization, a civilization that began the century fully believing itself superior, had descended. Harry J. Herder, Jr., one of the soldiers first to arrive at Buchenwald on April 11, 1945, writes about it, including his first encounter with the dead:

> The bodies of human beings were stacked like cord wood. All of them dead. All of them stripped. The inspection I made of the pile was not very close, but the corpses seemed to be all male. The bottom layer of the bodies had a north/south orientation, the next layer went east/west, and they continued alternating. The stack was about five feet high, maybe a little more; I could see over the top. They extended down the hill, only a slight hill, for fifty to seventy-five feet. Human bodies neatly stacked, naked, ready for disposal. The arms and legs were neatly arranged, but an occasional limb dangled oddly. The bodies we could see were all face up. There was an aisle, then another stack, and another aisle, and more stacks. The Lord only knows how many there were.
>
> Just looking at these bodies made one believe they had been starved to death. They appeared to be skin covering bones and nothing more. The eyes on some were closed, on others open. Bill, Tim, and I grew very quiet. I think my only comment was, "Jesus Christ."
>
> Then, passed the stacks of bodies, through the survivors and other soldiers. The living dead had more to reveal:
>
> They pointed to a long building which was about two stories high, and butted up tightly to the chimney. It had two barn-like doors on either end of the building we were looking at, and the doors were standing open. We turned and walked back to the building where we found others from our company, along with some of the prisoners milling around in the space between the bodies and the building. We moved gently through those people, through the doors and felt the warmth immediately. Not far from the doors, and parallel to the front of the building, there was a brick wall, solid to the top of the building. In the wall were small openings fitted with iron doors. Those doors were a little more than two feet wide and about two and a half feet high; the tops of the doors had curved shapes much like the entrances to churches. Those iron doors were in sets, three high. There must have been more than ten of those sets, extending down that brick wall. Most of the doors were closed, but down near the middle a few stood open. Heavy metal trays had been pulled out of those openings, and on those trays were partially burned bodies. On one tray was a skull partially burned through, with a hole in the top; other trays held partially disintegrated arms and legs. It appeared that those trays could hold three bodies at a time. And the odor, my God, the odor. (214)

The most famous of the eye-witness reports comes from Edward R. Morrow:

There surged around me an evil-smelling stink, men and boys reached out to touch me. They were in rags and the remnants of uniforms. Death already had marked many of them, but they were smiling with their eyes. I looked out over the mass of men to the green fields beyond, where well-fed Germans were ploughing....

[I] asked to see one of the barracks. It happened to be occupied by Czechoslovaks. When I entered, men crowded around, tried to lift me to their shoulders. They were too weak. Many of them could not get out of bed. I was told that this building had once stabled 80 horses. There were 1200 men in it, five to a bunk. The stink was beyond all description.

They called the doctor. We inspected his records. There were only names in the little black book — nothing more — nothing about who had been where, what he had done or hoped. Behind the names of those who had died, there was a cross. I counted them. They totaled 242 — 242 out of 1200, in one month.

As we walked out into the courtyard, a man fell dead. Two others, they must have been over 60, were crawling toward the latrine. I saw it, but will not describe it.

In another part of the camp they showed me the children, hundreds of them. Some were only 6 years old. One rolled up his sleeves, showed me his number. It was tattooed on his arm. B-6030, it was. The others showed me their numbers. They will carry them till they die. An elderly man standing beside me said: "The children — enemies of the state!" I could see their ribs through their thin shirts....

We went to the hospital. It was full. The doctor told me that 200 had died the day before. I asked the cause of death. He shrugged and said: "tuberculosis, starvation, fatigue and there are many who have no desire to live. It is very difficult." He pulled back the blanket from a man's feet to show me how swollen they were. The man was dead. Most of the patients could not move.

I asked to see the kitchen. It was clean. The German in charge.... showed me the daily ration. One piece of brown bread about as thick as your thumb, on top of it a piece of margarine as big as three sticks of chewing gum. That, and a little stew, was what they received every 24 hours. He had a chart on the wall. Very complicated it was. There were little red tabs scattered through it. He said that was to indicate each 10 men who died. He had to account for the rations and he added: "We're very efficient here."

We proceeded to the small courtyard. The wall adjoined what had been a stable or garage. We entered. It was floored with concrete. There were two rows of bodies stacked up like cordwood. They were thin and very white. Some of the bodies were terribly bruised; though there seemed to be little flesh to bruise. Some had been shot through the head, but they bled but little.

I arrived at the conclusion that all that was mortal of more than 500 men and boys lay there in two neat piles. There was a German trailer,

> which must have contained another 50, but it wasn't possible to count them. The clothing was piled in a heap against the wall. It appeared that most of the men and boys had died of starvation; they had not been executed.
>
> But the manner of death seemed unimportant. Murder had been done at Buchenwald. God alone knows how many men and boys have died there during the last 12 years. Thursday, I was told that there were more than 20,000 in the camp. There had been as many as 60,000. Where are they now?
>
> I pray you to believe what I have said about Buchenwald. I reported what I saw and heard, but only part of it. For most of it, I have no words.
>
> If I have offended you by this rather mild account of Buchenwald, I'm not in the least sorry.... (215)

Perhaps a comment he made outside of the report, expresses it best: "A French newsman, imprisoned there throughout the war, had summed it up: "To write about this, you must been here at least two years. And after that—you don't want to write anymore." (216)

Such conscious implementation of the most horrific possibilities of the human mind by the very center of civilization, by the very people who thought of themselves as the leading edge of morality and spirituality, leaves us who were not there stunned. We cannot know the reality of it, but even at a distance, we can sense the feelings of hopelessness and meaningless and depression that had to have entered into any discourse. Why, indeed, would anyone want to bother to "write" or do anything to give human existence some positive perspective when the "real" world has proven that if there is such a thing as an ethos, humans must be condemned as monsters. Or, even if the need to find a meaning still existed, and perhaps even if the human brain demanded meaning, certainly the culture that had been supplying the maps was at fault, was not to be trusted, certainly the very heritage that had underpinned the culture needed to be re-evaluated. But how? All of the tools of both logical thinking and artistic expression, all of the understandings of religion and spirituality had not done it.

And to simply write it off as a non-sequitur, an aberration required a blindness. During this same war, the Japanese had concentration camps where they were using humans for experiments with chemical weapons. The Americans had moved all Japanese citizens into concentration camps (perhaps not as horrific as the Nazi ones, but certainly denying them any dignity, taking away all of their possessions, separating them from their families, for no reason other than their bloodlines), and, of course, the frantically developed atomic bombs killed and maimed hundreds of thousands of non-military people. The list could be continued. Regardless how argued, how much justification can be found, it is hard to deny the darkness.

In a world such as this, *Waiting for Godot* might be the way to suggest meaning and value do exist. Why write, perform or view a play in a world gone so wrong? How can one find meaning after the loss of innocence? Perhaps a parody is the place. Perhaps by purposely trying to undermine all meaning and value, and yet ending up finding it cannot be denied is exactly the way to support it. Perhaps facing the ultimate evil in men's souls is where we must go for a chance at something more, for a chance at meaning, for the possibility of salvation.

"Poo-tee-weet." Kurt Vonnegut gives us a Dadaist-like comment on it all through the voice of a bird. Having been a prisoner of war in Dresden during the bombings there, bombings that it has been claimed killed more people than the nuclear bombing of Hiroshima, he wanted to find a way to express what he had experienced, but as the initial chapter to his book, *Slaughterhouse-Five* gives the reader, he is stuck, caught in a paradox. (217) How does one give meaning and value to that which is beyond meaning? In what has become designated a technique of *metafiction* a type of fiction which consciously, self-consciously highlights the devices of fiction, systematically drawing attention to its status as an artifact in order to raise questions about the relationship between fiction (illusion) and reality, by necessity creating an ironic perspective, similar to Theatre of the Absurd, that does not let the reader or audience forget they are reading a work of fiction or viewing a play, Vonnegut immediately confuses realities with the simple opening "All this happened, more or less, and brings the reader into a mutual evaluation of the work by apologizing for the fact that the novel is "so short and jumbled and jangled" and seeming to echo the thoughts of those who first witnessed the Buchenwald concentration camps at the end of the war, explaining that this is because "there is nothing intelligent to say about a massacre." As the novel unfolds and Vonnegut presents us with the fictional Billy Pilgrim's experiences, he refuses to let us fall completely into the fictional world, breaking it by intruding himself, and demanding we stop and take notice, stating: "That was I. That was me. That was the author of this book."

He also mixes in several science fiction devices, including time travel. Billy Pilgrim becomes "unstuck in time" and experiences his life in a disjointed journey where even his death is not the end but just one point to be revisited randomly. This frame, which has connections to "stream of consciousness" and "interior monologue" techniques made famous by James Joyce, gets explored in more depth by having Billy Pilgrim meet, and then be kidnapped by aliens from the planet Tralfamadore, who exhibit him in a Tralfamadorian zoo with Montana Wildhack, a pornographic movie star. The Tralfamadorians experience life in four dimensions, the fourth dimension being time. Though the Tralfamadorians have seen every instant of their lives already; they cannot choose to change anything about their fate.

What free will there is consists in the ability to choose to focus on any moment in their lives that they wish. Thus, outside of the individual brain/mind, there is no free will. Vonnegut states this through one of the Tralfamadorians, "I've visited thirty-one inhabited planets in the universe... Only on Earth is there any talk of free will." To the Tralfamadorians, everything always exists at the same time, and for them everyone is therefore always alive. They too have wars and tragic events (they destroy the universe testing spaceship fuels), but when asked by Billy what they do about wars, the Tralfamadorians tell him they simply ignore them.

It seems likely that, in the Tralfamadorians, Vonnegut is demonstrating that life is only worthwhile if there is an unknown. Obviously, for the Tralfamadorians, nothing can be changed, there is no chance of influencing existence, and nothing new can be learned. Furthermore, even inside the human brain/mind what free will there is has no value beyond self-gratification or pacification, for even one's experiencing of an event cannot be changed. One can choose to focus on any moment, but one cannot choose to change anything about that moment. At most one can only choose to dwell on the pleasant experiences and ignore the unpleasant ones.

However, again we are led to simply deeper or other contexts, rather than a dead-end. The mere fact that the Tralfamadorians can choose when to exist opens the door to free will. Outside of their minds, they cannot change the happenings, but they can travel through both time and space within their minds. Thus, they have a huge number, perhaps an infinite number of experiences to exist in, and since they can feel good or bad, they can experience emotions, and can make ethical judgments. So, again, even in a seemingly fatalistic worldview, meaning and value for humans exists. Existence simply needs to be understood as taking place in the human mind, at least if it is to have meaning and value as humans understand such things.

The next step in this sequence of logic is that the human brain is constructed so that only certain thoughts can take place within it. Thus, ultimately, it is all pre-determined by the scientific properties involved with the human brain. Ultimately, whatever logical path is taken from this leads to the edges of what logic and thus science can explain and comprehend, and we end up struggling with the underlying mathematic attempts to explain existence, which, at the moment, are saying we end up with the logic of illogic (as in Godel's famous theory of infinity squared, which proves under his mathematical system that there are at least two mathematical universes). In other words, at least at this time humans cannot comprehend everything through logic and explanation, cannot explain away either infinity or eternity, at least as these terms are generally used to refer to human existence. Ironically, this inability to explain is the key to meaning and value. It gives an opening to the kinds of knowing beyond logic, all of the possibilities of world of expression (of faith and spirituality).

Viktor Frankl did experience the horror of Auschwitz, satisfying Edward R. Murrow's requirement, and it can safely be said, living through an even more horrific imprisonment than Kurt Vonnegut. Yet, Frankl did want to write about it, and found a very different path to giving life, even in the most degrading and "inhuman" circumstances, meaning and value.

Born in Vienna to a Jewish servant family in 1905, and graduating from the Gymnasium in 1923, he went on to study medicine at the University of Vienna, eventually specializing in neurology and psychiatry, especially depression and suicide. In 1924 he became president of Sozialistische Mittelschuler Osterreich, a program to counsel students and prevent suicides, and his success (not a single student committed suicide during his tenure) resulted in Wilhelm Reich bringing him to Berlin. During the mid-thirties, he ran the "suicide pavilion" of Vienna's General Hospital, treating over 30,000 women at risk of suicide. However, in 1938 his Jewish bloodlines began getting him censored, and he was prevented from treating any more Aryan women. In 1940, he moved to Rothschild Hospital, the only place left where Jews were still admitted, where he continued both psychiatric and neurological work, including brain surgery, and using what influence he had to prevent euthanasia of as many Jews as possible.

In 1942, he, his wife, and his parents were deported to a concentration camp in Theresienstadt, where his father would die in 1943. In addition to his regular assignments, his background put him in a position to continue to work with other inmates and help them through the horrific conditions that caused a great deal of depression and suicide. In a dark irony the "Nazis sought to prevent Jewish suicides. Wherever Jews tried to kill themselves - in their homes, in hospitals, on the deportation trains, in the concentration camps - the Nazi authorities would invariably intervene in order to save the Jews' lives, wait for them to recover, and then send them to their prescribed deaths." (218)

October 19, 1944, Frankl was transported to Auschwitz, where his other died. His wife was transported to Bergen-Belsen, where she died. October 25, he was transported to Turkheim, a camp near Dachau. Finally, on April 27, 1945, he was liberated.

Viktor Frankl, thus, was in a unique position. He was a highly trained and experienced psychologist and neurologist who was thrust into the most inhumane environment, where he witnessed both personally and all around him the cruelest and darkest actions of humans and the deepest depressions of the victims.

It would seem he had to become the poster-child for the most negative views of the 20th century. Certainly he would find life to be meaningless, filled with at best a cruel, depressing endurance of pain and suffering.

However, instead he becomes a central figure in the argument against such negative views, offering a powerful confirmation for the belief in the strength of the human spirit to give life meaning and value no matter how extreme the tests. In fact, the more extreme the test, the more chance to prove the strength of the human spirit (or mind or heart or whatever term is used for the human self).

In his best known book, *Man's Search for Meaning: An Introduction to Logotherapy,* he first describes the standard experiences of a prisoner of a concentration camp, and then, in part two presents his central concepts. (219) He writes:

> Let me explain why I have employed the term "logotherapy" as the name form my theory. *Logos* is a Greek word that denotes "meaning"! Logo-therapy or, as it has been called by some authors, "The Third Viennese School of Psychotherapy," focuses on the meaning of human existence as well as on man's search for such a meaning. According to logotherapy, the striving to find a meaning in one's life is the primary motivational force in man. That is why I speak of a *will to meaning* in contrast to the pleasure principle (or, as we could also term it, the *will to pleasure*) on which Freudian psychoanalysis is centered, as well as in contrast to the *will to power* stressed by Adlerian psychology. (220)

Arguing that humans have not only physical and psychological dimensions, but a spiritual dimension, not external but "within the reference of logotherapy," he is careful to clarify "'spiritual' does not have a primarily religious connotation but refers to the specifically human dimension." (221) Putting his theories into the general philosophy of existentialism, he writes: "Man's will to meaning can . . . be frustrated, in which case logotherapy speaks of "existential frustration." The term "existential" may be used in three ways: to refer to (1) *existence* itself, i.e., the specifically human mode of being; (2) the *meaning* of existence; and (3) the striving to find a concrete meaning in a personal existence, that is to say, the *will* to meaning."

Thus, he continues, "Existential frustration can . . . result in neurosis. For this type of neurosis, logotherapy has coined the term "noogenic neurosis" in contrast to neurosis in the usual sense of the word, i.e., psychogenic neurosis. Noogenic neuroses have their origin not in the psychological but rather in the "noological" (from the Greek "noos" meaning mind) dimension of the human existence. This is another logotherapeutic term which denotes anything pertaining to the "spiritual" core of man's personality." (222)

As with the other theorists we've encountered, Viktor Frankl must deal with the source of *meaning*. Sounding much like Albert Camus, Viktor Frankl writes:

> We have to beware of the tendency to deal with values in terms of the mere self-expression of man himself. For *logos*, or "meaning," is not only an emergence from existence itself but rather something confronting existence. If the meaning that is waiting to be fulfilled by man were really nothing but a mere expression of self, or no more than a projection of his wishful thinking, it would immediately lose its demanding and challenging character; it could no longer call man forth or summon him. This holds true not only for the so-called sublimation of instinctual drives but for what C. G. Jung called the *"archetypes"* of the "collective unconscious" as well, inasmuch as the latter would also be self-expressions, namely, of mankind as a whole. This holds true as well for the contention of some existentialist thinkers who see in man's ideals nothing but his own inventions. According to Jean-Paul Sartre, man invents himself, he designs his own "essence"; that is to say, what he essentially is, including what he should be, or ought to become. However, I think the meaning of our existence is not invented by ourselves, but rather detected. (223)

Here Viktor Frankl is clearly separating himself from the views that separate humans from the external environment. "Wishful thinking" cannot help but bring Sigmund Freud's *pleasure principal* to mind, and Viktor Frankl wants to distinguish the "higher" forms of thinking, those of meaning and value from the "instinctual drives" that serve as a basis for both Sigmund Freud and Carl Jung's theories. Furthermore, he wants to separate his views from such existential thinkers as Sartre who claim that "ideals" are nothing more than human inventions.

Just what, then, is Viktor Frankl trying to claim. He attempts to clarify:

> Psychodynamic research in the field of values is legitimate; the question is whether it is always appropriate. Above all, we must keep in mind that any exclusively psychodynamic investigation can, in principle, only reveal what is a driving force in man. Values, however, do not drive a man; they do not *push* him, but rather *pull* him. This is a difference, by the way, of which I am constantly reminded whenever I go through the doors of an American hotel. One of them has to be pulled while the other has to be pushed. Now, if I say man is *pulled* by values, what is implicitly referred to is the fact that there is always freedom involved: the freedom of man to make his choice between accepting or rejecting an offer, i.e., to fulfill a meaning potentiality or else to forfeit it. (224)

Now comes the clarification he wants to make between a moral or religious *drive* and a moral or religious *choice*. He continues:

> . . . it should be made quite clear that there cannot exist in man any such thing as a *moral drive*, or even a *religious drive*, in the same manner as we speak of man's being determined by basic instincts. Man is never driven to moral behavior; in each instance he decides to behave morally. Man does not do so in order to satisfy a moral drive and to have a good conscience; he does so for the sake of a cause to which he commits himself, or for a person whom he loves, or the sake of his God. If he actually did it for the sake of having a good conscience, he would become a Pharisee and cease to be a truly moral person. (225)

Here Viktor Frankl is confronting the key distinction between *a meaningful and a meaningless existence*. It can be, and has been put forth many different ways. If life is predetermined, if there is some form of predestination, if the fates are in control, if, as Viktor Frankl focuses on, our actions are all determined by heredity and instincts, then life is meaningless. Life can only have meaning and value if there is some form of individual responsibility and free will.

Ever since Soren Kierkegaard phrased the human condition in terms of "Fear and Trembling" and "*Sickness Unto Death*," and Friedrich Nietzsche proclaimed the death of God, philosophers have gleefully plunged into a dark, meaningless world, overlooking the simple truth that, as long as humans exist, meaning and value exist, for the very definition of human is the giver of meaning and value. To separate humans from meaning and value is the error that lies at the basis of the dialogue of philosophy for at least the past 150 years. However one wishes to postulate the resulting world of meaning and value when the human brain and mind come together, that is the place it happens. And as much as clever intellectuals are able to play endless intricate language games to dodge around it, humans are a part of the physical and spiritual worlds, and give it meaning (whether as receptacles of some God or scientific process, the result is the same). It seems very likely that the reason this final click does not take place is simply the western world's interpretation of the human condition coming out of the Judeo/Christian tradition that condemns humans and demands they bow before the Jewish/Christian God. Hubris becomes the ultimate sin in this view, but, unfortunately, hubris is misapplied to such an extreme that blindness replaces it.

Furthermore, the simple breaking away from the huge hold of the Christian view that ultimately denies humans any true meaning in life, by denying them any true self-responsibility, and true freedom to be either Good or Bad, has made the outlaw, the condemned one so attractive that once the door got opened to deny God all power, the dark world became just too enticing to ignore. For only in going against Good could Self be established. This is the same thing as the embracement of the Shadow over the Ego in Jungian terms. The Shadow is where the Self separates from the social, the cultural, the legal, etc., and the individual assumes self-responsibility. Thus, even being Evil is better than being Good by default, for Evil equals existence and meaning in life. Good means being obedient and giving up all chance at a meaningful life.

The rebel, the maverick, the outlaw become attractive because they demand self-responsibility, which is a definition of human existence. Without it, human existence has no meaning or value. By tautology, the outsider is not ethically good in terms of the particular system he rebels against. That does not make him Evil in any universal sense. Rather it means the system itself is being called into account, the very basis of the system is brought into question, and the established ethos is denied or at least not absolute. Whether or not the New Ethos of the Rebel is a better one or not isn't the point. The ability to bring a new ethos into existence is what matters. It means that if humans have self-responsibility then by definition, they do create meaning and value. Thus, not only are the standard categories of the outsiders justified, but the artist is embraced as well, for the artist is the one bringing, by definition, something new into being, the one remapping or thickening the map of meaning and value. This is human existence at its highest level. And it is not meaningless, not nothingness, not mere physical existence. Rather, it is frightening, terrifying, because there is so much responsibility involved. Humans become, not the weak recipients of some huge, powerful, unknown force, some God, but, rather, the voice, the conscience, the meaning of existence.

Again, by definition, this does not mean humans always do wonderful things. If humans can only be Good then there is no choice involved, and we are back to zero, back to a meaningless existence. Furthermore, not only are humans capable of error, but great errors are to be expected, as humans are born into a mysterious world and ultimately have to trust their own minds, which they must struggle to comprehend with perhaps nothing outside of them to help.

Arthur Schopenhauer sees life as the stage necessary for the discovery of the *self*. He writes:

> [Life] is played that a man may come to understand himself, that he may see what it is that he seeks and has sought to be; what he wants, and what, therefore, he is. *This is a knowledge which must be imparted to him from without*. Life is to man, in other words, to will, what chemical re-agents are to the body: it is only by life that a man reveals what he is, and it is only in so far as he reveals himself that he exists at all. Life is the manifestation of character, of the something that we understand by that word; and it is not in life, but outside of it, and outside time, that character undergoes alteration, as a result of the self-knowledge which life gives.
>
> Life is only the mirror into which a man gazes not in order that he may get a reflection of himself, but that he may come to understand himself by that reflection; that he may see *what* it is that the mirror shows. Life is the proof sheet, in which the compositors' errors are brought to light. How they become visible, and whether the type is large or small, are matters of no consequence. Neither in the externals of life nor in the course of history is there any significance; for as it is all one whether an error occurs in the large type or in the small, so it is all one, as regards the essence of the matter, whether an evil disposition is mirrored as a conqueror of the world or a common swindler or ill-natured egoist. In one case he is seen of all men; in the other, perhaps only of himself; but that he should see himself is what signifies.
>
> Therefore if egoism has a firm hold of a man and masters him, whether it be in the form of joy, or triumph, or lust, or hope, or frantic grief, or annoyance, or anger, or fear, or suspicion, or passion of any kind - he is in the devil's clutches and how he got into them does not matter. What is needful is that he should make haste to get out of them; and here, again, it does not matter how.
>
> I have described *character* as *theoretically* an act of will lying beyond time, of which life in time, or *character in action*, is the development. For matters of practical life we all possess the one as well as the other; for we are constituted of them both. Character modifies our life more than we think, and it is to a certain extent true that every man is the architect of his own fortune. No doubt it seems as if our lot were assigned to us almost entirely from without, and imparted to us in something of the same way in which a melody outside us reaches the ear. But on looking back over our past, we see at once that our life consists of mere variations on one and the same theme, namely, our character, and that the same fundamental bass

> sounds through it all. This is an experience which a man can and must make in and by himself. (226)

Part XIII

Let us return to *Oedipus Rex*. In our rush to interpret and elaborate on it, we have forgotten something very important.

First, here is a quick overview of Freud's use of the tragedy to designate his theories of the Oedipus complex. In "classic" psychoanalytic theory, the Oedipus complex denotes the emotions and ideas that the mind keeps in the unconscious, through the dynamic of the preconscious, involving a child's desire to sexually possess the parent of the opposite sex (it's worth highlighting that this takes place in a child long before puberty, so sexual desires need to be put into this context; however real, they are certainly not those of an adult human). Initially, Freud used the Oedipus complex to refer to both male and female versions, but later he adopted Jung's use of Electra complex for females. According to Freud (and subsequently Jung) the Oedipus complex occurs in the third statge, the phallic stage (ages 3–6), of the five psychosexual development stages: (i) the oral, (ii) the anal, (iii) the phallic, (iv) the latent, and (v) the genital; in each stage the source of libidinal pleasure is in a different erogenous zone of the infant's body.

In classical Freudian theory, a child's identification with the same-sex parent is the successful resolution of the Oedipus and Electra complexes; key psychological experiences that are necessary for the development of a mature sexual role and identity. He also proposed that males and females experience the complexes differently, boys in a form of castration anxiety, girls in a form of penis envy; and that unsuccessful resolution of the complexes might lead to neurosis, paedophilia, and homosexuality. Men and women who are fixated in the Oedipal and Electra stages of their psychosexual development might be considered "mother-fixated" and "father-fixated," resulting in choosing an adult partner resembling that parent.

While these original, "classical" Freudian views have now been attacked and completely discounted or modified by some, especially in reference to homosexuality and numerous variations of sexual orientation, and by feminists not wanting to say they have "penis envy." Unfortunately, to use an old cliche, popular misrepresentations of Freud's views have tended to cause us to "throw out the baby with the bath water," similar to thinking that Jung's claims we all have anima (female) and animus (male) psychic energies getting translated into simplistic misrepresentations and condemned.

Perhaps, however, Freud is at least partially responsible for the misrepresentation of his point. He did, after all, choose Oedipus as his example, and that leads to a very literal, adult representation of the complex—the resulting "tragedy" of such an unresolved complex. On the level of id or libido, Freud's ideas work. The male child (in the phallic stage) competes with his father for his mother, and wishes to replace his father in her desires. Thus, Oedipus kills his father and marries his mother. However, at least in general, popular use, people forget or never knew that this is put forth as a flawed adult, an adult with an unresolved "complex." And the literal playing out of this is not the point. It is a psychological drama.

According to Freud's theories, the normal desires for this are a brief stage in a young child's life, and they are lived out in the psyche, are symbolic. A male child aged 3-6 does not plan or ever intend to carry out the murder of his father. Nor does he consider having real, physical sex with his mother (something boys that age could not even comprehend).

While Freud's theories concerning the Oedipus complex are more complicated than this, and such oversimplification is often done by his detractors to distort his point, this general straight-forward interpretation remains, and the fact that he chose this particular story to use for the ultimate example of the theory at the very least means he set himself up for this misinterpretation. Furthermore, his main discussion of a child replacing his father in *Totem and Taboo*, comparing humans to apes, also suggests exactly this. (227)

Furthermore, his focus on human psychology based on sexual drives tends to misrepresent the total human. Whether or not everything humans do can be followed to ever deeper levels of the body, of literal, physical existence, until, at the most basic level, they do, in fact, come from that drive to life, that initial desire to procreate, that drive to life of the libido or id, the drive to life is not the total being of a human. Nor was Freud saying it was, but his emphasis on its importance and influence over "higher" aspects of being human tends to misrepresent him and his theories by giving the sexual drive too much emphasis.

And that has happened to the very tragedy he used for his example of a psychological complex. A closer look at the play itself offers a much different view of the human condition. First of all, what gets forgotten, purposely or out of a careless viewing or reading of the play, is that Oedipus killed his father and married his mother out of ignorance (certainly not out of any kind of conscious intent). He did not do it willingly or knowingly (and it is really difficult to even suggest he did it out of some kind of unconscious knowledge and drive). In fact, when he was told by Apollo's oracle at Delphi he was destined to do it, he very purposedly determined to avoid it. Yet, of course, he did do it.

Once he discovered what he had done, through the gaining of *self*-knowledge, the knowledge of whom and what he was, he was empowered to think for him*self*, to have *self*-responsibility.

It is no accident that answer to the famous Riddle of the Sphinx that he solves is man, for the entire play is about what it means to be a human—it is a play about the riddle of the human condition.

And Oedipus is the one who solved it.

And he was given a horrible decision.

Either he could deny the responsibility for his actions, and in so doing deny the chance of saving his city (i.e., the entire human community). He, after all, was not the one who initiated the problem. His father, Laius had raped a young boy, and the boy's father, Pelops, had asked the gods to deny Laius a son or to be destroyed by that son. And the gods in the form of Apollo agreed, saying that Oedipus would be born fated to murder his father and have sex with his mother. This was determined before Oedipus was even conceived, and it certainly seems wrong to accuse him of being responsible for these actions. But now, once Oedipus has learned the truth, Apollo, still looking for an admission of guilt for sins Oedipus has been fated to commit because of his father's actions, offers Oedipus a hard choice. If Oedipus chooses to remain as king of Thebes, his city will continue to be punished for what his father has done.

Or, and here is the harsh alternative, Oedipus can take on the responsibility of redeeming his father (thus, human existence) through self-sacrifice. Of course, this is what he does. He sacrifices himself to save his city. He gives up his personal needs and desires to a system of judgment, to a demanding spiritual world, even if its values are not clear, do not seem right. This is another version of Abraham's God demanding that Abraham kill his only son for no reason other than God's demand. It is a human being asked to sacrifice for no reason other than faith, Kierkegaard's belief in the "absurd."

Here, then, comes the crucial realization. The drive to kill one's father is not of the world of higher knowledge, or of the higher functions of the human mind; rather it is of the lower, animal instinctual drives (and a stage of human psychological development, which Freud and certainly Jung saw in the sense of Darwin's theoris of evolution). This fits exactly with Freud's connection of the entire concept with the actions of apes (suggesting that the literal killing of the father ape by his sons is being done by humans psychologically and symbolically caught in the Oedipus complex, a mental sickness, a situation where a human has not developed as he should—has not developed out of the lesser, animal state of existence).

Instead of saying that *Oedipus Rex* represents a normal (in truth, a superior) adult human playing out the Oedipus complex, the Oedipus drama suggests that by gaining higher knowledge of what it means to be human (by opening the mind to decisions beyond the ignorant drives of mere flesh or a child's undeveloped psyche), humans can in fact overcome, gain control over, or at least accept responsibility for these drives. This, then, is the challenge, which is to both have the desire (the instinctual drive to life), and to gain control over it (to subjugate it to a system of meaning and value) with the higher ethical and spiritual psychic forces, what might be called in Freudian terms the qualities of the ego and super-ego (to put the ego or "I" into the maps of meaning and value of the super-ego, thus giving the ego or "I" a context, an ethical and spiritual landscape where it can judge itself and have an "other" to live within).

Indeed, this is closer to the focus of Freud's Oedipus complex than the simple desire to replace the father. What Freud emphasized is the *ambivalence* of the two desires during a particular stage of development, and that a normal, healthy psyche passes through this stage on the way to its full, healthy adult self.

In the end, *Oedipus Rex* is not a play about patricide but about redemption. The father is not knowingly killed by the son, but he is knowingly redeemed.

Remember, it all begins with a *fall* from *grace*, from *innocence,* represented in the Judeo/Christian mythos as a *sin* against God (the Father). Thus, the original human (son of God) and father (of Cain and Able) Adam needs redemption, which gets accomplished in the death of another *son* of God, Jesus Christ, who willingly sacrifices himself (allows himself to be crucified) to redeem the entire human race.

This all gets complicated. Certainly God is of that other reality, that invisible, spiritual existence outside or within or . . . somehow related to physical existence and yet other than physical existence. And yet, somehow He is both the original creator, the entity that created physical existence, including humans and yet He doesn't have total control over it.

The fall from grace is a fall into the knowledge of Good and Evil and Wisdom. This has to include Free Will, at least particial Self Determination. One cannot be either Good or Evil if one has no choice, no Free Will. And this is emphasized by God's reaction to humans gaining this level of existence, kicking them out of the Garden of Eden (of innocence, and of eternal life), and the strong suggestion is that all that God has over humans at this point is Eternal Life. Yet, of course, in reality, humans could never do the things God can do. Or, perhaps, somday they can. Perhaps, it's a matter of realizing their potential, the God within them (that breath God blew into them to give them life).

This fall, then, if viewed without or with the fear of the Wrath of God, is a fall up, a fall into meaning and value and Self.

An interesting discussion could take place at this point about the image of God as father, perhaps a father also needing to be both killed and resurrected on whatever levels of existence; and Nietzsche's dramatic claim that God is dead because humans have killed him comes immediately to mind. Does God exist or not depending on his creation? Must God have an *other* to exist? Is it possible for existence to exist without an *other*? Is nothingness something? If so, or not, can the other be nothing? Can God exist without an other? Can I exist without an other? Or does Descartes' simple claim, *Cogito, ergo sum* require a *Puto igitur esse aliud*, I think therefore an other exists?

What is redemption? If *Oedipus Rex* is about redemption, what does that mean? There are many mentions of it in the Bible, mainly in reference to being redeemed by Jesus Christ:

> In him we have redemption through his blood, the forgiveness of our trespasses, according to the riches of his grace. (Ephesians 1:7)

> In whom we have redemption, the forgivness of sins. (Colossians 1:14)

> And are justified by his grace as a gift, through the redemption that is in Chrsit Jesus, whom god put forward as a Propitiation by his blood, to be received by faith. This was to show God's righteousness, because in his divine forbearance he had passed over former sins. It was to show his righteousness at the present time, so that he might be just and the justifier of the one has faith in Jesus. (Romans 3: 24-26)

> Who gave himself for us to redeem us from all lawlessness and to purify for himself a people for his own possession who are zealous for good works. (Titus 2:14)

There are many more such quotes, but the gist is clear here, for Christians, *redemption* means being *forgiven* for past sins of some sort through *grace*, whether it come in the form of Jesus Christ or directly from God.

Of course, these definitions require a definition of *grace* and then an explanation of how it all works in terms of Jesus.

The many Biblical mentions of grace emphasize that it is a gift of God:

> For by grace you have been saved through faith. And this is not your own doing; it is the gift of God, not a result of works, so that no one may boast. (Ephesians 2:8-9)

> For sin will have no dominion over you, since you are not under law but under grace. (Romans 6:14)

> But if it is by grace, it is no longer on the basis of works; otherwise grace would no longer be grace. (Romans 11:6)

> But he gives more grace. Therefore it says, "God opposes the proud, but gives grace to the humble." (James 4:6)

> But by the grace of God I am what I am, and his grace toward me was not in vain. On the contrary, I worked harder than any of them, though it was not I, but the grace of God that is with me. (1 Corinthians 15:10)

> But God shows his love for us in that while we were still sinners, Christ died for us. (Romans 5:8)

Again, this is but a small sampling of the many mentions of *grace* in the Bible, and this last quote is the one that is the key. In other words, *salvation* is the *gift* of God's *grace*, and his main gift is his only son, Jesus Christ, whom he allows to die that humans might be *saved* through *faith*.

Faith has already been discussed at length. It is an absolute belief in that which cannot be proven, that which is beyond logic, even beyond ethics and love. It demands that one be willing to reject everything else that seems true and right. As Kierkegaard explained, it is an embracement of the absurd.

It cannot be earned or achieved in any sense except through faith. It is a gift whose only requirement (and this is a must) is absolute faith, absolute embracement of God and his demands without question.

This certainly seems to be a demand to reject the self and self-responsibility, to deny the self by giving it completely to God. If the *fall* from grace took place through the eating of the forbidden fruit of the Tree of the Knowledge of Good and Evil, then the *salvation* from that fall consists of denying that knowing, somehow stopping the mind from thinking, from being a participant in the making of meaning and value.

Sin, then, ultimately works back to that original sin, which consists of separating the self from God.

In terms of self and other, the other absorbs the self, and the human mind becomes the same as the animal mind, capable of immediate, practical thoughts, but not capable of symbolic or ethical thoughts. Creativity does not exist, either mentally or physically, for humans.

However, if humans do not give up that fall into knowledge and self as separate from God, then they need to participate in the making of meaning and value, need to decide for themselves what is right, wrong, good, bad and all of the shades of meaning for the world they live in. This, in turn, requires that they comprehend dual aspects of existence, self and other, and they develop a system of rules of how the two should interact, a world that is differientiated into ever more complex dualities, not just man and woman, day and night, right and wrong, but more meaningful and demanding levels of judgment, ways not just to live with the other but to gain meaning and value (in other words, understandings of what is Good and what is Bad).

If somehow they can connect with God, then God can be a part of this, perhaps helpful, perhaps not, depending on how one defines God and his creation, humans. For example, it is easy to claim that God exists within each human, that God, in some sense is a part of each human, perhaps the essence of each human, the spirit within the flesh (He did breath life into the first human, did turn the dead clay into a living thing with his own breath, so maybe God is the spiritual, the invisible, mental part of each human and each human is the physical container of God.) The possible variations are numerous, but perhaps some form of them is the point.

God is one of those words that no two people will ever completely understand in the exact same way. Yet there is the general acceptance that God is that which is beyond logic, reason, science and math, that something beyond explanation.

However, whether or not God is outside of humans, a part of humans, or perhaps even a creation of humans, self realization means there must be an other, or there is no boundary for self, and self does not exist. Given this, humans must be willing to place their selfs into a relationship with the other, and in doing this must be willing sacrifice their selfs to a map of rules. In other words, there cannot be self without the acceptance of nonself, and this means some form of self sacrifice. The self cannot do anything at anytime without any consequences, because the other is affected by what the self does and the other must respond, either willingly (perhaps as a God) or without any consciousness of its own self (in some automatic manner).

If a human goes against the best interests of the *other*, then he is doing something wrong, is not respecting the other, is putting himself before the better good of all (including himself). In simple language, he is being selfish or egotistical. It is easy to understand why being self centered can and most likely will hurt the other, but is it important to realize it also hurts the self. If the other is damaged in some way, that weakens the relationship of the self to the other. If the environment, however defined, is made less than it can be, then self must define itself as lessened, must be in a weaker totality (a musical composition is weakened if any of its parts is poor, out of key, off-beat, or simply poorly composed; or, even if it is in itself an extraordinary piece, it is pointless if the audience for it cannot hear or does not have the necessary knowledge to understand it). Two people waltzing only make their dance work if they are both able to perform at their highest level. If one does not know how to waltz, or if one is so overpowering that the other cannot perform at a high level, then both suffer the consequences. Thus, not only does an "I" require an "other," but the "I" and the "other" are dependent on each other.

With this in place, then, each human needs a world, an *other*, to exist in, and each human needs to sacrifice his *self* to the totality, must put his *self* in the context of the *other*.

Humans can only conceive of the other in terms of what humans can conceive (an obvious tautology), and thus, humans, wanting to achieve their highest potential, wanting to exist in a world of meaning and value, are driven to *create, to conceive of* an *other* that allows for that.

If the other were some god that humans could not conceive, then humans could not have an other with that god. Every conception humans have of the *other* is a conception that the human mind is able to have (I know, continued tautology going on here). But it is an important point. It is why Friedrich Neitzche is able to claim that God is dead because humans have killed him. God can only exist for humans because they allow for such an existence. But, of course, God is a vague, vague, vague entity. Ultimately, God is the other or the life force, or energy force of the other; and by believing that there is some other humans believe they exist, the self exists because there can only be an other beyond the logic of math and science if humans have a self, and there can only be a self if there is an other beyond those automatic happenings. As I continue to emphasize, there can only be a self if there is some kind of free will, some way the self can exist beyond predestination and predetermination. If each human is merely living out the patterns of a life that some kind of logic has put in place before the human was even born, then a human is nothing more than a complex machine, an organic computer.

How humans conceive of the other, then, becomes the key. This is where the religions and their mythologies come into play. These are attempts by humans to know, to create, to have an other that gives them meaningful lives. And, then, it becomes clear that the Christian God's demands focus on believing in God. That is the key. Otherwise that belief system doesn't work.

In terms of the Garden of Eden story, when humans bite that forbidden fruit, they are partaking of the other, of the realm of God, and they become co-creators, gain a self. Thus, it is a fall *up* into the birth of what humans are today, and they are able to join in the making of meaning because they have realized the duality, and a duality can be shaped and given form. It is not so much what the duality consists of, but that duality exists. Once duality exists for them, humans have gained the power to participate in their own creation.

In a sense each generation redeems the previous one on several levels: certainly on the simple physical level of the continuation of the species; but also on the higher levels of connecting with the previous generation (through compassion and empathy) and what ethical, spiritual realizations are passed on. It is, as just mentioned, in the willing sacrifice of Christ that all previous generations of humans going all the way back to Adam and Eve, who represent the birth of human consciousness, are redeemed; and, at the same time, Christ is redeeming his father (God), is reconnecting God with human existence. Each is necessary to the other. The Chinese symbol yin/yang (that circle with two fish like shapes in it) has a dot (an eye) from the other in it; in other words each does partake of the other. If humans and God were totally separate, then God would not exist for humans (back to those previous tautologies). So Christ, in redeeming humans, also redeems God. His suffering on the Cross connects humans with God in a form of mutual empathy, and coming together in passion, the Passion of the Cross.

Christ is the Savior because he is the means of connection with that other, with the Christian other, God. If one believes in Christ, believes he is the son of God, then one gains salvation. This, of course, leads back to the previous discussion about believing in something more than what can be proven through logic and reason, belieiving there is more to life an just going through predetermined motions. If one believes, then God exists. And then, if one has a map of life, a mythology, one can have a fully developed other to give one meaning and value.

Salvation, a saving of one from some kind of sin and its consequences (which ultimately leads back to faith), then, comes from redemption, not from patricide. And *Oedipus Rex*, rather than suggesting that the most important drive in humans is that of survival, or the drive to life, suggests it is the drive to meaning and value.

Again, to emphasize, salvation is not a one-way connection. Not only are humans reconnected to the spiritual, other world of God, but God is reconnected to the physical world of humans. The world beyond physical existence and the world of physical existence complete each other. The *self* and the *knowledge of the self* are equated.

Let me repeat: In Sophocles' *Oedipus Rex* Oedipus is born *fated* to kill his father and marry his mother. He has no choice in this. And, as much as everyone in the drama attempts to outwit fate, it cannot be done, and Oedipus fulfills his horrific fate. This is the realm of the body, of the physical universe outside of the human mind. But then comes salvation. By seeking knowledge (which is, in fact, knowledge of who and what he is, or *self*-knowledge) Oedipus empowers his *self*, because he puts himself into an ethical and spiritual position, a position to give his seemingly fated and meaningless life both meaning and value on the ethical level of right and wrong, and on the spiritual level of salvation. He chooses (free will) to *sacrifice himself* to save his city. It is a heroic decision. He could have chosen to remain as king of Thebes, in which case the plague would have remained. Instead, he condemned himself in order to have the plague lifted and save his city, his people--human existence. This self-sacrifice is the ultimate gift given to humans. It is the chance to have a life of meaning and value, a chance to outwit the meaningless plane of physical existence and put a pattern, an ethical map and a spiritual map, on the universe.

While the Chorus, upon the revelation to Oedipus of the truth of his crimes insists that he "were better dead than alive" (228) he, while fully realizing and accepting the horror of what he has done, without hesitation replies "Do not counsel me anymore. This punishment / That I have laid upon myself is just." (229)

Here the *self* is confirmed without denying the *other*, the environments of the visible world of physical existence and of the invisible world of fate. *Self* exists, but not separate from an environment. Furthermore, meaning and value comes into existence through sacrificing the *self* to the *relationship* of the *self* to the *environment*. Neither *self* nor *fate* is totally in control. One cannot exist without the other.

It amounts to sacrificing the *self* as more important than the *other*, in which case ethos would be denied, because life without a context would simply become self-gratification. This is a denial of ethos and spirituality because there is no truth beyond physical existence; only the selfish desires of the individual on a physical level would be recognized. Ethos and spirituality only exists, or exists for humans if humans allow it. If there is no judgment beyond gratifying the body desires, or the desires of the individual psyche, then there is no universal ethos or spiritual world, at best only the separate ethos of each individual human, and that ethos is at best simply an instinctual response to the environment, in which case ethos, if it can be called that, has no choice, only a blind following of one's instincts. And since ethos is by definition a matter of judgment, of distinguishing Good and Bad, Right and Wrong, it does not exist.

If ethos does exist, then it takes shape in the human mind/brain (certainly for human existence; this becomes another tautology). The question here is not if it originates there or arrives there from some outside source (a God, which can lead to everything being fated, and thus life being a meaningless following out of this God's desires, i.e., no self-responsibility, or not, depending on whether or not the God is all-powerful), or a particular structuring of the brain that results in humans having no choice but to consider certain things as Good and other things as Evil (ultimately such a situation can also lead to life being meaningless, as the structures of science and logic can be said to predetermine everything, and thus humans have no choice but to live out pre-programmed scientific and logical principals, or not, depending on whether or not science and logic can reduce all of existence to patterns that must be followed).

If, rather than saying that ethos does not exist because there is no free will or choice involved in human existence, ethos is denied by going beyond it, by saying that there is something more important or at least more powerful then even the values of an ethos, i.e., Soren Kierkegaard's teleological suspension of the ethical, that does not deny ethos in the same sense of it not existing. Rather, it says ethos does exist, but that there are higher truths than the ethical truths, truths that have to do with the *numinous*, the *sublime,* the realizations of *faith*, ultimately, the truths of *salvation*—the highest truths of meaning and value in human existence, truths that *humans can experience but never fully comprehend*, or, perhaps can comprehend in the manner of *sensing* them, but cannot fully explain.

The goal in life is not to avoid such truths, not to avoid a terrifying confrontation with the sublime or the numinous, but to have the chance to face them with human nobility, for they are the moments to experience the epiphanies that give human life a special place in the cosmos.

Indeed, it is not the final lines of Creon or the Chorus that define a man. They are but the landscape. It is Oedipus rising above this barren condemnation when he says "But let me go Creon! Let me purge my father's Thebes of the pollution" that resonates, that stays with the audience, that makes everyone relate, feel the desolation and admire the man capable of rising above it. (230)

This is it, the center of any great tragedy, a man facing the *other*, facing fate, and obtaining meaning and value through self sacrifice.

In William Shakespeare's *Hamlet*, the same realization of the harshness of human existence, the demands of "fate" are not denied, but instead become the chance for meaning, the key to what is best in a human. In his introduction to *Hamlet*, Edward Hubler writes:

> In a tragedy, the hero normally comes to the realization of a truth of which he had been hitherto unaware. This is, as Aristotle has it, "a change from ignorance to knowledge" . . . . In Hamlet and King Lear, for instance, there is a transformation in the character of the hero. Toward the end of his play Lear is the opposite of what he had been at the beginning. He has been purged of his arrogance and pride, and the pomp and circumstance of kingship, on which he had placed great store, is to him no more than an interesting spectacle. What matters now is the love of the daughter he had rejected in the first scene. When we first meet Hamlet he is in a state of depression. The world to him is "an unweeded garden" from which he would willingly depart. He has found corruption not only in the state but in existence itself . . . . By the final scene, however, his composure has returned. He no longer appears in slovenly dress; he apologizes to Laertes, and he treats Claudius with courtesy up to the point at which Gertrude's death discloses the king's treachery and compels him to the act of vengeance.
>
> All this is not simply a return to Hamlet's former self. In the course of the action he has grown in stature and wisdom. He is no longer troubled by reasoning doubts, for he knows now that reason is not enough. An over reliance on reason and a belief in untrammeled free will are hallmarks of the Shakespearean villain; the heroes learn better. In the beginning of the final scene Hamlet is still beset from without and within—"thou wouldst not think how ill all's here about my heart; but it is no matter." And it does not matter, because he has now come to put his trust in providence. Earlier in the scene he had said,
>
> > Our indiscretion sometime serves us well
> > When our deep plots do pall, and that should learn us
> > There's a divinity that shapes our ends,
> > Rough-hew them how we will.
> > (V.ii.8-11)
>
> This is not, as has been said, "a fatalist's surrender of his personal responsibility." It is the realization that man is not a totally free agent. With this realization Hamlet faces the fencing match and the king's intrigues without concern for self. What matters at the end of an important tragedy is not success or failure, but what a man IS. Tragedy of the first order moves into the realm of the human spirit, and at the close we contemplate the nature of man. In this respect Shakespeare and the Greeks are the same, but they reach the end by widely divergent paths . . ..

> Melodrama such as TITUS uses horror and grief as entertainment, bringing them as close to the spectator as it can. Tragedy uses them as truth. These, it says, are part of our human heritage, and we must face them. And in the end, partly because they are faced, they lose their terror, and the tragedy passes beyond them. It is not surprising, then, that the greatest tragedies are those involving the greatest horrors, for facing a great horror demands greatness of spirit. This greatness of spirit is what we contemplate at the end of a Shakespearean tragedy. At the close of the tragedy we are not so much concerned with Hamlet or Othello as individuals as with the spirit of man triumphant in defeat . . . .
>
> What matters now is not so much what a man DOES but what he IS. . . It might be the voice of Hamlet before the fencing match: "If it be now, 'tis not to come. . ."
>
> Hamlet's remarks on the bones [while talking to the skull of Yorick] are his last comment on the discrepancy between appearance and reality. He is coming to accept reality for what it is. Then as the generalization continues, a funeral procession enters, and Hamlet learns who is to be Buried today. He has seen the body of an old friend dug up to make room for the body of the woman he loves. He has looked on death at what is for him its worst. It is after the graveyard scene that the man who had continually brooded on death is able to face it. It seems axiomatic that any horror becomes less horrible once we have looked squarely at it When we see Hamlet again he can defy augury, for the augurs can foretell only such things as success or failure; but there is nothing, except himself, to prevent a man from facing his own private horror and rising above it. And so it is with Hamlet. When Horatio offers to cancel the fencing match "if your mind dislike anything," he is able to reply, "Not a whit, we defy augury. There is special providence in the fall of a sparrow. If it be now 'tis not to come; if it be not to come, it will be now; if it be not now, yet it will come. The readiness is all. Since no man of aught he leaves knows, what is't to leave betimes? Let it be." "Readiness" here means both submitting to providence and being in a state of preparation. It is not that death does not matter; it matters very much indeed, but readiness matters more. Shakespeare's tragic heroes do not renounce the world. The dying Hamlet is concerned about the welfare of the state and his own worldly reputation. Such values are never denied, but at the end of the tragedies they are no longer primary values. At such moments the central thing is the spirit of man achieving grandeur. (231)

And we suffer with him, are attracted to him, somehow understand the greatness of his decision, for it is the potential greatness in each of us.

But it is even more complex than this. For Soren Kierkegaard, Oedipus's self-sacrifice, the sacrifice of a Tragic Hero (Soren Kierkegaard uses Agamemnon's sacrificing of his daughter Iphigenia to appease the gods as the example) is a subordination of the individual to the universal (certainly on the level of a universal ethos), a denial of the individual (ethos). Soren Kierkegaard's Abraham, on the other hand, goes against the universal, the ethos of his community for a unique, personal relationship to his God. He is, in fact, denying everything else, logic, compassion, the entire history and mythology of his world, most importantly his own human nature, all of which condemn his actions, and he is not doing it for some benefit for either himself or his community (which would deny Kierkegaard's whole point). This is the opposite of the rebellion of the Garden of Eden, a denial of that gaining of the knowledge of Good and Evil, a return to pure innocence and obedience, carrying with it the ability to completely believe in God, and in that belief the belief that the killing of Isaac is not going to result in Isaac's death, but that, contrary to all logic and all conscious knowing, Isaac will regain his same living self after the act is completed or certainly that it is a possible outcome (faith here carrying with it a belief in the impossible).

This is more of a leap of faith then even the Biblical story gives us, as in the Biblical version, God intercedes before the act is completed, and there is no proof of the miracle of faith, only an assumed proof that Abraham had true faith in the belief that he was not going to fail his existential test at the moment of realization.

But here we encounter more complications. Are we not putting thoughts into Abraham's mind? Are we not guessing at his reasons? How do we know he believed Isaac would not die? Soren Kierkegaard actually begins his discussion with four different scenarios, four different versions of the thoughts of Abraham.

Does not Soren Kierkegaard give us more of a reason not to embrace faith than to embrace it? Are Abraham's actions not too pure, too innocent to be possible, or desirable? Can they carry any self-responsibility? Can they not be viewed as a denial of *self* through an absolute obedience to God? If all of the *self* is denied for faith, isn't *self* denied? It's one thing to go beyond logic and all of practical and rational knowing in favor of intuition and the kind of knowing beyond reason, another to go beyond all forms of knowing, including compassion for no reason other than obedience. Oedipus and Agamemnon sacrificed themselves, went against their deep desires for something more important than them, for meaning and value as humans comprehend it (for a "universal order"). For them, all of the knowledge of their mind, all of human mythology and history came into play, including the fall from innocence (represented for them in such stories as that of Prometheus giving fire to humans). They were doing their best to act in accordance with their understanding of what was best, and they were using all of their knowledge and intelligence (both logical and intuitive) to make their decisions. It is, in fact, through the insistence on gaining self-knowledge that Oedipus is able to gain meaning and value and stand up to the gods, deny them complete power.

It is clear that Soren Kierkegaard sets up Abraham as taking self-responsibility in terms of denying all influences from everything beyond pure faith, including both human nature and ethical considerations. However, the question becomes whether there is anything left.

If one says that this voice of God that he hears is some form of knowing and thinking (whether it be explicable or inexplicable), then that form becomes the key to self-responsibility, but that, in turn, denies pure faith, or, rather, becomes a definition of faith. Faith becomes listening to that inexplicable voice within (or that part of the brain/mind that some voice from without communicates with), and immediately the various disciplines of inquiry begin to identify just what that voice is, because that voice, then, becomes the key to a definition of human. And it is an existence beyond all other forms of knowing. Thus, all of human knowing, conscious and unconscious, other than that irrational voice, is at best worthless, and human existence becomes simply obedience, and thus meaningless. The paradox is clear. Either the value of human existence comes from complete obedience (which is a denial of self-responsibility, and thus a contradiction), or complete Faith is ultimately a denial of self-responsibility, and thus needs to be considered in the sense of one end of a continuum that needs to be balanced by some personal "I," which in turn involves forms of knowing and makes decisions for itself. Thus, it is the tension between believing in that which is beyond knowing and knowing itself that becomes the key to human existence.

A person of pure faith, what Soren Kierkegaard calls a knight of faith would be pure only if he denied all else, if the universal were completely eliminated in the relationship of I to God. Soren Kierkegaard, however, admits, not being able to do this himself, and in opposition to what Existentialism becomes, still maintains an impure relationship to the social world around him in addition to the essential relationship to God.

But wait. Perhaps this is a misinterpretation of Soren Kierkegaard. Is it not the absolute obedience after all, but rather the struggle that matters. It is a "sickness," a "fear" and a "trembling" that Soren Kierkegaard emphasizes. And isn't "existentialism" defined in terms of the I/Other relationship, the encounter? Reality exists not in the I nor in the Other, but in the encounter of the two. In the act of living. But again, we come to something of a tautology here. Existence is existence. It does, however, open the door to the idea that the meaning of life is in the journey of life. And thus, "existence" might be replaced on one side of the equation with meaning. Again, here, we end up defining human existence in terms of meaning and value.

We return to Richard Dadd, the man being imprisoned for killing his father at the same time Soren Kierkegaard was publishing his views about killing a loved one for some higher purpose. How aware was Richard Dadd of the intellectual, spiritual, and cultural world he lived in? He most certainly knew the classics, knew *Oedipus Rex*, even traveled to modern day Thebes, knew *Hamlet* and Shakespeare well. Was, in fact, a member of what might be called the intellectual elite of Europe. He also understood, was consciously aware that he had split into two people, lived in both sane and insane worlds, and even wrote about his loss of sanity. He knew he had entered a nightmarish world, faced some horrific evil, had encountered the demands of a god, demands that went against all of the values he had lived by, and amazingly he was able to give it a context, to express it in a work of art. How much he consciously realized the depths of the human condition he had entered is impossible to know. But certainly, he did realize some other world, the *numinous*, the *sublime*, the *absurd*, the *spiritual*, the *other*. It might be claimed he had experienced the shamanic break-up, might be claimed he managed to work through it, to return with his insights and express them in his painting, might be claimed he was one with Oedipus and Hamlet, perhaps with the Abraham of Soren Kierkegaard. A case certainly can be made that he was living out the theoretical "knight of faith," regardless if his God was not The God of Christianity. Perhaps he is the ultimate "existential man of faith." Or, perhaps, it is more honest to say he never did reach the end of his schizophrenic journey, was still stuck in his nightmare when he died, and his greatest work of art is not the result of coming to an enlightened conclusion but rather something of a brilliant flash of insight reaching us from the darkest forest of the psychological nightmare, perhaps a wondrous message in a bottle from a man lost on some psychic island unable to cross the great ocean of the unconscious and return to the hard, dry land of reason and civilization, the world of waking consciousness.

Perhaps a man to be pitied.

Perhaps, however, it is Dadd who is the enlightened one reversing all of this, and the painting is something of a joke, a commentary on the rest of the world, perhaps his line beginning his description of the characters in the painting, "We'll now advance these folks displayed as in a trance," is meant to comment on all of us, all of the world outside of the world of Richard Dadd, the world of merely meaningless, physical existence, an existence lived as if in a trance, a world waiting, poised, frozen, without life, a world in need of a master-stroke from an enlightened artist.

Two popular rock songs come to mind, the first, *The Thin Man* by Bob Dylan, might be seen as either a commentary on Dadd or as his commentary on those who do not understand the world he has come to know:

You walk into the room
With your pencil in your hand
You see somebody naked
And you say, "Who is that man?"
You try so hard
But you don't understand
Just what you'll say
When you get home
Because something is happening here
But you don't know what it is
Do you, Mister Jones

You raise up your head
And you ask, "Is this where it is?"
And somebody points to you and says
"It's his"
And you say, "What's mine?"
And somebody else says, "Where what is?"
And you say, "Oh my God
Am I here all alone?"
Because something is happening here
But you don't know what it is
Do you, Mister Jones

You hand in your ticket
And you go watch the geek
Who immediately walks up to you
When he hears you speak
And says, "How does it feel
To be such a freak?"
And you say, "Impossible"
As he hands you a bone
Because something is happening here
But you don't know what it is
Do you, Mister Jones

You have many contacts
Among the lumberjacks
To get you facts
When somebody attacks your imagination
But nobody has any respect

Anyway they already expect you
To just give a check
To tax-deductible charity organizations
You've been with the professors
And they've all liked your looks
With great lawyers you have
Discussed lepers and crooks
You've been through all of
F. Scott Fitzgerald's books
You're very well read
It's well known
Because something is happening here
But you don't know what it is
Do you, Mister Jones

Well, the sword swallower, he comes up to you
And then he kneels
He crosses himself
And then he clicks his high heels
And without further notice
He asks you how it feels
And he says, "Here is your throat back
Thanks for the loan"
Because something is happening here
But you don't know what it is
Do you, Mister Jones

Now you see this one-eyed midget
Shouting the word "NOW"
And you say, "For what reason?"
And he says, "How?"
And you say, "What does this mean?"
And he screams back, "You're a cow
Give me some milk
Or else go home"
Because something is happening here
But you don't know what it is
Do you, Mister Jones

Well, you walk into the room
Like a camel and then you frown
You put your eyes in your pocket
And your nose to the ground
There ought to be a law
Against you comin' around
You should be made

To wear earphones
Because something is happening here
But you don't know what it is
Do you, Mister Jones (232)

The second, *I Started a Joke* by the Bee Gees, has fewer words, less imagery, but contains a similar ironic, paradoxical understanding of truth:

I started a joke, which started the whole world crying,
But I didn't see that the joke was on me, oh no.

I started to cry, which started the whole world laughing,
Oh, if I'd only seen that the joke was on me.

I looked at the skies, running my hands over my eyes,
And I fell out of bed, hurting my head from things that I'd said.

Till I finally died, which started the whole world living,
Oh, if I'd only seen that the joke was on me.

I looked at the skies, running my hands over my eyes,
And I fell out of bed, hurting my head from things that I'd said.

Til I finally died, which started the whole world living,
Oh, if I'd only seen that the joke was one me. (233)

Does Dadd at some point come to see himself in these oxymorons, these contradictory truths? At the very least, he shows us that even a man who has both committed a horrible crime and knows it is horrible on many levels, perhaps all, a man condemned by his community with no hope of being released from his imprisonment back into the community, a man with little reason to continue living took it upon himself to devote years to complete an extremely detailed and complex work of art that certainly is an attempt to give a map of meaning and value to existence is *compelled* to do so. And it is very hard to see this as anything other than strong support for the innate need within humans to do just that for no other reason than that *they must*.

In *La Condition Humaine*, Andre Malraux writes:

> The great mystery is not that we should have been thrown down here at random between the profusion of the earth and the galaxy of the stars, but that in this prison we can fashion images of ourselves sufficiently powerful to deny our nothingness. (234)

Notes

1. Plato, Phaedrus. This translation based on translations of B. Jowett, *Symposium and Phaedrus*, Dover, 1994 (ccat.sas.upenn.edu/jod/texts/phaedrus.html); and *Phaedrus*, translated by Alexander Nehamas & Paul Woodruff, Indianapolis: Hackett, 1995.
2. These quotes and the following referring to witches in this exchange with but a few added phrases come from the beginning passages of Shakespeare's *MacBeth*, the Folio of 1623; rpt. Harcourt Brace Jovanovich, New York, p. 1234.
3. *Richard Dadd*, Patricia Allderidge, St. Martin's Press: New York, 1974, p.22.
4. Three of Robert Dadd's seven children by his first wife, Richard, George, and Maria, were to die in an insane institution, and another, Stephen, was to have a personal attendant to help him with his mental problems, strongly suggesting that there is some inheritance aspect to this, that at the very least one's genetic make-up has a tendency toward this form of insanity.
5. While Richard Dadd's paintings have not received as much notice as some of the more famous artists of the 1800s, he has gained more attention and is one of the artists movements such as Surrealism have turned to as a precursor to 20th century artistic theories coming out of psychology and moving art ever more into an expression of inner realities and truths beyond the literal world. Worthwhile discussions of him and his work include two books by Patricia Allderidge, Archivist to the Bethlem Royal Hospital and the Maudsley Hospital, *Richard Dadd,* Academy Editions, St. Martin's Press, New York, 1974, and *The Late Richard Dadd*, The Tate Gallery, 1974; a more in depth study of his work in relation to his life by David Greysmith, *Richard Dadd: The Rock and the Castle of Seclusion*, Studio Vista, London, 1973; and a chapter on him by John M. MacGregor, *The Discovery of the Art of the Insane,* Princeton University, 1989.
6. William Wood, *Remarks on the Plea of Insanity*, 1851, pp. 41-42. Wood records that at his request Dadd furnished him with a "long and rambling account of the ideas that had, from time to time, occurred to him," including a description of the murder.
7. p. 122, reprinted in *The Discovery of the Art of the Insane*, John M. MacGregor, Princeton University, 1989, p. 119, and reported to come from a letter to Dadd's family from French officials of the asylum. Also included in *The Late Richard Dadd,* Patricia Allderidge, The Tate Gallery, 1974, p. 28.
8. *The Catholic Encyclopedia* (New Advent edition, www.newadvent.org).
9. *The trial of Ebenezer Haskell, in lunacy, and his acquittal before Judge Brewster, in November, 1868, together with a brief sketch of the mode of treatment of lunatics in difference asylums in this country and in England,: with illustrations, including a copy of Hogarth's celebrated painting of a scene in old Bedlam, in London, 1635*; rpt online free for public use:
http://quod.lib.umich.edu/cgi/t/text/text.
10. May, 1845, p. 137.
11. His involvement is discussed by Jack Lindsay, *Turner,* New York, 1966.
12. London, 1976, by F. J. G. Jeffriss.

13. The Report of the Commissioners in Lunacy to the Secretary of State on Bethlem Hospital, 1852. For a detailed discussion of the scandal, and a report on the conditions in the hospital prior to 1852, see "The Commissioners in Lunacy Report on Bethlem Hospital," *The Journal of Psychological Medicine and Mental Pathology* 6 (1853), pp. 129-45, and "Bethlem Hospital in 1852," p. 62. Andrew T. Scull, *Museums of Madness*, Harmondsworth, 1982, also has a historical study of these reports.
14. Patricia Allderidge, *The Late Richard Dadd*, St. Martins Press, 1974, p. 30.
15. Patricia Allderidge, *The Late Richard Dadd*, St. Martins Press, 1974, p. 26.
16. Vol. 101, 1857, pp. 361-2.
17. Casebooks, Bethlem Royal Hospital.
18. Sala, "A Visit to the Royal Hospital of Bethlehem," pp. 304-5.
19. *Some Reminiscences of William Michael Rossetti,* 1906, I, pp. 269-70.
20. Dec. 26, 1877, "Her Majesty's Pleasure, The Parricide's Story, pp. 13-14.
21. Found in a letter from Mary Ann Dad dot Catherine (Carter Dadd), Galveston, Texas, Jan. 31, 1886, Dadd family papers, private collection.
22. http://www.getty.edu/art/gettyguide/artMakerDetails?maker=281.
23. *The Discovery of the Art of the Insane*, John M. MacGregor, Princeton, 1989, p. 127.
24. 1, 1892, pp. 16-18.
25. According to the online site www.Everything2.com/index.pl?node_id=1106011 "Though only 54cm by 39.4cm, the layers of paint are so thick that it has been said 'There is no possible way that Fairy-Feller can be properly reproduced in any medium. It is literally three dimensional'. It is also very detailed, for much of it Dadd worked with a magnifying glass and one of his own hairs."
26. It is much smaller in size, currently in the Fitzwilliam Museum, Cambridge, and appears to have been originally given to Dr. John Meyer, a later superintendent of Broadmoor.
27. *The Late Richard Dadd*, 1817-1886, Tate Gallery, 1974; ISBN: 0-900874-79-1, p. 127.
28. Patricia Allderidge, *The Late Richard Dadd*, 1817-1886, Tate Gallery, 1974, p. 127.
29. *The Discovery of the Art of the Insane*, Princeton, 1989, p.128.
30. *Richard Dadd: the Rock and the Castle of Seclusion*, David Greysmith, Studio Vista: London, 1973, p. 121.
31. *The Late Richard Dadd*, 1817-1886, Tate Gallery, 1974, p. 128.
32. *The Forgotten Language*, London, 1952
33. *Archetypes and the Unconscious*, trans., R. F. C. Hull, pp. 214.
34. *Archetypes and the Unconscious*, trans., R. F. C. Hull, pp. 215.
35. *Archetypes and the Unconscious*, trans., R. F. C. Hull, pp. 215.
36. *Elimination of a Picture & Its Subject—Called the Feller's Master Stroke*; rpt. p. 122, Greysmith.
37. *Archetypes and the Unconscious*, trans., R. F. C. Hull, pp. 215-216.
38. *Archetypes and the Unconscious*, trans., R. F. C. Hull, pp. 216-217.
39. *Holy Bible*, King James Version, Genesis 22, i-ii.
40. *Holy Bible*, King James Version, Genesis 22, iii-x.

41. *Journal* IV A 108; rpt. *Fear and Trembling*, 1843; trans. Walter Lowrie, Princeton, 1953, p. 10.
42. *Dionysus: Myth and Cult*, translated by Robert B. Palmer, Indiana University Press, 1965, p. 78.
43. *The Power of Myth: Joseph Campbell with Bill Moyers*, ed., Sue Flowers, Doubleday, 1988, pp. 216-220.
44. *Dionysus: Myth and Cult*, translated by Robert B. Palmer, Indiana University Press, 1965, p. 73.
45. *Dionysus: Myth and Cult*, translated by Robert B. Palmer, Indiana University Press, 1965, pp. 79-80.
46. *Dionysus: Myth and Cult*, translated by Robert B. Palmer, Indiana University Press, 1965, p. 80.
47. Euripides, "The Bacchae," *Greek Tragedies, Volume 3*, edited by David Grene and Richmond Lattimore, Phoenix Books, University of Chicago Press, 1968, lines 142-145.
48. Euripides, "The Bacchae," *Greek Tragedies, Volume 3*, edited by David Grene and Richmond Lattimore, Phoenix Books, University of Chicago Press, 1968, line 699.
49. Euripides, "The Bacchae," *Greek Tragedies, Volume 3*, edited by David Grene and Richmond Lattimore, Phoenix Books, University of Chicago Press, 1968, line 767.
50. Euripides, "The Bacchae," *Greek Tragedies, Volume 3*, edited by David Grene and Richmond Lattimore, Phoenix Books, University of Chicago Press, 1968, line 1109.
51. Euripides, "The Bacchae," *Greek Tragedies, Volume 3*, edited by David Grene and Richmond Lattimore, Phoenix Books, University of Chicago Press, 1968, lines 724-768.
52. *Dionysus: Myth and Cult*, translated by Robert B. Palmer, Indiana University Press, 1965, pp. 105-106.
53. *Dionysus: Myth and Cult*, translated by Robert B. Palmer, Indiana University Press, 1965, p. 116.
54. *Dionysus: Myth and Cult*, translated by Robert B. Palmer, Indiana University Press, 1965, p. 115.
55. *Psyche*, Vol. II, p. 13, note; p. 45, note; referred to in *Dionysus: Myth and Cult*, p. 117.
56. *Dionysus: Myth and Cult*, translated by Robert B. Palmer, Indiana University Press, 1965, p. 117.
57. *Dionysus: Myth and Cult*, translated by Robert B. Palmer, Indiana University Press, 1965, p. 118.
58. *Dionysus: Myth and Cult*, translated by Robert B. Palmer, Indiana University Press, 1965, p. 118.
59. *Dionysus: Myth and Cult*, translated by Robert B. Palmer, Indiana University Press, 1965, p. 119.
60. *The Iliad of Homer*, translated by Richmond Lattimore, Phoenix Books, 1970, "Book Five," lines 132-137.

61. *The Iliad of Homer*, translated by Richmond Lattimore, Phoenix Books, 1970, "Book Five," lines 132-137, p. 156; Walter Otto, *Dionysus: Myth and Cult*, translated by Robert B. Palmer, Indiana University Press, 1965, p. 127.
62. A Philosophical Enquiry into the Origin of Our Ideas of the Sublime and Beautiful, 1757; Routledge and Kegan Paul, London, 1958
63. *Beobachtungen über das Gefühl des Schönen und Erhabenen, 1764; Observations on the Feeling of the Beautiful and Sublime*. Trans. John T. Goldthwait. University of California Press, 1961, 2003.
64. *Die Welt als Wille und Vorstellung*, vol. one, 1819, vol. two, 1844. *The World as Will and Representation*, translated as *The World as Will and Idea,* AMS Press, 1975.
65. The book was written by B.R.Haydon (no relationship to G.H. Haydon). G.H. Haydon owned it, and it was extensively annotated by Dadd. G.H. Haydon later copied the annotations in ink to preserve them.
66. *Journal*; rpt. *Fear and Trembling*, 1843; trans. Walter Lowrie, Princeton, 1953, p. 14.
67. *A Little Book on the Human Shadow*, HarperCollins, 1988, p. 7.
68. *The Raven and Other Poems*, 1845; rpt., *Edgar Allen Poe*, Octopus Books, London, 1981.
69. "On the Psychology of the Unconscious," 1912; rpt., "Two Essays on Analytical Psychology, *The Collected Works of Carl Jung*, vol. 7, p.35.
70. "New Paths in Psychology," 1912; rpt., "Two Essays on analytical Psychology, *The Collected Works of Carl Jung*, vol. 7, p. 425.
71. "Archetypes of the Collective Unconscious," *The Collected Works of Carl Jung*, trans., R. F. C. Hull, vol. 9, part 1, pp. 39-40.
72. *The Delight Makers*, 1890; rpt., Bibliobazaar, 2006.
73. "On the Psychology of the Trickster-Figure," *The Collected Works of Carl Jung*, trans., R. F. C. Hull, vol. 9, part 1, p. 255.
74. "On the Psychology of the Trickster-Figure," *The Collected Works of Carl Jung*, trans., R. F. C. Hull, vol. 9, part 1, p. 256.
75. "On the Psychology of the Trickster-Figure," *The Collected Works of Carl Jung*, trans., R. F. C. Hull, vol. 9, part 1, p. 262
76. "On the Psychology of the Trickster-Figure," *The Collected Works of Carl Jung*, trans., R. F. C. Hull, vol. 9, part 1, p. 263.
77. "On the Psychology of the Trickster-Figure," *The Collected Works of Carl Jung*, trans., R. F. C. Hull, vol. 9, part 1, p. 266-7.
78. "On the Psychology of the Trickster-Figure," *The Collected Works of Carl Jung*, trans., R. F. C. Hull, vol. 9, part 1, 2nd ed, 1969, p. 266.
79. *The World of Primitive Man*, p. 3-5.
80. "On the Psychology of the Trickster-Figure," *The Collected Works of Carl Jung*, trans., R. F. C. Hull, vol. 9, part 1, Princeton Univ., 2nd ed., 1969, p. 268.
81. "On the Psychology of the Trickster-Figure," *The Collected Works of Carl Jung*, trans., R. F. C. Hull, vol. 9, part 1, Princeton Univ., 2nd ed., 1969, p. 269-70.
82. "On the Psychology of the Trickster-Figure," *The Collected Works of Carl Jung*, trans., R. F. C. Hull, vol. 9, part 1, Princeton Univ., 2nd ed., 1969, p. 270.

83. "On the Psychology of the Trickster-Figure," *The Collected Works of Carl Jung*, trans., R. F. C. Hull, vol. 9, part 1, Princeton Univ., 2nd ed., 1969, p. 272.
84. *Owning Your Own Shadow: Understanding the Dark Side of the Psyche*, Harper, 1991, pp. 4-5.
85. *Owning Your Own Shadow: Understanding the Dark Side of the Psyche*, Harper, 1991, pp. 7-8.
86. *Owning Your Own Shadow: Understanding the Dark Side of the Psyche*, Harper, 1991, p. 8.
87. *A Little Book on the Human Shadow*, ed., William Booth, Harper, 1988, pp. 17-18.
88. *A Little Book on the Human Shadow*, ed., William Booth, Harper, 1988, p. 18.
89. *The holy: an inquiry into the non-rational factor in the idea of the divine and its relation to the rational*, 1917; rpt., trans., John W. Harvey, London: Oxford University Press, 1926, p. 10.
90. *Partison Review*, 1957; rpt., *Going to Meet the Man*, Vintage Books, 1965.
91. *Holy Bible*, King James Version.
92. *On psychic energy, 1928*; rpt., *The Collected Works of Carl Jung,* Vol. 8, trans., R. F. C. Hull, Princeton Univ., 2nd ed., 1969, p 233.
93. *On psychic energy, 1928*; rpt., *The Collected Works of Carl Jung,* Vol. 8, trans., R. F. C. Hull, Princeton Univ., 2nd ed., 1969, p. 186.
94. *A psychological approach to the dogma of the Trinity*, 1942; rpt., *The Collected Works of Carl Jung,* Vol. 11, trans., R. F. C. Hull, Princeton Univ., 2nd ed., 1969, p. 184.
95. Psychology and religion, 1937-40; rpt., *The Collected Works of Carl Jung,* Vol. 11, trans., R. F. C. Hull, Princeton Univ., 2nd ed., 1969, p. 7.
96. Peter Tomory, *The Life and Art of Henry Fuseli*, New York: Praeger, 1972, pp. 100- 109.
97. William Shakespeare, *Romeo and Juliet*. Scene IV, 49-103, 1594-96; rpt., New American Lib.: New York, 1964.
98. Damon, S. Foster, *A Blake Dictionary: The Ideas and Symbols of William Blake*. Hanover, NH: University Press of New England, 1988.
99. Richard Schindler, www.victorianweb.org/painting/fairy/ras2.html.
100. Butlin, Martin. *The Paintings and Drawings of William Blake*. 2 vols. New Haven: Yale University Press, 1981. Also see Katherine Briggs, *An Encyclopedia of Fairies*, New York: Pantheon Books, 1976; and John Adlard, *The Sports of Cruelty: Fairies, Folk-Songs, Charms, and Other Country Matters in the Works of William Blake,* London: Cecil and Amelia Woolf, 1972.
101. Patricia Allderidge, *Richard Dadd*, St. Martin's Press: New York, 1974, p. 16.
102. Patricia Allderidge, *The Late Richard Dadd*, St. Martins Press, 1974, p. 13.
103. Henry Howard, *A Course of Lectures on Painting Delivered at the Royal Academy of Fine Arts*, ed., Frank Howard,1848.
104. W.P. Firth, *My Autobiography and Reminiscences*, III, 1888, Ch. IX, "Richard Dadd," pp. 177-195.
105. S.C. Hall, *Retrospect of a Long Life*, 1883, pp. 328-329; rpt. *The Late Richard Dadd,* St. Martins Press, 1974.

106. Oct., 1843, pp. 267-8; rpt., *The Late Richard Dadd* St. Martins Press, 1974, p. 15.
107. Patricia Allderidge, *Richard Dadd*, St. Martin's Press: New York, 1974, p. 16.
108. Richard Ormond, *Daniel Maclise, 1806-1870*, exhib. cat., London: Arts Council of Great Britain, 1972. p. 28.
109. *Richard Dadd*, St. Martin's Press: New York, 1974, p. 16.
110. *Richard Dadd*, St. Martin's Press: New York, 1974, p. 16
111. Archives of the Royal Academy, Art Union, Dec. 1839, p. 177, and Dec. 1840, p. 194; rep. *The Late Richard Dadd*, pp. 15-16.
112. Frith, *My Autobiography and Reminiscences, III*, 1888, Ch. IX, "Richard Dadd," pp. 177-195.
113. Frith, *My Autobiography and Reminiscences, III*, 1888, Ch. IX, "Richard Dadd," pp. 177-195.
114. *The Late Richard Dadd*, St. Martins Press, 1974, p. 23.
115. William Wood, *Remarks on the Plea of Insanity*, 1851, pp. 41-2; rpt., *The Late Richard Dadd*, St. Martins Press, 1974p. 22-23.
116. *Victorian Painting*, Macmillan, 1966, p. 40.
117. *Richard Dadd: The Rock and the Castle of Seclusion*, Studio Vista, London, 1973, p. 122.
118. *The Power of Myth: Joseph Campbell with Bill Moyers*, ed., Sue Flowers, Doubleday, 1988, pp. 55-59.
119. *The Power of Myth: Joseph Campbell with Bill Moyers*, ed., Sue Flowers, Doubleday, 1988, p. 59.
120. *The Courage to Create*, Norton/Bantam, 1975, p. 11.
121. *The Power of Myth: Joseph Campbell with Bill Moyers* ed., Sue Flowers, Doubleday, 1988, p. 59.
122. First serialized in *The Egoist*, 1914-1915; published as a book, 1916; rpt., Penguin, 2003, concluding passages.
123. *The Courage to Create*, Norton/Bantam, 1975, p. 21.
124. *The Courage to Create*, Norton/Bantam, 1975, p. 22.
125. *Poetry and Experience*, Boston, 1961, pp. 8-9.
126. *The Courage to Create*, Norton/Bantam, 1975, p. 23.
127. "Psychology and Alchemy," *Collected Works,* 12, London, 1953.
128. *La danza de espadas y la tarantela*, Barcelona, 1948; rpt., *A Dictionary of Symbols*, J. E. Cirlot, trans., Jack Sage, Routledge & Kegan Paul, 1962, p. 304.
129. mayoclinic.org.
130. James Russell Lowell's *The Pioneer*, January, 1843; Poe republished it in his periodical *The Broadway Journal,* August 23, 1845; rpt., *Tales of Mystery and Imagination.*
131. March 21, 1854; rpt., *Richard Dadd: The Rock and the Castle of Seclusion*, David Greysmith, Studio Vista: London, p. 66.
132. *Myths to Live By*, Bantam, 1973.
133. Vol. 96, Article 3, Jan. 27, 1962, pp. 853-876.
134. *Myths to Live By*, Bantam, 1973, p. 208.
135. *Myths to Live By*, Bantam, 1973, pp. 208-209.
136. *The Cry for Myth*, Bantam, 1991, pp. 17-19.

137. *American Anthropologist*, vol. 69, no 1, 1967.
138. *Myths to Live By*, Bantam, 1973, p. 225.
139. Ballantine Books, "Chapter Seven," 1967.
140. *Myths to Live By*, Bantam, 1973, p. 237.
141. *Myths to Live By*, Bantam, 1973, p. 237.
142. *The Power of Myth: Joseph Campbell with Bill Moyers*, ed., Sue Flowers, Doubleday, 1988, p. 122-124.
143. *The Power of Myth: Joseph Campbell with Bill Moyers*, ed., Sue Flowers, Doubleday, 1988, p. 107.
144. *The Power of Myth: Joseph Campbell with Bill Moyers*, ed., Sue Flowers, Doubleday, 1988, pp. 204-205.
145. Written May, 1819; rpt., *The Norton Anthology of Poetry*, ed. Arthur M. Eastman, 1970, p. 698.
146. "Plastic Art and Pure Plastic Art," *Documents of Modern Art,* vol. 2, New York: Witterborn and Company, 1945, p. 10.
147. From an interview with Marius de Zayas, translated with Picasso's approval and published as "Picasso Speaks," *The Arts*, New York, May 1923, pp. 315-326.
148. *The Structure and Dynamics of the Psyche* (1916); rpt., *Jung on Active Imagination,* ed. by Joan Chodorow, Princeton Univ., p. 56.
149. *The Power of Myth: Joseph Campbell with Bill Moyers*, ed., Sue Flowers, Doubleday, 1988, p. 277.
150. *The Power of Myth: Joseph Campbell with Bill Moyers*, ed., Sue Flowers, Doubleday, 1988, p. 277.
151. *The Interpretation of Dreams*, trans. By James Strachey, Avon Books, pp. 295-296.
152. *From Instruction to Delight: An Anthology of Children's Literature To 1850*, Patricia Demers & Gordon Moyles, Oxford Univ. Press, 1982, pp. 31-32, contains a reproduction of one of the illustrated versions of this popular alphabet poem, this one from *The New England Primer* (c. 1683-1830).
153. From *The Works of Anne Bradstreet*, ed. J. Hensley (Cambridge, Mass: Harvard Univ. Press, 1967, p. 54.
154. *From Instruction to Delight: An Anthology of Children's Literature To 1850*, by Patricia Demers & Gordon Moyles, Oxford Univ. Press, 1982, p. 42.
155. "Genesis," 3:6-24. *King James Bible: Red Letter Version*. Riverside: Iowa Falls, Iowa, 1971. pp. 3-2.
156. "Genesis," 2:25. *King James Bible: Red Letter Version*. Riverside: Iowa Falls, Iowa, 1971, p. 3.
157. "Genesis," 3:1-6. *King James Bible: Red Letter Version*. Riverside: Iowa Falls, Iowa, 1971, p. 3.
158. "The Journey Inward," *The Power of Myth: Joseph Campbell with Bill Meyers*. Betty Sue Flowers, ed. Doubleday: New York, 1988, p. 55.
159. *Il Simbolismo dei Sogni*, Milan, 1951; trans. J. E. Cirlot, Diccionario De Simbolos Tradicionales; trans, Jack Sage, A Dictionary of Symbols, Routledge & Kegan Paul: London, 1962, p. 287.
160. *Myths and Symbols in Indian Art and Civilization*. New York, 1946).
161. *Psychology and Alchemy* (collected works, 12). London, 1953.

162. *A Dictionary of Symbols*, 2nd ed. J. E. Cirlot. Trans. by Jack Sage, Routledge & Kegan Paul: London, 1976, p. 287.
163. *La langue sacree*. Paris, 1931; *A Dictionary of Symbols*, 2nd ed. J. E. Cirlot. Trans. by Jack Sage, Routledge & Kegan Paul: London, 1976, p. 287.
164. *Tratado de historie de las religions*. Madrid, 1954; *A Dictionary of Symbols*, 2nd ed. J. E. Cirlot. Trans. by Jack Sage, Routledge & Kegan Paul: London, 1976, p. 286.
165. *Le Symbolisme dans la mythologie grecque*, Paris, 1952; *A Dictionary of Symbols*, 2nd ed. J. E. Cirlot. translated by Jack Sage, Routledge & Kegan Paul: London, 1976, p. 286.
166. *Mytische Reste in der Paradieserzahlune*, Archive f. Rel. X, 345.
167. *Tratado de historia de las religions*. Madrid, 1954.
168. *The Alphabet of Ben Sira* is the source most refer to for the origin of this story. It didn't appear until Medieval times, but various sources for the elements in it have been found dating back to Jewish beliefs and to passages in the *Bible*. There are many sources discussing Lilith, including the following: Hurwitz, Siegmund. *Lilith*. Switzerland: Daminon Press, 1992. *Lilith's Cave: Jewish tales of the supernatural* by Howard Schwartz (San Francisco: Harper & Row, 1988) Patai, Raphael. *The Hebrew Goddess*, 3rd enlarged edition New York: Discus Books, 1978. *The Witch Book*, by Raymond Buckland, Visible Ink Press (November 1, 2001).
169. *A Dictionary of Symbols*, 2nd ed. J. E. Cirlot. Trans. by Jack Sage, Routledge & Kegan Paul: London, 1976, p. 288.
170. *The Further Reaches of Human Nature*. NY: Viking Press, 1971, pp. 45-49.
171. *The Power of Myth: Joseph Campbell with Bill Moyers*, ed., Sue Flowers, Doubleday, 1988, p. 276.
172. *The Power of Myth: Joseph Campbell with Bill Moyers*, ed., Sue Flowers, Doubleday, 1988, p. 286.
173. *Le Mythe De Sisyphe*, 1942; rpt., *The Myth of Sisyphus & Other Essays by Albert Camus*, trans. by Justin O'Brien, NY, Random House, 1955.
174. The Iliad, Book VI, pp. 155-157.
175. *The Immense Journey: An imaginative Naturalist Explores the Mysteries of Man and Nature, Vintage,* 1955, pp. 120-121.
176. *Man's Search for Himself*, Norton: New York, 153, pp. 74-76.
177. The Power of Myth with Bill Moyers, ed., Betty Sue Flowers. Doubleday, p. 140, 1988.
178. *The World as Will and Representation* 1859, *Die Welt als Wille und Vorstellung* (*The World as Will and Representation*) [third edition, two volumes], section 68.
179. *The World as Will and Representation* 1859, *Die Welt als Wille und Vorstellung* (*The World as Will and Representation*) [third edition, two volumes], sections 63-64.
180. *Über die Freiheit des menschlichen Willens,* 1839; rpt., trans. Konstantin Kolenda, Dover, May, 2005.
181. *Kritik der reinen Vernunft*, 1781, 2nd ed., 1787; trans., Norman Kemp Smith, Palgrave Macmillan, 2003.

182. *The World as Will and Representation* 1859, *Die Welt als Wille und Vorstellung* (*The World as Will and Representation*) [third edition, two volumes], section 22.
183. *The World as Will and Representation* 1859, *Die Welt als Wille und Vorstellung* (*The World as Will and Representation*) [third edition, two volumes], section 55.
184. *To Deny Our Nothingness: Contemporary Images of Man,* Delta, 1967, p.334.
185. *The Power of Myth: Joseph Campbell with Bill Moyers*, ed., Sue Flowers, Doubleday, 1988. p. 227.
186. *Le Mythe De Sisyphe*, 1942; rpt., *The Myth of Sisyphus & Other Essays by Albert Camus*, trans. by Justin O'Brien, NY, Random House, 1955, p. 91.
187. *To Deny Our Nothingness: Contemporary Images of Man,* Delta, 1967, p. 334.
188. *To Deny Our Nothingness: Contemporary Images of Man,* Delta, 1967, p. 336.
189. *To Deny Our Nothingness: Contemporary Images of Man,* Delta, 1967, p. 336.
190. "Letters to a German Friend," *Resistance, Rebellion, and Death*, trans. Justin O'Brien, Alfred A. Knopf, 1961, p. 28.
191. "Letters to a German Friend," *Resistance, Rebellion, and Death,* trans. Justin O'Brien, Alfred A. Knopf, 1961, p. 28-9.
192. "The Night of Truth," *Resistance, Rebellion and Death*. trans. Justin O'Brien, Alfred A. Knopf, 1961, p. 39.
193. *The Power of Myth: Joseph Campbell with Bill Moyers*, ed., Sue Flowers, Doubleday, 1988. p. 172-173.
194. *The Rebel; An Essay on Man in Revolt,* Alfred A. Knopf, 1956, p. 22.
195. *The Rebel; An Essay on Man in Revolt,* Alfred A. Knopf, 1956, p. 22
196. "Letters to a German Friend," *Cahiers de Liberation*, No 3, 1944; rpt., *Resistence, Rebellion, and Death*, Trans., Justin O'Brien, Alfred A. Knopf, 1961, p. 19.
197. *Concluding Unscientific Postscript to Philosophical Fragments,* Princeton Univ. Press, 1992.
198. Quoted in *The Way of Men,* by Martin Buber, pp. 38-9.
199. Martin Buber, *I and Thou*. trans., Ronald Gregor Smith. New York: Charles Scribner's Sons, 1958.
200. Ida Postma, *Sunrise magazine*, Theosophica Univ., December 1975.
201. *Slavery and Freedom*, trans. R. M. French, New York: Charles Scribner's Sons, 1944, p. 43.
202. *The Destiny of Man*, trans. by N. Duddington, New York: Scribner's, 1937, p. 378.
203. *Slavery and Freedom*, trans. R. M. French, New York: Charles Scribner's Sons, 1944.
204. *Slavery and Freedom*, trans. R. M. French, New York: Charles Scribner's Sons, 1944.
205. See "Tales of Love and Marriage," *The Power of Myth: Joseph Campbell with Bill Moyers*, ed., Sue Flowers, Doubleday, 1988.

206. *The Courage to Be*, Yale Univ., 1952.
207. *The Myth of Sisyphus and Other Essays*, trans., Justin O'Brien, Vintage, 1960, p. 21.
208. *The Myth of Sisyphus and Other Essays*, trans., Justin O'Brien, Vintage, 1960, p. 38.
209. *To Deny Our Nothingness: Contemporary Images of Man,* Delta, 1967, p. 310.
210. *To Deny Our Nothingness: Contemporary Images of Man,* Delta, 1967, p. 311.
211. *To Deny Our Nothingness: Contemporary Images of Man,* Delta, 1967, p. 311.
212. *To Deny Our Nothingness: Contemporary Images of Man,* Delta, 1967, p. 311.
213. *To Deny Our Nothingness: Contemporary Images of Man,* Delta, 1967, p. 313.
214. *Liberation of Buchanvald*, www.remember.org.
215. "They Died 900 a Day in 'the Best' Nazi Death Camp," *PM,* April 16, 1945.
216. *Murrow: His Life and Times*, Ann M. Sperber. Fordham, 1998, p. 252.
217. *Slaughterhouse Five; or the Children's Crusade, a Duty Dance with Death*, Delacore, 1968. Interestingly, Beckett read and was impressed by the novel in 1972; see James Knowlson, *Dammed to fame: The life of Samuel Beckett*, pp. 594-595.
218. Kwiet, K.: "Suicide in the Jewish Community," *Leo Baeck Yearbook*, vol. 38. 1993.
219. trans. Ilse Lasch, Washington Square Press (Simon & Schuster), 1959; rpt, ISBN 0-671-02337-3 (Softcover, January, 1997).
220. *Man's Search for Meaning*, pp. 153-154.
221. *Man's Search for Meaning*, p. 160.
222. *Man's Search for Meaning,* p. 159.
223. *Man's Search for Meaning,* pp. 156-157.
224. *Man's Search for Meaning,* pp. 157-158.
225. *Man's Search for Meaning,* p. 158.
226. "On Human Nature: Character: Part 1," *The Essays of Arthur Schopenhauer*, Kensinger Publ., 2004.
227. *Totem and Taboo: Some Points of Agreement between the Mental Lives of Savages and Neurotics*, translated by James Strachey, W. W. Norton, 1950.
228. *Oedipus Rex*, Final Scene, line 139.
229. *Oedipus Rex*, Final Scene, line 140-141.
230. *Oedipus Rex*, Final Scene, lines 222-223.
231. *The Complete Signet Classic Shakespeare*, ed. Sylvan Barnet, Harcourt, Brace, Jovanovich, 1972.
232. *Highway 61 Revisited,* Columbia Studios, 1965.
233. Composed by Barry, Robin and Maurice Gibb, IBC Studios, 1968.
234. 1933; rpt., *Man's Fate,* Vintage, February 19, 1990.

CHAPTER TWO

CHRIST OF THE COAL MINES

Every artist dips his brush in his own soul, and paints his own nature into his pictures.

—Henry Ward Beecher
Proverbs from Plymouth Pulpit

I have tried to express the terrible passions of humanity by means of red and green.

—Vincent van Gogh
Letter to Theo, September 8, 1888

Windmill at Montmartre, Vincent Van Gogh, Oil on Canvas, 46.5 x 38.0 cm, Paris, autumn, 1886, destroyed by fire, 1967
The original Windmill on Montmartre (in color) was destroyed in a fire. This copyright free copy was obtained from http://www.vggallery.com.

Part I

Sunday at Le Crau

No one heard the shot.

No one ever found the gun.

It was Sunday, July 27, 1890. Vincent had recently finished *Wheatfield with Crows*, thought to be his final painting, one that he described as representing “vast fields” of wheat beneath troubled skies.” One where he said in a letter he meant to send to Theo “I did not need to go out of my way to try to express cheerlessness and extreme loneliness.” The letter never got sent, but was found stuffed in his pocket.

That morning, as usual, he walked out into the wheat fields with his easel, brushes, tubes of color and folding stool, perhaps hoping to reach his destination before the gang of local boys and girls were up and able to tease him and throw tomatoes. *Le Crau*, a wide plain of ripe grain, fields of citron, yellow, tan, and ochre, spread out beneath the bright Provencal sun. It’s safe to assume he heard the cicadas singing loudly, the swiping swishes of the farmers’ scythes already cutting through the rich wheat stalks, the gusts of wind whispering through the olive branches.

Driven and filled with energy for months, he had been quickly, with an assurance that overcame and perhaps even came from his doubts and struggles, putting his own dramatic visions on canvas after canvas. But today he did not go into the fields to paint, or, perhaps, in the beginning he did, perhaps in the morning that was his intention.

No one will ever know.

He said he brought the revolver to frighten off the crows. Possibly that was his original intention when he included it with his lunch of bread and milk. In the end it’s probably not relevant, except for the endless attempts to analyze him, to dig into his complex psyche, at once brilliant and yet impelled to self-destruction.

The Ravoux family were sitting on the terrace of their café when he returned, a bit concerned because he was late, but not overly so. When he finally appeared, his walk was more uneven than usual, and he held his hand over his stomach. “Monsieur Vincent,” Mrs. Ravoux said, “we were worried, we are glad to see you come. Has anything bad happened?”

“No, but I . . .” he left his reply unfinished as he passed inside.

“Mr. Ravoux followed him upstairs, where he found him sitting on his bad, facing the wall.

“I wanted to kill myself.”

Part II

Christ of the Coal Mines

A brutal coal mining village located in the center of Borinage, Petit Wasmes fit the harsh environment Vincent wanted. Hadn't Christ walked among, lived with and shared the suffering, the harshest indignities of the most downtrodden people of his time, even the most unclean, the lepers. Vincent had read in the gospel of Luke how Jesus had not turned away in disgust, but had healed them, had promised them acceptance into God's kingdom. Yes, Vincent thought, here, in the real world of poverty and want is where a true preacher, an honest servant of Christ must live.

His letter to Theo in April, 1879, some three months after having accepted the position, expresses both the rough, hard existence of his new assignment and his embracement of it:

> Dear Theo,
>
> It is time that you heard from me again. From home I heard that you had been in Etten for a few days and that you were on a business trip. I certainly hope you had a good journey. I suppose you will be in the dunes some of these days and occasionally in Scheveningen. It is lovely here in spring, too; there are spots where one could almost fancy oneself in the dunes, because of the hills.
>
> Not long ago I made a very interesting expedition, spending six hours in a mine. It was Marcasse, one of the oldest and most dangerous mines in the neighbourhood. It has a bad reputation because many perish in it, either going down or coming up, or through poisoned air, firedamp explosion, water seepage, cave-ins, etc. It is a gloomy spot, and at first everything around looks dreary and desolate.
>
> Most of the miners are thin and pale from fever; they look tired and emaciated, weather-beaten and aged before their time. On the whole the women are faded and worn. Around the mine are poor miners' huts, a few dead trees black from smoke, thorn hedges, dunghills, ash dumps, heaps of useless coal, etc. Mans could make a wonderful picture of it.
>
> I will try to make a little sketch of it presently to give you an idea of how it looks.
>
> I had a good guide, a man who has already worked there for thirty-three years; kind and patient, he explained everything well and tried to make it clear to me.
>
> So together we went down 700 meters and explored the most hidden corners of that underworld. The maintenages or gredins [cells where the miners work] which are situated farthest from the exit are called *des caches* [hiding places, places where men search].

This mine has five levels, but the three upper ones have been exhausted and abandoned; they are no longer worked because there is no more coal. A picture of the maintenages would be something new and unheard of--or rather, never before seen. Imagine a row of cells in a rather narrow, low passage, shored up with rough timber. In each of those cells a miner in a coarse linen suit, filthy and black as a chimney sweep, is busy hewing coal by the pale light of a small lamp. The miner can stand erect in some cells; in others, he lies on the ground (***** tailles à droit, *** tailles à plat). The arrangement is more or less like the cells in a beehive) or like a dark, gloomy passage in an underground prison, or like a row of small weaving looms, or rather more like a row of baking ovens such as the peasants have, or like the partitions in a crypt. The tunnels themselves are like the big chimneys of the Brabant farms.

The water leaks through in some and the light of the miner's lamp makes a curious effect, reflected as in a stalactite cave. Some of the miners work in the maintenages; others load the cut coal into small carts that run on rails, like a street-car. This is mostly done by children, boys as well as girls. There is also a stable yard down there, 700 meters underground, with about seven old horses which pull a great many of those carts to the so-called accrochage, the place from which they are pulled up to the surface. Other miners repair the old galleries to prevent their collapse or make new galleries in the coal vein. As the mariners ashore are homesick for the sea, notwithstanding all the dangers and hardships which threaten them, so the miner would rather be under the ground than above it. The villages here look desolate and dead and forsaken; life goes on underground instead of above. One might live here for years and never know the real state of things unless one went down in the mines.

People here are very ignorant and untaught--most of them cannot read--but at the same time they are intelligent and quick at their difficult work; brave and frank, they are short but square-shouldered, with melancholy deep-set eyes. They are skillful at many things, and work terribly hard. They have a nervous temperament--I do not mean weak, but very sensitive. They have an innate, deep-rooted hatred and a strong mistrust of anyone who is domineering. With miners one must have a miner's character and temperament, and no pretentious pride or mastery, or one will never get along with them or gain their confidence.

Did I tell you at the time about the miner who was so badly hurt by a firedamp explosion? Thank God, he has recovered and is going out again, and is beginning to walk some distance just for exercise; his hands are still weak and it will be some time before he can use them for his work, but he is out of danger. Since that time there have been many cases of typhoid and malignant fever, of what they call la sotte fièvre, which gives them bad dreams like nightmares and makes them delirious. So again there are many sickly and bedridden people - emaciated, weak, and miserable.

In one house they are all ill with fever and have little or no help, so that the patients have to nurse the patients. "Ici c'est les malades qui soignent les malades" [here the sick tend the sick], said a woman, like, "Le pauvre est l'ami du pauvre." [The poor man is the poor man's friend.]

Have you seen any beautiful pictures lately? I am eager for a letter from you. Has Israëls done much lately and Maris and Mauve?

A few days ago a colt was born here in the stable, a pretty little animal that soon stood firm on his legs. The miners keep many goats here, and there are kids in every house; rabbits are also very common here in the miners' houses.

I must go out to visit some patients, so I must finish. When you have time, let me have a word from you soon, as a sign of life. My compliments to the Roos family, and to Mauve when you meet him. Many good wishes, and believe me always, with a handshake in thought,

Your loving brother, Vincent

Going down into a mine is a very unpleasant sensation. One goes in a kind of basket or cage, like a bucket in a well, but in a well from 500 - 700 meters deep, so that when looking upward from the bottom, the daylight is about the size of a star in the sky.

It feels like being on a ship at sea for the first time, but it is worse; fortunately it does not last long. The miners get used to it, yet they keep an unconquerable feeling of horror and fear which reasonably and justifiably stays with them.

But once down, the worst is over, and one is richly rewarded for the trouble by what one sees.

My address is - Vincent van Gogh, c/o Jean Baptiste Denis,
Rue de petit Wasmes,
Wasmes (Borinage, Hainaut) (1)

Others would have thought this assignment a punishment, a rejection. Certainly Vincent had not impressed the church hierarchy with his studies. However, no one could doubt his sincerity, his faith. If anything, it was too strong--that of a fanatic; and that was the problem.

There could be no doubt it was sincere, not a momentary or shallow impulse, certainly not driven by practical or economic concerns. Rather, it was a sincere desire that had taken root somewhere in his childhood, as his correspondence to Theo indicates time and again. On March 22, 1877, some two years previous to this assignment here, he wrote:

Father was unable to preach last Sunday and the Reverend Mr. Kam stood in for him. I know that his heart burns for something to happen that will allow me to follow in his footsteps, not just some of the way, but all the way. Father has always expected it of me, oh, may it come about and blessings be upon it.

The print you gave, "Heaven and earth shall pass away, but my words shall not pass away," and the portrait of the Reverend Mr. Heldring are already up in my little room, oh, how glad I am to have them, they fill me with hope.

Writing to you about my plans helps me to clarify and settle my thoughts. To begin with, I think of the text, "It is my portion to keep Thy word." I have such a craving to make the treasures of the Bible's word my own, to become thoroughly and lovingly familiar with all those old stories, and above all with everything we know about Christ.

In our family, which is a Christian family in the full sense of the term, there has always been, as far as one can tell, someone from generation to generation who was a preacher of the Gospel. Why should there not be a member of our family even now who feels called to that ministry, and who has some reason to suppose that he may, and must, declare himself and look for means of attaining that end? It is my prayer and fervent desire that the spirit of my Father and Grandfather may rest upon me, that it may be granted me to become a Christian and a Christian laborer, that my life may come to resemble, the more the better, those of the people I have mentioned above—for behold, the old wine is good and I do not desire new. Let their God be my God and their people my people, let it be my lot to come to know Christ in his full worth and to be impelled by his charity.

It is so beautifully put in the text, "As sorrowful, yet always rejoicing," what that charity is, and in Cr. 13 she "beareth all things, believeth all things, hopeth all things, endureth all things. Charity never faileth."

My heart is filled today with the text about those on the way to Emmaus, when it was toward even and the sun was going down: "But they constrained him, saying, Abide with us."

It is dear to you, too, that "sorrowful, yet always rejoicing," keep it in mind, for it is a good text and a good cloak to wear in the storm of life, keep it in mind at this time now that you have been going through so much. And be careful, for though what you have been through is no small thing, yet as far as I can see there is something still greater ahead and you too will be put in mind of the Lord's word. I have loved you with an everlasting Love, as one whom his mother comforteth, so will I comfort you. I shall comfort you as one who comforteth his Mother. I shall give you another Comforter, even the Spirit of truth. I will make a new covenant with you. Depart, touch no unclean thing, and I will take you to me for a people, and I will be to you a God. And I will be a Father unto you, and ye shall be my sons and daughters. Hate the evil and the places where it is rife, it draws you with its false splendor and will tempt you as the devil tried to tempt Christ by showing Him "all the kingdoms of the world, and the glory of them"; and saying, "All these things will I give thee, if thou wilt fall down and worship me." There is something better than the glory of the things of this world, namely the feeling when our heart burns within us upon hearing His word, faith in God, love of Christ, belief in immortality, in the life hereafter.

Hold on to what you have, Theo, my boy, brother whom I love, I long so fervently for the goal you know of, but how can I attain it? If only everything were already behind me as it is behind Father, but it takes so much hard work to become a Christian laborer and a preacher of the Gospel and a sower of the Word. You see, Father can count his religious services and Bible readings and visits to the sick and the poor and his written sermons by the thousand, and yet he does not look back, but carries on doing good.

Cast your eye up on high and ask that it be granted to me, as I ask if for you. May He grant your heart's desire, He who knows us better than we know ourselves, and is above prayer and above thought, since His ways are higher than our ways and his thoughts higher than our thoughts, as high as heaven is above earth. And may the thought of Christ as a Comforter and of God as a lofty dwelling be with you. (2)

Vincent was driven to preach the gospel. In May, 1877, he moved to Amsterdam to study theology, and with the full support of his family, who hired a tutor for him, he began to prepare in earnest for the theology entrance exams. However, his emotional, personal connection with God didn't match the more formal, dispassionate studies required of him. He did poorly in his studies and dropped out in July, 1878. Perhaps, he thought, a different program, a shorter program of study in Brussels to prepare him to be an *evangelical missionary* rather than a minister would suit him better.

A letter to his brother the following November suggests he thought he had made the right move and indicates he was looking forward to his potential assignment:

Dear Theo,

On the evening of the day we spent together, which passed only too quickly for me, I want to write to you again. It was a great day for me to see you again and to talk with you, and it is a blessing that such a day, that passes in a moment, and such a joy that is of so short duration, stays in our memory and will never be forgotten. When we had taken leave I walked back, not along the shortest way but along the tow-path. Here are workshops of all kinds that look picturesque, especially in the evening with the lights, and to us who are also labourers and workmen, each in his sphere and in the work to which he is called, they speak in their own way, if we only listen to them, for they say: Work while it is day, the night cometh when no man can work.

It was just the moment when the street cleaners came home with their carts with the old white horses. A long row of these carts were standing at the so-called Ferme des Boues, at the beginning of the tow-path. Some of these old white horses resemble a certain old aquatint engraving, which you perhaps know, an engraving that has no great art value, it is true, but which struck me, and made a deep impression upon me. I mean the last from that series of prints called "The Life of a Horse." It represents an old white horse, lean and emaciated, and tired to death by a long life of heavy labour, of too much and too hard work. The poor animal is standing on a spot utterly lonely and desolate, a plain scantily covered with withered dry grass, and here and there a gnarled old tree broken and bent by the storm. On the ground lies a skull, and at a distance in the background a bleached skeleton of a horse, lying near a hut where lives a man who skins horses. Over the whole is a stormy sky, it is a cold, bleak day, gloomy and dark weather.

It is a sad and very melancholy scene, which must strike everyone who knows and feels that we also have to pass one day through the valley of the shadow of death, and "que la fin de la vie humaine, ce sont des larmes ou des cheveux blancs." [the end of human life is tears or white hairs.] What lies beyond this is a great mystery that only God knows, but He has revealed absolutely through His word that there is a resurrection of the dead.

The poor horse, the old faithful servant, is standing there patiently and meekly, yet bravely and unflinchingly; like the old guard who said, "la garde meurt mais elle ne se rend pas," [the guard dies, but never surrenders] it awaits its last hour. Involuntarily I was reminded of that engraving, when I saw tonight those horses of the ash carts.

As to the drivers themselves with their filthy dirty clothes, they seemed sunk and rooted still deeper in poverty than that long row or rather group of paupers, that Master de Groux has drawn in his "Bench of the Poor." It always strikes me, and it is very peculiar, that when we see the image of indescribable and unutterable desolation - of loneliness, of poverty and misery, the end of all things, or their extreme, then rises in our mind the thought of God. At least this is the case with me and does not Father also say: "There is no place where I like better to speak than in a churchyard, for there we are all on equal ground; not only that, there we always *realize* it." I am glad that we had time to see the museum together and especially the work of de Groux and Leys, and so many other interesting pictures, like that landscape of Cooseman's for instance. I am very pleased with the two prints you gave me, but you ought to have accepted from me that small etching, "The Three Mills." Now you have paid it all yourself, and not allowed me to pay half as I wished to do. But you must keep it for your collection, for it is remarkable, even though the reproduction is not so very good. In my ignorance, I should ascribe it rather to Peasant Breughel than to Velvet Breughel.

I enclose the little hasty sketch, "Au Charbonnage." I should like to begin making rough sketches from some of the many things that I meet on my way, but as it would probably keep me from my real work, it is better not to begin. As soon as I came home I began a sermon about the "barren fig tree," Luke xiii 6-9.

That little drawing "Au Charbonnage" is nothing specially remarkable, but the reason I made it is that one sees here so many people that work in the coal mines, and they are rather a characteristic kind of people. This little house stands not far from the road; it is a small inn attached to the big coal shed, and the workmen come there to eat their bread and drink their glass of beer during the lunch hour.

When I was in England I applied for a position as Evangelist among the miners in the coal mines, but they turned me down, stating that I had to be at least twenty-five years old. You know how one of the roots or foundations, not only of the Gospel, but of the whole Bible is, "Light that rises in the darkness," *from darkness to light*. Well, who will need this most, who will be open to it? Experience has taught that those who walk in the darkness, in the centre of the earth, like the miners in the black coal mines for instance, are very much impressed by the words of the Gospel, and believe it too. Now there is in the south of Belgium, in Hainault, in the neighborhood of Mons, up to the French frontiers, aye, even far across it, a district called the Borinage that has a special population of laborers who work in the numerous coal mines. In a little handbook of geography I found the following about them: "The Borins (inhabitants of the Borinage, situated west of Mons) find their work exclusively in the coal mines. These mines are an imposing sight, 300 metres underground, into which daily descend groups of working men, worthy of our respect and our sympathies. The miner is a special Borinage type, for him daylight does not exist, and except on Sunday he never sees the sunshine. He works laboriously by a lamp whose light is pale and dim, in a narrow tunnel, his body bent double and sometimes he is obliged to crawl along; he works to extract from the bowels of the earth that mineral substance of which we know the great utility; he works in the midst of thousands of ever-recurring dangers; but the Belgium miner has a happy disposition, he is used to that kind of life, and when he descends the shaft, carrying on his hat a little lamp that is destined to guide him in the darkness, he trusts himself to God, Who sees his labour and Who protects him, his wife and his children."

So the Borinage is situated south of Lessines, where one finds the stone quarries.

I should very much like to go there as an Evangelist. The three months' trial demanded of me by the Rev. de Jong and the Rev. Pietersen is almost over. St. Paul was three years in Arabia before he began to preach, and before he started on his great missionary journeys and his real work among the heathen. If I could work quietly for about three years in such a district, always learning and observing, then I should not come back from there without having something to say that was really worth hearing. I say so in all humility and yet with confidence. If God wills, and if He spares my life, I would be ready about my thirtieth year, starting out with a peculiar training and experience, being able to master my work better, and riper for it than now.

I write you this again although we have already spoken about it many a time.

There are already in the Borinage many little Protestant communities and certainly schools also. I wish I could get a position there as Evangelist in the way we spoke about, preaching the Gospel to the poor, that means those who need it most and for whom it is so well suited, and then during the week devoting myself to teaching.

You have certainly visited St. Gilles? I too made a trip there, in the direction of the Ancienne Barrière. Where the road to Mont St. Jean begins there is another hill, the Alsemberg. To the right is the cemetery of St. Gilles, full of cedars and evergreen, from where one has a view over the whole city.

Proceeding further one arrives at Forest. The neighbourhood is very picturesque there, on the slope of the hills are old houses, like those huts in the dunes that Bosboom has sometimes painted. One sees all kinds of field labour performed there, the sowing of corn, the digging of potatoes, the washing of turnips, and everything is picturesque, even the gathering of wood, and it looks much like Montmartre. There are old houses covered with evergreen or vines, and pretty little inns; among the houses I noticed one was that of a mustard manufacturer, a certain Verkisten, his place was just like a picture by Thijs Mans for instance. Here and there are places where stone is found, so they have small quarries, through which hollowed out roads pass, with deeply cut wagon ruts, where one sees the little white horses with red tassels, and the drivers with blue blouses; shepherds are to be found there too, and women in black with white caps, that remind one of those of de Groux. There are some places here, thank God one finds them everywhere, where one feels more at home than anywhere else, where one gets a peculiar pristine feeling like that of homesickness, in which bitter melancholy plays some part; but yet its stimulation strengthens and cheers the mind, and gives us, we do not know how or why, new strength and ardour for our work. That day I walked on past Forest and took a side path leading to a little old ivy-grown church. I saw many linden trees there, still more interwoven, and more Gothic so to say, than those we saw in the Park, and at the side of the hollowed road that leads to the churchyard there were twisted and gnarled stamps and tree roots, fantastical like those Albert Dürer etched in “Ritter, Tod and Teufel.”

Have you ever seen a picture or rather a photograph of Carlo Dolci’s work “The Garden of Olives”? There is something Rembrandt-esque in it; I saw it the other day. I suppose you know that large rough etching on the same theme after Rembrandt, it is the pendant of that other, “The Bible Reading,” with those two women and a cradle? Since you told me that you had seen the picture by Father Corot on that same subject, I remembered it again; I saw it at the exhibition of his works shortly after his death and it deeply appealed to me.

How rich art is, if one can only remember what one has seen, one is never empty of thoughts or truly lonely, never alone.

A Dieu, Theo, I heartily shake hands with you in thought. Have a good time, have success in your work, and meet many good things on your road, such as stay in our memory and enrich us, though apparently we possess little. When you see Mauve greet him for me and believe me,

Your loving brother, Vincent

I kept this letter for a few days; the 15th of November is passed, so the three months have elapsed. I spoke with the Rev, de Jong and Master Bokma, they tell me that I cannot attend the school on the same conditions as they allow to the native Flemish pupils; I can follow the lessons free of charge if necessary - but this is the only privilege, - so in order to stay here longer I ought to have more financial means than I have at my disposal, for they are nil. So I shall perhaps soon try that plan involving the Borinage. Once I am in the country I shall not soon go back to a large city.

It would not be easy to live without the Faith in Him and the old confidence in Him; without it one would lose one's courage. (3)

A letter from his father to Theo supports Vincent's belief that the anticipated move is going well.

20 December 1878

Eight days ago we received a rather nice letter from Vincent; nice in so far as he seems to have found a good boarding house with farmers in Pâturages for thirty francs a month, the reason that it is not much being that he teaches the children in the evening. He was soon accepted with goodwill by many people, and working with these people seems to give him satisfaction. Furthermore, the Reverend Péron has promised him his cooperation. In the middle of January there will be another meeting of the committee of with Mr. Péron is the secretary, and he promised him that he would try to do something for him.

In the meantime, he spends his free moments drawing big maps of Palestine which can be used for talks and catechism, and with which he tries to make some money. I received one and ordered him to make four more copies for which I gave him ten francs each. (4)

On December 26, Vincent's letter to Theo suggests a good beginning:

My dear Theo,

It is time I wrote to you again, to wish you, firstly, all the best at the start of the New Year. May many good things be your lot and may God's blessing rest on your work in the year on which we are now embarking.

I very much long for a letter from you, to hear how things are going and how you are, and also if you have *seen anything beautiful and remarkable of late*. As far as I am concerned, you'll be aware that there are no paintings here in the Borinage, that by and large they do not even know what a painting is, so obviously I have not seen anything in the way of art since my departure from Brussels. But that does not alter the fact that the country here is very special and very picturesque, everything *speaks*, as it were, and is full of character. Lately, during the dark days before Christmas, snow was lying on the ground. Everything reminded one then of the medieval paintings by, say, Peasant Brueghel, and by so many others who have known how to depict the singular effect of red and green, black and white so strikingly. And often the sights here have made me think of the work of, for example, Thijs Mans or Albrecht Dürer. There are sunken roads here, overgrown with thornbushes and gnarled old trees with their freakish roots, which resemble perfectly that road on Dürer's etching, "Death and the Knight."

Thus, a few days ago, the miners returning home in the evening towards dusk in the white snow were a singular sight. These people are quite black when they emerge into the daylight from the dark mines, looking jut like chimney sweeps. Their dwellings are usually small and should really be called huts; they lie scattered along the sunken roads, in the woods and on the slopes of the hills. Here and there one can still see moss-covered roofs, and in the evening a friendly light shines through the small-paned windows.

Much as we have coppices and shrubby oaks in Brabant and pollard willows in Holland, so one sees blackthorn hedges around the gardens, fields and meadows here. Lately, with the snow, the effect is that of black lettering on white paper, like pages of the Gospel.

I have already spoken several times here, both in a fairly large room especially designed for religious meetings and also at the meetings they usually hold in the evenings in the workmen's cottages, and which may best be called Bible classes. Among other things, I have spoken on the parable of the mustard seed, the barren fig tree and the man born blind. On Christmas, of course, on the stable in Bethlehem and Peace on earth. If, with God‘s blessing, I were to get a permanent position here, I should welcome that with all my heart.

Everywhere round here one sees the large chimneys and the tremendous mountains of coal at the entrance to the mines, the so-called charbonnages. You know that large drawing by Bosboom, "Chaudfontaine" - it gives a good impression of the countryside in these parts, except that here everything is coal while to the north of Hainaut there are stone quarries and in Chaudfontaine they have iron.

I still keep thinking of the day you came to Brussels and of our visit to the Museum. And I often wish you were a bit nearer and we could be together more often. Do reply soon, I keep looking at that etching of "A Young Citizen" over and over again.

The miner's talk is not very easy to make out, but they understand ordinary French well, provided it is spoken quickly and fluently enough, for then it automatically sounds like their patois, which comes out with amazing speed. At a meeting this week, my text was Acts 16 9, "And a vision appeared to Paul in the night; There stood a man of Macedonia, and prayed him, saying, Come over into Macedonia, and help us." And they listened attentively when I tried to describe what the Macedonian was like who needed and longed for the comfort of the Gospel and for knowledge of the Only True God. That we should think of him as a workman, with lines of sorrow and suffering and fatigue on his countenance, without pomp or glory but with an immortal soul and needing the food that does not perish, namely God's word, because man liveth not by bread alone, but by all the words that flow from God's mouth. How Jesus Christ is the Master who can strengthen and comfort and enlighten one like the Macedonian, a workman and labourer whose life is hard. Because He Himself is the great Man of Sorrows who knows our ills, Who was called the son of a carpenter, though He was the Son of God and the great Healer of sick souls. Who laboured for thirty years in a humble carpenter's shop to fulfill God's will. And God wills that in imitation of Christ, man should live and walk humbly on earth, not reaching for the sky, but bowing to humble things, learning from the Gospel to be meek and humble of heart.

I have already had occasion to visit some of the sick, since there are so many of them here. Wrote today to the President of the Committee of Evangelization, asking him if my case could be dealt with at the next meeting of the committee.

It is thawing tonight. I can't tell you how picturesque the hilly country looks in the thaw, with the snow melting and now that the black fields with the green of the winter wheat can be seen again.

For a stranger, the villages here are real rabbit warrens with the countless narrow streets and alleyways of small worker's houses, at the foot of the hills as well as on the slopes and the top. The nearest comparison is a village like Scheveningen, especially the back streets, or villages in Brittany as we know them from pictures. But you have travelled through these parts by train on your way to Paris and may have fleeting memories of them. The Protestant churches are small, like the one at De Hoeve though a little larger, but the place where I spoke was just a large bare room which could hold a hundred people at most. I also attended a religious service in a stable or shed, so everything it is simple and original enough.

Write soon if you can find the time, and know that you are again and again, indeed constantly, in my thoughts. Wishing once more that God's blessing be yours in the New Year, and shaking your hand in my thoughts, believe me, always,

Your very loving brother, Vincent

> My regards to everyone at the Roos's, and wish them all the very best for the New Year, as well as anyone who may ask after me.
>
> When you write, please address your letter care of M. van der Haegen, Colporteur, à Pâturages, près de Mons, Borinage, Hainaut.
>
> I have just visited a little old woman in a charcoal-burner's home. She is terribly ill, but full of faith and patience. I read a chapter with her and prayed with them all. The people here have something unique and attractive about them thanks to their simplicity and good nature, not unlike the Brabant people in Zundert and Etten. (5)

Note the key beliefs he expresses here in the portion of the sermon he shares with his brother. They emphasize how Christ is "the great Man of Sorrow" . . . "who laboured for thirty years in a humble carpenter's shop to fulfill God's will, and God wills that in imitation of Christ man should live and walk humbly on earth, not reaching for the sky, but bowing to humble things, learning from the Gospel to be meek and humble of heart." The stress is to be *like* Christ, to be Christ*like*.

However, the letters from Vincent's parents to Theo during this time quickly shift from simple, positive support for a son to reveal their concerns about him struggling to maintain his sanity, concerns about a son fully committed to his mission, to his relationship to God, but also a son unable to take care of himself, much less administer to others.

They begin upbeat. On January 20, 1879, Vincent's father writes:

> We are glad to be able to tell you that Vincent has been accepted for the evangelization in the Borinage--provisionally for six months. He gets fifty francs a month--surely not much, but his boarding house costs him thirty francs. It seems he works there with success and ambition, and his letters are really interesting. He devotes himself to that job with all his heart and an eye for the needs of those people. It is certainly remarkable what he writes; he went down, for instance, in a mine, 635 meters. (6)

However, suggestions of the underlying fears already appear in his father's next letter, dated February 12, 1879:

> We are beginning to worry about him again. I am afraid he is wholly absorbed by the care for the sick and the wounded and by sitting up with them. . . .
>
> He also spoke about a plan of renting a workman's house and living there alone. We have tried to dissuade him from it. We are afraid he would not keep it in good shape and it would again lead to eccentricities. (7)

The single word "again" indicates the problems are long standing. *Eccentricities* also sounds like a euphemism, something that might better or more honestly be stated as suggesting possible *insanity*, a psychological unbalance, at the very least a troubling past.

Letters from his mother to Theo continue to reveal fears he is not taking care of himself, is ignoring his own well being to the point of harming himself. On February 27, 1879, she writes:

> Verhaegen, a colporteur, to whom Pa also sent his letters in the beginning, where Vincent had been lovingly received during the first eight days; he was the one who had found that good boarding house at Denis. . . .
>
> And now I have to tell you that Pa has gone to Vincent this week. We were worried about all the bad weather he had, and especially because while I was away, there had been a very unpleasant letter from him, confirming what we had already suspected, that he had no bed, and that there was nobody to watch his things but far from complaining he said that that was nobody's concern, etc. We were preparing a parcel for him, but we both thought that it would be so much better if Pa himself would take it to him. (8)

On July 2, 1879, her letter to Theo is more ominous:

> This week a letter from Vincent; we are always thinking about him with anxiety; poor boy, shortly after my visit to him he wrote that he had such a melancholy feeling when we said goodbye, as if it could have been for the first, but also for the last time. But now there has been a meeting, but that they hadn't said anything to him; before, they had always found fault with him. We have the idea they still want to wait and see for some time, but if he doesn't suit himself to their wishes and adopt the behaviour they demand of him, they can't accept him. He could still achieve so much, if only he knew how to control himself. Poor boy, what a difficult, unrewarding, much missing young life, and what is he going to do next? (9)

His father's letter to Theo on July 19 indicates that what began as a promising new start has already collapsed:

> You know, don't you that Vincent's situation in Wasmes does not become any clearer? They have given him three months to look for something else. He does not comply with the wishes of the Committee and it seems that nothing can be done about it. It is a bitter trial for us. We literally don't know what to do. There is so much good in him, but he simply doesn't want to cooperate. (10)

The 1879-1880 report from the Synodal Board of Evangelization of the Union of Protestant Churches of Belgium gives the official view:

> The experiment of accepting the services of a young Dutchman, Mr. Vincent van Gogh, who felt himself called to be an evangelist in the Borinage, has not produced the anticipated results. If a talent for speaking, indispensable to anyone placed at the head of a congregation, had been added to the admirable qualities he displayed in aiding the sick and wounded, to his devotion to the spirit of self-sacrifice, of which he gave many proofs by consecrating his night's rest to them, and by stripping himself of most of his clothes and linen in their behalf, Mr. Van Gogh would certainly have been an accomplished evangelist.
>
> Undoubtedly it would be unreasonable to demand extraordinary talents. But it is evident that the absence of certain qualities may render the exercise of an evangelist's principal function wholly impossible.
>
> Unfortunately this is the case with Mr. Van Gogh. Therefore, the probationary period—some months—having expired, it has been necessary to abandon the idea of retaining him any longer. (11)

On August 7 his father writes Theo:

> Last Friday [25 July], Vincent writes, he started on a trip to Maria Hoorebeeke in Flanders; he arrived there—on foot—on Sunday afternoon, intending to meet the Reverend Pietersen, who was in Brussels. Thereupon he went to that city, and he met him on Monday morning. After consultation with him, he is now in Cuesmes again, where he has found shelter; he hopes to find a small room there to stay for the time being. At present his address is: Chez M. Frank, Evangéliste Ã Cuesmes (prés de Mons) au Marais. In Brussels, he visited the families he had met there earlier - what impression will he have made? (12)

On August 19, his mother writes Theo:

> But now I must tell you something new, which is that Vincent, after much pressure from our side to visit us at home because we were worrying so much about him and he had nothing to do there, suddenly stood before us last Friday [August 15th]. The girls were boating with the Gezink family, and all at once we hear, "Hello father, hello mother," and there he was. We were glad; although seeing him again we found he looked thin; that is over now; it must have been the walking and bad food etc.--things, by the way, he does not talk about, but he looks well, except for his clothes. Pa immediately gave him his cherished new jacket. We bought him a pair of boots, and he now wears the little summer coat that I made for Pa's birthday every day. Some of your old underwear came in useful too, and as far as stockings, etc., are concerned, I had prepared them in

> advance, so that now he is quite well taken care of. He is reading books by Dickens all day long, and does not speak apart from giving answers--sometimes correct, sometimes strange ones; if only he adopted the good things from these books. For the rest, about his work, about the past or the future, not a word…Tomorrow, he and Pa will go to Prinsenhage, where CM's boys will come to see the paintings; they are going by train. Pa and Vincent will go on foot, maybe he will talk a little bit then. (13)

The letters are sparse over the next year, and reveal that Vincent was emotionally, psychologically lost. He continued to work in Wasmes without official position or pay, living in poverty, nursing the victims of a mining disaster, giving away whatever food and clothing came his way. The Decrugs, a family he stayed with, reported that they felt powerless to help him though they could hear him weeping alone at night. (14)

The following March 11, his father writes Theo:

> Vincent is still here--but alas! It is nothing but worry. Now he is talking about going to London in order to speak with the Reverend Jones. If he sticks to that plan, I'll enable him to go, but it is hopeless. (15)

On July 5 his father writes:

> Indeed that letter Vincent wrote to you gave me some pleasure. But oh! What will become of him, and isn't it insane to choose a life of poverty and let time pass by without looking for an occasion of earning one's own bread - yes, that really is insane. But we have to put up with it. None of all the things we tried has helped in any way. Maybe you should write back to him; in the last days of June I sent him 60 francs, which he acknowledged; some time later we sent him some clothes. Thinking of him always hurts, and we do think so continuously of him.
>
> [Lines added by Mrs. Van Gogh] We can agree with what you write about Vincent, but if reading books gives such practical results, can it then be called right? And for the rest, what kind of ideas his reading gives him. He sent us a book by Victor Hugo, but that man takes the side of the criminals and doesn't call bad what really is bad. What would the world look like if one calls the evil good? Even with the best of intentions that cannot be accepted. Did you answer him? If not, do so in any case; we were so glad that he thought of you, and we were so sad that he didn't want to have anything to do with anybody when he was here. We haven't heard from him for a long time now and shall write to him again. (16)

Books indeed were important to Vincent, and one in particular had become his guide. On September 4, 1877, some two years before he began pushing to work as an evangelist at Wasmes, he wrote Theo:

> I am also copying the whole of the *Imitation of Christ* from a French edition which I borrowed from Uncle Cor; the book is sublime, and he who wrote it must have been a man after God's own heart. A few days ago such an irresistible longing for that book came over me--perhaps because I so often look at the lithograph after Ruyperez--that I asked Uncle Cor to lend it to me; now I am copying it in the evening: it means much work, but I have finished most of it, and I know no better way to study it. I also bought Bossuet's Oraisons Funebres once more, for 40 cents. I feel it is necessary to work as hard as I can. I often think of that phrase, "The days are full of evil"; one must arm oneself and try to be filled with as much goodness as possible in order to be prepared and be able to resist. As you know, it is no small undertaking, and we do not know the result; but at all events I will try to fight the good fight.
>
> Thomas a Kempis' book is peculiar; in it are words so profound and serious that one cannot read them without emotion, almost fear - at least if one reads with a sincere desire for light and truth--the language has an eloquence which wins the heart because it comes from the heart. You have a copy, haven't you? (17)

Brief references to the book appear in Vincent's letters as early as 1875, and according to biographers Lawrence and Elisabeth Hanson, he was so taken by it he destroyed books by Michelet, Renan, and Heine, books he had previously held up to Theo as favorites, and even suggested to Theo that he also should burn all books except for the *Bible* and *The Imitation of Christ.* (18)

On December 2, 1910, *Het Algemeen Handelsblad* (an important newspaper in Amsterdam) included the following memoir written by Dr. M. B. Mendes da Costa describing his time as Vincent's tutor, both foreshadowing Vincent's inability to work for the established church and indicating how attracted Vincent was to Thomas a Kempte's book:

> It was probably in the year 1877 or thereabouts that the Reverend Mr. J. P. Stricker, a preacher universally respected in Amsterdam, asked me whether I was willing to give lessons in Latin and Greek to his cousin Vincent, son of the Reverend Mr. T. van Gogh, clergyman at Etten and De Hoeven, to prepare him for his matriculation. I was warned that I would not be dealing with any ordinary boy, and was apprised of his ways, so different from ordinary human behavior. However, this did not discourage me in the least, particularly as the Reverend Mr. Stricker spoke with much love of Vincent himself as well as of his parents.
>
> Our first meeting, of so much importance to the relationship between master and pupil, was very pleasant indeed. The seemingly reticent young man - our ages differed but little, for I was twenty-six then, and he was undoubtedly over twenty - immediately felt at home, and notwithstanding his lank reddish hair and his many freckles, his appearance was far from

unattractive to me. In passing, let me say that it is not very clear to me why his sister speaks of his "more or less rough exterior"; it is possible that, since the time when I knew him, because of his untidiness and his growing a beard, his outward appearance lost something of its charming quaintness; but most decidedly it can never have been rough, neither his nervous hands, nor his countenance, which might have been considered homely, but which expressed so much and hid so much more.

I succeeded in winning his confidence and friendship very soon, which was so essential in this case; and as his studies were prompted by the best of intentions, we made comparatively good progress in the beginning - I was soon able to let him translate an easy Latin author. Needless to say, he, who was so fanatically devout in those days, at once started using this little bit of Latin knowledge to read Thomas a Kempis in the original.

So far everything went well, including mathematics, which he had begun studying with another master in the meantime; but after a short time the Greek verbs became too much for him. However I might set about it, whatever trick I might invent to enliven the lessons, it was no use. "Mendes," he would say - we did not mister each other any more - "Mendes, do you seriously believe that such horrors are indispensable to a man who wants to do what I want to do: give peace to poor creatures and reconcile them to their existence here on earth?"

And I, who as his master naturally could not agree, but who felt in my heart of hearts that he - mind, I say *he*, Vincent van Gogh! - was quite right, I put up the most formidable defense I was capable of; but it was no use.

"John Bunyan's *Pilgrim's Progress* is of much more use to me, and Thomas a Kempis and a translation of the Bible; and I don't want anything more." I really do not know how many times he told me this, nor how many times I went to the Reverend Mr. Stricker to discuss the matter, after which it was decided again and again that Vincent ought to have another try.

But before long the trouble would start afresh, and then he would come to me in the morning with an announcement I knew so well, "Mendes, last night I used the cudgel again," or, "Mendes, last night I got myself locked out again." It should be observed that this was some sort of self-chastisement resorted to whenever he thought he had neglected a duty. In fact, during those days he lived in his uncle's house, Rear Admiral J. van Gogh, director and commander of the naval base at Amsterdam; the house was a big building inside the naval dockyard. Well, whenever Vincent felt that his thoughts had strayed further than they should have, he took a cudgel to bed with him and belabored his back with it; and whenever he was convinced he had forfeited the privilege of passing the night in his bed, he slunk out of the house unobserved at night, and then, when he came back and found the door double-locked, was forced to go and lie on the floor of a little wooden shed, without bed or blanket. He preferred to do this in winter, so that the punishment, which I am disposed to think arose from mental masochism, might be more severe.

He knew quite well that I was displeased by such announcements on his part, and therefore, to appease me as much as possible, he would, either before his confession or the day after, go to the park which was then the Oosterbegraaf-plaats [East Cemetery], his favorite walk, in the early morning, and pick some snowdrops for me there, preferably from under the snow. At the time I was living in Jonas Daniel Meyer Square and had my study on the third floor. In my mind's eye I can still see him come stepping across the square from the Nieuwe Herengracht Bridge, without an overcoat as additional self-chastisement; his books under his right arm, pressed firmly against his body, and his left hand clasping the bunch of snowdrops to his breast; his head thrust forward a little to the right, and on his face, because of the way his mouth drooped at the corners, a pervading expression of indescribable sadness and despair. And when he had come upstairs, there would sound again that singular, profoundly melancholy, deep voice: "Don't be mad at me, Mendes; I have brought you some little flowers again because you are so good to me."

As far as I can see, to be angry under such circumstances would have been impossible for anybody, not just for me, who had soon discovered that in those days he was consumed by a desire to help the unfortunate. I had noticed it even in my own home, for not only did he show great interest in my deaf and dumb brother, but at the same time he always spoke kindly to and about an aunt of ours whom we had taken in, an impecunious, slightly deformed woman who was slow-witted, and spoke with difficulty, thus provoking the mockery of many people. This aunt tried to make herself useful by "minding the bell," and as soon as she saw Vincent approach, she would run as quickly as her short old legs would carry her to the street door in order to welcome him with a "*Good* morning Mister Van Gort." [1]

"Mendes," Vincent used to say, "however much that aunt of yours may mutilate my name, she is a good soul, and I like her very much."

As I was not so very busy in those days, he often stayed talking for a while after the lesson, and naturally we often discussed his former profession, the art dealing business. He had kept quite a number of the prints which he had collected in those days, little lithographs after paintings, etc. He brought them to show me repeatedly, but they were always completely spoiled: the white borders were literally covered with quotations from Thomas a Kempis and the Bible, more or less connected with the subject, which he had scrawled all over them. Once he made me a present of *De Imitatione Christi,* without any intention of converting me, only to acquaint me with the deep humanity of it.

In no way could I guess in those days - no more than anyone else, himself included - that in the depths of his soul lay dormant the future visionary of colour.

I remember only the following incident. Proud of the fact that I could do it with the money I had earned myself, I had exchanged my Smyrna carpet, at least fifty years old and nearly threadbare, which had covered the floor of my room, for a very modest but brightly coloured cowhair one. "Mendes!" Vincent said when he saw it, "I hadn't expected this of you! Do you really think this finer than those old faded colours which had so much in them?" And Mendes was ashamed of himself, for he felt that this queer boy was right.

Our intercourse lasted for less than a year. By then I had come to the conclusion that he would never be able to pass the required examination; so what Mrs. Du Quesne tells us, namely that he had mastered Latin and Greek within a few months, is incorrect, as well as the statement that Vincent stopped at the very moment he was to start on his academic course proper. No, at least a year before he could have reached this point, even with the utmost exertion on his part, I advised his uncle, wholly in conformity with Vincent's own wishes, to let him stop. And so it happened.

After our cordial leave-taking before he went to the Borinage, I never saw him again. From there one letter from him to me, and an answer from me to him, and . . . after that, nothing...

Amsterdam, 30 November 1910 (19)

Thomas a Kempis was born in 1380, and at age 15 went to Deventer, a school run by the Bretheren of the Common Life, a Dutch association of monks and priests dedicated to living a simple life of devotion to Jesus Christ. The Bretheren had been founded in the 14th century by Geert Groote, a reformer in the Roman Catholic community with a worldly reputation as an educator who opposed what he considered a corrupt clergy. Eventually, he won enough followers to his brotherhood and it was approved by the Pope. After his death in 1384, Florence Radewyns took over his position and founded the monastery of Windesheim, near Zwolle, which then became the central location for the group.

Members took no vows, and neither asked for nor received alms. They wished for no worldly rewards, worked for their daily bread, and focused their lives on cultivating the inner self. Education was stressed, and they became the most important teachers and publishers of spiritual writings throughout Germany and the Netherlands by the end of the 15th century, stressing love of God, at first mainly teaching literacy at the elementary levels, but later expanding into humanities, philosophy, and theology.

Since they were, technically, outside of the formal hierarchy of the Catholic Church, they were generally not well thought of by the monks and priests and other official dignitaries, but they received support and protection from the Popes. And they claimed connections with a number of important spiritual and intellectual leaders, including Dierick Maertens,

Gabriel Biel, Jan Standonck, Master of the Collège de Montaigu in Paris, and the Dutch Pope Adrian VI.

Thomas a Kempus rose in the group hierarchy, eventually becoming a priest and demonstrating his abilities as a copyist. As a part of his position, he collected, edited, translated from Netherlandish into Latin, and revised a collection of devotionals written by a small group of earlier Brothers, publishing the result under the title *The Imitation of Christ*, a book that has since become second only to the Bible in Christian readership. (20) The stress of the book is to live one's life in the manner of Christ, especially following Christ's example of humility, lowliness, discipline, prudence, inward consolation, and the acceptance, even embracement of misery.

It begins:

Book One

Thoughts Helpful in the Life of the Soul

The First Chapter

Imitating Christ and Despising All Vanities on Earth

HE WHO follows Me, walks not in darkness," says the Lord. By these words of Christ we are advised to imitate His life and habits, if we wish to be truly enlightened and free from all blindness of heart. Let our chief effort, therefore, be to study the life of Jesus Christ.

The teaching of Christ is more excellent than all the advice of the saints, and he who has His spirit will find in it a hidden manna. Now, there are many who hear the Gospel often but care little for it because they have not the spirit of Christ. Yet whoever wishes to understand fully the words of Christ must try to pattern his whole life on that of Christ.

What good does it do to speak learnedly about the Trinity if, lacking humility, you displease the Trinity? Indeed it is not learning that makes a man holy and just, but a virtuous life makes him pleasing to God. I would rather feel contrition than know how to define it. For what would it profit us to know the whole Bible by heart and the principles of all the philosophers if we live without grace and the love of God? Vanity of vanities and all is vanity, except to love God and serve Him alone.

This is the greatest wisdom—to seek the kingdom of heaven through contempt of the world. It is vanity, therefore, to seek and trust in riches that perish. It is vanity also to court honor and to be puffed up with pride. It is vanity to follow the lusts of the body and to desire things for which severe punishment later must come. It is vanity to wish for long life and to care little about a well-spent life. It is vanity to be concerned with the present only and not to make provision for things to come. It is vanity to love what passes quickly and not to look ahead where eternal joy abides.

Often recall the proverb: "The eye is not satisfied with seeing nor the ear filled with hearing." Try, moreover, to turn your heart from the love of things visible and bring yourself to things invisible. For they who follow their own evil passions stain their consciences and lose the grace of God.

The Second Chapter

Having a Humble Opinion of Self

EVERY man naturally desires knowledge; but what good is knowledge without fear of God? Indeed a humble rustic who serves God is better than a proud intellectual who neglects his soul to study the course of the stars. He who knows himself well becomes mean in his own eyes and is not happy when praised by men.

If I knew all things in the world and had not charity, what would it profit me before God Who will judge me by my deeds?

Shun too great a desire for knowledge, for in it there is much fretting and delusion. Intellectuals like to appear learned and to be called wise. Yet there are many things the knowledge of which does little or no good to the soul, and he who concerns himself about other things than those which lead to salvation is very unwise.

Many words do not satisfy the soul; but a good life eases the mind and a clean conscience inspires great trust in God.

The more you know and the better you understand, the more severely will you be judged, unless your life is also the more holy. Do not be proud, therefore, because of your learning or skill. Rather, fear because of the talent given you. If you think you know many things and understand them well 4 enough, realize at the same time that there is much you do not know. Hence, do not affect wisdom, but admit your ignorance. Why prefer yourself to anyone else when many are more learned, more cultured than you?

If you wish to learn and appreciate something worth while, then love to be unknown and considered as nothing. Truly to know and despise self is the best and most perfect counsel. To think of oneself as nothing, and always to think well and highly of others is the best and most perfect wisdom. Wherefore, if you see another sin openly or commit a serious crime, do not consider yourself better, for you do not know how long you can remain in good estate. All men are frail, but you must admit that none is more frail than yourself.

In truth, this is a kindly track, espousing a simple, sincere faith, not requiring extreme hardships. Christian Classic Ethereal Library prefaces it: "For five hundred years, this gentle book, filled with the spirit of the love of God, has brought understanding and comfort to millions of readers in over fifty languages, and provided them with a source of heart-felt personal prayer. These meditations on the life and teachings of Jesus, written in times even more troubled and dangerous than our own, have become second only to the Bible as a guide and inspiration." (21)

Vincent related to this personal form of self-discipline and took it to extremes, seeking out the miners of Wasmes because he saw them as being the honest, hard-working, suffering poor that Christ walked among.

Then he went against the Church doctrine, just as Christ had gone against the Church Elders of his time, and determined to live the life of Christ, purposely seeking out the most run-down hut in the village, giving away his clothes, even his bed, living on scrapes of food, and personally going down into the mines to experience the hard lives of these noble wretches. (22)

Many biographies and other studies of Vincent have focused on his seeming need for self-punishment. Albert J. Lubin put the following frame over it:

> His identification with Christ and immortal artists of the past generated a secondary self-confidence in him. (A "secondary" self-confidence is intended to contrast with the more solid self-confidence built on a satisfactory parent-child relationship.) These grandiose identifications also justified the expression of a self-righteous anger that helped neutralize depression. He was also able to replace depression with a hyperactive, hypersensitive state, similar in some ways to the manic phase of a manic-depressive state.
>
> Finally, he had the ability to convert the nonfunctioning misery of depression into the "active" suffering of masochism, and alternative way of having relations with other people.
>
> In the light of Bernhard Berliner's convincing evidence that masochism is a form of depression, ["Psychodynamics of the Depressive Character," *Psychoanalytic Forum,* I (1966), pp. 244-264)] the masochist can be seen as a depressed person who has attempted to preserve or restore hope through a display of suffering that appeals for love. I have suggested that a continuum exists between theoretically "pure" depression and "pure" masochism:
>
>> Broadly speaking, the depressive suffers and the masochist exhibits suffering. Though he complains bitterly of his misery, the masochist tends to minimize or deny to himself the affect of depression. The emotion is distilled out of the depressive mixture, leaving behind a residue of depressive ideation. Depression is infectious; gloom spreads in its wake. But, partly because of this dissociation between affect and thought, masochism stirs resentment and recrimination, and the masochist is sometimes condemned as a

> fraud who exaggerates his woes. Analysis, however, discloses profound unhappiness.
>
> The depressive feels the misery inside himself and blames himself for it. Using provocative behavior, the masochist manages to cause others to bring it on him. This helps him to deny the inner unhappiness. Through externalizing it, he can exhibionistically appeal for help and sympathy, righteously express anger toward his persecutors and absolve himself of guilt. These safety valves are not available to the depressive. ["Psychodynamics of the Depressive Character," *Psychoanalytic Forum,* I (1966), pp. 254-256]

> Vincent is a good example of this relationship between depression and masochism. He took the suffering of depression and, instead of be crushed by it, glorified it—first in the name of Christ and then in the name of art. He exhibited it to his parents, his brother, and to all the world. He swung back and forth between paralyzing depressive states of relatively short duration and productive masochistic-creative states. In his identification with the crucified Jesus, the masochistic use of depression enabled him to accept unhappiness as a means of obtaining the approbation of his fellow man as well as eternal joy in heaven. With the glorification of suffering, the thought of suffering remained but the feeling diminished. He appeared to accept the cruel demands of the rejecting, punishing, shaming world by rejecting, punishing, and shaming himself; but, at the same time, he rejected this world and asserted his intimacy with God and heaven. He made a compact with the sadists, but he also defied them. By becoming a martyred hero, he turned guilt into innocence and shame into pride. (23)

His letters, those of his family, and the report from the Synodal Board of Evangelization of the Union of Protestant Churches of Belgium reveal a man simply not able to preach the word of God in the formal, organized, ceremonial manner of the Church, a man desperately wanting to experience and share the world of Christ, that spiritual existence beyond mere physical existence where life gains meaning and value, wanting to sacrifice himself, to be as Christ had been, yet unable to do it in the approved manner, unable to do it within a *system*. By following Christ's example to an extreme, he found only the Church's removal of his ministry. His letter to Theo in July of 1880, expresses it well. I include the entire letter, because the details help to clarify the total Vincent, but for the moment, focus on the section that begins "You should know that it is the same with evangelists as it is with artists."

> My Dear Theo,
>
> I am writing to you rather reluctantly because, for a good many reasons, I have kept silent for such a long time. To some extent you have

become a stranger to me, and I to you perhaps more than you think. It is probably better for us not to go on like that. It is probable that I would not have written to you even now, were it not that I feel obliged, compelled, to do so - because, be it noted, you yourself have compelled me to.

I heard in Etten that you had sent 50 francs for me. Well, I have accepted them. With reluctance, of course, with a feeling of some despondency, of course, but I have reached a sort of impasse, am in trouble, what else can I do? And so

I am writing to thank you.

As you may know, I am back in the Borinage. Father said he would prefer me to stay somewhere near Etten, but I refused and I believe I was right to do so. To the family, I have, willy-nilly, become a more or less objectionable and shady sort of character, at any rate a bad lot. How then could I then be of any use to anyone? And so I am inclined to think the best and most sensible solution all round would be for me to go away and to keep my distance, to cease to be, as it were. What the moulting season is for birds - the time when they lose their feathers - setbacks, misfortune and hard times are for us human beings. You can cling on to the moulting season, you can also emerge from it reborn, but it must not be done in public.

The thing is far from amusing, not very exhilarating, and so one should take care to keep out of the way. Well, so be it.

Now, though it is a fairly hopeless task to regain the trust of an entire family, one which has perhaps never been wholly weaned from prejudice and other equally honourable and respectable qualities, I am not entirely without hope that, bit by bit, slowly but surely, the good relationship between one and all may be restored. In the first place I should be glad to see this good relationship - to put it no more strongly than that - restored at least between Father and me, and further, I set great store by seeing it restored between the two of us. A good relationship is infinitely preferable to a misunderstanding.

Now I must trouble you with certain abstract matters, hoping that you will listen to them patiently. I am a man of passions, capable of and given to doing more or less outrageous things for which I sometimes feel a little sorry. Every so often I say or do something too hastily, when it would have been better to have shown a little more patience. Other people also act rashly at times, I think.

This being the case, what can be done about it? Should I consider myself a dangerous person, unfit for anything? I think not. Rather, every means should be tried to put these very passions to good effect.

To mention just one by way of an example, I have a more or less irresistible passion for books and the constant need to improve my mind, to study if you like, just as I have a need to eat bread. You will understand that. When I lived in other surroundings, surroundings full of pictures and works of art, I conceived a violent, almost fanatical passion for those surroundings, as you know. And I do not regret that, and even now, *far from home, I often feel homesick for the land of pictures*

You may remember that I knew very well (and it may be that I know it still) what Rembrandt was or what Millet was or Jules Dupré or Delacroix or Millais or Matthijs Maris.

Well, today I am no longer in those surroundings, yet they say that what is known as the soul never dies but lives on for ever, continuing to seek for ever and again.

So instead of succumbing to my homesickness I told myself: your land, your fatherland, is all around. So instead of giving in to despair I chose active melancholy, in so far as I was capable of activity, in other words I chose the kind of melancholy that hopes, that strives and that seeks, in preference to the melancholy that despairs numbly and in distress. I accordingly made a more or less serious study of the books within my reach, such as the Bible and Michelet's La révolution Française, and then last winter Shakespeare and a little Victor Hugo and Dickens and Beecher Stowe and recently Æschylus and then various less classical writers, a few great minor masters. You know, don't you, that Fabritius and Bida are counted among the minor masters?

Now anyone who becomes absorbed in all this is sometimes considered outrageous, `shocking,' sinning more or less unwillingly against certain forms and customs and proprieties. It is a pity that people take that amiss.

You know, for example, that I have often neglected my appearance. I admit it, and I also admit that it is `shocking.' But look here, lack of money and poverty have something to do with it too, as well as a profound disillusionment, and besides, it is sometimes a good way of ensuring the solitude you need, of concentrating more or less on whatever study you are immersed in. One essential study is that of medicine. There is scarcely anybody who does not try to acquire some knowledge of it, who does not at least try to grasp what it is about (and you see, I still know absolutely nothing about it). And all these things absorb you, preoccupy you, set you dreaming, musing and thinking.

Now for the past five years or so, I don't know how long exactly, I have been more or less without permanent employment, wandering from pillar to post. You will say, ever since such and such a time you have been going downhill, you have been feeble, you have done nothing. Is that entirely true?

What is true is that I have at times earned my own crust of bread, and at other times a friend has given it to me out of the goodness of his heart. I have lived whatever way I could, for better or for worse, taking things just as they came. It is true that I have forfeited the trust of various people, it is true that my financial affairs are in a sorry state, it is true that the future looks rather bleak, it is true that I might have done better, it is true that I have wasted time when it comes to earning a living, it is true that my studies are in a fairly lamentable and appalling state, and that my needs are greater, infinitely greater than my resources. But does that mean going downhill and doing nothing?

You might say, but why didn't you go through with university, continue as they wanted you to? To that I can only reply that it was too expensive, and besides, the future then looked no better than it does now, along the path I am now taking.

And I must continue to follow the path I take now. If I do nothing, if I study nothing, if I cease searching, then, woe is me, I am lost. That is how I look at it - keep going, keep going come what may.

But what is your final goal, you may ask. That goal will become clearer, will emerge slowly but surely, much as the rough draught turns into a sketch, and the sketch into a painting through the serious work done on it, through the elaboration of the original vague idea and through the consolidation of the first fleeting and passing thought.

You should know that it is the same with evangelists as it is with artists. There is an old academic school, often odious and tyrannical, the `abomination of desolation', in short, men who dress, as it were, in a suit of steel armour, a cuirass, of prejudice and convention. When they are in charge, it is they who hand out the jobs and try, with much red tape, to keep them for their protégés and to exclude the man with an open mind.

Their God is like the God of Shakespeare's drunken Falstaff, "the inside of a church." Indeed, by a strange coincidence, some evangelical (???) gentlemen have the same view of matters spiritual as that drunkard (which might surprise them somewhat were they capable of human emotion). But there is little fear that their blindness will ever turn into insight.

This is a bad state of affairs for anyone who differs from them and protests with heart and soul and all the indignation he can muster. For my part, I hold those academicians who are not like these academicians in high esteem, but the decent ones are thinner on the ground than you might think.

Now, one of the reasons why I have no regular job, and why I have not had a regular job for years, is quite simply that my ideas differ from those of the gentlemen who hand out the jobs to individuals who think as they do. It is not just a question of my appearance, which is what they have sanctimoniously reproached me with. It goes deeper, I do assure you.

I am telling you all this not to complain, not to make excuses for matters in which I may perhaps have been somewhat at fault, but simply to tell you the following: during your final visit last summer when we were walking together near that abandoned mineshaft which they call "La Sorcière," you reminded me of another walk we once took at another time near the old canal and the mill at Rijswijk, and, you said, we used to agree about many things, but, you added, "You have changed since then, you are no longer the same." Well, that is not entirely true. What has changed is that my life then was less difficult and my future seemingly less gloomy, but as far as my inner self, my way of looking at things and of thinking is concerned, that has not changed. But if there has indeed been a change, then it is that I think, believe and love more seriously now what I thought, believed and loved even then.

So you would be mistaken should you continue to think that I have become less keen on, say, Rembrandt, Millet, or Delacroix or whoever or whatever, for the reverse is the case, but there are many different things worth believing and loving, you see - there is something of Rembrandt in Shakespeare, something of Correggio or of Sarto in Michelet and something of Delacroix in Victor Hugo, and there is also something of Rembrandt in the Gospel or, if you prefer, something of the Gospel in Rembrandt, it comes to much the same thing, provided you understand it properly, do not try to distort it and bear in mind that the elements of the comparisons are not intended to detract in any way from the merits of the original individuals.

And in Bunyan there is something of M. Maris or of Millet, a reality that, in a manner of speaking, is more real than reality itself, something hitherto unknown that, if only you can read it, will tell you untold things. And in Beecher Stowe there is something of Ary Scheffer.

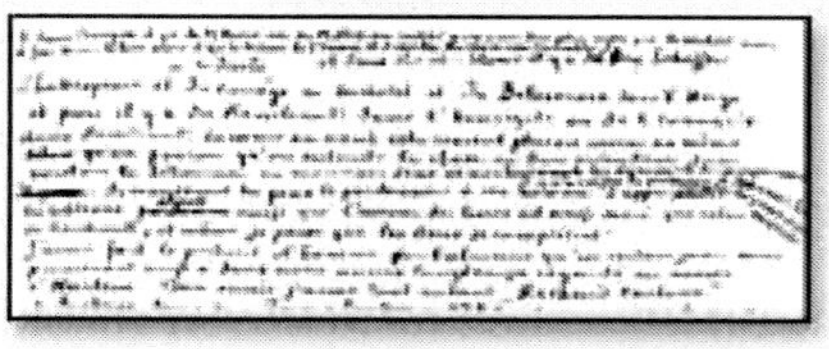

Now, if you can forgive someone for immersing himself in pictures, perhaps you will also grant that the love of books is as sacred as that of Rembrandt, indeed, I believe that the two complement each other.

I very much admire the portrait of a man by Fabritius that we stood looking at for a long time in the gallery in Haarlem one day when we took another walk together. Admittedly, I am as fond of Dickens's 'Richard Cartone' [Sydney Carton] in his Paris & Londres in 1793 [A Tale of Two Cities], and I could point to other particularly gripping characters in other books with a more or less striking resemblance. And I think that Kent, a character in Shakespeare's "King Lear," is as noble and distinguished a man as that figure by Th. de Keyser, though Kent and King Lear are reputed to have lived much earlier.

Let me stop there, but my God, how beautiful Shakespeare is, who else is as mysterious as he is; his language and method are like a brush trembling with excitement and ecstasy. But one must learn to read, just as one must learn to see and learn to live.

So please don't think that I am renouncing anything, I am reasonably faithful in my unfaithfulness and though I have changed, I am the same, and what preys on my mind is simply this one question: what am I good for, could I not be of service or use in some way, how can I become more knowledgeable and study some subject or other in depth? That is what keeps preying on my mind, you see, and then one feels imprisoned by poverty, barred from taking part in this or that project and all sorts of necessities are out of one's reach. As a result one cannot rid oneself of melancholy, one feels emptiness where there might have been friendship and sublime and genuine affection, and one feels dreadful disappointment gnawing at one's spiritual energy, fate seems to stand in the way of affection or one feels a wave of disgust welling up inside. And then one says "How long, my God!"

Well, that's how it is, can you tell what goes on within by looking at what happens without? There may be a great fire in our soul, but no one ever comes to warm himself by it, all that passers-by can see is a little smoke coming out of the chimney, and they walk on.

All right, then, what is to be done, should one tend that inward fire, turn to oneself for strength, wait patiently - yet with how much impatience! - wait, I say, for the moment when someone who wants to comes and sits down beside one's fire and perhaps stays on? Let him who believes in God await the moment that will sooner or later arrive.

Well, right now it seems that things are going very badly for me, have been doing so for some considerable time, and may continue to do so well into the future. But it is possible that everything will get better after it has all seemed to go wrong. I am not counting on it, it may never happen, but if there should be a change for the better I should regard that as a gain. I should rejoice, I should say, at last! So there *was something after all!*

But, you will say, what a dreadful person you are, with your impossible religious notions and idiotic scruples. If my ideas are impossible or idiotic then I would like nothing better than to be rid of them. But this is roughly the way I see things. In Le Philosophe sous les Toits by Souvestre you can read what a man of the people, a simple craftsman, pitiful if you will, thinks of his country: "Tu n'as peut-être jamais pensé à ce que c'est la patrie, reprit-il, en me posant une main sur l'épaule; c'est tout ce qui t'entoure, tout ce qui t'a élevé et nourri, tout ce que tu as aimé. Cette campagne que tu vois, ces maisons, ces arbres, ces jeunes filles qui passent là en riant, c'est la patrie! Les lois qui te protègent, le pain qui paye ton travail, les paroles que tu échanges, la joie et la tristesse qui te viennent des hommes et des choses parmi lesquels tu vis, c'est la patrie! La petite chambre où tu as autrefois vu ta mère, les souvenirs qu'elle t'a laissés, la terre où elle repose, c'est la patrie! Tu la vois, tu la respires partout! Figure toi, tes affections et tes besoins, tes souvenirs et ta reconnaissance, réunis tout ça sous un seul nom et ce nom sera la patrie." [You may never have thought what your country really is, he continued, placing his hand on my shoulder; it is everything around you, everything that has raised and nourished you, everything that you have loved. This countryside that you see; these houses, these trees, these young girls laughing as they pass, that is your country! The laws that protect you, the bread that rewards your labour, the words you speak, the joy and sorrow that come from the people and things in whose midst you live, that is your country! The little room where you used in days gone by to see your mother, the memories she left you, the earth in which she rests, that is your country! You see it, you breathe it, everywhere! Imagine your rights and your duties, your affections and your needs, your memories and your gratitude, gather all that together under a single name, and that name will be your country.]

In the same way I think that everything that is really good and beautiful, the inner, moral, spiritual and sublime beauty in men and their works, comes from God, and everything that is bad and evil in the works of men and in men is not from God, and God does not approve of it.

But I cannot help thinking that the best way of knowing God is to love many things. Love this friend, this person, this thing, whatever you like, and you will be on the right road to understanding Him better, that is what I keep telling myself. But you must love with a sublime, genuine, profound sympathy, with devotion, with intelligence, and you must try all the time to understand Him more, better and yet more. That will lead to God, that will lead to an unshakeable faith.

To take an example: one man will love Rembrandt, genuinely, and that man will surely know that there is a God, he will really believe it. Another will make a thorough study of the French Revolution - he will not be an unbeliever, he will see that there is a supreme authority that manifests itself in great affairs. Yet another has recently attended a free course of lectures at the great university of sorrow and has heeded the things he saw with his eyes and heard with his ears, and has reflected upon them. He too will come to believe in the end and will perhaps have learned more than he can tell.

Try to grasp the essence of what the great artists, the serious masters, say in their masterpieces, and you will again find God in them. One man has written or said it in a book, another in a painting. Just read the Bible and the Gospel, that will start you thinking, thinking about many things, thinking about everything, well then, think about many things, think about everything, that will lift your thoughts above the humdrum despite yourself. We know how to read, so let us read!

Now then, you may well have bouts of being a little absent-minded, a little dreamy, indeed there are some who become too absent-minded, a little too dreamy. That may indeed have happened with me, but all in all that is my own fault, maybe there as a reason for it, perhaps I was lost in thought for one reason or another, anxious, worried, but one gets over that in the end. The dreamer sometimes falls into the doldrums, but is said to emerge from them again. And the absent-minded person also makes up for it with bouts of perspicacity. Sometimes he is a person whose right to exist has a justification that is not always immediately obvious to you, or more usually, you may absent-mindedly allow it to slip from your mind. Someone who has been wandering about for a long time, tossed to and fro on a stormy sea, will in the end reach his destination. Someone who has seemed to be good for nothing, unable to fill any job, any appointment, will find one in the end and, energetic and capable, will prove himself quite different from what he seemed at first.

I am writing somewhat at random, writing whatever flows from my pen. I should be very happy if you could see in me something more than a kind of fainéant [idler]. For there is a great difference between one idler and another idler. There is someone who is an idler out of laziness and lack of character, owing to the baseness of his nature. If you like, you may take me for one of those. Then there is the other kind of idler, the idler despite himself, who is inwardly consumed by a great longing for action who does nothing because his hands are tied, because he is, so to speak, imprisoned somewhere, because he lacks what he needs to be productive, because disastrous circumstances have brought him forcibly to this end. Such a one does not always know what he can do, but he nevertheless instinctively feels, I am good for something! My existence is not without reason! I know that I could be a quite a different person! How can I be of use, how can I be of service? There is something inside me, but what can it be? He is quite another idler. If you like you may take me for one of those.

A caged bird in spring knows perfectly well that there is some way in which he should be able to serve. He is well aware that there is something to be done, but he is unable to do it. What is it? He cannot quite remember, but then he gets a vague inkling and he says to himself, "The others are building their nests and hatching their young and bringing them up," and then he bangs his head against the bars of the cage. But the cage does not give way and the bird is maddened by pain. "What an idler," says another bird passing by - what an idler. Yet the prisoner lives and does not die. There are no outward signs of what is going on inside him; he is doing well, he is quite cheerful in the sunshine.

But then the season of the great migration arrives, an attack of melancholy. He has everything he needs, say the children who tend him in his cage - but he looks out, at the heavy thundery sky, and in his heart of hearts he rebels against his fate. I am caged, I am caged and you say I need nothing, you idiots! I have everything I need, indeed! Oh! please give me the freedom to be a bird like other birds!

A kind of idler of a person resembles that kind of idler of a bird. And people are often unable to do anything, imprisoned as they are in I don't know what kind of terrible, terrible, oh such terrible cage.

I do know that there is a release, the belated release. A justly or unjustly ruined reputation, poverty, disastrous circumstances, misfortune, they all turn you into a prisoner. You cannot always tell what keeps you confined, what immures you, what seems to bury you, and yet you can feel those elusive bars, railings, walls. Is all this illusion, imagination? I don't think so. And then one asks: My God! will it be for long, will it be for ever, will it be for eternity?

Do you know what makes the prison disappear? Every deep, genuine affection. Being friends, being brothers, loving, that is what opens the prison, with supreme power, by some magic force. Without these one stays dead. But whenever affection is revived, there life revives. Moreover, the prison is sometimes called prejudice, misunderstanding, fatal ignorance of one thing or another, suspicion, false modesty.

But to change the subject - if I have come down in the world, you have in a different way come up in it. And if I have forfeited sympathy, you have gained it. I am glad of that, I say that it in all sincerity, and it will always give me pleasure. If you lacked seriousness or consideration, I would be fearful that it might not last, but since I think that you are very serious and very considerate, I tend to believe it will!

But if you could see me as something other than a idler of the bad sort, I should be very happy.

For the rest, if I can ever do anything for you, be of some use to you, know that I am at your disposal. Now that I have accepted what you have given me, you are, should I be able to render you some service, in a position to ask me. It would make me happy, and I should take it a sign of trust. We have moved rather far apart and may in certain respects have perhaps different views, but some time, some day, one of us may be of service to the other.

For now I shake your hand, thanking you once again for having been so good to me. If, one of these days, you feel like writing, my address is, chez Ch. Decrucq, Rue du Pavillon 8, Cuesmes, near Mons, and know that it will do me good to hear from you.

Yours, Vincent (24)

Vincent was unable to perform in the role of a minister, a preacher, an evangelist because his *faith* came from a different place, a different part of his psyche. Though he tried very hard during this time to make *faith* and *religion* correspond or at least overlap, he could not give up his *faith* for his *religion*, and for him *faith* demanded *suffering and sacrifice*, the same *suffering and sacrifice* exhibited by Christ himself. That, as Vincent saw it, was the only way to salvation, the only way to a connection with the world of the spirit, the world of meaning and value, the only way to an existence beyond the mere physical, animal existence of the body.

He failed as an evangelist, but he was so intensely dedicated to his *faith*, so determined to live as Christ had lived, to *suffer* as Christ had *suffered*, that he became known as Christ of the Coal Mines. (25)

Part III

The Replacement Child Syndrome

In 1852 in the small village of Groot-Zundert, Holland, Calvinist minister Theodorus van Gogh and artisticly inclined Anna Cornelia Carbentus van Gogh are delighted to learn they are about to have a baby, a boy they name Vincent William van Gogh. And on March 30 of that year he is born. However, their joy turns to sorrow because he is born dead.

In March 30 precisely one year later, they have a second child, and give him the same name, Vincent William van Gogh. He doesn't die in birth, but according to several biographers, his parents, especially his mother, apparently unable to accept the death of the previous child, never grow close to him. This is called the Replacement Child Syndrome, and unfortunately becomes a major psychological approach to analyzing him. Whatever the reasons, according to Theo's wife, Johanna van Gogh-Bonger, he feels rejected and has a moody, melancholy personality, is seldom happy and is often combative and prone to arguments. In a letter to Theo later in life, one filled with attempts by him to justify the current discord in his relationship with the family, especially with their father, he writes "My youth was gloomy and cold and sterile under the influence of the rayon noir." (26)

Most of what is assumed about Vincent's childhood comes from the memoirs of him by this same Johanna, a good but not definitive source. For example, she gives the wrong name for a girl with whom he later is suspected of having a failed love affair (mistakenly gives her mother's name). Also, such quotes as the above one taken from his letters have been used to support a wide range of theories about his later psychological make-up, but again, they are far from definitive, and are often taken out of context to support often interesting but wildly speculative theories. The quote above, for example, is meant by him to indicate he did not agree with the rayon noir approach to art, not necessarily that all of his youth was "gloomy and cold and sterile." Perhaps it was, but other passages from him contradict such a dark upbringing and instead clearly indicate a longing to return to it later in life.

His parents will have five more children, and as is clear from the letters above, Vincent will form an unusually strong bond with his brother Theo, held together by the many letters they will exchange through the years, and later by Theo's support for Vincent's art and constant attempts to help him, both emotionally and practically.

Vincent's mother is given credit for introducing him to two of her own interests, nature and drawing, and he spends much of his time wandering about the countryside, catching insects, watching the birds, and doing drawings and watercolors of nature, though at the time he exhibits no special artistic talent. In 1860 he attends the local Zundert village school, which consists of one Catholic teacher and about 200 students. This lasts for but one year. From 1861-1864 he and his sister Anna are taught at home by a governess. In October of 1864, he is sent to the elementary boarding school of Jan Provily in Zeyenbergen, about twenty miles away. He is unhappy about being sent away from his home, and the experience is a sad one. September 18, 1866, he is sent to the new middle school, Willem II College in Tilburg, where Constantijn C. Huysmans, who had achieved a bit of success himself in Paris, instructs him in drawing, advocating a systematic approach Vincent thinks is wrong. In March, 1868 Vincent abrupt leaves school and returns home.

One thing worth noting before gettingt to the theories to come about his childhood is that whenever he leaves home he quickly becomes homesick (suggesting that his family life is not as negative as they are often portrayed, unless the negative aspects are turned upside down, and result not from a lack of love and attention, but rather, from too much of it, smothering his own independence).

Whatever the reasons, and they can only be guessed at because there is so little to go on, although he is an intelligent child, at age fifteen, his uncle Cent (nickname for Vincent; it is a standard name in the family) arranges for him to work at The Hague in the Goupil and Cie Gallery, a well-known firm of art dealers with establishments in Hague, London, Paris, New York, and Brussels. This movement into the art world is neither surprising nor indicative of artistic talent or interest, as it is simply the result of his father having three brothers working in the profession.

On August, 1872, three years after he begins this employment, the first of his letters to his younger brother Theo that remains gets written.

My Dear Theo,

Many thanks for your letter. I was glad you arrived back safely. I missed you the first few days & it felt strange not to find you there when I cam home in the afternoons.

We have had some enjoyable days together, and managed to take a few walks & see one or two sights between the spots of rain.

What terrible weather. You must have sweltered on your walks to Oisterwijk. There was harness racing yesterday for the exhibition, but the illuminations & the fireworks were put off because of the bad weather, so it's just as well you didn't stay on to see them.

Regards from the Haanebeek & Roos families.
Always your loving
Vincent (27)

While there is not much to it, it clearly suggests a "normal" young man. The letters from his time at The Hague that follow continue in a similar light, "happy" tone.

In June, 1873, his training completed, he is transferred by Goupil to a gallery in London, and he finds lodgings at 87 Hackford Road, Brixton. He is twenty years old.

Dear Theo,

My address is c/o Messrs. Goupil & Co., 17 Southampton Street, Strand, London. You must be eager to hear from me, so I will not keep you waiting any longer for a letter.

I hear from home that you are living with Mr. Schmidt now and that Father has been to see you. I certainly hope this will please you better than your former boardinghouse, and I'm sure it will.

I am very anxious for a letter; write me soon, and tell me how you spend your day, etc. You must tell me especially what pictures you have seen lately and also if any new etchings or lithographs have been published. Let me know as much as you can about these things, for I do not see much of them here as it is only a wholesale house.

Considering the circumstances, I am doing pretty well. So far the boardinghouse where I am staying pleases me. There are also three German boarders who are very fond of music, they play the piano and sing, so we spend very pleasant evenings together. I am not so busy here as I was in The Hague; I work only from nine in the morning to six in the evening, and on Saturdays we close at four o'clock. I live in one of the suburbs of London, where it is relatively quiet. It reminds me of Tilburg or some such place.

I spent some very pleasant days in Paris, and, as you can imagine, I enjoyed all the beautiful things I saw at the exhibition and in the Louvre and the Luxembourg. The house in Paris is splendid and much bigger than I had thought, especially the one in the Place de l'Opera.

Life is very expensive here, my accommodation alone costs me eighteen shillings by week, washing excepted, and then I still have to take my dinner in the city. Last Sunday I went to the country with Mr. Obach, my principal, to Boxhill; it is a high hill about six hours by road from London, partially chalky and overgrown with box and on one side a wood of high oak trees. The country is beautiful here, quite different from

Holland or Belgium. Everywhere you see charming parks with high trees and shrubs. Everyone is allowed to walk there. At Easter, I made an interesting excursion with the Germans, but these gentlemen spend a great deal of money and I shall not go out with them in the future.

I was glad to hear from home that Uncle Hein's health is good. Give him and Aunt my best and tell them something about me. Give my compliments to Mr. Schmidt and Eduard and write to me soon. À Dieu, best wishes,

Vincent (28)

Ronald De Leeuw writes:

For the time being, it seemed, London continued to please him. Vincent made excursions with his German friends, went rowing on the Thames and discovered the joys of gardening. He urged Theo to read the *Gazette des Beaux-Arts* and William Burger's book on French and Dutch museums and galleries. He himself visited gallery after gallery in London, from the Royal Academy to the Dulwich Picture Gallery. (29)

A few things can and have been noticed about these letters. First, he doesn't have a lot of money, though he doesn't seem to be starving either. In other words, he is in a typical young man situation economically, just getting started in his career and needing to watch his finances. More importantly, he is happy or at least content with his life at the moment, and has positive relations with his family and friends.

In another letter, this one to the Van Stockurn-Haanebeek family, on July 2, 1873, he writes:

Dear friends,

I should have liked to write sooner, and now I will not postpone it any longer. How are you? I heard that your house has been smartened up, and that all is well with you. I hope very much you will drop me a line when you have a moment to spare.

All is well with me. I see much that is new and beautiful, and have been fortunate in finding a good boardinghouse, so that on the whole I feel quite at home already. Yet I do not forget The Hague, and should very, very much like to spend an evening in the Poten, and look in on you, too.

The business here is only a stockroom, and our work is quite different from what it is in The Hague; but I shall probably get used to it. At six o'clock my work is already done for the day, so that I have a nice bit of time for myself, which I spend pleasantly - taking walks, reading and letter writing.

The neighbourhood where I live is quite beautiful, and so quiet and intimate that you almost forget you are in London. In front of every house there is a small garden with flowers or a few trees, and many houses are built very tastefully in a sort of Gothic style. Still, I have a good half-hour's walk to get to the country.

We have a piano in the sitting room, and there are also three Germans living here who are very fond of music, which is very pleasant.

One of the finest sights I have seen is Rotten Row in Hyde Park, where hundreds of ladies and gentlemen ride on horseback.

In all parts of the town there are beautiful parks with a wealth of flowers such as I have never seen anywhere else.

Enclosed I am sending you a copy of a poem by Van Beers, which you possibly do not know. Our Elisabeth copied it for me on my last evening in Helvoirt because she knew I thought so much of it. It is genuine Brabant: I thought you would read it with pleasure, and therefore I copied it for you.

It was very considerate of your sister Marie to send me the announcement. I long to hear something of the wedding, and I congratulate you all.

Will you kindly let me have a list of your birthdays some time? I did have one, but lost it.

And now good-by; remember me to everybody in the Poten, and good luck to you all. Excuse the bad handwriting; it is late and time to go to bed.

Sleep well.
Vincent

DE AVONDSTOND

Langzaam galmde `t gesamp der beelok over de velden,
Die, volzalig, in `t goud van de avondzonne zich baadden…
[Literal translation]

THE EVENING HOUR

The toll of the curfew calling to prayer resounded lazily across the fields,
which blissfully lay bathed in the gold of the evening sun.
Right in front of him lay the village, with hills to the north and to the south,
between whose ridges the sun, sinking in the west with a crimson blush,
poured forth its whole wealth of colours and the magic of its rays.
Now the little bell in the grey steeple veiled in dark green
was silent. The brown sails of the mills, on yonder height,
hung motionless; the foliage was still; and over the cottages
little puffs of peat-smoke, tinged with blue, rose so straight
from the chimneys that they too seemed to hang motionless in the tingling

air.
After the sun's good night kiss it was as if this hamlet, this field, these hills,
everything around, silent and grateful, once more recalled
the richness and peace they had enjoyed, before wrapping themselves in the cloak of evening dew to sleep.
Farther on…but just beside the narrow footpath
followed by the Painter, the sudden loud peal of cheering met his ears.
Swaying to and fro, a wagon came rumbling toward him, piled high with the harvest of buckwheat.
Horse and freight were decked with fluttering ribbons and flowering
branches; children, each with a wreath of flowers around his little flaxen head, sat atop it, brandishing alder twigs,
or raining down a shower of leaves and flowers,
whilst below, around the wagon, a crowd of servant lads and lasses
leaped and sang, so as to startle the whole slumbering plain.
Behind the shrubs, the silently smiling Painter watched
the noisy throng wind its way along the bumpy road.
And thus, pondering the calm and deep delight the soul
savours in the country, or with his artist's mind reconstructing
in silent rapture the whole glorious scene of a short while ago,
he came, without perceiving it himself, sauntering into the hamlet.
In the west the purple and yellow had already faded to grey;
and in the east, quite close to the little church, the full copper-coloured
disk of the moon, lightly shrouded in the haze of the gloaming, had risen
when he entered "The Swan," the inn where he was lodging.

—Jan Van Beers, "The Pauper" (30)

Once again, the letter suggests a happy young man enjoying his new life in London, a man getting out and seeing the new world, a man able to be social, even quite good at being social. While the poem is titled "The Pauper," and refers to a "Painter," what probably attracted Vincent to it is the description of the world in gentle, pleasant images. Similar positive feelings are expresses in the following letter:

Dear Theo,

Thanks for your letter, which was very welcome. I am glad you are doing well and that you like living with Mr. Schmidt, Mr. Obach was very pleased to have met you. I hope that in the future we shall do much business with each other. That picture of Linder's is very beautiful.

As to the photo engravings, I have never seen them being made; I know a little about how they are done, but not enough to explain.

At first English art did not appeal to me; one must get used to it. But there are clever painters here, among others, Millais, who has painted: "The Huguenot," "Ophelia," etc., of which I think you know the engravings; his things are beautiful. Then there is Boughton, whose "Puritans Going to Church" is in our Galerie Photographique; I have seen wonderful things by him. Among the old painters, Constable was a landscape painter who lived about thirty years ago; he is splendid - his work reminds me of Diaz and Daubigny. Then there are Reynolds and Gainsborough, whose forte was very beautiful ladies' portraits, and Turner, whose engravings you must have seen.

Some good French painters live here, including Tissot, of whose work there are several photographs in our Galerie Photographique; and Otto Weber and Heilbuth. The latter is at present painting exquisitely beautiful pictures in the manner of Linder.

Sometime you must write me if there are any photographs of Wauters's work other than "Hugo Van der Goes" and "Mary of Burgundy," and if you know about any photographs of pictures by Lagye and De Braekeleer. I don't mean the elder Braekeleer, but, I think, a son of his who had three beautiful pictures called "Antwerp," "The School" and "The Atlas" at the last exhibition in Brussels.

I am quite contented here; I walk a lot and the neighborhood where I live is quiet, pleasant and fresh - I was really very lucky to find it. Still, I often think with regret of the delightful Sundays at Scheveningen and other things, but what's the use of worrying?

Thanks for what you wrote me about pictures. If you happen to see anything by Lagye, De Braekeleer, Wauters, Maris, Tissot, George Saal, Jundt, Zeim, or Mauve, you must not forget to tell me; those are the painters I am very fond of, and whose work you will probably see something of.

Enclosed is a copy of the poem about the painter who "entered `The Swan,' the inn where he was lodging," which I am sure you remember. It is typical Brabant, and I am fond of it. L. copied it for me the last evening I was home . [L is an abbreviation of Lies, a nickname for their sister Elisabeth.]

How I should like to have you here. What pleasant days we spent together at The Hague; I think so often of that walk on the Rijswijk road, when we drank milk at the mill after the rain. When we send back the pictures we have from you, I will send you a picture of that mill by Weissenbruch; perhaps you remember him, his nickname is Merry Weiss. *That Rijswijk road holds memories for me which are perhaps the most beautiful I have. If we meet again, maybe we shall talk about them once more.*

And now, boy, I wish you well. Think of me from time to time and write me soon, it is such a delight to get a letter.

Vincent (31)

Of course, everyone latches onto whatever Vincent says about art and artists in his letters, and the fact that he is not particularly drawn to English art gets highlighted here. It is more interesting in terms of this to put all of his framework for art into larger categories, because his early art is dark and colorless and not really indicative of the genius to come, and then, later in life, when he finds his palette of brilliant colors, thick texture, energetic short brush-strokes and a kind of semi-realism infused with a unique form of impressionism, the change is dramatic, and though it is obviously partially a matter of him getting introduced to the new art of the Impressionists and their use of color and atmosphere, and other obvious influences, such as the "in-the-air" embracement of various arts outside of Europe, most importantly oriental and primitive art, there is ultimately something more, and this something more is the reason for his appeal.

Again, the main thing to emphasize is how much this letter continues to suggest a normal, contented young man, one who is very capable of getting along well with others, who, in fact, seems to enjoy social situations, and it is even possible to suggest that he is the family member most at the middle of the family exchanges, the one who most wants to keep in contact with the rest of the family and share positive hopes for all. Of course, this last suggestion must be speculative, as we simply don't have available any longer many of the letters of the other members, so we cannot know for sure whether they were prone to write as much or in as positive terms as Vincent, though the letters we do have suggest the entire family was close at this time.

In a letter to the Van Stockum-Haanebeek family, Vincent writes:

> Dear friends,
>
> It was a pleasant surprise to me to receive Caroline's letter. Thanks. With all my heart I hope she is quite well again; a good thing it is over now!
>
> In your next letter I should like to hear more about that last play you wrote. I was really amazed: for ten characters - it must be the biggest you have done.
>
> These last days I have greatly enjoyed reading the poems of John Keats; he is a poet who, I think, is not very well known in Holland. He is the favourite of all the painters here, and so I started reading him. Here is something by him. His best-known piece is "The Eve of St. Agnes," but it is a bit too long to copy.
>
> I have visited neither Crystal Palace nor the Tower yet, nor Tussod; I am not in a hurry to see everything. For the present I am quite satisfied with the museums, parks, etc.; they interest me more.

Last Monday I had a nice day. The first Monday in August is a holiday here. I went with one of the Germans to Dulwich, an hour and a half outside L., to see the museum there, and after that we took about an hour's walk to another village.

The country is so beautiful here; many people who have their businesses in London live in some village outside L. and go to town by train every day; perhaps I shall do the same shortly, if I can find a cheap room somewhere. But moving is so horrible that I shall stop here as long as possible, although everything is not so beautiful as it seemed to me in the beginning. Perhaps it is my own fault, so I shall bear with it a little longer.

Pardon me if this letter is not as I should like it to be, for I am writing in a hurry. I wanted to congratulate you on Willem's birthday and wish you many happy returns.

I was most pleased to learn that you have renewed your acquaintance with the Tersteeg family. I have been hoping you would for a long time.

When you have a chance, please let me know what photographs you have received - I am curious to know. I have had a letter from Marinus, from which I learned that he is going to Amsterdam. This will mean a great change for him; I hope he will do well. I was very glad he wrote me.

A few days ago a brother of Iterson's paid me a call, and for the first time since May I had a chance to speak Dutch. We live far apart, much to my regret.

Good luck to you. Remember me to all in the Poten. Good luck!

Yours truly, Vincent

Gladden my heart with a letter as soon as you can find time.

[Enclosed]

THE EVE OF SAINT MARK (Unfinished)
Upon a Sabbath-day it fell;
Twice holy was the Sabbath-bell,
That call'd the folk to evening prayer;
The chilly sunset faintly told
Of unmatur'd green vallies cold,
Of the green thorny bloomless hedge,
Of rivers new with spring-time sedge,
Of primroses by shelter'd rills,
Of daisies on the aguist hill.
Bertha was a maiden fair,
Dwelling in the old Minister-square;
From her fire-side she could see,
Sidelong its rich antiquity,
Far as the bishop's garden-wall;

Where sycamores and elm-trees tall,
Full-leav'd, the forest had outstript,
By no sharp north-wind ever nipt,
So shelter'd by the mighty pile.
All was silent, all was gloom,
Abroad and in the homely room;
Down she sat, poor cheated soul!
And struck a lamp from dismal coal;
Lean'd forward, with bright drooping hair,
And slant book, full against the glare.
Untir'd she read, her shadow still
Glower'd about, as it would fill
The room with wildest forms and shades,
As though some ghostly queen of spades
Had come to mock behind her back,
And dance, and ruffle her garments black;
Untir'd she read the legend page,
Of Holy Mark, from youth to age,
On land, on sea, in pagan chains,
Rejoicing for his many pains…

JOHN KEATS (1818)

The imagination may be compared to Adam's dream: "He awoke and found it truth."
[Written on the back of the same page]

AUTUMN
Season of mists and mellow fruitfulness,
Close bosom-friend to the maturing sun;
Conspiring with him how to load and bless
With fruit the vines that round the thatch-eaves run;
To bend with apples the moss'd cottage-trees,
And fill all fruit with ripeness to the core.
Where are the songs of Spring? Ay, where are they?
Think not of them, thou hast thy music too, -
While barred clouds bloom the soft-dying day,
And touch the stubble-plains with rosy hue…
Madame Tussaud's Wax Works. (32)

It is worth nothing that he is quoting John Keats, whose final couplet from *Ode on a Grecian Urn,* "Beauty is truth, truth beauty,"—that is all / Ye know on earth, and all ye need to know," is often used as the most straight-forward expression of the Romantic Movement's embracement of the artist (the poet) as the prophet, the person able to grasp the invisible truth within or behind the visible world and express it in a work of art (a poem), thus leading those less sensitive to those higher truths. It is unlikely that Vincent has grasped the idea of art as a means to connect with God at this point in his life, but his reading of Keats foreshadows that likelihood later in life.

Again, these letters suggest an upbeat man, neither the unhappy child nor depressed man driven to self-condemnation and punishment of later years, certainly not the depressed or even deranged young man depicted in many biographies. Theo's wife Johanna says it was a good time for him, a time when he found the people and places and activities constantly "pleasant," and reveals an active mind, eager to learn and grow, a mind filled with ideas from the books he reads, the paintings he seeks out, and the museums he visits. (33) Frank Milner writes "Vincent's first year in England was among the happiest of his life and his letters to Theo are full of his curiosity about London's landscape and life. He walked a great deal, visited galleries, read Dickens, Longfellow and George Eliot and worked on the expansion of the London branch of Goupil, from trade-only sales to public picture gallery. (34)

In August, he moved to new lodgings with the Loyer family in Hackford Road, Braxton, South London. At first, the positive tone remains.

Dear Theo,

In my letter to Uncle Hein I enclose a little note for you. I wonder if you were in Helvoirt for Mother's birthday and how you enjoyed it.

Did you get my letter and the lithograph after Weissenbruch which I put in the box with the pictures? Oh! Old man, I very much want you to come here to see my new lodgings, the ones I have already spoken about to you. I now have a bedroom such as I always longed for, without a sloping ceiling and without blue wallpaper with green fringes. I lodge with some charming people now; they keep a school for little boys.

One Saturday some time ago, I went boating on the Thames, in the company of two Englishmen. It was glorious.

Yesterday I saw an exhibition of Belgian art, where I noticed many of the same pictures that were at the Brussels exhibition. There were several beautiful things by Alb. and Julien de Vriendt, Cluysenaer, Wauters, Coosemans Gabriel, De Schampheleer, etc. Have you ever seen anything by Terlinden? If so, tell me about it. It was a real pleasure to see those Belgian pictures; the English ones are with a few exceptions very bad and uninteresting. Some time ago I saw one which represented a kind of fish or dragon, six yards long. It was awful. And then a little man, who came to kill the above-mentioned dragon. I think the whole represented "The Archangel Michael, Killing Satan."

> Adieu, boy, best wishes and write soon,
> Vincent
>
> Another English picture is "Satan Possessing the Herd of Swine at the Lake of Gadarena." It represented about fifty black pigs and swine running helter-skelter down the mountain, and skipping over one another into the sea. But there was a very clever picture by Prinsep.
>
> I just received your letter. Going to The Hague will be a great change for you. I imagine it will be hard to leave beautiful, pleasant Brussels, but you will enjoy The Hague, too. Thanks for what you wrote me about the pictures. That picture by Millet must have been splendid. À Dieu, I will write soon again. (35)

Life continues well for him into March, 1874, when he is delighted to welcome his sister Anna, who has come to London to find a job. In July, he visits his parents at their new address in Helvoirt, Brabant, and then returns to London with Anna, who also moves in with the Loyers.

And then he falls in love with Eugenie Loyer, his landlady's daughter. At least that's the general consensus. The truth is not clear, but the theories are wide-ranging and interesting, and if taken as different facets of a textured canvas reveal why Vincent is an endlessly intriguing study.

A good place to begin is Ronald de Leeuw's prosaic account of it:

> Michelet's dictum in *'Les aspirations de l'automne',* as quoted by Van Gogh, that "a woman is a 'quite different being' from a man, & a being we do not yet know" became harsh reality for Van Gogh during this period, when he expressed his feelings for Eugenie Loyer, the nineteen-year-old daughter of his landlady, by proposing to her. There are no letters extant in which Van Gogh refers to being in love, but the situation may be inferred from the family correspondence. We have only one letter from Vincent himself mentioning Eugenie. In it he describes her as "a girl with whom I have agreed that we should be as brother and sister to each other." When Van Gogh asked for her hand, it transpired that Eugenie was secretly engaged to someone else and that there could be no question of a serious relationship between them. When he nevertheless continued to press Eugenie to call off her engagement, the situation became intolerable and Vincent and his sister were obliged to move out. (36)

There is also a letter from Anna to Theo mentioning Eugenie, and it suggests Vincent both knew she was already engaged to another man and that he saw her as a friend rather than a love interest. Not to mention that Vincent is happy about her engagement, not at all jealous and distraught. Here it is:

24th February 1874

> I also got a very kind letter from Eugénie; she seems to be a natural and amiable girl. Vincent wrote that she was engaged to a good natured youth who would know to appreciate her…We two are just [like] old people who try to know all about persons who are in love. But I am very glad for Vincent that he found such a kind family to live [with], you know yourself how agreeable it is. He seems to be always in good spirits. In the last letter he writes to me: "I fear that after all the sunshine I enjoy from there could be very soon rain - but I will only enjoy as long as possible the sunshine and have my umbrella in the neighbourhood for the rain that could come." (37)

In other words, De Leeuw's seemingly straight-forward summary, while attempting to avoid speculation (and much more even-handed than most), still carries a common suggestion of more support for this romantic love than there is. In truth, it becomes difficult to believe Vincent would not have written something, at least to his brother Theo, about his feelings, both at the time and later in thinking back on it, especially since they had such a close, personal correspondence, so the fact that there is only one mention of her in his correspondence, and it suggests a friendship rather than a deep love does undermine the theories that have emerged based on the assumption of a rejected love here. There is only one real source for all of the stories of Vincent's failed love affair, and that is his sister-in-law Johanna van Gogh-Bonger's *Memoir* of him. It is, nevertheless, a fairly convincing source for many reasons. First, there is no reason for her to have fabricated it. She had nothing to gain by so doing, and in fact would most likely rather have not exposed this negative aspect of him. Second, she states it all so matter-of-factly and without fanfare that it "feels" honest. Third, she is the biographer with both the life interactions and the correspondence to know. All of the other biographies turn to her for most of their materials, and thus, it makes sense to bypass them and go directly to the source. What follows is the major passage of her recollections of this time period, offering a sincere interpretation of Vincent's life at the time and of what is generally considered his first mental breakdown:

> In 1873 the latter had been transferred to the firm in London. When leaving The Hague, he got a splendid testimonial from Mr. Tersteeg, who also wrote to the parents that at the gallery everybody liked to deal with Vincent—art lovers, clients, as well as painters—and that he certainly would succeed in his profession. "It is a great satisfaction that he can close the first period of his career in that way, and withal he has remained just as simple as he was before," wrote Mother. At first everything went well with

him in London; Uncle Vincent had given him introductions to some of his friends, and he threw himself into his work with great pleasure. He earned a salary of £90 a year, and though the cost of living was high, he managed to lay by some money to send home now and then. He bought himself a top hat like a real businessman—"You cannot be in London without one"—and he enjoyed his daily trips from the suburbs to the gallery on Southampton Street in the city.

The first boardinghouse he stayed in was kept by two ladies who owned two parrots. The place was nice, but somewhat expensive; therefore, he moved in August to the house of Mrs. Loyer, a curate's widow from the south of France, who with her daughter Ursula ran a day school for little children. There he spent the happiest year of his life. Ursula made a deep impression upon him. "I never saw or dreamed of anything like the love between her and her mother," he wrote to one of his sisters; and, "Love her for my sake."

He did not mention it to his parents, for he had not even confessed his love to Ursula herself—but his letters home were radiant with happiness. He wrote that he enjoyed his life so much—"Oh fullness of rich life, your gift O God."

In September an acquaintance was going over to London and undertook to bring a parcel for Vincent. Characteristically, it contained, among other things, a bunch of grass and a wreath of oak leaves made at home during the holidays by Theo, who had meanwhile been transferred from Brussels to the House of Goupil at The Hague. Vincent had to have something in his room to remind him of the beloved fields and woods.

He celebrated a happy Christmas with the Loyers. He would send home now and then a little drawing, from his house and the street and from the interior of his room, "so that we can imagine exactly how it looks, it is so well drawn," wrote his mother. In this period he seems to have weighed the possibility of becoming a painter; later he wrote to Theo from Drenthe, "... how often I stood drawing on the Thames Embankment, on my way home from Southampton Street in the evening and it came to nothing. If there had been somebody then to tell me what perspective was, how much misery I should have been spared, how much further I should be now!"

At that time he occasionally met Matthijs Maris, but was too bashful to speak out freely to him and shut all his longings and desires within himself—he still had a long road of sorrow to go ere he could reach his goal.

In January his salary was raised, and until spring his letters remained cheerful and happy. He intended to visit Holland in July, and before that time he apparently spoke to Ursula of his love. Alas, it turned out that she was already engaged to the man who boarded with them before Vincent came. He tried everything to make her break this engagement, but he did not succeed.

With this first great sorrow his character changed; when he came home for the holidays he was thin, silent, dejected—a different being. But he drew a great deal. Mother wrote, "Vincent made many a nice drawing: he drew the bedroom window and the front door, all that part of the house, and also a large sketch of the houses in London which his window looks out on; it is a delightful talent which can be of great value to him."

Accompanied by his eldest sister, who wanted to find a situation, he returned to London. He took furnished rooms in Ivy Cottage, 39 Kensington New Road; there, without any family life, he grew more and more silent and depressed, and also more and more religious.

His parents were glad he left the Loyers. "His living at the Loyers' with all those secrets has done him no good, and it was not a family like others...but not realizing his hopes must have been a great disappointment to him," Father wrote. Mother complained, "The evenings are so long already and his work finishes early; he must be lonely. If only it does not harm him."

They felt uneasy and worried about his solitary, secluded life. Uncle Vincent also insisted on his mixing more with other people: "That is just as necessary as learning your business." But the depression continued. Letters home grew more and more scarce, and Mother began to think that the London fog depressed him and that even a temporary change might do him good: "Poor boy, he means so well, but I believe things are very hard for him just now."

In October, 1874, Uncle Vincent did indeed effect a short removal to the firm in Paris. Vincent himself was little pleased by this, in fact, he was so angry that he did not write home, to the great grief of his parents. "He is only in a bad temper," his sister said; and Theo comforted, "He is doing all right."

Toward the end of December he returned to London, where he took the same rooms and led the same solitary life. For the first time he was described as eccentric. His love for drawing had ceased, but he read a great deal. The quotation from Renan which closes the London period clearly shows what filled his thoughts and how high he aimed even then: "... to sacrifice all personal desires...to realize great things...to attain nobility and to surmount the vulgarity of nearly every individual's existence." He did not know yet how to reach his goal.

In May, 1875, he was transferred permanently to Paris and assigned especially to the picture gallery, where he felt quite out of place. He was more at home in his "cabin," the little room in Montmartre where, morning and evening, he read the Bible with his young friend, Harry Gladwell, than among the mundane Parisian public.

His parents inferred from his letters that things were not going well. After he had come home at Christmas and everything had been talked over, Father wrote to Theo, "I almost think that Vincent had better leave Goupil within two or three months; there is so much that is good in him, yet it may be necessary for him to change his position. He is certainly not happy." And they loved him too much to persuade him to stay in a place where he would be unhappy. He wanted to live for others, to be useful, to bring about something great; he did not yet know how, but not in an art gallery. On his return from Holland he had a decisive interview with Mr. Boussod (the son-in-law and successor of Mr. Goupil) that ended in his dismissal as from April 1, and he accepted it without offering any excuses for himself. One of the grievances against him was that he had gone home to Holland for Christmas and New Year's, the busiest time for business in Paris.

In his letters he seemed to take it rather lightly, but he felt how gloomily and threateningly the clouds were beginning to gather around him. At the age of twenty-three he had been thrown out of employment, without any chance of a better career; Uncle Vincent was deeply disappointed in his namesake and had washed his hands of him; his parents were well-meaning, but they could not do much for him, as they had been obliged to touch their capital for the education of their children. (The pastor's salary was about 820 guilders a year.) Vincent had had his share, now the others had to have theirs. It seemed that Theo, who was soon to become everybody's helper and adviser, had already at that time suggested Vincent‘s becoming a painter; but for the moment he would not hear of it. His father suggested a position in a museum or opening a small art gallery for himself, as Uncle Vincent and Uncle Cor had done before him; he would have then been able to follow his own ideas about art and have been no longer obliged to sell pictures which he considered bad. But his heart again drew him to England, and he planned to become a teacher.

Through an advertisement, in April, 1876, he got a position in Ramsgate at Mr. Stokes's, whose school moved in July to Isleworth. He received only board and lodging, no salary. He soon accepted another position at the somewhat richer school of Mr. Jones, a Methodist preacher, where Vincent finally acted as a kind of curate.

His letters home were gloomy. "It seems as if something were threatening me," he wrote. His parents perceived full well that teaching did not satisfy him. They suggested his studying for a French or German college certificate, but he would not hear of it. "I wish he could find some work in connection with art or nature, wrote his mother, who understood what was going on within him. With the force of despair he clung to religion, in which he tried to satisfy his craving for beauty as well as his longing to live for others. At times he seemed to become intoxicated with the sweet, melodious words of the English texts and hymns, the romantic charm of the little village church, and the lovely, holy atmosphere that enveloped the English service. His letters in those days contained an almost morbid sensitivity. Over and over he spoke about a position connected with the church—but when he came home for Christmas, it was decided that he would not go back to Isleworth because there was absolutely no prospect for the future. He remained on friendly terms with Mr. Jones, who later came to stay a few days at the Etten parsonage, and whom he subsequently met in Belgium.

Once again Uncle Vincent used his influence and procured a place for him in the bookshop of Blussé and Van Braam in Dordrecht. He accepted it, but without great enthusiasm. The words written to Theo by one of the sisters were characteristic: "You think that he is something more than an ordinary human being, but I think it would be much better if he thought himself just an ordinary being." Another sister wrote, "His religion makes him absolutely dull and unsociable."

To preach the Gospel still seemed to him the only desirable thing, and at last an attempt was made to enable him to begin the study of theology. The uncles in Amsterdam promised to give their aid. He could live with Uncle Jan van Gogh, Commandant of the Navy Yard, which would be a great saving; Uncle Stricker found the best teacher in the classical languages, the well-known Dr. Mendes da Costa, and gave Vincent some lessons himself; he could satisfy his love for pictures and prints in Uncle Cor's art gallery. Everybody tried to make it easy for him, all except Uncle Vincent, who was strongly opposed to the plan and would not help promote it—in which he proved to be right after all. Vincent set to work full of courage; first, he had to prepare himself for a State examination before he could he admitted to the university, and then it would take seven years to become fully qualified. His parents anxiously asked themselves whether he would have the strength to persevere, and whether he, who had never been used to regular study, would be able to force himself to it at the age of twenty-four.

That period in Amsterdam, from May, 1877, to 1878, was one long tale of woe. After the first half year Vincent began to lose ardour and courage. Writing exercises and studying grammar was not what he wanted to do; he wanted to comfort and cheer people by bringing them the Gospel—and surely he did not need so much learning for that! He really longed for practical work, and when at last his teacher also perceived that Vincent would never succeed, he advised him to give up his studies. In the Handelsblad of December 2, 1910, Dr. Mendes da Costa wrote his personal recollections of the pupil who later became so famous. He recorded many characteristic particulars: Vincent's nervous, strange appearance that yet was not without charm; his fervent intention to study well; his peculiar habit of self-discipline, self-chastisement; and finally, his total unfitness for regular study. Not along that path was he to reach his goal! He confessed openly that he was glad things had gone so far and that he could look forward to his future with more courage than when he devoted himself hopelessly to his theological studies, which period he afterward called "the worst time of my life."

He would remain "humble," and now wanted to become an evangelist in Belgium; for this no certificates were required, no Latin or Greek—only three months at the School of Evangelization at Brussels. There the lessons were free, the only charges being board and lodging, and he could get his nomination. In July he traveled thither with his father, accompanied by Mr. Jones, who on his way to Belgium had spent a few days with them at Etten. Together they visited the members of the Committee of Evangelization: the Reverend Mr. Van den Brink from Rousselaere; the Reverend Mr.

Pietersen from Malines; and the Reverend Mr. De Jong from Brussels. Vincent explained his case clearly and made a very good impression. His father wrote: “His stay abroad and that last year at Amsterdam have not been quite fruitless after all, and when he takes the trouble to exert himself, he shows that he has learned and observed much in the school of life.” Vincent consequently was accepted as a pupil.

But the parents regarded this new experiment with fresh anxiety: “I am always so afraid that wherever Vincent may be or whatever he may do, he will spoil everything by his eccentricity, his queer ideas and views on life,” his mother wrote. His father added, “It grieves us so to see that he literally knows no joy of life, but always walks with bent head, whilst we did all in our power to bring him to an honorable position! It seems as if he deliberately chooses the most difficult path.”

In fact, that was Vincent’s aim—to humble himself, to forget himself, to sacrifice himself, mourir à soi-même (to mortify himself)—that was the severe ideal he tried to reach as long as he sought his refuge in religion, and he never did a thing by halves. But to follow the paths trodden by others, to submit to the will of other people, that was not in his character; he wanted to work out his own salvation. Toward the end of August be arrived at the school in Brussels which had been opened only recently and had but three pupils. He certainly was the most advanced in Mr. Bokma’s class, but he did not feel at home at the school, he was “like a fish out of water,” he said, and was ridiculed for his peculiarities in dress and manners. He also lacked the ability to extemporize, and was therefore obliged to read his lectures from manuscript. But the greatest objection against him was, “He is not submissive”; and when the three months had elapsed, he did not get his nomination. Though he wrote (in letter 126) in an offhand way to Theo, he seems to have been greatly upset by it. His father received a letter from Brussels, probably from the school, saying that Vincent was weak and thin, did not sleep, and was in a nervous and excited state, so that it would be best to come and take him home.

He immediately traveled to Brussels and succeeded in arranging everything. At his own risk Vincent went to the Borinage, where he boarded at 30 francs a month with M. Van der Haegen, Rue de L’Église 39, at Pâturages near Mons. He taught the children in the evening, visited the poor, and held Bible classes; when the Committee met in January, he would again try to get a nomination. The intercourse with the people pleased him very much, and in his leisure hours he drew large maps of Palestine, of which his father ordered four at 10 francs apiece. At last, in January, 1879, he got a temporary nomination for six months at Wasmes at 50 francs a month, for which he would have to give Bible classes, teach the children, and visit the sick—the work of his heart. His first letters from there were very contented, and he devoted himself heart and soul to his work, especially the practical part of it; his greatest interest was in nursing the sick and wounded. Soon, however, he fell back into the old exaggerations—he tried to practice the doctrines of Jesus, giving away

> everything—his money, clothes and bed—leaving the good Denis boardinghouse in Wasmes, and retiring to a miserable hut where every comfort was wanting. Already his parents had been notified of it, and when, toward the end of February, the Reverend Mr. Rochelieu came for inspection, the bomb exploded. So much zeal was too much for the Committee, and a person who neglected himself so could not be an example to others. The Church Council at Wasmes held a meeting, and it agreed that if he did not listen to reason, he would lose his position. He himself took it rather coolly. "What shall we do now?" he wrote. "Jesus was also very calm in the storm; perhaps it must grow worse before it grows better." Again his father went to him and succeeded in stilling the storm; he brought him back to the old boardinghouse and advised him to be less exaggerated in his work. (38)

One immediate problem with the claim here is that Johanna names not Eugenie but her mother as the object of Vincent's love. This, of course, opens the door to debate. Was this substitution simply an oversight? This is the most likely possibility. However, if so, at the very least, it suggests Johanna was not as careful a biographer as she might have been and justifies questioning similar mistakes in all of what she has written. On the other hand, perhaps she meant exactly what she wrote. Did she really intend to say that Vincent loved the mother and not the daughter? That is unlikely; because she immediately points out that the woman rejects him partially because she is already promised to another man, which means it would be the daughter, not the mother. Perhaps she has simply mistaken Vincent's comments about his love for both of them as romantic, rather than as the love of a friend. Vincent constantly wrote about loving others in a Christian or humanitarian sense, and seems to have thought of both of them, at least at first, as wonderful people he loved in that way and wished well. Unlikely as it is, it is also possible that he was romantically and even sexually attracted to both at the same time. While these speculations are unlikely, they are, nevertheless, what serious scholars have suggested.

Vincent's own letters, while not offering much, if any support for the failed love theories, confirm his shift from a happy, intelligent, dynamic young man about to make himself a good life as an art dealer to an unlikable, discontented man, prone to arguments and determined to withdraw from the world into religion and faith.

There is no guessing here. This shift is obvious in his letters. While they continue to discuss art, literature, and the beauty of nature, they move more and more into passages from the Bible, and the confident young man now strives to be humble, to be a sincere servant of God. A few examples should suffice:

Dear Theo,

I am glad you've been reading Michelet and that you understand him so well. If that kind of book teaches us anything it is that there is much more to love than people generally suppose. To me, that book has been both a revelation and a Gospel.

'Il n'y a pas de vielle femme!' [There are no old women.] (That does not mean that there are no old women, only that a woman does not grow old as long as she loves and is loved.) And then a chapter like "The Aspirations of Autumn," how rich that is … That a woman is a 'quite different being' from a man, and a being that we do not yet know, or at best only superficially, as you put it, yes, that I am sure of. And that a man and a woman can become one, that is to say, one whole and not two halves, I believe that too.

Anna is bearing up well, we go on marvellous walks together. It is so beautiful here, if one just has a good and single eye without too many beams in it. And if one does have that eye, then it is beautiful everywhere.

Father is far from well, although he and Mother say that he's better. Yesterday we received a letter with all sorts of plans (wouldn't we just try this and that) which will prove to be unworkable and certainly useless and at the end Father said once again that he leaves it all to us, etc., etc. Rather petty and disagreeable, Theo, and it reminded me so much of Grandfather's letters, but qu'y faire [What can you do?]. Our beloved Aunts are staying there now and are no doubt doing much good! Things are as they are and what can a person do about it, as Jong Jochem said.

Anna and I look at the newspaper faithfully every day and reply to whatever advertisements there are. On top of that we have already registered with a Governess agency. So we are doing what we can. More haste less speed.

I'm glad that you go round to see the Haanebeeks so often, give them all my kindest regards and tell them some of my news.

The painting by Thijs Maris that Mr. Tersteeg has bought must be beautiful, I had already heard about it and have myself bought and sold one quite similar.

My interest in drawing has died down here in England, but maybe I'll be in the mood again some day or other. Right now I am doing a great deal of reading

On 1st of January 1875 we shall probably be moving to another, larger shop. Mr. Obach is in Paris at the moment deciding whether or not we should take that other firm over. Don't mention it to anybody for the time being.

Best wishes and write to us again soon. Anna is learning to appreciate paintings and has quite a good eye, admiring Boughton, Maris and Jacquet already, for instance, so that is a good start. *Entre nous,* I think we are going to have a difficult time finding something for her, they say everywhere that she is too young, and they required German, too, but be that as it may, she certainly has a better chances here than in Holland.

Goodbye,
Vincent

You can imagine how delighted I am to be here together with Anna. Tell H. T. [Herman Tersteeg] that the pictures have duly arrived and that I shall be writing to him soon. (39)

The statement "'Il n'y a pas de vielle femme!'" [There are no old women.], which does not mean that there are no old women, only that a woman does not grow old as long as she loves and is loved, has been taken to suggest Vincent was indeed in love with Ursula (the mother), and used as evidence for various claims about his relationship with his own mother.

Another line that jumps out is "My interest in drawing has died down here in England, but maybe I'll be in the mood again some day or other."

The next letter immediately brings in Biblical sayings (ones that have since been interpreted as Vincent's references to his desires for Eugenie).

Dear Theo,

"Ye judge after the flesh; I judge no man."

"He that is without sin among you, let him cast a stone at her."

So keep to your own ideas, and if you doubt whether they are right, test them with those of Him who dared to say, "I am the truth," or with those of some very human person, Michelet, for instance…

Virginity of soul and impurity of body can go together. You know the "Margaret at the Fountain," by Ary Scheffer, is there a purer being than that girl "who loved so much"?

"Leys n'est pas un imitateur mais un semblable" [Leys is not an imitator but a similar one] is a true saying that struck me too. One might say the same of Tissot's pictures, of his "Walk in the Snow," "Walk on the Ramparts," "Marguerite in Church," etc.

With the money I gave you, you must buy Alphonse Karr's Voyage autour de mon jardin. Be sure to do that - I want you to read it.

Anna and I walk every evening. Autumn is coming fast and that makes nature more serious and more intimate still. We are going to move to a house quite covered in ivy; I will soon write more from there.

Compliments to anyone who may inquire after me.
Vincent (40)

The letters continue in this mixed vein of past positive views and increasingly dark and religious centered views. On May 8, 1875, he writes:

Dear Theo,

Thanks for your last letter. How is the patient? [referring to their cousin Annette Haanebeek, who died soon after.] I had already heard from Father that she was ill, but I did not know it was as bad as that. Please write me soon about her.

Ay, boy, "What shall we say?" C. M. and Mr. Tersteeg have been here and left again last Saturday. In my opinion they went too often to the Crystal Palace and other places where they had nothing particular to do. I think they might just as well have come to see the place where I live. I hope and trust that I am not what many people think I am just now. We shall see, some time must pass; probably they will say the same of you a few years hence, at least if you remain what you are: my brother in both senses of the word.

Farewell and my compliments to the patient.
With a handshake,
Vincent

Pour agir dans le monde il faut mourir à soi-même. Le peuple qui se fait le missionnaire d'une pensée religieuse n'a plus d'autre patrie que cette pensée.

L'homme n'est pas ici-bas seulement pour être heureux, il n'y est même pas pour être simplement honnête. Il y est pour réaliser de grandes choses pour la société, pour arriver à la noblesse et dàpasser la vulgarité où se traîne l'existence de Presque tous les individus.

[To act well in this world one must sacrifice all personal desires. The people who become the missionary of a religious thought have no other fatherland than this thought.

Man is not on this earth merely to be happy, or even to be simply honest. He is there to realize great things for humanity, to attain nobility and to surmount the vulgarity of almost everybody.] RENAN (41)

The change is confirmed by his sister Anna:

28 April 1875

Do you sometimes hear from Vincent? I never do. It seems to me he has illusions about people and judges them before he knows them, and then, when he finds out how they really are and that they don't come up to the expectations he had formed *too soon* about them, he is so disappointed that he throws them away like a bouquet of wilted flowers without looking whether among those wilted flowers there would not be some that are not "quite rubbish" if only they would be treated with some care. I am really sorry I went to stay with him during the school holidays and was a burden to him. If I would have had any reason to foresee that that would be the case I certainly would have found some way to arrange things differently. I did not tell this to the people at home; they think he is a great support to me and that seems to give Father and Mother a great sense of relief. Well, the sun shines into my little room too beautifully and joyfully to think about unpleasant things and still less to write about them; if, however, you know why he is like that to me, I should be very glad if you would tell me. (42)

His father's correspondence also demonstrates a concern for the changes:

9 July 1875

If with regard to Vincent, we sometimes were worried about something strange in him, this does not mean - you do know that, don't you?--that we overlooked all the good qualities he has.

There is a kind of naturalness that is blamable. Someone who yields to low passions, follows *nature*, that is to say bestial nature, but *human nature* teaches him to dominate those passions.

Now don't misunderstand me; I don't want to say that I suspect you or Vincent of yielding to those ignoble passions, no indeed!

But this is the course of my reasoning: a person can sometimes be *not natural enough*. Youth is allowed to be lively, gay, cheerful; a youthful person is allowed to enjoy meeting people who are also youthful, gay and cheerful. In those years it is even a good thing if one doesn't go against one's nature, for there is in a cheerful mood a beneficial force. Melancholy can be harmful, and to indulge in melancholy does not help to produce energy.

My dear Theo! You should really think about that; I see that recently your liveliness has diminished, your cheerfulness is no longer what it was before. (43)

31 December 1875

How Vincent is going to go on, we don't know yet - he certainly is *not* happy. I believe it is not the right place for him there. We talked quite openly and discussed possibilities. Yesterday he went to Uncle Cor to consult him too; he is also a businessman. I tend to believe that I must advise Vincent to ask for his resignation in two or three months. (I tell you this *confidentially*!) Don't think I act hastily; I have noticed the signs of the times, seriously noticed them! In the meantime, these are only deliberations; it is not a definite decision. We also keep an eye on God in this matter. May His light give us wisdom and courage to act, when we see it necessary […] There is so much good in Vincent. That is why it may be necessary to make a change in his position. (44)

Vincent's letters continue to shift more and more into religion and take on an ever more preachy quality.

Dear Theo,

I wanted to write you earlier. I am happy that Father has accepted the nomination at Etten; in the circumstances, I think it right that W. and A. [their sisters] are leaving together. I would have liked to be with you all that Sunday when you went to Helvoirt - did I write you that I spent the day with Soek and his family at Ville d'Avray? I was surprised to find three pictures by Corot in a little church there.

Last Sunday and the previous Sunday, I went to church to listen to Dr. Bercier; I heard him preach on "Toutes choses prouveront le bien de ceux qui aiment Dieu" [All things work together for the good of those that love God] and on "Il fit l'homme à son image" [God created man in his own image.] It was beautiful and noble.

You should go to church every Sunday if you have the time; even if the preaching is not good, it is better to go; you will not regret it. Have you been to church to hear the Reverend Mr. Zubli?

On the list of engravings that hang in my room, I forgot the following:

N. Maes – The Nativity
Hamon – If I Were the Sombre Winter
Francois – Last Summer Days
Ruyperez – The Imitation of Jesus Christ
Bosboom – Cantabimus and Psallemus

I am trying to get another engraving of Rembrandt's "Bible Reading" for you; if possible, I will send it to you in the first box of pictures. Did I send you a lithograph of Troyon's "Morning Effects" and François' "Last Summer Days"? If not, tell me. I have duplicate copies of them.

And now, be as happy as you can; do well and don't look back if you can help it.

Believe me, always your loving brother,
Vincent

My compliments to Mr. and Mrs. Tersteeg and Mauve. Also to the Van Stockums, Haanebeeks, Aunt Fie and the Rooses! (45)

And again:

Dear Theo,

Thanks for your last letter and the poem by Rückert.

Last Sunday, I again went to hear M. Bercier. He had chosen this text: “Il ne vous est pas permis” [It is not permitted] and he concluded with “Heureaux ceux pour qui la vie a toutes ses épines” [Happy is he whose life is all thorns]

I know that uncle Vincent really likes this phrase, “Young man, rejoice in your youth and that your heart is made content by the days of your youth, and live like your heart guides itself according to your eyes, but understand that for everything God will make judgment. Take away the chagrin of your heart, and put away malice, for youth and adolescence is not in vain. But remember your Creator during the days of your youth, before the bad days come and the years arrive when you will say `I don’t have any pleasure’.”

For me, I find the following phrase better:

Fear God and keep his commandments, for this will give pleasure for all men, and lead us not into temptation, but deliver us from evil.

Enclosed is a note for Mr. Tersteeg. It is to ask him if he will have two engravings framed for me, “Good Friday” and “St. Augustine.” You will find them in the next box I send you. And will you be so kind as to have them sent to Helvoirt by September 10? I should like it to be a present [for his mother’s birthday] from both of us, so will you pay 2.50 towards the frames? I told Mr. Tersteeg that you would write me how much it cost and that I would then remit the money to him; the 2.50 you can give me when we meet. That will probably be before Christmas; I think it would be better not to ask for leave before that time.

Tonight I am invited to dine at Mr. Hamman’s. [a French painter, friend of Uncle Vincent.] À Dieu, write me soon and believe me,

Your affectionate brother,
Vincent (46)

And again:

Dear Theo,

A feeling, even a fine feeling, for the beauties of Nature is not the same as a religious feeling, though I believe these two are connected. *

Nearly everyone has a feeling for nature, some more, some less, but there are some who feel: God is a Spirit, and they that worship Him must worship Him in spirit and in truth. Father is one of those few, Mother too, and Uncle Vincent as well, I think.

You know that it is written: “The world passeth away and the lust thereof”, and that on the other hand we are also told about “that good part which shall not be taken away”, and about “a well of water springing up into everlasting life.” Let us also pray that we may grow rich in God. Still, do not dwell too deeply on these matters - in the fullness of time they will become clearer to you of their own accord - and just take the advice I have given you.

Let us ask that it may fall to us to become the poor in the kingdom of God, God's servants. We are still a long way from that, however, since there are often beams in our eye that we know not of. Let us therefore ask that our eye may become single, for then we ourselves shall become wholly single.

Regards to Roos and to anybody who may ask after me, and believe me, always,

Your loving brother, Vincent

You are eating properly, aren't you? In particular eat especially as much bread as you can. Sleep well, I must go and polish my boots for tomorrow.

*The same is true of the feeling for art. Do not succumb too much to that either. Above all, save some love for the business and for your work, and respect for Mr. Tersteeg. One day you will appreciate, better than now, how much he deserves it. No need to overdo it, though. (47)

And again:

Dear Theo,

The path is narrow, therefore we must be careful. You know how others have arrived where we want to go, let us take that simple road too.

Ora et Labora, [Pray and work] let us do our daily work, whatever the hand finds to do, with all our strength and let us believe that God will give good gifts, a part that will not be taken away, to those who ask Him for it.

"Therefore if any man be in Christ, he is a new creature: old things are passed away; behold all things are become new!" [2 Cor. v.17.]

I am going to destroy all my books by Michelet, etc. I wish you would do the same.

How I am longing for Christmas, but let us have patience, it will come soon enough.

Courage, lad; my compliments to all the friends, and believe me,

Your loving brother,
Vincent

As soon as possible I will send the money for the frames. When I write to Mr. Tersteeg, I will tell him that for the moment I am rather short of cash; I asked our cashier to hold back every month a part of my salary as I shall want a lot of money around Christmas for my journey, etc., however I hope to send it before long. (48)

What a different Vincent appears here than the one that came through in earlier letters. Now he has decided to destroy all of his books by Michelet (and other similar books), and advises Theo to do the same. And his religious preaching moves ever further into religious fanaticism.

Paris, 27 September 1875

Dear Theo,

"The Kingdom of God cometh not with observation; the kingdom of God is within you." "The son of Man is not come to be served, but to serve," and we who want to become His followers, Christians, we are not greater than our Lord [see Like 22:26-27; John 13:16]. Blessed are the poor in spirit, blessed are the pure in heart.

Narrow is the path which leadeth unto life, and those that find it are few. Struggle to enter by the narrow gate, for many will seek to enter, and will not be able [see Matt. 7:14].

My brother, let us be prudent; let us ask of Him Who is on high, Who also prayeth for us, that He take us not away from the world, but that He preserve us from evil. Yea, let us be sober, and watch, let us trust in God, and not lean upon our intellect. Let us ask of Him that He force us to come in; that He enable us to fulfill a Christian's life; that He teach us to deny ourselves, to take our cross every day and follow after Him; to be gentle, long-suffering and lowly of heart.

A part that shall not be taken away, a spring of living water, springing into Life eternal [see Luke 10:42; John 4:13], these are the good gifts that the Hearer of prayers, the Giver of all perfect gifts, will give unto those that pray for them unto Him.

And over and above this the assurance that there is "a Father's house in which are many mansions," and that, when He that hath gone and prepared a place for us, He will receive us all unto Himself. And for our comfort in life on the road to that "Father's house," the Comforter, the Spirit of Truth, who shall lead us in truth [see John 14:2-6, 16-17].

Yet a Christian's life has its dark side too, it is principally man's work.

For those that walk with God, God's friends, God's pious ones, those who worship him in the Spirit and in Truth, are tried and purified, and often have received from God a thorn in the flesh; blessed shall we be, if we can repeat after our father, Paul: "When I was a child, I spake as a child, I understood as a child, but now that I have become a man, I have put away childish things, and am I become, and God hath made me sorrowful yet always rejoicing."

Write soon and give my regards to all acquaintances, and believe me

Your loving brother,
Vincent (49)

Certainly, something has changed:

Dear Theo,

Although I have written you recently, I send you the following because I know that sometimes rough obstacles arise on our path. Courage, old son, after the rain, good weather: keep hoping for that.

The rain and good weather alternate on *The road that goes uphill all the way, yes to the very end, and one rests also from time to time during the journey that takes the whole day long, from morn till night.*

Think on this, now and in later years, "This also will pass away."

Jules Dupré liked to repeat: 'One has one's beautiful days'. Let us believe it too.

Today, I have the opportunity to send a parcel to A. and W. in England. I have sent them notably The Imitation of Jesus Christ and some separate fragments of the Bible published in a collection of the Psalms which I have sent you.

Read them with assiduity. Do you want also the four Evangelists and some Epistles, which are published separately?

I would like to have a Hymnbook in Dutch. If the opportunity arises, send me the least expensive edition than you can find. I already have the Psalms. There are some very beautiful English hymns, for instance:

Thy way, not mine, O Lord,
However dark it be:
Lead me by Thine own hand
Choose out the path for me…
I dare not choose my lot;
I would not, if I might;
Choose Thou for me, my God:
So shall I walk aright.
The kingdom that I seek,
Is Thine, so let the way
That leads to it be Thine,
Else I must surely stray.
Choose Thou for me my friends,
My sickness or my health;
Choose Thou my cares for me,
My poverty or wealth.
Not mine, not mine the choice,
In things or great or small
Be Thou my guide, my strength
My wisdom, and my all.

And the following:

Nearer, my God, to Thee
Nearer to Thee.

E'en though it be a cross,
That raiseth me;
Still all my song shall be,
Nearer, my God, to Thee,
Nearer to Thee.

And also:

Oft in sorrow…oft in woe
Onward, Christians, onward go:
Fight the fight, maintain the strife,
Strengthened with the Bread of life.
Let your drooping hearts be glad:
March in heavenly armour clad:
Fight, nor think the battle long,
Soon shall victory tune your song.
Let not sorrow dim your eye,
Soon shall every tears be dry;
Let not fears your course impede,
Great your strength, if great your need.

Best regards to all the friends. How is Caroline Van Stockum? Give her my kindest regards and believe me, your loving brother,

Vincent

Does the road go uphill all the way?
Yes, to the very end.
Will the journey take the whole long day ?
From morn till night, my friend.[1]
Christina Rossetti. Inaccurately quoted by Vincent in the original. (50)

Vincent has changed. That cannot be denied. And those who know him best and love him most are seriously worried about this change. He is exhibiting qualities they do not like and they can clearly see are hurting him. In fact, he seems bent on self-destruction. However, it is possible to put a positive spin on this from a religious perspective (and some scholars do just that). Vincent has found God, and this is not a superficial conversion, but a deep one demanding that he give up many of the more pleasant things in life, that he become strident about his views, and that he work to save others, especially those closest to him. Nevertheless, it is, even if a positive movement into a religious and/or spiritual existence, a dramatic change. And his father, certainly a religious man, finds it worrisome. In their very impressive work *Van Gogh: The Complete Paintings*, Ingo F. Walther and Rainer Metzger quote Vincent and write:

> "Who sees that once our first life, the life of youth and young manhood, the life of worldly pleasure and vanity, has withered away, as wither away it must—and it will do so, just as the blossoms fall from the trees—who sees that then a new life is ours, in all its strength, a life filled with love of Christ and with a sadness that causes sorrow to no one, a divine sadness?"
>
> Most of the products of van Gogh's pen up to 1880 read like this passage from Letter 82a, dating from November 1876. He had developed what can only be described as a religious mania, and it involved excessive mortification of the flesh: van Gogh cudgelled his back, went around wearing only a shirt in winter, and slept on the stone floor beside his bed. It was as if he wanted to catch up on his forefathers' piety—and was doing it at double the normal pace. (51)

This passage is careless when it claims that "most of the products of Van Gogh's pen up to 1880" have such strong spiritual concerns. As demonstrated above, Vincent's letters exhibit a definite shift to this quality from previous letters. The part to highlight here, however, is that concerning his "excessive mortification of the flesh." It's not Vincent's move into religion that bothers his father and the other members of his family. That would be ridiculous, as his father is a minister. It's the form of it, the desire for self-condemnation and very real physical denial and punishment. However, again, this can be given a positive, or at least a legitimate spin, as such Christianity has a long history, and a fully developed theology. Flagellants have existed and held a great deal of influence throughout the centuries, and Opus Dei continues such practices to what most would consider an extreme today (beyond the degree and forms Vincent practiced).

But more on that later. For the moment, consider the plethora of theories concerning how much of this change was the result of a failed love interest (and especially how this is tied to theories about his own childhood and relations with his mother).

First, as an example of the many books that make large assumptions based on little evidence and actually embarrass themselves by demonstrating a lack of knowledge of simple facts, consider a very impressive looking book by D. M. Field, a man whose credentials include having "read history at Cambridge University, in England, taught history at a boys' school in New York, and later worked for several years for an international publisher of art books before becoming a full-time writer in London. . ." and to have authored "many popular books on social history and the arts." (52) His spin on Vincent's experience in London is that "he had fallen in love," that he always wanted to be a part of a community, but that he was unable to maintain friendly relationships because of "the difficulties of his temperament," which meant he "could not manage intimate relations with an individual." (53) With this unsubstantiated frame in place, Field is ready to plunge into some fun but unfounded speculation. He writes:

> When the powerful element of sex was added, disaster threatened. He was so thoroughly enveloped in his own emotions that he could not conduct even normal human relations. It was, for instance, quite characteristic that he was in love for the best part of a year with 19-year-old Ursula Loyer [here is the standard factual mistake, as Ursula was Eugenie's mother, and even though this confusion goes all the way back to Johanna van Gogh-Bonger, Ursula was certainly not 19, and the confusion has long since been clarified, so D. M. Field should have realized the mistake; others have suggested that, in fact, it was the mother that Vincent had fallen in love with, not the daughter, but only poor scholarship would make the mother 19.], his landlady's daughter, who lived in the same house and whom he saw every day, without ever betraying to her a sign of his feelings. Nor did he ever wonder, as he silently admired Ursula across the supper table, if the frequent presence at the evening meal of a former lodger, a young, good-looking, unmarried engineer, had any significance. [as we've already seen from a letter above, he was aware of this relationship well before the date of the suggested proposal.] (54)

Then Field pushes even further into unfounded speculation about Vincent's understanding of sexual attraction:

> As we are often told, people were more reticent about sex in the nineteenth century, but Vincent cannot have been entirely ignorant. It would be odd if he had never discussed the matter with boyhood friends. He had seen plenty of sensual pictures. He had known and had lived under the same roof as married couples, and had witnessed courtships, such as that of Anton Mauve with his cousin Jet (they had just become officially engaged). He had read a little book, *L'Amour*, by the historian Jules Michelet which discoursed on contemporary notions of femininity (and incidentally remarked that an Englishwoman made the ideal spouse).
>
> He seems to have supposed that feelings between a man and a woman were necessarily mutual even though unstated, therefore that Ursula felt the same for him as he did for her. Vincent's inability to look at a situation from another person's point of view seems extraordinary to us, and suggest that a degree of autism may have been amoung his problems. (55)

Tempting speculation, but filled with endless flaws or at best guesses, including not only assumptions about sexual views in the nineteenth century and Vincent's own knowledge and views on sex without any support, but claims that Vincent was unable to "look at a situation from another's person point of view," something his letters contradict (though it seems he was an argumentative person, at least later in life), and the careless claim that this indicates possible autism. At the very least, large assumptions are being made here, and unfortunately, they feed into what has become a general popular view of Vincent as the ideal example of an insane artist on a very superficial level. Field continues:

> The first that Ursula knew of Vincent's feelings was his announcement, in early summer 1874, that it was about time they thought about getting married. Horrified, she flatly rejected him, explaining that she was already secretly engaged to the young engineer.
>
> Overwhelmed, Vincent sank into severe depression, which developed into what might be regarded as his first serious mental breakdown. (56)

Rather than being a singularly poor representation of the facts, D. M. Field's book presents the standard view. In another biography, Pierre Cabanne claims that "She admitted she had lead him on, making fun of him; and as he persisted, continued to plead with her, burst again into laughter, cutting him to the heart." (57) Sounds like a good bit of dramatic writing, directions for a theatrical performance. In another, Frank Elgar writes "He soon fell in love with his landlady's daughter Ursula . . . proposed to her, was rejected, and became so depressed that he returned to his family in Holland." (58)

In another, Philip Callow, while dealing with Vincent's childhood in a much more even-handed manner than most, suddenly loses his objectivity in assuming Vincent's failed love for Eugene, perhaps because he loves the "fictional" potential of it all:

> Towards the end of July, 1874, about to go home for his summer holidays, he made a move that must have seemed violent, petrified as he was by the audacity of his love. Its consequences shattered him for months to come, stayed with him for years, and changed the course of his life.
>
> All at once he came to the end of his fantasizing. Dutch tradition and his own upbringing had taught him to fear and distrust imaginings and prefer the real thing. Worshipping Eugenia from a distance was an indulgence, a sin. Vincent blundered up to her and declared himself. Perhaps the glorious weather had incited him. They were alone in the garden. No one knows how she reacted, whether she frowned, looked contemptuous, was astounded, or simply horrified. What she told him made him recoil in shocked disbelief. She was secretly engaged to the man who had boarded with them before Vincent arrived.
>
> The paralysing fear of rejection which had kept him in check for nearly a year now became an absolute refusal to accept what his pain was telling him. Desperately this future devotee of suffering begged to be let off. He demanded to know why he hadn't been told, when it was obvious to

> anyone how he felt about her. It was apparently not evident at all, but how could he be expected to believe that, when the very skin of his face proclaimed it to the world, as her presence poured enriching rays straight into his view? Where was this fiancé? Had he visited her since Vincent had taken over his room? If she hadn't seen him for over a year, what kind of love was that? How could it be compared with his burring passion, seething now to anguish and falling back shamed in his blood. . . .
>
> Vincent was distraught. He 'tried everything' to make her break off her engagement, but in vain. At the gallery he lost interest in and was brusque to the point of rudeness with customers. If some chose to buy bad rather than good art he could hardly contain his disgust. Even worse, he objected to the practice of selling worthless pictures to uninformed buyers. In his rage and despair he took up religion, endlessly reading and studying the Bible. If he had been "strange" in his childhood, now he was seen as eccentric rather than just willful. . . .
>
> In the house, Eugenia went out of her way to avoid him. His last few days there dragged by. Vincent had lost all his illusions. When he went home in defeat, the mysterious fiancé from Wales, the rival he had done his best to supplant, took over his room. His disgrace was total. (59)

The creative writer in Philip Callow has taken over here, and endless assumptions are stated as fact, including not only the certainty that Vincent had fallen deep in love and been rejected, but many details that have no source, along with several that have shaky sources. And the claim that he turned to the Bible and his religion in "rage and despair" as a result of this failed love affair almost certainly misrepresents his years deeply devoted to his religion and faith.

In the very impressive work *Van Gogh: The Complete Paintings,* Ingo F. Walther and Rainer Metzger, while retaining the assumed love and affect it had on Vincent, put it more factually:

> Vincent lacked experience with women, and one unfortunate love seems to have given him a considerable jolt. Six months previously he had nursed a secret love for Eugenie Loyer, the daughter of his landlady in London, and when at last he dared to tell his adored young woman of his feelings he proved to be too late—another man had got there first. Thwarted relations with women were to devastate Vincent's emotional life at later periods too. At the time of his rejection, about the end of 173, van Gogh's behaviour underwent a complete change. Hitherto an open, entertaining and liberal-minded man, he became an eccentric, taciturn loner who substituted late-night Bible reading for contact with his fellow-beings. He now reviewed all the books he had previously read so voraciously to check their usefulness for pious purposes. In the tone of a preacher he instructed Theo [letter 36a] "Do not read Michelet or any other book but the Bible till we meet again at Christmas." (60)

Humberto Nagera, with his huge credentials in Freudian child psychology, especially in the field of obsessional neurosis, and strong support from Freud's daughter Anna, who wrote a brief introduction to his book, boldly offers the first important Freudian analysis of Vincent's childhood, basing it, as he emphasizes, mainly on *The Complete Letters of Vincent Van Gogh* published by Thames and Hudson, and deciding it is the result of the Replacement Child Syndrome that will dominate future discussions.

It is quickly apparent that Nagera is going to make some huge assumptions based on little evidence, as on the first page of the book, he states that "It was Vincent's fate to come into the world without an identity of his own but as a substitute for his dead brother; witness the fact that he was given exactly the same name by his parents. This dead brother was buried near the entrance to his father's chapel in Zundert. Vincent probably saw the grave at least every Sunday. It must indeed have made a strange impression on him, to see his name on the tombstone of his brother." (61)

Nagera immediately takes this into a favorite topic of his studies, that of the psychological problems developing out of parents having a dead child and how they relate that to future children:

> The replacement of a dead child by another has been a subject of study by a number of psychoanalysts. [Reference is given here to Albert C. Cain and Barbara S. Cain, "On Replacing a Child," *Journal of the American Academy of Child Psychiatry,* Vol. 3, No. 3, July, 1964]. They have shown that in some families, the parents, after the death of the child become the victims of important psychopathological changes that will profoundly affect and interfere with their relationship to the child whose fate it is to become a substitute for the dead sibling. These parents tend to impose the identity of the child,—who in the meantime becomes highly idealized and the compendium of all their phantasies and hopes—upon his substitute, unconsciously identifying the two siblings. Furthermore the parents are usually unable to accept the substitute as 'the same' with the result that he becomes an increasing disappointment to the parents and naturally, always falls short of the highly idealized image of the dead sibling. The personality development of many such children becomes in consequence greatly affected and distorted. In addition such parents, especially the mothers, are not infrequently panic-ridden by phantasies of the substitute child dying too, which leads to an abnormal concern and oversolicitous attitude towards him. The child, on the other hand, acquires a conviction of his inadequacy and vulnerability in a world of constant unpredictable dangers (Cain). Vincent's anxiety when away from home at some points in his life was no doubt partly determined by these circumstances. Their impact on his personality left clear imprints. (62)

Nagera will return to this assumption much later in his book, when he discusses Vincent's views about the birth of his brother Theo's son, again making claims from some lines taken out of context in letters leading up to the birth:

> Vincent's unconscious hostility towards Jo and her son yet to be born found expression on a conscious level towards the end of her pregnancy in the form of a tormenting preoccupation about her safety and that of the baby. He referred repeatedly to this in different letters. To Jo herself he wrote, "How I am longing to get the news that you have come safely through, and that your child is living." (Letter 624) On February 1st he received the news that the baby was born and all was well. Consciously he was pleased and relieved, but the dreaded rival had finally arrived safely. From his letters it is possible to reconstruct the following sequence, which is of great significance in respect of the tensions that may have led to the new crisis. On February 15th he wrote to his mother stating that he would have preferred Theo's child to be called not Vincent after himself, but Theo in memory of their father. It was, after all, his brother's name as well. He continued saying that he started "a picture for him, to hang in their bedroom, big branches of white almond blossom against blue sky." Letter 622) It is not surprising that it was while working at this painting for his nephew that he first felt ill again. He wrote "I felt ill at the time I was doing the almond blossoms" (letter 629), and "My work was going well, the last canvas of branches in blossom—you will see that it was perhaps the best, the most patiently worked thing I had done, painted with calm and with a greater firmness of touch. And then the next day, down like a brute." (Letter 628) (63)

He is reaching for his predetermined conclusions here, and the claims can easily be interpreted as the opposite of what the passages suggest. Furthermore, to tie Vincent's physical and psychic sickness to his thoughts about Theo's baby is unjustified. However, this is just the set up. With it in place, Nagera makes his big claim:

> In some ways this child was like a reincarnation of the dead brother whose absence-presence had haunted him all his life. And the boy was named Vincent like his brother, and himself. The first one was dead, and the second mad; thus the name was a bad omen. For the child to have this name of his dead brother must have increased his anxiety over his unconscious death wishes against it. This did not show directly but appeared in the form of a reaction formation, that is, as an intense preoccupation for the health of the child and extreme anxiety when the child became ill. Added to all this, to have given the boy his name and not that of the boy's grandfather was an offence to his dead father's memory and we know that he had been thinking a great deal about his father at

> some points during his illness. Thus it is not surprising that he himself traced the beginning of the crisis to the picture he was painting for his nephew, though he may not have been fully aware of the implications of this link. Clearly, the positive aspects of his ambivalence towards the child, the only ones allowed into his conscious mind, gave him not only the idea of making the picture for his nephew's room but led him to try his best in this picture; as he said it was the most patiently worked thing he had done. The negative aspects of the ambivalence—of which he had little if any conscious awareness—created a situation of tension and conflict that was much reinforced by the other elements referred to above. They led to a new and prolonged attack of insanity. This was the fifth acute psychotic episode he had suffered in little more than a year. (64)

As mentioned, Nagera returns to the dead child syndrome implications near the end of his book:

> I have already many times referred to an element in Vincent's history which played an essential role in his development and in the distortion of his personality. Indeed it looms heavily everywhere and is an essential part to the background of Vincent's character. I refer to the impact on his family, and through them on him, of the stillbirth of his older brother for whom he was meant to substitute and whose name he was given. We have mentioned some psycho-analytic studies of families where a child has been conceived to take the place of a dead one. In Vincent's case the brother, being stillborn, had never had an identity of his own in reality, but for this very reason an ideal one had been created in the phantasy life of the parents. He would have been the perfect child, the compendium of all virtue, ability and kindness. He would always have done everything right, and, especially where Vincent failed, the other, the dead Vincent, would have been successful. The extreme degree of idealization of a dead child is by no means uncommon and such were the ideals of behaviour and accomplishment which his parents offered unconsciously to Vincent for emulation. This explains the high ego-ideals which he set himself, his dread of failing (exhibiting his painting always implied the risk of failing), and his fear of success (quite apart from the factors mentioned already), because there was always the risk that success would be transitory, in which case the later feelings of failure would become all the more painful. We have seen that he always dreaded this possibility. On the other hand, once successful, one is forced to perform at that very high standard all the time, or face great losses in terms of one's self-esteem. It is on a similar basis that we can understand his constant search for an identity, but always one of "heroic proportions" such as that of a great Christian, a great painter, etc.
>
> Against such high ego-ideals he would, of course, nearly always fall short. His badly batter ego needed to be shown some appreciation and admiration but the constant fear of not being up to these high standards

> forced him to withdraw from people and activities precisely out of the fear of criticisms hat he felt certain he was to receive.
>
> A further important aspect of these conflicts was the unconscious dread of competing with the *idealized dead Vincent*. Unconsciously he must have felt that his success was an attack on the memory of the dead one, an attempt to take his place in the affection of the parents. Such phantasies are highly conflictive since, as Cain points out, siblings of a dead child feel in some form responsible for their death even "in the face of their not having even been born during the child's lifetime." Cain further found that even in the phantasies of the parents, the substitute child was frequently felt to be responsible for the death of the dead one and that they could never accept the substitute sibling as being as good as the idealized dead one; in fact they tended to become increasingly disappointed from the moment of the birth onwards.
>
> Furthermore, it seems possible that under these circumstances Vincent came to associate death and success. To be recognized as good as the brother or better even it was necessary to be dead like him, an association that may have made no small contribution to his fear of exhibiting and meeting success and would justify and explain his feeling of impending disaster and punishment as expressed to his sister after Aurier's article, his successful exhibitions and so on. It is indeed not surprising that under the impact of such unconscious pressures his mind collapsed.
>
> Painting was in a way a most suitable area of activity for this type of conflict since as Vincent well knew and frequently stated, it is a profession where one can be considered an utter failure during one's lifetime only to rise to fame after death; a possibility well adapted to his unconscious requirements. In this sense his recognition as an outstanding painter came perhaps too soon. (65)

It is curious to find such extreme claims based on so little evidence and on what appears to be such a clear misrepresentation of what little there is in a book that is a generally well reasoned discussion by a well respected psychologist.

Nagera does give a sensible view of Vincent's relationship with Ursula (Eugenie, again, we see the standard misnaming here):

> Though in *The Complete Letters* and in most biographical studies this episode of his disappointed love for Ursula Loyer is given as a turning point in Vincent's life, and though there is no question as to its importance, I am rather inclined to think that a number of other factors too came together at this time to determine Vincent's change of character. First there was the fact that Vincent had been for about a year away from home and this enforced separation has placed a tremendous strain in his psychological resources. The Ursula episode has to be seen against this background. Vincent was in distress and lonely, away from home and was

> immediately attracted by the close relationship that existed between this daughter and her mother. Vincent longed for such a relationship, as we will have opportunity to verify later on. Since, on the other hand, there was no question of his been able to stay at home in Holland as he would have wanted, the possibility of a close relation to Ursula and her mother was a second best choice open to him. (66)

Interestingly, probably because he is so trained to notice such things, Nagera quickly focuses on the later period of Vincent's collapse, and on his brief friendship with an eighteen-year-old boy named Gladwell. The two of them meet because they are living in the same house and walk home together from work. By this time Vincent is deep into his religious fervor, and reads to his new friend from the Bible. Though he doesn't describe it in the terms of "projection," what Nagera finds in their relationship is just that, Vincent projecting his own psychic needs onto Gladwell, and what Nagera finds him projecting is his recent love for and admiration for his own father: "This conscious and overt admiration of his father is the expression of an unconscious and rather conflictive homosexual passive surrender to the all powerful father. This passivity, Vincent says to Gladwell, can be acceptable only if expressed in the relationship to God but not to the father." (67). Nagera continues, "Vincent is indeed moving fast in this direction, his obsession with religion is increasing to the point of a mania. It will lead him finally to attempt to follow on the footsteps of his father by trying to become a parson. Thus he will be surrendering his whole self to the Supreme Being, to God (the father). This is an attempt at sublimation of his strong bisexual conflict and passive homosexual striving towards his father." (68)

The homosexual aspects of this discussion carry suggestions of literal homosexual desires that would be hard to support. Furthermore, Freud's own views on homosexuality and much of early psychology based on those views have long since been discounted. (Later, however, in his discussion of Vincent's relationship with Gauguin, Nagera presents an excellent interpretation of an aspect of it in these terms, using the symbolism of two chairs Vincent painted, one to represent himself and one to represent Gauguin, to demonstrate Vincent's confusing understanding of his own dynamics in relation to "fathers" and other males). In this section of the book, they get in the way of the discussion of the psychological father/son dynamics, which otherwise works well. Indeed, the letters of the time do indicate a struggle to both please and rebel against his father, and the whole desire to become a more religious person in relation to Freudian father/son psychological theory fits.

Shortly, Nagera moves even more into Vincent's ever increasing religious fanaticism, continuing to couch it in terms of his need for acceptance from his father, and offering several passages from Vincent's letters demonstrating both this desire and at times the belief that he was indeed being successful satisfying his father's desire. Nagera quotes Vincent:

> "It is good to think of Jesus in all places and circumstances . . . You do not know how I am drawn to the Bible; I read it daily, but I should like to know it by heart and to view life in the light of that phrase, They word is a light unto my path and a lamp unto my feet [sic. Ps. 119:105]. I hope and believe that my life will be changed somehow, and that this longing for him will be satisfied. I too am sometimes lonely and sad, especially when I am near a church or parsonage." (Letter 88). (69)

Then Nagera elaborates:

> "Him" in this context can be seen as a condensation of the idealized "quasi-God" image of his father and of God as he conceived of it. The longing for him is a sublimated expression in religious terms of unconscious passive homosexual striving in relation to his father. The identification in his unconscious mind between his father and God needs no elaboration here. It is quite evident for anyone who reads the letters. At the end of this same letter there appears another example of such a connection. He refers now to his father's words as he did above those of God. His father's word is too "a light unto my path and a lamp unto my feet." He writes to Theo, "Hate sin; remember how Father prayed every morning, "Deliver us from evil, especially from the evil of sin," and well he knows." This letter contains—as do many of his letters—a condensed version of his oedipal struggles and defensive attitudes. At the end of it, and without any apparent connection with the rest of the text, there appears the following sentence, "The photograph "Mater Dolorosa" which you sent me is hanging in my room. Do you remember, it was always hanging in Father's study at Zundert?" The unconscious, perhaps conscious connection between this "Mater Dolorosa" and his mother, who had lost her firstborn child and was consequently a true "Mater Dolorosa" herself is clear. Furthermore this "Mater Dolorosa" was the same as the one that his father always had hanging in his study; it was the "Mater Dolorosa" so to say, that belonged to the father. It is, of course, by no means extraordinary that a man of Vincent's background attempted to deal at this point in his life with his unconscious conflicts by the means of the rich symbolic imagery that religion offers. It presents, as is well known, the unconscious aspects of the human mind, the id, with the right kind of elements on which to displace and express otherwise forbidden conflicts. And religion is often equally as welcome to other aspects of the human mind that is, to the ego,

> because of the multiple possible ego identifications made available and the innumerable possibilities of sublimation that it opens up. It has been said, and I believe rightly so, that Van Gogh's many unconscious identifications includes an identification with Christ. His behaviour with the miners in the Borinage district supports this view. His mother, like Christ's mother, was a true "Mater Dolorosa"; his father, "the handsome parson," from being a man of God, became God himself, the Father, the possessor of the "Mater Dolorosa." Who else could Vincent be then than Christ himself, how could he fail to fulfill his "Father's" purposes for him? (70)

In the end, Nagera's biography is a frustrating read because it has some interesting, worthwhile insights and yet continues to wander in-and-out of assumptions (stated as truths) that are obviously based on preconceived conclusions given very little support. Certainly, Vincent saw himself as Christ*like* or as wanting to be Christ*like* during this time period. And it is legitimate to claim he was idealizing his father, at least trying to place his father-as-minister into the role of a good man. In fact, it is plausible to claim that this role of minister representation of the father ended up causing or representing Vincent's struggle to both please and rebel against his father, a struggle that can be represented as Vincent's struggle between *religion* and *faith*.

Just how huge a door Nagera opens here might or might not be clear to him, or, in truth, might not be what he would claim, though he certainly is aware he has found a Freudian explanation. Vincent's ambivalence towards his father at this time is obvious, and strong. And Vincent most definitely blurs the reality of his father's religious beliefs and his own struggles to be completely sincere, to an extreme, a perfect Christ for God. Suffering and submission are essential. This cannot be a pretend form of religion or faith. Both failure and the fear of failure get intertwined. Vincent cannot succeed in the formal process. He must fail. At the same time, he must succeed in his failure. He must rewrite the rules, be a more perfect Christian than the Church itself. It works to see his life, certainly his time devoted to religion and faith from this perspective. He cannot win, because he must both win and lose at the same time. And he ends up realizing that formalized religion and faith become the two poles. From a Freudian perspective, he cannot give into the formalized religion, the establishment father, but must escape it in order to realize his own being. Still he must recognize this higher power, this God/father. Faith is the key. The Father both exists and has the highest qualities possible (in faith), but the formal Father, the established Father of religion who judges in this world (his own father in terms of family and society and theology) must be overcome in order to realize the *self*. Ultimately, he cannot please his own father, something he believes deep inside, but he can satisfy his father without his father realizing it, through faith.

Vincent's attempt to imitate or be as Christ, this father/son/ religion/faith conflict is evident throughout Christian history. Jesus Christ can be claimed to have been the most successful rebel against established religion and the ruling powers of all time. He went against both the powerful Roman Empire (at a time when others similar to him were being crucified daily for preaching against the government and the established religion) and the religious leaders of his own religion, Judaism, who condemned him and demanded Pilot crucify him. In other words, this man, with no power structure of any kind, boldly stood up to both the Roman Empire and the Judaic elders (the religious leaders of his own faith), and in the end became the most well known and influential rebel of all time. There is, of course, an irony here, because the faith of this man, which contains endless statements against the wealthy and powerful, the establishment, became an established religion, wealthy and powerful itself, and endless questions can be raised about whether faith and religion are compatible or, at least in some ways, opposites.

Joseph Campbell quotes Carl Jung: "'Religion is a defense against the experience of God." (71) The point here is that religion (theology) is the logical frame, the established set of rules and concepts, the kind of thinking associated with the left cerebrum, but faith is that which both admits and denies them, involving an experience of the mystery of life beyond explanations, the kind of knowing that takes place in the right cerebrum. An established religion can be the support system, can be a frame through which the mystery is experienced, but it also can be a barrier. This is the same for the entire realm of explanation, both needed and valuable for expression but also often the final barrier to it.

Rollo May discusses this concept at length in *The Courage to Create*. He writes:

> In these notes I shall explore the hypothesis that limits are not only unavoidable in human life, they are also valuable. I shall also discuss the phenomenon that *creativity itself requires limits,* for the creative act arises out of the struggle of human being with and against that which limits them.
>
> To begin with, there is the inescapable physical limitation of death. We can postpone our death slightly, but nevertheless each of us will die and at some future time unknown to and unpredictable by us. Sickness is another limit. When we overwork we get ill in one form or another. There are obvious neurological limits. If the blood stops flowing to the brain for as little as a couple of minutes, a stroke or some other kind of serious damage occurs. Despite the fact that we can improve our intelligence to some

> degree, it remains radically limited by our physical and emotional environment.
>
> There are also metaphysical limitations which are even more interesting. Each of us was born into a certain family in a certain country at a certain historical moment, all with no choice on our part. If we try to deny these facts—like Jay Gatsby in Fitzgerald's *The Great Gatsby*—we blind ourselves to reality and come to grief. True, we can surpass to some extent the limitations of our family backgrounds or our historical situation, but such transcendence can occur only to those who accept the fact of their limitation to begin with. (72)

The limits are the rules that result from forms, from structure. In the sense of religion, the limits are the forms put over the spiritual experience. Rollo May continues:

> *Conscious itself is born out of the awareness of these limits.* Human consciousness is the distinguishing feature of our existence; without limitations we would never have developed it. Consciousness is the awareness that emerges out of the dialectical tension between possibilities and limitations. Infants begin to be aware of limits when they experience the ball as different from themselves; mother is a limiting factor for them in that she does not feed them every time they cry for food. Through a multitude of such limiting experiences they learn to develop the capacity to differentiate themselves from others and from objects and to delay gratification. If there had been no limits, there would be no consciousness.
>
> Our discussion so far may seem, at first glance, to be discouraging, but not when we probe more deeply. It is not by accident that the Hebrew myth that marks the beginning of human consciousness, Adam and Eve in the Garden of Eden, is portrayed in the context of a rebellion. Consciousness is born in the struggle against a limit, called there a prohibition. Going beyond the limit set by Yahweh is then punished by the acquiring of other limits which operate inwardly. In the human being—anxiety, the feeling of alienation and guilt. But valuable qualities also come out of this experience of rebellion—the sense of personal responsibility and ultimately the possibility, born out of loneliness, of human love. Confronting limits for the human personality actually turns out to be *expansive*. *Limiting* and *expanding* thus go together.
>
> Alfred Adler proposed that civilization arose out of our physical limitations, or what Adler called inferiority. Tooth for tooth and claw for claw, men and women were inferior to the wild animals. In the struggle against these limitations for their survival, human beings evolved their intelligence.
>
> Heraclitus said, "Conflict is both king of all and father of all." He was referring to the theme I am here stating: conflict presupposes limits, and the struggle with limits is actually the source of creative productions. The limits are as necessary as those provided by the banks of a river, without

> which the water would be dispersed on the earth and there would be no river—that is, the river is constituted by the tension between the flowing water and the banks. Art in the same way requires limits as a necessary factor in its birth.
>
> Creativity arises out of the tension between spontaneity and limitations, the latter (like the river banks) forcing the spontaneity into the various forms which are essential to the work of art or poem. Again listen to Heraclitus: unwise people "do not understand how that which differs with itself is in agreement: harmony consists of opposing tension, like that of the bow and the lyre." In a discussion of how he composed his music, Duke Ellington explained that since his trumpet player could reach certain notes beautifully but not other notes, and the same with his trombonist, he had to write his music within those limits. "It's good to have limits," he remarked. (73)

May even usse Van Gogh's art as an example:

> Sooner or later the growing person's art must relate itself to the dialectic tension that comes out of confronting limits and is present in all forms of mature art. Michelangelo's writhing slaves; Van Gogh's fiercely twisting cypress trees; Cezanne's lovely yellow-green landscapes of southern France, reminding us of the freshness of eternal spring—these works have that spontaneity, but they also have the mature quality that comes from the absorption of tension. This makes them much more than "interesting"; it makes them great. The controlled and transcended tension present in the work of art is the result of the artists' successful struggle with and against limits. (74)

Certainly the case can be made that artistic form replaces religion for Vincent. He realizes that religion isn't going to work as a structure for his need to give his life meaning and value, and turns (or returns) to art for that frame. He is expressing his faith through his art.

Several biographies about him have stressed this. Meyer Shapiro in his book *Van Gogh* writes "that art was for him . . . a choice made for personal salvation. . . . a deeply lived means of spiritual deliverance or transformation of the self. . . . an alternative to older moral-religious man. . . . His career as an artist is a high religious-moral drama." (75) Julius Meier-Graefe, in his biography, *Vincent Van Gogh: A Bibliographical Study*, writes that he could just as well have titled the work *Van Gogh and God.* (76)

Albert J. Lubin writes: "The story of Vincent van Gogh is a never-ending struggle to control, modify, glorify, or deny a deep-rooted melancholy and loneliness. Religion and art were simply different means he employed for this purpose." (77)

Lubin also focuses on the father/son relationship, and offers a similar view to that of Nagera concerning Vincent's overly high assessment of his father (connections of his father to the Christian God as father) at this time:

> He decided to abandon the art business and follow in his father's footsteps as a minister. As one who helped others while looking down on them from the pulpit, perhaps he would feel less guilty, less inferior, and less ugly. And as one who related to people in a spirit of Christian love, his distrust and anger could be attenuated. His father became an idealized model in this venture; such a man was "more beautiful than the sea." Not even the greatest artist was permitted to outshine such a guide; his father's life resembled that of his favorite artists, Rembrandt and Millet, but was even ore valuable. By trying to be like this kind of father as well as his other religious antecedents, he could carry on their work. "As far as one can see," he wrote, ". . . there has always been someone in our family who preached the Gospel. . . . It is my fervent prayer and earnest desire that the spirit of my father and grandfather may also rest upon me . . . and that my life may resemble more and more [their] lives."
>
> Idealization of one's father is ubiquitous among children, and nowhere is this stronger than among ministers' children. The minister is a holy father endowed with magical qualities found in no one else. Set off from other parents by special dress and special demeanor, he carries on elaborate rituals, during which he becomes the intermediary between a mysterious God in heaven and the ordinary members of the congregation. Idealization is usually replaced with a more realistic appraisal, or turns into outright antagonism, during adolescence. This disillusioning experience, painful though it is, serves a useful function. Part of the price of independence, the youth creates a gulf between himself and those who tie him to his dependent childhood. This no doubt happened to Vincent. But now, in his mid-twenties, a melancholic hunger for love and a desperate search for a better life caused him to return once again to an early awe-inspiring image. (78)

Lubin's view supports the general context of Nagera's claims, though in much less depth. It might also make more sense to suggest that Vincent had not yet gone through the initial stage of separation, of disillusionment and was circling back, but rather that this time in his life was the struggle to both idolize and separate (not unusual for a rite of passage from the psychological dependency of a child to the psychologically independent adult).

Later on in his book, Lubin will also try to tie Vincent's embracement of religion (and/or faith) to his mother:

> Exercising a selective vision for Bible passages concerned with sorrow, suffering, and self-denial, he wrote, "Let us ask of Him that He teach us to deny ourselves, to take our cross every day and follow after Him; to be gentle, long-suffering and lowly of heart." [Letter 39b] Vincent had hoped to achieve a relationship with Ursula that had been denied him by his mother; when she failed him, he renounced worldly desires and displaced his frustrated longings to God. Through helping the unfortunate and directing his suffering into the service of God, he hoped that God in turn would make up for the failure.
>
> Vincent followed Christian tradition when he referred to God as the Father, but the image of God he portrayed in his letters (and in his art as well) was more often that of an idealized, loving mother who, through devoted care, brought joy to humanity. To obtain this love, however, earthly pleasures had to be disowned and suffering glorified. These ideas grew out of his unhappy childhood, which, coupled with his Calvinistic heritage, helped set the pattern for his personal understanding of The New Testament. A rambling letter, written at the height of his religious fervor, shows how God became Mother: "This Charity is Life in Christ, this charity is our Mother; all the good things of the earth belong to Her, for all is good if enjoyed with thankfulness, but She extends much further than those good things of the earth. To her belongs the draught of water from a brook on a hike or from a fountain in the hot streets of London or Paris, to Her belong also "I shall make thy bed in sickness," "as one whom his mother comforteth, so I will comfort you," and to Her belongs: "'Constancy unto death toward Christ, who giveth us the strength to do all.'" When the real mother has failed to do her job, the Mother God takes her place: "Can a woman forget her suckling child, that she should not have compassion on the son of her womb? Yea, they may forget, yet I will not forget thee.'" The Lord also takes the place of those later loves, who have disappointed him: "'All thy lovers have forgotten thee. I shall restore health unto thee, and take the plagues away from thee.'" He repeated for the second, and then for the third time: "'As one whom his Mother comforteth, so I will comfort you, saith the Lord.'" [Letter 82a] (79)

This insistent psycological discussion based on his early childhood relationships with his parents spreads into sometimes interesting, often frustrating, and even contradictory claims that tend to be confusing and miseading rather than helpful. Nagera, for example, intertwines his father/son analysis with other Freudian odds-and-ends. He can't help but highlight a standard representation of a retreat by Vincent into Freud's anal-masochistic stage, the wearing of dirty clothes, the argumentative attitude, the need for oral gratification (exemplified by taking up smoking a pipe), and so on. Such insights, in and of themselves, are well done Freudian analysis, even if a bit wandering and, though elaborated upon at different times throughout the text, tending to get in the way.

However, Nagera's sometimes insightful, though rambling Freudian annotating of the letters cannot let go of the initial Replacement Child Syndrom:

> We referred earlier to some psychoanalytic studies which demonstrated how the death of a child affected the family and the substitute brother. In Vincent's case the situation had been further complicated by the fact that he was conceived only three months after the death of his brother so that his mother's mental representation of him was developing while she was in deep sorrow and mourning for her first son. Here again the psycho-analysis of mothers in their second pregnancy whose first child was alive had shown the existence of innumerable conflicts of loyalty and of other types in relation to the first-born which had to be controlled and mastered if the relationship to the second child or to both was to develop within normal limits. These difficulties increase one hundredfold and may well become insuperable if it happens that the first has died as was the case here. From the insight gained from such studies one can attempt to reconstruct the psychological state of the painter's mother while she was pregnant with him. She must have felt the usual turmoil of inner contradictions, the usual conflicts of loyalty, particularly because all the hopes and plans for her first-born child had been shattered by fate. She may have felt at times that this new life was that of a usurper. Was this new child so soon to take the place in her heart of the dead one? Was he to fulfill the same expectations, phantasies and hopes that were aroused in relation to the first conceived? But how could he possibly do this, becoming somebody who was already dead, that he was not really meant to be? Was not this an intrusion in her pain, in the intimacy of feelings, of expectations, of hopes, of plans and longings between a mother and her first conceived child? And how could she think, hope and plan any differently for this second child at such a close range from the death of the first without feeling that she was insulting his memory and was in fact a traitor to it? Similar phantasies even if in a lesser degree must have haunted at least the unconscious mind of the father. It was against this conflictive background of emotional uncertainty, instability, conflictive loyalties, ambivalence, sadness and withdrawal that Vincent was to be born. Perhaps taking this into account, we will not be so surprised at his constant struggle to find an identity for himself in life, an identity in his art belonging only to him, a unique style, that, as he was to say would distinguish him from everybody else and would make his work recognizable even when unsigned. The unique identity he most certainly found through his art in the end. (80)

Again, though this body of theory is, at least at first, interesting, it is so speculative that it loses creditability. One might say yes, it *would be* an interesting possibility or group of possibilities, but it's hard to accept that it *could be* from the paucity of evidence, and from the fact that what little we have of his relationship to his parents as a child can more easily be interpreted as presenting a *positive* childhood, perhaps even a spoiled childhood than as consisting of some dark psychological, nightmarish drama.

At the very least, it is suspect to suggest that Vincent's *assumed* failed love affair with Ursula/Eugenie was the result of *assumed* psychological problems in childhood because of *assumed* psychological damage to both Vincent and his parents as a result of him having a brother born dead a year earlier. Interesting yes, but . . . !

Basing his views on those of Nagera, Marc Edo Tralbaut, in his thick, well respected biography, writes at length about Vincent's foiled love with Ursula (Eugenie) and how "it was suggested earlier [in his book] that Vincent might have suffered in his earliest childhood from an unsuspected mental depression; in which case Ursula's refusal may have aroused dormant neurotic tendencies, in particular an inferiority complex in regard to women." (81) and concludes this portion of Vincent's life: "Unless we consider it from this psychoanalytical point of view, it is hard to understand this complete change of course, in which Vincent abandoned a career that seemed so promising and embarked on an adventure into new and unknown territory." (82)

Lubin, in his psychological study of Vincent, *Stranger on the Earth*, refers to Johanna van Gogh-Bonger's *Memoir of Vincent van Gogh*, in his claims about the relationship with Eugenie: "Refused, he tried everything to convince her to give up her fiancé in favor of himself but failed. The extreme sensitivity to rejection that heretofore prevented him from approaching her now precipitated a prolonged state of melancholy. This experience, he wrote later, caused him 'many years of humiliation.'" (83)

Again, Lubin puts Vincent's study into terms of the Replacement Child Syndrome, again without solid support, but rather on assumptions of its correctness. The fact that Vincent never mentions the first brother once in his letters in-and-of-itself suggests such an assumption is probably wrong.

In truth, Lubin takes this whole concept much deeper, making it the central basis for his entire book. In Chapter 5, titled appropriately "The first Vincent and the sad mother," he puts forth the standard support and assumptions, beginning:

> Vincent's deep misery seems to have been present from his earliest years. "My youth," he wrote, "was gloomy and cold and sterile. . . ." Even then, as his sister Elizabeth noted, he was a stranger to his family, even as he was a stranger to the world later on. Although he sometimes fancied that his childhood was happy, second thoughts caused him to suspect that he only imagined it. [letters 347, 388a]

> He had a theory about the origins of his chronic unhappiness, and he voiced it in a language that came easily to him—the pictorial language of nature: "The germinating seed must not be exposed to a frosty wind—that was the case with me in the beginning." [letter 341] He was saying, in effect, that during his earliest years he was deprived of those ingredients that comprise the mysterious entity called mother-love: the freely-given, cuddling, cooing, nourishing, protecting, reassuring behavior of a maternal figure. He was like a scraggy, stunted, deformed plant, constantly struggling for its life because it had been neglected by an unfriendly Mother Nature when it was beginning to root and grow. (84)

These are the identical passages Nagera uses to support the same conclusion, and it's not surprising, since it's clear Lubin is well acquainted with Nagera's book. But again they are taken out of context (seem to have been taken directly from Nagera's book) and if they do suggest the possibilities of a sad childhood, they most certainly do not prove it, nor, when taken in context, do they stand up well against the numerous positive statements by Vincent about his childhood and his obviously strong desires to recapture it throughout his life. Furthermore, the passages are to be found in Vincent's endless attempts to convince his brother he is right in the dispute he is having at the time with his family (literally the entire family) after his strange actions towards Kee Vos and subsequent relationship with Sien. The truth is that, if one considers the overall context of these comments, they suggest strong family ties, ties that are important to Vincent, ties that make it hard for him to break away and establish his own identity.

Fortunately, Lubin does suggest this context as well, immediately going into a superficial explanation of the process of "self-differentiation" that takes place between a child and a mother, and how it might be negatively affected if the mother is not under emotional distress. Throughout his discussion, Lubin is even-handed enough to mention, even stress that little is really known about Vincent's childhood and any psychological conclusions based on it can only be conjecture. However, he follows this with the theory of the dead child's possible affects on both mother and son, and gives a few quotes from a few psychiatrists about such a possibility, but again rightfully points out how any claims about "Vincent's unconscious fixation concerning the circumstances of his earliest years may not have been based on fact." (85) Making the necessary admission that "His mother may not have been depressed and unloving, and her mind may not have constantly dwelled on the first Vincent during the second's early years." Rather, Lubin points out, "The idea could have been reconstructed retrospectively in order to rationalize later childhood conflicts that cause him to hate his mother and, in turn, feel hated by her." (86)

While reading through these passages of Lubin, I found myself constantly flip-flopping from saying "Yes, I'm pleased you're seeing the contexts of your claims" to saying "Wait, where did that claim come from." The bold claims here that Vincent hated his mother and the ones Lubin attaches to them in other passages that she hated him are far from obvious.

Once Lubin has considered this mother/son relationship, stressing the likelihood that it was a failure, he applies it to Vincent's other failed relationships with women. Certainly Vincent's relationships with women didn't work out well. However, Lubin's same mixture of convincing truths gets undermined by numerious unsupported assumptions. For example, in the middle of his discussions of Vincent's early love interests, Lubin writes, "In falling in love with Ursula and Kee, Vincent was choosing partners who had committed their love to someone else, a situation reminiscent of his mother's attachment to her bereaved son." (87) The unsubstantiated assumption here are so intrusive that is seems patronizing to go into them, but in an attempt to not duplicate Lubin's flaws, I want to highlight a few of them. First, there is little proof that he fell in love with Ursula (Eugenie). Second, if he did, the standard view at the time of Lubin's book (and still today) was that he did not know that she was already in love with and promised to another man (in which case Lubin's theory collapses in terms of her as an example). Second, a similar mixture of conjectures are evident in such claims about Vincent's relationship to Kee, who supposedly again was alreadly taken, perhaps even more suspect, since these are based on the idea that she was still committed to her dead husband. Even if that were true, it does not lead necessarily to him purposely reliving an impossible relationship with his mother. More likely, he thought he could win her away from this attachment, or his empathy was the main attraction (a kind of empathy that he exhibited throughout his life).

Lubin, however, takes these initial assumptions to: "As an intruder into these loves, Vincent re-experienced the childhood situation. When he complained that Kee's love for her husband caused her to bury herself in the past and that her guilt made it impossible to transfer this love to another person, he might have been paraphrasing an explanation of his mother's attachment to the first Vincent. And when he fantasized that he was united with Kee "forever and ever," he was repeating an unfulfilled childhood wish to be united with a loving mother." (88)

Then, trying to tie all together, Lubin turns Vincent's relationship with Sien upside down. "As an adult with Sien, however, he was the rescuer, not the helpless, unloved, ignored child. The hopelessness that followed his failure with two women who represented the 'good' aspects of his religious mother provoked him to turn to a social outcast." (89) Here, I become even more frustrated, because, to a certain extent I agree, but the larger Replacement Child Syndrome confuses it all. It is possible that Vincent's embracement of Sien was both the result of his natural empathy for those in suffering and need (similar to his embracement of the poor coal miners and others throughout his life, including his own parents) and a rebellion against his parents' condemnation of his actions towards Kee. If he saw his love for Kee within the context of an empathy for her suffering following her husband's death (and it is likely that to some extent he did), and he also saw this compassion in the context of Christianity (which he probably also did), then on one level it would represent him living up to the values of his parents and the Christian community; it could be seen by him as an attempt at being a good Christian, though, of course, it's far more complicated than this, especially in terms of his previous conflicts between religion and faith. In this sense, then, he might well, on some level, have thought he was being a good person to pursue Kee, at very least might have used it as a justification for his seemingly fanatical pursuit of her. This being the case, then, when his actions were condemned by everyone, Kee, Kee's family, Vincent's own mother and father, and even his confidant Theo, he might well have rebelled, have basically said, "Well, the Hell with it! I've had it with all of you! I've tried and tried and tried to please you! Now I'm going to show you that you can no longer tell me what to do!" And, at the same time as he is likely to have had an emotional meltdown (something the records suggest he was prone to do, and the circumstances suggest he was likely to do) he was also wanting a woman, wanting not only a loving kind of relationship, but wanting sex.

Sien fit all the requirements here. He could not have found a woman more opposite of his family's Christian views. She was as down-and-out as one could get, and she was as sexual a woman as one could get. It is very easy to suggest that Vincent thought at the time, "Well if I am to be damned, then I'll be damned!"

Lubin, however, is not going to let go of the Replacement Child Syndrome:

> Sien also reminded Vincent of his childhood nurse. "Do you remember our old nurse a Zundert, Leen Veerman?" he asked Theo. "If my memory does not deceive me, Sien is that kind of person." [Letter 201] Both the

> nurse and the prostitute were depreciated, paid substitutes, with analogous functions. During childhood, a nurse is paid to give the child the care a mother would otherwise give freely. During adulthood, a prostitute is paid to give the sexual satisfaction that a wife would otherwise give freely. Vincent learned to accept a second-best substitute early in life, even though he longed for the original. As an adult, his longing for a mothering wife encouraged him to repeat an unhappy childhood experience with the forlorn hope of transforming it into a happy experience. When he failed with a woman who resembled his mother, he turned to one who resembled her paid substitute. (90)

The problems are obvious here. First, having a nurse or a nanny was common (for various groups, still is) and in no way suggests his mother was neglecting him or that she was where he found love as a child. Second, what comparison Lubin wants to make between Sien and the nurse is unclear, and there is little reason to believe Vincent saw his nurse as someone in need of love and compassion as Lubin has already claimed he saw Sien. Also, to compare giving sexual satisfaction with giving love and care is a dangerous claim. Then to turn it all upside down by claiming that he was looking for a mothering wife (rather then he being the care-giver suggested earlier) suggests Lubin is simply reaching for clever comparisons and suggestions.

Lubin briefly claims that by the time Vincent had his final relationship with a woman, the suggested one with Margot, "he had accepted his fate as a man who could not even succeed in having a permanent relationship with a wretched stand-in. Margot, not he, was the one who longed to be married." (91) This final relationship is a hard one to prove, and what little support there is suggests it was simply an unspoken desire on her part.

Then Lubin sums up this entire discussion by making the very claim he previously couched in terms of a possible but perhaps unlikely possibility:

> The contrasting themes in this art—sorrow and joy, isolation and togetherness, death and rebirth, darkness and light, earth and heaven—arose, I suggest, out of the persistent but buried memory of his childhood. He was the unhappy outside, ignored and rejected by a grieving mother who affection lay with the dead brother who was buried in the earth but had ascended into heaven. In defense, the second Vincent developed an envious identification with the first, made almost inevitable by the mysterious coincidence of name and birthdate. In the fantasy of his artistic life, he alternated between depicting the depressed, unloved outsider, living in darkness, whose salvation lay in death, and the adored child, reborn on the earth or ascended into the light of heaven. (92)

Lubin then goes on to point out that most of the women in Vincent's art are "marked by sadness and tragedy," (93) giving a few examples, before he returns to his by now assumed truth:

> Vincent perceived himself as a burden his depressed mother was forced to tolerate. His unsatisfied craving for maternal care continued to make him feel like a burden for the rest of his life, and he was forever divided between attempts to satisfy this craving and contradictory attempts to free himself from it or to take on other people's burdens. After he became an artist, his brother Theo took his mother's place as the one who tolerated him, and Vincent repeatedly accused himself of being a burden to his brother." (94)

Most psychological interpretations, including Nagera, suggest Theo as a substitute father, not mother.

Extending his claims, Lubin writes:

> It has been found that psychiatric patients who have been reared as replacements for a dead child tend, like Vincent, to be preoccupied with death, illness, and body-mutilating accidents. Like Vincent, too, they are apt to believe that they will die at an early age and have an inordinate interest in cemeteries. (95)

Lubin then gives many examples of Vincent's art that fit into these macabre categories. But Lubin has more support, claiming that even Vincent's writings "hint that he retained an image of his dead brother." (96) And he gives the following passage as proof. In it, Vincent is explaining to Theo that a portrait of Alfred Bruyas by Delacroix, painted in 1853, the year of Vincent's birth, looked like he and Theo, so Vincent wrote "one must boldly believe that what is is, and Delacroix's portrait of Brais [sic] resembles you and me as if he were a new brother." (97) Hard to connect this to Vincent's dead brother.

Ken Wilkie in his engaging biography *In Search of Van Gogh* contains the following mixture of autobiography and biography:

> The gaunt house where Vincent was born stands directly opposite the old town hall in the village square. I walked the short distance from the manse to the churchyard—the route Vincent must have followed every Sunday as a child.
>
> Near the graveyard gate I stumbled on a little gravestone, much smaller than the rest. Inscribed on it was the name *Vincent van Gogh.* This was the Van Gogh's family's first child, who was stillborn on March 30, 1852. Their second, whom they also called Vincent, was born a year—to the day—after his baby brother.

Here I was, at the beginning of my journey, beside a forgotten gravestone, already contemplating life and death. What a profound effect it must have had on Vincent every Sunday, as he passed that tombstone with his name engraved on it. A constant confrontation with the idea of death.

When he revisited this cemetery in 1877, at the age of 24, Vincent described the sunrise he witnessed there as "reminiscent of the Resurrection." It made me think that the initial grief suffered by Vincent's mother following the death of Vincent I must have persisted as a state of melancholy into the early years of Vincent II.

The warmth and intimacy of a happy mother's loving care certainly appear to have been absent from Vincent II's childhood. He was described as introverted, self-willed, intelligent, difficult, melancholy, not like other children, extraordinarily serious. Deprived of the love his mother could not give him, the foundation for his later depressive tendencies may have been laid at that time.

In simple terms, death became synonymous with being loved and cherished, while being alive was identified with rejection. Did the contrasting themes in his art—sorrow and joy, isolation and togetherness, death and rebirth, darkness and light, earth and heaven—have their roots in the buried memory of his childhood?

As Albert and Barbara Cain pointed out in a relevant article, "On Replacing a child," in the *Journal of the American Academy of Child Psychiatry*, people who have been reared as replacements for a dead child tend, like Vincent, to be preoccupied with death, illness and self-mutilation. Like Vincent, too, they tend to believe that they will die at an early age and have an obsessive interest in places of burial. Cemeteries were the goal of many of Vincent's walks, and from his letters we see that he often chose graveyards as rendezvous of reunion. He seemed to regard graveyards not as repositories for decaying corpses but as beautiful places where living things grow out of the ground [a note: this seems to contradict the negative thrust being put forward here].

Vincent's unhappy alliances with women continued throughout his life and they often ended in disaster. Were the seeds of his malaise before me at this grave of his stillborn namesake? I wondered who his first love had been. This must surely have had a crucial effect on him. Had he ever experienced human warmth and affection from a woman? From childhood, was he in the grip of irreversible alienation?

As I walked past the willows of the graveyard, a wave of sadness mingled with the anticipation I felt at the start of my journey. In spite of the bizarre adventures to come, it was a sadness that would deepen as I came closer to the person whom fate had led me to follow. (98)

Later, as Wilkie is getting into Vincent's relationship with Ursula and Eugenie Loyer, he continues to stress this negative relationship between Vincent and his mother:

> During his early years he had been deprived of all the elements of mother-love. The deficiency in this relationship created an impairment which impeded the boy's development into a stable character. Indeed, he felt inferior, lonely, unloved and hypersensitive. The threat of not being loved caused him great anxiety, while rejection resulted in depression that continued into adulthood. Moreover, as Albert Lubin pointed out in one of the post perceptive character studies ever written about Vincent, *Stranger on the Earth*, "the combination of low self-esteem and high expectation is most apt to produce adults who are highly susceptible to depressed states." (99)

Completely convinced Vincent had this love affair, Wilkie takes the reader on his own exciting journey of rediscovery of the Loyer's house, and even an assumed drawing of it by Vincent found in some old, old photographs of Enid Dove-Meadows, a surviving daughter of Eugenie, who is surprised to hear of Wilkie's suggestion her mother was once proposed to by Vincent, suggesting her mother was much too strict and "severe" a woman to allow such a thing.

Wilkie's account is one of the more interesting and valuable, partially because it involves visiting the sites and actively seeking out the people who might have known or had parents who knew Vincent. It is, in fact, Wilkie who sorts out the factual reasons for why there is so much confusion about Ursula and Eugenie as the proposed object of Vincent's love, involving mixed up names in correspondence between Vincent and his sister Anna. However, even with his research, it still remains speculative, guessing that they "accidentally" mixed up the names.

Nevertheless, even such a seemingly careful and sincere researcher simply assumes the romantic love without much proof of it, stating:

> Eugenie's rejection of Vincent caused him intense suffering that was rooted in the melancholic atmosphere and emotional deprivation of his childhood. It precipitated a prolonged state of depression and a four-month gap in his published correspondence. It is presumed he did write during this time but for some reason these letters have not come to light. (100)

Unfortunately, Wilkie is making a number of his own presumptions here beyond this. Undoubtedly, in spite of his own research, he is being influenced by Lubin, whom he quotes, and by Tralbaut, who it has already been noted was influenced by Nagera. Wilkie certainly considers Tralbaut THE biographer of Vincent, and discusses his pleasure at having the chance to visit with and seek leads from Tralbaut in his own biography. The more the various biographies are studied, the more it is obvious the

the early ones are strongly influencing those that followed, and their claims go from guesses to truths.

Derek Fell's book *Van Gogh's Women: Vincent's Love Affairs and Journey Into Madness* also focuses on the Replacement Child Syndrome, specifically applying the views of Dr. Decker, a psychologist who specializes in family counseling and served as the psychological consultant for the book, and some articles written about this concept, mainly "The Ghost in the Mother: Strange Attractors and Impossible Mourning," by Edward Dmery (101)

As he discusses Vincent's forbidden love for Kee, he immediately puts the standard "out of context"quotes in place, along with the standard assumptions about his mother's sorrow over her previous miscarriage and her poor treatment of Vincent as a result. It is worth quoting this portion of the book to get a full view of the claims and the means to be employed supporting them. After a summary of Vincent's affair with Kee and his "thirty-five pages of letters in which he asserts and explores his love for Kee" which are "so farcical in their expressions of profound despair and absurd optimism that they are worthy of Shakespeare at his most comical and romantic," Fell contradicts himself, suggesting a deeper psychological interpretation. It begins with an excerpt from one of these letters:

> Let this my weakness be my strength. I will be dependent on "her, and no other": even if I could, I should not want to be independent of her. But she has loved another and her thoughts are always in the past: and her conscience seems to bother her even at the thought of a possible new love. But there is a saying, and you know it, "one must have loved, then unloved, then loved again."
>
> I saw that she was always thinking of the past and buried herself in it with devotion. Then I thought, though I respect the feeling and though that deep grief of hers touches and moves me, yet I think there is some fatalism in it.
>
> So it must not weaken my heart, but I must be resolute and firm, like a steel blade. [November 7, 1881.] (102)

Then comes the psychological speculation:

> The word *mother* might be substituted for all reference to Kee in this passage, as Lawrence Decker, Ph.D, clinical psychologist, points out to illustrate Vincent's perception of Kee as a mother substitute. "*I will be dependent on her and not other*" recalls Vincent's psychological history of living with an emotionally remote mother. *"But she loved another and her thoughts are always in the past"* could apply as much to his mother's chronic mourning over a dead child as to Kee's widowly bereavement. *"I saw that she was always thinking of the past and buried herself in it with*

> *devotion"* reads like a subconscious evocation of his mother's pious withdrawal into the past, while *"That deep grief of hers touches and moves me, yet I think there is a fatalism in it"* echoes his mother's long grieving over her dead infant son. These allusions provide the first clear clue to the deep hurt Vincent experienced in his mother's grief over the death of her firstborn child—a boy she had named Vincent—and to the profundity of pain he felt as a "replacement child," to use the term preferred by psychologists. Indeed, he was even named after the dead child. Vincent constantly craved to be loved for himself, yet he bore his dead brother's name—a circumstance that did not enable him to ease his mother's mourning or to gain her love for himself. Vincent must have suffered early from a feeling of abandonment that few of us could ever imagine, and his extreme sensitivity to emotional rejection appears to have exacerbated the innate personality disorders that would eventually drive him to the depths of despair. (103)

As emphasized earlier, there is no proof and little evidence that Vincent's mother suffered this Replacement Child Syndrome, that she was distant and in mourning, or that she took out her sorrow on Vincent. Nor is there any proof or evidence that Vincent suffered from it. Instead, the words and lines used to give the proof are taken out of context and misrepresented. The obvious claim to make in terms of Vincent's relationship with Kee is that he wanted to comfort her, rather than that he wanted her to comfort him, and though these particular lines can be interpreted as him wanting a mother to be dependant upon, in the context of all of his letters, it is clear he does not mean the word dependent in this manner. More accurate would be to say he is juxtaposing dependent with independent in the sense of being connected to, promised to, and related to, rather than the opposite. And the constant attempts to prove there was some strange psychological situation between Vincent and his dead brother because he was given the same name are hard to support; rather it is most likely he was give the same name because it was a commen name throughout his family and was most likely given both his dead born brother and him because it's the name his parents wanted their first son to have. In truth, it makes more sense to believe they had put their dead son out of their lives, and as such had no problem using this name again. Suggestions that they gave Vincent the name as some kind of punishment or sick reminder of his dead brother suggest his seemingly normal parents were really psychotic beneath what is known about them. Also, the suggestion that Vincent was gloomy and always thinking about this dead brother because he was buried in a nearby graveyard make little sense. A baby does not even register this. And once he is old enough to even realize it, there is no reason for him to feel some morbid connection to a child he never knew (unless his parents were beating it into him, a claim for which there is no support and what evidence there is suggests was not the case, except for the brief "out-of-context" passages that "might" be referring to it).

Fell continues:

> Vincent described his childhood as "gloomy and cold and sterile." He wrote that his mother did not love him enough; that Theo was the one who comforted her and the one she found worthy to be comforted by her. (104)

This is it, the main support for a claim that Vincent had an unhappy childhood. Of course, it is taken out of context, especially out of the huge context of all of his letters, where there are many passages that can be referanced suggesting he had a happy childhood, including him feeling lonely and sad whenever he had to leave home, constantly returning home, and wanting to recreate his childhood. The fact that some phrases written at times he was trying to defend himself against his parents' concerns suggest they didn't always agree on everything is not indicative of an overwhelming childhood conflict. As Fell pointed out, much of Vincent's writing in these exchanges can be seen as comical justification for his actions. In other words, they fall more into the category of a young man acting like a young man trying to justify doing some things that his parents believe are not in his best interests—typical adolescent rebellion from parents. Next Fell goes into his experts' discussion of just what the psychosis is:

> Anne Stiles Wylie, writing about Vincent's infancy in the bulletin of the Van Gogh Museum in 1975, explained that the psychological factors put in motion by this so-called Replacement Child Syndrome can assume a variety of dynamic position. "The mother's guilt may be displaced on the new baby, or he may assume the melancholy into which a mother's mourning is often transformed," states Wylie. "From his first hours the new baby may be challenged to live up to unrealistic standards of behavior and success: dirtying his diapers is hardly part of a mother's fantasy of how the dead child would have pleased her had he live." Whatever the weight of Vincent's psychological burden, Wylie concludes that the first Vincent's grave was a daily symbolic reminder to his namesake of the special forces that had shaped, and would continue to shape, his destiny.
>
> Dr. Decker summarizes the problem at Vincent's core as follows: "The tragedy of Vincent is that he was shrouded under the force this dead brother exerted on him and not free to become his own person. In fact, he was in an impossible situation, for whenever he tried to be himself, he would betray the perfect image of his deceased brother, for whom he was nominated by his mother to replace. Poor Vincent felt condemned to be some approximation of himself and not the object of his mother's true

> affection. Vincent could never be good enough in his own right, and would always sense that something was wrong with him that couldn't be identified. There was an impossible longing to know himself, which could never happen because he was carrying around the ghost of his brother. He was condemned to be the longed-for other." (105)

An interesting theory. By now it should be obvious there is little to back it up. But Fell is not yet through with his expert's views:

> Decker identifies a second major problem for Vincent: "His early life set up an impossible longing for the kind of unconditional and boundless love that only a mother could provide. He would forever be incapable of awakening the part of his mother who was lost in a mournful swoon over her beloved stillborn child. He could never repair the damage. He would always be the exiled third, the endless witness to others' successes. He would always choose impossible love objects, and idolize them." (106)

Not only, once again, is there the claim that he experienced this replacement child syndrome, but added to it is the claim he would always choose impossible love objects and idolize them. Fell continues:

> A third problem for Vincent, according to Decker, was born of his impossible longing for unconditional love. "Rejected over and over again, Vincent tried to come to grips with the anger he felt whenever threatened with abandonment. He was sensitized to this, of course, by not having a chance to be nurtured unconditionally by his mother. His acts of violence were all in response to threatened rejection, and were the internalization of aggression felt towards those who threatened him. He roasted his hand over a flame, cut off his ear, and shot himself in the stomach, all following anger at the rejection he received from people he loved." (107)

Perhaps his acts of violence were in response to a felt rejection from people he loved, but taking it back to the Replacement Child Syndrome undermines a better approach to this. Nevertheless, this is where Fell wants to take it:

> Edward Emery, Ph.D., training and supervising analyst at Westchester Institute for Training in Psychoanalysis and Psychotherapy, in a paper entitled *The Ghost of the Mother: Strange Attractors and Impossible Mourning,* wrote that "ghostin" by the mother—treating a second child as a replacement—damages the replacement child by burdening it with the mother's *impossible mourning,* her refusal to let go of the loss of the first child, and *impossible longing,* her desire that the second child replace her loss. Vincent's mother, Anna, became pregnant with him barely three months after delivering a stillborn child, her first, and Vincent was born

> exactly one year after her dead baby. She saw Vincent as the second coming, the incarnation of the dead child she idolized and preserved in her subconscious.
>
> Anna probably became disappointed in Vincent around the time he turned two, when his demands as a normal, healthy child would have spoiled her fantasy of Vincent as the embodiment of her perfectly imagined firstborn. As Vincent failed to ease the pain of her loss, she descended deeper into depression and left young Vincent confused about his identity. Abandoned by his mother as she withdrew from him into melancholia, Vincent could neither redeem her nor esteem himself. He could only sense his failure to assuage her impossible mourning and to fulfill her impossible longing. His rejection by his mother appears to have engendered his lifelong quest for the perfect, nurturing love that he ardently desired. (108)

Fell's theoretical set-up for his book continues in this manner, but I am going to stop the explication of it here, as my comments have already become redundant and anything more is unnecessary.

Fell offers similar weak interpretations of passages in Vincent's letters to support theories about Vincent's desires for a mother substitute:

> In a letter to his sister Anna, dated January 6, 1874, written following Eugenie's rejection and shortly before he left the Loyer household, Vincent reveals that he love *both* women. Anna wanted to work in England, and he urged her to visit the Loyers to see for herself that they bore him no grudge, that they were a respectable family and not a household that Vincent's father considered "full of secrets." Vincent said "I'll say nothing more than that I never heard or dreamed of anything like the love between her and her mother," and he cautioned Anna not to put the wrong meaning on his affections for both women, "to read nothing more than that he valued them as friends." He obviously envied their close relationship and the emotional support they afforded each other, an element that was still missing from his own life. Subconsciously, it seems, his love for Eugenie's mother was more important to him than his love for Eugenie, and so Eugenie's rejection was a crushing disappointment because it thwarted Vincent's desire for Michelet's woman in black. It was undoubtedly this lost hope for an idealistic mother substitute in the person of Mrs. Loyer that damaged his personality the most, and made him all the more introverted and argumentative with customers at the London branch of art dealers, Goupil & Company. (109)

These claims are easy to turn upside down. Why is it not likely that he compared the relationship between Ursula and Eugenia to a similar feeling about his own family that he now missed, since he had to live away from home, and it is obvious he missed his home life, instead of jumping to the conclusion that they represented some happy family life that Vincent had never had but deeply desired? And where do these lines that actually stress not to make stupid assumptions suggest that he loved the mother more than the daughter? And why would Eugenie's rejection be seen by him as a rejection by her mother? Whose love was more important to him than hers? On and on and on.

Fell is aware of, and refers to Lubin in the book, though he is apparently unaware of Nagera, as he misrepresents Lubin as being the first to suggest the Replacement Child Syndrome concept, though it is clear Lubin has taken it from Nagera. In truth, Fell is little more than the latest reincarnation of their theories, the most recent version of a psychological approach working hard to apply a rather strict Freudian interpretation of Vincent based on negative childhood experiences. Nagera was the first to apply the Replacement Child Syndrome theories. Lubin repeated Nagera's theories, and then Fell updated them, though the update of this particular portion of the book is weaker than the original discussion by Nagera. The Replacement Child Syndrome theories remain the weakest portion of a sequence of psychological books that have other, defensible insights.

Interestingly, Fell's book, after setting itself up with the beginning materials on Replacement Child Syndrome following its suggestive title, especially its suggestive subtitle, *Van Gogh's Women: Vincent's Love Affairs and Journey into Madness*, seldom refers back to these theories throughout the rest of the book. And, in many ways, that is also the situation for both Nagera and Lubin. They put the theories in place, and then simply move on. Nagera interweaves them the most throughout his book, and Lubin returns to them enough to keep them a part of his total discussion. Fell only briefly mentions them again after his initial discussion, which is mainly in reference to the failed love affair with Kee. In fact, Fell's title and initial set-up are misleading. While it is true that he focuses on Vincent's letters with his mother Anna, his sister Wil, and his sister-in-law Jo, while he does discuss Vincent's love affairs, and while he does deal with Vincent's "journey into madness," it's hard to claim these all come together or that any of them really is the center. The main title *Vincent's Women* only generally encompasses the materials in the book, and the suggestions that somehow everything is brought together in the dual subtitle are misleading. Just as Nagera and Lubin did, Fell wanders off into other matters and offers more of a general biography than a focused interpretation. In the case of all three, some specific discussions of Vincent's works of art in the context of psychology are the best portions. Nagera's discussion of Vincent's two paintings of chairs representing him and Paul Gauguin is especially interesting, though the suggested psychological reasoning behind it is suspect, and one might wonder if Vincent wasn't consciously playing a symbolic game rather than unconsciously expressing latent desires. Lubin's analysis of *The Potato Eaters* also works well on the level of symbolic criticism, though it is similarly flawed in its reference to the underlying psychological assumptions about Vincent in the Replacement Child Syndrome theories.

Jan Hulsker, one of the most respected biographers of Vincent, gives a straight-forward commentary on this:

> Vincent Willem van Gogh was born on 30 March 1853 in the Brabant village of Zundert, near the Belgian border. He was the eldest child of Theodorus van Gogh, the village parson, and Anna Cornelia Carbentus. A curious coincidence about the date was that exactly one year earlier, on 30 March, 1852, his mother had given birth to a still-born child. This child was buried in Zundert in the graveyard on the side of the small Protestant church. It has often been assumed that this fact made an indelible impression on Vincent. Jo van Gogh's son, Vincent W. van Gogh, who edited a revised edition of Vincent's letters in 1952 to 1954 titled *Verzamelde brieven van Bincent van Gogh*, wrote in the last volume of this edition, "Near the entrance to the little church at Zundert there is the grave of the little boy who was born precisely a year before Vincent, and who was also called Vincent; he lived for only a few weeks. Until he left his parent's home, Vincent saw this little grave at least once a week when he went to church, but he saw it too when he came home on weekends, holidays, etc. Besides, it is certain that he heard the little boy mentioned continually. This must have made a deep impression on the baby as well as on the growing child." It was probably this passage that induced many authors to elaborate on the influence this death had on Vincent's personality. Psychoanalysts have often stressed how detrimental it can be for the emotional development of a person to be a "replacement child." [he refers specifically to Humberto Nagera, Albert Lubin and Viviane Forrester in a footnote here] However, the facts in this passage are not altogether correct. The idea that the first child had lived for a few weeks was an error. The child was still-born, as is confirmed by the birth certificate and by Jo van Gogh-Bonger in her 1914 edition of the letters. In any case, it is doubtful that a young village boy would be overly affected by the sight of a gravestone with which he had been acquainted for years. It is also doubtful that in a parson's family, with all its numerous "taboos," the still-born first child would have been mentioned often, let alone "continually." Infantile deaths were surely not a rare phenomenon in nineteenth-century Holland. (110)

It can work to interpret art from a Freudian psychological, symbolic perspective—can thicken the meanings of the art tremendously. And it can work to interpret a person's thoughts and actions from a Freudian psychoanalytic perspective, can help to clarify how and why people behave in the ways they do. However, there are many dangers. Humberto Nagera, Albert J. Lubin, Derek Fell, and other Freudian biographers of Van Gogh, while they include valuable insights, also fall prey to some of the dangers inherent in a discipline that must guess at human behavor, even if the guesses are sophisticaed, educated guesses. At worst these psychologists get so excited about their claims and the wonderful symbolic maps of meaning they're able to suggest that they lose perspective and get lost in their own landscapes. In the case of Van Gogh, they have especially gotten blinded by the Replacement Child Syndrome, and, as Hulsker points out, even their beginning assumptions that the first Vincent was born alive are wrong. Unfortunately, that causes them to lose creditability for some of their more insightful claims.

Van Gogh, The Life, a more recent biography by Steven Naifeh and Gregory White Smith (Random House, 2011) dismisses this entire Replacement Child Syndrome discussion: "The coincidence that Vincent Willem van Gogh was born a year to the day after the stillborn buried under a marker inscribed 'Vincent van Gogh' would prove of far greater interest to later commentators than to the Van Goghs." (p. 23)

However, in doing this, it doesn't dismiss some similar psychological guesswork. Rather, it puts a strong frame immediately in place by emphasizing the mental make-up of Vincent's mother Anna:

> Anna grew up with a dark and fearful view of life. Everywhere forces threatened to cast the family back into the chaos from which it had just recently emerged, as suddenly and finally as the sea swallowing up a village. The result was a childhood hedged by fear and fatalism: by a sense that both life and happiness were precarious, and there fore could not be trusted. By her own telling, Anna's world was "a place full of troubles and worries [that] are inherent to it"; a place where "disappointments will never cease" and only the foolish "make heavy demands" on life. Instead, one must simply "learn to endure," she said, "realize that no one is perfect," that "there are always imperfections in the fulfillment of one's wishes," and that people must be loved "despite their shortcomings." Human nature especially was too chaotic to be trusted, forever in danger of running amok. "If we could do whatever we wanted," she warned her children, "unharmed, unseen, untroubled—wouldn't we stray further and further from the right path?" (p. 16)

This passage sounds a lot like the ones those earlier biographers were writing, filled with psychological analysis based on scattered phrases from her writings that can easily be interpreted in far different ways then they are. It is also crouched in a brief overview of the social, political, religious world she grew up in, agaon written in a flourid, dramtic style that makes for a more dynamic read, but that undermines how seriously it can be accepted. Here is but a brief example: "But neither war nor these paroxysms of communal rage posed the greatest danger to the Carbentus family. Like many of his countrymen, Gerrit Carbentus lived his entire life on the edge of extinction by flood. . . . It was a precarious life, and Gerrit Carbentus, like all his countrymen, inherited an acute sense, a sailor's sense, of the imminence of disaster." (p. 14) Not only can this brief passage be amplified, but the pages surrounding it are, if anything, even more caught up in the larger-than-life dangers that generations of Anna's family lived through, all resulting in her having a humorless, dark vision of life. If this biography is to be believed, she married out of disparation of ending up a spinster, and accepted what was viewed by her and her family as a husband and situation beneath her, had it not been that she was getting old and needed a spouse. In the end, what drove her was a desire to be "respectable" and "normal," which meant "conforming to the conventions of her class." (p. 25) Thus, "Anna stamped this fearful, insular view on all her children. Neither physical nor affectionate by nature, she waged instead a relentless campaign of words: affirmations of family ties, invocations of filial duty, professions of parental love, and reminders of parental sacrifice, endlessly knitted into the fabric of everyday life," a form of "family totalitarianism." (p. 25)

As a result, Vincent (and all of her children) "suffered the pain of withdrawal" (p. 25) when they were separated from her. This view has already been mixed in with the privious theories, some supporting it, more claiming the opposite, all with the same scattered quotes from the letters used as support, and none offering strong evidence or psychological theory. It's clear that Vincent's relationship with his mother was more fractuous than his syblings, yet when Naifeh and Smith try to explain this, instead of offering convincing insights, they fall into a discussion of how Vincent and his mother had similar looks and traits (again adding in her own artistic qualities—for which there is little more than that "for a while, she *may* (my emphasis) have kept up her own amateur artwork," and that "her daughters learned to play the piano," and "everyone took singing lessons." If this is the best support there is, the claims should be changed to a more honest one that there is little to demonstrate her having anything beyond an average interest in the arts (especially since a woman wanting to maintain a "normalcy" and "respectablilty" in her family, especially one married to a minister) would very likely have her children taking both piano and singing lessons. And, of course, there is the more certain fact that she showed no interest in Vincent's art, was actually dismissive of it.

As they push forward in their stress that Vincent was much the same as his mother and that "Vincent's attachment to his mother was profound," (p. 37) they fall even further into the same over-the-top claims of previous biographers, now, instead of the more common claims about Vincent's relationship with his father, wanting to put Vincent's personality into his relationship with his mother: "Despite this special attachment, or perhaps because of its inevitable disappointment, Vincent hardened into an obstreperous, ill-tempered child." (p. 37)

One thing they mention in passing that should be brought more into a psychological analysis is that Anne's father Willem is thought to have died of a mental disease, and her sister Clara was considered to be an "epileptic." Certainly such mental instability in Vincent's family heritage figures into his own mental make-up, and professionals in both psychology and neurology will to consider it in attempts to unfold his own psychological picture.

Part IV

Sigmund Freud

Whereas the fields of neurology and psychology have scattered beginnings, psychoanalysis has a more definite parent. However, that doesn't mean it's a more precise beginning, because, for good or ill, this theorist is as enigmatic and difficult to sort through as the multiple strands of neurology and psychology, and his theories, though in many ways far more imaginative and enticing, also remain less provable, because both their strength and their weakness result from the fact that his science is less scientific, which, in turn, allows his concepts to range beyond the limitations of the more traditionally rigorous science of neurology (and more scientifically based forms of psychology) and still wait for science to catch up, wait for it to refine and redefine its truths, because, curiously, his ideas are of the very right brain (right lobe of the cerebrum) world of truths that the left brained (left lobe of the cerebrum) world of science is still struggling to rephrase in its own language.

Perry Meisel concludes his discussion of Sigmund Freud, "Just as Shakespeare uses traditions at will in a mingled discourse that appeals to countless regimens while submitting, in the end, to none in particular, so Sigmund Freud contaminates science with literature, literature with science, to produce a prose-poetry whose only real boundaries are those of his own imagination… And just as John Milton chooses the most authoritative of anterior myths in a gamble to assert his priority over the past, so Sigmund Freud chooses for his equivalent purposes the most authoritative of anterior nineteenth-century myths, the myth of science." (111)

Lest it be overlooked, it is worth noting that, long before he developed his psychoanalytical theories, Freud studied neurological theories, and even wrote a draft for a paper titled "Project for a Scientific Psychology," where he offered a beginning discussion of a neurological explanation for neurotic behavior. (112)

Freud even admits that in the very case studies he begins to build his theories upon "the neurotic symptoms were not related to actual events but to wishful *phantasies*, and that as far as the neurosis was concerned psychical reality was of more importance than material reality." (113) In other words, Freud's deepest insight, which he never completely understood himself, is that the truths of the mind are of two worlds, the visible and the invisible, both valid, but, if anything, the more real, more important, and more powerful are those of the invisible world.

In *The Interpretation of Dreams*, he postulated a psyche consisting of the conscious, the unconscious (holder of deep, hidden memories, and unacceptable thoughts and feelings), and the preconscious (which is, in actuality, a part of the unconscious). Initially, all psychic thoughts, feelings, and desires exist as impulses in the unconscious. These thoughts, feelings, and desires must first pass through the preconscious, a link between unconscious and conscious bounded by a censor on each side. If either censor deems the material unfit, it is either repressed or transformed (though, in either case, it will still influence the conscious mind). This preconscious, then, gives artistic works a position as the transformed expressions of the deeper meanings buried in the unconscious. Freud called it a topographical approach to mental processes. (114)

After struggling with it for some time, he later shifted his emphasis to what he called the structural theory, and, while not denying the earlier landscape, now moved most of the ego and super-ego into the unconscious with the id. The id (completely in the unconscious) is pure energy, pure instinct, the reservoir of the libido (basic instinctual, sexual drive, in more noble terms, the drive to life), both destructive and creative, amoral; the superego (in unconscious and preconscious) is pure conscience, the church, the police force, the parent; and the ego (in all three parts of the psyche) is the "I" within, the conscious sense of self, and acts as a mediator or compromiser, working to satisfy both the drives of the id and the censoring of the superego. (115)

At first, it all seems rather simple, if difficult to prove. But there is some proof in the very case studies the theories are derived from, and even if the details are slightly off, the concepts "feel" right, if only they can be formulated more clearly, and tested, and refined.

And if ultimately their truth is more of the world of the arts, just maybe it is possible to postulate that the truths of this world are more true than the truths of the scientific world, are, in fact, more defensible than science as non-myth. Certainly during a time when mathematical systems are being judged by their aesthetic qualities it is possible to speculate that an interpretation of the psyche with a strong aesthetic footing has some merit. And intuitively, aesthetically and artistically, the ideas run deep and rich.

"Thoughts of Freud, of course," Franz Kafka writes in reference to his nightmarish story "The Judgment." (116)

It should be of no surprise that literature embraced Freud. Leonard Woolf wrote, "Whether one believes in his theories or not, one is forced to admit that he writes with a great subtlety of mind . . . a broad and sweeping imagination more characteristic of the poet than the scientist or the medical practitioner." (117)

Upon the publication of Freud's *Beyond the Pleasure Principle* and *Group Psychology and the Analysis of the Ego*, John Crowe Ransom pointed out that Freud accommodates both biological and mythical truths, thus avoiding both a "rampant vitalism" and a "sociological determinism":

> Freudian psychology, if it keeps that name, will be far more than one man's work before it is completed. It will be like a mediaeval Gothic cathedral, for whole generations of scholars will have helped to put it together; and we could delimit offhand a dozen or so separate fields of labor, such as ethnology, biology, comparative religions, primitivism, language, the "lost knowledge" of symbols, the biography of genius and poetry. And when the grand edifice is completed, the result will be a complexity and yet a unification of doctrine, perhaps as imposing a structure as the world has seen. (118)

In fact, Freud was acknowledged for his literary abilities officially in 1930 with the award of the annual Goethe Prize in Frankfurt, an award he said was "the climax" of his life as a citizen (119), crediting Goethe's "beautiful essay on Nature read aloud" as the decisive moment in his own decision to become a medical student instead of a lawyer. (120) And it is not surprising that Freud should be so connected to Johann Wolfgang von Goethe, as Freud's own writing has strong foundations in the literary world from Greek drama (the Oedipus complex) to William Shakespeare (and Hamlet's struggles).

Perhaps it was Thomas Mann who was the first to connect Freud directly with the literary world of Romanticism and the Romantic quest, linking him directly to the "romantic-biologic fantasies" of Novalis (Georg Philipp Friedrich Freiherr von Hardenberg), the philosophic writings of Friedrich Nietzsche, the moral attitude of Arthur Schopenhauer, the psychological extremes of Soren Kierkegaard, and the moral love for the ideal of Henrik Ibsen. Finally, and at some length, Thomas Mann linked his own writings to Sigmund Freud:

> Indeed, it would be too much to say that I came to psychoanalysis. It came to me. Through the friendly interest that some younger workers in the field had shown in my work, from *Little Herr Friedemann* to *Death in Venice*, *The Magic Mountain*, and the Joseph novels, it gave me to understand that in my way I "belonged"; it made me aware, as probably behoved it, of my own latent, preconscious sympathies; and when I began to occupy myself with the literature of psychoanalysis I recognized, arrayed in the ideas and the language of scientific exactitude, much that had long been familiar to me through my youthful mental experience. (121)

In making this connection, Thomas Mann pushed Freud's id and the unconscious into the prime position, the driving force, the irrational, instinctual, primitive energy that has the power, and the ego is beset with having to struggle to keep it under control. Mann didn't continue into this extreme form of romanticism but rather turned to the connection of Freud with myth, finding Freud's struggle of the ego with the world of the id, the unconscious, not a tragedy, not a romantic battle of ego versus id, but rather a mythic union, partaking of the eternal, connecting with the same patterns that have been encountered by countless humans before. Myth, then, is the naturalistic, eternal expression of eternal human rhythms. It is the cultural representation of instinct, speaking directly to man's biological core. He wrote:

> The mythical interest is as native to psychoanalysis as the psychological interest is to all creative writing. Its penetration into the childhood of the individual soul is at the same time a penetration into the childhood of mankind, into the primitive and mythical. Freud has told us that for him all natural science, medicine, and psychotherapy were a lifelong journey round and back to the early passion of his youth for the history of mankind, for the origins of religion and morality—an interest which at the height of his career broke out to such magnificent effect in *Totem and Taboo*. The word Tiefenpsychologie ("deep" psychology) has a temporal significance; the primitive foundations of the human soul are likewise primitive time, they are those profound time-sources where the myth has its home and shapes the primeval norms and forms of life. For the myth is the foundation of life; it is the timeless schema, the pious formula into which

> life flows when it reproduces its traits out of the unconscious. Certainly, when a writer has acquired the habit of regarding life as mythical and typical there comes a curious heightening of his artist temper, a new refreshment to his perceiving and shaping powers, which otherwise occurs much later in life; for while in the life of the human race the mythical is an early and primitive stare, in the life of the individual it is a late and mature one. What is gained is an insight into the higher truth depicted in the actual; a smiling knowledge of the eternal, the ever-being and authentic; a knowledge of the schema in which and according to which the supposed individual lives, unaware, in his naïve belief in himself as unique in space and time, of the extent to which his life is but formula and repetition and his path marked out for him by those who trod it before him." (122)

It is worth noting that these connections with eternal human rhythms and repetitive motifs connect Thomas Mann with Carl Jung's theories of archetypes as much as with Freud's more personalized symbols. (123) Others at the time, however, focused in more on the need to separate or moderate the instinctual prominence. As a result of the heightened discussions of the importance of the id's instinctual drives underlying all actions, especially creative and destructive ones, by the 1930s it had become popular to equate the artist, the madman, and the neurotic, and artists especially wanted to clarify some distinctions. W. H. Auden, "Psychology and Art To-Day," attempted to separate the neurotic and the poet by saying that the neurotic gets stuck in his nightmare world, but the poet finds a way back out of it, mainly based on the his "ability to mould his particular material." He claimed that even a dream has a bit of poetry in it, as it has a "balance" of interests, and is perhaps even "moral." The instinctual or "automatic element" is but the beginning of the work of art; it is the "inspiration," but the real work of art still needs the labor of poetry, the fashioning and crafting of the poet. (124)

W. H. Auden also moved away from "symbols" as the basic building blocks and instead used "words." This distinction became one of the central areas of confusion and discussion.

Edmund Burke offered a solution to the division represented by Thomas Mann and W. H. Auden, placing the symbolic act in context, and suggesting that Sigmund Freud himself struggled with "a distinction between . . . an essentializing mode of interpretation and a mode that stresses proportion of ingredients." On the first of these, the essentializing mode, Edmund Burke pointed out how Sigmund Freud tended to emphasize the sexual manifestation as the cause of most of a man's motivations, just as any scientist would work toward a simplification of a problem by placing one aspect at the center and looking for the others as lesser variations of that. On the second aspect, the proportion of ingredients, he pointed out that Freud did consider the context, i.e., symbols in psychoanalysis depend on the context, and vary in different contexts. Freud, of course, was aware of this, and while at times claiming certain meanings for certain symbols, also stressed that each patient had his own spin on these symbols, was unique.

Edmund Burke then applied this concept to poetry, pointing out how each poet's symbol system varies from others, and it is necessary to note the context of the symbols, the images in use to arrive at the deeper meanings. Thus, Edmund Burke claimed, it is at the level of the dream, not word or myth or symbol, that Freud comes closest to mapping out the logic of poetry through two functions of a dream that Sigmund Freud calls "condensation" and "displacement," which are the same as "metaphor" and "metonymy." Thus, the dream and the semantics of language become the same, and "something can be only in so far as it is not something else." And the something else is ever-changing. (125)

Lionel Trilling worked to combine Thomas Mann's views with Edmund Burke's. In "Art and Neurosis," Lionel Trilling embraced the idea incumbent in Sigmund Freud's biological determinations of human consciousness as a way out of the trap of culture determination. He wrote, "It is so far from being a reactionary idea that it is actually a liberating idea. It proposes to us that culture is not all-powerful. It suggests that there is a residue of human quality beyond the reach of cultural control, and that this residue of human quality, elemental as it may be, serves to bring culture itself under criticism and keeps it from being absolute." (126)

In attempting to clarify the problems of "reality" in the biological versus cultural determination of man, he, in the end, placed both in the poetic and social reality rather than the scientific and universal reality, claiming that the reality that Freud appeals to is "the reality of social life and of value, conceived and maintained by the human mind and will. Love, morality, honor, esteem—these are the components of a created reality. If we are to call art an illusion then we must call most of the activities and satisfactions of the ego illusions: Sigmund Freud, of course, has no desire to call them that."

Then he connected art to the thinking process, saying that, just as there is no one correct meaning for any work of art, there is no one correct meaning for a thought. This allowed him to state that "of all mental systems, the Freudian psychology is the one which makes poetry indigenous to the very constitution of the mind." Thus, he said,

> Freud has not merely naturalized poetry; he has discovered its status as a pioneer settler, and he sees it as a method of thought. Often enough he tries to show how, as a method of thought, it is unreliable and ineffective

> for conquering reality; yet he himself is forced to use it in the very shaping of his own science, as when he speaks of the topography of the mind and tells us with a kind of defiant apology that the metaphors of space relationship which he is using are really most inexact since the mind is not a thing of space at all, but that there is no other way of conceiving the difficult idea except by metaphor. In the eighteenth century Vico spoke of the metaphorical, imagistic language of the early stages of culture; it was left to Freud to discover how, in a scientific age, we still feel and think in figurative formations, and to create, what psycho-analysis is, a science of tropes, of metaphor and its variants, synecdoche and metonymy." (127)

These ideas are, at bottom, leading to a duality of meaning in psychic thoughts, one based on universal meanings, and one based on meanings in context. Just as the literal meanings of words are but one level of meaning, the universal meanings of symbols are but one level of meaning. It exists. However, at the same time, the context influences this meaning. Words in context take on different, and, at times, even opposite meanings. For example, the word "bad" can actually mean "extra good" in certain contexts, i.e., "He's a really bad dude," or "When she's good she's good, but when she's bad she's better." This contextual meaning does not deny the original, universal meaning, but adds to it, builds on it, gives is a new perspective. In fact, if the original meaning were removed, the contextual meaning would be less, and maybe even meaningless.

Julia Kristeva put forth a theory of meaning she calls "semanalysis," based on a division of language into semiotic (a feminine side rooted in a child's preverbal, pre-Oedipal stage and connected to the "chora," the intuitive both physical and psychic bond of a child with its mother, the irrational, sensual, feminine qualities, the unconscious), and the symbolic (the male side, associated with conscious order and reason). She then claimed that the semiotic has been devalued and repressed, only to surface in dreams and poetic language, the language of myths, ritual and the arts (also in puns, rhyme, rhythm, and *jouissance* (playfulness)). Each individual experiences a tension between these two aspects of language. It is an interesting idea, and one, that, if the feminist slant is removed (she claims not to be a feminist), resolves easily into both psychological and neurological models of the dual forms of human thinking. (128)

Jacques Derrida attempted to separate the workings of the unconscious from both Thomas Mann's concept that it connects up with a mythic expression of instinctual drives and needs, and a view of Sigmund Freud as claiming that it is an "ideal" language, a "nonphonetic writing," (i.e., ideograms or hieroglyphs). Rather, said Jacques Derrida, Sigmund Freud should be seen as giving us an unconscious that consists of "memory traces," a form of psychic writing that results from some sort of psychic trauma that sets the psyche in motion, a traumatic moment when it starts to resist stimuli, when it experiences some form of stimulation and resists it.

Jacques Derrida, then, said that Sigmund Freud searches to explain how the psyche both continues to receive new impressions (in the conscious) and yet still store them away (in the unconscious), and that Freud finally comes to the solution in his vision of memory, those traces or pathways and their intensity. In other words, it is in context that both memory and language work.

Thus, Jacques Derrida theorized, there is no fixed meaning, but, rather, an ever-changing psychic map of meaning, a continual response or resistance to stimulation. Thus, thought and language both are deferred responses. He found Sigmund Freud coming to a solution of the new thought/stored memory conflict in what Jacques Derrida called a "writing machine," a two sided writing-pad that both erases and writes at the same moment. At the same time this allows for the needed protection from oneself, the idea that the unconscious does not always let itself be known by the conscious. (129)

Sigmund Freud offered an overview of this in his *Autobiographical Study*:

> How had it come about that the patients had forgotten so many of the facts of their external and internal lives but could nevertheless recollect them if a particular technique was applied? Observation supplied an exhaustive answer to these questions. Everything that had been forgotten had in some way or other been painful; it had been either alarming or disagreeable or shameful by the standards of the subject's personality. The thought arose spontaneously: that was precisely why it had been forgotten, i.e., why it had not remained conscious. In order to make it conscious again in spite of this, it was necessary to overcome something that fought against one in the patient; it was necessary to make an expenditure of effort on one's own part in order to compel and subdue it. . . . The expenditure of force on the part of the physician was evidently the measure of a *resistance* on the part of the patient. It was only necessary to translate into words what I myself had observed, and I was in possession of the theory of *repression*. . . . The ego drew back, as it were, after the first shock of its conflict with the objectionable impulse; it debarred the impulse from access to consciousness and to direct motor discharge, but at the same time the impulse retained its full charge of energy. I named this process *repression*. . . . the repressed impulse, which was now *unconscious*, was able to find means of discharge and of substitutive gratification by circuitous routes and thus to bring the whole purpose of the repression to nothing. . . . The theory of repression became the foundationstone of our understanding of the neuroses. . . . I showed my recognition of the new

> situation by no longer calling my method of investigation and treatment catharsis but *psychoanalysis.* (130)

This, then, Jacques Derrida said, is the difference between literary writing and positivistic science, where Sigmund Freud uses the literal language of science to demonstrate its real position as metaphor, as literary language in its own right. Thus, says Jacques Derrida, Sigmund Freud actually recreates the working of the psychic in his own writing, because, just as the psyche cannot comprehend its own beginnings (before the initial repression), writing cannot uncover its own beginnings. Jacques Derrida then expressed Sigmund Freud's suggestion that in the beginning there can only have been repression, what Sigmund Freud called "primal repression" (131); the idea being that the initial exposure to stimuli from the external world sets the repression in place. However, since logic requires both repression and stimuli at the same time, neither can have come first, and thus, we are driven back to the realization that both the language of the psyche and the language of science are in truth literary languages.

Harold Bloom brought Jacques Derrida's claims to a clearer focus (132), pointing out that, if, as Jacques Derrida claimed, Sigmund Freud's very language is the subject of the psychic writing it describes, then we have reached a tautology, and Sigmund Freud is, in the end, a poet, since he "cannot invoke the trope of the Unconscious as though he were doing more (or less) than the poet or critic does by invoking the trope of the Imagination, or than the theologian does by invoking the trope of the Divine."

Harold Bloom mainly focused on Sigmund Freud's later writings, especially on "Beyond the Pleasure Principle," and attempted to clarify Sigmund Freud's own struggles with "anxiety over stimuli," where individuals are purposely creating an anxiety (or fear) in order to gain mastery over it. This original anxiety, then, is what motivates the "primal repression," the "theoretical fiction" that sets the ego in motion. Thus, the original repression, the primal repression, comes from a created anxiety, and it, in turn, creates a force, so the resistance has a force to resist. This force, then, is the drive, which is put into place after the fact, as a scenario by which the consciousness can imagine its beginnings as a sudden shock, as the instant drive surprised it. In other words, drive emerges as a resistance to its elimination. It is, then, in an attempt to avoid stimuli, that the ego establishes means of defense against them, and it does this through "tropes," and, thus, the ego creates the very thing it is trying to avoid, the drives. Thus, drives are fictions. And the drives are not only the psyche's originating fictions, they are also the enabling fictions for Sigmund Freud's own writings.

Thus, Sigmund Freud's structure of the psyche and his own language are a match. Just as there is the delayed emergence of the drive in the rhetoric of the psyche (i.e., the unconscious), there is a delayed effect in Sigmund Freud's own writing, an inferred division of the psyche where the unconscious affects our thoughts and actions behind the scenes, only to be known to consciousness by inference, not by direct knowledge. Primal repression, then, is the central Freudian trope, since it is the model for literary reference itself, retroactively creating the reference, through rhetoric, which language establishes outside of itself, just as primal repression creates the drive retroactively as a sudden shock to begin one's life.

Furthermore, if the psychic and literary texts are the same, then the psyche should provide the answers to why we are pushed to create. Harold Bloom suggests that, if the reason for the repetition-compulsion is to "master a stimulus retroactively by first developing the anxiety," then the desire to create is also affected by the threat of "anteriority," some previous force that hangs over the artist as a rearguard catastrophe, which Harold Bloom linked to a literary notion of the *sublime*, the creative moment that he equated with an encounter with a previous creative moment of another, a moment of negation involving the realization of a previous writer of equal power. Harold Bloom, then, equated Sigmund Freud's drive as his own earlier achievement, an achievement now looming behind him that he must defend himself against by revising his entire theory of drives. Now, if the force of drive actually threatens the very emergence of the self in the beginning, then drive must also be associated with death. However, this seems, at first, a contradiction, as the drive is Eros, that drive to life of the libido. Sigmund Freud theorized the death drive, a drive beyond the pleasure principle, and ended up with Eros uniting with repression to struggle against the death drive. Harold Bloom then identified the death instinct with literal meaning and the life instinct with figurative meaning. Here the bond of Eros and repression represents Sigmund Freud's sublime moment, his self-conscious rhetorical achievement, knowing without denial that his drives are fictions, products of his own rhetorical devices, his own figurative language, and Eros is drive as fiction, poetic figure bound to culture because it is literary invention. In contrast, the looming, earlier Sigmund Freud meant himself literally, scientifically, measurable by science, Thanatos, and in disavowing himself of his earlier literal connection of libidinal or biological energy for the drive, he reached the *sublime* moment of negation of his previous writings, the moment of defending his previous self by disowning it.

In these later writings, Sigmund Freud explicitly disavowed his earlier presumption of the psyche functioning through an energy that can be measured by science, to a theory of psychic relationships that can only be measured proportionally, the earlier a literal language of empiricism, the later a figurative language of poetry. Thus, the later Sigmund Freud presented truths that are not bound to external reference but only to their own inter-relationships. He became the poetic of figurative truths by a purposeful misrepresentation of science.

Perry Meisel writes:

> Though Freud's language swerves, often wildly, from the regularities of literary and scientific discourse alike, each swerve is nonetheless lawful from the point of view of the other—what is literary is precisely that which cannot be vouchsafed in the name of science, and vice versa. After all, the trope of biology, for example, in a late visionary work like *Beyond the Pleasure Principle*, stands out as a poetic figure only at the moment it transgresses what biology as a science is privileged to say, that among the instincts there is one that wishes for death. The boundaries of poetry and science, in other words, are in each case an effect of the violation of one by the other. Freud's double language of science and vision, then, is an apparatus or machine, to use Derrida's vocabulary, that allows Freud to employ the rhetoric of each tradition even as it simultaneously frees him from the obligation to stay bound by either one. . . . (133)

It is the later Sigmund Freud, the one who has moved fully into meta-science, who writes:

> In France the interest in psychoanalysis began among men of letters. To understand this, it must be borne in mind that from the time of the writing of *The Interpretation of Dreams* psychoanalysis ceased to be a purely medical subject. Between its appearance in Germany and in France lies the history of its numerous applications to departments of literature and of aesthetics, to the history of religions and to pre-history, to mythology, to folklore, to education, and so on. . . .
>
> A number of suggestions came to me out of the Oedipus complex, the ubiquity of which gradually dawned on me. The poet's choice, or his invention, of such a terrible subject seemed puzzling; and so too did the overwhelming effect of its dramatic treatment, and the general nature of such tragedies of destiny. But all of this became intelligible when one realized that a universal law of mental life had here been captured in all its emotional significance. Fate and the oracle were no more than materializations of an internal necessity; and the fact of the hero sinning without his knowledge and against his intentions was evidently a right

expression of the *unconscious* nature of his criminal tendencies. From understanding this tragedy of destiny it was only a step further to understanding a tragedy of character—*Hamlet*, which had been admired for three hundred years without its meaning being discovered or its author's motives guessed. It could scarcely be a chance that this neurotic creation of the poet should have broken down, like his numberless fellows in the real world, at the Oedipus complex; for Hamlet was faced with the task of taking vengeance upon another for the two deeds which are the subject of the Oedipus desires, and before that task his arm was paralyzed by his own obscure sense of guilt. Shakespeare wrote *Hamlet* very soon after his father's death [in a note at the bottom of the page, Freud claims to no longer wish to emphasize this point]. The suggestions made by me for the analysis of this tragedy were fully worked out later on by Ernest Jones. And the same example was afterwards used by Otto Rank as the starting-point for his investigation of the choice of material made by dramatists. In his large volume upon the incest theme he was able to show how often imaginative writers have taken as their subject the themes of the Oedipus situation, and traced in different literatures of the world the way in which the material has been transformed, modified, and softened.

It was tempting to go on from there to an attempt at an analysis of poetic and artistic creation in general. The realm of imagination was evidently a 'sanctuary' made during the painful transition from the pleasure principle to the reality principle in order to provide a substitute for the gratification of instincts which had to be given up in real life. The artist, like the neurotic, had withdrawn from an unsatisfying reality into this world of imagination; but, unlike the neurotic, he knew how to find a way back from it and once more to get a firm foothold in reality. His creations, works of art, were the imaginary gratifications of unconscious wishes, just as dreams are; and like them they were in the nature of compromises, since they too were forced to avoid any open conflict with the forces of repression. But they differed from the asocial, narcissistic products of dreaming in that they were calculated to arouse interest in other people and were able to evoke and to gratify the same unconscious wishes in them too. Besides this, they made use of the perceptual pleasure of formal beauty as what I have called an 'incitement-premium'. What psychoanalysis was able to do was to take the inter-relations between the impressions of the artist's life, his chance experiences, and his works, and from them to construct his constitution and the impulses at work in it—that is to say, that part of him which he shared with all men. With this aim in view, for instance, I made Leonardo da Vinci the subject of a study, which is based upon a single memory of childhood related by him and which aims chiefly at explaining his picture of 'St. Anne with the Virgin and Child'. It does not appear that the enjoyment of a work of art is spoiled by the knowledge gained from such an analysis. The layman may perhaps expect too much from analysis in this respect, for it must be admitted that it throws no light upon the two problems which probably interest him the

> most. It can do nothing towards elucidating the nature of the artistic gift, nor can it explain the means by which the artist works—artistic technique.
>
> I was able to show from a short story by W. Jensen called *Gradiva*, which has no particular merit in itself, that invented dreams can be interpreted in the same way as real ones and that the unconscious mechanisms familiar to us in the 'dream-work' are thus also operative in the processes of imaginative writing. (134)

It becomes obvious that Freud was aware of what he was doing, or, at least later in life, of what he had done; and if we can grasp his genius, we can see that, in creating the unconscious, the preconscious, and the conscious, he gave us a powerful landscape for the mapping of meaning and value in human existence. And he even began to populate this world, uneasy with his characters of ego, id, and super-ego, yet sensing they were at least the beginnings of important entities in this as yet uncharted or incompletely charted landscape.

The key is the unconscious, a place where meanings beyond the literal, physical world dominate, where the truths of the metaphor and the symbol can be explored, where connotative and intuitive truths are expressed and embraced. These are the truths beyond explanation, beyond logic, beyond reason, beyond math and science, the truths that finally reach into the world of the spirit, the truths of the heart, where logic and reason often are simply unable to function. And without these truths life is meaningless.

In the preconscious, Sigmund Freud gave us the gate-keeper, the artist within each of us, capable of giving the truths of the unconscious form and shape, of transforming the dream into a physical entity (an art work), of placing a pattern, a map of meaning a value over the wild landscape of the unconscious, capable of taking the logic of the conscious world and using it as a beginning template to be shaped and bent to the needs of the dream. It is the place where the world of logic comes together with the world of imagination and faith.

One aspect Sigmund Freud perhaps did not realize, at least not strongly enough, is that the passage through the preconscious is a two way street, the conscious supplying the unconscious the needed raw data to be molded and shaped by the powerful tools of figurative meanings.

Only then can they return to the preconscious to pass to the conscious, and the ever changing mapping of meaning and value continue through its endless circle, or, perhaps, spiral, for as long as the human mind remains alive.

After we study the complex logic of the Freudian critics (positive and negative), what we realize is that they cannot finally reduce his ideas to a fixed logic, but actually end up suggesting by the very great efforts of great minds failing to reduce it that it must be beyond logic. Actually, this is Sigmund Freud's greatness, that he is giving support for a huge, meaningful world of the mind beyond logic. Whether his specific "logics" and geography of that world is the best is ultimately not the point at all.

Rather, ultimately, it comes down to language (symbol systems) and thought being the same, and that there are several languages (art forms, mathematical systems, etc.). Language can be separated into literal aspects and figurative aspects (the literal understood by the left hemisphere, the figurative by the right). The figurative aspects open up the world of meaning and knowing beyond explanation, the world of meaning beyond the literal, the world of meaning of the metaphor. This is where the "mystery" of life resides, where meaning and value exist. If there is only literal reality, then there is no meaning or value in existence. Then it is all reducible to math and science, and human choice and self-responsibility do not exist.

Thus, while the discussions are interesting and spark endless possible debates, they also demonstrate how Freud has opened a door to a world of meaning and value that ultimately is beyond logic and explanation. This world is the world of metaphor and pun and art, and in the end, of faith. It is, finally, the same world as that opened up by neurologists, that right hemisphere of the brain, where we sense and feel and know without being able to explain and prove it. It is the world where we cry at a movie or laugh at a pun. It is the world of meaning and value. And it is the center of all human existence. For humans are the givers of meaning and value.

Part V

The Religious Context

Marc Edo Tralbaut in one of the better biographies of Vincent writes:

> If we read between the lines of the uninterrupted confession which Vincent's letters provide, we find that his life was governed by various processes of repression, and that, moreover, fantasies of guilt caused by the death of the first child frequently appear right up to the end of his life. This feeling of guilt showed itself for the last time, at a subconscious level, in the suicide which liberated him from the world, and indeed was one of the motives for it. (135)

Erwin van Meekeren, in his interesting study *Starry starry night: Life and psychiatric history of Vincent van Gogh*, claims that Vincent not only had the rejected love experience with Eugenie, but wrote about it in future letters, stating "He later called this declaration of love 'senseless' since someone who 'surrenders to his feelings catches too much wind and his boat will ship water.'" (136) Unfortunately, Erwin van Meekeren doesn't give the needed information to find this statement and confirm its existence and context. Further along he states that Vincent "later confessed his clumsiness, calling this love 'senseless, wrong, overdone, arrogant, audacious,' since he only wanted to 'give, but not receive (…). Who deviates from this, on the right or left side, he will fall (…) and it was a wonder that I recovered from it.' (Letter 181)" (137) The context of the quote again is not included or discussed, and it is even difficult to know if the quote is meant to be referring to Eugenie or Vincent's later failed affair with Kee.

In fairness to Erwin van Meekeren, or in a demonstration of contradictions in his claims, he later admits to the paucity of evidence for this, stating: "Vincent seemed to go downhill from the moment he declared his love, in vain, to his landlady's daughter. The connection is only speculation in view of the lack of data." (138)

While such assumptions and at least seeming contradictions hurt his study, Van Meekeren opens a legitimate door into a psychological study of a man who certainly seemed to exist in some twilight zone between genius and insanity. If not completely avoiding the temptation to a Freudian interpretation of Vincent's childhood based on very little evidence, he does keep it to a minimum:

> At first sight it appears that there was little wrong with Vincent up until the age of 20. He was a good student and model employee, and no major calamities or negative experiences were reported. The main conspicuous points are his Protestant descent and his unfinished school education. Vincent grew up in a small Protestant community living in a predominantly Catholic area. The lack of information about why he left the 'HBS' and stayed at home for over a year forms a remarkable gap in his history. It is not known how he spent his time. Could this have been the first sign of mental instability? It is striking how the lives of the Van Gogh children played out: two suicides at a young age (Vincent and Cor), a son who died young from a venereal disease (Theo), a daughter who had a child out of wedlock (Lies), insanity (Vincent, Theo and Wil), being unmarried (Vincent, Wil) and marrying late or for a very short period (Theo, Cor). It may, of course, be bad luck or randomness but it could have been related to the atmosphere in the Van Gogh residence. One can imagine emotionally cold surroundings in which religious and bourgeois morality prevailed. The starting points were good schooling, looking neat and tidy, conforming and doing one's duty rather than striving for one's own dreams and desires, particularly if such dreams did not match the

> 'status' of a minister's family. Relatively little is known about how Vincent experienced his childhood." (139)

Nevertheless, as Erwin van Meekeren points out, directly following his experience in the Loyer residency, Vincent has a dramatic change. Fortunately, while Meekeren assumes the failed love relationship, he downplays it: "Although much has been written about this affair and many psychological interpretations have been given, little is really known about it." (140)

Let's take a different perspective on Vincent's childhood and all of the elaborate attempts to explain his adult psychological problems by some form of Freudian interpretation. First, the views suggesting he had an unhappy childhood have little to support them. This form of reverse psychology that assumes he must have had an unhappy childhood because he was psychologically unbalanced as an adult puts the assumption before the facts and causes those assuming it to bring an unreasoned perspective to their studies. The main theory they have put forth is the Replacement Child Syndrome, a theory that in-and-of itself is suspect. Whenever I encounter it, I can't help but think of how Bruno Bettelheim's career came tumbling down because of his theories about Refrigerator Moms in connection with autism. (141) Furthermore, even if the general theory has some merit, there is little to support that it applies to this particular situation, little reason to suspect either of Vincent's parents was cold toward him because of the death of a previous child. In truth, what evidence remains, mainly the letters, suggests that they went out-of-their-way to try and help him, that the conflicts between them and him later in life were much, much more the result of his actions than theirs. Furthermore, there is also little to support claims he had a sad childhood in any sense. While some passages in his letters refer to negative experiences as a child, especially during the time he had a falling out with his father, there are no indications of any terrible hardships or punishments he suffered as a child, and there are more references to happy aspects (not to mention the obvious dislike of having to be away from home and family as an adult and the desire to recreate what he once had as a child in later life).

Ronald de Leeuw summarizes this more even-handed view:

> Little is known of the young Vincent other than that he was a rather trying, sometimes troublesome boy—probably because his mother tended to spoil her children—and that he loved animals and flowers. Vincent and Theo kept each other company a good deal and their childhood against the background of 'the wheat fields, the heath and the pine forests, in that

> peculiarly intense atmosphere of a village 'parsonage' was later described by Johanna van Gogh-Bonger as a poetic age, a Brabant idyll. (142)

Ingo F. Walter and Rainer Metzger suggest that Vincent's childhood not only was not unhappy and cold but that it was so friendly and comfortable his later violence was a reaction against this overly idyllic world:

> His family lived a quiet life in the modest vicarage at Zundert near Breda, in Dutch Brabant. Theodore's father had been a pastor too; indeed, so had generations of the van Goghs. They were not strict Calvinists in belief, but adherents of the Groniger party, a liberal branch of the Dutch Reformed Church. Vincent was profoundly influenced by the hard-working and pious atmosphere of his parental home; and the eruptive violence with which he expressed himself strikes us as a necessary strategy for ridding himself of the cozy image of the world that was imposed on him in childhood. . . . The childhood and youth of the siblings seem to have been much what we would expect in a petit bourgeois household, and later, after a fit of madness, Vincent was to long for the unruffled happiness of his Zundert home: "During my illness I saw every room in the Zundert house, every path, every plant in the garden, the surroundings, the fields, the neighbours, the graveyard, the church, our kitchen garden at the back—down to the magpie nest in a tall acacia in the graveyard.: (Letter 573). In that last year of his life, longing for his childhood home never relaxed its grip on Vincent van Gogh. (143)

Others have suggested that Vincent's sudden change of direction while working in London was the result, not so much of a psychotic reaction to a failed love, but rather, a positive move into religion and spirituality. After all, Vincent had reached the age common for people to begin seeing life from a larger, more philosophical perspective, from an ethical and thus spiritual perspective. And Vincent's childhood had to have given him the seeds of understanding life from a Christian perspective. Perhaps, now, when he has been thrust out of his protective home and family, has been given the task of making his rite of passage from child to adult, has had a brief taste of what his proposed adult life will consist of, has experienced how the "real" world of business and social interaction "really" work, has found that the "cozy" world he grew up in with probably even stricter embracements of such traditional "Puritan" values as honesty and the value of hard work than most doesn't match the practical world, perhaps, being overly idealistic and sincere in his desires to be a good person, which for him, with his childhood programming, had to have meant being a good Christian, perhaps realizing that he was but a small cog in a large world that swirled around him and took little notice of him, far different from a childhood where he held a central position, perhaps all of these ideas came together (and *perhaps* some form of relationship with the Loyers; *perhaps* falling in love with Eugenie and/or her mother, *perhaps* simply seeing the positive *family* relationship they had and relating to what he also once had but recently had been pushed to move beyond), perhaps all or some of these things came together in him. A lot of *perhapsing* here, but all legitimate, all based on similar feelings young men have at this age in this kind of situation.

Overly idealistic, prone to take his ideas to the extreme, needing to figure out how to both embrace his parents and yet establish himself as a self-responsible adult, and quite likely having been given an overly high image of himself, Vincent's dramatic and extreme turn to religion makes sense.

Some of the critics suggesting this approach couch it in the Replacement Child Syndrome theories. Jean Leymarie writes:

> The eldest of two brothers and three sisters, he was marked by fate at birth. He was named after a previous child who, born exactly a year before him, to the very day, had not lived; and, a further coincidence which shows the stability of the population in this rural area, his birth was recorded at the registry office under the same number, twenty-nine. His dead brother's grave stood right at the entrance of the churchyard at Zundert, near the little church where the pastor, his father, read the service; so that Vincent, as soon as he could read, had the odd experience of seeing his own name on a tombstone.
>
> Such cases are familiar to psychoanalysts, and it is certain that, born as he was to parents in mourning who in effect substituted him for their dead child, his relations with his mother were distorted from the start, and so gave rise, with his attachment to Theo, the second boy four years his junior, to the sense of anguish and guilt that grew upon him in later years. Vincent stepped into another's place, he was himself another. Who was he then and how could he redeem himself from this usurpation?
>
> "How then can I make myself useful, what end can I serve? There is something within me, what is it then?" Such was the source of the soul-searchings that tormented him all his life, the quest for a *lost identity*, the craving to *serve* even to the point of self-sacrifice. (144)

Obviously, the same assumptions are being made here, and since Nagera is referred to later in the book, it is clear that his theories have been an influence. The difference is the stress on the connections between Vincent's religious fervor and his art. Jean Leymarie prefaces the book:

> At the age of twenty-seven, in the black winter of the Borinage, among the most sorely disinherited workers in the world whose destitution he shared, the decisive soul-struggle took place within him.
>
> There, exacerbated by the revelation of human misery and social injustice, his religious frenzy rose to fever pitch and, without losing its momentum or changing its nature, reversed its direction and turned into an artistic vocation. It was at the bottom of the mine, where his passionate sense of human brotherhood led him to descent, that at last the *painter* within him was stirred to life and discovered, smoldering under the coals before blazing forth, what was to be his consubstantial element, fire. Thrown back on painting as on a last hope of salvation, but ever faithful to the letter and the spirit of the Bible's central message, which epitomizes his own struggle for fulfillment, he worked his way up "from darkness to light" and went from the mists of the North with fresh eyes and to immerse himself in it "without ever letting the fire of his spirit go out." (145)

For Jean Leymarie, Vincent's life was a search for self-realization, a search to give his life meaning, and his initial move into religion, whatever the spark might be for it, was a sincere attempt at salvation, one driven by compassion, the desire for a human community, and the belief in and perhaps need for self-inflicted suffering and martyrdom resulting from a "sense of anguish and guilt." Even here, in a biography wanting to stress the spiritual aspects of Vincent's life, the Replacement Child Syndrome lurks in the shadows.

Marc Edo Tralbaut writes:

> If we read between the lines of the uninterrupted confession which Vincent's letters provide, we find that his life was governed by various processes of repression, and that, moreover, fantasies of guilt caused by the death of the first child frequently appear right up to the end of his life. This feeling of guilt showed itself for the last time, at a subconscious level, in the suicide which liberated him from the world, and indeed was one of the motives for it. (146)

We just cannot seem to get away from the psychological theories based on the Replacement Child Syndrome and claims that it resulted in a sense of guilt, a need to suffer, and the drive to give life meaning and value through religion and then through art.

Cliff Edwards has noticed this same situation and writes:

> My own view is that the Ursula love story and the replacement child theory have too often been utilized in place of a careful examination of the artist's own letters, which themselves indicate a complex network of religious tradition, nineteenth-century social forces, and personal concerns

> that led to his so-called crisis and period of "religious fanaticism." Certainly some romantic considerations played a role when twenty-year-old Vincent pondered what course his life might take; that much he himself tells us. [See especially letters 156-157] True, there may have been both sadness and over-solicitousness by Vincent's parents in view of the loss of a first child twelve months before Vincent's birth. But the letters themselves indicate that Vincent's reading of Michelet and Renan, his experiencing the deaths of friends and relatives, and his deepening disillusionment with a commercialized art trade played a significant part in his search for an alternative vocation worthy of self-sacrifice. Further, Vincent's sense of guilt, unworthiness, anxiety, and melancholy, shared in part by his brother Theo, had obvious roots in the tradition of his religion, which would not be connected to either Ursula or the replacement child theory. (147)

This sounds promising, a new perspective, and the title of his book, *Van Gogh and God; A Creative Spiritual Quest,* makes it clear he is going to connect Vincent's life and work to religion and faith.

Not surprisingly given this slant, Edwards begins his biography with an interpretation of a passage from Vincent's first sermon:

> Vincent van Gogh found deep, personal meaning in the biblical description of life as pilgrimage, and devoted the first sermon he ever preached to that theme. In November 1876, he mounted the pulpit of a small Methodist church in Richmond, England, and affirmed that "our life is a pilgrim's progress":
>
>> We are pilgrims on the earth and strangers—we have come from afar and we are going far—The journey of our life goes from the loving breast of our Mother on earth to the arms of our Father in heaven. Everything on earth changes—we have no abiding city here—it is the experience of everybody. That it is God's will that we should part with what is dearest on earth—we ourselves change in many respects, we are not what we once were, we shall not remain what we are now. (148)
>
> Vincent, at age twenty-three, convinced that he would follow in his father's footsteps as pastor, missionary, or teacher of the Bible, seems to have drawn together in this one sermon the experiences of his youth and marked out directions his spiritual quest would take during the remaining fourteen years of his life.
>
> A sense of passing through life in loneliness and alienation permeate the sermon, and "pilgrimage" provides a spiritual interpretation for this experience of estrangement. The related theme of impermanence, "we have no abiding city here," endows with religious meaning the bewildering

> changes of late nineteenth-century Europe and the lonely and often unsympathetic rooming houses and workplace far from home with which Vincent was forced to cope. The "great storms of life" described further on in the sermon soon become storms in "the heart of man," and Vincent applies the theme of change directly to the self, a description of personal transformation: "we are not what we once were, we shall not remain what we are now." Alienation, impermanence, and transformation become key words within the greater religious drama of pilgrimage toward "the arms of our Father in heaven." Both life's problems and the spiritual resources for interpreting and transforming those problems appear deeply rooted in Vincent van Gogh's recollections of his early years and religious environment." (149)

These are some large claims based initially on but one small passage from a letter written by a still young man, and as such are similar to the problems with previous claims about the Replacement Child Syndrome.

He immediately builds on this passage by offering a number of quotes collected from those who knew Vincent and from Vincent himself, offering us a view of Vincent both physically and psychologically (in terms of his personality traits). While other biographies have some of the exact same sources and quotes, Edwards brings more of them into play and states them in a more straight-forward manner, removing them from assumptions built on the Replacement Child Syndrome.

First there is the description of him by his sister Elisabeth Huberta van Gogh:

> Turning around, one of them saw approaching their older brother Vincent, a seventeen-year-old boy as square as he was tall, with a slight stoop for he had the bad habit of walking with his head down. His close-cropped red-blond hair was hidden under a straw hat. He had a strange, not young looking, face, the forehead full of lines, the eyebrows on the large, noble brow drawn together in deepest thought. The eyes, small and deep-sel, were now blue, now green, according to the impression of the moment. But in spite of all awkwardness and the ugly exterior, one was conscious of greatness, through the unmistakable sign of the deep inner life. (150)

Some of the previous descriptions of Vincent as *different* get support here, since his sister sees him as "awkward, unattractive, and yet having some quality of a "deep inner life."

S. Aertsen-Honcoop, a maid in the Van Gogh parsonage, describes him as follows:

> There was something strange about him. He did not seem like a child and was different from the others. Besides, he had queer manners and was

> often punished. He was covered with freckles. His hair was read as fire. (151)

This quote (also in Lubin's book) suggests Vincent was "different" and "queer" and was "often punished," supporting the negative views of his childhood; it should be noted in the context of him having a difficult personality.

Hendrik Hoppenbrouwers, a classmate of Vincent states:

> He was an ugly red-headed boy who liked to go by himself on many long walks across the fields. . . . Vincent was a good pupil and read a great deal. We were beaten from time to time, but on other occasions we also got up to mischief together. All the same, as I said before, Vincent went off on his own for most of the time and wandered for hours together around the village, and quite a long way from it. (152)

Thus, Vincent both hung around with other boys his age and apparently got into the same kinds of mischief as they did, was, in other words, fairly normal, and got the same punishment as they did, again fairly normal, and yet he separated himself from them, was something of a loner. It is also worth mentioning that it was not uncommon for children to be physically punished for misbehaving at the time. It was standard.

Mr. Gorlitz, an assistant teacher Vincent shared a room with at age twenty-three, describes him as follows:

> He was a singular man with a singular appearance into the bargain. He was well made, and had reddish hair which stood up on end; his face was homely and covered with freckles, but changed and brightened wonderfully when he warmed into enthusiasm, which happened often enough. Van Gogh provoked laughter repeatedly by his attitude and behavior—for everything he did and thought and felt, and his way of living, was different from that of others of his age. At table he said lengthy prayers and ate like a penitent friar; for instance, he would take no meat, gravy, etc. And then his face had always an abstracted expression—pondering, deeply serious, melancholy. (153)

Again, some of the standard traits get support. Vincent is "different" in some way. He has a "singular" appearance, both because of his inherited looks which must have been unattractive, as description after description of him call him "ugly" or "homely" (though his father is considered to have been strikingly handsome), because he dressed in ways found comical or unusual, and because of his mannerisms. Also, time and again, he is described as having some kind of serious, deep-in-thought quality about him. At the same time, he is not so different that it is fair to claim his is mentally unbalanced, or he has had a terrible or unusual upbringing. There is, for example, nothing to suggest he is epileptic, a claim that has constantly been made.

Mendes da Costa, Vincent's tutor when he was twenty-five, offers the following:

> In my mind's eye I can still see him come stepping across the square from the Nieuwe Herengracht Bridge, without an overcoat as additional self-chastisement: his books under his right arm pressed firmly against his body, and his left hand clasping the bunch of snowdrops to his breast; his head thrust forward a little to the right, and on his face, because of the way his mouth drooped at the corners, a pervading expression of indescribable sadness and despair. And when he had come upstairs, there would sound again that singular, profoundly melancholy, deep voice: "Don't be mad at me, Mendes; I have brought you some little flowers again because you are so good to me. (154)

Again, the serious, somber, profound qualities are mentioned. Two other qualities are also worth noting: first, the "self-chastisement," which is constantly brought up in reference to him in other places, the fact that his form of religion and spirituality was far more acetic and brutal than his family's (especially important in placing him within the context of the religion of his family and upbringing), and second, the kind, sweet quality suggested in the final sentence.

Cliff Edwards includes two descriptions of Vincent later in life that contrast in several ways with these. First in one from Archibald Hartrick, a Scottish painter who knew Vincent in Paris:

> I can affirm that to my eye Van Gogh was a rather weedy little man, with pinched features, red hair and beard, and a light blue eye. He had an extraordinary way of pouring out sentences, if he got started, in Dutch, English and French, then glancing back at you over his shoulder, and hissing through his teeth. In fact, when thus excited, he looked more than a little mad; at other times he was apt to be morose, as if suspicious. (155)

The physical appearance here contradicts, or at least modifies earlier descriptions, where Vincent is pictured as "well made," "square," and "tall," never as good looking, but also not in the same category as a "weedy little man." The "pinched features" can fit with earlier descriptions of him looking older than his age. The "morose" continence fits with the earlier depictions of him having a "serious," "deep," "sad" appearance. It is worth noting that this is someone not especially attracted to Vincent, someone wanting to represent him in a negative light, and also that, as Vincent grew older, he became less robust, less "square" and "tall," and his features became more "pinched," at least partially because he did not eat well.

The second of these later descriptions included by Edwards comes from Dr. Felix Rey, the surgeon at the Arles hospital who attended Vincent when he mutilated his ear:

> First and foremost Vincent was a miserable, pitiful man, small of stature . . ., lean. He always wore a sort of overcoat, smeared all over with colors—he painted with his thumb and then wiped it on his coat—and an enormous straw hat without a hatband, of the type usually worn by the shepherds of the Camargue as a protection against the scorching sun. (156)

Vincent's own letters confirm also suggest his negative appearance:

> I shall have to suffer much, especially from those peculiarities which I cannot change. First, my appearance and my way of speaking and my clothes . . . (Letter 190) (157)

> All the anxiety and worry cannot but make me nervous and flurried in speech and manner. When Mauve imitated me, saying "This is the sort of face you make, this is the way you speak," I answered, My dear friend, if you had spent rainy nights in the streets of London or cold nights in the Borinage—hungry, homeless, feverish—you would also have such ugly lines in your face and perhaps a grating voice too. (Letter 191) (158)

These written records of Vincent's appearance are not the only, and not the best source for his true physical looks. There are few people ever to have as many portraits painted of themselves, both self-portraits and those done by others, to offer visual evidence. And, even better, there are *actual photographs*. Interestingly what is revealed is *not* that Vincent was ugly, but, rather, that he was well within the range of average, even better-than-average looking. Thus, what is revealed in the descriptions is more the general response to certain aspects of his personality, and these are not as one-sided as many biographers want to make them. While he was prone to arguing, he was also capable of great kindness and compassion. While he was prone to being a loner, to going off on his own for walks, he was also capable of socializing, wanted to be around people, to have a community of friends, or, at times, just a single friend. While he apparently got into trouble for various childhood pranks, he also took his studies seriously. While he was filled with some kind of guilt and need for self-punishment, he also thought himself a person capable of great things. And while he was selfish in constantly asking for money from his borhter Theo, he was generous with whatever belongings and food he might have, giving away the very shirt on his back.

Edwards follows his passages about Vincent's physical appearance by putting them into the general views of phrenology at the time:

> The effects of Vincent's appearance on others may have been further influenced by the public's fascination in his day with physiognomic classification, phrenology, pantomime, and caricature. The increased mobility of large numbers of people had contributed to the growing alienation of urban workers, who no longer knew their neighbors, and had apparently led many to look to some *science* of physical appearance for help in the evaluation of strangers. The population of Paris, for example, had become avid viewers of the caricatures drawn for the daily papers by Daumier, Gavarni, Grandville, and Monnier, who all employed a vocabulary of facial features and postures that were intended to represent the class, temperament, and quality of their characters and, by extension, their counterparts in society. As one of the many popular physiognomic manuals of the day put it: *Le Dedans juge par le Dehors* ("The Inside judged by the Outside"). Vincent's bristling red hair, small deep-set eyes, grating voice, and tendency to hiss and to hang his head may have marked him as a "low class" and suspect person to be avoided. (159)

After Edwards provides the collection of passages about Vincent's looks and general personality traits, he wants to move quickly into the religious and faith based facets. In order to do this, he compares Vincent and his brother Theo with the biblical story of Esau and Jacob, which he finds "both poignant and revealing when applied to Vincent, for they focus upon external appearance as revelatory of internal character." (160) As a doorway into his theories on this, he quotes Vincent:

> Father used to ponder over the story of Jacob and Esau with regard to you and me—not quite wrongly—but fortunately there is less discord, to mention only one point of difference, and in the Bible itself there are plenty of examples of better relations between brothers than existed between the venerable patriarchs mentioned above. (Letter 338) (161)

Edwards goes into a summary of the story, and after offering the standard disclaimer that we shouldn't push the idea too far, he ignores his own advice, moving into various passages from Vincent's letters where Vincent uses animal imagery to describe some aspect of himself, and suggesting that these images are meant to refer to Vincent seeing himself as Esau, who "first came forth red, all his body like a hairy mantle." (Genesis, 25:25) (162) Enamored with his analogy, he then ties the animal imagery to coarse looks and behavior, and "coarse" with being of the peasants, those hard working, labor worn, poor and sorrowing people Vincent championed. Having put this in place, he then uses it to claim that Vincent consciously saw himself as Esau (even though the passage it all comes from indictes Vincent didn't agree with the anology) and then pushes this to Vincent seeing Esau representing him as the outsider who is being unfairly treated by the father, the community and the church, and who is actually more spiritual because he is less polished, less likeable, and less refined. Then, stretching this anology even further, he claims that Vincent connect all this with the peasants, also seen as animals, lesser beings, but in reality higher beings, more spiritual because they are poor and uneducated and dirty and "coarse," animals just like Esau.

As with previous theories, this one undermines itself by claiming too much from too little to support it. It is an interesting suggestion, and there might be some facets of it that have some truth, but the passages from the letters used to support them are too weak, and the claims too extreme. For example, to suggest that Vincent was consciously using animal analogies to represent himself as Esau in his letters has no basis and is almost certainly wrong, undermining other possible insights. It is easy to support the claim thatVincent saw himself as an outsider and thought of his Christlike form of religion as more spiritual, more sincere than his established religion, and in turn that he consciously and purposely lived among the poor, the coarse peasants to emulate Christ, but there isn't enough evidence for the particulars used to support the Esau claims, and since those are what gets stressed, all of the rest loses creditability.

The other biblical connection Edwards highlights comes from Isaiah 53, the passage known as the "Suffering Servant Song," the very passage Vincent opened the Bible to in a still life he painted. Edwards suggests, since Vincent purposely chose this particular passage, it represents how he saw himself, which is likely, especially in the context of his other actions during his "religious" period. However, what Edwards wants to highlight is the ugly physical appearance represented in it, stressing a physically ugly person as the savior and that Vincent specifically chose this because of his own sensitivity about his physical appearance.

Again, there is no support for this view, and it might be mentioned that he painted many, many self-portriats, an unlikely activity for someone ashamed of his looks. And to create a convoluted argument that he turned ugly into something to be proud to be, so he could find a way to turn something he was ashamed to be into something he was proud to be only serves to discount serious consideration.

If we put aside the stress on physical appearance, it is easy to support claims that Vincent saw himself, at least at this time in his life, as being similar to this despised servant who was "smitten by God," "wounded for our transgressions" (Isaiah 53; 4-5), "a lamb that is led to the slaughter," a sacrificial "offering for our sin" (Isaiah 53:7, 10), a despised man whose mission was to "pour out his soul to death," and bare "the sin of many." (Isaiah 53: 13), and who saw himself "preaching the gospel to the poor," and comforting "all who mourn (Isaiah 61: 1-2), and the many references in his letters to this passage certainly confirm he knew the passage and it was an important one for him (see, for example, letters 71, 77, and 82a)

To bring his discussion in this chapter to a close, Edwards offers better support for his theories from outside the Bible:

> As we will see later, Vincent's religious resources were not limited to the Bible, and so we will close this chapter with a moving passage Vincent treasured from a book by Jules Michelet, the French historian and social critic. Almost a dozen of Michelet's books are mentioned in over fifty references in Vincent's letters from 1873 to 1889, a period of seventeen years. According to Vincent, "Michelet even expresses completely and aloud things which the Gospel whispers only the germ of" (letter 161), and Michelet's book *L'Amour* became to him, "a revelation and a Gospel" (letters 19, 20). In *L'Amour*, Michelet asserted that surface beauty, "youthful complexion," was of little consequence compared to the deeper beauties that can be developed in the human soul. An anecdote used at that point in Michelet made a lasting impression Vincent, an anecdote concerning "Socrates, ugly as a satyr" (letter 572). In 1889 Vincent applied the anecdote to his friend Roulin, but as early as 1883 he made clear that he applied the story directly to himself:
>
>> Socrates was born as a true satyr, but by devotion, work and renouncing frivolous things he changed so completely that on the last day before his judges and in the face of death, there was in him something, I do not know what, of a god, a ray of light from heaven that illuminated the Parthenon. (Letter 306) (163)

Let's make a couple of tentative claims from what we've got so far. Vincent's letters are filled with contradictions. One passage or a few passages suggest he is concerned about his looks and concerned about what others think of his looks, but then other passages suggest he is not concerned about such things, and at times is even purposely taking on the appearance of an outsider or unattractive person (in the strange clothes he chooses to wear, in the neglect of his general hygiene, and in the dark, gloomy facial expressions and slouching posture he assumes).

Rather than jump to conclusions about him being highly affected psychologically by his looks, it seems clear here that Vincent is putting forth what many others have put forth, especially young men who have reached an age when they begin thinking about ethical and spiritual matters, which is that there are more important things in life than physical beauty, that there is an inner beauty, a spiritual beauty, which refers to the higher truths of meaning and value.

Just how much Vincent connected concepts of physical ugliness to his own looks remains guesswork, and more importantly just how literally he applied these views of physical beauty to his theoretical views of beauty and spirituality can easily lead to a tangled wandering down theoretical paths to little purpose. When all is said and done, does it not seem a bit shallow on the part of the critics to claim that Vincent turned so fanatically to religion and then to art simply or mainly to justify his own ugliness? Even if one were to say, well, that wasn't the only reason, it was just a major one makes me want to vent my own frustrations, makes me want to say, listen, humans do have deeper drives, deeper desires, and we're discussing a man who, though he does have some shallow traits (and, by the way, what human hasn't has some thoughts about his or her looks?), is obviously driven by something more than his sorrow over being unattractive.

It is clear that, whatever the reasons, Vincent turned to religion and spirituality in the manner of a young man while he was living in London. Certainly, the religious foundation had been given him throughout his childhood, but now, at this point in his life, he was ready to embrace it from an adult perspective.

This, as quoted earlier, is where Edwards offers a more even-handed discussion than those centering on an emotional collapse resulting from a failed love affair. Edwards continues his discounting of the Ursula/Eugenie theories:

> A closer look at the primary evidence in Vincent's letters regarding his crisis and consequent "religious obsession" reveals a complex of *triggering* factors among which we might choose. The fact that Michelet's *L'Amour* was to him "both a revelation and a Gospel" (letter 20) may indicate that he was seriously considering the options of young love or self-improvement, particularly as he was struck by Michelet's example of Socrates, who "by the sculpture of reason, virtue, and self-sacrifice," transformed his "ugliness" into a face that shone with divine light." [Michel, *Love*, p. 301] Or perhaps Ernest Renan, whose *The Life of Jesus* he had just read, was his chief inspiration for a new life-direction. (164)

Edwards also suggests that the death of Annette Haanebeek, a young relative Vincent and Theo were close to, combined with the deaths of a thirteen-year-old girl at Vincent's boardinghouse, his uncle Carbentus, and an acquaintance named Weehuizen, all within a six month period, probably had an affect on him. After offering a few passages from Vincent's letters relating death to his religious concerns, Edwards makes the obvious point that death is central to religious concerns. I agree. An immediate experience of the reality of death, generally by experiencing the death of someone close, is likely to be the most powerful stimulus to transform from a child to an adult. Like it or not, hardships change people, cause people to grow into richer, more complex, deeper thinkers. And it really does make sense that the various hardships Vincent was facing came together to cause the dramatic change, a change resulting in a more sincere exploration of religion and faith, that occurred in Vincent at the time.

Another kind of hardship, and one that Edwards immediately brings up, is the practical, very important aspect of life that has to do with one's job. Neither Edwards nor I can figure out why other biographers ignore the obvious here. They all want to write off Vincent's job at the time as being positive, perhaps simply because Vincent begins it on such positive terms with a glowing letter from his boss, and then assume that Vincent's fall from grace at work is the result of other factors, when it is just as likely that the job itself was one of the negative factors causing his change. In fact, if we don't "read between the lines," it is obvious on the surface that Vincent's work turns negative for him. Perhaps that's why the critics ignore or discount it; perhaps it's too obvious, too simple, and just not glamorous enough.

One thing to consider here, as Edwards notes, is that Goupil and Company, even in just the short time Vincent has been working for them, have noticeably changed. Now, with a new generation of owners, its name is Boussod, Valadon and Company, and it is rapidly expanding. Vincent's boss in London, as Vincent indicates in Letter 20, is making rapid changes to increase profits. Eight years later, Vincent will air his feelings in a letter to Theo, expressing his sadness over the "triomphe de la mediocrite" of the art trade in general and his former company and manager in particular:

> And now everything is gone—once again materialism instead of moral principle. Do you know what I think of the copy I'm sending you? [Vincent sent along a copy of *Graphic* as proof of his points] It is like Obach's kind of talk, for instance, the manager of Goupil and Company in London. And it has success, *yes, that has success, yes, that is listened to and that is admired.* . . . But it makes me sad, it takes away my pleasure, it upsets me, and personally I am absolutely at a loss about what to do. What

> sometimes makes me sad is this: formerly, when I started, I used to think, if only I make so or so much progress, I shall get a job somewhere, and I shall be on a straight road and find my way through life. . . In short, instead of meeting with an opinion, a sentiment, an aim like Dickens' . . . one is confronted with a philosophy like Obach's. It makes me sad, and then I feel helpless. One can only undertake a thing if one has sympathy and co-operation. (165)

While such ramblings later in life do not necessarily mean Vincent quite his job because he was so upset over the state of the business of art, especially since Vincent did go through changes in his views (and I stress that it's dangerous to take views expressed later in life and use them to support his earlier actions). However, with that caution, it is still fairly easy to establish that Vincent was upset over how inferior art was being pushed by the company he worked for. In fact, one of the main reasons he got into trouble at his job was that he became so honest in telling customers that he did not think various works for sale were worth purchasing, discouraging them from buying instead of manipulating them into buying.

A reasoned approach to viewing Vincent at this time, instead of assuming some kind of insane genius is being revealed, suggests the following. Vincent was a young man, recently forced to leave the security of his family, his childhood, forced to begin to grow up, get a job, support himself, move to a city where he was without friends, didn't even speak the language, and had little of what might be called a support system beyond his family. This is the standard rite-of-passage time in human life. For some it is less harsh and demanding, for some more. It is the time when adolescents begin reassessing their lives. Vincent certainly had a serious side, was well read, and was prone to think about the important truths of life. It's not at all difficult to comprehend why the profession of business, and in particular this business world of selling inferior art would not appeal to him. He never demonstrated at any point in his life that he had any interest at all in the world of business. Whether or not he had a failed love affair (and if it did exist it certainly was mainly in his own head with little reality to it), he no doubt saw the Layors family as a happy one, and related this to his own childhood, which he was now being forced to leave behind. It might even be possible that he compared them to a happy childhood that he now saw as gone.

The first of Vincent's letters at this time that has any religious content is dated August, 1874 (Letter 21), and it is anything but fanatical or demonstrating a complete, dramatic change. It simply demonstrates Vincent knew the Bible (which, of course, he must have from a young age), and that he is now beginning to use passages from it to help clarify points he wants to make to his brother. Unfortunately, a five month gap now takes place in his correspondence. When the letters begin again, the biblical references continue in much the same manner, as examples or references to support various points he wants to make. But the religious perspective increases rapidly as Vincent is transferred to Paris and heads quickly toward the end of his job as an art dealer. By September, 1875, he has made the decision to shift his life from the world of business to religion. It is a conscious decision, and though it is evident his religious views are taking a turn away from the more moderate views of his family and religious upbringing (certainly influenced in part by his reading of the *Imitation of Christ*), there is little to support a sudden "rebirth" kind of revelation or dramatic emotional breakdown. What seems more likely is that Vincent had decided he just could not live the life his family had set out for him, and, prompted by his own desire to find meaning and value in life, the obvious alternative, the obvious choice open to him, was religion.

Furthermore, this shift would necessarily involve a reassessment of his father, driven by the normal revising of a boy's relationship with his father during a rite of passage from child to adult, and by the impossible to avoid simple fact that Vincent's father was, after all, a minister, a man of God. It is obvious that Vincent now began to re-image his father in the dual role of father and of spiritual guide, and, as discussed earlier, to struggle with the love/hate relationship common at this stage of a boy's/man's life.

However, this new path, while certainly at first it has a lot of simple logic, *does* move dramatically to the extreme. As Vincent's currently idolized parents are doing their best to help him move into this new career, and it is easy to imagine them as fairly typical parents frustrated by their son, whom they are trying to help get started in life, but who just cannot get his act together, Vincent is unable to follow the standard path that will opens doors to a ministry or an evangelical career. He has demonstrated his intellectual abilities, has in fact shown he is adept at language and the learning of foreign languages, yet he has decided he simply doesn't want to learn the necessary languages of Greek and Latin. In fact, though he is frankly given a rather spoiled set up, spending a year (May 1887 to July 1878) living in Amsterdam with Uncle Jan van Gogh, Commandant of the local Navy Yard, studying with Uncle Stricker, an establish Dutch pastor (of the more liberal Groningen School of Calvinism and a later even more liberal form of Calvinism promoted by Cornelius Willem Opzoomer), and with a hired tutor, Mendes da Costa, Vincent refuses to study, finds endless excuses that are going to obviously result in him failing, and adopts an interpretation of the Bible that allows him to attack the very religious establishment he is trying to enter. The reasons for this can be written off as a standard, perhaps a bit more than most, but still within the norm, rebellion against growing up and accepting self-responsibility, or can, perhaps, indicate some kind of abnormal psyche. The refusal to put in the work necessary to obtain one's goals might be driven by a spoiled child (opposite of what all of the psychological approaches are claiming, yet as supportable as their claims), or it might be driven by something more noble. And as is always the case with Vincent, the contradictions in him make opposite claims plausible. He certainly is willing to work hard and to sacrifice. In fact, the path he chooses is one of extreme sacrifice.

One other element needs to be put in here, one that lurks in everything he does, and that is his refusal to let others tell him what to do. He is most certainly a rebel, a maverick, a man determined to have his own way. This quality in-and-of-itself might be so strong that it is the basis for everything else in his life. And it is wrong to equate it with too much pride or hubris. That just confuses the issues, as he constantly criticizes himself, constantly punishes himself, seeks out a form of religion based on sacrifice and self-abuse. It is not a matter of him seeing himself as "Godlike," though he does see himself as similar to Christ or as trying to be "Christlike." But again, we need to be very careful here. This is not a matter of conceit, but a more sincere attempt to be a good person. Christlike is not, for him, some kind of superior or proud position. Rather it embraces such Christian values as humility and servitude. And it is not meant to be taken in the sense of being better than others because he is more Christian than them, though that aspect is hard to eliminate. Vincent is doing this for himself, not to grandstand for others. His sympathy, compassion, and connections with the poor, the suffering and the peasants is sincere.

At the same time he is standing up to what he sees as a hypocritical establishment, both in the business world and in the church. This view makes him a romantic hero, much the same as a Christian martyr, or even a Christ, a man who also stood up to the Jewish elders of his religion and to the business, political establishment of his time, the all powerful Roman Empire.

Rites-of-passage by necessity are difficult and especially difficult for a man refusing to conform and give up his quest of meaning and value. Vincent's time in London and later in Borinage, if nothing else, demonstrates a man trying to figure out life, willing to suffer if his beliefs demand it, and a man unable to live in the established systems of business and religion. By the summer of 1878, Vincent is ready to move forward, ready to take his spirituality out of religion, his art out of business, and plunge into the task of mapping out the values of his spirituality in his artist vision. He writes Theo:

> Dear Theo,
>
> Your letter has done me good and I thank you for having written to me in the way you have.
>
> The roll with a new selection of etchings and various prints has just arrived. First and foremost the masterly etching, “Le Buisson” by Daubigny and Ruysdael. Well! I propose to make two drawings, in sepia or something else, one after that etching, the other after “Le Four dans les Landes” by Th. Rousseau. Indeed, I have already done a sepia of the latter, but if you compare it with Daubigny’s etching you will see that it contrasts feebly, although considered on its own the sepia may betray some tone and sentiment. I shall have to return to it and tackle it again.
>
> I am still working on Bargue’s *Cours de Dessin,* and intend to finish it before I go on to anything else, for both my hand and my mind are growing daily more supple and strong as a result, and I cannot thank Mr. Tersteeg enough for having been so kind as to lend it to me. The models are outstanding. Meanwhile I am reading one book on anatomy and another on perspective, which Mr. Tersteeg also sent me. These studies are demanding and sometimes the books are extremely tedious, but I think all the same that it’s doing me good to study them.
>
> So you see that I am working away hard, though for the moment it is not yielding particularly gratifying results. But I have every hope that these thorns will bear white blossoms in due course and that these apparently fruitless struggles are nothing but labour pains. First the pain, then the joy.
>
> You mention Lessore. I think I remember some very elegant watercolour landscapes by him in a blonde tone, worked with an apparent ease and a light touch, yet with accuracy and distinction, and a somewhat decorative effect (that is not meant badly, but on the contrary, in a favourable sense). So I know a little about his work and you mention someone not entirely unknown to me.
>
> I admire the portrait of Victor Hugo. It is done very conscientiously with the evident intention of portraying the truth without straining after effect. That is precisely what makes it so effective.
>
> Last winter I pored over some of Hugo’s works, *Le Dernier Jour d’un Condamné* and an excellent book on Shakespeare. I first started studying this writer long ago. He is just as splendid as Rembrandt. Shakespeare is to Charles Dickens or Victor Hugo what Ruysdael is to Daubigny, and Rembrandt to Millet.
>
> What you say in your letter about Barbizon is perfectly true and I can tell you one or two things that will make it clear how much I share your view. I haven’t been to Barbizon, but though I haven’t been there, I did go to Courrières last winter. I went on a walking tour in the Pas-de-Calais, not the English Channel but the department, or province. I had gone on this

trip in the hope of perhaps finding some sort of work there, if possible - I would have accepted anything - but in fact I set out a bit reluctantly, though I can't exactly say why. But I had told myself, You must see Courrières. I had just 10 francs in my pocket and because I had started out by taking the train, that was soon gone, and I was on the road for a week, it was a rather gruelling trip. Anyway, I saw Courrières and the outside of M. Jules Breton's studio. The outside of the studio was a bit of a disappointment, seeing that it is in a brand-new studio, recently built of brick, of a Methodist regularity, with an inhospitable, stone-cold and forbidding aspect, just like C. M.'s Jovinda, which, between ourselves, I am none too keen on either, for the same reason. If I could have seen the inside, I am quite certain that I should have given no further thought to the outside, but there you are, I could not see the inside because I dared not introduce myself and go in. Elsewhere in Courrières I looked for traces of Jules Breton or any other artist. All I was able to find was a portrait of him at a photographer's and a copy of Titian's *Entombment* in a corner of the old church which looked very beautiful to me in the darkness and masterly in tone. Was it by him? I don't know because I was unable to make out any signature.

But of any living artist, no trace, just a cafe called Cafe des Beaux Arts, also of new, inhospitable, stone-cold, repulsive brick - the café was decorated with a kind of fresco or mural depicting episodes from the life of that illustrious knight, Don Quixote.

To tell the truth, those frescoes seemed to me rather poor consolation, and fairly mediocre at the time. I don't know who did them.

But anyway I did seen the country around Courrières then, the haystacks, the brown farmland or the marled earth, almost coffee-coloured (with whitish spots where the marl shows through), which seems somewhat unusual to people like us who are used to a blackish soil. And the French sky looked to me much finer and brighter than the smoky, foggy sky of the Borinage. What's more, there were farms and barns that, God be praised, still retained their mossy thatched roofs. I also saw the flocks of crows made famous by the pictures of Daubigny and Millet. Not to mention, as I ought to have done in the first place, the characteristic and picturesque figures of all manner of workmen, diggers, woodcutters, a farmhand driving his wagon and a silhouette of a woman in a white cap. Even in Courrières there was still a coal mine or pit, I saw the day shift come up at nightfall: but there were no women workers in men's clothes as in the Borinage, just the miners looking tired and careworn, black with coal dust, dressed in ragged miners' clothes, one of them in an old army cape.

Although this trip nearly killed me and I came back spent with fatigue, with crippled feet and in more or less depressed state of mind, I do not regret it, because I saw some interesting things and the terrible ordeals of suffering are what teach you to look at things through different eyes.

I earned a few crusts here and there en route in exchange for a picture or a drawing or two I had in my bag. But when my ten francs ran out I tried to bivouac in the open air the last 3 nights, once in an abandoned carriage which was completely white with hoarfrost the next morning, not the best accommodation, once in a pile of faggots; and once, and that was a slight improvement, in a haystack, that had been opened up, where I succeeded in making myself a slightly more comfortable little hideaway, though the drizzle did not exactly add to my enjoyment.

Well, and yet it was in these depths of misery that I felt my energy revive and I said to myself, I shall get over it somehow, I shall set to work again with my pencil, which I had cast aside in my deep dejection, and I shall draw again, and from that moment I have had the feeling that everything has changed for me, and now I am in my stride and my pencil has become slightly more willing and seems to be getting more so by the day. My over-long and over-intense misery had discouraged me so much that I was unable to do anything.

I saw something else during the trip - the weaver's villages.

The miners and the weavers still form a race somehow apart from other workers and artisans and I have much fellow-feeling for them and I should consider myself fortunate if I could draw them one day, for then these as yet unknown, or virtually unknown, types would be brought out into the light of day.

The man from the depths, from the abyss, *de profundis*, that is the miner. The other, with the faraway look, almost daydreaming, almost a sleepwalker, that is the weaver. I have been living among them now for nearly 2 years and have learned a little of their special character, in particular that of the miners. And increasingly I find something touching and even pathetic in these poor, humble workers, the lowest of the low in a manner of speaking, and the most despised, who, owing to a possibly widely held but quite baseless and inaccurate presumption, are usually considered a race of knaves and scoundrels. Knaves, drunkards and scoundrels may be found here, of course, just as elsewhere, but the real type is nothing at all like that.

You refer vaguely in your letter to my coming sooner or later to Paris or its environs, if it were possible and if I wanted to. It is of course my eager and fervent wish to go either to Paris or to Barbizon, or somewhere else, but how can I, when I do not earn a cent and when, though I work hard, it will be some time before I reach the point at which I can give any thought to something like going to Paris. For honestly, to be able to work properly I need at least a hundred francs a month. You can certainly live on less, but then you really are hard up, much too hard up in fact!

"Poverty stops the best minds in their tracks" the old Palissy saying goes, which has some truth in it and is entirely true if you understand its real meaning and import. For the moment I do not see how it could be feasible, and the best thing is for me to stay here and work as hard as I can, and, after all, it is cheaper to live here.

At the same time I must tell you that I cannot remain very much longer in the little room where I live now. It is very small room indeed, and then there are the two beds as well, the children's and my own. And now that I am working on Bargue's fairly large sheets I cannot tell you how difficult it is. I don't want to upset these people's domestic arrangements. They have already told me that I couldn't have the other room in the house under any circumstances, not even if I paid more, for the woman needs it for her washing, which in a miner's house has to be done almost every day. In short, I should like to rent a small workman's cottage. It costs about 9 francs a month.

I cannot tell you (though fresh problems arise and will continue to arise every day), I cannot tell you how happy I am that I have taken up drawing again. I had been thinking about it for a long time, but always considered it impossible and beyond my abilities. But now, though I continue to be conscious of my failings and of my depressing dependence on a great many things, now I have recovered my peace of mind and my energy increases by the day.

As far as coming to Paris is concerned, it would be of particular advantage to me if we could manage to establish contact with some good and able artist, but to be quite blunt about it, it might only be a repetition on a large scale of my trip to Courrières, where I hoped to come across a living example of the species Artist and found none. For me the object is to learn to draw well, to gain control of my pencil, my charcoal or my brush. Once I have achieved that I shall be able to do good work almost anywhere and the Borinage is as picturesque as old Venice, as Arabia, as Brittany, Normandy, Picardy or Brie.

Should my work be no good, it will be my own fault. But in Barbizon, you most certainly have a better chance than elsewhere of meeting a good artist who would be as an angel sent by God, should such a happy meeting take place. I say this in all seriousness and without exaggeration. So if, sometime or other, you should see the means and the opportunity, please think of me. Meanwhile I'll stay here quietly in some small workman's cottage and work as hard as I can.

You mentioned Méryon again. What you say about him is quite true. I know his etchings slightly. If you want to see something curious, then place one of his meticulous and powerful sketches next to a print by Viollet-le-Duc or anyone else engaged in architecture. If you do, then you will see Méryon in his true light, thanks to the other etching which will serve, whether you like it or not, as a foil or contrast. Right, so what do you see? This. Even when he draws bricks, granite, iron bars, or the railing of a bridge, Méryon puts into his etchings something of the human soul, moved by I do not know what inner sorrow. I have seen Victor Hugo's drawings of Gothic buildings. Well, though they lacked Méryon's powerful and masterly technique, they had some-thing of the same sentiment. What sort of sentiment is that? It is akin to what Albrecht Dürer expressed in his "Melancholia," and James Tissot and M. Maris (different though these two may be) in our own day. A discerning critic once rightly said of James Tissot, "He is a troubled soul." However this may be, there is something of

> the human soul in his work and that is why he is great, immense, infinite. But place Viollet-le-Duc alongside and he is stone, while the other, that is, Méryon is Spirit.
>
> Méryon is said to have had so much love that, just like Dickens's Sydney Carton, he loved even the stones of certain places. But in Millet, in Jules Breton, in Jozef Israëls, the precious pearl, the human soul, is even more in evidence and better expressed in a noble, worthier, and if you will allow me, more evangelical tone.
>
> But to return to Méryon, in my view he also has a distant kinship with Jongkind and perhaps with Seymour Haden, since at times these two artists have been extremely good.
>
> Just wait, and perhaps you'll see that I too am a workman. Though I cannot predict what I shall be able to do, I hope to make a few sketches with perhaps something human in them, but first I must do the Bargue drawings and other more or less difficult things. Narrow is the way and straight the gate and there are only a few who find it.
>
> Thanking you for your kindness, especially for "Le Buisson," I shake your hand,
>
> Vincent
>
> I have now taken your whole collection, but you will get it back later and in addition I've got some very fine things for your collection of wood engravings, which I hope you will continue, in the two volumes of the Musee Universel, which I am keeping for you. (166)

The entire letter, filled with the typical rambling of a letter, also indicates a man able to step back and give a reasoned analysis of himself, and a man consciously moving in this new direction. The key passage begins: "Well, and yet it was in these depths of misery that I felt my energy revive and I said to myself, I shall get over it somehow, I shall set to work again with my pencil, which I had cast aside in my deep dejection, and I shall draw again, and from that moment I have had the feeling that everything has changed for me, and now I am in my stride and my pencil has become slightly more willing and seems to be getting more so by the day."

Kathleen Powers Erickson also wants to approach an understanding of Vincent in terms of religion and spirituality, and, as with Edwards, to tie Vincent's religion to both the Bible and to outside sources, to what she calls the modernist movement taking place at the time. Having read Edwards book, she wants to offer a much more in-depth context for Vincent's religious background, and she does. Interestingly, and surprisingly for an author determined to put a more religious spin on Vincent's life, Edwards does not provide a well researched landscape of the form of Calvinism

(Gronigen Calvinism, a form of rebellion to more strict forms of Calvinism) Vincent grew up in, but instead simply emphasizes the harsh Calvinism that he assumes Vincent's father preached and would have naturally emphasized in his home, stating that Vincent's sense of alienation and pilgrim status got established and reinforced by the "Calvinist atmosphere of a home in which the pastor-father uttered daily warnings regarding the sinfulness of the outside world, or where parents considered the local school "too rough" and so withdrew their parsonage children." (167)

Erickson begins with the following claim, one that, even in its initial brevity and with its need for more clarification (which she will provide), is still more accurate than what most of the biographies offer:

> Initially, van Gogh embraced the theology of his father, whom he admired and adored. Known as the Groningen School, named for the Groningen region in the northern Netherlands where it originated, it is often called "the mediating school" because its proponents shared tenets of faith embraced by both the most orthodox and the most liberal thinkers of the time. On the one hand it insisted on the verbal inspiration of Scripture and its ultimate authority; and on the other hand, it denied the deity of Jesus. Vincent intended to follow in his father's footsteps and become a pastor of the Dutch Reformed faith, but he lacked formal training and chose instead to become a missionary to the Belgian coal miners in the Borinage district. (168)

Let me elaborate on the religious frame provided by Erickson. It will help us to understand some things about Vincent.

The split from traditional Calvinism formally began in 1610, when the beliefs of Jacob Arminius, professor of theology at the University of Leiden from 1603 to 1609, put forth objections to two central beliefs of Calvinism, predestination and limited atonement, the belief that only a few chosen people from the beginning of time would ultimately receive salvation.

Interestingly, Arminius had been taught his theology by Theodore Beza, Calvin's personally chosen successor. Of course, he is not the first student to push past his teacher to new ideas. And just as certain, the new views were not easily received. His main opponent was Franciscus Gomarus, and a national synod was needed to resolve the conflict. Holland's State General requested a paper from Arminius clarifying his theology. Unfortunately, Arminius died before he could argue his beliefs.

In January 14, 1610 some forty followers of Arminians met at Gouda and came up with a petition, what they called the "Remonstrance," stating that: 1. God's preordained plan was to save the fallen human race through

Christ's death on the cross, not just those who have been election, but all those who believe in Christ; 2. Christ died for all who believed in him, not just a select few; 3. Man cannot earn his own salvation but must be regenerated by the Holy Spirit in order to be saved; 4. Man is free to accept or reject God's saving grace; and 5. Christians with the help of the Holy Spirit are able to resist evil, but may also fall from grace.

"Religious Politics" were in full swing. The Dutch Calvinists convinced Prince Maurice of Nassau to take charge of the conflict. He systematically removed Armenian magistrates from office and called for a nation synod at Dordrecht (where the Arminians were excluded).

The Synod of Dordrecht met from 1618-19 to come up with a response to these five points, and the result was a demand that all Calvinists subscribe to the Heidleberg Catechism, the Belgic Confession, and the Five Points of Calvinism (TULIP) which upheld the established doctrines of Calvinist doctrine: (1) Total depravity, also referred to as "total inability," asserts that, as a result of the fall of man in the Garden of Eden, each person is born in sin, inclined toward evil and unable to resist it on their own because of their very nature [Arminius agreed with this, though the Remonstrants rejected it]; (2) Unconditional election (predestination), asserts that God chose from the beginning or prior to the beginning which people he will save, and it is not based on how they live but simply on God's mercy; (3) Limited atonement, also referred to as "particular redemption" and "definite atonement," asserts the Christ died, not for the sins of all mankind, but just to atone for the sins of the select; (4) Irresistible grace, also referred to as "efficacious grace," asserts that God's grace, God's decision to save someone cannot be refused; and (5) Perseverance (or preservation) of the saints, also known as "eternal security," asserts that all of the chosen are saints, all are set apart by God, and if any seeming saints fall into sin that simply means they never truly were chosen or they will return to it.

In other words, by affirming the Calvinist doctrines of original sin and the total depravity of man, justification by the grace of God's act of salvation through Christ's death on the cross, and redemption limited to the predestined elect, the Synod of Dort categorically rejected the Arminian doctrines of the Remonstrants. Arminians across Holland were removed from office, imprisoned, banished, and sworn to silence. Twelve years later Holland officially granted Arminianism protection as a religion, although animosity between Arminians and Calvinists continued

This would remain the situation until the early 19th century, when Dutch philosopher P. W. van Heusde, a professor of literature and history at Utrecht from 1804-1839, began preaching more liberal, more "humanistic" views, putting forth a dynamic view of history and emphasizing progress through education. Three of his followers, Petrus Hofstede de Groot, J. F. van Oordt, and I. G. Pareau, joined the theological faculty at Groningen, and originated the Groningen Richting (The Evangical Party). They appealed to the rising nationalism in the nineteenth century, rejected the Calvinism of Dort as a foreign intrusion, and espoused a humanistic, mystical, pietistic form of Christianity, embracing the Remonstrants, Arminius, Erasmus, and notable in terms of Vincent's own embracement Thomas a Kempis. Though they retained the traditional insistence that Scripture is grounded in real history, accepting the various miracles and other incidents reported in the Bible as fact, they replaced the doctrines of election, predestination, and limited atonement with doctrines of universal atonement and all-inclusive perfection. (169)

In opposition to the traditional Calvinist doctrines, the Van Gogh family, going all the way back to Cornelious Van Gogh, a Remonstrat preacher at Boskoop, embraced Arminiam doctrines on these matters. (170) Kathleen Powers Erickson states:

> Vincent van Gogh's grandfather received his theological degree at the University of Leiden, which held the least doctrinally confessional views of the three state universities, conferring even to this day a Remonstrant degree. Van Gogh's father, Theodorus, was the first to break with the family tradition in obtaining his degree from Utrecht, but even there he declared his alliance with the Groningen theology, which was also Arminian. (171)

Though he was not Dutch, and this new form of Calvinism was centered on nationalism, the other main source of Groningen theology was Friedrich Schleiermacher, an early 19th century German theologian and philosopher, often referred to as the "Father of Modern Protestant Theology," whose attempt to combine traditional Protestant orthodoxy with the views of the Enlightenment was imbued with the ideas of Schlegal, Leibniz, Lessing, Fichte, and Jacobi, views providing him a mystical passage into the depths of the human psyche, views put forth most completely his book titled appropriately *The Christian Faith*. (172)

To a large extent it was Schleiermacher who gave Groningen beliefs their strong embracement of inward piety, and the belief that each human should devote his life to creating in himself the consciousness and character of Jesus. This, of course, fits Vincent's views exactly. Christ is in all of us, and the way to salvation is to be as Christ was, to suffer as Christ suffered, to sacrifice as Christ sacrificed, to in effect live as Christ lived. Vincent's Christianity is driven by this central belief. And once that is clear, much of what he did that seems negative, perhaps even insane, makes sense. Yes, he wanted to be Christ*like*. That was the way to salvation in the religion he knew.

For Schleiermacher, the person, represented as the ego, is an individualization of universal reason (the underlying logic or pattern of existence), and self-consciousness or realization of self is first found when the universal and individual, reason incarnate, come together. In this way each person is a specific and original representation of the underlying structure of the universe, a miniature totality of humanity, a micro cosmos of the cosmos. Rejecting the idea that we can obtain a complete understanding of the unity of thought and being through cognition or volition, we can find it in our own personality, our own self-awareness, self-consciousness, our own *feelings*. By this Schleieracher does not mean the feelings of the physical senses, but, rather, an abolishment of the sense of separateness between the individual and the universal. In other words, we can feel a unity with that which is other. Singular and universal can merge in consciousness of self or feeling. This is a religious feeling, what Schleiermacher also referred to as a "feeling or intuition of the universe," a "consciousness of the unity of reason and nature," and a "consciousness of the infinite and the eternal within the finite and the temporal," the feeling of absolute dependence, the consciousness of being in relation to God. In 1799, he wrote:

> Religion is the outcome neither of the fear of death, nor of the fear of God. It answers a deep need in man. It is neither a metaphysic, nor a morality, but above all and essentially an intuition and a feeling. ... Dogmas are not, properly speaking, part of religion: rather it is that they are derived from it. Religion is the miracle of direct relationship with the infinite; and dogmas are the reflection of this miracle. Similarly belief in God, and in personal immortality, are not necessarily a part of religion; one can conceive of a religion without God, and it would be pure contemplation of the universe; the desire for personal immortality seems rather to show a lack of religion, since religion assumes a desire to lose oneself in the infinite, rather than to preserve one's own finite self. (173)

This, then, is the more accurate picture of the religious world Vincent grew up in, not the misleading form of Calvinism most biographers apply. However, it is also worth noting that it was not nearly so liberal and humanistic as it might seem in the context of this discussion. The fact that human existence is released from predestination and that God is in man, rather than separate, does not deny a strict, strong stress on suffering, demonstrated clearly in the work of Thomas a Kempus that Vincent embraced, and even an interpretation of it demanding physical self-abuse.

Erickson claims that the family was moving to even more liberal beliefs, which she details out mainly in a discussion of Uncle Stricker, the uncle Vincent lived with and studied under while preparing for his exams, who, according to her, was very influential on him, and she provides passages from Vincent's letters of the time to support that Vincent was attracted to him, attended and was influenced by his sermons, and was a good student determined to be successful; in spite of passages others have quoted demonstrating the opposite.

It is interesting to consider the two sides of Vincent's religion and faith. On the one hand, he is obviously influenced by both his father and his Uncle Stricker, who espouse a Groningen embracement of the individual and piety, clearly expressed in Uncle Stricker's final sermon, *Een Laaste Woord bij het Nederleggen zijner Evangeliebediening* (*A Last Word upon Retiring as Servant of the Gospel*, 1884), where he puts forth that "Ethics is not religion, but religion is ethics" and that "the religion of love is the root from which ethics grows," that true religion is to be a "servant of God," and that "religion is not the fruit of rational contemplation and scientific investigation, but especially, the understanding of the human heart of its kindred sense of a higher world than the purely sensual and material world." (174)

On the other hand, in applying these views to an extreme, Vincent breaks away from his father and his uncle, and it seems that the falling out they all have is a demonstration that Vincent is the more sincere spiritually, as he would claim the rest of his life. However, of course, it's not that simple. A close understanding of the religious views reveals some other differences, which I'm pleased to find Erickson pointing out. A major one has to do with Vincent's move into an evangelical faith that stressed being "born again," this based on the idea that we are all born in sin and need to be redeemed through the cross, a view rejected by Groningens, as it goes against the belief in the intrinsically divine nature of man. Furthermore, the letters of the time reveal Vincent's father continually concerned about the practical aspects, about Vincent earning a living. Of course, that might seem petty in relation to being a man of faith and holding high ethical and spiritual values, but it is also a reality of life, and it also does figure into self-responsibility, something Vincent fails to accept in any kind of practical sense.

Erickson puts a great deal of time into establishing Vincent's actions during this religious period of his life in the context of his religion, concluding:

> Far from being peculiar or eccentric, Van Gogh was actually a traditional disciple of the cross. In choosing to center his spiritual life on

> *The Imitation of Christ* and *Pilgrim's Progress,* moreover, van Gogh chose two of the most popular and influential devotional works in the Christian tradition. Like many Christians before him, he found them profoundly insightful and exemplary guides. In his self-denial, in his commitment to spiritual pilgrimage, in his active ministry to the poor, van Gogh sought to follow Jesus and imitate his life of humility, to the letter. Rather than pathological or even irrationally fanatical, he was entirely consistent with the tradition he chose, the imitatio Christi. Although thwarted in his efforts to live out his religious commitment and forced in disgust to abandon the institutional church, van Gogh did not bury his piety. As profoundly as these ideas had influenced him in the first twenty-eight years of his life, he did not forget them but took them with him on his pilgrimage to make art his next all-consuming work. (175)

This sounds good. He is not *insane* or eccentric, rather simply a sincere Christian in a common tradition of Christianity. But wait. Aren't his actions extreme? Isn't it abnormal to starve oneself, to purposely give up all comforts, even one's clothes, to find as unfriendly a hut to sleep in as possible? Isn't there a sense of being outside of the norm when a person seeks out the poorest, most unfriendly environment possible to live in? Is this completely driven by religion or faith, or is it at least partially a form of masochism justified by faith? Perhaps it is in a grand tradition of martyrs, but doesn't it go against natural human needs and desires?

Is salvation really the result of self-inflicted suffering?

These questions lead to other questions. Are martyrs insane? Was Christ insane? Take away the religious interpretation of Christ for a moment; see him as no different than any other man. Can it be claimed he suffered some form of psychosis?

Perhaps *insanity* is *necessary* to salvation.

What does it mean to be *insane* after all?

Part VI

Penance

Remember the memories quoted early on from Vincent's former language tutor Mendes da Costa. They included not only references to Vincent embracing Thomas a Kempis, but references to his interpretation of it in terms of his own self inflicted suffering. Here are some of the key passages:

I succeeded in winning his confidence and friendship very soon, which was so essential in this case; and as his studies were prompted by the best of intentions, we made comparatively good progress in the beginning - I was soon able to let him translate an easy Latin author. Needless to say, he, who was so fanatically devout in those days, at once started using this little bit of Latin knowledge to read Thomas a Kempis in the original. . . .

"John Bunyan's *Pilgrim's Progress* is of much more use to me, and Thomas a Kempis and a translation of the Bible; and I don't want anything more." I really do not know how many times he told me this, nor how many times I went to the Reverend Mr. Stricker to discuss the matter, after which it was decided again and again that Vincent ought to have another try. . . .

But before long the trouble would start afresh, and then he would come to me in the morning with an announcement I knew so well, "Mendes, last night I used the cudgel again," or, "Mendes, last night I got myself locked out again." It should be observed that this was some sort of self-chastisement resorted to whenever he thought he had neglected a duty. In fact, during those days he lived in his uncle's house, Rear Admiral J. van Gogh, director and commander of the naval base at Amsterdam; the house was a big building inside the naval dockyard. Well, whenever Vincent felt that his thoughts had strayed further than they should have, he took a cudgel to bed with him and belabored his back with it; and whenever he was convinced he had forfeited the privilege of passing the night in his bed, he slunk out of the house unobserved at night, and then, when he came back and found the door double-locked, was forced to go and lie on the floor of a little wooden shed, without bed or blanket. He preferred to do this in winter, so that the punishment, which I am disposed to think arose from mental masochism, might be more severe. . . .

As I was not so very busy in those days, he often stayed talking for a while after the lesson, and naturally we often discussed his former profession, the art dealing business. He had kept quite a number of the prints which he had collected in those days, little lithographs after paintings, etc. He brought them to show me repeatedly, but they were always completely spoiled: the white borders were literally covered with quotations from Thomas a Kempis and the Bible, more or less connected with the subject, which he had scrawled all over them. Once he made me a present of *De Imitatione Christi,* without any intention of converting me, only to acquaint me with the deep humanity of it. . . .
Amsterdam, 30 November 1910 (176)

If Mendes da Costa is to be believed, and there is little reason not to believe him, it is clear that Vincent was practicing self abuse, including flagellation (a mortification of the flesh through some form of whipping it). However, before simply labeling Vincent as masochistic, this needs to be placed in context. Flagellation has been practiced as a form of religious ritual from the beginnings of recorded history, includes followers of Isis during the glorious Dynasties of Ancient Egypt and Dionysus during the Golden Age of Greece, and begins gaining a foothold in Christianity towards the end of the 10th century *as a form of penance*, supported by the idea that one needs to put aside fleshly concerns in order to become spiritually pure, which in turn gets connected with putting aside self-concerns, egotism. One is to worship God, not man. One is to become humble. Remember, it is thought that humans are born in sin. This all gets tied to the idea that in purifying the body one purifies the soul. (177)

The main Christian support for flagellation comes from Christ's own punishments and humiliation before he was crucified by the Roman soldiers. Interestingly, it is thought that such physical abuse as flagellation was not an established Roman punishment (maybe not flagellation, but they certainly were enamored with endless forms of physical punishment), but, rather, a Jewish one (the Old Testament is filled with examples of it), and that Pilate's hope in demanding this of Christ was to satisfy the Jewish Elders so it would not be necessary to Crucify Christ, a plan that obviously failed. Ultimately, whatever the reasons behind it and for it, the fact that Jesus accepted his humiliation, both the flagellation and ultimately the crucifixion, has been used ever since to sanctify the need for an acceptance of physical suffering as necessary to salvation. And remember, Vincent wanted to become Christlike.

It is recorded that in the 11th century Dominicus Loricatus accompanied a repetition of the entire Psalter twenty times in a week by giving himself 100 lash strokes for each psalm. At about the same time, St. Peter Damian wrote a treatise in praise of self-flagellation that appears to have been highly influential. By the 12th century Flagellants had become organized into groups and began demonstrating their piety in public spaces, combining flagellation with elaborate ceremonies, including processions, hymns, and costumes. There are several incidents reported throughout the 13th and 14th centuries, the main one sparking a movement taking place in Perugia in 1259 (thought perhaps to have been sparked by the preaching of Gioacchino da Fiore). Thousands of people joined in precessions, singing, carrying crosses and banners, and whipping themselves. Interestingly, especially in relation to Vincent, there are recorded acts of charity in conjunction with the violence. But the violence dominated, and those not joining in were accused of being in league with the devil. These, the Jews, and any priests opposing this uprising were killed. This uprising spread rapidly through Modena, Bologna, Reggio and Parma, becoming so disruptive that the Catholic Pope condemned and banned the movement January, 1261, and it lost its momentum in Italy.

However, it continued to spread into Austria and Germany, peaking in the mid-fourteenth century during the Black Plague (also referred to as the Black Death or Great Death). The Brothers of the Cross movement in Germany is well documented. They moved across Germany, wearing white robes, stopping each day at a new place, and holding two rituals before moving on. These rituals consisted of reading a letter supposedly delivered by an angel that justified Flagellantism, which included falling to their knees, scourging themselves, making gestures to indicate their sin, and, of course, striking themselves rhythmically to songs until blood flowed.

As the movement grew and spread, the Catholic Church became less tolerant. Clement VI condemned them in a bull, October 20, 1349, and ordered Church leader to suppress them. Gregory XI reaffirmed this in 1372 and suggested Flagellants were connected to other heretical groups, such as the Beghards. They were also accused of such heresies as questioning the need for the sacraments, denying ecclesiastical jurisdiction, and claiming to work miracles. One of the Flagellant leaders in Germany actually claimed to be the reincarnated emperor Frederick II. He ultimately got burned by the Inquisition. In 1399, the White Penitents or Bianchi movement arose in northern Italy, the result of a peasant who saw a vision, eventually gathering 15,000 followers for a march on Rome. But, as the Church was wont to do, on the order of Boniface IX, it ordered one of the leaders burned at the stake, and the movement quickly lost its momentum.

The Inquisition was now moving rapidly across Europe. In 1414, two groups of Flagellants totaling over 100 members were burned in Germany. In 1416, three hundred were burned in Thuringia in a single day. Such actions ended the large public movements. However, kings and queens such as Henry III of France and Catherine de Medici continued to support them, and the Catholic Church accepted flagellation as a form of penance, so long as it was done within the frame of the Church.

Also, various groups continue, such as the Hermanos Penitentes in colonial Spanish America (currently Colorado and New Mexico), continued against the Church's orders. And many contemporary Flagellant ceremonies continue to take place, especially during Lent in Spain, Portugal, and Italy

Perhaps the most important of these groups today is Opus Dei (also known as The Prelature of the Holy Cross), an organization sanctioned by the Roman Catholic Church that stresses that everyone is called to holiness and ordinary life is the means to sanctity.

The organization was founded by Josemaria Escriva, a Spanish Roman Catholic priest (later canonized by Pope John Paul II) on October 2, 1928, the result of him seeing a vision, and sanctified by Pope Pius XII in 1950;

and it considered by many to be the most controversial group in the Catholic church with serious questions being raised about it recruiting methods, its strict rules, its secretiveness, its elitism, its very conservative politics, and its embracement of *mortification of the flesh*.

It has some 87,000 members in over 80 countries, with about 30 percent of them leading celibate lives in official Opus Dei establishments, and the rest living traditional Catholic lives and perusing secular careers. Originally just open to men, in 1930, it also took in women. In 1939 Escirva published *The Way*, giving its major maxims about spirituality. The 1940s found involved in a great deal of controversy from such people as Jesuit leader Wlodimir Ledochowski, who considered it "dangerous," as a result of it "secretive" qualities. But it continued to gain support from the Catholic Popes, moved its center to Rome in 1946, and was granted definitive approval by Pope Pius XII in 1950, meaning married people could join the organization. Escriva died in 1975 and was succeeded by Alvaro del Portillo. In 1982, Opus Dei became a personal prelature. In 1994 Javier Echevarria became the Prelate. The Catholic Church has continued to support and glorify Opus Dei, including the September, 2005 blessing by Pope Benedict XVI of a newly installed statue of Escriva located in an outside wall niche of St. Peter's Basilica.

However, in spite of the support of the Catholic Popes, it remains a very controversial organization, and it was represented negatively in *The Da Vinci Code* by Dan Brown (a film both Opus Dei and the Catholic Church condemned).

The general thrust of the doctrines of Opus Dei embraces the individual, espousing the idea that each of us should aspire to be a saint, that ordinary life is to be sanctified, that God is to be found in daily life, that "God created man to work" and that the Christian duty is to follow the example of Christ, for Jesus "has done everything well (Mark, 7:37). Other beliefs include man's "divine filiation" (we are the children of God), "freedom" (personal choice and responsibility), and "charity" (love of God and others). All members are trained to follow a "plan of life" or the "norms of piety," also called the Catholic devotions that include the following: the Heroic Minute (waking up at a specific time and uttering "serviam"; I will serve); the Morning Offering (a focus on doing everything for the glory of God); Spiritual reading (daily reading of the New Testament each day); Mental prayer (talk with God); Mass, Communion, and Thanksgiving after Communion; Rosary; The Preces (the Opus Dei's common prayer); Angelus (the prayer that recalls God becoming man); Memorare prayer to the Blessed Virgin Mary (to help the member most in need at the moment); the visit to the Blessed Sacrament

(Catholic practice of greeting Jesus in the Eucharist); Examination of conscience to end each day; Three Hail Marys before bed; and other spontaneous prayers throughout the day; weekly confession, Marian antiphon each Saturday; and the application of Psalm 2 for mental prayers on Tuesdays. Furthermore, each member needs to make a day pilgrimage and recite three 5-decate rosaries each May to honor Mary, and participate in a one week (supernumeraries) or three week (numeraries) spiritual retreat seminar each year.

These rules, while they probably seem a bit "over-the-top" for outsiders, are not so controversial, and really not much different than other Christian religions, especially those that focus on personal salvation. However, in addition to the secrecy of the organization and its support of very right-wing politics, what has caused the controversy is its embracement of the "mortification of the flesh."

In Christianity, as stated earlier, this mainly comes in as a result of Christ's own acceptance of punishment and his words for any who would follow him, "Let him deny himself, take up his cross daily and follow me." (Luke 9:23) Those who embrace mortification of the flesh claim that those opposing them have lost their Christianity, have lost the realization of the enormity of sin, the evilness of offense against God, have lost the truths of how sinful human nature is, what a huge battle is necessary to drive it out of human nature, and have lost the spirit of sacrifice, the need to move beyond physical comfort and desire for spiritual love.

For the average person, Opus Dei demands lesser sacrifices, but for celibate members such sacrifices include sleeping without a pillow or on the floor, fasting, remaining silent for certain hours each day, and more controversial, the wearing of a cilice, a metal chain with inward-pointing spikes worn around the upper thigh (generally for two hours daily).

Escriva is supposed to have performed more extreme mortification, including flailing himself over a thousand times, and such maxims as "No ideal becomes a reality without sacrifice. Deny yourself. It is so beautiful to be a victim!" (*The Way*, 175) and "Blessed be pain. Loved be pain. Sanctified be pain. . . . Glorified be pain!" (*The Way*, 208)

Critics of Opus Dei include Opus Dei Awareness Network (ODAN), Maria Carmen del Tapia (a former member), Fr. James Martin (a Jesuit writer), Penny Lernoux (and other supports of Liberation Theology), and Michael Walsh (a former Jesuit), David Clark (anti-cult activist). The general attacks on the group include such recruiting practices as "love bombing" (showering potential members with praise), programmed forming of friendships for the purpose of recruitment (including written reports on the progress in such activities), the strong control the organization holds over its members, even requiring members to submit all incoming and outgoing mail, forbidding members to read certain books, pressure to severe contacts with non-members, including their own family, supporting the governments of Francisco Granco, Augusto Pinochet, and Alberto Fujimori, and even Adolf Hitler.

Supporters deny all of these accusations or say that they were only in place briefly, and basically claim that the detractors are misrepresenting Opus Dei. (178)

The reminiscences of Mendes da Costa are the only place there is any mention of Vincent going to the extreme of flagellation. However, it seems likely to be true for several reasons. First, there is no reason for Mendes da Costa to make it up. It is put forth in a reasoned manner. It also fits with Vincent's general penchant for self-denial, for self-punishment, including such things as going without food and sleeping outside to discipline himself. It fits his form of Christianity at the time, perhaps to the extreme, but not to a contradiction, rather to a confirmation. It also fits his personality that he took it to an extreme, something he did with everything in his life. And it fits his general tendency to self-punishment to the extreme, evident in such actions as holding his hand over a burning candle, cutting off his ear, and ultimately committing suicide.

In giving Vincent *credit* for self-abuse by claiming it is a sincere form of Christianity, we must in turn support it, at least in principle in its various forms throughout the history of Christianity (and if not support it certainly not condemn it in its appearances in other religions and cultures). Either we say something was psychologically wrong with Vincent, perhaps simply such a desire to be a good Christian that he couldn't see the error of his ways, perhaps something more serious, or we say he was correct, and those of us not practicing such self-abuse are wrong.

At the very least, some interesting and important questions can be raised: Are *suffering* and *sacrifice* necessary for *salvation*? What is *salvation*? Does this have anything at all to do with *the mapping out of meaning and value in human existence?*

The key is to separate the *literal* aspects of *physical* existence with the *symbolic expressions* of *spiritual* existence. This is a major confusion made by many sincere people. The brain is the physical holder of the mind. The mind is the invisible reality brought into physical reality in the brain. It is in the mind, specifically in the right hemisphere of the cerebrum, *beyond explanation*, where meaning and value exist, where the spiritual and the physical merge. What Vincent believed during the time he was trying to be *Christlike in religious terms* was that one needed to be physically Christlike, needed to suffer physically as Christ had suffered physically. His great insight would come during the year after he *failed in religious terms*. Literal physical suffering might lead to psychological suffering, and thus to mental insights, to a *realization of the spiritual*, but it can also be a barrior to the connection of physical and spiritual worlds in that it can be seen as the same thing, thus denying that there is a spiritual world at all. This was Vincent's mistake. His realization of this mistake would result in his great *insight*.

Part VII

Transformation

There are no letters exchanged by Vincent and Theo from August 1879 to July 1880, the longest break since their correspondence began, and few letters or documents of any kind remain to detail Vincent's life and feelings during this time. Not only do Vincent's letters to Theo and the rest of his family end, but so do his desires to be an evangelist, to embrace religion, and to pursue a career as a servant of God, at least under the umbrella of formalized religion. It is clear that by August, 1880 Vincent's extreme self-denial practices have ended, at least in terms of Christianity. His religion has not given him the map of meaning and value he wants.

Then, after months of silence, he writes what might be his most endearing letter to Theo, a letter that clearly reveals him as lost and feeling alone:

> My Dear Theo,
>
> I am writing to you rather reluctantly because, for a good many reasons, I have kept silent for such a long time. To some extent you have become a stranger to me, and I to you perhaps more than you think. It is probably better for us not to go on like that. It is probable that I would not have written to you even now, were it not that I feel obliged, compelled, to do so - because, be it noted, you yourself have compelled me to.
>
> I heard in Etten that you had sent 50 francs for me. Well, I have accepted them. With reluctance, of course, with a feeling of some despondency, of course, but I have reached a sort of impasse, am in trouble, what else can I do? And so I am writing to thank you.
>
> As you may know, I am back in the Borinage. Father said he would prefer me to stay somewhere near Etten, but I refused and I believe I was right to do so. To the family, I have, willy-nilly, become a more or less objectionable and shady sort of character, at any rate a bad lot. How then could I then be of any use to anyone? And so I am inclined to think the best and most sensible solution all round would be for me to go away and

to keep my distance, to cease to be, as it were. What the moulting season is for birds - the time when they lose their feathers - setbacks, misfortune and hard times are for us human beings. You can cling on to the moulting season, you can also emerge from it reborn, but it must not be done in public.

The thing is far from amusing, not very exhilarating, and so one should take care to keep out of the way. Well, so be it.

Now, though it is a fairly hopeless task to regain the trust of an entire family, one which has perhaps never been wholly weaned from prejudice and other equally honourable and respectable qualities, I am not entirely without hope that, bit by bit, slowly but surely, the good relationship between one and all may be restored. In the first place I should be glad to see this good relationship - to put it no more strongly than that - restored at least between Father and me, and further, I set great store by seeing it restored between the two of us. A good relationship is infinitely preferable to a misunderstanding.

Now I must trouble you with certain abstract matters, hoping that you will listen to them patiently. I am a man of passions, capable of and given to doing more or less outrageous things for which I sometimes feel a little sorry. Every so often I say or do something too hastily, when it would have been better to have shown a little more patience. Other people also act rashly at times, I think.

This being the case, what can be done about it? Should I consider myself a dangerous person, unfit for anything? I think not. Rather, every means should be tried to put these very passions to good effect.

To mention just one by way of an example, I have a more or less irresistible passion for books and the constant need to improve my mind, to study if you like, just as I have a need to eat bread. You will understand that. When I lived in other surroundings, surroundings full of pictures and works of art, I conceived a violent, almost fanatical passion for those surroundings, as you know. And I do not regret that, and even now, *far from home, I often feel homesick for the land of pictures*

You may remember that I knew very well (and it may be that I know it still) what Rembrandt was or what Millet was or Jules Dupré or Delacroix or Millais or Matthijs Maris.

Well, today I am no longer in those surroundings, yet they say that what is known as the soul never dies but lives on for ever, continuing to seek for ever and again.

So instead of succumbing to my homesickness I told myself: your land, your fatherland, is all around. So instead of giving in to despair I chose active melancholy, in so far as I was capable of activity, in other words I chose the kind of melancholy that hopes, that strives and that seeks, in preference to the melancholy that despairs numbly and in distress. I accordingly made a more or less serious study of the books within my reach, such as the Bible and Michelet's *La révolution Française*, and then last winter Shakespeare and a little Victor Hugo and Dickens and Beecher

Stowe and recently Æschylus and then various less classical writers, a few great minor masters. You know, don't you, that Fabritius and Bida are counted among the minor masters?

Now anyone who becomes absorbed in all this is sometimes considered outrageous, `shocking,' sinning more or less unwillingly against certain forms and customs and proprieties. It is a pity that people take that amiss.

You know, for example, that I have often neglected my appearance. I admit it, and I also admit that it is `shocking.' But look here, lack of money and poverty have something to do with it too, as well as a profound disillusionment, and besides, it is sometimes a good way of ensuring the solitude you need, of concentrating more or less on whatever study you are immersed in. One essential study is that of medicine. There is scarcely anybody who does not try to acquire some knowledge of it, who does not at least try to grasp what it is about (and you see, I still know absolutely nothing about it). And all these things absorb you, preoccupy you, set you dreaming, musing and thinking.

Now for the past five years or so, I don't know how long exactly, I have been more or less without permanent employment, wandering from pillar to post. You will say, ever since such and such a time you have been going downhill, you have been feeble, you have done nothing. Is that entirely true?

What is true is that I have at times earned my own crust of bread, and at other times a friend has given it to me out of the goodness of his heart. I have lived whatever way I could, for better or for worse, taking things just as they came. It is true that I have forfeited the trust of various people, it is true that my financial affairs are in a sorry state, it is true that the future looks rather bleak, it is true that I might have done better, it is true that I have wasted time when it comes to earning a living, it is true that my studies are in a fairly lamentable and appalling state, and that my needs are greater, infinitely greater than my resources. But does that mean going downhill and doing nothing?

You might say, but why didn't you go through with university, continue as they wanted you to? To that I can only reply that it was too expensive, and besides, the future then looked no better than it does now, along the path I am now taking.

And I must continue to follow the path I take now. If I do nothing, if I study nothing, if I cease searching, then, woe is me, I am lost. That is how I look at it - keep going, keep going come what may.

But what is your final goal, you may ask. That goal will become clearer, will emerge slowly but surely, much as the rough draught turns into a sketch, and the sketch into a painting through the serious work done on it, through the elaboration of the original vague idea and through the consolidation of the first fleeting and passing thought.

You should know that it is the same with evangelists as it is with artists. There is an old academic school, often odious and tyrannical, the `abomination of desolation', in short, men who dress, as it were, in a suit of steel armour, a cuirass, of prejudice and convention. When they are in charge, it is they who hand out the jobs and try, with much red tape, to keep them for their protégés and to exclude the man with an open mind.

Their God is like the God of Shakespeare's drunken Falstaff, "the inside of a church." Indeed, by a strange coincidence, some evangelical (???) gentlemen have the same view of matters spiritual as that drunkard (which might surprise them somewhat were they capable of human emotion). But there is little fear that their blindness will ever turn into insight.

This is a bad state of affairs for anyone who differs from them and protests with heart and soul and all the indignation he can muster. For my part, I hold those academicians who are not like these academicians in high esteem, but the decent ones are thinner on the ground than you might think.

Now, one of the reasons why I have no regular job, and why I have not had a regular job for years, is quite simply that my ideas differ from those of the gentlemen who hand out the jobs to individuals who think as they do. It is not just a question of my appearance, which is what they have sanctimoniously reproached me with. It goes deeper, I do assure you.

I am telling you all this not to complain, not to make excuses for matters in which I may perhaps have been somewhat at fault, but simply to tell you the following: during your final visit last summer when we were walking together near that abandoned mineshaft which they call "La Sorcière," you reminded me of another walk we once took at another time near the old canal and the mill at Rijswijk, and, you said, we used to agree about many things, but, you added, "You have changed since then, you are no longer the same." Well, that is not entirely true. What has changed is that my life then was less difficult and my future seemingly less gloomy, but as far as my inner self, my way of looking at things and of thinking is concerned, that has not changed. But if there has indeed been a change, then it is that I think, believe and love more seriously now what I thought, believed and loved even then.

So you would be mistaken should you continue to think that I have become less keen on, say, Rembrandt, Millet, or Delacroix or whoever or whatever, for the reverse is the case, but there are many different things worth believing and loving, you see - there is something of Rembrandt in Shakespeare, something of Correggio or of Sarto in Michelet and something of Delacroix in Victor Hugo, and there is also something of Rembrandt in the Gospel or, if you prefer, something of the Gospel in Rembrandt, it comes to much the same thing, provided you understand it properly, do not try to distort it and bear in mind that the elements of the comparisons are not intended to detract in any way from the merits of the original individuals.

And in Bunyan there is something of M. Maris or of Millet, a reality that, in a manner of speaking, is more real than reality itself, something hitherto unknown that, if only you can read it, will tell you untold things. And in Beecher Stowe there is something of Ary Scheffer.

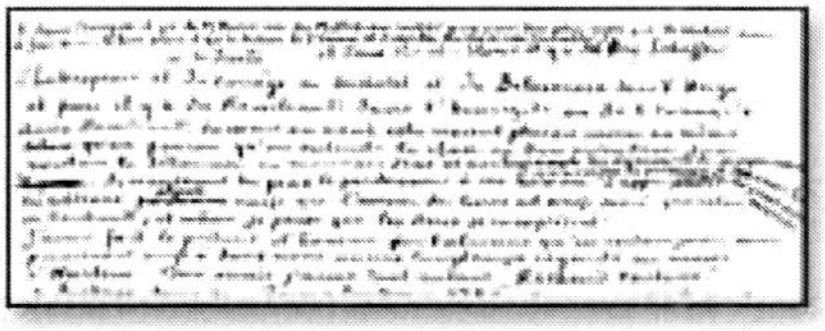

Now, if you can forgive someone for immersing himself in pictures, perhaps you will also grant that the love of books is as sacred as that of Rembrandt, indeed, I believe that the two complement each other.

I very much admire the portrait of a man by Fabritius that we stood looking at for a long time in the gallery in Haarlem one day when we took another walk together. Admittedly, I am as fond of Dickens's 'Richard Cartone' [Sydney Carton] in his Paris & Londres in 1793 [*A Tale of Two Cities*], and I could point to other particularly gripping characters in other books with a more or less striking resemblance. And I think that Kent, a character in Shakespeare's "*King Lear*," is as noble and distinguished a man as that figure by Th. de Keyser, though Kent and King Lear are reputed to have lived much earlier.

Let me stop there, but my God, how beautiful Shakespeare is, who else is as mysterious as he is; his language and method are like a brush trembling with excitement and ecstasy. But one must learn to read, just as one must learn to see and learn to live.

So please don't think that I am renouncing anything, I am reasonably faithful in my unfaithfulness and though I have changed, I am the same, and what preys on my mind is simply this one question: what am I good for, could I not be of service or use in some way, how can I become more knowledgeable and study some subject or other in depth? That is what keeps preying on my mind, you see, and then one feels imprisoned by poverty, barred from taking part in this or that project and all sorts of necessities are out of one's reach. As a result one cannot rid oneself of melancholy, one feels emptiness where there might have been friendship and sublime and genuine affection, and one feels dreadful disappointment gnawing at one's spiritual energy, fate seems to stand in the way of affection or one feels a wave of disgust welling up inside. And then one says "How long, my God!"

Well, that's how it is, can you tell what goes on within by looking at what happens without? There may be a great fire in our soul, but no one ever comes to warm himself by it, all that passers-by can see is a little smoke coming out of the chimney, and they walk on.

All right, then, what is to be done, should one tend that inward fire, turn to oneself for strength, wait patiently - yet with how much impatience! - wait, I say, for the moment when someone who wants to comes and sits down beside one's fire and perhaps stays on? Let him who believes in God await the moment that will sooner or later arrive.

Well, right now it seems that things are going very badly for me, have been doing so for some considerable time, and may continue to do so well into the future. But it is possible that everything will get better after it has all seemed to go wrong. I am not counting on it, it may never happen, but if there should be a change for the better I should regard that as a gain, I should rejoice, I should say, at last! So there *was something after all!*

But, you will say, what a dreadful person you are, with your impossible religious notions and idiotic scruples. If my ideas are impossible or idiotic then I would like nothing better than to be rid of them. But this is roughly the way I see things. In *Le Philosophe sous les Toits by Souvestre* you can read what a man of the people, a simple craftsman, pitiful if you will, thinks of his country: "Tu n'as peut-être jamais pensé à ce que c'est la patrie, reprit-il, en me posant une main sur l'épaule; c'est tout ce qui t'entoure, tout ce qui t'a élevé et nourri, tout ce que tu as aimé. Cette campagne que tu vois, ces maisons, ces arbres, ces jeunes filles qui passent là en riant, c'est la patrie! Les lois qui te protègent, le pain qui paye ton travail, les paroles que tu échanges, la joie et la tristesse qui te viennent des hommes et des choses parmi lesquels tu vis, c'est la patrie! La petite chambre où tu as autrefois vu ta mère, les souvenirs qu'elle t'a laissés, la terre où elle repose, c'est la patrie! Tu la vois, tu la respires partout! Figure toi, tes affections et tes besoins, tes souvenirs et ta reconnaissance, réunis tout ça sous un seul nom et ce nom sera la patrie." [You may never have thought what your country really is, he continued, placing his hand on my shoulder; it is everything around you, everything that has raised and nourished you, everything that you have loved. This countryside that you see; these houses, these trees, these young girls laughing as they pass, that is your country! The laws that protect you, the bread that rewards your labour, the words you speak, the joy and sorrow that come from the people and things in whose midst you live, that is your country! The little room where you used in days gone by to see your mother, the memories she left you, the earth in which she rests, that is your country! You see it, you breathe it, everywhere! Imagine your rights and your duties, your affections and your needs, your memories and your gratitude, gather all that together under a single name, and that name will be your country.]

In the same way I think that everything that is really good and beautiful, the inner, moral, spiritual and sublime beauty in men and their works, comes from God, and everything that is bad and evil in the works of men and in men is not from God, and God does not approve of it.

But I cannot help thinking that the best way of knowing God is to love many things. Love this friend, this person, this thing, whatever you like, and you will be on the right road to understanding Him better, that is what I keep telling myself. But you must love with a sublime, genuine, profound sympathy, with devotion, with intelligence, and you must try all the time to understand Him more, better and yet more. That will lead to God, that will lead to an unshakeable faith.

To take an example: one man will love Rembrandt, genuinely, and that man will surely know that there is a God, he will really believe it. Another will make a thorough study of the French Revolution - he will not be an unbeliever, he will see that there is a supreme authority that manifests itself in great affairs. Yet another has recently attended a free course of lectures at the great university of sorrow and has heeded the things he saw with his eyes and heard with his ears, and has reflected upon them. He too will come to believe in the end and will perhaps have learned more than he can tell.

Try to grasp the essence of what the great artists, the serious masters, say in their masterpieces, and you will again find God in them. One man has written or said it in a book, another in a painting. Just read the Bible and the Gospel, that will start you thinking, thinking about many things, thinking about everything, well then, think about many things, think about everything, that will lift your thoughts above the humdrum despite yourself. We know how to read, so let us read!

Now then, you may well have bouts of being a little absent-minded, a little dreamy, indeed there are some who become too absent-minded, a little too dreamy. That may indeed have happened with me, but all in all that is my own fault, maybe there as a reason for it, perhaps I was lost in thought for one reason or another, anxious, worried, but one gets over that in the end. The dreamer sometimes falls into the doldrums, but is said to emerge from them again. And the absent-minded person also makes up for it with bouts of perspicacity. Sometimes he is a person whose right to exist has a justification that is not always immediately obvious to you, or more usually, you may absent-mindedly allow it to slip from your mind. Someone who has been wandering about for a long time, tossed to and fro on a stormy sea, will in the end reach his destination. Someone who has seemed to be good for nothing, unable to fill any job, any appointment, will find one in the end and, energetic and capable, will prove himself quite different from what he seemed at first.

I am writing somewhat at random, writing whatever flows from my pen. I should be very happy if you could see in me something more than a kind of fainéant [idler]. For there is a great difference between one idler and another idler. There is someone who is an idler out of laziness and lack of character, owing to the baseness of his nature. If you like, you may take me for one of those. Then there is the other kind of idler, the idler despite himself, who is inwardly consumed by a great longing for action who does nothing because his hands are tied, because he is, so to speak, imprisoned somewhere, because he lacks what he needs to be productive, because disastrous circumstances have brought him forcibly to this end. Such a one does not always know what he can do, but he nevertheless instinctively feels, I am good for something! My existence is not without reason! I know that I could be a quite a different person! How can I be of use, how can I be of service? There is something inside me, but what can it be? He is quite another idler. If you like you may take me for one of those.

A caged bird in spring knows perfectly well that there is some way in which he should be able to serve. He is well aware that there is something to be done, but he is unable to do it. What is it? He cannot quite remember, but then he gets a vague inkling and he says to himself, "The others are building their nests and hatching their young and bringing them up," and then he bangs his head against the bars of the cage. But the cage does not give way and the bird is maddened by pain. "What a idler," says another bird passing by - what an idler. Yet the prisoner lives and does not die. There are no outward signs of what is going on inside him; he is doing well, he is quite cheerful in the sunshine.

But then the season of the great migration arrives, an attack of melancholy. He has everything he needs, say the children who tend him in his cage - but he looks out, at the heavy thundery sky, and in his heart of hearts he rebels against his fate. I am caged, I am caged and you say I need nothing, you idiots! I have everything I need, indeed! Oh! please give me the freedom to be a bird like other birds!

A kind of idler of a person resembles that kind of idler of a bird. And people are often unable to do anything, imprisoned as they are in I don't know what kind of terrible, terrible, oh such terrible cage.

I do know that there is a release, the belated release. A justly or unjustly ruined reputation, poverty, disastrous circumstances, misfortune, they all turn you into a prisoner. You cannot always tell what keeps you confined, what immures you, what seems to bury you, and yet you can feel those elusive bars, railings, walls. Is all this illusion, imagination? I don't think so. And then one asks: My God! will it be for long, will it be for ever, will it be for eternity?

Do you know what makes the prison disappear? Every deep, genuine affection. Being friends, being brothers, loving, that is what opens the prison, with supreme power, by some magic force. Without these one stays dead. But whenever affection is revived, there life revives. Moreover, the prison is sometimes called prejudice, misunderstanding, fatal ignorance of one thing or another, suspicion, false modesty.

But to change the subject - if I have come down in the world, you have in a different way come up in it. And if I have forfeited sympathy, you have gained it. I am glad of that, I say that it in all sincerity, and it will always give me pleasure. If you lacked seriousness or consideration, I would be fearful that it might not last, but since I think that you are very serious and very considerate, I tend to believe it will!

But if you could see me as something other than a idler of the bad sort, I should be very happy.

For the rest, if I can ever do anything for you, be of some use to you, know that I am at your disposal. Now that I have accepted what you have given me, you are, should I be able to render you some service, in a position to ask me. It would make me happy, and I should take it a sign of trust. We have moved rather far apart and may in certain respects have perhaps different views, but some time, some day, one of us may be of service to the other.

For now I shake your hand, thanking you once again for having been so good to me. If, one of these days, you feel like writing, my address is, chez Ch. Decrucq, Rue du Pavillon 8, Cuesmes, near Mons, and know that it will do me good to hear from you.

Yours, Vincent (179)

The letter begins with an apology, revealing how well Vincent could observe and judge himself. If, as he says, he has sinned against social conventions, the sins were minor and perhaps necessary for his studies and self-assessment. He then discusses the image of himself as an "idler," revealing his feeling of simply not knowing where to take his life, of not being an idler on the inside, but struggling with how to translate he inner qualities to the outer world. A later passage reveals that he has expanded his views of Christianity, of how to know God beyond a strict embracement of only *religion* (remember how previously he had given up Michelet and all other books except the *Bible*). Now he writes "I think that everything which is really good and beautiful—of inner, spiritual and sublime beauty—in men and their works, comes from God, and that all which is bad and wrong in men and their works is not of God." And he uses a love of Rembrandt as an example. In truth Vincent had not stopped reading all other books, had always continued his reading of literature outside of the Bible and even outside of strictly religious writings, but still the perspective he puts on it now has shifted.

Joann van Gogh-Bonger writes of it in her *Memoir*:

Now, in the days of deepest discouragement and darkness, at last the light began to dawn. Not in books would he find satisfaction, nor find his work in literature, as his letters sometimes suggested; he turned back to his old love: "I said to myself, O I will take up my pencil again, I will go on with my drawing, and from then on everything has seemed transformed for me." It sounds like a cry of deliverance, and once more, "do not fear for me. If I can only continue to work, it will set me right again." (130)

By September 24, he has moved so much into the light he is able to make a clear statement that he has decided to become an artist. His letter to Theo begins,

> Your letter has done me good and I thank you for having written to me in the way you have.
>
> The roll with a new selection of etchings and various prints has just arrived. First and foremost the masterly etching, "Le Buisson" by Daubigny and Ruysdael. Well! I propose to make two drawings, in sepia or something else, one after that etching, the other after "Le Four dans les Landes" by Th. Rousseau. Indeed, I have already done a sepia of the latter, but if you compare it with Daubigny's etching you will see that it contrasts feebly, although considered on its own the sepia may betray some tone and sentiment. I shall have to return to it and tackle it again.
>
> I am still working on Bargue's *Cours de Dessin,* and intend to finish it before I go on to anything else, for both my hand and my mind are growing daily more supple and strong as a result, and I cannot thank Mr. Tersteeg enough for having been so kind as to lend it to me. The models are outstanding. Meanwhile I am reading one book on anatomy and another on perspective, which Mr. Tersteeg also sent me. These studies are demanding and sometimes the books are extremely tedious, but I think all the same that it's doing me good to study them.
>
> So you see that I am working away hard, though for the moment it is not yielding particularly gratifying results. But I have every hope that these thorns will bear white blossoms in due course and that these apparently fruitless struggles are nothing but labour pains. First the pain, then the joy. . . . (181)

From this point on, Vincent's fanatical devotion to religion has been shifted to art. Do not believe that he suddenly is a great artist. Up to this point, he has, truthfully, neither received the necessary training, nor demonstrated any special artistic ability. But he is now going to focus on art, and as always, that means taking it to an extreme.

Part VIII

Love and Syphllus

It is the summer of 1881.

This time the drama is real, not fictional, not guesswork.

Vincent holds his hand over the flame of an oil lamp. "I will keep my hand here until I am allowed to see Kee!"

Pastor Stricker, the uncle who once befriended him and whom he once wrote glowing praise of, is not impressed. He blows out the flame. "Now leave!"

Vincent has fallen in love with Kee Vos-Stricker, a cousin he first meet while she was still married to her now diseased husband Christoffel Martinus Vos, who has left her with Johannes Paulus, an eight year old son.

At the time, he told no one of his love, but afterwards, he poured out his heart to Theo in twelve long letters, clearly indicating the he and Kee had spent many hours together in such direct statements as "she and I had walked together and spoken together for days and weeks. (182) Several things are obvious. First, he has most certainly fallen in love. On November 3, he first confirmed it to Theo: "There is something in my heart that I must tell you; perhaps you know about it already and it is not new to you. I want to tell you that this summer such a deep love has grown in my heart for Kee Vos that I cannot find other words for it than: it is as if Kee Vos is my nearest and I the nearest of Kee Vos, and—I have told it her in these words." Second, Kee has given him no illusions. She does not return his love. The following sentence leaves no doubt: "But when I said this, she replied that to her, past and future remained one, so that she could never return my feelings." By the time Vincent begins pouring out his feelings to Theo, the rest of the family has already been made aware of them, and all have let him know they are strongly against it (except apparently his Uncle Vincent who treated it all as something of a joke).

Such a dramatic declaration of love. But Uncle Stricker is not impressed. Kee has stated it clearly, left no doubt. She is not interested in Vincent as a lover or future husband. And as the drama unfolds, the letters reveal that, eventually Vincent begins to accept this, and even begins to see Kee in less emotional ways, but with his usual compassion for her suffering over the loss of her husband.

Before it calms down for Vincent, a lot of anger and disillusionment with his family fills the air.

As Vincent's religious period, his attempt to find his own salvation through religion comes to an end, once again in a sudden way, his relationship to his father matches it, and not surprisingly the "ideal" father to please becomes an overly demanding father to rebel against.

Humberto Nagera, probably correctly, writes:

> During this period there was an increasing deterioration of Vincent's relationship with his home and especially of the relationship with his father. The very old conflicts with his father, especially his hostility and negative feelings so far expressed covertly and strongly reacted against,

> were to find an increasingly overt form of expression. The previous idealization of the father was thus coming to an end, transforming itself, if anything, into a sort of "negative idealization." This transformation allowed Vincent to split his ambivalent feelings in such a way that the father became the recipient of most of his negative feelings and hostility which was thus largely directed away from Theo who had become in the meantime a father substitute for him and the recipient of his positive feelings. (183)

For the group of critics led by Nagera, it is not surprising that a woman is involved, and in this case there can be no doubt about Vincent falling in love and being rejected.

Ken Wilkie summarizes it, with the usual assumptions about Eugenie, and connects into his next relationship with a woman (one where Vincent, as seems to be his nature, dramatically responds to his failure by embracing its opposite):

> Vincent now abandoned all hope of following his father's footsteps through religion and was on the point of following his mother's—into art. For even in the most unlikely situations in the Borinage, he would always find time to draw the miners plodding wearily to and from the pit. He actually wrote to Theo at this time that devotion to Rembrandt was as sincere as his devotion to Christ. Drawing, he said, liberated him. And he had begun to make copies after Millet, the French painter of labourers. . . .
>
> This was in April 1881. The spring began peacefully enough till Vincent's cousin Kee Vos and her little son came to stay with the Van Goghs for the summer. Kee Vos's husband had just died and Vincent felt a sympathy for her that developed into a deep passion—a feeling that had been smouldering in him since his unsuccessful love affair with Eugenie Loyer in London.
>
> Till now, he admitted to Theo, he had been living in the emotional wake of Eugenie's refusal. [There is no proof of his claim here, though of course it has been endlessly guessed at through some of the passages in his letters.] And his cousin Kee seemed to appeal to his combination of sexual need and human compassion. But when he declared his love, she, too, firmly rejected him. She was still in mourning and left immediately for Amsterdam. Vincent was in torment. He bombarded her with letters which she apparently refused to read. Theo even gave him money to go to Amsterdam to meet her, but she refused to see him. "Kee left the house as soon as she heard you were at the front door," said his uncle at the doorstep of Keizersgracht 453. Vincent was admitted to the parlour and was convinced Kee was still in the house. There followed the scene, borrowed from one of Multatuli's characters, where Vincent held his hand above the flame of an oil lamp, insisting he would keep it there for as long as he

> could not see Kee. But Pastor Stricker blew the lamp out and told his eccentric nephew to leave the house.
>
> On Christmas Day, 1881, things came to a head at Etten when Vincent and his father had a violent quarrel, mainly over Vincent's refusal to attend church. The minister told his son to get out within the hour and Vincent took the train to The Hague to take lessons with the painter Anton Mauve. While there his thoughts turned away from religion. He wrote to Theo: "I am a man and a man with passions, I must go to a woman, otherwise I shall freeze or turn to stone."
>
> The woman he found was a prostitute called Clasina Maria Hoornik [usually referred to as Sien], and not long after meeting her, he set up home with her and her daughter Maria. He spent several weeks in [the] hospital being treated for gonorrhea and meanwhile Clasina bore another child, Willem, several months after they had begun living with each other. When Vincent began to talk of marrying her, his father contemplated having he son confined in a mental asylum. Instead, the family threatened to withdraw Vincent's allowance unless he left her: he was faced with the choice between losing Clasina and losing his sole means of support. Unwillingly, he chose the former. (184)

As expected, Nagera places this follow-up relationship in the contexts of his views mentioned earlier:

> Whatever one may conclude about this relationship it still remains that it is the closest to a home life, to a family of his own—for which he so longed all his life—that Vincent ever got.
>
> Furthermore there is the fact that the massive anal regression that had taken place in his personality made the relationship to a prostitute a particularly suitable form of object choice, allowing for the gratification of important unconscious elements and phantasies belonging to the anal stage of development, a stage that played such a significant role in Vincent's character structure. In addition this relationship was, in terms of his anal regression, an aggressive, obstinate and provocative piece of behaviour.
>
> Vincent's masochistic needs, unconscious guilt and need for punishment found an excellent opportunity of fulfillment in a relationship that was bound to bring much disapproval, rejection, isolation, suffering and difficulty into his life. (185)

It is easy to believe Vincent's relationship with Clasina (Christine), whom he generally refers to as Sien, was driven by some fairly straightforward circumstances. He was now in his late twenties, and the assumption is that he had not had sex with a woman. Whether or not that is true, it would certainly be normal for a man his age to want to satisfy sexual desires, something he states more than once in his letters. The one most often quoted was written December 21, 1881, a lengthy one revealing his struggles dealing with his love for Kee Vos and the whole rejection of it by both her and his parents, and apparently everyone except him. In it he discusses rather openly his sexual desires, including the following passage: "And whether I do right or wrong, I have no choice, that damned wall is too cold for me, I need a woman, I cannot, will not, may not, live without love. I am only a man and a man with passions, I must have a woman, otherwise I shall freeze or turn to stone or, in short, I shall have let things browbeat me." (186)

In addition, it is clear Vincent is feeling bitter about the rejection by Kee Vos, and about everyone, including even Theo, but most especially his parents siding against him. It seems to him at the time that the entire world is against him. He has tried and failed in his religion/faith struggles. He could not win given the scenario he had created. He had tried to be Christ*like*, to such an extreme that he would live only for agape, and when that had failed, he had tried a more conventional form of Christianity, one that allowed for amore and Eros, but he had been rejected there as well. It is not surprising that he fells emotionally distraught. And it is also not a large claim to say that, in this sense, Clasina is a punishment, a self-rejection, especially since she represents such a strong rejection of the Christianity he has just embraced and now considers himself to have failed at exemplifying. Embracing her is, in other words, a way of rejecting the entire world of his parents, which he feels has just rejected him, and mixed into this rejection is an honest feeling of having failed. It's easily possible to imagine him shouting out to the night "All right, then, if I am a failure, I will prove to everyone that's what I am!"

Moving in with a down-and-out prostitute is about as perfect a solution as can be found. Along with these aspects, there are also the less dramatic but real desires throughout his life to seek out poor people, and to live with such people, carrying with it an honest desire to help them. I think it fair to say his empathy is real.

He doesn't reveal his new relationship to Theo until the following May:

> Last winter I met a pregnant woman, deserted by the man whose child she was carrying. A pregnant woman who walked the streets in the winter—she had her bread to earn, you'll know how. I took that woman on as a model and have worked with her all winter. I couldn't pay her a model's full daily wages, but I paid her rent all the same, and thus far, thank God, I have been able to save her and her child from hunger and cold by sharing my own bread with her.
>
> When I first came across this woman, she caught my eye because she looked ill. I made her take baths and as many restoratives as I could manage, and she has become much healthier. I have been with her to Leiden, where there is a maternity hospital in which she will be confined.

> [there is reference to the following footnote here: Small wonder she wasn't well, the child was in the wrong position and she needed an operation, that is, the child had to be turned round with forceps. But there is a good chance that she will pull through. She is due to give birth in June.]
>
> It strikes me that any man worth his salt would have done the same in a case like this. I consider what I did so simple and natural that I thought I could keep it to myself. She found posing difficult, yet she has learned, and I have made progress with my drawing because I have a model. The woman is now attached to me like a tame dove. For my part, I can only get married once, and when better than now, and to her, because it is the only to go on helping her and she would otherwise be sent back by want on to the same old path which leads to the abyss. She has no money, but she is helping me to earn money with my work. (187)

The previous Christmas Vincent's disputes with his family had resulted in a violent quarrel, and he had moved out of their house to The Hague, where he had reconnected with his former boss Tersteeg and Mauve, both at first very helpful, but both turning against him when they found out about his relationship with Sien.

Some biographers, including Wilkie, have suggested that Vincent met Sien in early December and was the father of the child she would birth, have suggested that the main reason he moved to The Hague was to return to her. However, as Jan Hulsker has pointed out:

> It can be said with absolute certainty that Wilkie's premise is wrong. To doubt Vincent's statement that he had met Sien only late in January would mean to deny completely his honesty and frankness; it would also mean that he spoke in an extremely unsympathetic and hypocritical way when he described Sien as "a pregnant woman, deserted by the man whose child she carried" (Letter 192). However, there is proof of a more objective nature that Willem van Wijk was not Vincent van Gogh's child. Even if one accepts for a moment that Vincent met Sien around 1 December 1881, and that the child had been conceived at that time, the child would have been two months premature at its birth in July. In reality, the child that Sien gave birth to on 2 July 1882 was a mature baby weighing roughly seven and a half pounds. Given all this, Vincent could not possibly have been the father. (188)

As biographers have pointed out, this family is the closest Vincent comes to having his own family, and it can easily be seen as an alternative to, an opposite of his own family. The correspondence between Vincent and his family during this time period is filled with bickering and anger. This can be demonstrated by endless passages. However, what biographers sometimes fail to note is that the evidence of negative relationships is intermixed with passages of love and care. In other words, love and the desire for acceptance are always just underneath the surface of the anger and rejection. In fact, a careful, and even a not so careful reading of the letters suggests that he had a rather close relationship with his family and that, even if they disagreed with many of the things he did, they still loved him and often only condemn him or go against him because they believe he is hurting himself (and often he is).

However much Vincent wanted this strange family to work out, by the summer of 1883, it is clearly falling apart. Though he had refrained from discussing Sien with Theo for several months, probably because the two brothers had come to some kind of agreement to disagree about her, Vincent begins to mention problems in his letters: "the difficulties with the woman," (189) and in more detail:

> As the professor who attended her confinement told me, it will take years before the woman has completely recovered her health. That is to say the nervous system remains very sensitive, and she possesses the changeability of women to a high degree.
>
> The great danger is - as you will understand - her falling back into former errors. This danger, though of a moral nature, is connected with the physical constitution. And what I should like to call oscillations between improvement and falling back into former bad habits worry me continually and seriously. At times her temper is such that it is almost unbearable even for me - violent, mischievous, bad. I can tell you, I am sometimes in despair. She comes round again, and she has often told me afterward: *I myself do not know what I am doing then.* Do you remember you wrote me last year that you were afraid the mother would become a burden on me? Sometimes I wish it had been so. The mother is so energetic when she wants to be, and might have done so much better than she did. Now she often is more of a trouble than a help. Well, when the woman does wrong, it is sometimes the mother's fault; and when the mother does wrong, sometimes the family is in back of it. These are things which are not so very bad in themselves, but they prevent improvement and destroy or neutralize all better influences.
>
> The woman has certain faults and shortcomings in her behavior - how could it be otherwise? - *but for all that she is not bad* in my opinion. But those errors must be corrected - habits of slovenliness, indifference, lack of activity and ability, oh, a lot of things. But they all have the same root: bad education, years of quite wrong views on life, fatal influence of bad company. (190)

Just two weeks later, his letter to Vincent reveals more ominous problems:

> Theo, do you know what the difficulties I had with the woman were when I wrote you last? - her family tried to draw her away from me; I have had nothing to do with any of them except the mother, because I did not trust them. The more I tried to analyze the history of that family, the more I was strengthened in that opinion. Now, just because I kept out of their way, they plot against me, and so a treacherous attack occurred. I told the woman my opinion of their intentions, and said she had to choose between her family and me, but that I did not want to have anything to do with any of them, primarily because I thought that relations with her family would lead her back to her former bad life. The family proposed that she, with her mother, should keep house for a brother of hers who divorced his wife and is rather an infamous scoundrel. The reason why the family advised her to leave me was that I earned too little, and I was not good to her, and did it only for the posing, but would certainly leave her in the lurch. Nota bene, she has hardly been able to pose the whole year because of the baby. Well, you can judge for yourself just how far these suspicions of me have any foundations. But all these things were secretly discussed behind my back, and at last the woman told me. I said to her, "Do just as you like, but I shall never leave you unless you turn back to your former life."
>
> The worst is, Theo, that if we are hard up now and then, they try to upset the woman in that way, and that rascal of a brother, for instance, tries to drive her back to her former life. Well, I can only say of her that I should think it sensible and loyal of her if she broke off all relations with her family. I myself dissuade her from going there, but if she wants to, I let her go. And the temptation to show off her baby, for instance, often drives her back to her family. And that influence is fatal, and makes a greater impression upon her because it comes from her family, who upset her by saying, He will certainly leave you someday. So they try to make her leave me. (191)

By September 2, Vincent is realizing the relationship cannot work. In a surprising instance of calm, rational decision making, Vincent and Sien discuss the situation and the decision is made that he will move to Drenthe alone. Perhaps they will reconnect, perhaps not, but for now they decide to part as friends. Subsequent letters suggest Vincent, as expected, still struggles with the decision to end the relationship, but they also reveal that he is realizing ever more that she has not been honest with him, which in turn makes it easier for him to accept the necessary parting.

His relationship with Sien almost had to fail. She was a prostitute with few redeeming qualities, and even though it seems at least possible that he would have continued to persevere, it becomes obvious that she has grown tired of him, sees him more as a mark than a loved one, and *wants* to return to prostitution. One difference in this failed relationship is that

Vincent now expresses his sorrow over it in the context of *art* rather than *religion.* Albert J. Lubin states:

> But whereas formerly he praised his unhappiness on religious grounds, now he conceived of it as a necessary adjunct to an artistic career. Why should he give it up? After all, he explained, "the history of great men is tragic. . . . For a long time during their lives they are under a kind of depression because of the opposition and difficulties of struggling through life." Relating himself to creative men, he felt less hopeless." (192)

On September 11, 1883, Vincent departs The Hague, ending his relationship with Sien.

His subsequent time in Drenthe can only be described as miserable, and lasts but three months. On December 5 he arrives at his parents' home in Nuenen. Obviously, the tensions are high, but matters improve fairly quickly, partially the result of his mother's illness, where he demonstrates his usual compassion for people in need by caring for her. In May he rents a study from Schafrat, the verger of the Catholic Church, and experiences his next love affair, once again demonstrating both his compassion and his inability to successfully consecrate a relationship.

Margot Begemann, the Van Gogh family's next-door neighbor, who is ten years older than Vincent, is one of the few women to fall deeply in love with him. But her father and sisters work to turn her against him, which drives her into a nervous collapse. Vincent is considering marriage until Margot tries to poison herself with strychnine. Vincent rushes her to the hospital where she is saved. But then she is sent to a sanatorium in Utrecht.

Lubin's discussion of this goes as follows:

> Less than a month after his arrival at Nuenen, his mother seriously injured her hip and became bedridden. Accusations ceased, and he cared for her with tenderness and devotion. Her pain and helplessness brought forth an outpouring of love that can be found no place else in his letters. [Letter 352] Sufferers always aroused his compassion, and he nursed his mother with the same intensity with which he had previously nursed sick miners and Sien.
>
> Another tragic incident occurred later in the year. Vincent had become attached to Margot Begemann, a lonely spinster who lived next to the parsonage and helped nurse his mother. Margot was the youngest of three sisters but ten years older than Vincent; she was described as "neither beautiful nor gifted," [Joanna van Gogh-Bonger, *Memoirs of Vincent van Gogh*, p. xxxvi.] and their mutual attraction seems to have been based on shared unhappiness. They discussed marriage, and she, at least, was eager for it. Vincent, ready to respond with a consoling interest toward an

> unhappy woman, would have married her, but out of a love that was based on pity.
>
> Vincent's description of Margot suggests that he was seeing his own past in hers: "[I]n her youth she let herself be crushed by disappointments, crushed in the sense that the orthodox religious family *thought they had to suppress the active, aye, brilliant* quality in her, and have made her utterly passive." Like her, he too "used to be very passive and very soft-hearted and quiet; I'm not anymore." [Letter 378] In Margot, Vincent became attached to an externalized image of himself, a sad person who was misunderstood and mistreated by an intolerant family, suffering what he had so often suffered. Saving her was like saving himself. (193)

It should be noted here that this assumption by Lubin needs to be understood in context. It probably fits at this time because Vincent is in the middle of a strong conflict with his parents over his recent relationships with Kee Vos and Sien, but it should not be taken as a life long relationship with his family. Also, in the middle of this family conflict, Vincent has demonstrated a strong loving care for his injured mother. Vincent's expression of having been very passive in his past also needs to be taken in context. It is not hard to see him as being forthright and combative throughout his life. If anything, he initiated many of the conflicts he suffered, and there are endless passages to indicate that people who liked and even loved him (such as his brother Theo) had a hard time putting up with him.

Lubin continues his description:

> Not wishing to part with a useful helper, Margot's family strenuously objected to her relations with this strange fellow. Unfortunately, Margot was a woman with formidable emotional difficulties, extremely vulnerable to her family's disapproval, and she became "*excessively melancholic*" and "felt deserted by everybody and everything." [Letter 377] When she told Vincent she wanted to die, he tried in vain to rescue her from her family, but this only stirred up more trouble. The drama reached its climax when Margot attempted suicide by swallowing strychnine; she had a convulsion while walking in the fields with her controversial friend. Finally, she was sent off to a sanitarium in Utrecht, and the relationship seems to have terminated. (194)

Thus, in rapid succession, Vincent experienced three failed love affairs. However, we need to be careful in trying to make each of them fit a preconceived psychological view of Vincent. This last affair almost certainly was mutual and involved two people who were both intelligent and respected each other. Unfortunately, it was similar in that everyone else condemned it. Margot's two sisters and her sister-in-law did their best to prevent it, and, at the time, Vincent was in such a strained relationship with Theo that he didn't discuss it in his letters until it was over.

Unfortunately for Vincent, shortly after this last failure, he was to be falsely accused of impregnating a young girl. Nagera describes it briefly as follows (Albert J. Lubin gives the same description, probably simply using Nagera):

> In early 1885, Vincent spent a lot of time at the home of a peasant family, the De Groots, who posed for *The Potato Eaters* and many other studies. Unfortunately, the unmarried daughter, Gordina, became pregnant while he worked there, and Vincent, of course, was an easy target for suspicion. The local Catholic priest even issued an edict forbidding Catholics to pose for him, thus cutting off his chief source of models. Vincent vehemently denied the accusation and Gordina herself claimed the father of her child was one of her nephews. Nevertheless all this only increased his conviction that he was misunderstood and victimized. (195)

Vincent's letter to Theo about it strongly suggests he was not to blame, but was being blamed because he was an easy target. Here is the central passage:

> These last two weeks I have had a lot of trouble with the reverend gentlemen of the clergy, who gave me to understand, albeit with the best intentions and believing like so many others that they were obliged to intervene - that I ought not to be too familiar with people below me in station. But while they put the matter to me in these terms, they used quite a different tone with the "people of lower station," namely, threatening them if they allowed themselves to be painted. This time I went straight to the Burgomaster and told him all about it, pointing out that it was no business of the priests, and that they ought to stick to their own sphere of more abstract concerns. In any case, for the moment I am having no more opposition from them and I hope it will stay like that.
>
> A girl I had frequently painted was about to have a baby and they suspected me, though I had nothing to do with it. But I heard what had really happened from the girl herself, namely that a member of the priest's own congregation in Nuenen had played a particularly ugly part in the affair, and so they could not get at me, at least not on that occasion. But you can see that it isn't easy to paint people in their own home or to draw them going about their business. (196)

Two years later, Vincent writes his sister Wil about it:

> You would please me greatly by letting me know how Margot Begemann is, and how things are with the De Groots: how did that affair

turn out—did Sien [referring to Gordina here] marry that cousin of hers? And did her child live? (197)

Further support that Vincent wasn't the father comes from the fact that the accusers dropped their charges, though they had, in effect, accomplished their goal of spreading distrust of him.

It is possible Vincent had one last love affair, and as was always the case for him, it was forbidden. In his research into Vincent's life, Ken Wilkie visited the daughter of Madame Liberge, and is told the following:

> "My mother was Marguerite Gachet's best friend. And she was the only person, as far as I am aware, who knew about the love affair between Vincent van Gogh and Marguerite. Marguerite was a proud girl but suppressed by her father. She confided to my mother that Vincent and she had fallen in love with each other and that Vincent wanted to marry her. But the thorn in the flesh was Marguerite's father, Dr. Gachet. Though an advocate of free love in theory, he was strongly against an association between Vincent, who was of course his patient, and his daughter. Gachet forbade Marguerite to see the painter." (198)

Ken Wilkie, much influenced by the theories of Nagera and Lubin, offers his conclusions about Vincent's love life:

> In fact, it had been Vincent's letters to Theo, and not his paintings, which initially had interested me back in 1968. I had been struck by their almost mesmeric intensity of feeling. Later, I realized that his deep melancholy was rooted in the birth of his stillborn namesake brother whose shadow he followed as a child and whose grave would reappear again and again to him in hallucinations.
>
> Repeatedly failing to establish lasting relationships with women, I could see him sink into depression following the inevitable rejection, then glorify his sorrow and channel it into first religion, then art, and express it in the most exceptional words and images. In both his letters and his art, Vincent's personality leaps off the page or painting in an amazing symbiosis of nature and human nature, always using his brother Theo whose birth must have relieved his childhood isolation, as a life-long link with the human world. (199)

In his research, Wilkie also comes up with support for the theory that Vincent and Theo both suffered from syphilis, which in turn caused Vincent's dramatic changes in temperament and by extension might have been a major factor in his artistic vision, in effect suggesting that this disease and the remedies for it were central to his genius. Though Wilkie doesn't elaborate on it, the suppositions here are huge. They offer support for claims that exceptional mental ability, including both insanity and genius, are the result of, or at least can be the result of a disease and the chemicals taken to fight against it. In other words, it is possible to alter the mind chemically to produce exceptionally brilliant and exceptionally deranged people. In today's world of neurology, the idea that chemicals can and probably should be used to help people with some kind of brain malfunction or even with standard brain loss is a generally accepted matter. This is especially being applied to various mental diseases that might be categorized under the vague term of *insanity* (though endless less dramatic and more precise terms such as schizophrenia, autism, and bipolar disorder are used). Theories of chemically induced genius are less supported, though the embracement of such drugs as LSD, promoted by a serious psychologist, Timothy Leary, as being mind expanding, certainly fall into this category, and although that entire movement resulted in a backlash from the culture, it also resulted in or at least became central to a huge counter-culture movement that remains.

In pursuing Vincent's time in Antwerp in 1885-6, Wilkie attempted to understand why Vincent went through one of his changes in temperament. He writes:

> His earliest letters there describe with delight the harbour area, where he spent time watching the bustle of daily life, talking to "various girls who seemed to take me for a sailor," and observing the light on the docks and buildings. . . .
>
> However, as the letters continued, Vincent's mood changed. He spoke increasingly of his health, which was not good. He confessed to Theo that he was afraid of dying before his talent was recognized, and that he was afraid of madness. He produced two macabre pictures: one a painting called *Skull with Cigarette,* which shows a skeleton smoking, and another drawing of a hanging skeleton. Moreover, his first self-portraits which date from this period seem to reflect a morbid introspection.
>
> Perhaps Vincent's ill health was connected with this self-awareness. Perhaps the key to the change in his personality was to be found in the illness itself. This was the line I decided to follow.
>
> I soon discovered that I was by no means the first to try to identify Vincent's disease. Among the numerous hypotheses that had been put forward were one or another form of epilepsy, schizophrenia, dementia praecox, meningo-encephalitis luetica, cerebral tumour, hallucinatory psychosis, chronic sunstroke and the influence of yellow, dromamania, turpentine poisoning, and hypertrophy of the creative forces. . . . (200)

Wilkie remembered Tralbaut had mentioned to him that the name "Cavenaile" was scrawled in the back of one of Vincent's Antwerp sketchbooks of the time. Wilkie decided to see if he could find what it referred to. And he did, fairly easily by looking up the name in a phone book and finding the grandson of the doctor who treated Vincent and who stated simply when asked if his grandfather had treated Vincent, "He said he treated Van Gogh for syphilis. He prescribed a treatment with mercury and sent him to the Stuyvenberg hospital for hip-baths. Van Gogh did not have a bathe in his own lodgings." (201)

In perusing this possibility, Wilkie found that, at the time, there was no cure for syphilis, and that, according to Dr. Cavenaile (grandson was also a doctor) "In its third, and final, stage it could have been, at the very least, a contributory cause of his ultimate madness, yes. (p. 146-7) Wilkie pursues this angle, finding especially Theo's son, "the engineer" reluctant to agree with it, perhaps especially since Wilkie also finds it likely Theo had the same illness.

Wilkie looked up the standard treatment of syphilis and prognosis of it at the time:

> I found a book, published in Paris in 1890, entitled *La Syphilis Aujourdhui et Chez Les anciens,* sub-titled *Nihil Sub Sole Novum* (Nothing New Under the Sun). Dr. F. Buret's opening statement reveals the social attitude to the disease at the time: "Daughter of prostitution, syphilis was born when commerce, chasing away love, presided over the exchange of kisses."
>
> He writes that in Paris of the 1880s, decades before penicillin (the present cure) was discovered, the basic treatment for the disease was mercury and iodine, with increasing doses of iodide of potassium as the illness progressed into its third stage.
>
> "In the second stage of syphilis," he says, "one of the best preparations is *Le Sirop de Gibert* which includes 0.5 grams of iodide of potassium combined with one centigram of biodure hydrargie." Another method practiced in 1890 was the combination of a dose of iodide of potassium of sodium and massage with Neapolitan ointment.
>
> "In the third stage," Dr. Buret continues, "larger doses of iodide of potassium are prescribed: one, two, three, and up to six grams per day for periods of two months."
>
> The most serious of the sexually transmitted diseases then, syphilis, could prove fatal. In its third stage, the brain and spinal cord were often affected, which caused insanity (general paralysis of the insane, or GPI) and loss of muscular co-ordination. Indeed, tertiary syphilis may involve any organ and mimic virtually and other chronic disease.
>
> As we know, Vincent suffered from gonorrhea as well as syphilis, and it seems probable that Theo also suffered from both diseases for he later suffered a structure which was a narrowing of the urinary passage and a

> late consequence of infection in the days before antibiotics. And gonorrhea could also spread to the joints, tendons, muscles, heart and brain. (202)

One of the problems with a chemical approach to the workings of the mind, while there is some obvious truth to it (which makes it even more dangerous) is that it can easily lead to a *deterministic* view of human existence that ultimately is exactly the same as traditional Calvinism, only now couched in the current belief systems of science. In other words, if our minds are the results of the chemical processes of the brain, then our thoughts, our hopes and fears, our achievements and failures can be said to be all *predetermined* by the laws of science and math. If this is so, then life is *meaningless*, because we ultimately have no choice about how we act or think. Everything has been put in place from the beginning, and each of us is merely living out a *predetermined* path in the complex but *meaningless* web of existence.

Part IX

The Potato Eaters

From August of 1879 to July of 1880, there are no letters between Vincent and Theo. Vincent is going through a struggle, and even though he lives with his parents part of this year, he is having tremendous battles with them. Perhaps because there are few records of it, it does not get a lot of attention, but it certainly can be seen as the year of Vincent's "time in the wilderness," the year he goes from an evangical imitation of Christ focused on a very narrow, literal interpretation of Christianity to an understanding of it in a larger context, one that allows him to step outside of his failed role as an evangilest and into a new role as an artist. It is a major transformation in his thinking, a realization that religion specific images are *not in-and-of-themselves-that-which-is-beyond-them*, but, rather, are images, symbols, metaphors *to that-which-is-beyond-them*.

Though his future biographers and art collectors have since dug into any possible artwork by Vincent prior to September 24, 1880, the date of his letter to Theo declaring he has made the commitment to become an artist, what little there is shows no genius. He exhibits more interest in drawing than the average child, possibly because of his family's connections to the world of art and his mother's encouragement, and, perhaps, he can be said to be a slightly better draftsman than the average child or young adult, but any claims that he is superior on the level of genius are not easy to support.

Having gone through his disillusionment and emerged with a higher knowledge, he enrolls in the Academy at Brussels, and works hard to improve his skills.

When he moves in with his parents the following April, 1881, he begins practicing landscapes, demonstrating a great deal of improvement over his previous work, and he also begins to draw from live models, though, as Jan Hulsker points out "On close inspection of the few dozen surviving drawings from this time, it becomes evident that, in spite of his confidence, Vincent's technical skill in rendering the human figure was developing much more slowly than his ability to realistically represent a landscape. Most of the figures from Etten look somewhat clumsy, and the proportions are often still awkward." (203) Perhaps the best of his portraits of the time is a caricature of an older man thought possibly to be a representation of his father. This caricature quality plugs into a common category of such drawings popular in Europe at the time, and also foreshadows similar such faces in his more famous work to be completed soon, *The Potato Eaters*.

At the end of this summer, he visits Anton Mauve, who gives him advice on his art, and then The Hague, where he has a chance to catch up on the latest art, including the chance to speak with such artists as Theophile de Bock. In January, 1882, he moves to The Hague and begins studying in earnest under Mauve, who, among other things introduces him to watercolors. For a brief time this relationship with Mauve goes well, but then, by April, everything changes and both Anton Mauve and H. G. Tersteeg turn against Vincent because of his relationship to Sien.

As discussed above, this relationship brought condemnation from everyone, most importantly from Vincent's entire family. Something not mentioned, however, concerns their views about his sanity. June 1 or 2, Vincent responded to Theo's suggestion of the family's considerations: "What you seem to be afraid of, namely the possibility that the family might put me under guardianship—this is what I want to say to all that. When you say, 'Only a few witnesses (and even false ones) would have to declare that you are unable to manage your own financial affairs: this would be sufficient to entitle Father to take away your civil rights and put you under guardianship'—if, I say, you really mean this, and think that nowadays this might be an easy thing to do, I take the liberty to doubt it." (204) In this same letter Vincent refers to "the case of Gheel." Gheel is a village in Belgium that housed the insane, and there are other references in Vincent's letters revealing that his father had threatened him with this, including Letter 158: "Father grew very angry, ordered me out of the room with a curse, at least it sounded exactly like one! Now this causes me much pain and sorrow, but I can't believe that a father is right who curses his son and (think of last year) wants to send him to a madhouse (something which I oppose with all my might), and calls the love his son untimely and indelicate." How serious these threats were can only be speculated. They are likely more meant to be threats than serious consideration. However, it is also likely that when Vincent refers to "last year" it is in reference to the time he was deep into his radical form of Christian deprivation, and that might well have pushed his father to consider institualizing him for his own good. Now, at this point, Vincent's taking up with Sien can hardly have been seen as anything other then the most egregious of actions, and from his father's perspective a form of masochism. Once again, whether because the family actually sees Vincent as insane or more likely simply because they think it in his best interest, it is possible the threats are more than just threats.

In the middle of all this, Vincent gets some disease (as discussed earlier, probably some form of venereal disease) and is hospitalized for three weeks. During the time he lived with Sien, he continued working on his drawings, and the improvement is obvious. The drawing that gets the most attention is one titled *Sorrow*, a nude drawing of Sien that well expresses its title. In March his Uncle Cor commissions twelve drawings, followed by an order for six more. It is safe to assume this was not an act of pity or simple encouragement from a relative, as Uncle Cor would not have done so. In fact, Vincent's family throughout was, if anything, overly critical of Vincent, rather than encouraging. It is clear Vincent has made rapid progress--most critics consider it amazing progress. His black-and-white drawings have reached a level that can be considered professional, and have moved from caricatures to serious depictions demonstrating a deep respect for the dignity of the poor, the old, and the suffering who are his subjects.

Though he purchases materials for oil paintings, nothing special results, as he probably realizes, because he puts oil painting aside and continues to concentrate on his drawings. The drawings continue to depict poor people in harsh environments, and begin to move more and more into expressive works purposely pushing against realistic representations. Vincent writes how he wants to express the inner beauty of the ordinary, working class people, farmers, carpenters, the homeless. By the summer of 1883, while he continues with his drawings, he returns again to oil paintings, but still, at this time, nothing special comes of this medium. While the drawings are now clearly well done, they are not selling, Vincent's financial condition worsens, and, along with the deteriorating relationship with Sien, results in him moving out of the city to Drenthe (along with all of the other reasons for the move is that he cannot afford to live there).

He constantly writes about how much he enjoys being in the country, near nature. However, that is about all the positive that can be given this time period. Little remains of whatever art he produced. Uncle Cor no longer demonstrates any interest in Vincent's art (probably at least partially the result of the backlash from the Sien affair). Vincent is, in effect, stuck between where he was and where he should go next. In a sense, the transition takes place without him really deciding, sparked by Theo's problems with work, which in turn ignite Vincent to invite Theo to quit his job and come live in Drenthe (which Vincent sees as an idyllic life for the two of them), but then, as a result of Vincent misrepresenting his concerns for Theo and the money he has been living off of from Theo, Theo declines. Vincent feels guilty and makes what must have been a difficult decision to move back into his parents' house so he won't continue to put Theo in the position of supporting him.

On December 3 or 4, 1883, Vincent returns home, a curious looking man for this small village. And he does not come home apologetically. A day after arriving, he writes Theo:

> I am sick at heart about the fact that, coming back after two years' absence, the welcome home was kind and cordial in every respect, but basically there had been no change whatever, not the slightest, in what I must call the most extreme blindness and ignorance as to the insight in our mutual position." (205)

About a week later, his views are expressed even more strongly (including the passage where he refers to himself as being treated as a dog):

> He would run into the room with wet paws—and he is so rough. He will be in everybody's way. And he barks so loud. . . . The dog is only sorry that he did not stay away, for it was less lonely on the heath than in this house, notwithstanding all the kindness." (206)

Theo's response isn't what Vincent wants. Rather than consoling him, Theo gives him "a good scolding." Hard not to agree with Theo here. Vincent's parents have taken him in their home, and as he himself says shown him kindness and love, even though they are obviously upset with his recent decisions and are in the middle of a major fight with him. Yet they have not suddenly changed their views about what they see as an endless series of wrong actions and apologized to him. Hmmmm . . . . So he responds by condemning them. In his response to Theo he states that he has decided to leave his parents as soon as possible, wants money from Theo, and suggests that he and Theo's relationship is probably ending. (207) Hard not to see Vincent as throwing a childish tantrum here.

Jan Hulsker writes: "All this sounds very definite, and if they had parted, probably not much would have become of the painter Vincent van Gogh." (208) However, as fate will have it, Vincent and his father have a talk that apparently resolves, or at least modifies their conflicting views, and it is agreed that Vincent will stay, and be allowed to continue much as he is. Seems as if his father gave into his demands, rather than there being much of a compromise. Then, again seeming very selfish, Vincent travels to The Hague to get his things, *visits Sien!*, and writes Theo about it, condemning his parents for not understanding! (209) This continues in the same vein in the next letter! (210) And the next, "So you have me at your mercy, you particularly, along with may others, none of whom can agree with me. And yet you will not be able to force me to renounce her, whatever your financial power. And because I shall make no concessions in the matter of the woman—and I will clearly declare it, loud enough for even ears that are most hard of hearing—I announce in advance that I have resolved to share with her all that is my property and I do not wish to accept any money from you, except that I may regard as my property without *arriere pensee."* (211)

Interestingly, this is the last he mentions her, and the assumption is that *she* broke off the relationship. On the other hand, especially since he is waging such a strong battle with everyone about his love for her, almost immediately he takes up a new relationship, this one with Margot Begemann. They first meet in January, 1884, and appeerently are immediately attracted to one-another.

Vincent pushes his relationship with Theo even more. "Now I want to make a proposal for the future. Let me send you my work, and keep what you like for yourself, but I insist on considering the money I receive from you after March as money I have earned." (212) On February 13, he sends Theo a parcel with three little panels and nine watercolors. (213) Theo does not respond as quickly as usual. When he does respond, it is negatively, including comments about Vincent's art not being good enough to sell. Vincent explodes back: "For my part I will also tell you frankly that I think it is true what you say, that my work must become much better, but at the same time, that your energy to do something with it might become much stronger too. You have *never sold a single one from me*—neither for much or for little—and in fact *have not even tried."* (214) The vicious letters continue, until Vincent writes: "That very idea of yours is sufficient proof to me that we have come to the point where more words won't do any good, and I think it better to let this question rest." (215) Of course, Vincent still wants his monthly check from Theo, and in the end the business relationship continues, though the letters take on a much colder relationship.

In the meantime, Vincent continues his drawings, and they continue to improve, becoming larger and retaining their solemn respect for the subjects. In May, 1884, he rents a new studio, larger than what he has had, at the house of Johannes Schafrat, the sexton of the Roman Catholic Church. He gets a commission from Antoon Hermans, an amateur painter who is decorating his ceilings and walls and wants six panels in the dining room filled with compositions of a Last Supper and saints. Vincent convinces him to change the subject matter to depictions of peasant life, and the arrangements are made. Vincent will paint six canvases (which he will get to keep), and Hermans will copy them onto his walls. In the end, this does not prove as lucrative as Vincent hopes, because Hermans does not come up with the money. Also, at this time Vincent is beginning to experiment with color, and indicates he has heard of the Impressionists in Paris, though it is obvious he knows very little about what they are doing.

During the fall of 1884, Vincent continues in his poor relationship with Theo and his family, works on the panels for Hermans, but does not appear to have done much other art work, though he mentions in letters of the time that a visit from his artist friend Anthon van Rappard has sparked new interest in his own work. (216) He also takes on some students, most notably Anton Kerssemakers, who later includes some comments about it in his own published recollections. (217) Vincent does not charge them, except for the cost of tubes of paint, because he cannot afford to buy his own paint. Indeed, reports from the time all indicate that Vincent lives in poverty, eating little more than dry bread and an occasional piece of cheese. He has painted some still-lives that he hopes to sell in Antwerp, but they are not of great interest.

Through the winter, Vincent and Theo continue to fight, and it seems very likely during this time that the only reason Vincent doesn't completely break away from Theo is his need for Theo's money, which Theo continus to send in spite of the quarrel. Then, March 26, 1885, their father dies of a stroke, and although Vincent writes very little about it, he and Theo have a chance to visit in person, resulting in an improvement in their relationship. The death of his father also prompts Vincent to decide to move.

This, then, is the situation when Vincent begins *The Potato Eaters*, the work considered to be the best of his from this period, and the first work by him worthy of serious consideration. By April, he has gone through several versions of it and sends the final work to Theo on May 4 (218), writing:

> Though the ultimate picture will have been painted in a relatively short time and for the greater part from memory, it has taken a whole winter of painting studies of heads and hands. And as to those few days in which I have painted it now, it has been a real battle, but one for which I feel great enthusiasm. (219)

Vincent obviously thinks highly of his work and gives some insights into it in the same letter:

> I have tried to emphasize that those people, eating their potatoes in the lamplight, have dug the earth with those very hands they put in the dish, and so it speaks of manual labor, and how they have honestly earned their food. I have wanted to give the impression of a way of life quite different from that of us civilized people. Therefore I am not at all anxious for everyone to like it or to admire it at once" (220)

He continues:

> All winter long I have had the threads of this tissue in my hands, and have sought for the ultimate pattern; and if it has become a tissue of rough, coarse aspect, nevertheless the threads have been chosen carefully and according to certain rules. And it might prove to be a real peasant picture. I know it is. But he who prefers to see the peasants in a sentimental way may do as he likes. I personally am convinced one gets better results by painting them in their roughness than by giving them a conventional charm. (221)

Excited about his work, Vincent immediately begins plans for reproducing it, and makes a number of lithographic copies of it, one of which he sends to his artist friend Anthon van Rappard, expecting an enthusiastic response. However, what he gets is the exact opposite, resulting in a break in their relationship. Part of Rappard's criticism follows:

> I hope I was mistaken in my opinion of your manner of working, and I hope so still; but for this very reason I was deeply sorry to see such a complete confirmation of my opinion in what you sent me that I myself was terrified by it. You will agree with me that such work is not meant seriously. Fortunately you can do better than that, but why then did you see and treat everything so superficially? Why didn't you study the movements? Now they are only posing. How far from true that coquettish

> little hand of the woman in the background is—and what connection is there between the coffee kettle, the table and that hand that is lying on top of the handle? Whatever is that kettle doing? It isn't standing, it isn't being held—so what then? And why can the man on the right not have a knee, no belly and no lungs? Or are they in his back? And why must his arm be a meter short? And why must one half of his nose be lacking? And why must the woman on the left have such a little pipe-stem with a die at the end for a nose? And dare you, working in such a manner, invoke the names of Millet and Breton? Come! Art stands in my opinion too high to be treated so carelessly." (222)

The following passage from Vincent's reply indicates, though the two of them had just recently shared a positive visit and seemingly similar views on art, Vincent has already moved beyond Van Rappard's perspectives:

> *The Potato Eaters* is a subject that I tried to paint, being carried away by the singular light of the grimy hut. . . . Now you call the ensemble of my work utterly weak and show at great length that the faults exceed the qualities. As with my work, so with myself. Well, I do not accept that, certainly not. The work in question, painting peasants, is such a hard job that the utterly weak left to themselves would not even start it. . . . And yet I believe that even if I keep on producing work in which one can, if one so desires, especially looking at it just from that side and with that purpose, point out mistakes, it will have a certain life and a raison d'etere of its own which will outdo those faults—in the eye of those who appreciate character and the passings of things through the mind. And with all my faults I cannot be so easily outdone as one would think. I know too well which purpose I am aiming for. I am too firmly convinced that after all I am on the right road—when I want to paint what I feel and feel what I paint—rather than that I bother much about what people say of me. Still it makes life very difficult for me at times, and I think it quite possible that some people will later be sorry, either of the things they said of me, or of the opposition or indifference with which they have pestered me. What I do against it is to withdraw so far that I literally don't see people any more other than the little peasants with whom I am directly concerned in order to paint them. And this will remain my system and it is quite possible that before long I shall give up my studio and go to live in a hut so that I don't hear or see any more of those who call themselves civilized people. (223)

Vincent makes a few changes for the final version. In the earlier version the woman behind the table seems to be the older, but in the final version the woman with the coffee kettle is obviously the older one, and the now younger woman is less of a caricature, more idealized. It seems that Vincent took the features of previous portraits he had done of Gordina de Groot to compose her face. In addition to revising these two women's looks, Vincent makes the young peasant's face more coarse, no doubt the result of him moving ever more out of representation to expressionism.

The entire painting demonstrates that Vincent is consciously trying to move passed a realistic representation towards an expressionist style, and into a form of symbolism. He writes Theo about his struggles with the color of the faces that he tried combining various colors, yellow ocher, red ocher, white, but "that was much too light and was decidedly wrong. What was to be done? All the heads were finished, and even finished with great care, but I immediately repainted them, inexorably, and the color they are painted in now is like the color of a very dusty potato, unpeeled of course." Because he wants a rough, earthy quality, he uses heavy brushstrokes in the peasant heads he is working with at the time, experimenting to find a way to connect his painting style up to his subject matter, to move beyond merely attempting to duplicate the physical appearance. As he works through the summer, he continues to clarify what he is now trying to do in his art, and the following passage is always highlighted because of how it foreshadows both his own artistic vision and the direction art is about to take:

> It is not yet well expressed. Tell Serret that I should be desperate if my figures were correct, tell him that I do not want them to be academically correct, tell him that I mean: If one photographed a digger, he certainly would not be digging then. Tell him that I adore the figures by Michelangelo though the legs are undoubtedly too long, the hips and the backsides too large. Tell him that, for me, Millet and Lhermitte are the real artists for the very reason that they do not paint things as they are, explored in a dry analytical way, but as they—Millet, Lhermitte, Michelangelo—feel them. Tell him that my great longing is to learn to make those very incorrectnesses, those deviations, remodelings, changes in reality, so that they may become, yes, lies if you like—but truer than the literal truth. (224)

The drawings he does during the following summer demonstrate a new confidence in what he is doing, most of them peasant women bending deeply down while working in the fields, each filled with a sureness of line and a control of light and shadow superior to what came before.

Theo apparently does not appreciate Vincent's new direction, and so, as usual, Vincent is left without the desired praise for his work and, as always, begging for more money. His time in Nuenen is about to end. And it will end with some important foreshadowings of the truly brilliant art to come. First, he will visit a new art museum in Amsterdam with his friend Kerssemakers, and he will realize in viewing the paintings of those he admires that "most of them painted quickly, that these great masters, such as Frans Hals, a Rembrandt, a Ruysdael and so many others—dash off a thing as much as possible *de premier coup* [from the first stroke] and did not retouch it so very much. . . . To paint in one rush, as much as possible in one rush. What joy to see such a Frans Halls, how different it is from those pictures—there are so many of them—where everything has been smoothed down in the same way." (225) Second, he will plunge into color.

The mature Vincent van Gogh style is about to come-together—quick, thick brushstrokes, an expressionism beyond realistic reproduction, even beyond Impressionism, energetic colors, all for the purpose of exploring the inner quality, the psychological reality. Just as he has realized that the images of religion are not the resulting spirituality but rather are the passages to it, he realizes that the images of art are not the resulting spirituality but are also passages to it. The idea is not to paint the physical, but to paint that which is beyond the physical.

Of course, since this is Vincent after all, there remains yet another incident with a woman, this the one mentioned above with Gordina De Groot, one where it seems very likely he was being falsely accused of impregnating her, but one that confirmed that the people of this rural world really did not want a strange artist among them, and one that would encourage him in his thoughts about leaving. Which he does. On November, 1885, Vincent leaves Nuenen and will never return.

However, before following him, let's take a closer look at *The Potato Eaters*, the major work he did to this point, and one that has received a great deal of commentary, with widely divergent conclusions.

First, let's note that neither Vincent's artist friend Van Rappard nor Vincent's brother Theo thought much of it. It is, in fact, somewhat surprising how vicious Van Rappard was in his letter to his friend Vincent about this work. These less than enthusiastic responses were, certainly in part, the result of Van Rappard's artistic embracement of realism and Theo's more surprising failure to move into this more expressionist, impressionist style of painting, something Theo had to have been exposed to in his art dealings.

Jean Leymarie provides the basics:

> It was by a determined effort of synthesis that in the *Potato Eaters*, the only large picture he ever painted, he combined the many studies of heads and hands he had accumulated during the winter. He got the idea for the picture while taking his meals in the de Groot family, and the son and daughter sat for him. It represents the consecration of manual labor by a meal in common under the lamp—a sort of rustic last Supper whose rugged truthfulness coveys a vehement grandeur. The impetus behind it

> came from Millet, Rembrandt, Delacroix and also Zola, but the direct sources for the choice and arrangement of the figures seem to have been Charles de Groux's *Grace before Meat* and Jozef Israels' *Frugal Meal.* In March 1885 he sketched out on cardboard a group of four figures silhouetted in a violent chiaroscuro, which he showed to Theo during his visit. His father's death, for which he unconsciously assumed responsibility, interrupted the work for several days, but it also gave him the necessary independence and boldness to tackle a large-scale composition. In April he painted "from life" an initial version on wood in where the grouping and setting are fully worked out. The addition of a fifth figure, the peasant woman with a large white cap, gave rise to the central group of three women, marking the ascendancy of the matriarchal tradition to which Van Gogh remained strongly attached.
>
> He then made a lithograph of it and sent a print to Van Rappart—who, however, criticized it with such unexpected asperity that Van Gogh was disgusted. Their friendship cooled and was broken off in September. In early May he painted in his studio, from memory, "by heart," as he expressly says, the final version on canvas, a little larger and with slight changes, which he presented to Theo as the summing up of his artistic and social preoccupations at Nuenen: "a true picture of the peasants," painted in the very colors of the soil they till and of the crops they live on. His concern with objectively symbolic truth thus coincided with the rugged expressiveness of his style. There is no communication between the five haggard figures assembled singly around the table, but the "russet yellow" glow of the lamp in the greenish penumbra gives the picture, optically and spiritually, a unified focus. (226)

It is important to re-emphasize that Vincent is moving into symbolic representation, seeing his art as expressing the inner values of the peasants, and connecting these with their own connection to the earth. It is also clear that Vincent purposely has the figures not connecting, and that the glow of the light is intended to be the unifying element in the painting.

Derek Fell quotes Bradley Collins as he pushes into the possible symbolic meanings:

> During this period Vincent completed what he considered his finest painting, *The Potato Eaters*, which shows a family of peasants, the de Groots, at an evening meal. Potatoes served as a symbol for Vincent of his love of the earth and connection to the soil. He told Theo: "I have tried to emphasize that those people, eating their potatoes in the lamplight, have dug the earth with those very hands they put in the dish, and so it speaks of *manual labor*, and how they have honestly earned their food." When Vincent lived in London, his fascination for soil and gardening shines through in a letter written to Theo: "I am busy gardening and have sown a little garden full of poppies, sweet peas and mignonette. Now we must wait

> and see what becomes of it." In a subsequent letter from London he described weeding a row of potatoes that later in the season would be plucked from the soil like nuggets of gold. Significantly, the color gold is the predominant color in *The Potato Eaters*. Gold tones are used to represent lamplight reflected in the faces of the family, in contrast to the black shadowy interior of their humble home.
>
> First and foremost, *The Potato Eaters*, represents a rebellion against the traditional teachings of the academy, which Vincent described as a mistress "who freezes you, who petrifies you, who sucks your blood." Art historian Bradley Collins sees it as Vincent's celebration of the "coarse and ugly," and discusses it poignant symbolism in his book *Van Gogh and Gauguin*. Describing the work as an homage to Rembrandt's *Supper at Emmaus* in the Louvre, Collins sees the mysterious child in the foreground as Christ revealing Himself to the apostles. "The solemnity of the figures and the ponderousness of their gestures also make the scene readable as a *Last Supper*. The older man in particular seems to hold his coffee cup so reverentially that he seems to offer it up as the body and blood of Christ. Vincent himself cues the viewer to his religious allusions by placing a small print of the Crucifixion behind the young man."" (227)

While it is fun to try and connect this work up to Vincent's religious views, it needs to be done in a more Jungian form of symbolic analysis, not one that works to turn the symbols into such a literal translation. In other words, it might be possible to find the painting symbolizing the same truths as the Last Supper, but to suggest pieces of it literally match pieces of the Last Supper undermines such considerations rather than confirming them and completely misunderstands how symbols work.

Albert J. Lubin provides a psychological interpretation of the symbols that has similar flaws:

> Vincent's work in the North had culminated in *The Potato Eaters*, a "composition of those peasants around a dish of potatoes." [Letter 398] More study went into this painting than any other in his career. After a long series of preliminary studies of heads and hands, he made a rough sketch in March 1885, a preliminary oil in April, and the definitive version in May. He had mastered the model and the scene so well that he painted this last version form memory; "[T]he thing is so fixed in my mind," he wrote, "that I can literally dream it." In contrast to Vincent's usual humility about his work, he had no qualms about praising *The Potato Eaters* Looking back on his achievements more than two years later, he called it "the best one after all." The scene so fascinated him that he returned to it again in Provence. *The Potato Eaters* was the focus of all his energy, talent, and hope—a situation suggesting that it depicted a central issue in his life.

> The painting was a sermon on the unjust treatment of the peasant by the Dutch ruling class and an ode to the farmer's endurance. It was a stirring commentary on the difficult role of the peasant, a man like himself who suffered, lived in darkness, bore heavy burdens, and received few rewards on earth. While the painting depicts a family group seated around a small table united by the rays of a single lamp above them, their closeness is only physical. Emotionally, they are remote from each other, unable to communicate. Vincent identified himself with these coarse peasants, with their suffering and their isolation. Meyer Schapiro observes that "each figure retains a thought of its own and two of them seem to be on the brink of an unspoken loneliness." [Meyer Schapiro, *Vincent van Gogh* (New York: Harry N. Abrams, 1950, p. 40)] He attributes this effect to the difficulties involved in copying the heads from earlier portrait studies. Yet this same sense of isolation is common to many of Vincent's Dutch works, where it cannot be explained away solely on technical grounds. A wall separates the woman on the right from the rest of the family—Professor Schapiro says that it "creates a strange partition of the inner space"; I have suggested that Vincent often used such partitions to portray a feeling of isolation. (228)

Up to this point, Lubin is on solid ground. He hasn't said much, but what he has said is easily supported. However, he is just setting the stage for what he really wants to deal with.

> Such are a few of the messages conveyed by *The Potato Eaters*. But these same messages are contained in other van Gogh pictures; they do not account for the elaborate detail. Preliminary studies were sketched at the hut of the de Groot family, and the family posed for them. The painting itself, however, was done in his studio, using his memory and imagination. The sketches drawn directly from the models were "food for one's imagination," he wrote, "but in the painting I give free scope to my own head." [Letter 430] Such freely elaborated thinking, like the content of dreams, has roots both in the present and in the distant past, and is derived from conscious as well as unconscious thoughts. Vincent began his work on the painting immediately after his father's death.* [*The first letter to Theo after the father's death is numbered 397. Vincent first mentions "those peasants around a dish of potatoes" in the following letter, which begins, "I am still greatly under the impression of what has just happened. . . ."] While he spent the winter preparing for a peasant picture, he nowhere indicated that he had chosen the subject of the potato eaters until this time. Perhaps the renewed confrontation with his mother's grief reactivated memories of the childhood situation. (229)

The general theory of how art expresses psychological truths again is on the mark, the unconscious working on some psychological struggles, "like the content of dreams," results in an artistic expression that puts a pattern on, that helps resolve the conflicts. However, Lubin is determined to slide into his favorite theory, that of the Replacement Child Syndrome, which as discussed earlier is on shaky ground. He continues:

> If we look at *The Potato Eaters* from this point of view, the younger man on the left represents Vincent: the name "Vincent" is inscribed on the top slat of the back of his chair, although it is difficult to make out. The older woman on the right represents his mother. Her head is bowed, seemingly because she is intent on pouring coffee. But it is modeled from the melancholic heads that Vincent had been drawing in preparation for the painting. The act of pouring coffee diverts attention from her downcast expression; at a deeper level, melancholy accounts for her inability to show interest in those around her. The goggle-eyed "Vincent" gazes at her, vainly trying to make contact.
>
> The enigmatic figure of a child in the foreground—faceless and ethereal—stands between Vincent and his mother. No chair is visible, and the child—a ghostly presence, not a flesh and blood human being—is obviously not partaking of the meal. Curiously, steam arises from only one side of the plate, forming an aureole that envelops the child's head and shoulders. It would seem that Vincent was portraying the grieving mother who could not mother him; her spirit remained with the dead but perfect child who stood between them, separating them in the painting as in life. (230)

This is the point, the moment in his discussion where the conclusions have moved beyond any real support. This enigmatic figure, seen only from the back, is a young female. I'm not sure how Albert J. Lubin can be so certain she is not partaking of the food. The steam rising from the plate appears more to be an awkward attempt to distinguish the dark head of the girl from the darkness and other heads near it. It does not seem to be rising from the plate at all. Whether or not she is sitting on a chair or some form of stool is also only to be guessed at. In truth, the table appears to be taking up the same space she is. If one wants to attribute this to the fact she is ephemeral, then I suppose she is to be considered a ghost, though even that would not offer a lot of support that she is the ghost of the younger mother. It is an arbitrary claim. While critics don't want to admit the possibility (the likelihood) that Vincent did not get his overall structure well worked out, it certainly is a legitimate claim. The table itself has an awkward tilt to it, the chair the man is sitting in has an awkward twist to it, and so on. Yes, these can be attributed to Vincent's move beyond realistic representation, but they certainly are not a mastered understanding of it. But, back to Lubin:

> The two other members of the family, seated in the rear, are secondary figures, not essential to the theme, perhaps added to focus attention on the two protagonists on whom their attention is fixed. Indeed, only the three foreground figures are present in a preliminary sketch. In a preliminary version in oil the younger woman is omitted and the father is but a disembodied head. In the definitive version the younger woman, gazing toward Vincent, is on his side; perhaps she represents his sister Wil, for Wil was the only sister who was on his side. The father is on the mother's side (indeed, Vincent often saw them on the same side), although he is isolated from her by the partition and his own vague remoteness in the background; he is making her an offering that she cannot acknowledge. (231)

The flaws in the discussion continue here, and seem so obvious I feel patronizing highlighting them, but of course I must. If two of the main figures have no real purpose or Lubin finds he must admit having no good explanation for them, then either he is saying that the work is flawed or that his ability to interpret it is flawed.

> The situation is sad and apparently hopeless. But a crucifix, similar to the one in *Bearers of the Burden*, hangs in the murky atmosphere above Vincent's head. Hope lies here, in this hint of his identification with Christ. Indeed, the iconography of *The Potato Eaters* may have been influenced by Vincent's frequently expressed admiration for Rembrandt's *Christ in Emmanus*. In it, the risen Christ sits at a table; the others gaze with awe upon him. Vincent praised the "deeply mysterious" quality of Rembrandt's painting. In it he was a "soul in a body," and a "tenderness of gaze." [Letter 435] and these are the qualities that he carried into his own picture.
>
> It is interesting to compare *The Pieta* of 1889 with *The Potato Eaters* of 1885. *The Pieta*, based on a painting of Delacroix, depicts the dead Christ lying in the entrance of a cave. His grieving mother stretches out her arms toward him in a "large gesture of despair"; [Letter W14] because he has suffered and died, she shows him her love. The contrast between the two heads causes them "to seem like one somber-hued flower and one pale flower, arranged in such a way as to intensify the effect." Death and grief were on Vincent's mind at the time he painted *The Pieta*. He had just painted *The Reaper*, the one in yellow that he called "the image of death," and a self-portrait in which he pictured himself "thin and pale as a ghost." [Letter 604] He was also concerned about his mother's grief, then occasioned by the loss of her last son, Cor, who had gone off to South Africa.
>
> The red-headed, red-bearded Christ of *The Pieta* shares the same brilliant colors, the same pathos, the same shadow effects as the ghostly self-portrait. While the crucifix above Vincent's head only hints at his identification with Christ in *The Potato Eaters* he has become Christ in *The*

> *Pieta*. Even as the somber hues of *The Potato Eaters* have been replaced by the brilliant blues and yellows, the downcast rejecting mother has been replaced by the adoring mother and the live rejected son had been replaced by the dead adored son. A haloed self-portrait painted shortly before *The Pieta* reveals that the red-haired Christ is also Vincent. Anticipating the death he is soon to bring upon himself, Vincent proclaims once again that it is better to be dead than alive, for death is the prerequisite for a state of perpetual bliss in which a good mother loves her adored child. (232)

While it is true that Vincent saw himself as wanting to be like Christ, and perhaps at times represented this in his paintings, there is a lot of reaching here to say the peasant in this painting was meant to symbolize Vincent, and much more to say it was meant to connect him to Christ. Lubin continues:

> The unconscious, according to psychoanalysts, conceives of death as the "good sleep" of a baby following a satisfying feeding or as a state of happiness in which the dead one is united with a good mother. [Bertram D. Lewon, *The Psychoanalysis of Elation* (New York: W. W. Norton, 1970), pp. 150-155)] Martin Grotjahn writes, "The unconscious may not know death but it knows peace and sleep and longs for reunion with mother." [Martin Grotjahn, "Ego Identity and the Fear of Death and Dying," *The Journal of the Hillside Hospital,* IX (1960), pp. 147-155)] This is the latent meaning behind much of Vincent's art, perhaps most obvious during the last year of his life when his brush strokes move ecstatically into the heavens. For it is in heaven where this happy reunion will occur. *Landscape with Olive Trees,* for instance, painted at about the same time as *The Reaper* and *The Pieta*, continues the theme of death and a other's love: The cloud in the sky, toward which anguished trees are reaching, resembles "a wraith-like mother and child," [Schapiro, p. 108] according to Professor Schapiro. In a picture sketched just before Vincent killed himself, a man stands on a roof-top as a breast-shaped cloud formation awaits him in the sky above. (*Thatched Roof with Man on Top*.) (233)

I like attempts to understand artistic expression in terms of psychological symbolism, and believe they work well in Vincent's paintings. Partially because of this, I am annoyed when it is poorly done. Here, Lubin has plugged onto the psychological interpretation of Nagera, added some rather unusual claims by Bertram D. Lewon and Grotjahn, thrown in some fairly highly respected art criticism from Schapiro, mixed well, and ended up with a frustrating attempt to give this early work more psychological meaning than is justified. To the extent *The Potato Eaters* works it works, as Vincent says, in presenting peasants as spiritual in their poverty. The darkness surrounding them is certainly an attempt to copy or incorporate Rembrandt's chiaroscuro, and isolation or loneliness is a possible but not necessary interpretation. The religious ideas work better in a more impersonal way, Vincent seeing himself as Christ-like in suffering with peasants, rather than as dealing with some kind of rejection by the mother. The connection of this painting with later paintings has problems, because Vincent changes dramatically before his explosion of brilliant works. It is much more likely he was thinking about his own death in later paintings, not necessarily but more likely. It is hard to say that in this one. Since it was done during the time his father died, that probably influenced it, but claims it represented Vincent's struggles with his own death are hard to support. The psychological theories presented are at best controversial; death in a dream is generally thought to represent not real death but a transformation. Also, the suggestions at the end that the clouds represent "a wraith like mother and child" and that the clouds in *Thatched Roof with Man on Top* represent breasts are rediculous—hard not to completely reject anyone so determined to make the pieces fit.

H. R. Graetz, a critic who deals much better with the symbolism in Vincent's work, offers the following discussion of *The Potato Eaters*

> His famous painting the "Potato Eaters" . . . , done in 1885, warrants close attention. At first the canvas seems very dark, but the longer one looks at it, the lighter it becomes. In a greenish-gray, stable-like room five people are sitting round a table, eating their supper of steaming potatoes. The evening meal has gathered the family together after a day of hard work, but if we observe the faces and bodies of these men and women we see that each one seems lonely and isolated and that there is no response, no relieving communication between them. They are not looking at one another. The two men are turned towards the elder woman on the right, but the approach of the old man and the mute appeal of the younger man on the left are alike disregarded. She looks down while pouring coffee into the cups. Her own isolation is underlined by a wall between her and the old man next to her. There is also no contact between any one of the adults and the young girl in front. Particularly sad and forlorn is the expression of the younger man. Inside the upper slat of his chair the name "Vincent" is painted in such dark color that it can hardly be detected. The young woman at his side looks at him, but he does not seem to notice her attention. His absentminded look is directed towards the old woman on the other side of the table and goes beyond her into infinity. (234)

After this objective description of the painting, Graetz presents his main claim about the symbolism being used:

> But there is one uniting feature in this somber atmosphere of isolation: the lamp. It throws equal light on everyone in the room and brings out the

> warming effect of the steam from the hot potatoes and coffee. The same light which Vincent had once taken into the huts of the miners, he now brings into the hovels of the peasants and weavers. The *burning lamp*, his symbol of love, is the light to console them in the loneliness from which he himself suffered so much in his own life. (235)

I agree that the lamp symbolizes some form of positive buffer against the surrounding darkness; however, I am not sure it can be particularized as love, and find it difficult to say it accomplishes anything in terms of consolation, as no consolation seems to be taking place. They, indeed, are not connecting in any way. If the light is offering consolation or love, they are not realizing it, not allowing it to happen. Thus, if anything, it is an ironic representation of it. Graetz continues:

> In the dark tone of this painting, he portrays the feelings he had expressed to Theo about the lack of understanding in his parents' house and his yearning for *a ray of light, or kindness.* In one of these letters he pointed to isolation as a *kind of prison* and added that he felt more at ease in being " . . .*with peasants and weavers who do not even know the word* [isolation] *than with educated people*." Nevertheless, he described the weavers as *"very miserable people in miserable little rooms.*" The frequent portrayal of their cage-like huts suggests a projection of his own isolation and imprisonment in his work. Vincent himself hinted at it by saying that he *wanted to be more himself in his studies.* He wrote this just before he did the "Weaver's Loom" . . ., a subject he drew and painted many times at that period. (236)

While there is some legitimacy in it, taking out-of-context phrases in Vincent's writing and applying them in a scattered way to his paintings is dangerous, as has already been pointed out in previous incidents. And again, if he is dealing with peasants as representing the opposite of the isolation and prisons of educated people, why is he presenting these peasants as being so isolated!? And demonstrating no compassion or interest in one-another?

Graetz follows up this discussion by plunging into the same contexts as the previous critics concerning the negative response from Van Rappard, and Vincent's reply that he is purposely breaking away from realistic presentation to try and give a more psychological perspective. And then Graetz puts the painting into an important context in terms of other works Vincent is doing at the time, and Vincent's own comments about how he is moving into a more symbolic form of expression:

> Later, in 1885, Vincent painted various still lifes with birds' nests Colleting them had been a hobby since his childhood, and from his lonely

> excursions he brought many home to keep in his room. In one of them . . . we see some eggs in the nest's almost black cavity. He did five different versions, one after the other, and when he wrote to Theo about them he added some thoughts on *"reality and at the same time symbolism in art."* (237)

He also painted *Open Bible* at this time, a harsh, powerful Bible opened to Isaiah with a small, yellow book, slightly torn, titled *La Joie de Vivre*, a novel by Emile Zola, in the foreground. The symbolism here is clear, the Bible representing Vincent's father and the established world of religion he represented, and the passage in Isaiah condemning joy in life, the smaller book representing Vincent, visibly damaged.

Certainly Vincent is beginning to grasp the means and to understand the techniques for how paintings can be more than realistic reproductions, can be expressions of spiritual truths through a form of symbolism, a means of connecting visible and invisible worlds. In this context *The Potato Eaters* has some value. However, it is clear that *The Potato Eaters* is not yet a mature expression of this new understanding. The combination of *realism* with hints of *impressionism, expressionism*, and *surrealism* results in an awkward painting, one that never achieves that desired click where the total becomes more than the sum of its parts. In fact, the painting splinters into parts, some well executed, some not so well executed. The faces of the four peasants are well done, and it can be claimed they manage to take us beneath the surface into an expression not so much of nobility or spirituality, but of loneliness, of being lost and isolated. The hands are also carefully drawn, and hands in general do have strong symbolism. Perhaps Vincent was purposely giving them their qualities and particular actions to symbolise something. Of course, on one level they do represent the rough hands of peasants who work with the earth, something Vincent obviously wanted to depict. The light might have indeed been meant by Vincent to symbolize some form of positive aspect, some sort of redeeming quality in the darkness, but even with the critics' claims to the contrary, it doesn't give the total painting a unity, a meaning. If anything its scattered patches bouncing in different directions off the peasants help emphasize the fragmented separation of them. The young girl might be said to have a ghost-like quality. She certainly doesn't fit realistically in the painting, and perhaps her less detailed qualities can be said to be purposely so in order to give her a spectral quality. In terms of a visual, aesthetic quality, she does not work well, but perhaps there is some point in that. The fact that the table is represented so awkwardly again might be attributed to a purposeful break with realism. The same might be said of the various literal flaws Van Rappard detailed out. Vincent, after all, did respond that they were *intentional.* At the same time, there are flaws that do not work, perhaps intentional flaws in terms of realistic representation, attempts at offering more, but flaws that don't work on the higher levels Vincent is trying to convey.

In other words, the painting is interesting because it is transitional, not because the final genius is realized in it. As a result, the unrealized painting reveals what Vincent is learning, how he is working at a unity of the visable and invisible, trying to connect the two realities. It is not surprising that he is excited about it. It is a moment of great insight, a moment when he has touched his genius, and he knows it. But it is still a clumsy genius.

Part X

Antwerp

Antwerp, Vincent's next stop, lasts only three months. It is a curious interlude between his previous art and what is to come. He arrives eager to explore the city and its museums, and begins visiting the *cafes-concerts*, where he can observe and draw the visitors. He even attends a popular sailor's ball, where he sketches some of the ladies. He writes Theo twenty-three lengthy letters describing the city and his feelings at the time. One will suffice:

> My dear Theo,
>
> I just wanted to send you a few more impressions of Antwerp. This morning I took a long walk in the pouring rain, the object of the outing being to fetch my things from the custom house. The various warehouses and storage sheds on the quays look splendid.
>
> I've walked in many different directions along the docks and quays several times already. The contrast is particularly marked for one who has just arrived from the sand and the heath and the tranquility of a country village and has been in quiet surroundings for a long time. It's all an impenetrable confusion.
>
> One of de Goncourt's sayings was, "Japonaiserie for ever." Well, those docks are one huge Japonaiserie, fantastic, peculiar, unheard of - or at any rate, that's one way of looking at them. I would love some day to take a walk there in your company, just to find out if we see things in the same way.
>
> Everything could be done there, townscapes, figures of the most diverse character, ships as the main subject with water and the sky a delicate grey - but, above all - Japonaiseries. The point I'm trying to make

is that there are always figures in motion there, one sees them in the strangest setting, everything looks fantastic, with interesting contrasts at every turn. A white horse in the mud in a corner where piles of merchandise lie covered with a tarpaulin - against the old, black, smoke-stained walls of the warehouse. Perfectly simple, but with a Black and White effect.

Through the window of a very elegant English public house one can look out on the filthiest mud and on a ship from which, say, such pleasing wares as hides and buffalo horns are being unloaded by docker types as ugly as sin, or by exotic sailors, while a very fair, very delicate English girl stands at the window looking out at this or at something else. The interior with figure wholly in tone, and for light--the silvery sky above the mud and the buffalo horns--again a series of fairly strong contrasts.

Flemish sailors with excessively ruddy faces and broad shoulders, lusty and tipsy, Antwerpers through and through, are to be seen eating mussels or drinking beer with a great deal of noise and commotion. In contrast - there goes a tiny little figure in black, small hands clasped close to her body, scuttling noiselessly past the grey walls. In an encadrement [frame] of jet-black hair, a small oval face. Brown? Orange-yellow? I'm not sure. For a moment she looks up and gives a slanting glance from a pair of jet-black eyes. She is a Chinese girl, quiet as a mouse, stealthy, small, naturally bedbug-like. What a contrast to the group of Flemish mussel-eaters!

Another contrast - one walks down a very narrow street between tremendously tall buildings, warehouses and storehouses. But at ground level in the street - alehouses for every nationality, with males and females to match, shops for food, for seamen's clothing, colourful and bustling. The street is long, at every turn one sees a typical scene, a commotion, perhaps, more intense than usual, as a squabble breaks out. For example, there you are walking along, just looking around - and suddenly cheers go up and there's a lot of yelling. A sailor is being thrown out of a brothel by the girls in broad daylight and is being pursued by a furious fellow and a string of prostitutes, of whom he seems to be terrified - anyway, I see him clamber over a pile of sacks and disappear through a window into a warehouse.

When one has had enough of this hullabaloo - with the city behind one at the end of the landing stages where the Harwich and Havre steamers lie, there is nothing, absolutely nothing to be seen in front except for an infinite expanse of flat, half-flooded pasture, immensely melancholy and wet, with undulating dry reeds, and mud - the river with a single small black boat, water in the foreground grey, sky misty and cold, grey - still as the desert.

As to the overall impression of the harbour, or of one of the docks - at one moment it is more tangled and fantastic than a thorn hedge, so chaotic that one finds no rest for the eye, grows giddy, and is forced by the "papillot-ering" [flickering. Vincent "Dutchified" the French *papillotement*] of colours and lines to look first here, then there, unable to distinguish one thing from another - even after looking at the same point for a long time. But if one moves on to a certain spot with an undefined stretch of land in the foreground, then one again encounters the most beautiful, most peaceful lines and those effects which Mols, for instance, so often achieves.

Here one may se a splendidly healthy-looking girl, who is, or at least seems, wholly honest and unaffectedly cheerful; there a face so slyly vicious, like a hyena's, that it frightens one. Not to forget faces ravaged by smallpox, the colour of boiled shrimps, with dull, grey little eyes, no eyebrows and sparse, greasy, thinning hair the colour of pure hog bristle, or a bit yellower - Swedish or Danish types.

I'd like to do some work around there, but how and where, for one would get into trouble exceedingly quickly. All the same I've roamed through quite a number of streets and alleyways without mishap, have even sat down to talk in a very friendly way with various girls, who seemed to take me for a bargee.

I think it not unlikely that painting portraits may help me to come by some good models. I got my gear today, and some materials, to which I'd been looking forward very eagerly. So now my studio is all ready. If I could come by a good model for a song, I'd be afraid of nothing. Nor do I mind very much that I haven't enough money to force the pace. Perhaps the idea of doing portraits and getting the subjects to pay for them by posing is a safer method. You see, in the city things aren't the same as when one deals with peasants.

Well, one thing is certain, Antwerp is a splendid and very remarkable place for a painter.

My studio isn't at all bad, especially now that I've pinned up a lot of small Japanese prints which I enjoy very much. You know, those small female figures in gardens or on the beach, horsemen, flowers, gnarled thorn branches.

I'm glad I came here - and hope not to sit still and do nothing this winter. Anyway, it's a relief to have a small hideaway where I can work when the weather is bad. It goes without saying that I won't be living in the lap of luxury.

Try to send your letter off on the first, for while I've enough to live on until then, I shall be getting the wind up after that.

My little room has turned out better than I expected and certainly doesn't look dreary.

Now that I have the 3 studies I took along with me here, I shall try to make contact with the marchands de tableaux [picture dealers], who seem, however, to live for the most part in private houses, with no display windows giving on to the street.

The park is beautiful too. I sat there one morning and did some drawing.

> Well - I've had no setbacks so far, and I'm well off as far as accommodation is concerned, for by sacrificing another few francs I've acquired a stove and a lamp. I shan't easily get bored, believe me.
>
> I've also found Lhermitte's *Octobre*, women in a potato field in the evening, splendid, but not his November yet. Have you kept track of that by any chance? I've also seen that there's a Figaro illustré with a beautiful drawing by Raffaelli.
>
> My address, as you know, is 194 Rue des images, so please send your letter there, and the second de Goncourt volume when you've finished with it. Regards,
>
> Ever yours, Vincent
>
> It's odd that my painted studies look darker here in the city than in the country. Is that because the light isn't as bright in the city? I'm not sure, but it might matter more than one might think at first sight. I was struck by it and can imagine that some of the things that are with you now also look darker than I thought they were in the country. Yet those I brought along with me don't seem the worse for it - the mill, avenue with autumn trees, a still life, as well as a few small things. (238)

At the same time, he is broke, eats almost nothing, and as a result begins experiencing stomach problems, coughing incessantly and losing his teeth. It is possible some of his discomfort is the result of more than just poor eating habits, perhaps syphilis, but whatever the cause, he is exhibiting serious health problems.

The letters contain endless begging for more money: On December 19, he writes, "I must tell you that I am at the end of my rope; with my last remaining five francs I had to buy two canvases . . ., and the laundry woman had just brought me my clean linen, so that for the moment I have only a few centimes left. (239) In late January, he writes, "Therefore try, as I asked you, to send me another fifty francs, then I can keep going till the end of the month, and could buy a new pair of trousers and a waistcoat at once, and the coat in February." (240) On February 18, he writes, "If you can send anything, even if it's only five francs, do so. There are still ten days left in the month, and how am I to get through them? For I have absolutely nothing left. (241)

And the lack of money leads into mention of his health problems: He writes, "For you must realize that in the whole time I have been here now, I've only had three warm meals, and for the rest nothing but bread. In this way one becomes vegetarian more than is good for one." (242) He assumes that the lack of food is the reason, writing that his stomach "having deteriorated to such a degree" it is the cause of his constant coughing and the rotting out of his teeth. (243). He elaborates, "I am having my teeth seen to, for instance; there are no less than ten teeth that I have either lost or may lose, and that is too many and too troublesome, and besides, it makes me look over forty, which is not to my advantage. So I must have that taken care of. It will cost me a hundred francs, but it can be done better now while I am drawing than at any other time, and I have had the bad teeth cut off and have just aid half the money in advance. They told me at the same time that I ought to take care of my stomach, for it is in a bad state. And since I have been here this has far from improved." (244)

He produces little art during his stay, and what he does is not of a high quality, but he is getting exposure to other artists. The one most critics focus on is Peter Paul Rubens, but others such as Henri de Braekeleer, Charles Mertens and Piet Verhaert also interest him. And there are indications he is beginning to move to a brighter, more colorful palette. He writes, "In the woman's portrait I have brought lighter tones into the flesh, white tinted with carmine, vermillion, yellow, an a light background of gray-yellow, from which the face is separated only by the black hair. Lilac tones in the dress." (245)

Though he did not go to Antwerp with the idea of attending the art academy there, by January, 1886, he has decided to check it out, mainly with the idea of being able to work with nudes and study under its director Karel Verlat. He applies, is accepted, and begins his studies with high expectations. However, things quickly go badly for him. Not unexpectedly he ends up quarreling with his instructors and disagreeing on their views about art. Also, the timing is wrong, as he joins the classes just before they are concluding. And he finds that he does not get to work with nudes, or very little. He apparently does find a solution to this by joining some sketching clubs in the evening that employ nudes, and some of his sketches of them remain. However, as Jan Hulsker points out, the "nudes, all quite large, clearly show Vincent's lack of skill in this field; most of them are quite clumsily done and not well proportioned, especially compared to the masterly drawings of peasants he had produced in Nuenen." (246) In the end, Jan Hulsker concludes, "that by sheer lack of materials it is impossible to get a firm grasp of the work Vincent achieved in the two drawing classes at the academy. (247)

During February Vincent is pushing Theo to let him come and live with him in Paris. Theo is not as enthusiastic, and the exchange of letters is confusing. Theo certainly wants Vincent to return to Nuenen first, to help their mother make a move to live with Wil in Breda. Vincent does not want to return to Nuenen, seems so set against even a short visit that one wonders if there is more to the refusal than simply a desire to come to Paris. In the end, Vincent gets on the train and descends on Theo, whether or not Theo wants it.

An aspect of Vincent's sudden move that does not reflect well on him is that he consciously leaves Antwerp without paying his bills, and is thus able to assure Theo he still has some money. However, it is probably best understood more in the sense of him simply leaving suddenly, without any plans, rather than him cheating others, as he also left whatever he owned behind. The exchange of letters debating when he would come to Paris was dragging on too long for him, and on the spur of the moment he decided to take action. It is assumed he arrived March 1 or 2, 1886.

Part XI

Paris

The beginning of March, 1886, Vincent moves in with Theo and immediately joins the dynamic, cutting edge of the art world. Even for one already knowledgeable about art and aware, though from a distance, of the new trends, he has to have been dazzled by the community of artists, outsiders, rebels, mavericks, people like him who just don't fit in the more "normal," conservative communities where he has spent his life. This is his world, a world with an electric atmosphere that embraces his own eccentricities.

Think of it. Within but a few months of his arrival, Vincent connects up with Edgar Degas, Paul Signac, Louis Anquetin, Emile Bernard, Arnold Koning, Henri deToulouse-Lautrec, and shortly after Paul Gauguin, Armand Guillaumin, Camille Pissarro, and Georges Seurat.

Vincent and Theo were two young men in their prime (even if Vincent was in poor health), living on their own in the middle of this wild bohemian life. In spite of their hardships with money and their health struggles, they had to have led, at least briefly, lifes filled with the excitement and some would say the debauchery of young adults. Certainly the West Bank of the Seine, and the hill district just north of Paris named Montmartre was a community of artists consciously adopting a bohemian life style. The term bohemian was first used in the nineteenth century to describe the untraditional lifestyles of marginalized and impoverished artists, writers, musicians, and actors in major European cities who gathered together to live lifes of poverty that flaunted the establishment, engaged in extra-marital sex, and questioned traditional, orthodox assumptions about morality. The use of the particular term referenced the lives of gypsies and reflected a belief in France at the time that the gypsies had come from Bohemia. While it became a popular term for such groups throughout France and the rest of Europe, and has since been used for similar communities throughout the world, this particular community in Paris has come down to us today as the classic example, the reference or touchstone for all of the others. (248)

Portraits of Vincent at the time suggest his health has improved (his teeth and stomach problems at least temporarily are less bothersome), and he is even depicted wearing a stylish hat, suggesting he was trying to fit into the persona of this new world. Of course, he never ate well, and it's safe to assume his poor eating habits continued. Also, while Jan Hulsker claims the evidence is sparse, it's true that he joined in the general consumption of wine and absinthe, a powerful drink to which some have since attributed his mental problems. While there are few references to it in the correspondence surrounding him, there are enough, one to Gauguin from Sept. 29, 1888, where he claims to have been "nearly an alcoholic," and there are paintings of him with glasses of wine, even a still life he did titled *Absinthe*.

Absinthe was the "in" drink of the group, and has since been closely associated with them. It is a distilled, anise-flavored liquor (60 – 80 %), derived from herbs, including the flowers and leaves of the medicinal plant *Artemisia absinthium*, also called "wormwood." Since it is typically of a natural green color, a result of the chlorophyll in the herbal ingredients present during maceration, it is often called "the Green Fairy." The usual way of drinking it involved pouring it through a sugar cube on a slotted spoon. It originated in Val-de-Travers, Switzerland, as a medical elixir, but gained popularity in the bohemian culture Vincent became a part of in Paris, and has been said to be both dangerously addictive and a psychoactive drug, thought to result in nervous and mental disorders, do to the chemical *thujone*, which is present in small quantities in the wormwood it comes from. In fact the Lanfray murders of 1906 in Switzerland were thought to be the result of it, and it was subsequently banned there, and in 1915 it was also banned in France. Vincent's friend Henri de Toulouse-Lautrec, who provides a painting of Vincent sitting at a table with a glass of wine, was addicted to it, and eventually was institutionalized and died from a combination of alcoholism and syphilis.

Throughout the nineteenth century, Europe was rapidly evolving, not just artistically but politically and economically as well. In France, the political movements, the conflict between the Monarchists and the Republicans that had dramatically sparked the French Revolution with the storming of Bastille in 1889, eventuated in the constitution of the French

Third Republic, and on December 27, 1848, Napoleon III, born Charles Louis-Napoleon Bonaparte and also known as Louis-Napoleon Bonaparte, became the first President of it, gaining the support of the monarchist right and the upper class as the best of what there was to choose from, and with the hope that he would end the disorder and instability of the country and prevent a communist styled revolution. The working class supported him with the hope he would implement more progressive economic policies that would improve their working conditions. However, it was the rural voters, the peasants who overwhelmingly supported him, mainly because he represented what they saw as the great Napoleon Bonaparte bloodlines that had brought France to its highest levels and established, at least briefly, some stability after the French Revolution.

In spite of his overwhelming victory, he was hand-cuffed by a Monarchist Parliament determined to restore either the Bourbon or Orleans family to power. He also had to deal with the religious situation, and satisfy both the Catholics, who demanded restoration of the Pope's rule in Rome, and the secularist demands that the Pope introduce liberal changes to its governance of the Papal State.

In the third of his four-year mandate, he asked the National Assembly for a revision of the constitution to enable him to run for re-election, arguing that four years were not enough to implement his political and economic program. Fearing that a President would abuse his power to transform the Republic into a dictatorship for life, the Constitution of the Second Republic had stated that the Presidency of the Republic was to be held for a single term of four years, with no possibility of re-election. With plans of restoring the monarchy, the National Assembly refused the amendment and changed the electoral law to place restrictions on universal male suffrage, imposing a three-year residency requirement thus preventing a large proportion of the lower class from voting. Although he had originally acquiesced to this law, Louis-Napoleon now used it as a pretext to break with the Assembly. He surrounded himself with lieutenants, such as Momy and Persigny, completely loyal to him, secured the support of the army, and traveled the country giving populist speeches condemning the assembly and presenting himself as the protector of universal male suffrage.

Finally, after months of stalemate, and funded by the wealth of his mistress, Harriet Howard, he staged a coup d'etat and seized dictatorial powers on December 11, 1853, the 54th anniversary of Napoleon II's crowning as Emperor, and also the 46th anniversary of the famous Battle of Austerlitz (gaining the nickname "The Man of December", "*l'homme de Décembre*"). In a controversial national referendum, the coup was declared to have been approved by the French. The coup alienated the Republicans, and some of Napoleon III's previous supporter, notably Victor Hugo, turned against him.

However, of course, with money and the military on his side, he had gained control, at least temporarily, and sought to solidify and justify his new position. He established himself as the Second Emperor of the French (since Napoleon II's reign has been very short and largely fictional) under the name Napoleon III, and ruled from December 2, 1852 to September 4, 1870. While the appearance of a Republic was somewhat maintained by reestablishing universal male suffrage and an elected Parliament, all power was now in Napoleon III's hands, and he began eliminating his competition by sending dissidents to either Devil's Island or New Caldonia.

Furthermore, following standard practices of the Monarchy, he immediately began to look for a wife to produce a legitimate heir. What he found was that his status among Europe's nobility was not what he might have hoped, and after being turned down by Princess Carola of Sweden and Queen Victoria's German niece Princess Adelheid of Hohenlohe-Langenbury, he married into the lesser bloodlines, but still wealthy and beautiful Countess of Teba, Eugenie de Alexandre, a Spanish noblewoman with some Scottish ancestry who grew up in Paris, who provided him with a son and heir in 1856..

However, as much as he manipulated to empower himself, life for Napoleon III was far from secure. On April 28, 1855, and again on January 1858, he survived a attempted assassinations.

No doubt, the precariousness of his position drove him to be authoritarian, censoring the press, manipulating elections, denying free debate in Parliament, and imprisoning his enemies. By the 1860s, however, perhaps feeling more secure, he began making concessions to placate his liberal opponents, now allowing free debates in Parliament and uncensored reports of them, even appointing the Liberal Emile Ollivier, a leader strongly opposing him in the past, as Prime Minister in 1870.

As with all political leaders, Napoleon III was concerned about his legacy, and one thing he wanted was to known as a social reformer. Of course, as always with political leaders, it's hard to know just how much their actions and policies are the cause of economic and social developments. Nevertheless, during his rule, the France economy made rapid strides, and the inevitable industrialization managed to keep both the owners and the workers relatively satisfied. Downtown Paris was renovated, many of the slums eliminated, the streets widened, and numerous parks constructed. Working class neighborhoods were moved to the outskirts of Paris, nearer the factories where they were employed. He was supported by the Saint-

Simonians, a group of bankers who established a new form of banking called the Credit Mobilier, which sold stocks to the public and used the money to invest in industry, sparking a rapid growth in economic development. Also important for economic development, developments outside of France were favorable for economic expansion. Discoveries of gold in California and later Australia increased the European money supply, which in turn encouraged company promotion and investment of capital. The railroads in France also increased from about 3,000 to 16,000 kilometers during the 1850s, providing access to mines and factories. At the same time, the many smaller rail lines merged into six major ones, and the disjointed transportation of materials became much more unified and efficient. More efficient iron steamships also began to replace the old majestic and beautiful but inefficient wooden ships, and France build the Suez Canal.

However, Napoleon III's Second French Empire was about to come crashing down. Germany was transforming even faster, and was about to defeat France in the Franco-Prussian War, bringing an end to Napoleon III's reign.

Prussia's Prime Minister Otto von Bismarck needed a war with France to unify the German states with Prussia, and he manipulated just that. Napoleon III was captured at Sedan, and the French surrendered in January of 1871, ending Napoleon III's reign, and beginning a messy transformation in French government. The new government under Adolphe Thiers was overshadowed by the settlement of peace terms with Prussia, resulting in a revolution in Paris known as the Paris Commune, which maintained a radical regime for two months until its bloody suppression by Thiers' government in May 1871, which in turn would be followed by firm repression of the labor movement.

The French people and National Assembly wanted to fill the political void with a return to a constitutional monarchy. The Legitimists supported the heirs to Charles X, and put forth his grandson Henri, Comte de Chambord (Henry V), and Orleanists, supported the heirs of Louis Philippe, putting forth his grandson, Louis-Philippe, Comte de Paris. Surprisingly, they did manage to compromise, naming the childless Comte de Chambord as king, with the Comte de Paris recognized as his heir. Thus, in 1871, the throne was offered to the Comte de Chambord. Chambord, however, had no wish to be a *constitutional* monarch but a semi-absolutist one like his grandfather Charles X, who, in 1830 had abdicated in favor of Chambord, or like the contemporary rulers of Prussia/Germany. Furthermore, he refused to reign over a state that used the Tricolor associated with the Revolution of 1789 and the July

Monarchy of the man who had seized the throne from him in 1830, the citizen-king, Louis Philippe, *King of the French*. This became the ultimate reason the restoration never occurred. As much as France wanted a restored monarchy, the nation was unwilling to abandon the popular *tricolore*. Instead a "temporary" republic was established, pending the death of the elderly childless Chambord and the succession of his more liberal heir, the Comte de Paris.

Ahhhh, the whims of the rich and famous! Such power, such arrogance, such stupidity!

Thus, in February 1875, a series of parliamentary Acts established the organic or constitutional laws of the new republic. At its apex was a *President of the Republic*. A two-chamber parliament was created, along with a ministry under a prime minister referred to as "President of the Council" who was nominally answerable to both the President of the Republic and parliament.

But the battle wasn't yet completely over. On May 16, 1877, with public opinion swinging heavily in favor of a republic, the President of the Republic, Patrice MacMahon, duc de Magenta, a *monarchist*, made one last desperate attempt to salvage the monarchical cause by dismissing the republic-minded prime minister Jules Simon and appointing the monarchist leader the Duc de Broglie to the office. Then, in October, he dissolved parliament and called a general election. But his clever plan backfired, and he was accused of having staged a *constitutional coup d'etat*, subsequently known as *le seize Mai* after the date on which it happened.

The Legitimists were defeated. Republicans (called Opportunist Republicans because they favored moderate changes to establish the new regime) returned to power, ending any chance of a restored French monarchy by gaining control of the Senate, January 5, 1879. MacMahon resigned on January 30, leaving a seriously weakened presidency in the name of Jules Grevy. It was not until Charles de Gaulle, eighty years later that another President of France would unilaterally dissolve parliament. In 1881 and 1882, the republican control of the Republic voted in the Jules Ferry laws establishing mandatory, secular education, thus removing education from the exclusive control of the Catholic Church. In 1883, Henri, Comte de Chambord died, and many of the remaining Orleanists gave up hope of returning to a monarchy, instead now turning their support to the Republic, as Adlophe Thiers had already done. The political battles weren't suddenly over, but Republic had been established and would remain the political frame within which Vincent's years at Paris and Arles took place.

While these dramatic political events were transforming the French world, the Academie des Beaux-Arts, determined to uphold traditional views of both content and style in French painting, controlled the French art scene. Historical subjects, religious themes, and portraits were valued (landscape and still life were not), and the Académie preferred carefully finished images that mirrored reality, somber and conservative colors, and a repression of brush strokes, all meant to conceal the artist's personality and techniques.

Each year the Academie held a juried art show, the Salon de Paris, and even just being invited to display one's work in the show brought prestige and garnered commissions. Winners at these shows indicate what the Academie was looking for, and included works that fit into what has been designated as Neo-classicism, such refined works as those of Jean-Leon Gerome and Alexandre Cabanel.

Younger artists painted in a lighter, less cold and more lively manner, extending the realism of Gustave Courbet and the Barbizon school. Their subjects tended to landscapes and contemporary life, rather than the classical subjects and historic scenes of Neo-classicism. But the Academie was not impressed, and year after year rejected their submissions. Finally a group of young realists featuring Claude Monet, Pierre Auguste Renoir, Alfred Sisley, and Frederic Bazille came together. Soon, they were joined by Paul Cezanne, Camille Pissarro, and Armand Guillaumin. Often they would meet at places such as Café Guerbois and share views with such recognized arts as Edouard Manet.

In 1863, Edouard Manet's *The Luncheon on the Grass (Le dejeuner sur l'herbe)* was rejected by the Salon, because it depicted a nude woman with two clothed men at a picnic, a contemporary setting. While nudes were routinely accepted by the Salon when featured in historical and allegorical paintings, it was considered bad taste to present a realistic nude in a contemporary setting. The jury's sharply worded rejection of Manet's painting, as well as the unusually large number of rejected works that year sparked the young French artists.

After seeing the rejected works, Napoleon III got into the fray, decreeing that the public be allowed to judge the work themselves, and the Salon des Refuses (Salon of the Refused) was organized. While many came only to ridicule, the Salon des Refusés drew attention to the existence of a new tendency in art, actually attracting more visitors than the established Academie Salon.

Nevertheless, artists' petitions requesting a new Salon des Refusés in 1867, and again in 1872, were denied. Frustrated and looking to promote their new views on art, Claude Monet, Pierre-Auguste Renoir, Camille

Pissarro, Alfred Sisley, Paul Cezanne, Berthe Morisot, and Edgar Degas organized their own exhibit at the studio of the photographer Nadar (Gaspard-Félix Tournachon). Though Johan Jongkind and Edouard Manet declined, thirty artists, including one of Edouard Manet's mentors, Eugene Boudin participated. Seven more exhibits would be held over the next twelve years.

As expected, the critical response was mixed. Paul Cezanne and Claude Monet were especially condemned, and, as is not unusual, the detractors gave the group its since famous name. Louis Leroy, an art critic and humorist, wrote a scathing review in the *Le Charivari* newspaper, where he displayed his clever wordplay by taking the title of Claude Monet's painting *Impression, soleil levant* (*Impression, Sunrise*) for the title of his article, *The Exhibition of the Impressionists*, and setting up his review in the form of a dialogue between viewers, including such passages as: "Impression—I was certain of it. I was jest telling myself that, since I was impressed, there had to be some impression in it . . . and what freedom what ease of workmanship! Wall paper in its embryonic state is more finished than that seascape." (249)

How could such a clever play on words not catch on, and it did. The term "Impressionists" quickly gained favor with both the artists and the general public. In terms of what Impressionism has come to represent, Claude Monet, Alfred Sisley, Berthe Morisot, and Camille Pissarro in their embracement of spontaneity, sunlight, and color, fit it the best. In truth, the group as a whole brought a somewhat diverse mixture of approaches to art to the exhibit, held together more by their position outside the established Academie. Edgar Degas was actually against many of the central views, believing in the primacy of drawing over color, and rejecting the practice of painting outdoors. Pierre-Auguste Renoir also turned against it in the 1880s. And the original leader of the group, Edouard Manet never moved away from his strong use of black in his work, and never did participate in any of the exhibitions, instead continuing to submit works to the Salon, where his painting *Spanish Singer* had won a second class medal in 1861, urging his followers to do the same in the belief that that was how to really establish a reputation. Some of them took his advice.

Paul Cezanne, Pierre-Auguste Renoir, Alfred Sisley, and Claude Monet, all withdrew from the group exhibitions in order to submit to the Salon. Jean Frédéric Bazille was killed in the Franco-Prussian War, 1870. Disagreements were common, resulting from issues such as Guillaumin's membership in the group, championed by Camille Pissarro and Paul Cézanne, but rejected by Claude Monet and Edgar Degas. Edgar Degas, in turn, wanted to included Mary Cassatt, Jean-Francois Raffaelli, Ludovic

Lepic and others that fit more under traditional realist categories, to which Claude Monet replied that the group had become nothing more than a place for "first-come daubers" to exhibit. The group also fought over whether or not to include Paul Signac and Georges-Pierre Seurat. In the end, only Camille Pissarro participated in all eight exhibitions.

Interestingly, while social, economic, religious, political and artistic transformations are evident beneath the surface, and sometimes not so submerged, the years between the Franco-Prussian War and WWI, are sometimes referred to as *belle epoque* (the beautiful era). For a brief time, though tensions remained high between France and Germany, especially the result of the French loss of Alsace-Lorraine to Germany in 1871, there was relative political stability throughout Europe. Diplomacy in such conferences as the Congress of Berlin in 1878 and the Berlin Congo Conference in 1884 kept military conflict off-stage. The wealthy, even middle-classes, traveled throughout Europe without passports. The International Workers' Movement reorganized itself in such transnational socialist organization as the Second International and reinforced a European, rather than national consciousness.

Nevertheless, ominous tensions were just below the surface. The working-class socialist parties, the bourgeois liberal parties, and the aristocratic conservative parties were gathering strength, and a confrontation was becoming inevitable. Also, Europe had created an impossible situation in its relations with the rest of the world. What has become known as the New Imperialism or High Imperialism, involving a struggle by the various European countries for overseas colonies, especially in Africa, was about to explode into the largest war yet.

In a sense, it is the calm before the storm, and within it is a rapid development in all arenas. Tremendous scientific and technological advances take place, ones that are still being unfolded today, including the invention of the automobile, the airplane, the phonograph, the telephone, motion pictures, and the subway. The entire field of bacteriology enters the world of medicine, and modern understanding of the origins results in a century of anti-biotics and subsequent developments in microbiology. Max Planck, Albert Einstein, and Niels Bohr give the world modern physics. Social Sciences explode on the scene, with such notables as Georg Simmel, Emile Durkheim, Max Weber, and Thorstein Veblen leading the field into improved research and interpretation techniques.

In Paris, *belle epoque* brought such popular new forms of entertainment as cabaret, the can-can, and the cinema. In 1889, the *Exposition Universelle* took place there, and the Eiffel Tower, meant to represent this exciting new world of transformation was built, and would become the symbol of Paris.

For those embracing the bohemian life, the restaurant-concert hall at *Moulin de la Galette* and other such gathering places in Montmartre seethed with this rich alternative life style of the times. His correspondence reveals that Vincent must have enjoyed discussions there, arguments about art. The *Moulin de la Galette*, often used in mistaken contexts was actually a restaurant-concert hall built around three windmills The *Blute-Fin*, the *Radet*, and the *Debray* or *Poivre*. Several of Vincent's paintings and drawings from this time are of this group or various parts of it and include: *Moulin de la Gallette, View of Montmartre with Windmills, Windmills at Montmartre, Windmill at Montmartre with Quarry,* and *View of Montmartre, Montmartre, the Windmills*. Not a surprising subject. Such gathering places were popular subjects for Vincent's current community of artists. August Renoir painted several, including *Dance Hall at the Moulin de la Galette*. Another one of these gathering places, *The Folies-Bergere,* featuring operettas, comic opera, popular song, gymnastics, and bawdy dancing, was the setting for Edouard Manet's appropriately named painting *A Bar at the Folies-Bergere* depicting a bar-girl, one of the *demimondaine*, standing before a mirror, completed in 1882. In 1889, Josep Oller, owner of the Paris Olympia, built the famous cabaret, *Moulin Rouge* (Red Windmill), so named because of a large, red windmill on its roof, in the "red-light" district of Pigalle, at the bottom of the hill that held the three windmills that made up *Moulin de la Galette,* and it became the subject of many paintings by Toulouse Lautrec. Vincent's paintings, however, are of the windmills from the outside and depict them more as traditional windmills, rough and without the excitement of the life that was going on inside them.

While Vincent couldn't have known him or his work, this same environment was about to include Alfred Jarry, known primarily for his Ubu plays. His first play, which he wrote at the age of fifteen with two other pupils at the Rennes *Lycee* in 1888, was a comic satire of their physics teacher, Monsieur Hebert--nothing more than a childish prank. In time, however, it would come to be considered the world's first absurdist drama.

The first and most famous of Alfred Jarry's Ubu plays *Ubu Roi* or *Ubu Rex* parodies Shakespeare's Macbeth, presenting the world the grotesque figure of Pa Ubu, a foul old man determined to conquer Poland by any means necessary, a personification of all that is foul and stupid in mankind. It premiered at the *Theatre de L'OEuvre* on December 10, 1896 to mixed results from an angry and violent crowd, some outraged, some intrigued. In either case, it created a stir. Two years later it was performed again, this time using marionettes at the *Theatre de Pantins*. About the same time, Jarry completed the second play in what he designated his Ubu trilogy, *Ubu Cuckolded*. He completed *Ubu Enchanted,* the final play of the trilogy, September of 1899. However, neither *Ubu Cuckolded* nor *Ubu Enchained* was ever performed during his lifetime.

Unfortunately, Alfred Jarry's died in 1907 at the age of 34, but what he initiated prefigured and influenced the works of Eugene Ionesco and Samuel Beckett.

Vincent's stated reason for coming to Paris is to study under Fernand Cormon, whom he has been told is a master of composition. Vincent has come to believe that Fernand Cormon's philosophy of learning by "drawing from the antique" will help him develop his own technique, which he considers weak. Upon his arrival, Vincent immediately goes to No. 104 boulevard de Clichy, the location of Fernand Cormon's studios and begins his studies. Little is known about his experiences there, though he became close friends with Emile Bernard, and their exchange of letters reveals some of what took place.

The painting done at that time that has garnered the most commentary is *Two Shoes.* H. R. Graetz writes:

> Shoes are coverings for man's feet, carrying his weight as he walks on the this earth. They bear the imprint of his movements and thus attest to his traveling along the road. In the common expression "to be in someone's shoes," the shoe stands for the wearer and, as in other still lifes, we have a portrait her, this time in the form of shoes. They seem to be a pair, though their shape could also suggest two left shoes. Time, wear, and weather have left their marks equally on them as on twin brothers, worn in *toil and sweat.* The lace on the left-hand shoe changes direction and suddenly turns to the upright shoe, again almost touching it—as if Vincent's thoughts were going from him to his brother. In such close touch with each other, the two shoes are like a symbolic expression of the two brothers together on their road. (250)

Vincent, indeed, is aware of symbolism on this level. He has clearly used more blatant symbolism in past works, and it is certain he meant these shoes to symbolize something. While it is a guess to say he consciously meant them to be Theo and him, it is a legitimate one.

Also worth noting in the style of the painting is how he manages to give the shoes both a personality and a feeling of life beyond mere physical representation. The jagged brushstrokes, foreshadowing what is to come in his work, the sharp bits of light, the luminous quality, all suggest that this is a painting meant to take the viewers out of mere physical existence into some numinous world beyond it. The painting both connects back to *The Potato Eaters* and looks forward to the masterpieces to come.

At the same time, he does a still life of six shoes, the first two on the left looking very similar to the two in the previous painting and very likely meant to represent the same two people. All of the shoes are worn. Whoever they belong to has lived a rough life. Perhaps, as H. R. Graetz suggests, they represent the friends Vincent has met through Theo, perhaps Vincent's family. There is no way to know. Certainly he is making some kind of statement in having the third shoe upside down.

As it turns out, Vincent only studies at Cormon two-four months, depending on what correspondence one believes, before Cormon suddenly decides to close down the studio. When he reopens it, Vincent does not return. However, during his brief time there, he is said to have worked very hard and demonstrated amazing improvement, two qualities that he has demonstrated throughout his life.

Three months after Vincent arrives in Paris, the brothers move into a new apartment on the third floor of a building at N. 54 rue Lepic, in Montmartre, an apartment that looks over much of Paris, and some of Vincent's paintings at the time are his interpretations of what he sees. It was perhaps the most comfortable place Vincent has lived in, the two brothers seem to get along better, and Vincent is in better spirits than usual, at least at first. It is also clear his health has much improved.

All of these positives, of course, need to be taken in context, and there are reports, such as a letter by Andries Bonger, that Vincent does not look so well at all, continues to quarrel with everyone, and is not getting along with Theo. Furthermore it becomes clear that Vincent isn't the only brother with health problems and on the edge of a mental breakdown: "It is quite possible I have missed the news, because owing to van Gogh's illness I have had little time to read. He has had such serious nervous complaints that he has been unable to move. To my great astonishment I found yesterday that he was as he had been before it occurred; he still felt stiff, as if he had had a fall, but it had left no other consequences. He has now decided at last that he must take care of his health. He is in great need of doing so. He has decided to leave Vincent; living together is impossible." (251) However, the parting does not take place. As usual, the brothers manage to work through their disagreements; their love/hate relationship is strong enough to keep them together, at least for a time.

Vincent's friend Emile Bernard will later write of him: "Red haired, with a goatee, rough moustache, shaven skull, eagle eye, and incisive mouth as if he were about to speak; medium height, stocky without being in the least fat, lively gestures, jerky step, such was van Gogh, with his everlasting pipe, canvas, engraving or sketch. He was vehement in speech, interminable in explaining and developing his ideas, but not very ready to argue. And what dreams he had: gigantic exhibitions, philanthropic communities of artists, colonies to be founded to the south, and the conquest of public media to re-educate the masses—who used to understand art in the past. . . ." (252)

At Cormon's studio Vincent makes friends with Henri Toulouse-Lautrec, and within a few months has become a central member of this *avant-garde*, spending daily (and nightly) social sessions with them arguing over their artistic beliefs. There can be little doubt but that it is a wearing life, filled with smoking, liquor, sex, and driven by an honest and demanding desire to take artistic expression into new territory. This is, in fact, the artistic community Vincent longed for, a world of people passionate about life where he can argue his views, certainly not all happy and peaceful, much the opposite. It a way, it can be said he blossoms here. The anti-social outsider, undernourished and pessimistic, even self-abusive, becomes a lively social man. Even his health seems to have improved, at least for a time, and he puts aside his ragged clothes for clothes matching the world he now lives in.

Marc Edo Tralbaut writes:

> Many of the women who sat to Vincent were prostitutes. Most of middle-class society had closed its doors to him when he was in Holland, and he had no choice. It was much the same in Belgium and France, and it was natural enough that Vincent, who had never enjoyed normal marital relations, should have sought in the brothels what he was denied by a serious of unfortunate circumstances.
>
> An echo of this low life and perhaps of their adventures together is to be found in an album of water-colours which Emile Bernard gave to him. It was entitled "At the Brothel" and it has never been published. It now belongs to Dr. V. W. van Gogh. (253)

Artistic influences surround him. He visits the Louvre regularly, where he sits and studies individual paintings for hours. He admires the work of Adolphe Joseph Thomas Monticelli, a superb colorist, who dies in June of 1886, and whose work influences Vincent to move out of his harsh colorless style. Of course, the artists surrounding him, whether or not they all fit the technical category of Impressionism, are exploring color, and now, though he once condemned them, Vincent is quickly being converted. He soon encounters Japanese art, even purchases Japanese prints and puts them on his wall. Cliff Edwards devotes an entire chapter of his biography to Vincent's "Oriental Connection," pointing out several references in Vincent's letters, including the following:

> If we study Japanese art, we see a man who is undoubtedly wise, philosophic and intelligent, who spends his time doing what? In studying the distance between the earth and the moon? No. In studying Bismarck's policy? No. He studies a single blade of grass.
>
> But this blade of grass leads him to draw every plant and then the seasons, the wide aspects of the countryside, the animals, then the human figure. So he passes his life, and life is too short to do the whole.
>
> Come now, isn't it almost a true religion which these simple Japanese teach us, who live in nature as though they themselves were flowers?
>
> And you cannot study Japanese art, it seems to me, without becoming more joyful and happier, and we must return to nature in spite of our education and our work in a world of convention. (254)

Vincent has already encountered Japanese art, but while in Paris sees it in new ways, and as Tralbaut writes:

> There is no record of subjects discussed on the Petit Boulevard, but one of the topics about which Vincent must certainly have often held forth to Bernard, Toulouse-Lautrec, Gauguin and the others was certainly Japanese art. Japanese prints were very popular at this time, and they had a considerable effect on the development of Impressionism and the trends that were to emerge thereafter. These decorative pictures, printed in clear flat colour from several wood-blocks, attracted Vincent greatly, and he pinned them up in his room wherever he went. Their influence upon his work at a superficial level is obvious: he copied some of them, and they appeared in the background of his painting. (255)

Tralbaut promises immediately after this to devote an entire book to this Japanese influence. And Cliff Edwards expands his discussion of Vincent's Oriental influence to an in-depth claim that Buddhist and Zen beliefs underlie all of Vincent's mature art. Just how central the influence is beyond giving him a different perspective on techniques, though interesting, must be speculative. However, it is obvious he embraces both Japanese artistic techniques and some of the Buddhist views of life. He and Theo purchase a number of Japanese prints from the Siegfried Bing dealership, and some of the copies he paints from them remain. Also, in March-April of 1887, he organizes an exhibition of Japanese prints in the Café Le Tambourin, a café owned by Agostina Segatori, with whom Vincent apparently has an affair, and where he exhibits a portrait he did of her that includes Japanese prints in the background. He also uses Japanese prints for the background of two portraits of Pere Tanguy, the kind art dealer who was one of the first to offer Vincent's work for sale.

Following this exhibit Vincent begins accompanying Paul Signac to the do paintings along the Seine at Asnieres. Paul Signac is a deciple of Georges Seurat, whose careful, systematic use of dots or points of pure color in juxtaposition to be combined in the viewer's eye will become known as "Post-impressionism" or more commonly simply "Pointillism," yet another technique Vincent experiments with, though he cannot bring himself to completely abandon his rapidly developing style of rough brushstrokes. Nevertheless, it seems clear that the short, energetic strokes of his mature paintings have been influenced by this experience.

Though his relationship with Theo, which when he first moved to Paris was, for a time, more positive, has by now become yet again almost unbearable, and he is, as usual, beginning to wear out his friendships with his artists friends, he is still participating and at the center of this alternative art scene, as is indicated by the fact that in November he organizes an exhibition of the Impressionists in the Restaurant du Chalet, showing his own work along with that of Louis Anquetin, Emile Bernard, Arnold Koning, and Henri Toulouse-Lautrec, and meeting Paul Gauguin, Jean-Baptiste Armand Guillaumin, Camille Pissarro, and Georges Seurat. A couple of months later, December-January, he exhibits a painting in the Theatre Libre d'Antoine.

However, Vincent's time in Paris is about to end. The endless activity, both professional and social, has worn him out. One evening he tells Emile Bernard he is leaving the next day for Arles in southern France. They shake hands, plan to connect up again, but never do. On February 20, 1888, Vincent climbs on a train and travels to Arles.

His work over the past six months has foretold it. His many studies of sunflowers, while forcefully painted, filled with dramatic energy, also carry obvious symbolism. In *Cut Sunflowers,* the once beautiful flowers are now dying, their cut stems facing the viewer, their flowers beginning to wilt. The small red strokes of blood on the lower left balance the flower lying face down on the right, almost certainly representing Vincent cut off from the rest. Another one, *Two Sunflowers*, again has the flowers cut off, here with no stems at all, just the two flowers left wilting, and noticeably one facing toward the viewer, the other face down. It doesn't take much to identify them as symbolizing Vincent and Theo, here the opposite of the earlier painting of two shoes.

Paris has been the place of transformation. The works before Paris, while perhaps suggesting what was to come, are not the works of an artist who has found his genius. The works done while in Paris are a mixed bag of experimentation, explorations of different styles and techniques, certainly now taking on a more mature understanding of symbolism and the means of connecting visible and invisible realities, some of them coming together well, but more important as a prefiguring of the mature works about to explode out of Arles.

Part XII

Arles

Upon arriving in Arles, February 20, 1888, Vincent takes a small attic-room at the Hotel-Restaurant Carrel. It is obvious that, even though they couldn't stand to live together any longer, Vincent and Theo immediately miss each other.

Vincent writes Theo: "I have thought of you at least as much as about the new country I saw." (256) He later writes to Gauguin: "When I left Paris quite heartbroken and ill, nearly an alcoholic and hardly able to keep my head up, and my strength gone—then I withdrew into myself and no longer dared to hope." (257)

Ironically, since a major reason for moving to Arles is the bright, warm landscape, when Vincent arrives the land is covered with snow, and Vincent paints a relatively unknown work, *Effect of Snow at Arles*. With his usual strenuous work habits, even in his depressed state, he plunges in, painting *The Charcuterie,* a painting that has more of the flat Japanese texture, mainly interesting in that he paints it directly with his brush without the preliminary drawing or study he always used in the past. In March, he paints *The Viaduct Near the Station,* a work that retains the flat, washed texture, but also has the some of the short brushstrokes and the uneven lines that give it a more luminous quality of his mature work.

By April, he is fully into his new world, and paints a number of scenes of orchards in bloom, his mastery now complete. He has taken in all of the different art movements of Paris, refused to simply commit to one, and now has combined elements of each into his own expressions. He has crossed over into the world of genius, that twilight zone between the normal world, the literal, physical world, a world that people with various levels of intelligence understand and map out in various degrees of complexity, and that other world, the numinous world that can be sensed lurking beneath or above the intelligible world.

His orchards contradict the gloomy, depression he has taken with him from Paris. Instead, they are filled with joy and life and energy. Marc Edo Tralbaut writes that "His own enthusiasm for life was greater than it had ever been before, and he seized it while he could. Such was his perception that his mind was not merely contemplating the outward forms of the things around him, but somehow sensing the mysteries of creation." (258) Meyer Schapiro writes, "If the somber peasant-painting in Holland pictured a man after the expulsion, doomed to a wearisome labour, the painting of Arles was a return to Eden. (259)

However, it is not all joy, and Vincent's firm control of technique includes a mastery of symbolism. Tralbaut rightly offers the *Pear-Tree in Blossom* as an example. In the middle of a painting filled with life and growth is a stump cut off about three feet above the ground, obviously denied the blossoming and life around it. Next to it stands the title pear tree, fully blossoming out with beauty and the promise of life. One branch of this tree reaches down to the dead stump. Certainly it refers to the two brothers, and Vincent's letter to Theo about it, containing a pen sketch of the painting excludes the branch reaching down. Furthermore, the stump is not painted as distinctly as the rest of the painting, and that cannot have been accidental.

On May 1, Vincent rents four rooms at the now infamous Yellow House, but cannot afford to furnish them. He moves from the Restaurant Carrel into a room at the Café de la Gare. He begins to prepare the Yellow House for what he originally hopes will become a community or cooperative of artists and later reduces to a hope that even one artist will come and live with him, finally, of course, focusing that goal on Paul Gauguin.

In the meantime, in June, he briefly visits Les-Saintes-Maries-de-la-Mer, a fishing village on the Mediterranean coast twenty-five miles south of Arles, where he becomes enchanted by the little, wooden fishing boats, and begins drawing them. Upon returning to Arles, he takes on Paul-Eugene Milliet as a student, and much later in life Paul-Eugene gives his impressions of Vincent at this time, matching what others have said throughout Vincent's life:

> "He was an odd fellow," he said. "His face was a bit sunburnt, as if he had been serving up-county in Africa. But of course he hadn't; he had none of the makings of a soldier, not one. And artist? Of course he was an artist, he drew very well indeed. He was a charming companion when he wanted to be—which didn't happen every day. We often went on good walks around Arles, and in the country we would make no end of sketches. Sometimes he would set up his canvas and start daubing away. After that there was no budging him. The fellow had talent in his drawing, but he became quite different when he picked up his brushes. As soon as he started painting I would leave him alone, otherwise I should either have to

> refuse to tell him what I thought, or we would start arguing. He hadn't got an easy nature, and when he lost his temper you'd think he'd gone mad. . . .
>
> Very nervous. Furious when I criticized his pictures. But that didn't last, and we always made it up in the end. He was exceptionally sensitive. Reacted like a woman sometimes. Aware that he was a great artist. He had faith in his talent, rather a blind faith. And pride. His health didn't seem very strong. He complained of his stomach. But on the whole he was a good friend and not a bad fellow." (260)

Paul-Eugene Milliet is also famous as the subject of one of the portraits Vincent painted during this summer and into fall, most of them done in a naïve style of large, almost caricature shapes with the energetic, striking colors, all of them exemplifying an easy confidence in his work, the confidence of someone who knows he has broken through, has connected with some higher plane. Those done of his local postal official Armand Roulin force the viewer to Armand Roulin's sad, distant eyes, the rest of each painting reflecting that sadness in not just the literal down turned mouth and heavy facial cheek, but in the drooping quality of the lines and brushstrokes. Another one done of his son, Camille Roulin surrounds the "lost" look in the blue eyes (a blue matched by the hat) with an ever-changing green yellow, all of the colors slipping into one-another in what by now has become Vincent's signature style of short, rapid strokes. The last of these portraits are being done in November, shortly after the anticipated arrival of Paul Gauguin takes place.

Vincent makes the move into the Yellow House on September 16. Paul Gauguin will arrive Oct. 23. Vincent is excited to have an artist he greatly admires join him, someone to share his enthusiasm for art, but the expected reunion with this leading member of the community of artists in Paris will not turn out well.

Eugene Henri Paul Gauguin was born June 7, 1848 to journalist Clovis Gauguin and half-Peruvian Aline Maria Chazal, daughter of socialist leader Flora Tristan. Just three years later Paul Gauguin's father died on a voyage from Paris to Peru, and he, his mother, and his sister lived for the next four years in Lima, Peru, with his uncle's family.

At the age of seven, he returned with his family to Orleans, France, to live with his grandfather, where he learned French and proved himself an exceptional student. At age seventeen, he fulfilled his required military service as a pilot's assistant in the merchant marine. Three years later, he joined the navy, where he remained for two years. In 1871 he returned to his birthplace, Paris, securing a job as a stockbroker, and two years later he married Mette Sophie Gad. Over the next ten years they would have five children.

While in Paris, he began painting as a hobby. He visited the galleries and purchased the work of the newer artists. Eventually he formed a friendship with Camille Pissarro, and through him became a part of the emerging Impressionist community, including his work with theirs in the 1881 and 1882 exhibitions. During this time he also grew close to Paul Cezanne and often painted with him. However, in 1884, he moved his family to Copenhagen to continue his business career as a stockbroker.

Then comes the commitment to pursue an artistic career. Just a year after moving to Copenhagen, he leaves his family and moves back to Paris. His wide-ranging life experiences around the world and his eloquent speaking abilities impress the emerging community of arts, and he quickly becomes a central figure. But, as with so many of them, he has bouts of depression, and at one time attempts suicide. He also wants to move beyond Impressionism, feeling that even that progressive movement is too imitative and without the depth and symbolic levels of meaning he finds in Japanese, African, and folk art. The cloisonné enameling technique of Emile Bernard also interests him, and he begins moving into what critic Edouard Dujardin would designate Cloisonnism styled art, of which Paul Gauguin's *The Yellow Christ*, painted in 1889, is considered the classic example. The subject has been reduced to a presentation of areas of pure color separated by heavy black outlines, purposely eliminating classical perspective and any gradations of color. Later his paintings evolve into what has since been designated as "Synthetism," an attempt to merge form and color equally.

By 1891, annoyed by his perceived lack of recognition of his work and broke, he sails to the tropics, claiming that everything about Europe is artificial and conventional. He has already spent brief periods in Martinique and worked as a laborer in the currently ongoing construction of the Panama Canal, which it should be no surprise to know he was fired from after but two weeks. This time he is to settle in Mataiea Village, Tahiti, where he paints such classic works as *Fatata te Miti* and *Ia Orana Maria.* In 1897 he moves to Punaauia, where he paints *Where do We come From*, and then lives the rest of his life in the Marquesas Islands, only returning briefly to France once. The works he produces during this final stage of his life are filled with quasi-religious symbolism and an exotic view of the natives of Polynesia, whom he identifies with and fights for against the Catholic Church and colonial authorities. He also writes *Avant et après* (*Before and After*), a memoir of his life with various observations on painting and literature. In 1903, as a result of his conflicts with the church and government, he is sentenced to three months in jail and a fine. However, appropriately enough, he dies of syphilis before the prison sentence begins. As with many of the Paris community of artists, he has taken such poor care of himself, led such an unhealthy life of alcohol and sex that it catches up with him and he dies at the age of 54.

He has since been given credit for being the first to employ the techniques and subject matter of Primitivism, works characterized by exaggerated body proportions, animal totems, geometric designs and bold contrasts resulting from Europe's current fascination with the art of Africa, Micronesia, and even Native American. He is also considered one of the central artists of the Post-impressionists, a large category of those artists, including Vincent, who were emerging out of the Impressionist movement, each with a unique style, but all intersecting mainly in that they all lived together in their bohemian world for about a decade.

What mainly brought Paul Gauguin and Vincent together artistically, and separated them from the Impressionists, was that they saw art as a movement beyond nature, more of a synthesis of experiences, rather than an immediate perceptual experience, taking artistic expression to deeper psychological levels. For Gauguin, the civilized world of Europe was a barrier to the deeper truths, and the alternative cultures not yet corrupted by accumulation of Western Tradition and Culture were where one could entire the higher realms of artistic meaning and value. Although, so far as I know, he did not have a strong knowledge of the theories of such philosophers as Jean Jacques Rousseau, this partakes of that "noble savage" idea that we are born pure and lose both that purity and the ability to see the higher truths as we become corrupted by civilization. Gauguin wanted his works to take us out of our normal, daily, practical, meaningless thoughts into another realm, that place where the invisible world, the numinous world exists. In works such as *Vision after the Sermon* he uses explosive, intense colors, colors beyond "descriptive" colors, and sinuous lines, again beyond "descriptive" lines, to represent, to recreate, a spiritual experience. This style he gave yet another name, "Synthetism," claiming it synthesized observation of the subject with the artist's feelings about the subject.

His famous painting *Mahana no atua (Day of God)* demonstrates how he tried to take the viewer from the surface levels to the deeper worlds of what he called the "mysterious center of thought." In it he divided the canvas into three large horizontal zones, the upper painted realistically (at least for him), and has the image of a god, behind which extends the naturalistic beach and ocean populated by the native Tahitians. The middle zone is dominated by three Tahitians on a beach, but one with unnatural, though not startling colors. One of these people is sleeping in a clear fetal position, suggesting prebirth or recent birth. The second person is slightly beyond this stage, born, but not yet mature and upright. The third occupies the dominant center of the painting. She is fully developed, sitting strong, and looking boldly out at the viewer. Her feet are submerged in the water that is the bottom third zone of the painting, and takes the form of a dazzling array of abstract colors and shapes, easily connected to that mysterious spiritual world beyond the literal. Clearly, the water is not giving us the expected literal mirror of the shapes above it, but is transforming those shapes into an enigmatic dream world. And water is the place of such a transforming reality, liquid, ever-changing, the source of life. It is the amniotic fluid out of which each of us is born and must someday return. This is the world Gauguin is trying to make his viewers enter.

It is the entrance into yet another artistic movement he is considered to have originated, Symbolism. And it takes us to the larger contexts of this entire manuscript. Marilyn Stokstad writes:

> Like the Romantics before them, the Symbolists opposed the values of rationalism and material progress that dominated Western culture and instead explored the nonmaterial realms of emotion, imagination, and spirituality. Ultimately the Symbolists sought a deeper and more mysterious reality than the one we encounter in everyday life, which they conveyed not through traditional iconography but through ambiguous subject matter and formal stylization suggestive of hidden and elusive meanings. Instead of objectively representing the world, they transformed appearances in order to give pictorial form to psychic experience, and they often compared their works to dreams. It seems hardly coincidental that Sigmund Freud, who compared artistic creation to the process of dreaming, wrote his pioneering *The interpretation of Dreams* (1900) during the Symbolist period. (261)

Unfortunately, Gauguin's ego was too large for a friendly reunion with Vincent. In truth, the only reason he came to live at The Yellow House was because he was broke. In fact, he did not think much of Vincent or his art. Tralbaut includes the following description of Gauguin's actions following Vincent's death as a demonstration of this:

> If both men had been equally sincere in the venture it might have succeeded. But they were not; and, whatever Gauguin may have said late, he had a very poor opinion of Vincent and was quite unaware of his talent, let alone his genius. After Vincent died he have ample proof of this which none of his subsequent assertions could contradict.
>
> Several months after Vincent's death Theo asked Bernard to organize a commemorative exhibition of van Gogh's work. When Gauguin learnt of this project from Bernard he did all in his power to stop this mark of

> respect to the dead man from ever taking place. He not only sent Meyer de Haan to Paris to try to prevent it, but he also wrote to Bernard and said that "it was not really very wise to exhibit the works of a madman" and that Bernard would risk everything by acting so thoughtlessly." (262)

The initial meeting and first few weeks go well, they seem to have mutual agreements about the cooking and housework, they paint together, and they even go together to the local brothels, "Maison Tutelle" and what was called the "Hot Quarter." They take an excursion to the museum in Montpellier together, and Paul Gauguin even paints a portrait of Vincent in the act of painting sunflowers. But by December Vincent is already expressing serious doubts. He writes Theo, "I think myself that Gauguin was a little out of sorts with the good town of Arles, the little yellow house where we work, and especially with me." (263) At the same time Gauguin writes Theo, "I would be greatly obliged to you for sending me part of the money for the pictures sold. After all, I must go back to Paris. Vincent and I simply cannot live together in peace because of incompatibility of temper." (264)

During this emotionally charged December, Vincent begins painting two pictures, *Van Gogh's Chair* and *Gauguin's Chair,* which perhaps indicate his views of their relationship better than his letters. At any rate, they have received a great deal of commentary. Alfred Nemeczek writes:

> The still lifes of items of furniture from early December 1888 are like coded portraits, forming a diptych despite being so consciously different from one another: one picture is in daylight, the other at night; the tiled floor of one suggests rural simplicity, while something much more intellectual is alluded to in the luxurious carpet and the books on the seat of the chair of the other. The two chairs belong together like Symbolism and Impressionism, as two strands of modern art still looked down upon, as were the friends Van Gogh and Gauguin. The significance of the empty chair as a symbol of death for Van Gogh has been well documented by researchers in the filed. The art historian Meyer Schapiro also point to the complex composition of the picture of the simple country chair, resulting from the geometry of the wooden struts, the tiles running in the opposite direction, and the right angle drawn on the door. The simpler of the two motifs is thus anything but simple in its execution. (265)

Especially interesting is the fact that Vincent begins the chairs before the dramatic break-up with Gauguin, but completes them after it takes place. After beginning them in mid-December, he writes Theo:

> I can at all events tell you that the last two studies are quite curious. Canvases of 30, a chair of wood and rush all yellow on red tiles against a wall (the day). Then Gauguin's armchair, red and green, effect of night, wall and floor also red and green, on the seat two novels and a candle." (266)

Upon completing the paintings a month later, after the break-up, he writes:

> I should like de Haan to see a study of mine of a lighted candle and two novels (one yellow, the other pink) but on an empty armchair (actually Gauguin's chair), canvas of 30, in red and green. I just worked again today at the companion piece, my own empty chair, a chair of white wood with a pipe and tobacco pouch. In these two studies, as in others, I have sought an effect of light with color: de Haan would probably understand what I am looking for, if you read to him what I am writing you on this subject. (267)

A year later, he will again refer to them in a letter to Albert Aurier, who has recently published a positive review of Vincent's work, that he owes a great deal to Gauguin, that: "A few days before we separated, when my illness forced me to enter a hospital, I tried to paint "his empty place. This is a study of his armchair in dark brown-red wood, the seat of greenish rush, and in the place of the seat a lighted candle and modern novels." (268)

As H. R. Graetz points out, "When Vincent used the words "his empty place" and put them in quotation marks, he not only suggested that the chair stood for his companion, he also pointed out that Gauguin's place had become empty even before their separation at the outbreak of his illness." (269)

The two paintings are, as Vincent states, "quite curious." Gauguin's chair, painted in dark, masculine colors, and having an elegance, a luxury to it, expresses a man of power with rich tastes and confidence. The books speak of a person of knowledge, a well-read man. Though he fails to mention it in his letters, Vincent has placed an important circular yellow light on the wall, a symbol he has constantly used to symbolize love (and perhaps meant to symbolize himself here), and it contrasts with the lesser light of the candle on the chair. The matching painting of Vincent's chair is dominated by a pale yellow, and the chair itself is that of an unsophisticated, country person, with little money, little social polish, a person of austerity. The pipe and tobacco clearly represent Vincent's own smoking habits, but even more the fact it is unlit and cold comments on his own feelings of inferiority an unfulfilled potential in comparison with his virile, lighted candle of Gauguin's chair. While Vincent's chair has a luminous quality, created by the yellow color with light blue-gray outlines, the wavering lines, and the uneasy planes, at the same time it is isolated, sparse, and alone; this is especially evident when contrasted to Gauguin's chair, which though also alone, does not project a sad, sparse loneliness but rather a comfortable though swaying ease. The one other object in the painting of Vincent's chair is what H. R. Graetz calls a "coffin-like" box, certainly a rough box suited more for a barn than a house. Interestingly, in the box are the sprouts of an onion, thus life contrasting to the isolation, a flicker of hope in the sadness.

Nagera goes through a detailed discussion of how chairs symbolized death for Vincent, connects this symbol especially to the death of Vincent's father, and then is in place to connect Vincent's views of Gauguin to his father, especially to the loss of his father:

> The reaction of Gauguin's possible departure can now be better understood. His attack on Gauguin is a break-through into actual behaviour of his hostility and death wishes against his father now reinforced by his hostility towards Theo and towards Gauguin himself. To further understand the pictures and the elements depicted on them we must remind ourselves of some of the episodes that took place during the Nuenen period. (270)

After reminding us of the bitter battles that took place between Vincent and his father at that time, Nagera gets to the point he wants to make, the painting Vincent did to commemorate his father's death, a painting that included his father's pipe and tobacco, now repeated in Vincent's own empty chair. Nagera is reaching here, as he often does, for his psychological insights. Yet there might be some truth. It is clear that Vincent saw the pipe and tobacco as symbolizing each of them, and that it was a unifying element in them. And it is true that Vincent saw an empty chair as symbolizing death. Nagera wants to take it further, to suggest Vincent had guilty feelings of "the crime he has symbolically committed," in other words, guilt about his falling out with his father, and about being on negative terms with his father when his father died. Nagero obviously thinks Vincent really did nearly kill Gauguin:

> Now he felt the full impact of his father's death and of his "responsibility" in bringing it about. The attack on Gauguin (in so far as he represented the father) constituted a re-enactment of the whole complex of Vincent's father's death years earlier, when he seemed to have successfully repressed any feelings of guilt. The association of events and multiple factors in his present life that led to the attack reactivated all sorts of earlier conflicts belonging to the relationship to his father. There is just one

> important difference: on this occasion the impulse to kill *did* break through into his conscious mind and *was* acted upon, though in the last instance he managed to turn it away from Gauguin and against himself. This is in sharp contrast to the situation at the time of his father's death when such impulses, undoubtedly present in his unconscious phantasies, were able to find expression only through provocative behaviors and verbal abuse of his father. In Gauguin's "empty chair" are thus condensed the conscious murderous intentions towards Gauguin and the unconscious wishes to kill the father—but these impulses have now been acted upon. Vincent has thus in symbolic terms committed the oedipal crime. Such a crime cannot go without punishment and so his own chair is painted at the same time as that of Gauguin. His chair is empty too, symbolizing his own death; his pipe and his tobacco lie on the empty chair, just as his father's pipe and tobacco pouch appear in the picture that commemorated his death. (271)

But Nagera isn't through yet. He is about to tie these paintings and the immediate symbolism in them to the events to take place leading all the way to Vincent's own suicide:

> Vincent must die, and die he will through his own hand not long after the incident. But before that final act is played he dies symbolically through becoming deranged. Such symbolic death postpones for a short while his actual physical death. It is necessary at this point to recall The Hague period. It was then that Theo suggested to him that the family (Vincent took this to mean his father) was considering having him placed under guardianship in a mental institution. They thought him unfit to manage his affairs because of his abnormal behaviour and his association with Sien the prostitute. He reacted then by warning Theo that he would not take this passively, uttering a veiled threat that in such circumstances one could even kill the offenders (in this case his father) and get acquitted by the law as had happened in the example he quoted. Yet in losing his reason and by being institutionalized he was complying with what he thought to be his father's wishes. The father was right after all, he was mad and dangerous. He should be under care, especially after the attack on Gauguin. He was imposing on himself the punishment he thought the father had ready for him when he rebelled against his authority. Further we have seen how he decided freely, during a lucid interval, that it would be best for himself and others for him to live in a mental institution for some time. This in fact he did by going to St. Remy. Finally, we should point out that form the time Vincent cut his ear lobe onwards there is a significative change in his signature, pointing to the symbolic self-castration (another unconscious equivalent of his death) that he has just carried out. His paintings were usually signed, if at all, with his Christian name, but the initial "V" was a character with a very sharp angle. This "V" was now with only a few exceptions changed into a rounded "V", all sharpness disappearing from it. This change had already started to appear

> occasionally after Gauguin's arrival in Arles—and is to be noted in the picture of his own empty chair—but it became a permanent feature of his signature at this point. (272)

While the details and perhaps the exact psychological struggles he claims might be slightly off or on-and-off the mark, there is an enticing sense that he has at least provided some clues, some flashes of light in the darkness. And there is no doubt but that Vincent is fully into symbolic expression at this point in his career. Furthermore, most critics believe Vincent saw Gauguin as someone to look up to, whether as a father figure or a worldly man or a great artist or simply a man with a great aura about him (Gauguin certainly held such a high opinion of himself and expected it of others). As much as Vincent believed in himself as an artist, he also constantly revealed insecurities, and it must also be remember that Vincent had spent most of his still short life living in the country, living outside of the great civilized world of Paris. He must have been intimidated by it all, especially by such a larger-than-life personality as Gauguin. And his letters to Theo, while containing the standard mixture of a love-hate relationship with Gauguin, constantly indicate this worshipful aspect. Thus, just how consciously or unconsciously Vincent was living out his love-hate relationship with his father, the possible similarities exist.

Just to confuse the mixture of the real life relationship and the symbolism taking place, it's also true that Vincent very consciously decorated Gauguin's room in the manner of a beautiful boudoir, a "woman's quarters"! There can be no doubt about this, as he writes Theo: "there will be a prettier room upstairs, which I shall try to make as much as possible like the boudoir of a really artistic woman . . . the sunflowers, 12 or 14 to the bunch. Crammed into this tiny boudoir with its pretty bed and everything else dainty." (273) Nagera considers this the result of Vincent's bisexual conflicts, a defense mechanism for his homosexual longings. It is true that the painting of Gauguin's chair, while it can be seen as symbolizing a powerful, intellectual, worldly man, can also be seen as being overly feminine, especially in contrast to Vincent's hard, workingman's chair.

As stated earlier, these two paintings straddled the big event. Vincent began them before it, and completed them after it. Let's take a closer look at the real life actions that are getting symbolized in the paintings.

With two powerful personalities, both prone to explosive fits, the drama has to come to a climax, and it takes but two months. Accounts vary. Even though he was one of the central characters, and partially because he was a central character, most critics todaydiscount the details of Gauguin's version; he had an agenda, which was to romanticize himself and justify his abrupt departure without even checking on Vincent at a time when he might have even been near death. But it is a place to begin.

In his version, he claims that the night previous to it, while he and Vincent were sharing a drink at a local bar, Vincent threw a glass at his head, suggesting Vincent's madness was at the bottom of things. Even if this happened, it was probably not an out-of-the-ordinary thing in the type of establishment they frequented, one of the local down-and-out brothels and bars where such seemingly "wild" actions were likely common. They were, after all, drinking absinthe, probably to excess, and their normal relationship was filled with arguments. This one is likely the one that meant the end of their friendship, the end of them living together; it is the final straw, certainly for Gauguin, something that probably, especially immediately afterwards, sends Vincent into a dramatic depression. It is also worth noting that records in the Arles archives indicate that nearly every week someone is being confined in an institution resulting from an act of madness, suggesting either that the small town is filled with mad people or that the category of madness is a loosely applied one, and given the atmosphere might be the result of the endless partying at these places of entertainment, and might be related to the drinking of absinthe, which has since been found to contain a drug that brings about a psychological experience beyond that of normal wine or alcohol. In other words, those committed to the institutions might well be nothing more than the standard cases of people getting intoxicated on drinks and drugs in a nightly atmosphere filled with such happenings. At any rate, this supposed argument the night before is just the prologue.

According to Gauguin, the following day, after having a light supper, he went outside and claims, "I had crossed almost the whole of the place Victor Hugo, when I heard behind me a light step. It was rapid and abrupt, and I knew it well. I turned round just as Vincent was coming at me with an open razor in his hand. I must have looked at him then with a very commanding eye, for he stopped, lowered his head, and turned round and ran back towards the house." (274)

This is certainly an exaggeration. Four days afterwards, Emile Bernard spoke with Gauguin and wrote to art critic Albert Aurier, "My dear, best friend Vincent is crazy. Since I heard that I have just about been crazy too. I rushed to see Gauguin and this is what he said: 'On the day before I left Arles Vincent ran up behind me—it was night—I turned round because Vincent had been rather strange for a while and I was on my guard. He said to me: 'You are silent, but I will be too.' I went to stay overnight in a hotel." (275)

Neither Vincent, who claimed to not remember, nor Gauguin suggested Vincent had threatened violence at this moment. In fact, Gauguin's immediate description makes Vincent out to be submissive.

The next day, the series of events reached its critical moment. *Le Forum Republicain*, the local news, reported on December 30, 1888, that Vincent had appeared at the *maison de tolerance n. 1,* a brothel and given his ear to a woman there named Rachel, telling her to "guard this object carefully." The police were informed, and the next morning they came to his house, found him unconscious, and took him to the hospital. (276) It has since been confirmed that Rachel was indeed a real woman, and one Vincent and Gauguin had often visited. These are the few certain facts of the incident.

Johanna van Gogh-Bonger described it as follows: "The day before Christmas—Theo and I had just become engaged and intended to go to Holland together . . . a telegram arrived from Gauguin which called Theo to Arles. On the evening of December 24 [has to have been a mistake on her part, as it was Dec. 23] Vincent had in a state of violent excitement . . . cut off a piece of his ear and brought it as a gift to a woman in a brothel. A big tumult ensued. Roulin the postman had seen Vincent home; the policemen had intervened, had found Vincent bleeding and unconscious in bed, and sent him to the hospital. There Theo found him in a severe crisis and stayed with him during the Christmas days. The doctor considered his condition very serious." (277) While both the policeman who was given the ear and the prostitute Rachel claimed the entire ear was cut off, Dr. Gachet, his son, the painter Signac, and, as quoted above, Johanna van Gogh-Bonger all modified this. It has since been accepted that what Vincent cut off was the lower half of the left ear.

Speculation is endless and sometimes laughable, but unfortunately this real-life drama is so rich that people want it to be larger-than-life, and sadly lose the larger-than-life truth of it.

According to A. J. Westerman Holstijn, this self-mutilation was the result of two things, the engagement of Theo, which seemed to have a strong effect on Vincent and the failed relationship with Gauguin. (278) This is a promising beginning theory, and in-and-of-itself works as an umbrella. However, Holstijn wants to take it into less secure arenas. He claims that the ear is a phallic symbol and that Vincent's cutting off of it was a symbolic act of castration, supporting this partially with the comparison of the two Dutch words *lul* (slang for penis) and *lel* (earlobe). He then claims that it all revolved around Vincent's homosexual tendencies toward Paul Gauguin. Subsequently, another psychoanalyst, Daniel Schneider, agreeing with Holstijn, claims that Vincent lived "under the constant overpowering threat and masochistic homosexual unconscious wish for castration." Thus, "when he slices off his ear and gives it to the prostitute who accepted Gauguin, he brings it about rather than face it any longer." (279) A slightly different twist is put on this by Jacques Schnier, who suggests that, in giving the ear to a prostitute, Vincent is fulfilling an unconscious wish to possess his mother following the assault upon his father substitute Gauguin. (280)

Art critic Frank Elgar speculates:

> We have to imagine his state of mind when Gauguin announced his departure. In this fresh defeat he had the appalling sensation of having been outlawed by that very humanity with which he had always longer, from childhood, to be united. This inferiority complex was reinforced by one of guilt. Like Orestes turning, in his mania, upon himself, he punishes his guilt by severing his own ear. Next, in a Christian spirit of self-sacrifice he carries this fragment of himself, his own living flesh, to the most fallen of human beings. (281)

J. Olivier suggests that the bullfights in Arles made a deep impression on Vincent, and that a part of them consisted of the matador receiving the ear of the bull his has just defeated as a trophy, which he then gives to his own lady or to some other woman in the crowd. He writes:

> I am absolutely convinced that Van Gogh was deeply impressed by this practice. . . . Van Gogh cut off the ear, his own ear, as if he were at the same time the vanquished bull and the victorious matador. A confusion the mind of one person between the vanquished and the vanquisher. This is often the case with all of us. It may easily have been the case with Van Gogh that same night when he was provoked to overexcitement by Gauguin, yet refused to be dominated by him. Personally, I see in it a collapse followed by a courageous and magnificent exaltation, which ends in a relief of tension and an appeasement. (282)

H. R. Graetz, however, puts a bit of a damper on this idea, pointing out that "when Monsieur Olivier drew his convincing conclusion he seemed to have ignored the fact that the bulls wer not killed in the fights at the time Van Gogh was in Arles. Consequently, he could not have witnessed that Spanish custom which was introduced there only later. He may have heard about it but there is no mention of it in his letters." (283) "On the other hand," Graeta suggests, "he had written to Bernard in June 1888: *Then . . . I feel quite an ox that will prevent me from being ambitious."* In earlier letters he had referred to St. Luke as the protector of the painters and to the ox as his symbol. Vincent's identification with an ox—the castrated animal—who admires the bull is symptomatic of his feeling of inferiority." (284)

And then gives the interpretation even more strength:

> Notwithstanding Monsieur Olivier's mistaken reference to the bull fights, he accurately perceives Vincent's inner condition, which had shown itself in aggression against Gauguin, in self-mutilation and submission to a woman by offering her the sacrificed ear. His impression of a magnificent exaltation brings back the symbolic meaning of the spirals in the self-portrait, of Vincent's rising above the level of bodily limitations. It is remarkable that one should say that cutting off the ear and giving it to a woman is a normal linking and that confusion between the vanquished and the victor is often the case with all of us. These words express Monsier Oliver's identification with Vincent's conflict, at a time when confusion marks the state ofmind of our whole age. His generalization is all the more interesting as it connotes a widespread sympathy in the world with Van Gogh, the painter, and to no less a degree with the man. (285)

It is clearly true that Vincent tended to put his own struggles into larger contexts, often quoting passages from literature or mentioning paintings to try and provide a context for understanding his own life, to try and give meaning to his own actions and the events of his own life, and he liked to put things into a symbolic context. There can be no doubt but that he was comprehending life on symbolic levels at this time. The symbolism of the bullfight would certainly have attracted him, and as he was prone to do, he might well have translated it into the literal reality of his own life in order to supply a symbolic map of meaning for it.

Just the previous April, he had written Theo: "Yesterday I saw another bullfight, where five men played the bull with darts and cockades. One toreador crushed one of his balls jumping the barricade. He was a fair man with grey eyes, plenty of san-groid; people said he'll be ill long enough. He was dressed in sky blue and gold, just like the little horseman in our Monticelli, the three figures in a wood. The arenas are a fine sight when there's sunshine and a crowd" (286)

At the same time he wrote Emile Bernard: "By the way, I have seen bullfights in the arena, or rather sham fights, seeing that the bulls were numerous but there was nobody to fight them. However, the crowd was magnificent, those great colourful multitudes piled up one above the other on two or three galleries, with the effect of sun and shade and the shadow cast by the enormous ring. (287)

It is easy to envision Vincent seeing the arena as symbolic of life, enthralled by the glamour of it, perhaps seen himself as the equivalent of the matador, an artist acclaimed by the world. As must be clear by now, he had both such grandiose thoughts and the opposite thoughts of inferiority struggling within him. And that struggle has to be considered in his actions here.

Yet another influence is suggested by Lubin, at first one hard to buy into, but then, upon reasoning, one that might be considered. Lubin goes to great pains to suggest that even the distant and somewhat tangential happenings in London might have been on Vincent's mind. At just this time, the infamous Jack-the-Ripper was mutilating prostitutes in London, and, as Albert J. Lubin details out, Vincent devoured the local newspapers, which carried fifteen articles about Jack the Ripper from September 8 through December 22, the final one, just a day before Vincent's mutilation. One of them included a quote from a letter from the assassin announcing that "in his next crime he would cut off the ears of his victim, and this was in fact done on one of the bodies found yesterday." (288)

Lubin goes into a great amount of detail concerning the possible connection of Vincent cutting off his ear with the Biblical scene in the Garden of Gethsemane, where Simon-Peter cuts off the ear of Malchus, a savant of the high priest who had come to imprison Christ. "Then," the Bible says, "Simon-Peter having a sword drew it, and smote the high priest's servant, and cut off his right ear. The servant's name was Malchus. Then said Jesus unto Peter, 'Put up thy sword into the sheath." (John 18:10, 11) The convoluted argument Lubin gives for this loses much of its force simply by being so complicated. It is true Vincent knew the Bible and saw himself as similar to Christ, not as Christ, but as similar to him, at least at one time. It is also true Vincent had stopped thinking in Christian terms, Christian symbolism, or, at the very least, had pushed such thinking to the background. At any rate, Albert J. Lubin, surrounding this passage with his elaborate, if convoluted, parallels, states:

> If we grant that Vincent identified himself with Jesus, that the ear mutilation was a reenactment of the Gethsemane scene, and that the Society [Vincent's hoped for community of artists] was equated with the group of Apostles, Gauguin becomes Judas Iscariot, the Apostle who betrayed Jesus in Gethsemane for thirty pieces of silver. There was another similarity between Gauguin and Judas: until a few years before, Gauguin had been a man of finance—a successful stock-broker—while Judas was in charge of the finances of the Apostles (John 12:6 and 13:29). Vincent also felt the same ambivalence toward Gauguin that has been described in the alliance between Jesus and Judas, in which the kiss of betrayal bears witness to the fusion of loving and destructive impulses. Nor was Vincent lacking in discernment in casting Gauguin in this role. The latter's behavior during their final days together and in later attempts to vindicate himself through condemning Vincent were hardly those of a friend. (289)

In conjunction with all of this, Lubin suggests the prostitute Rachel works in several psychological roles, including both those of Mary Magdalene and Virgin Mary, and even of the Rachel from the Old Testament. Furthermore, he suggests, that the fact it took place on Christmas was not merely a coincidence, especially since so many dramatic events for Vincent took place on Christmas, a time Vincent seemed to see as a time for the family to come together.

Two other suggestions come together from opposite directions. One is that he had a physical, medical problem with his hearing. This diagnosis of Tinnitus, a form of Meniere's disease suggests that he cut off his ear because of the constant ringing in it. It has largely been dismissed, and it seems rightly so. Another is that he heard voices, had aural hallucinations, this caused by some form of psychosis. This has much more to justify it. Marc Edo Tralbaut writes that it is supported by Vincent's own letters written while he was in the asylum at Saint-Remy: "While arguing with Gauguin he heard a voice in his ear whispering, 'Kill him!'" (290)

While the above suggestions carry various degrees of truth, some truths stand out as more likely than others. First, it is clear Vincent saw Paul Gauguin as a powerful, important artist, a man he respected and wanted respect from in return, and Gauguin saw himself as deserving such admiration. In return, however, Gauguin saw Vincent as inferior, and almost certainly moved in with him for purely selfish reasons. It is likely that the relationship with the prostitute Rachel fit into this dynamic, probably with Gauguin having the upper hand, perhaps with both of them sharing her, perhaps with Vincent wanting to share her, perhaps with Gauguin having, in a sense, taken her from Vincent, or some other scenario similar to these. And she quite possibly was the reason for their argument in the bar, or at least figured into it. Vincent also saw Gauguin moving in with him as a beginning of the Society or Community of Artists he envisioned, which, in turn, was a hoped for relief from his own loneliness, as well as a dream of accomplishment through art. Finally, Vincent's physical condition was terrible. He was most likely suffering from syphilis, badly undernourished, smoking a great deal, and perhaps an alcoholic, not to mention the possible dangers specifically related to absinthe. Others have also suggested that the chemicals in the paints Vincent used and perhaps even consumed were poisenous. Psychological interpretations are always dangerous, especially when applied to someone who has been dead for over half a century, but some of those applied carry both an interest and some possibilities. Vincent's relations with his parents, his siblings, especially Theo, and others are filled with unusually deep emotions. By the time all of these various aspects are brought together the question becomes not so much why did he have his breakdown, but how did he avoid it for as long as he did.

The diagnosis by Dr. Felix Rey, the doctor at the hospital who treated Vincent, was that he suffered an attack of epilepsy, and the cure was to lead a healthier life—eat better food and cut down on smoking and drinking.

One other thing that gets lost in all of this is that Vincent is rapidly producing large numbers of brilliant works of art at the time. Genius and insanity are partnered in a frantic dance through a living nightmare. That thin skin between the prosaic, mundane world of physical existence and the frightening, beautiful world of the sublime has been shredded, and is flowing across his canvases in thick streams of rich color.

Part XIII

Entering the Sublime

On December 26, 1888, while Vincent is recovering in the hospital, his friend Roulin writes Theo, "Not only his mind is affected, but he is very weak and downcast. . . . I believe that they are taking the necessary measures to have him admitted to a lunatic asylum." On December 28, he writes, "they have had to put him in a specially isolated room. Since he has been locked up in it, he has not eaten anything and has been completely silent." (291)

Then, suddenly, much everyone's surprise, Vincent recovers. On December 30, Reverend Salles writes, "His condition did not appear abnormal in any way . . . he calmly spoke to me, without the least of nonsense." (292) During the first week of January, Vincent is freely walking throughout the hospital, and on January 3, Roulin writes that his friend has "made a full recovery." On January 7, Vincent is discharged from the hospital and returns to The Yellow House.

The next eighteen months will consist of one of the most interesting intertwinings of *insanity and genius* ever recorded. Vincent's letters are filled with lucid, intelligent passages, including some amazingly clam and honest remarks about his bouts of insanity. Others offer positive comments about his sanity and both physical and mental health. However, the sanity is clearly interspersed with insanity. And while this psychic battle is taking place, he produces an explosion of brilliant paintings! It is as if he has entered into a psychic existence where he touches that which is beyond explanation, a state of knowing, of experience of what is called *genius*, but he cannot maintain it and must collapse back into the literal world in a state of psychic emptiness.

On January 9, just two days back from the hospital, he writes Theo:

> Physically I am well. The wound is healing very well, and the great loss of blood is righting itself, as I am eating well and my digestion is good. The thing I dread most is insomnia, but the doctor hasn't mentioned it to me, nor have I mentioned it to him as yet. But I am fighting that myself.
>
> I fight it with a very, very strong dose of camphor in my pillow and my mattress, and if ever you're unable to sleep, I recommend this to you. I very much dreaded the idea of sleeping alone in the house, and I've been worried about not being able to fall asleep, but all that's quite over now and I dare say it won't recur.
>
> In the hospital I suffered terribly from it, and yet during it all, when it was worse than losing consciousness, I can tell you as an odd fact that I continued to think about Degas. Gauguin and I had been talking about Degas beforehand, and I had pointed out to Gauguin that Degas had said . . . "I am saving myself for the Arlesiennes."
>
> Now you know how discerning Degas is, so on your return to Paris, just tell Degas that I confess that up to now I have been powerless to paint the women of Arles without venom, and that he mustn't believe Gauguin if Gauguin is too quick to speak well of my work, since it is nothing more than that of a sick man.
>
> Now if I recover, I must start afresh, but I shall never again be able to reach the heights to which the illness to some extent led me. (293)

Vincent is aware of his illness, and sees one aspect of it as enabling him to reach higher levels of artistic achievement. Also worth noting is that he is not visualizing Paul Gauguin negatively, not even supposing that Gauguin is going to see him negatively. Rather, he is believing Gauguin will over-estimate him. In a letter to Gauguin on January 22, he will again see Gauguin in very positive terms, beginning the letter "My dear friend Gauguin" and then going into a friendly, personal discussion, including a reasoned comment on his illness: "In my mental or nervous fever, or madness—I am not too sure how to put it or what to call it—my thoughts sailed over many seas. I even dreamed of the phantom Dutch ship and of Le Horla, and it seems that, while thinking what the woman rocking the cradle sang to rock the sailors to sleep, I, who on other occasions cannot even sing a note, came out with an old nursery turn, something I had tried to express in an arrangement of colours before I fell ill, because I don't know the music of Berlioz." (294) Indeed, during the remainder of Vincent's life, he and Gauguin will exchange "friendly" letters, even amazingly a suggestion from Vincent that he join Gauguin in Brittany and subsequently one from Gauguin that the two of them form a studio in

Antwerp, making Gauguin's actions towards Vincent even more confusing. While it is clear that he moved in with Vincent largely because of his own money problems, clear he thought of himself as the superior artist, clear that his actions at the time of Vincent's serious possibility of dying from cutting off his ear (completely abandoning him), and clear that he spoke against Vincent and his artistic abilities, it is also clear that all of these negatives were contradicted by numerous positive exchanges, numerous positive comments, and signs of friendship. One might use the old cliché here and say that "with friends like Gauguin who needs enemies"; however, as with most things about Vincent, this relationship was complicated, and the complications here get even more convoluted because of Gauguin's personality, where it can easily be claimed that he treated Vincent as well as any of his other "friends." Important critics such as Tralbaut who have strongly condemned Gauguin, depicting him as coldly using Vincent, only present part of the prism of the relationship. (295)

On February 3 Vincent writes Theo again a lucid letter with matter-of-fact comments about his illness, including: "When I came out of the hospital with good old Roulin, I fancied there'd been nothing wrong with me; it was only afterwards I felt I'd been ill. Well, that's only to be expected, I have moments when I as twisted with enthusiasm or madness or prophecy, like a Greek oracle on his tripod. I display great presence of mind then in my words, and speak like the Arlesiennes, but in spite of all that, my spirits are very low, especially when my physical strength returns. But I've already told Rey that at the first sign of a serious symptom I would come back and submit myself to the alienists [psychitriasts or neurologists] in Aix, or to himself." (296) Vincent also offers his own diagnosis: "the explanation of all this is probably not Ricord's [syphilis] but Raspail's [some form of parasite]."

Vincent also mentions visiting Rachel, and the response suggests what was stated earlier in the discussion about the whole event: "Yesterday I went to see the girl to whom I had gone when I was off my head. They told me that there's nothing surprising about things like that in this part of the world." It seems Vincent's calm remarks here are nothing more than an accurate description of this, and it gives the whole relationship a much more mundane quality than the biographers and critics want, but then, of course, it is the report of a man who apparently doesn't remember the whole event and who is obviously in a weakened state both physically and mentally.

During this time, he continues his painting of still lifes and portraits, including the famous self-portrait with his head bandaged. H. R. Graetz offers an excellent analysis of it:

> We are at once caught by the gripping sadness of his eyes. The distance between them seems narrower than usual, as if strain and pain had drawn them closer together, thus increasing the expression of suffering that his grave, straight look transmits to us. Contrasted by the dark blue fur cap which is deep-set on his forehead and the white bandage over the wound, the face is very pale. The mouth firmly holding a pipe between the lips shows tense determination. He wears a green jacket outlined in yellow and black. The background is divided in two, the lower part read and the upper orange; the line of the horizon is exactly at the level of his eyes. The tension in this portrait is greatly increased by the extraordinary rendering of Vincent's breath and of the smoke from his pipe. They are in yellow spirals which rise above the horizon. His recurrent motif of struggle is apparent in the dark strokes from the fur cap in front and above the bandaged ear, stabbing against the sky.
>
> The line of the horizon separating earth and heaven, the infinite life line of circular movement, so prominent in some of Vincent's other paintings, acquires a rare significance by running precisely at the height of his eyes. As if going through them, it becomes a symbol of transcending lucidity, of his deep perception of the incident; an insight arming him with almost superhuman strength and courage to present himself in this condition. (297)

After the initial paragraph here, which mainly consists of a careful description of the literal aspects, the suggestions about the line of the horizon beginning with a wonderful claim, about it representing the "infinite life line of circular movement," which, of course, "acquires a rare significance by running precisely at the height of his eyes." Yes, yes, yes, the eyes are the window to the soul, and the line of the horizon connecting the physical world with the world beyond it works excellently. The portrait is of Vincent, but it forces us to realize it is of his inner self, not his physical self, and his inner self connects up to the infinite world beyond physical existence, as H. R. Graetz says, "a symbol of transcending lucidity, of his deep perception of the incident." This is great insight, and can lead to a very interesting discussion. However, I'm not convinced of the final conclusion here that this insight arms him "with almost superhuman strength and courage to present himself in this condition." The two worlds are coming together here, but the union is not so confident, empowering and well understood. This is the entrance into that "other world," that world of the numinous, and while Vincent is aware he has entered this inexplicable world, he is not confident, just the opposite. It might be said he is similar to Christ in that he is experiencing what Christ experienced, what Christ has come to represent, the union of the spiritual with the physical, the Christhood.

Back on track, Graetz continues:

> Whenever Vincent painted a self-portrait, it was in order to place before his own eyes, and thus to make more objective, his search into himself. In this portrait he renders an uncompromising and merciless account of what he saw, and so documents this momentous experience on his road of life: the appalling affliction which had befallen him. It is an image of his suffering—his own Ecce Homo.
>
> The spiral symbol appears in this painting for the first time. From now on it is often repeated in curls of waves, most strikingly in the *Starry Night*. One may perceive the meaning of the spiral in Vincent's portrait if it is kept in mind that in several basic languages the words meaning breath and soul are the same: in Greek psyche, in Hebrew neshama, in Indian atman. As his breath crosses the line of the horizon, his soul rises above the life line to the sky. It us of great importance that this takes place at the level of his eyes, of sight and consciousness, at the very level where the blood-red earth threatens to submerge him. This red contrasted by the green of the coat, recalls an earlier letter saying that *by red and green he sought to express the terrible human passions*. But against the negative act of destruction imparted by the bandage is the positive reaction of the rising spirals. Their bright yellow, Vincent's color of light and love, lightens the blood-red of the earth and changes it into the orange of the sky. Yet the spirals have one more function. They link the lower and upper parts of the divided background. They symbolize his spiritual force, and their uniting movement signifies the beginning of a tenacious battle against the disintegration which the divided background conveys. They emphasize Vincent's rising into his spiritual sphere, with the life of the soul becoming ever more predominant. (298)

By February 9, shortly after his friend (some would claim substitute father figure) Roulin leaves for Marseilles, Vincent is not normal. The Reverend Salles writes: "Since three days he believes he is being poisoned and he sees nothing but poisoners or poisnees (. . .) he is absolutely silent, hides under his blankets and starts crying without saying a word (…) and refuses all food." (299)

By February 17 Vincent is well again and released from the hospital. However the town people fear him and petition the mayor to have him locked up, which he does, and on February 26, Vincent agrees to be readmitted. While he is there, the police seal up The Yellow House. He writes: "Anyhow, here I am, shut up in a cell all the livelong day, under lock and key and with keepers, without any guilt being proved or even open to proof. . . . So you understand what a staggering blow between the eyes it was to find so many people here cowardly enough to join together against one man and that man ill." (300)

According to Albert J. Lubin:

> This episode was representative of the way Vincent set himself up as a martyr and then exploited the reaction for the benefit of his self-esteem. He was jeered and threatened by a crowd, even as Christ was jeered and threatened in Gethsemane, and Christ, too, was seized and falsely convicted. As a result of this misfortune, Vincent reaffirmed what Christ had taught him long before: "[T]o suffer without complaining is the one lesson that has to be learned in this life." [Letter 579] By channeling self-defeating behavior into the Christ identification, he put himself above his tormentors. This forestalled the depression that would otherwise have followed such rejection; and intense feelings, compounded of righteousness and anger, became available for artistic purposes, rather than being wasted away in a paralyzing state of unhappiness." (301)

While one would assume anyone in Vincent's situation would naturally feel sorry for himself and want to present the situation as one where he has been unfairly treated, the truth is that Vincent's letters throughout these years reveal much the opposite of what Albert J. Lubin claims here. Vincent constantly accepts his condition in fair, evened tones, sees himself during this time as unable to handle life and needing to be, in fact, locked up (though no doubt not in such an unfriendly situation as the current one), even states that he does not want to be considered a martyr in them. In other words, especially when this brief passage in this one letter is put into the larger contexts of the situation (an immediate response to a very unfriendly situation) and of his other letters and actions throughout his life and especially during this time of his life, it is hard to defend that he saw himself is such grandiose terms. Perhaps it can be claimed that his once strong desires to live as Christ had lived, to sacrifice as Christ had sacrificed, to suffer as Christ had suffered can be brought to bear, but that ultimately goes against what Albert J. Lubin claims here, for Christ would never have desired pity, and Vincent's actions throughout his life in reference to being Christlike would not have sought to manipulate his situation for the sake of self-pity, just the opposite, unless one wants to interpret his time as "Christ of the Coalyards" as one of seeking pity instead of seeking to be an exceptionally sincere Christian. In fact, if we follow this line of reasoning, Vincent's entire life becomes a very shallow, self-serving one, and the truth is that he is one of the most difficult people to have ever lived to force into that mold.

The truth is much deeper, and worth exploring. First, even though he was humiliated and aware that this particular forced restraint was not necessary as he was in full control at the time, he did not respond irrationally or emotionally. Instead, he accepted his situation with complete silence for three full weeks before writing to Theo. Here is the full letter:

My dear brother,

I seemed to see so much brotherly anxiety in your kind letter that I think it my duty to break my silence. I write to you in the full possession of my faculties and not as a madman, but as the brother you know. This is the truth. A certain number of people here (there were more than 80 signatures) addressed a petition to the Mayor (I think his name is M. Tardieu), describing me as a man not fit to be at liberty, or something like that.

The commissioner of police or the chief commissioner then gave the order to shut me up again.

Anyhow, here I am, shut up in a cell all the livelong day, under lock and key and with keepers, without my guilt being proved or even open to proof.

Needless to say, in the secret tribunal of my soul I have much to reply to all that. Needless to say, I cannot be angry, and it seems to me a case of qui s'excuse s'accuse.

Only to let you know that as for setting me free - mind, I do not ask it, being persuaded that the whole accusation will be reduced to nothing - but I do say that as for getting me freed, you would find it difficult. If I did not restrain my indignation, I should at once be thought a dangerous lunatic. Let us hope and have patience. Besides, strong emotion can only aggravate my case. That is why I beg you for the present to let things be without meddling.

Take it as a warning from me that it might only complicate and confuse things.

All the more because you will understand that, while I am absolutely calm at the present moment, I may easily relapse into a state of overexcitement on account of fresh mental emotion.

So you understand what a staggering blow between the eyes it was to find so many people here cowardly enough to join together against one man, and that man ill.

Very good - so much for your better guidance; as far as my mental state is concerned, I am greatly shaken, but I am recovering a sort of calm in spite of everything, so as not to get angry.

Besides, humility becomes me after the experience of the repeated attacks. So I am being patient.

The main thing, I cannot tell you this too often, is that you should keep calm too, and let nothing upset you in your business. After your marriage we can set ourselves to clearing all this up, and meanwhile I beg you to leave me quietly here. I am convinced that the Mayor as well as the commissioner is really rather friendly, and that they will do what they can to settle all this. Here, except for liberty and except for many things that I could wish otherwise, I am not too badly off.

Besides, I told them that we were in no position to bear the expense. I cannot move without expense, and here are three months that I haven't been working, and mind, I could have worked if they had not vexed and worried me.

How are our mother and sister?

As I have nothing else to distract me - they even forbid me to smoke - though the other patients are allowed to - I think about all the people I know all day and all night long.

It is a shame - and all, so to speak, for nothing.

I will not deny that I would rather have died than have caused and suffered such trouble.

Well, well, to suffer without complaining is the one lesson that has to be learned in this life.

Now with all this, if I am to take up my task of painting again, I naturally need my studio, and some furniture, and we certainly have nothing to replace them with in case of loss. You know my work would not permit being reduced to living in hotels again. I must have my own fixed niche.

If these fellows here protest against me, I protest against them, and all they have to do is to give me damages and interest by friendly arrangement, in short, only to pay me back what I have lost through their blunders and ignorance.

If - say - I should become definitely insane - I certainly don't say that this is impossible - in any case I must be treated differently, and given fresh air, and my work, etc.

Then - honestly - I will submit.

But we have not got to that, and if I had had peace I should have recovered long ago.

They pester me because of my smoking and eating, but what's the use? After all, with all of their sobriety, they only cause me fresh misery. My dear boy, the best we can do perhaps is to make fun of our petty griefs and, in a way, of the great griefs of human life too. Take it like a man, go straight to your goal. In present-day society we artists are only the broken pitchers. I so wish I could send you my canvases, but all of them are under lock and key, guarded by the police and keepers. Don't try to release me, that will settle itself, but warn Signac[1] not to meddle in it, for he would be putting his hand into a hornets' nest - not until I write again. I shake your hand in thought. Give my kind regards to your fiancée, and to our mother and sister.

Ever yours, Vincent

I will read this letter just as it stands to M. Rey, who is not responsible, as he was ill himself. Doubtless he will write to you himself as well. My house has been closed by the police.

If, however, you have not heard from me direct for a month from now, then take action, but as long as I go on writing you, wait.

I have a vague recollection of a registered letter from you which they made me sign for, but which I did not want to take because they made such a fuss about the signature, and I have heard nothing about it since.

Explain to Bernard that I have not been able to answer him. It's quite a production to write a letter, there are as many formalities necessary now as if one were in prison. Tell him to ask Gauguin's advice, but give him a handshake for me.

Once more kind regards to your fiancée and Bonger.

I would rather not have written to you yet for fear of dragging you into it and upsetting you in what is before you. Things will settle down, it is too idiotic to last.

I had hoped that M. Rey would have come to see me so that I could talk to him again before sending off this letter, but though I sent word that I am expecting him, no one has come. I beg you once more to be cautious. You know what it means to go to the civil authorities with a complaint. At least wait till after you've been to Holland.

I am myself rather afraid that, if I were at liberty outside, I should not always keep control of myself if I were provoked or insulted, and then they would be able to take advantage of that. The fact remains that a petition has been sent to the Mayor. I answered roundly that I was quite prepared, for instance, to chuck myself into the water if that would please these good folk once and for all, but that in any case if I had in fact inflicted a wound on myself, I had done nothing of the sort to them, etc.

So cheer up, though my heart fails me sometimes. For you to come just now, honestly, would precipitate everything. I shall move out, of course, as soon as I see how to manage it.

I hope this will reach you all right. Do not be afraid of anything, I am quite calm now. Let them alone. Perhaps it would be well if you wrote once more, but nothing else for the time being. If I have patience, it can only strengthen me so as to leave me in less danger of a relapse. Of course, since I really had done my best to be friendly with people, and had no suspicion of it, it was rather a bad blow.

Good-by, my dear boy, for a little while, I hope, and don't worry. Perhaps it is a sort of quarantine they are forcing on me, for all I know.

1. Theo had heard from Signac that he was going to the South, and had asked him to visit Vincent. (302)

Vincent's *sane* assessment of the situation here can be wonderfully contrasted with the *sane* actions of the community, as expressed in this report from M. Jullian, who was seventeen at the time, and would later become the municipal librarian at Arles:

> I was one of the "flash lads" of the time. We were a gang of young people between sixteen and twenty, and like a lot of young imbeciles we used to amuse ourselves by shouting abuse at this man when he went past, alone and silent, in his long smock and wearing one of those cheap straw hats that you could buy everywhere. But he had decorated his with ribbons, sometimes blue, sometimes yellow. I remember—and I am bitterly ashamed of it now—how I threw cabbage-stalks at him! What do you expect? We were young, and he was odd, going out to paint in the country, his pipe between his teeth, his big body a bit hunched, a mad look in his eye. He always looked as if he were running away, without daring to look at anyone. Perhaps this was why we used to pursue him with our insults. He never made any scandal, except when he had been drinking, which happened often. We only became afraid of him after he had maimed himself, because then we realized that he really was mad! I have often thought about him. He was really a gentle person, a creature who would probably have liked us to like him, and we left him in his terrifying isolation, the terrible loneliness of genius." (303)

In a following letter, Vincent writes: "As far as I can judge I am not mad, properly speaking. You will see that the canvases I have done in the intervals are calm and not inferior to the others. I miss the work more than it tires me. . . If they should continue, these repeated and unexpected emotions may change a passing and momentary mental disturbance into a chronic disease. I am sure, if nothing intervenes, I should now be able to do the same and perhaps better work in the orchards than I did last year."

Vincent realizes that he is going in-and-out of another realm, one that he guesses most would designate as "madness" or a state of insanity, yet also knows those aren't quite the correct terms, though he cannot figure out how to explain the difference. He writes Theo:

> What consoles me a little is that I am beginning to consider madness as an illness like any other and accept the thin as such, while during the attacks themselves, it seemed to me that everything I imagined was real. . . . (304)

Realizing he cannot help but to continue being pulled into this psychic twilight zone between conscious control, or, perhaps in neurological terms, the logical, rational world of the left hemisphere of the brain, and the imaginary (but very real) world of the unconscious, the world that works as do *dreams*, he knows that the only form of *sanity* open to him is his art, for this is where he can give *insanity* form and shape and meaning. I stress that *insanity* and *madness* are dangerous terms here because they have such negative connotations.

Graetz correctly writes:

> He undertook this fight against disintegration in full consciousness and with the greatest determination. The three hundred paintings and many hundred drawings and sketches he did in the nineteen months form the first attack of his illness until his death are proof of this fight. Many of these works have had extraordinary influence and have attained wide fame in the generations which followed his own. There can be no doubt that the import and greatness of his art was not diminished but fully maintained, if not increased, after the outbreak of the illness. Vincent's spiritual and physical control of his creative capacity, which was indispensable for achieving his magnificent work, certainly makes his case an uncommon one; it still challenges medical diagnosis. (305)

Vincent has entered the world of the *shaman*, and this connects up to the world of the savior. Carl Jung writes:

> There is something of the trickster in the character of the shaman and medicine-man, for he, too, often plays malicious jokes on people, only to fall victim in his turn to the vengeance of those whom he has injured. For this reason, his profession sometimes puts him in peril of his life. Beside that, the shamanistic techniques in themselves often cause the medicine-man a good deal of discomfort, if not actual pain. At all events the "making of a medicine-man" involves, in many parts of the world, so much agony of body and soul that permanent psychic injuries may result. His "approximation to the savious" is an obvious consequence of this, in confirmation of the mythological truth that the wounded wounder is the agent of healing, and that the sufferer takes away suffering. (306)

Joseph Campbell explains it in connection with myths, which he defines as those stories which transform our consciousness, put it into a realm that is spiritual (brings the mind into the invisible world that underlies and gives meaning to the visible world):

> Myth must be kept alive. The people who can keep it alive are the artists of one kind or another. The function of the artist is the mythologization of the environment and the world. . . .
>
> The mythmakers of earlier days were the counterparts of our artists . . . shamans. The shaman is the person, male or female, who in his late childhood or early youth has an over-whelming physiological experience that turns him totally inward. It's a kind of schizophrenic crack-up. The whole unconscious opens up, and the shaman falls into it. This shaman experience has been described many, many times. It occurs all the way from Siberia right through the Americas down to Tierra del Fuego. (307)

John G. Neihardt first met Black Elk in August of 1930. He had come to Pine Ridge Reservation in search of finding some of the old medicine men who had been active in the Messiah Movement, the Messianic *Dream* that had come to the Sioux while they were being defeated by the white men, ultimately leading up to the famous massacre at Wounded Knee, South Dakota, December 29, 1980. He was told by the field agent there he should seek out Black Elk, told Black Elk was a *wishasha wakon* (a holy man, a priest, more accurately, a shaman). He drove to Manderson, where he met up with Flying Hawk, an interpreter he knew, and was taken to Black Elk. When he arrived at Black Elk's home, a one-room log cabin with weeds growing out of the roof, Black Elk indicated he had sensed the arrival. Black Elk was old and nearly blind, and only spoke Sioux, so all of the exchanges had to come through Flying Eagle, and what Flying Eagle told John G. Neihardt was that Black Elk felt he was a man sent to learn of the "other world." Black Elk told John G. Neihardt to return the following spring and he would teach him of his "great vision."

When the snows melted and the grass began to grow, John G. Neihardt did just that, and this *shaman* begin his story:

> I was four years old then, and I think it must have been the next summer that I first heard the voices. . . . Maybe it was not this summer when I first heard the voices, but I think it was, because I know it was before I played with bows and arrows or rode a horse, and I was out playing alone when I heard them. It was like somebody calling me, and I thought it was my mother, but there was nobody there. This happened more than once, and always made me afraid, so that I ran home.
>
> It was when I was five years old that my Grandfather made me a bow and some arrows. The grass was young and I was horseback. A thunder storm was coming from where the sun goes down, and just as I was riding into the woods along a creek, there was a kingbird sitting on a limb. This was not a dream, it happened. And I was going to shoot at the kingbird with the bow my Grandfather made, when the bird spoke and said: "The clouds all over are one-sided." Perhaps it meant that all the clouds were looking at me. And then it said: "Listen! A voice is calling you!" Then I looked up at the clouds, and two men were coming there, headfirst like arrows slanting down: and as they came, they sang a sacred song and the thunder was like drumming. I will sing it for you. The song and the drumming were like this:
>
> "Behold, a sacred voice is calling you;
> All over the sky a sacred voice is calling."
>
> I sat there gazing at them, and they were coming from the place where the giant lives (north). But when they were very close to me, they wheeled about toward where the sun goes down, and suddenly they were geese. Then they were gone, and the rain came with a big wind and a roaring.
>
> I did not tell this vision to any one. I liked to think about it, but I was afraid to tell it. (308)

Such experiences today would be explained neurologically as a form of autistic savant, where a young child, generally even younger than Black Elk was, goes beyond the logic of literal existence into a *dream*, a world considered *unreal*, only *imagined*. There are many examples: Jessy Park, the autistic artist whose works have a luminous quality similar to Vincent's though they are done in a very different style, comes immediately to mind. Also important immediately in this passage is that Black Elk did not have any control over the appearance of the visions, and was both comforted and frightened by them.

For the next few years, Black Elk was occasionally visited by voices, but nothing major. Then, one day, when he was nine, the voice came to him and said: "It is time; now they are calling you." Black Elk got up and exited the tepee to follow where the voice would take him, but his thighs and his legs hurt, so he returned to the tepee.

The next day this "sickness" came back, now worse, causing legs, arms and face to swell. That night, lying in his tepee next to his mother and father, he saw out the opening the same men on horse back coming down from the clouds. This time they stopped, looked at him and said: "Hurry! Come! Your Grandfathers are calling you!"

Then they left. Black got up, his legs no longer hurting, and followed. A small cloud gathered him in and took him rapidly into the sky. In this world of the clouds, Black Elk watched as a bay horse gathered together other horses, until horses everywhere danced about him. Soon they began to morph into other animals, which eventually fled into the four corners of the world from whence the horses had come. Then the cloud became a tepee with six men sitting in a row inside it. Then the two men with spears joined them, and the oldest spoke kindly and invited Black Elk to join them, saying: "Your Grandfathers all over the world are having a council, and they have called you here to teach you."

> His voice was very kind, but I shook all over with fear now, for I knew that these were not old men, but the Powers of the World. And the first was the Power of the West; the second, of the North; the third, of the East; the fourth of the South; the fifth, of the Sky; the sixth, of the Earth. I knew this, and was afraid, until the first Grandfather spoke again: "Behold them yonder where the sun goes down, the thunder beings! You shall see, and have from them my power; and they shall take you to the high and lonely center of the earth that you may see; even to the place where the sun continually shines, they shall take you there to understand." (309)

Then Black Elk saw a rainbow come over him. He was given a cup of water, which is the sky, and a bow, which is the power to destroy, and his spirit, called Eagle Wing stretched. Then the old Grandfather got up, ran to the setting sun, turned into a sick, old horse, and looked at the boy. Then the Grandfather of the North gave the horse a herb, which healed him, and he again became the first Grandfather.

Then the Grandfather of the North said, "On earth a nation you shall make live, for yours shall be the power of the white giant's wing, the cleansing wing." The he ran to the north, where he turned into a goose, and all of the horses of the west became thunders and the horses of the north became geese. And this Grandfather sang two songs, one about the "thunder nation appearing," the other about the "geese nation appearing."

Then the third Grandfather spoke: "Take courage, younger brother, for across the earth they shall take you!" And he pointed to two men flying beneath a shinning star to the east. "From them you shall have power, from them who have awakened all the beings of the earth with roots and legs and wings." Then he held out a peace pipe and said: "With this pipe, you shall walk upon the earth, and whatever sickens there you shall make well." Then he pointed to a red man, and the man rolled on the earth and changed into a bison, and the bison got up and ran to the horses of the east, which now also became bison, healthy and fat.

Then the fourth Grandfather spoke: "Younger brother, with the powers of the four quarters you shall walk, a relative. Behold, the living center of a nation I shall give you, and with it many you shall save." And he held up a red stick, which immediately began to transform, to grow and become a living tree with singing birds, and beneath it Black Elk saw villages of people, a happy world. Then the Grandfather said, "It shall stand in the center of the nation's circle, a cane to walk with and a people's heart; and by your powers you shall make it blossom."

> Then when he had been still a little while to hear the birds sing, he spoke again: "Behold the earth!" So I looked down and saw it lying yonder like a hoop of peoples, and in the center bloomed the holy stick that was a tree, and where it stood there crossed two roads, a red on and a black. "From where the giant lives (the north) to where you always face (the south) the red road goes, the road of good," the Grandfather said, "and on it shall your nation walk. The black road goes from where the thunder beings live (the west) to where the sun continually shines (the east), a fearful road, a road of troubles and of war. On this you shall walk, and from it you shall have the power to destroy a people's foes. In four ascents you shall walk the earth with power." (310)

Then the fifth Grandfather stretched out his arms and a spotted eagle appeared. "Behold, all the wings of the air shall come to you, and you and the winds and the stars shall be like relatives. You shall go across the earth with my power." Then the sky filled with wings.

Then the sixth Grandfather, the Spirit of the Earth, began to morph, seeming to grow backwards into his youth, until he became and boy, and Black Elk knew the boy was himself. Then he said: "My boy, have courage, for my power shall be yours, and you shall need it, for your nation on the earth will have great troubles. Come."

With that Black Elk followed the Grandfather out the rainbow door, and found he was riding a bay horse. Then all of the horses of the various directions came together behind him and men rode them, and they all rode the "fearful road" to the east. And below the earth was silent and sick.

Then he saw the images of the great battle with the white men. And he led his tribes to war. Then he rode around his villages, filled with sick and dying people, and as he did they began to rise and come forth, smiling and happy. "And a Voice said: "Behold, they have given you the center of the nation's hoop to make it live." So he rode to the middle of the nation's hoop and thrust the red stick into it, and there was peace and joy throughout the nation. Then the Voice offered a great insight: "It shall be a relative to them; and who shall see it, shall see much more, for thence comes wisdom; and those who do not see it shall be dark."

This is the great insight! This is the invisible world of meaning and value beyond the visible world. Black Elk experienced it. It is the shamanic experience.

Then the Voice said:

> "Behold the circle of the nation's hoop, for it is holy, being endless, and thus all powers shall be one power in the people without end. Now they shall break camp and go forth upon the red road, and your Grandfathers shall walk with them." (311)

Then the spirits lead Black Elk up four ascents, which represent four generations, and he saw the future, and it was a harsh one, but once he passed through them a Voice said to him:

> "Behold this day, for it is yours to make. Now you shall stand upon the center of the earth to see, for there they are taking you."
>
> I was still on my horse, and once more I felt the riders of the west, the north, the east, the south, behind me in formation, as before, and we were going east. I looked ahead and saw the mountains there with rocks and forests on them, and from the mountains flashed all colors upward to the heavens. Then I was standing on the highest mountain of them all, and round about beneath me was the whole hoop of the world. And while I stood there I saw more than I can tell and I understood more than I saw; for

> I was seeing in a sacred manner the shapes of all things in the spirit, and the shape of all shapes as they must live together like one being. And I saw that the sacred hoop of my people was one of many hoops that made one circle, wide as daylight and as starlight, and in the center grew one might flowering tree to shelter all the children of one mother and one father. And I saw that it was holy. (312)

This is the key, and fortunately, John C. Neihardt realized the importance of adding the following footnote: "Black Elk said the mountain he stood upon in his vision was Harney Peak in the Black Hills. "But anywhere is the center of the world," he added. This is key. This is the intersection of the physical and the spiritual. This is that passage between the two worlds.

This is what Pablo Picasso means when he says: "We all know that Art is not truth. Art is a lie that makes us realize truth, at least the truth that is given us to understand. . . . And from the point of view of art there are no concrete or abstract forms, but only forms which are ore or less convincing lies. That those lies are necessary to our mental selves is beyond any doubt, as it is through them that we form our aesthetic point of view of life. . . To me there is no past or future in art. If a work of art cannot live always in the present it must not be considered at all. The art of the Greeks, of the Egyptians, of the great painters who lived in other times, is not an art of the past; perhaps it is more alive today than it ever was." (313)

Black Elk has been entered by a world he does not completely comprehend, cannot completely explain, even to himself, because it is beyond explanation. He knows what he has experienced is important, an encounter with the spiritual world, the world beyond logic and reason, and he knows that it has chosen him, and that he has been given a responsibility. Unfortunately, there is a lot he doesn't know. While he realizes the meanings must be understood symbolically rather than literally, he cannot be sure what all of the symbols mean, must at times make guesses, and he cannot be certain what he is expected to do with the knowledge given him. He is both amazed at what he has experienced and frightened of it, wants it and fears it.

This is the kind of world Vincent has entered, and it is why he is so fascinating. We cannot help but be drawn into it to try and comprehend his experiences as his is pulled in-and-out of sanity and insanity, and as he takes us through windows of higher perception, where we, too, need to struggle with the images and symbols to understand, and we are aware that something important is taking place, if only we can grasp it. At least those of us who are open it, for remember Black Elk's insight, "It shall be a relative to them; and who shall see it, shall see much more, for thence comes wisdom; and those who do not see it shall be dark."

H. R. Graetz writes:

> As Vincent himself had been an intermediary of *a great hidden force*, so has his work remained a powerful medium, which he had prophesied *would not be outdone.* Indeed, it revives the earliest tradition of the symbolon, the urge fro brotherhood and friendship. *"More soul, more love, more heart"* he wanted in the artist's work and he saw a *similarity between art and love.* One will recall his affirmation: *"As y work is, so am I myself."* His relations with Theo and others—Christine, Madame Roulin, the postman, Milliet, Gauguin, Gachet—were projected in his work—symbolically and realistically. As *a link in the chain of artists*, he always dreamed of his idea, the working community of painters, and never did he cease to aver that his work was done by Theo and him together. Thus it was a means of union with his friends and brother, a bond of friendship as told by the early legend of the symbolon. *"To live, to work, to love are really one,"* he had written to Theo at the time he was in love with Kornelia, and later: "*You are good to the painters, and you should know, the more I think about it the more I feel that there is nothing more really artistic than to love the people."* (314)

The Reverend Salles reports to Theo confirm that Vincent's condition continues to alternate between lucidity and insanity. On March 2, he writes: "They say that children surround him and that he in turn tries to catch them and could harm them;—that he drinks much (the inn-keeper, his next door neighbour has confirmed it while he has just told me the opposite) and finally the women are afraid of him since he has taken some by the waist and has allowed himself to touch them." On March 18, he writes that Vincent is "completely clear and fully aware of his situation" but "his state has something that defies definition and the sudden and complete changes in him cannot be explained."

On March 23, Vincent's friend Paul Signac visits and writes Theo that Vincent is in perfect health, both physically and mentally (though later he will also write that Vincent tried to drink a bottle of the essence of turpentine). (315)

On March 24, Vincent tries to explain his condition:

> Am I to suffer imprisonment or the madhouse? Why not? Didn't Rochefort and Hugo, Quinet and others give an eternal example by submitting to exile, and the first even to a convict prison? But all I want to say is that this is a thing above the mere question of illness and health. (316)

Graetz catches the importance of this final line, and writes: "His approach to his inner condition as to something *beyond the question of illness or health* in fact went in the direction of today's psychological thinking in extending beyond the regular framework of medicine where answers to psychological problems have so far remained palliative. (317)

Alfred Nemeczek writes:

> There is something uncanny about the clarity with which he recognized his hopeless position, and the calm with which he accepted the inevitable was superhuman in its aspect: "I am absent-minded and could not direct my own life now." And yet in those last days in Arles he was by no means dull or apathetic—on the contrary, he was actively involved in winding up his studio. He rented storage for his furniture and took the pictures down in the Yellow House—some of them damaged: "the house itself had no fires in it during my absence, so when I cam back, the walls were oozing water and salt-peter." He saw the moldy canvasses as symptomatic for the "pitiful, painful failure" of his whole idea: "not only [was] the studio wrecked, but even the studies which would have been a souvenir of it ruined; it is so final, and my enthusiasm to found something very simple but lasting was so strong. It was a fight against the inevitable." He packed the pictures into two crates and sent them off to Paris, but his "feelings of profound remorse" continued to torment him. (318)

On March 29, Vincent gives one of his many insightful comments on the "madness" he is experiencing: "These last three months do seem so strange to me. Sometimes moods of indescribable mental anguish, sometimes moments when the veil of time and the fatality of circumstances seemed to be torn apart for an instant." (319)

In his next letter, he writes: "I am well just now, except for a certain undercurrent of vague sadness difficult to define--but anyway–I have rather gained than lost in physical strength, and I am working." (320)

Albert J. Lubin writes:

> During the attacks, people seemed to be at a great distance, voices to come from afar, and things to be changing before his eyes. He did not always recognize people; they looked "*quite different* from what they are in reality," he explained, "so much do I see in them pleasant or unpleasant resemblances to persons I knew in the past and elsewhere." He complained of dizziness, although this is difficult to evaluate in view of the fact he had suffered from it for many years. In the course of recovering, his mind remained foggy for a while, nightmares replaced with terrifying hallucinations, and he complained that his eyes were "very sensitive." The doctors said that his intellectual functions were unimpaired between attacks, and the content of his letters testifies to this. He was unable to

> paint during the acute psychotic phase of his illness; he returned to his easel only when the disorientation, the hallucinations, and the delusions disappeared. Vincent's case does not aid those who wish to prove that a psychotic man can create great pictures. (321)

These descriptions of Vincent's experiences match up with shamanic experiences. However, Lubin's conclusion miscomprehends the coming together of the dual mental states necessary. Those times Vincent is completely in his psychotic state, he has lost the necessary connection to the sensible world, the world of the senses, and thus, though he has walked through the door of perception to the other world, it is shut behind him. If Vincent never came back through that door into the physical world of sensibility, then there would be no communication, no coming together of the two worlds.

Lubin continues:

> To the clinician, the presence of delirium accompanied by disorientation, hallucinations, delusions, terror, and violent behavior suggests a syndrome commonly called a "toxic psychosis." Caused by various toxic or other organic disturbances of the brain, it is differentiated from the so-called "functional" psychoses—schizophrenia and the manic-depressive states. It seems likely that Vincent's emotional problems and poor diet added their share to these primary causes, but it is unlikely that either factor in itself would cause such a disturbance.
>
> Delirium tremens is the best known toxic psychosis. It appears when the alcoholic has been drinking heavily or after he stops abruptly. Although there is no substantiating evidence, the three attacks that followed Vincent's visit to Arles might have been due to resumption of heavy drinking on returning to the city, and the attack that occurred after being forcibly confined in a hospital cell in Arles might have been due to abrupt withdrawal. On the other hand, the Christmas attack in 1899 occurred in the sanitarium where, presumably, his intake of alcohol was controlled. Other factors also argued against the diagnosis of delirium tremens: While visual hallucinations are typical, auditory hallucinations are unusual; fainting spells are not observed; and the characteristic tremor of the hands that contributes to its name does not appear in descriptions of Vincent's attacks. The fact that his physicians were opposed to this diagnosis is perhaps the damaging evidence against it; in Provence, they undoubtedly would have been well-versed in the pathological manifestation of alcohol. (322)

On April 17, Theo marries Johann Gesina Bonger in Amsterdam. On April 19, Reverend Salles writes "It sometimes appears that there is no trace of his illness left," but also writes, 'I am not able, he said the day before yesterday, to take care of myself and to control myself; I feel rather a different man than before.'"

At the end of April, Reverend Salles accompanies Vincent to St-Remy, a private psychiatric institution nearby, and writes, "The journey has gone excellently. Mister Vincent was perfectly calm and he himself has explained his case to the directing physician like a man fully aware of his condition." Theo has to fill out an official request for Vincent to be admitted, and he requests that Vincent be allowed to paint outside and be served a half a liter of wine with each meal. Vincent arrives on May 8. During the next year, Vincent will complete nearly 300 drawings and paintings, including the brilliant *Starry Night.*

The directing physician at St-Remy is Theophile Peyron, who has once been a naval doctor and then an ophthalmic doctor before taking over his current position. Dr. Rey sends along his views that Vincent is suffering from epilepsy, and Theophile Peyron, who appears to have no background in neurology or psychiatry, simply accepts it as the cause. The treatment for it consists of a two-hour bath twice a week.

Lubin writes:

> Without evidence of an actual convulsion, this diagnosis seems remarkable for those days, even for an expert. During the 1870s and 1880s an argument was raging among neurologists concerning "masked epilepsy" and most—the celebrated John Hughlings Jackson among them—had concluded that episodic abnormal mental states in epileptics occur only after fits. [John Hughlings Jackson, "On Post-Epileptic States," 1888-1889, *Selected Writings of John Hughlings Jackson, London: Hodder and Stoughton, Ltd., 1931. I, pp. 366-384.*] Today, however, some episodic mental disturbances, including delirium, are generally accepted as bona fide manifestations of epilepsy. The triggering of attacks by external stimuli and the illusionary distortion that Vincent described have, in recent years, been found to be characteristic symptoms in epileptics with an abnormal focus of electrical activity in the temporal lobe of the brain. Perhaps Dr. Rey and Dr. Peyron were ahead of their times. Perhaps they were influenced by Vincent's statement that there were several epileptics in his family. Perhaps they had knowledge that the fainting attacks that Vincent mentioned were in fact convulsive seizures but did not enter it into their brief records. (323)

May 26, Theophile Peyron writes Theo that Vincent has been fine since his arrival, calm and becoming more healthy. However, he includes a note of caution, "I have every reason to think that the attack he has had is caused by an epileptic condition, and if this is confirmed we must fear for the future."

July 16, following a visit to Arles, shortly after Johanna mentions she is pregnant and tells him Theo is having serious health problems, Vincent suffers another attack while out painting in the fields, this one lasting 45 days.

By August 22, Vincent is writing Theo that he has come out of his latest attack and "You will perhaps do well to write a word to Dr. Peyron to say that the work at my paintings is somewhat necessary for me to recover. (324) and continues in the vein in the very next letter, "The work distracts me infinitely better than anything else and if I could once throw myself really into it with all my energy, that would possibly be the best remedy." (325) By September 5, he is putting it more forcefully: "I am struggling with all my energy to master my work, saying to myself that if I win, that will be the best lightning conductor against the illness. . . And what's to be done, there is no cure, or if there is one, it is to work with ardor. . . And altogether I would rather have a downright illness like this than to be as I was in Paris when this was brewing." (326) Vincent clearly realizes that it is in his art that he can find a way to gain control over his "illness," to in truth both have it and make sense out of it, to give it shape and form and meaning, to map out in the symbols and images of art the insanity that is beyond explanation. But he cannot completely sort it all out. In the next letter he writes: "Well, with the mental disease I have I think of so many other artists who suffer morally and I say to myself that this does not prevent one from exercising a painter's profession as if nothing were amiss. When I see that here the attacks tend to take an absurd religious turn." (327)

In September, two of his paintings are shown at the fifth exhibition of the Societe des Artistes independents in Paris. During the following autumn, he "translates" paintings by Millet, Delacroix and Rembrandt into color. This makes him feel guilty and fear he will be accused of plagiarism. He also claims he cannot copy a painting of Virgin Mary because "My illness makes me very sensitive now and for the moment I do not feel capable of continuing these "translations" when it concerns such masterpieces." (328)

December 24, he has another attack, this one lasting seven days. It appears he has been eating his own paints, as he writes Theo afterwards about it, and brief remarks are exchanged about whether or not Vincent should be allowed to continue painting since the paint is poisonous.

In January, the seventh annual exhibition of the Bigtistes in Brussels includes six of his paintings, leading to the only sale of one of his works, *Red Vineyard*, for the going rate of 400 francs, while he is alive.

January 21, the day after writing Wil a lucid letter suggesting nothing suspicious or worrisome, Vincent has another attack, again lasting seven days. While in this state, on January 25, he receives Albert Aurier's laudatory article entitled "*Les Isoles: Vincent van Gogh.*" Upon reading it Vincent is both pleased and bothered, feeling he doesn't deserve the praise.

On January 31 Theo and Johanna name their son after him. Vincent expresses mixed emotions, wishing the son were not named after him, worrying incessantly over whether Johanna will survive the birth, working on but having trouble finishing a picture to be hung in the baby's bedroom. Humberto Nagera ties this to his theories of the Replacement Child Syndrome:

> In some ways this child was like a reincarnation of the dead brother whose absence-presence had haunted him all his life. And the boy was named Vincent like his brother, and himself. The first one was dead, the second mad; thus the name was a bad omen. For the child to have this name of his dead brother must have increased his anxiety over his unconscious death wishes against it. This did not show directly but appeared in the form of a reaction formation, that is, as an intense preoccupation for the health of the child and extreme anxiety when the child became ill. Add to all this, to have given the boy his name and not that of the boy's father was an offence to his dead father's memory and we know that he had been thinking a great deal about his father at some points during his illness. Thus it is not surprising that he himself traced the beginning of the crisis to the picture he was painting for his nephew, though he may not have been fully aware of the implications of this link. Clearly, the positive aspects of his ambivalence towards the child, the only ones allowed into his conscious mind, gave him not only the idea of making the picture for his nephew's room but led him to try his best in this picture; as he said it was the most patiently worked thing he had done. The negative aspects of the ambivalence—of which he had little if any conscious awareness—created a situation of tension and conflict that was much reinforced by the other elements referred to above. They led to a new and prolonged attack of insanity. The was the fifth acute psychotic episode he had suffered in little more than a year. (329)

He visits Arles on February 22, something of a test to see if he can leave St. Remy and live on his own. He finds he cannot.

February 23, he has an attack that lasts for 65 days. Dr. Peyron writes on February 24: "He suffers from an attack that happened to him after he went to Arles. It appears to me that the crises follow one another with shorter intervals and happen whenever he goes out of doors. I do not think that he abandons himself to excess, since I know him as a sober and modest man (…) I have been compelled to send two men with a carriage to collect him from Arles."

On April 1, Dr. Peyton writes: "This attack takes more time to recede than the last one. At certain moments one would say that he is getting himself together: he explains what he is feeling and then, just hours later, his condition changes, the ill man becomes sad and despondent again and does not answer questions one puts to him."

Since Vincent is not writing him, on April 24 Theo writes: "Your silence is a proof that you are still ill," (330) to which Vincent responds: "What can I say about the last two months? I am not doing well at all, I feel more downcast and ill than I can say and I don't know what the matter is at all." (331)

This is to be his last attack before his suicide. In total, there were seven major attacks beginning with the incident of cutting off his ear. While that was the dramatic moment when Vincent's insanity clearly had entered his life, there can be no doubt that it was foreshadowed. After his fourth major attack, he wrote Wil: "I have had in all four great crises, during which I didn't in the least know what I said, what I wanted and what I did. Not taking into account that I had previously had three fainting fits without any plausible reason, and without retaining the slightest remembrance of what I felt." (332)

Having been declared fit to travel, he leaves the institution for Auvers-sure-Oise on May 16, visits Theo and Johanna in Paris on the way, having a chance to see his new nephew. Johanna describes the scene: "Silently the two brothers looked at the quietly sleeping baby—both had tears in their eyes. Then Vincent turned smiling to me and said, pointing to the simple crocheted cover on the cradle, 'Do not cover him too much with lace, little sister.'" (333)

Vincent arrives in Auvers on May 20, where he is immediately put under the care of Dr. Paul Gachet, a physician (not a psychiatrist but a cardiologist with a practice in Paris) and amateur artist, who is friends with several of the same artists Vincent hung out with when he lived in Paris. Vincent's first letter from Auvers sets in place subsequent views of Paul Gachet as being perhaps as insane as Vincent himself:

> My dear Theo and dear Jo,
>
> After having made Jo's acquaintance, henceforth it will be difficult for me to write only to Theo, but Jo will allow me, I hope, to write in French, because after two years in the Midi, I really think that I shall say what I have to say more clearly this way.

Auvers is quite beautiful, among other things a lot of old thatched roofs, which are getting rare. So I should hope that by settling down to do some canvases of this there would be a chance of recovering the expenses of my stay - for really it is profoundly beautiful, it is the real country, characteristic and picturesque.

I have seen Dr. Gachet, who made the impression on me of being rather eccentric, but his experience as a doctor must keep him balanced while fighting the nervous trouble from which he certainly seems to me to be suffering at least as seriously as I.

He piloted me to an inn where they asked 6 francs a day. All by myself I found one where I will pay 3.50 fr. a day. And until further notice I think I will stay there. When I have done some studies, I shall see if it would be better to move, but it seems unfair to me, when you are willing and able to pay and work like any other labourer, to have to pay almost double because you work at painting. Anyway, I am going to the inn at 3.50 first.

Probably you will see Doctor Gachet this week - he has a very fine Pissarro, winter with a red house in the snow, and two fine bouquets by Cézanne.

Also another Cézanne, of the village. And I in my turn will gladly, very gladly, do a bit of brushwork here. I told Dr. Gachet that for 4 francs a day I should think the inn he had shown me preferable, but that 6 was 2 francs too much, considering the expenses that I have. It was useless for him to say that I should be quieter there, enough is enough.

His house is full of black antiques, black, black, black, except for the impressionist pictures mentioned. Nevertheless, he is a strange fellow. The impression he made on me was not unfavorable. When he spoke of Belgium and the days of the old painters, his grief-hardened face became smiling again, and I really think that I shall go on being friends with him and that I shall do his portrait.

Then he said that I must work boldly on, and not think at all of what went wrong with me.

In Paris I felt very strongly that all the noise there was not for me.

I am so glad to have seen Jo and the little one and your apartment, which is certainly better than the other one.

Wishing you good luck and health and hoping to see you again soon, good handshakes,

Vincent (334)

He rents a room in Ravoux's Inn and plunges into his painting, resulting in seventy paintings and thirty drawings in the less than the seventy days he is there. One of them is the brilliant *Thatched Roofs at Montcel*, which he does during his first month there. He has now replaced earlier patterns of dots with his famous mature brushstrokes and expressive lines and colors. The naturalistic clouds become energized mysteries of another world beyond mere physical existence. The trees, the bushes, the fields, even the rocks and buildings come alive as if in motion. Another, *The Church at Auvers* depicts the church as an austere building that seems to be melting into the swift river of yellow and red and green life that rushes forward, pulling it out of the majestic blue-black sky with the flaming shapes, as if a dark fiery night is in pursuit.

On June 4, Vincent writes his mother and his sister Wil expressing concerns about Theo's health, and revealing that he is becoming friends with Dr. Paul Gachet, not surprising because Gachet is a kind man with eccentricities and interests matching Vincent's own.

In fact, it is Dr. Gachet who arranges for the two brothers to visit, inviting Theo and family down for a day, and on Sunday, June 8, Theo, Johanna and their son arrive at the doctor's house. Vincent presents his nephew with a bird's nest, gets a chance to hold him and to show him the various animals in the yard. Later that same month Vincent will write his mother: "There my little namesake made the acquaintance of the animal world for the first time, for at that house there are 8 cats, 8 dogs, besides chickens, rabbits, ducks, pigeons, etc., in great numbers. But he did not understand much of it all as yet, I think." Later in the afternoon, they all went for a long walk through the neighborhood. For the moment, life was as happy as it could be for Vincent. He later wrote to Theo: "Sunday has left me a very pleasant memory; in this way we feel that we are not so far from one another, and I hope that we shall often see each other again."

However, the joy does not last. Soon Theo writes Vincent that his nephew has taken ill and that Theo has had a bad disagreement with his employers and is considering quitting his job. Even though Theo couches the bad news in as much of a positive frame as he can, Vincent is disturbed.

Arrangements are made for Vincent to visit, and on July 6 he travels to Paris for the final time in order to discuss Theo's problems at Boussod & Valadon. Theo plans a full day of pleasant surprises for Vincent. Albert Aurier visits, and even his old friend Henri de Toulouse-Lautrec. His sister Wil also plans to visit. But suddenly Vincent can take no more of the excitement, the socializing. Suddenly, he quickly leaves. There is no concrete explanation, only the guesses that he became overwhelmed by so much excitement and interaction.

Dr. Gachet shares the same medical views of Dr. Rey and Dr. Peyton about Vincent's illness, some kind of non-convulsive epilepsy, complicated by periods of reactive depression. His treatment includes the standard use at the time of bromide therapy, which we know today might actually have hurt more than it helped. (335)

Throughout this period of Vincent's boughts of insanity, there are indications that Vincent has been thinking about suicide. March 24, 1889, he writes: "Let's guard against unwise deeds—you by marrying, me by getting old," and later, "I try to keep everything that's similar to being a hero and a martyr away from me" On April 11, he writes Wil, "Every day I take the medicine that the marvelous Dickens prescribes against suicide. It consists of a glass of wine, a piece of bread and cheese and a pipeful of tobacco." Later, he writes Theo, "If I did not have your friendship, they'd send me on the road to suicide without remorse and even when I am the coward I am, eventually that is what would happen." (336) On September 5 or 6, he writes: "I am writing you this letter bit by bit in the intervals when I am worn out with painting. The work is going pretty well. I'm struggling with a canvas I started a few days before my illness - a reaper. The study is all yellow, extremely thickly painted, but the subject was beautiful and simple. For I see in this reaper - a vague figure toiling away for all he's worth in the midst of the heat to finish his task - I see in him the image of death, in the sense that humanity might be the wheat he is reaping. So it is, if you like, the opposite of the sower which I tried to do before. But there's no sadness in this death, this one takes place in broad daylight with a sun flooding everything with light of pure gold." (337) On September 10, he writes: "During the attack I feel a coward before the pain and suffering—more of a coward than I ought to be, and it is perhaps this very moral cowardice which whereas I had no desire to get better before, makes me eat like two now, work hard, limit my relations with the other patients for fear of relapse—altogether I am now trying to recover like a man who meant to commit suicide and finding the water too cold, tries to regain the bank." (338) At the end of December, 1889, he writes "I often told myself that I would prefer if there was nothing anymore, that all would be over. But then again we are not masters of our existence and it appears that one has to learn to want to live on." (339)

July 27, 1890, Vincent shoots himself in the chest. He dies on July 29. Theo is present. The funeral is held in Auvers on July 30, and is attended by many friends.

On January 25, Theo also dies.

Most everyone agrees that Vincent's death is a suicide. However, the details are blurred and disputed. Marc EdoTralbaut interviewed Adeline Ravoux, daughter of Arthur Gustave Ravoux, owner of the inn where Vincent was staying.

She was but seventeen at the time Vincent was there, and remembers Vincent amusing her at night by drawing various versions of a little old man with a piece of chalk, and says she realized instinctively when she saw him playing with the children that he was a good man. She sat five or six times for him as he painted her portraits, and as everyone has said about him, found that when he worked he became so absorbed in his painting that he did not talk or even notice what was going on around him. She confirms the general view that he was not physically attractive, and says she was frightened at the time by the paintings he made of her, thinking them not a good likeness, though in later years she says she was able to see how he had projected the woman she was to become into them. Of all of the artists that stayed at the inn, he is the one she remembered the best because he was unique. She did not know at the time he had been declared insane or that he had been in an asylum.

Paul and Marguerite Gachet related an interesting, perhaps important incident to Marc Edo Tralbaut that took place shortly before the suicide. According to them, the first time Vincent entered Dr. Paul-Ferdinand Gachet's living-room, Vincent noticed an unframed picture by Guillaumin sitting among the various canvases Gachet had purchased and simply left unarranged. This angered Vincent and he demanded Gachet show the painting more respect and immediately get it properly framed. Upon returning a couple of days later and finding it still not framed, Vincent put his hand in his pocket, and the suggestion is that he had a gun there and was threatening to use it, though when he took his hand out of his pocket it was empty. Then Gachet gave Vincent a stern look and Vincent slunk away defeated. Paul and Marguerite watched this drama unfold and thought it would be the last time they ever saw Vincent. However, the next day he reappeared at their home as if nothing had happened. (340)

This has been compared to Vincent's encounter with Paul Gauguin, both preceding Vincent's insanity attacks and perhaps indicating some kind of common pattern.

Much has been made about the possibility that Vincent was in love with Marguerite, that her father, Dr. Gachet was firmly against this, and that it was the ultimate rejection that caused the suicide. However, there is little evidence for this view, and Marguerite's close friend Madame Liberge suggests the opposite, that Marguerite was in love with Vincent, though she never made it known. After Vincent's death Marguerite went into a depression, became a recluse, and lived the rest of her life in isolation. It is, of course, impossible to know how much her feelings about Vincent figured into all of this, but it is possible they were a factor.

Sunday, July 27, begins with Vincent eating lunch with the Gachet family, then abruptly once again threatening Dr. Gachet over the unframed picture, throwing down his napkin, and leaving to get his paints.

Adeline Ravoux relates what happened next. Towards the end of the afternoon the Ravoux family watched Vincent leave, not unusual for he came and went whenever he pleased. However, they began to wonder about him when he did not return in time for supper, as he had a consistent schedule about showing up for the meal and going to bed early.

Then, just at dusk, they saw him approaching, but moving in a strange way, staggering in big strides, as if he had been drinking. Since the sun was nearly down, it was hard to make out details, but Adeline relates that her mother was the first to notice Vincent was bolding his stomach over a buttoned up jacket. He passed them, leaned on their billiard table, finally responded to their queries with “I am wounded,” and slowly climbed the steps to his attic room. They heard him groaning, and Gustave Ravoux went up to check on him. He found Vincent lying on his bed, face turned to the wall, and asked what was wrong. Vincent turned, revealed his wound at the bottom of his chest and said, “I shot myself . . . I only hope I haven’t botched it.”

They immediately sent for the local doctor, Dr. Mazery.

The general belief is that Vincent walked into the wheat fields and shot himself there. However, another story, told by Madame Liberge, daughter of a friend of Marguerite Gachet, is that Vincent, instead of walking into the wheat fields, walked toward Chaponval, entered a small farmyard, hid behind its dunghill, and shot himself there. This is obviously a less romantic version, but has as much support as the more accepted one.

Vincent did not know Dr. Mazery, and requested Dr. Gachet, who arrived with his son shortly. According to the son, the wound was “level with the edge of the left ribs, a little in front of the axillary line,” and “formed a little circle of dark red, almost black, surrounded by a purple halo, and bled with a thin stream of blood.” Apparently no vital organs had been hit. The two doctors decided it was not possible to extract the bullet, so the best thing to do was nothing, and hope the wound would heal on its own.

Vincent apparently was not in pain, asked for and was given his pipe to smoke, told Dr. Gachet that if he healed he would simply repeat the act, and refused to give out Theo’s address. Shortly Dr. Gachet left, his son staying behind to keep watch. Anton Hirschig also went up to see Vincent and wrote: “I can still see him, with his maimed ear and his wild eyes which had a touch of madness in them, so that I did not like to look at them, as he sat on a bench in front of the care window. And I shall never forget him coming in, with his hand on his stomach, after we had waited for him at supper. I can see him in his little bed in his little attic, in the grip of terrible pain. *I couldn’t stick it any longer, so I shot myself,* he said. *But will nobody cut my belly open?* It was shelteringly hot up there under the roof." (341)

Dr. Gachet did write Theo through Goupil.

The police arrived. Vincent was uncooperative, and though Ravoux told them he had loaned Vincent the gun because Vincent had told him he needed it to scare off the crows that were bothering him when he painted, Vincent refused to confirm it. The gun was never found.

Theo received the letter the following morning, and immediately made the trip to be by Vincent's side. He found him in better shape then he expected, and wrote Johanna a hopeful note about it. But Vincent would not live long. At 1:00 AM, July 29, he passed away. Theo was at his bedside. According to Theo, shortly before he died, Vincent said, "I wish I could pass away like this."

Naifeh and Smith lean towards Rene Secretan's narrative of the event. He and his brother had been two of the local boys who tormented Vincent, and then many years later, 1957, as he was about to die, related his unkind actions towards Vincent to Victor Doiteau. While Rene did not admit to killing Vincent, Naifeh and Smith suggest it is likely he did, pointing out that their scenario solves the mysteries the best. However, yet again, they have to make a leap to explain why Vincent would not turn the Secretan brothers in to the police. He had many chances, was pushed to explain what had happened, and they had humiliated and even physically harmed him on a daily basis. To explain this, Naifeh and Smith, in the end, have to make as large an assumption as any of the others, claiming that Vincent wanted to die, wanted to commit suicide but couldn't because he thought it a cowardly thing to do, and as a result was thankful they had done it for him.

Yet again, it makes for a good story about a larger-than-life person. Rather than the death being accidental, a botched suicide, or some other less than noble act, it becomes the act of a man so extraordinary that, as they quote from Wilfred Arnold's account, he "decided to protect them (the Secretan boys) and to be a martyr." (*Van Gogh: The Life,* p. 879)

Part XIV

Starry Night

Starry Night, painted at Saint-Remy, June, 1889, stuns viewers with its explosive collision of the two planes of being, the solid, physical world of the senses, of logic and reason, and the invisible world beyond logic and reason, beyond the senses, the world of meaning and value, of the spirit and of the soul.

It takes place at night, the time when life becomes mysterious and dreams hang in the air, tempting, as if waiting to be grasped, the entire world of the unconscious waiting, quietly waiting, wanting to enter consciousness, but then always sliding away, like a shadow whenever consciousness is about to touch it, feel it, hold it, know it, the time when the world of Titania and Oberon hold court, and Puck creates mischief, the time of the nightmare, filled with the terror of the unknown, the time when Queen Mab's carriage is completed and the characters of *The Fairy-Feller's Master Stroke* come to life. It is the world of the woman, given the name "mother of the gods" by Hesiod for it is believed to be of the time before the creation of the world. (342) It is, like water, a liquid, flowing potentiality, a place of transformation, carrying with it both birth and death. (343) It is the time of primordial darkness, pre-existing the divisions of matter into the dualities of life, the *obscurum per obscurius*, the path to the profound mystery of the Creation, the ultimate mystery of the birth life from nothingness. (344)

For the great Greek civilizations before the coming of the Romans, Night (Nyx) is the daughter of Chaos, the time before the beginning of shape and form. She is the mother of the Sky (Ouranos) and the Earth (Gaia), and moves through the sky in a chariot drawn by four black horses, followed by the Fates and the Furies, an Underworld goddess who receives black ewe-lamb sacrifices. The Mayans also saw her as the goddess of the Underworld and death. The Celtics also saw her as that which came before the birth of the year and the day, a creature related to winter, the time of the winter solstice, when the days begin to grow longer, the same symbolism used by Christians for the birth of Christ, a new beginning, a resurrection.

Jean Chevalier and Alain Gheerbrant write:

> In mystical theology, Night symbolizes the disappearance of all knowledge which may be defined, analyzed or expressed and, further still, the state of being deprived of all proof and psychological support. With such other terms as "obscurity," Night applies to the wiping clean of the intellect, "emptiness" or "nakedness" applies to that of the memory and "dryness" or "aridity" to that of all longings, sensual emotions and even the highest aspirations." (345)

Most importantly, night is the time of the world of the dream, the maker of symbols and the meanings they encompass. For Sigmund Freud, they are "the royal road to knowledge of the soul," the expressions and perhaps fulfillment of repressed wishes. (346). For Carl Jung they are the "helpful powers slumbering in the deeper strata of man's nature "that" can come awake and intervene." (347) For Joseph Campbell they are "realizations of some kind that have then to find expression in symbolic form." (348) The Ancient Egyptians thought they were created by the gods to reveal the future. The Bantu of Kasai, a tributary of the Congo River, believed that some dreams come from souls of the dead. The Negritos of the Andaman Islands believed that they were the dark side of the human soul. North American Indians thought dreams were the highest truths, the source of the powers of the shaman. (349)

Within this *Starry Night*, this world of the dream, a moon and a sun are combined, as if the moon were taking on the bright, flaming powers of the sun. Always associated with the cycle of the tides, the moon is a life force, both of the unconscious world of the water, and of the universal unconsciousness of the cosmic ocean. It is connected to the female menstrual cycle, and thus to birth and transformation, as depicted in the lunar goddesses, Ishter, Hathor, Anaitis, and Artemis, which in turn always demands death. It's soft light both illuminates and yet leaves unlit, a twilight world between perception and deception, a place for the imagination, the world beyond the bright light of the sun.

Or, perhaps, in this painting, it is the sun swallowing the moon. Perhaps the bright sun behind the moon is burning through it, the world of consciousness, that world of brilliant, blinding, hot, yellow fire of a violent transformation about to sear into this "other" world. The stars are all shinning through the blue-black lines of the night, circles, spheres of life, symbolizing self and woman and completeness all at the same time. That which is beyond explanation is shattering the deceptions of physical sight. The higher sight of the invisible world has entered visible existence.

Beneath the blazing sky, the controlled world of logic and science and math rests motionless, or nearly so, in the geometric shapes of houses built by human reason, representing a world where human logic has provided some small amount of control over the environment. This is the hard, dry earth of the conscious mind, the place where humans have molded nature to their physical needs. The most prominent building, the church, suggesting a union of or a rising out of the conscious mind with the unconscious, the spiritual ocean raging above it, thrusts its masculine tower towards the fertile sky, but it is not powerful thrust, not strong enough to deliver its seed and join in the energetic birth taking place; no this church is not partaking of the spiraling, mysterious life swirling about it; rather, it is more of a commentary on the failure, the sterility of such a building (and possibly with it of organized religion).

Everything else is in motion. The mountains appear as if huge waves of an ocean, and the bushes in front of them, as blue as they are green, as waves crashing against the shore, a flood about to overflow the weak logic of the buildings. Above, the mountainous waves, the yellow energy of the sun and moon rushes s across the top of the earth, as if the cosmic ocean of the sky had descended and is now merging with the earth. The large dark green cypress on the left reaches into the sky, a living plant in the shape and with the energy of a raging fire.

This is the numinous realm of existence become visible. This is the shamanic experience.

Vincent writes: "when shall I ever get round to doing the *starry sky*, that picture which is always in my mind? Alas, alas, it is just as the excellent fellow Cyprien says in J. K. Huysman's "En ménage": the most beautiful paintings are those which you dream about when you lie in bed smoking a pipe, but which you never paint. Yet you have to make a start, no matter how incompetent you feel in the face of inexpressible perfection, of the overwhelming beauty of nature" (350)

W. Jos de Gruyter writes about it: "From inner suffering this artist has expressed with overwhelming power the mysticism which was the supreme mark of his genius." (351)

Tralbaut writes:

> When Vincent created THE STARRY NIGHT he bridged the gap between reality and dream. If one compares what Vincent actually saw, and which is still to be seen at Saint-Remy, with the poetic flight that he derived from it, one sees that reality, without any loss of truth, has given way to a new dimension, the dimension of the cypress on the left of the picture belongs to the real Provencal landscape. Vincent placed it where he did in order to form a sort of corridor and to push the roofs of the houses and the church tower back as far as possible into the distance. He experimented with this device in Paris in the views out of the window of his studio in the rue Lepic, influenced no doubt by the perspectives in Japanese prints; and he used it in composing his VIEW OF THE ARENA at Arles.
>
> Beyond the objects in the foreground of THE STARRY NIGHT rise the mountains. But they are not the Alpilles, which are really on the other side, they are imaginary mountains. The landscape is no more real than the sky that Vincent has imagined above it.
>
> The centre of the firmament is dominated by an eddying spiral nebula surrounded by eleven brilliant stars, and in the top right-hand corner there is a shining globe that sets a riddle. Is it the moon or the sun—or both at once? Meyer Schapiro, who considers that this picture is inspired by a religious mood, says it is a "confused memory, perhaps, of an eclipse" and recalls that Vincent had quoted Hugo's dictum, "God is a lighthouse in eclipse."

Graetz maintains that the crescent moon is merely a confirmation of the two clouds in the center. Moreover the strange shape of the moon recalls an old Chinese symbol, known as the Yin and the Yang, which represents the union between opposites.

Be that as it may, it was undoubtedly at this moment of Vincent's career, during his "visionary madness" that this whirling firmament of suns and wandering tumbled landscape reveals most clearly the extraordinary penetration of his art. A freedom of spirit is joined to a freedom of expression so daring that it would never be surpassed. . . .

In this picture Vincent has exalted real elements and carried them beyond our experience. He has thrown matter and light into the crucible and revealed the fast movement of the birth of a universe. He was undertaking something which most of his fellow-artists could not have conceived, and in doing so he became one of the greatest painters of all time. (352)

Schapiro isn't the only one wanting to put the painting into an expression of Christianity. Kathleen Powers Erickson writes:

Starry Night, as the preeminent expression of van Gogh's religious experience, is an autobiographical landscape, which we can divide into three separate areas, illustrating three of the most significant ideas in van Gogh's art and life. The village scene, the cypress three, and the sky are all representative of specific religious beliefs van Gogh held. The church provides both a focal point and vertical accent in the village scene. Art historians point out that van Gogh's rendition of this church is imaginary, since the steeple is typical of the Dutch landscape, but not the landscapes of Provence. In addition to being a Dutch church in style, van Gogh's rendition of the church is curious in another way. While every house glows with yellow light under the brilliance of the starry sky, the church remains completely dark. This is also true of his *Church in Auvers* . . . , in which the foreground is brightly lit by the sun, but the church neither reflects not [nor] emanates any light of its own. The darkness of the "inside of a church" is van Gogh's symbol of the empty and unenlightened preaching of the clergy which left him embittered and alone when he was forced to leave the ministry in 1880. In *Starry Night*, van Gogh reveals, however, that he did not close the door on religious faith, just the church. *Starry Night* shows van Gogh's journey from the darkness of the inside of a church, with its reference to his Dutch past, to the triumph of the mystic's communion with God through nature. While many have argued that the painting indicates van Gogh's rejection of Christianity and the supernatural, his observation (often quoted in conjunction with essays on *Starry Night*) that "When all sounds cease, God's voice is heard under the stars" actually comes from the heart of his "evangelical period," 1877, [Letter 100] and reflects a lifelong spiritual conviction. (353)

This part of her argument goes along with my own claims, especially the stress on "mystic," though the context of "God" is strongly Christian, and the suggestions are that it is couched in "religion" as opposed to a "break through" experience beyond the walls of religion.

She continues by suggesting that the cypress "which shoots up into the firmament like a giant flame, represents van Gogh's own as well as the universal striving for ultimate release from the sufferings of this world and ultimate union of the soul with the infinite." (354). Sounds good, but then she struggles to put it all into a more literal Christian frame. After pointing out that he painted a number of works featuring a cypress, she quotes a passage in a letter to Theo: "'You need a certain dose of inspiration, a ray from above which is not ours, to do the beautiful things. When I had done these sunflowers, I looked for the contrary and yet the equivalent, and I said this is the cypress. . . . It is as beautiful as the Egyptian obelisk." [Letter 596] (355) In other words, she is pointing out that Vincent used the "sunflower" to represent "inspiration," a "ray from above," interpreted as God or rather Christ, and "beauty," and the cypress as the opposite of this, still beautiful, as in an Egyptian obelisk, interpreted as representing death. Next she refers us to her earlier argument that "the sunflower, as we have already noted, appeared with increasing regularity in the landscapes of van Gogh's St. Remy period as symbols of devotional piety and love of God." In referring back to that "note," what we find is that she has quoted Tuskasa Kodera: "The sun as a symbol of God or Christ and the sunflower as the symbol of a pious soul is, of course, not new. In the tradition of emblem books and visual arts, sunflowers were often used as symbols of faith or love. In his *Emblemata* published in 1625, Zacharias Heyns depicted a sunflower facing the sun with the motto '*Christi action imitation nostra*' [Let us imitate Christ]. . . ." (356) Then she writes: "Given the traditional appropriation of the sun as an image of the divine presence, then, it is surely unnecessary and misleading to suggest that van Gogh perceived the sun itself as god. More likely, van Gogh simply used the sun to represent Christ, as others had done since the third century. This interpretation is far more consistent with Gogh's oeuvre and life. It is consistent, too, with his repeated depiction of sunflowers which, as noted earlier, are traditionally associated with piety and devotion." (357) In other words, what she is doing is taking a claim by Tuskasa Kodera, which offers one example to support itself, and turning it into a "obvious" claim that Vincent used sunflowers to symbolize piety to God or Christ, the argument seems to flip-flop between them, but perhaps meant to be understood as to God through the sun which symbolizes Christ. Thus, I think she is saying that Vincent saw himself as the sunflower, nourished by Christ (the Sun). With this in place, then, the cypress, which Vincent saw as a contrast yet equivalency, becomes symbolic of death, which is interpreted positively because it means the soul's desire to reunite with God through death.

It is best to point out that what she is trying to do here is to make *Starry Night* an expression of Vincent's Groningen theology, of his "Christ of the Coalyards" period. Immediately after stating that *Starry Night* is a visionary masterpiece, recounting the story of Van Gogh's ultimate triumph over suffering, and exalting his desire for a mystical union with the divine," she emphasizes that "In many ways, it recalls the mysticism of the religion of van Gogh's youth, the Groningen theology and the piety of his uncle Stricker." (358)

She is working hard to have both the visionary, the mystic experience here, and still to have the Christian reduction of it to literal existence, doing much the same thing the psychological interpretations did, which is to mix in wonderful, insightful statements with the flaws of a narrow, determined literal focus.

Her analysis moves next into a defense that Vincent saw the sky as a symbol of infinity, and that this in turn evoked an ecstatic, mystical mood in Vincent. Okay. Wish she had taken this more into the meaning of a mystical experience, more into what it consisted of, represented, expressed, beyond her easy tying of it to Vincent seeing himself as a "traveler," a "pilgrim" in relation to the pilgrim in John Bunyan's book who is traveling to the Celestial City, and thus, ultimately as Vincent believing his journey is coming to and end and is expressing his desire to unit with God in death. She leads it to: "The passing of day into night, the imposing image of the cypress as it soars from the earth to the heavens, and the stars, which represent van Gogh's longing for ultimate union with the Infinite Being, evoke thoughts of death and immortality." (359) I must admit, that this statement in-and-of-itself has a lot of potential, but the surrounding insistence on the narrow religious context makes it less powerful than it might be.

This is not to say, Vincent didn't have death and what comes next on his mind (as indeed he seemed to most of his life). Marc Edo Tralbaut speaks for most biographers in terms of this:

> In his contemplation of the starry skies, which he had attempted to fathom just as he had tried to penetrate every other natural object, he had begun to feel that there was some intangible association between stars and death.
>
> In a letter to Theo written the previous July he had been considering how all artists in all the arts were unfortunate in material things, and he

went on: *That brings up again the eternal question: Is the whole of life visible to us, or isn't it rather that this side of death we see only one hemisphere? Painters—to take them alone—dead and buried speak to the next generation or to several succeeding generations through their work. Is that all, or is there more to come? Perhaps death is not the hardest thing in a painter's life. For my own part, I declare I know nothing whatever about it, but looking at the stars always makes me dream, as simply as I dream over the black dots representing towns and villages on a map. Why, I ask myself, shouldn't the shining dots of the sky be as accessible as the black dots on the map of France? Just as we take the train to Tarascon or Rouen, we take death to reach a star.*

Looking at the stars made him dream, not that man might soon reach some of the heavenly bodies as prosaically as if he had taken a train, but mystical, metaphysical dreams. These had their repercussion in the first version of THE STARRY NIGHT, which he painted only a few weeks after he wrote this letter, and even more in the Saint-Remy version.

The Arles picture was Vincent's first venture into a new territory and led the way to the second version, which would have been inconceivable if Vincent had not suffered visions of death after his attack on Christmas Eve 1888. (360)

Erickson also brings in an interesting connection with Walt Whitman, mentioned by many critics. Marc Edo Tralbaut writes:

Yet the origin of this "apocalyptic fantasy," as Meyer Schapiro called it, may have been due to a literary influence. In a letter to Wilhelmien written in September or early October 1888, shortly before Gauguin arrived at Arles, Vincent wrote: *Have you read the American poems by Whitman? I am sure Theo has them, and I strongly advise you to read them, because to begin with they are really fine, and the English speak about them a good deal. He sees in the future, and even in the present, a world of healthy, carnal love, strong and frank—of friendship—of work—under the great starlit vault of heaven a something which after all one can only call God—and eternity in its place above this world.*

Vincent could hardly have provided a better description of his picture than that key phrase: *under the great starlit vault of heaven something which after all one can only call God.* This also seems to show that Whitman may have provided the germ of inspiration for this painting, a view that I first discussed with Paul van Ostaijen, the poet, in 1918, forty years before I first had a chance to read Vincent's letter to Wilhelmien. (361)

No doubt Vincent was struggling with spiritual matters, and the terms spiritual and religious often get intertwined, making it difficult to distinguish, but it is important to emphasize that Vincent is in the spiritual, not the religious world when he paints *Starry Night*.

As briefly mentioned in a quote from Marc Edo Tralbaut above, H. R. Graetz points out the easy connection of the central spiraling lights in the painting with the yin-yang symbol, symbol of being-non-being, a symbol Vincent must have known from his studies of Japanese art and Buddhism, a topic Cliff Edwards go into in great depth, supporting his claims that "Vincent not only saw himself as a monk worshiping Buddha, he saw his Yellow House as a "refuge for many," a Buddhist monastery under the direction of an abbot, whom he identified as Gauguin (Letters 544a, 544). In a letter to Wil about a painting he did of himself, he wrote: "I have a portrait of myself, all ash-colored. . . . But as I also exaggerate my personality I have in the first place aimed at the character of a simple bronze worshiping the Eternal Buddha." (Letter 544a)

Vincent must have realized the Yin-Yang resemblance and known at least the basic symbolism.

> It stands for the unity of dark and light, female and male, of negative and positive, non-being and being. The two parts enter into each other in a synchronous motion. They are inseparable and together they become one in a perfect circle of wholeness. In each part is a nucleus as in a cell, the equivalent of the kernel in the seed recalling the point in the star that symbolizes man's original oneness and individuality. The black nucleus in the center of the white field and the white nucleus in the black indicate the basic existence of Yin within Yang and Yang within Yin, of darkness in the light and light in darkness. There fore, each element also possesses the quality of its opposite and thus at the same time contrasts with and resembles the other. (362)

Carl Jung, who did in-depth studies of Chinese symbolism, explicitly used the yin-yang symbol as a symbol of *self*. For Carl Jung the *ego* is similar to but not quite the same as it is for Sigmund Freud, because for Carl Jung the *ego* is completely in the *conscious* world and responds to *conscious needs only*. Thus, it is imperfect, not connecting to the *unconscious*. It's mechanism for this is the *persona*, a social façade, a mask, a role each of us puts on to meet the particular demands of the external environment, the *other*. Beneath or behind this *conscious persona* this *ego* in the depths of the *unconscious* other forces are at work. These forces are *not* socially acceptable. They are the *dark side*, but not necessarily *evil side* of human behavior, and as they come together in the unconscious, they become the *shadow* or *alter-ego*.

In *Memories, Dreams, Reflections,* Jung discussed how he came to understand this archetype through a dream:

> It was night in some unknown place, and I was making slow and painful headway against a mighty wind. Dense fog was flying along everywhere. I had my hands cupped around a tiny light which threatened to go out at any moment. Everything depended on my keeping this little light alive. Suddenly I had the feeling that something was coming up behind me. I looked back and saw a gigantic black figure following me. But at the same moment I was conscious, in spite of my terror, that I must keep my little light going through night and wind, regardless of all dangers. (363)

He awoke with the realization that this frightening stalker was his own shadow cast in the mist by the flickering candle. While the *ego* is in charge of the *conscious* personality, the *shadow* and the *anima* (female archetype, the opposite sex characteristics) hold sway in the *unconscious*. The conscious and the *unconscious* together make up the *self.*

For Jung, the *libido* is the drive to life (more than just a sexual drive), the impetus for all of our thoughts and intentions, and it operates on the principle of opposites, for every good thought a corresponding bad one. Thus, every time the *conscious* good takes place, a corresponding bad gets stored in the *unconscious*. Archetypes contain *stored libido,* what Carl Jung calls *numen*. If the *archetypes* in the *unconscious* cannot find a way to release their stored up energy, there is a build up of energy, a *one-sidedness*, what Carl Jung says is the foundation of *neurotic disorders*, such as an unresolved Oedipal complex. In other words, either the *conscious ego* is so dominant in its desire to maintain a *persona* that it will not accept any contributions from the *unconscious* (i.e., a wall gets put up between the *conscious* and *unconscious*) or there are not enough *conscious symbols* for the archetypes to be expressed and thus the imbalance relieved.

Self individuation or realization is the goal to a healthy psyche, and this can only be accomplished through the union of opposites which make up the whole. Thus, Carl Jung embraces the sacred circle, the *mandala* as the strongest symbol of this union of opposites representing the total human. The yin-yang design takes the form of a circle containing the two opposites, yin the masculine *p'o*, the higher, the breath, the soul, yang the feminine *hun*, the lower, the earth. (364) Thus, the yin-yang symbol is the symbol of *individuation*, which is the process of the *realization* of *self.* The first step is simply realizing, admitting an unconscious element, the *shadow*, not to destroy it but to incorporate it, complete the totality of the *self.* A key in bringing the *ego* and the *shadow* together is the *anima.* The *anima* is located just behind the *shadow* in the *unconscious*, and it must emerge into *consciousness* as a personification to be confronted and brought under control. Once the *anima* and *shadow* have been realized and the *persona* has thus been weakened, the *self* can emerge, can be realized, and a balanced psyche results.

From the perspective of Jung's theories, and from the mythology of China the symbol of yin-yang in Vincent's painting becomes extremely powerful, a representation of the union of opposites on many levels, both the personal psychic levels and the universal psychic levels. Carl Jung, it must be remembered, laid the foundation for the *universal psyche*, the concept that the same *symbols* are understood universally by all humans. *Archetypes are universal mental patterns.* For humans, the entire world works as a dream, understood symbolically as if each individual dream were intertwining in an ever-changing web of light and meaning to join in the universal creation of human existence.

And this is precisely what Vincent's *surrealistic* painting presents, the invisible world of the unconscious of the human psyche breaking through the visible world of the human body. Here are the two largest opposites of all existence in collision. The yin-yang *mandala* is not contained. There is no circle surrounding the two forces. They have burst out of their circumspect realities. And the circles that surround them threaten to do the same.

Furthermore, though it cannot be as certain, Vincent might also have been aware of the Chinese symbolism of the moon within the sun. In *The Secret of the Golden Flower* (based on *I Ching, the Book of the Changes*), "Li is the sun, K'an the moon; the marriage of K'an and Li is the secret magical process which produces the child, the new man." (365) If so, then he is duplicating the explosive union of opposites expressed in yin-yang, and extending it into a birthing of a new union, a *transformation*, a movement beyond the dual realities into a reality that contains them both. This is the reality of the *spirit* ripping through the thin veneer of physical existence. This is the logic and theology of religion dwarfed by the extraordinary realization of the higher existence of the *spiritual, spiritus mundi!*

H. R. Graetz offers an interesting spin on the "lightless" and thus "spiritless" church depicted by Vincent:

> Could the remarkable choice of "Starry Night" as a religious painting mean that the church too, in a dark urge from within, is searching for light? Should the church itself be destined to fulfill Van Gogh's hope that this painting might be *consoling*? [Letter 595] And should the clergy who helped to select this work have been inspired by it to become instrumental in spreading his faith: the belief in God and eternity, lived and practiced by creative men and women in equality and love of soul and body. In this *badly reputed humane* [Letter 219] sense the "Starry Night" can be called

> religious. It is Vincent's spiritual testament, his message of a conception that *does not pass in the passing,*[Letter 253] communicated in a change of form, a metamorphosis, as he wrote: " . . . *as necessary as the renewal of the green in spring."*[Letter 253] (366)

Starry Night is not a *religious* expression or experience, but a *spiritual* one. This is the conflict that centers all of Vincent's life, how to take people from the mundane literal world, the explanations of theology, into a realization of that which is beyond explanation.

Part XV

Diagnosis

The people who knew Vincent while he was alive were not qualified to diagnose his psychic condition. Not only were the fields of psychology and neurology still in their infancy, but the doctors who treated Vincent were not trained in it. It appears that Vincent supplied them with their original diagnosis of epilepsy in telling Dr. Felix Rey his mother's sister had it. This got passed on to Vincent's next doctor, Theophile Peyron, who had no knowledge of psychiatric diseases, and he, in turn, passed it along to Paul Gachet, who became an eccentric friend with medical credentials, but was anything but a practicing or even knowledgeable psychiatrist or neurologist.

However, since it was the original diagnosis, it has since been reiterated again and again through the years by such people as John Rewald, Boiteau and Leroy, who claim it was "epileptoid psychosis," Thurler, who calls it a form of "suppressed epilepsy," and Birnbaum, Evensen, Koopan, Minkowska and Mayer-Grosz. (367)

Nevertheless, it has its detractors, some of whom are highly trained in their particular fields, and articles continue to appear in such respected publications as the *Journal of the American Medical Association* with such other difficult to believe assertions as that "Van Gogh had Meniere's disease and not epilepsy," (368) or that he had "polysurgical addiction," a "compulsion to submit to surgical operations." (369)

While such unlikely reasons for rejecting epilepsy are difficult to believe, most experts today agree that epilepsy should be considered a wrong analysis, at least in terms of how epilepsy is defined today. Neurologist and Van Gogh expert P. H. A. Voskuil writes that "Van Gogh certainly did not suffer from epilepsy as it is currently defined according to the usual and most frequent neurological categories." Rather, he continues,

"The attack phenomena are not, according to recent notions, typical of epilepsy. In general we are dealing with consciousness disorders and psychotic phenomena which occur in attacks that last for days, up to weeks. Some elements in this symptomatology could fit in with the partially motor or partially complex type of epileptic attack. Based on recent developments in the views concerning the overlap between epilepsy and psychiatry it seems adequate not to reject the epileptic phenomena in Van Gogh's diagnostics, but to place them in a larger biological/psychiatry reference frame in which depression, psychotic phenomena and impulse control disorders may find a place." (370)

Probably, Vincent and subsequently his doctors, in using this term, were meaning it as a larger umbrella than it is currently used to cover various mental diseases. It is clear that Vincent's family was prone to mental illness. His brother Cor committed suicide, and his sister Wil spent the final four decades of her life in an asylum. Frank Milner writes that "recent evidence surrounding Theo's death, connected with the disappearance of all medical records, has led to suggestions that in a state of mental collapse, he too committed suicide." (371) This is interesting in connection with the similar situation in Richard Dadd's family, and suggests that perhaps there is some close affinity between insanity and genius, some inherited quality.

In addition to heredity, there are some obvious contributing factors in terms of Vincent's poor health habits. His chronic stomach problems, dizziness, and very bad teeth were at least partially the result of his terrible eating habits. His letters constantly reveal he is eating almost nothing, living mainly on coffee, perhaps a piece of bread or cheese, and to various degrees wine and liquor and abstenthe. Furthermore, he pushed his body to exhaustion, at times punishing himself physically to satisfy his spiritual beliefs, at times because he had a strong work ethic.

His consumption of various kinds of alcohol needs to be emphasized. While some of his biographers want to downplay this, to almost deny it, the evidence is that he drank a good deal, was perhaps, even likely, an alcoholic. And the absenthe he drank has been proven to have dangerous properties, an ingredient called "thujone" that can cause psychotic experiences, bring attacks similar to epileptic attacks, and result in serious health problems, even death. Furthermore, he was constantly ingesting or breathing various toxic substances. The very paints he used and both breathed and even ate had lead, zinc, chromium, cobalt, various oils, and turpentine. He also used large quantities of camphor. Such toxins can cause "organic psycho syndrome" or "chronic toxic encephalopathy," a weakening of the central nervous system resulting in both physical and mental fatigue, strong irritability, changes in personality and emotions, loss of learning abilities, and dementia.

Vincent's two portraits of Doctor Paul Gachet, both including a branch of foxglove (digitalis), and the fact that it was being used to treat epilepsy at the time have led to speculation that Dr. Gachet was prescribing it to Vincent and that it contributed to his condition, though its effects would not have explained most of Vincent's symptoms.

Biochemical and molecular expert W. N. Arnold suggests that Vincent suffered from Acute Intermittent Porphyria (AIP), a disease of the blood where Haem, one of the building blocks of hemoglobin, is not being properly produced in the system, resulting in vomiting, constipation, weakness, shaking, trembling, high blood pressure, a fast pulse, and symptoms of psychosis. Attacks of it can be brought on by several of the things Vincent was exposed to, such as camphor, lead, and alcohol. A key symptom is a purple discoloration of urine, something never mentioned by Vincent or anyone, and thus one reason this diagnosis is not taken seriously by others.

Whether or not he had Syphilis is also controversial. As discussed previously, Ken Wilkie presents a strong argument for it, and for the fact that both the disease and the treatment might well have slowly eaten away at Vincent going all the way back to the time he was with Sien, resulting in his final insanity.

A. J. Westerman Holstijn, who is considered to have done the first comprehensive psychological study of Vincent, feels that two frustrations combined to cause Vincent to cut off his ear, Theo's engagement and the failure of the relationship with Gauguin. He also suggests the ear was a phallic symbol and that the self-punishment was partially the result of his homosexual impulses." (372) Daniel Schneider agrees, stating that Vincent lived "under the constant overpowering threat and masochistic homosexual unconscious wish for castration." And that, "when he slices off his ear and gives it to the prostitute who accepted Gauguin, he brings it about rather than face it any longer." (373), Art and psychiatry professor Jacques Schnier suggests Vincent's giving his ear to a prostitute was a way of fulfilling a wish to possess his mother following his imagined attack of his father upon a father substitute (Paul Gauguin). (374) This leads back to the lengthy discussions of the views of Humberto Nagera, Albert J. Lubin, and their followers and reinterpretors, especially their focus on the Replacement Child Syndrome.

These psychological approaches mesh with the many other claims he suffered some form of schizophrenia put forth by such people as Walter Riese, who refers to "episodic twilight states," and Hans Prinzhorn, famous for his collections and studies of art of the insane and found their works, especially those of schizophrenics, exhibited qualities similar to those of serious artists, including Vincent, who he says had a singular case of "dementia." (375) The list can be extended and includes numerous variations, some combining schizophrenia and epilepsy, some moving into other psychosis. Boiten calls it a form of psychopathy, Hutter a psychosis of degeneration, Kahn "schizoform reation," Rader a cerebral tumour. Lange Eichbaum an active luetic schizoid and epileptoid disposition, Rose and Mannheim a phasic schizophrenia, Bychowske a dementia praecox, Dupinet a meningo-encephalitis luetica, Vinchon a psychotic-exhaustion caused by great creative effort, Storch an atypical psychosis heterogeneously compounded of elements of epileptic and schizoid disposition, Stertz a phasic hallucinatory psychosis, Kerschbaumer a schizophrenia, probably of a paranoid type, Fels and Grey a neurasthenia, Grey a chronic sunstroke with the influence of yellow, Gastaut a psychomotor epilepsy. (376)

Existentialist Karl Jaspers is the most respected psychiatrist and philosopher to put forth this view that Vincent suffered from a form of schizophrenia. In 1910 he published an important paper providing a more comprehensive approach to paranoia then was currently in place, one that involved biological changes in people, and he stressed including patients' self-diagnosis in clinical psychiatry, an approach that became known as the *biographical method.* He emphasized a focus on the *form* of a patient's symptoms rather than the content. In other words, how a patient experiences a hallucination is more important than the content of it. Thus, hallucinations and delusions should be understood not on their content but on how a patient comes to have them, why a patient has hallucinations and delusions. He also divided delusions into primary, those arising without apparent cause, and secondary, those influenced by the patient's background and current mental state. Furthermore, he believed that some delusions are simply beyond comprehension, a very controversial view, but one that opens the door to perhaps considering them beyond explanation, but understandable if one puts the need to understand in terms of explanation (of left brain) aside.

This belief in an unexplainable aspect of psychology also infused his philosophy, which embraces Soren Kierkegaard and Friedrich Nietzsche's underlying views about the value of the individual, and says that empirical (or scientific) approaches to reality cannot transcend, resulting in despair and resignation, and one must move beyond them, must make Soren Kierkegaard's *leap of faith* to *Transcendence*. This is the only way to face humans' limitless freedom (*existenz*) and experience authentic existence. *Transcendence*, paired with the term *Encompassing* is that which exists beyond time and space, the ultimate non-objectivity or no-thing-ness. This opens the door to argue that Karl Jaspers is a monist, though he stressed the necessity of both subjectivity and objectivity. (377)

In *Strindberg and Van Gogh: An Attempt at a Pathographic Analysis with Reference to Parallel Casses of Swedenbory and Holderlin,* Jaspers attributed Vincent's art after 1888 to schzophrenia, claiming that the art took place during a time of "great tension" and "emotional excitement," resulting in "impoverishment and unsureness" in the drawings and a "lack of discipline" in the paintings resulting in a "smearing without a sense of form," which "represents energy without content, or doubt and terror without expression." (378)

The National Institute of Mental Health (NIMH) summarizes current views of schizophrenia, and includes the following:

> The symptoms of schizophrenia fall into three broad categories:
>
> *Positive symptoms* are unusual thoughts or perceptions, including hallucinations, delusions, thought disorder, and disorders of movement.
>
> *Negative symptoms* represent a loss or a decrease in the ability to initiate plans, speak, express emotion, or find pleasure in everyday life. These symptoms are harder to recognize as part of the disorder and can be mistaken for laziness or depression.
>
> *Cognitive symptoms* (or cognitive deficits) are problems with attention, certain types of memory, and the executive functions that allow us to plan and organize. Cognitive deficits can also be difficult to recognize as part of the disorder but are the most disabling in terms of leading a normal life.
>
> ***Positive symptoms*:**
>
> *Positive symptoms* are easy-to-spot behaviors not seen in healthy people and usually involve a loss of contact with reality. They include hallucinations, delusions, thought disorder, and disorders of movement. Positive symptoms can come and go. Sometimes they are severe and at other times hardly noticeable, depending on whether the individual is receiving treatment.
>
> *Hallucinations.* A hallucination is something a person sees, hears, smells, or feels that no one else can see, hear, smell, or feel. "Voices" are the most common type of hallucination in schizophrenia. Many people with the disorder hear voices that may comment on their behavior, order them to do things, warn them of impending danger, or talk to each other (usually about the patient). They may hear these voices for a long time before family and friends notice that something is wrong. Other types of

hallucinations include seeing people or objects that are not there, smelling odors that no one else detects (although this can also be a symptom of certain brain tumors), and feeling things like invisible fingers touching their bodies when no one is near.

Delusions. Delusions are false personal beliefs that are not part of the person's culture and do not change, even when other people present proof that the beliefs are not true or logical. People with schizophrenia can have delusions that are quite bizarre, such as believing that neighbors can control their behavior with magnetic waves, people on television are directing special messages to them, or radio stations are broadcasting their thoughts aloud to others. They may also have delusions of grandeur and think they are famous historical figures. People with paranoid schizophrenia can believe that others are deliberately cheating, harassing, poisoning, spying upon, or plotting against them or the people they care about. These beliefs are called delusions of persecution.

Thought Disorder. People with schizophrenia often have unusual thought processes. One dramatic form is disorganized thinking, in which the person has difficulty organizing his or her thoughts or connecting them logically. Speech may be garbled or hard to understand. Another form is "thought blocking," in which the person stops abruptly in the middle of a thought. When asked why, the person may say that it felt as if the thought had been taken out of his or her head. Finally, the individual might make up unintelligible words, or "neologisms."

Disorders of Movement. People with schizophrenia can be clumsy and uncoordinated. They may also exhibit involuntary movements and may grimace or exhibit unusual mannerisms. They may repeat certain motions over and over or, in extreme cases, may become catatonic. Catatonia is a state of immobility and unresponsiveness. It was more common when treatment for schizophrenia was not available; fortunately, it is now rare.[2]

Negative symptoms:

The term "negative symptoms" refers to reductions in normal emotional and behavioral states. These include the following:

Flat affect (immobile facial expression, monotonous voice), lack of pleasure in everyday life, diminished ability to initiate and sustain planned activity, and speaking infrequently, even when forced to interact.

People with schizophrenia often neglect basic hygiene and need help with everyday activities. Because it is not as obvious that negative symptoms are part of a psychiatric illness, people with schizophrenia are often perceived as lazy and unwilling to better their lives.

Cognitive symptoms:

Cognitive symptoms are subtle and are often detected only when neuropsychological tests are performed. They include the following:

Poor "executive functioning" (the ability to absorb and interpret information and make decisions based on that information), inability to sustain attention, and problems with "working memory" (the ability to keep recently learned information in mind and use it right away)

Cognitive impairments often interfere with the patient's ability to lead a normal life and earn a living. They can cause great emotional distress. (379)

Derek Fell states:

Psychiatrists disagree about schizophrenia, a psychosis characterized by withdrawal from reality and by highly variable, irrational emotional or behavioral disturbances, as Vincent enjoyed long periods of clear thinking between his devastating lows. Indeed, his letters reveal that he was well aware of his erratic behavior, keenly self-analytical about his disorder, and completely rational about his goals in life. Furthermore, Vincent's seizures were clearly connected to his intense anxiety and melancholia either over emotional rejection or from vivid memories of it; and indeed the seizures and anxiety both—i.e., the physiological ramifications of the psychological disorder—may have been rooted in the same neurological cause. (Also, while schizophrenics are known to have a distorted sense of color that heightens their color perception, Vincent's color choices were deliberate exaggerations.) (480)

Firmly refuting claims that Vincent suffered any form of schizophrenia, University of Missouri biochemist Wilfred Niels Arnold claims ("Vincent van Gogh: Chemicals, Crises, and Creativity") that Vincent suffered from "acute intermittent porphyria," a form of liver enzyme deficiency, complicated by his abuse of absinthe. Wilfred Niels Arnold points out that the symptoms of this disease closely match many of Vincent's symptoms, including abdominal pain, delirium, seizures and paralysis. However, in truth, there is little to support it, including the fact that several of the symptoms, such as paralysis, actually don't fit Vincent, and the fact that urine turns red, something again never suggested in any of Vincent's correspondence.

Psychiatry professor Kay Redfield Jamison also denies Vincent had schizophrenia, and claims he was simply and obviously a "manic depressive," also referred to as "bipolar," his condition again complicated by absinthe abuse, a diagnosis R. E. Hemphill had put forth as long ago as

1961. A lot has happened since then to help confirm his views. Derek Fell writes:

> Psychiatric medicine and research in the past century have evolved a much better understanding of mental illness. The evidence in Vincent's case today would seem to indicate that he suffered from bipolar syndrome, or manic-depression, a mental illness characterized by alternating periods of euphoria and depression as well as by suicidal tendencies. Of course, in Vincent's day little was known about mental illness, and manic-depression had not yet been identified as a disease. In the end, the culmination of such intense emotional disappointment, combined with his implacable manic-depression, proved to be too much for Vincent to bear.
>
> The connection between creative genius and manic-depression has been well documented by biographers of numerous artists—among them, other Impressionist painters like Pissarro, Renoir, and Cezanne, all of whom sought treatment from Dr. Gachet. Add to them such notable literary figures as Lord Byron, Edgar Allan Poe, Ernest Hemingway, Winston Churchill, F. Scott Fitzgerald, and Virginia Woolf. (381)

The National Institute of Mental Health (NIMH) has an excellent online site dealing with the current knowledge of Bipolar Disorder or Manic-Depressive Illness. Here are some pertinent parts of it:

> Bipolar disorder, also known as manic-depressive illness, is a brain disorder that causes unusual shifts in a person's mood, energy, and ability to function. Different from the normal ups and downs that everyone goes through, the symptoms of bipolar disorder are severe. They can result in damaged relationships, poor job or school performance, and even suicide. . . . Like diabetes or heart disease, bipolar disorder is a long-term illness that must be carefully managed throughout a person's life.
>
> "Manic-depression distorts moods and thoughts, incites dreadful behaviors, destroys the basis of rational thought, and too often erodes the desire and will to live. It is an illness that is biological in its origins, yet one that feels psychological in the experience of it; an illness that is unique in conferring advantage and pleasure, yet one that brings in its wake almost unendurable suffering and, not infrequently, suicide." . . .
>
> Bipolar disorder causes dramatic mood swings—from overly "high" and/or irritable to sad and hopeless, and then back again, often with periods of normal mood in between. Severe changes in energy and behavior go along with these changes in mood. The periods of highs and lows are called *episodes* of mania and depression.
>
> Signs and symptoms of *mania* (or a *manic episode*) include: increased energy, activity, and restlessness, excessively "high," overly good, euphoric mood, extreme irritability, racing thoughts and talking very fast, jumping from one idea to another, distractibility, can't concentrate well,

little sleep needed, unrealistic beliefs in one's abilities and powers, poor judgment, spending sprees, a lasting period of behavior that is different from usual, increased sexual drive, abuse of drugs, particularly cocaine, alcohol, and sleeping medications, provocative, intrusive, or aggressive behavior, denial that anything is wrong, a manic episode is diagnosed if elevated mood occurs with three or more of the other symptoms most of the day, nearly every day, for 1 week or longer. If the mood is irritable, four additional symptoms must be present, signs and symptoms of *depression* (or a *depressive episode*) include: lasting sad, anxious, or empty mood, feelings of hopelessness or pessimism, feelings of guilt, worthlessness, or helplessness, loss of interest or pleasure in activities once enjoyed, including sex, decreased energy, a feeling of fatigue or of being "slowed down," difficulty concentrating, remembering, making decisions, restlessness or irritability, sleeping too much, or can't sleep, change in appetite and/or unintended weight loss or gain, chronic pain or other persistent bodily symptoms that are not caused by physical illness or injury, thoughts of death or suicide, or suicide attempts.

A depressive episode is diagnosed if five or more of these symptoms last most of the day, nearly every day, for a period of 2 weeks or longer.

A mild to moderate level of mania is called *hypomania*. Hypomania may feel good to the person who experiences it and may even be associated with good functioning and enhanced productivity. Thus even when family and friends learn to recognize the mood swings as possible bipolar disorder, the person may deny that anything is wrong. Without proper treatment, however, hypomania can become severe mania in some people or can switch into depression.

Sometimes, severe episodes of mania or depression include symptoms of *psychosis* (or psychotic symptoms). Common psychotic symptoms are hallucinations (hearing, seeing, or otherwise sensing the presence of things not actually there) and delusions (false, strongly held beliefs not influenced by logical reasoning or explained by a person's usual cultural concepts). Psychotic symptoms in bipolar disorder tend to reflect the extreme mood state at the time. For example, delusions of grandiosity, such as believing one is the President or has special powers or wealth, may occur during mania; delusions of guilt or worthlessness, such as believing that one is ruined and penniless or has committed some terrible crime, may appear during depression. People with bipolar disorder who have these symptoms are sometimes incorrectly diagnosed as having schizophrenia, another severe mental illness.

It may be helpful to think of the various mood states in bipolar disorder as a spectrum or continuous range. At one end is severe depression, above which is moderate depression and then mild low mood, which many people call "the blues" when it is short-lived but is termed "dysthymia" when it is chronic. Then there is normal or balanced mood, above which comes hypomania (mild to moderate mania), and then severe mania. . . .

In some people, however, symptoms of mania and depression may occur together in what is called a *mixed* bipolar state. Symptoms of a mixed state often include agitation, trouble sleeping, and a significant change in appetite, psychosis, and suicidal thinking. A person may have a very sad, hopeless mood while at the same time feeling extremely energized.

Bipolar disorder may appear to be a problem other than mental illness—for instance, alcohol or drug abuse, poor school or work performance, or strained interpersonal relationships. Such problems in fact may be signs of an underlying mood disorder. . . .

Like other mental illnesses, bipolar disorder cannot yet be identified physiologically—for example, through a blood test or a brain scan. Therefore, a diagnosis of bipolar disorder is made on the basis of symptoms, course of illness, and, when available, family history. The diagnostic criteria for bipolar disorder are described in the *Diagnostic and Statistical Manual for Mental Disorders, fourth edition (DSM-IV).*

Descriptions offered by people with bipolar disorder give valuable insights into the various mood states associated with the illness:

> *Depression:* I doubt completely my ability to do anything well. It seems as though my mind has slowed down and burned out to the point of being virtually useless.... [I am] haunt[ed]... with the total, the desperate hopelessness of it all.... Others say, "It's only temporary, it will pass, you will get over it," but of course they haven't any idea of how I feel, although they are certain they do. If I can't feel, move, think or care, then what on earth is the point?
>
> *Hypomania:* At first when I'm high, it's tremendous... ideas are fast... like shooting stars you follow until brighter ones appear.... All shyness disappears, the right words and gestures are suddenly there... uninteresting people, things become intensely interesting. Sensuality is pervasive, the desire to seduce and be seduced is irresistible. Your marrow is infused with unbelievable feelings of ease, power, well-being, omnipotence, euphoria... you can do anything... but, somewhere this changes.
>
> *Mania:* The fast ideas become too fast and there are far too many... overwhelming confusion replaces clarity... you stop keeping up with it—memory goes. Infectious humor ceases to amuse. Your friends become frightened.... everything is now against the grain... you are irritable, angry, frightened, uncontrollable, and trapped.
>
> *Suicide:* Some people with bipolar disorder become suicidal. . . . Signs and symptoms that may accompany suicidal feelings include: talking about feeling suicidal or wanting to die, feeling hopeless, that nothing will ever change or get better, feeling helpless, that nothing

> one does makes any difference, feeling like a burden to family and friends, abusing alcohol or drugs, putting affairs in order (e.g., organizing finances or giving away possessions to prepare for one's death), writing a suicide note, putting oneself in harm's way, or in situations where there is a danger of being killed. . . .

While some suicide attempts are carefully planned over time, others are impulsive acts that have not been well thought out. . . .

Episodes of mania and depression typically recur across the life span. Between episodes, most people with bipolar disorder are free of symptoms, but as many as one-third of people have some residual symptoms. A small percentage of people experience chronic unremitting symptoms despite treatment.

The classic form of the illness, which involves recurrent episodes of mania and depression, is called *bipolar I disorder*. Some people, however, never develop severe mania but instead experience milder episodes of hypomania that alternate with depression; this form of the illness is called *bipolar II disorder*. When four or more episodes of illness occur within a 12-month period, a person is said to have *rapid-cycling* bipolar disorder. Some people experience multiple episodes within a single week, or even within a single day. Rapid cycling tends to develop later in the course of illness and is more common among women than among men.

People with bipolar disorder can lead healthy and productive lives when the illness is effectively. Without treatment, however, the natural course of bipolar disorder tends to worsen. Over time a person may suffer more frequent (more rapid-cycling) and more severe manic and depressive episodes than those experienced when the illness first appeared.[4] But in most cases, proper treatment can help reduce the frequency and severity of episodes and can help people with bipolar disorder maintain good quality of life. . . .

Both children and adolescents can develop bipolar disorder. It is more likely to affect the children of parents who have the illness.

Unlike many adults with bipolar disorder, whose episodes tend to be more clearly defined, children and young adolescents with the illness often experience very fast mood swings between depression and mania many times within a day.[5] Children with mania are more likely to be irritable and prone to destructive tantrums than to be overly happy and elated. Mixed symptoms also are common in youths with bipolar disorder. Older adolescents who develop the illness may have more classic, adult-type episodes and symptoms.

Bipolar disorder in children and adolescents can be hard to tell apart from other problems that may occur in these age groups. For example, while irritability and aggressiveness can indicate bipolar disorder, they also can be symptoms of attention deficit hyperactivity disorder, conduct disorder, oppositional defiant disorder, or other types of mental disorders more common among adults such as major depression or schizophrenia. Drug abuse also may lead to such symptoms. . . .

Scientists are learning about the possible causes of bipolar disorder through several kinds of studies. Most scientists now agree that there is no single cause for bipolar disorder—rather, many factors act together to produce the illness.

Because bipolar disorder tends to run in families, researchers have been searching for specific genes—the microscopic "building blocks" of DNA inside all cells that influence how the body and mind work and grow—passed down through generations that may increase a person's chance of developing the illness. But genes are not the whole story. Studies of identical twins, who share all the same genes, indicate that both genes and other factors play a role in bipolar disorder. If bipolar disorder were caused entirely by genes, then the identical twin of someone with the illness would *always* develop the illness, and research has shown that this is not the case. But if one twin has bipolar disorder, the other twin is more likely to develop the illness than is another sibling.

In addition, findings from gene research suggest that bipolar disorder, like other mental illnesses, does not occur because of a single gene. It appears likely that many different genes act together, and in combination with other factors of the person or the person's environment, to cause bipolar disorder. . . .

Often people with bipolar disorder do not realize how impaired they are, or they blame their problems on some cause other than mental illness.

A person with bipolar disorder may need strong encouragement from family and friends to seek treatment. Family physicians can play an important role in providing referral to a mental health professional. (382)

In an interesting article, "Vincent Van Gogh from a Clinical Psycho-art Therapy Viewpoint," Heike E. Stucke agrees with this Bipolar view, and offers an overview of it, some application of it to some of Vincent's paintings, and quotes from various theorists connecting it to creativity, at one point writing:

Monroe (1991) pointed out that Vincent had a low threshold of tolerating unusual stressors and the lack of being independent e.g. the anxiety of losing Theo's support due to his impending marriage, the sexual dominance-submission conflict with Gauguin, and the impact of heavy drinking and poor diet made him more vulnerable to excessive neuronal discharges. When an individual repeatedly fails to achieve adapt ional ens, [ends?] there is often extreme uncertainty and doubt as to the validity of his or her goals and means of achieving them. His or her identity is strongly threatened.

Jamison Redfield (1993) suggests that the evidence includes the nature of van Gogh's psychiatric symptoms (extreme mood changes, including long periods of depression and extended episodes of highly active, volatile and exited states, altered sleep patterns, hyper religiosity, extreme irritability, visual and auditory hallucinations, violence, agitation and alcohol abuse), the age of onset of his symptoms (late adolescence, early twenties), his pre morbid personality, the cyclic nature of his attacks, which were interspersed with long periods of highly lucid functioning, the lack of intellectual deterioration over time, the increasing severity of his mood swings, the seasonal exacerbation's in his symptoms, and his quite remarkable family history of suicide and psychiatric illness as explained with the charts in the next two pages.

Characteristics of art in bipolar depression have not been studied systematically. Descriptions in the literature are generally congruent with the disorganized hyperactivity of manic behavior. Content described includes sexual symbols (Zimmermann and Garfinkle, 1942), and euphoric themes (Enachescu, 1971). Style is described as having wild, vivid, or hot colors (Dax, 1953; Enachescu, 1971; Plokker, 1965; Reitman, 1954) as well as a lack of color variety (Dax, 1953). Other stylistic characteristics are deterioration in composition (Plokker, 1965), carelessness (Dax, 1953), distorted lines (Reitman, 1954), scant detail and excitement (Zimmerman and Garfinkle, 1942), and indication of activity (Schube and Cowell, 1939).

Vincent's art exuded depression and mania. During the manic depressive stage he was able to paint 70 canvases in 70 days. He made those statements as "ideas for my work are coming to me in swarms" (Letter 535) or "Continual fever to work" (Letter 474), or "an extraordinary feverish energy" (Letter 544A), or, "...terrible lucidity..." (Letter 543), and finally "...the pictures come to me as in a dream" (Letter 543)

Ferguson (1973) noted in regard to the perception of color, such comments from maniacs are made as, "it was as if we were surrounded by a golden glow"; "I emerged alone in the radiant white light"; "it began to glow a dull purple which turned to a deep cherry and the heat of it was overwhelming"; "colors seemed to flow"; colors seemed to hold great and uncanny significance"; "fields had a kind of luminescence"; "whirling colors into color, angle into angle"; there was no glow in them but only revolution." (383)

Erwin van Meekeren shifts the diagnosis to another current category of psychosis, and argues that Vincent suffered *Borderline Personality Disorder* (BPD). The National Institute of Mental Health provides the following current views on this:

Borderline personality disorder (BPD) is a serious mental illness characterized by pervasive instability in moods, interpersonal relationships, self-image, and behavior. This instability often disrupts family and work life, long-term planning, and the individual's sense of self-identity.

Originally thought to be at the "borderline" of psychosis, people with BPD suffer from a disorder of emotion regulation. While less well known than schizophrenia or bipolar disorder (manic-depressive illness), BPD is more common, affecting 2 percent of adults, mostly young women. There is a high rate of self-injury without suicide intent, as well as a significant rate of suicide attempts and completed suicide in severe cases. Patients often need extensive mental health services, and account for 20 percent of psychiatric hospitalizations. Yet, with help, many improve over time and are eventually able to lead productive lives. . . .

While a person with depression or bipolar disorder typically endures the same mood for weeks, a person with BPD may experience intense bouts of anger, depression, and anxiety that may last only hours, or at most a day. These may be associated with episodes of impulsive aggression, self-injury, and drug or alcohol abuse. Distortions in cognition and sense of self can lead to frequent changes in long-term goals, career plans, jobs, friendships, gender identity, and values. Sometimes people with BPD view themselves as fundamentally bad, or unworthy. They may feel unfairly misunderstood or mistreated, bored, empty, and have little idea who they are. Such symptoms are most acute when people with BPD feel isolated and lacking in social support, and may result in frantic efforts to avoid being alone.

People with BPD often have highly unstable patterns of social relationships. While they can develop intense but stormy attachments, their attitudes towards family, friends, and loved ones may suddenly shift from idealization (great admiration and love) to devaluation (intense anger and dislike). Thus, they may form an immediate attachment and idealize the other person, but when a slight separation or conflict occurs, they switch unexpectedly to the other extreme and angrily accuse the other person of not caring for them at all. Even with family members, individuals with BPD are highly sensitive to rejection, reacting with anger and distress to such mild separations as a vacation, a business trip, or a sudden change in plans. These fears of abandonment seem to be related to difficulties feeling emotionally connected to important persons when they are physically absent, leaving the individual with BPD feeling lost and perhaps worthless. Suicide threats and attempts may occur along with anger at perceived abandonment and disappointments.

People with BPD exhibit other impulsive behaviors, such as excessive spending, binge eating and risky sex. BPD often occurs together with other psychiatric problems, particularly bipolar disorder, depression, anxiety disorders, substance abuse, and other personality disorders.

Although the cause of BPD is unknown, both environmental and genetic factors are thought to play a role in predisposing patients to BPD symptoms and traits. Studies show that many, but not all individuals with BPD report a history of abuse, neglect, or separation as young children. Forty to 71 percent of BPD patients report having been sexually abused, usually by a non-caregiver. Researchers believe that BPD results from a

> combination of individual vulnerability to environmental stress, neglect or abuse as young children, and a series of events that trigger the onset of the disorder as young adults. Adults with BPD are also considerably more likely to be the victim of violence, including rape and other crimes. This may result from both harmful environments as well as impulsivity and poor judgment in choosing partners and lifestyles.
>
> NIMH-funded neuroscience research is revealing brain mechanisms underlying the impulsivity, mood instability, aggression, anger, and negative emotion seen in BPD. Studies suggest that people predisposed to impulsive aggression have impaired regulation of the neural circuits that modulate emotion. The amygdala, a small almond-shaped structure deep inside the brain, is an important component of the circuit that regulates negative emotion. In response to signals from other brain centers indicating a perceived threat, it marshals fear and arousal. This might be more pronounced under the influence of drugs like alcohol, or stress. Areas in the front of the brain (pre-frontal area) act to dampen the activity of this circuit. Recent brain imaging studies show that individual differences in the ability to activate regions of the prefrontal cerebral cortex thought to be involved in inhibitory activity predict the ability to suppress negative emotion.
>
> Serotonin, norepinephrine and acetylcholine are among the chemical messengers in these circuits that play a role in the regulation of emotions, including sadness, anger, anxiety, and irritability. Drugs that enhance brain serotonin function may improve emotional symptoms in BPD. Likewise, mood-stabilizing drugs that are known to enhance the activity of GABA, the brain's major inhibitory neurotransmitter, may help people who experience BPD-like mood swings. Such brain-based vulnerabilities can be managed with help from behavioral interventions and medications, much like people manage susceptibility to diabetes or high blood pressure. (384)

Erwin van Meekeren's study is an interesting combination of strengths and weaknesses. He attempts to, as he says, "sketch" in Vincent's "personality" and offer enough background in terms of biography, environment, and other diagnosis so there are some general frames. Unfortunately, he tends to dismiss "other diagnostic considerations" too casually, for example dismissing schizophrenia by incorrectly saying "this possibility was dropped long ago by everybody," and syphilis by saying "no serious author is in favor of this diagnosis." Nevertheless he offers a good possibility in terms of the current state of neurology and psychology. This is his conclusion:

> To summarize, the conclusion of this book is that the primary diagnosis is borderline personality disorder, combined with an organic psycho syndrome in the last 18 months of Vincent's life which was expressed by psychotic and epileptic symptoms. A miltifactorial analysis which includes

> the factors acting upon a person's life gives a fuller picture. The abandonment by Gauguin can be considered to be the moment of dislocation, the point at which the unstable mental equilibrium—caused by family heredity, malnutrition, intoxication, exhaustion and personality pathology—was shifter. An organic psycho syndrome also arose, causing psychotic and epileptic phenomena. The progression of the disorders, the mental stress (social isolation, being a mental patient, bad outlook) and the problems surrounding Theo caused a downward spiral that terminated in suicide.
>
> Vincent's symptoms and the unstable thinking, feeling and acting during his adulthood make borderline personality disorder a likely primary diagnosis. Vincent's relatively good recovery between crises can probably be explained by the effects of him being admitted: rest, structure, better food and less intake of toxic substance. It is possible that he also felt safer there than in a society which did not understand of recognize him and even rejected him—albeit only for some time.
>
> A number of the behaviours and circumstances of his life must be considered in the light of the times in which he lived.
>
> Vincent had very strong personality traits at his disposal, such as an iron will and much endurance. His constitution must have been very strong. He was able to perform at a top level while undermining his body, sometimes to extremes. Vincent van Gogh had a unique gift: he was a great artist, a sensitive man. He could love only in his paintings. In daily life he remained an unquiet man, a "rolling stone that gathers no moss." (385)

Lab experiments based on theories of *Brain Localization* or *Cognitive Neuropsychological Models* are also providing insights:

Allen W. Snyder and his co-workers at Centre for the Mind, Australian National University, Canberra, Australia, have been and are continuing to explore what happens to their thinking process when the left fronto-temporal lobe in normal people is suppressed. He writes:

> The astonishing skills of savants have been suggested to be latent in everyone, but are not normally accessible without a rare form of brain impairment. We attempted to simulate such brain impairment in healthy people by directing low-frequency magnetic pulses into the left fronto-temporal lobe. Significant stylistic changes in drawing were facilitated by the magnetic pulses in 4 of our 11 participants. Some of these "facilitated" participants also displayed enhanced proofreading ability. Our conclusions are derived from eleven right-handed male university students, eight of whom underwent placebo [pretend] stimulation. We examined performance before, during and after exposure to the stimulation. (386)

In other words, the experiment was meant to test whether at least some forms of savant skills, of artistic genius are the result, not of some "extra" or "special" thinking ability, but rather of some suppression of normal thinking that allows the extraordinary forms of thinking to appear. If true, then this suggests we all have these abilities, and they can be produced in each of us by simply eliminating or drastically cutting down on other forms of thinking that inhibit them. In less precise but basically correct language, it means that by blocking left brain thinking, the form of thinking of the right brain is increased. This is similar to psychoanalytical claims that the form of thinking of the conscious suppresses the form of thinking of the unconscious (in other words, when our conscious minds go to sleep, our unconscious is freed up to emerge in the form of a dream). In even more informal terms, it's the idea that, if we work on some project or concept for a while, then take a break from it to allow our mind to rest, we will suddenly have the flash of insight that gives us the solution to the problem.

To test this theory, Snyder and his co-workers gave their subjects repetitive transcranial magnetic stimulation (rTMS) to see if savant-like skills would appear. Low frequency rTMS inhibits brain activity, creates "virtual lesions," duplicating the condition found in young autistic savants. Specifically, for these experiments, the left fronto-temporal lobe was repressed.

The tests focused on two abilities or tasks. The first was Drawing Ability. For this test, the participants completed two drawing tasks. For the first one, participants were asked to draw either a dog or a horse. One week later, they were asked to draw the second animal. A committee judged the initial drawings randomly, looking to see which were the best artistically, and did not come to any consensus.

Then participants were asked to repeat their drawings, some given only a placebo, some given differing amounts of magnetic stimulation. And a subsequent committee was asked to look for changes. What they found was that the participants given the placebo did not show any differences, but those given the stimulation demonstrated a dramatically changed style. Allan W. Snyder states:

> After 10 minutes of stimulation, participants N.R. and A.J. radically changed their schema for dogs from their initial two drawings before stimulation. The drawings completed before the stimulation were typical caricatures of an animal standing up and facing to the left. During and after stimulation, the style dramatically changed, with the drawings becoming more complex.
>
> The horses of D.C. [one of the participants] changed their direction becoming more life-like, even flamboyant, compared to the drawing he completed before stimulation. D. C. attended a placebo stimulation session the week later, and no style change was observed.

The second drawing task involved seeing a series of faces and then drawing them from memory. Snyder discusses participant R.Y.:

> During and after real stimulation, R. Y. changed his convention for drawing faces. In his placebo stimulation session, the week prior to the real test, and in the drawings completed before real stimulation a distinct schema is present in the drawing of R. Y. However, during and after real stimulation, this style changed and R. Y. became preoccupied by the details of eyes.

Furthermore, Snyder reports:

> Three of the four "facilitated" participants experienced altered psychological states after stimulation. For example, N.R. said he was more "alert" and "conscious of detail" and that we had "taught him how to draw doges." He wished he had been asked to "write an essay", something he previously disliked, because when stimulated he became acutely aware of detail in his surrounds. Furthermore, the drawings of these three participants had not reverted to their original convention 45 minutes after stimulation had ceased. It is possible that the altered psychological states persisted beyond this time frame or that the newly acquired schema was preserved once learnt under magnetic stimulation.

The second test involved proofreading in order to demonstrate that, unlike autistic savants, our propensity to impose meaning blocks us from seeing the details, the literal truths.

In this test, each participant was presented with a series of 10 proverbs on a computer screen. Each proverb was presented on the screen for two seconds, during which time the participant read the proverbs aloud. Two of the proverbs in the set contained an error where a word was duplicated. Following the proverbs, a paragraph appeared on the screen, which again the participant read aloud. The paragraph also contained two duplicated word errors as well as some spelling and grammatical errors. Participants were instructed to read all the text *including* the errors, i.e., to identify any errors they saw in the text.

This test was designed to demonstrate that, unlike autistic savants, our propensity to impose meaning and concept blocks our awareness for the detail making up the concept.

As with the drawing experiments, those given a placebo showed no improvement the second time this was done. However, those given stimulation did dramatically improve during the stimulation, but did poorly both before and after it.

Snyder's conclusion is that magnetic stimulation mimics certain kinds of autistic brain impairment by shutting down part of the left fronto-temporal lobe allowing other portions of the brain to function at a higher level.

The results of a subsequent experiment resulted in the following comment in *Discover: Science Technology and the Future*:

> What if you could become a savant for a few minutes? Australian National University neuroscientist Allan Snyder says you can. Snyder induced instantaneous "jelly-bean counting" skills in normal people using transcranial magnetic stimulation. He used the powerful magnetic pulses to zap the brains of 12 volunteers, disrupting activity in the left anterior frontal lobe, an area where damage is known to cause sudden savant syndrome. For an hour or so after treatment, 10 of the volunteers were twice as accurate in guessing at a glance the number of dots flashed on a computer screen. "This is an extraordinarily good result—I'm amazed it came out so well," Snyder says.
>
> When Snyder used the same technique three years ago to induce episodes of savant like artistic ability in ordinary people, skeptics objected to his claims, saying that artistic abilities were too subjective to judge. But numbers don't lie, says Eric Wassermann of the National Institute of Neurological Disorders and Stroke. "There's nothing that looks bogus about it." (387)

Perhaps Vincent was an autistic savant, perhaps genius is the result of blocking restricting forms of brain activity, a form of subtraction rather than of addition.

There is also a possible fit here in terms of his genius appearing later in life. As much as researchers want to find it in him from the beginning, his early art work does not demonstrate anything special. Even his better young adult works would not be valued today if it weren't for what came later. Whatever happened to him, whatever changed him is debatable, as the previous discussions demonstrate. But something did. And it happened during times he was being exposed to numerous chemicals and diseases, any of which might have, almost certainly did, change his mental make-up. Furthermore, it is as likely, even more likely, that the change was a result of a subtraction than an addition to his mental processes.

Part XVI

Wheatfield with Crows

Starry Night is Vincent's most well known and popular painting. However, close behind it, perhaps even more attractive and enigmatic to the critics, is *Wheatfield with Crows* (has several titles, including *Crows over a Wheatfield, Crows in Wheatfield,* and even *Crows on the Cornfield*). Though there is no proof, it is often said to be his final painting. Whether or not it was the final one, it was one of his final paintings done shortly before he committed suicide, and it is a brilliant expression both foreshadowing the suicide and giving it a wonderfully complex expression filled with meaning.

The green streams of three roads carry life to into the sky. Three symbolizes a totality, if one wants to use the Christian representative of it, the Trinity. This works well, as the sky is the spiritual world, the cosmic ocean, represented in energetic strokes of blue, black, and holding two blue-gray-white clouds, possibility symbolizing Vincent himself and perhaps his brother, both about to entire this world beyond the hard dry earth of physical existence. It's interesting that the flock of crows is mainly going together in one direction, but the one crow caught in one of the clouds seems to be headed off on its own, which could symbolize many things, perhaps Vincent himself. For Freud, three symbolized the family, father, mother, and child, and so much has been written about Vincent's relationship with his parents, that the fact the three roads are diverging might suggest feelings of a family not all on the same path. The yellow-green fields of wheat are the fiery yellow of violent transformation. Both filled with life, growing, living, nurturing nature, the world beyond the conscious control of the city and civilization, and the always positive connections to the sun that Vincent saw as representing love, and yet also consuming it in a rushing rage, this is the unstoppable movement to the ocean of the sky. The field of wheat and the sky are the two worlds, visible and invisible, colliding. The black birds are the birds of knowledge and of death. They are the Jungian Shadow circling, similar to the blackbirds of Alfred Hitchcock's movie *The Birds*, frightening awakened Shadows symbolizing the dark side of the universal human psyche entering into and threatening the daylight world of consciousness, tricksters to overturn sanity and bring about the creative turmoil of insanity, and at the same time messengers of the Savior descending upon the earth, the one meant to take into himself all of the suffering, to free the world of sin and offer hope for a better life in the next world.

Pierre Cabanne says that *The Wheat Field with Crows* is a exactly this, a world where art and psychosis are inseparable, the movement of the corn and that of the crows oppose and contradict each other, rising in one direction, sinking in the other, creating the dual rhythm of the painter's psychopathic manifestation. (388)

Yes, yes, yes. I cannot know for sure how much Pierre Cabanne realized the extensions of what he said, but they are implied. The invisible world of the mind has become visible, and Vincent's own manifestation is the universal human manifestation, just as Carl Jung would claim. In it, the artist has reached into the depths of the human unconscious and brought them to light, has given all of us the expression of that dark, nightmarish world of the Shadow, and at the same time has shown us how it is actually the world of the Savior.

Frank Elgar asserted that a great poet, also on the threshold of madness, had attached a poignant significance to the crows, expressed in the "Nevermore" of Edgar Allan Poe's *The Raven*. In Vincent's painting, crows fly across corn tinged with red towards a lowering sky, inscribing it with omens of coming disaster, that dark side that has always been near the surface in Vincent's life, but has now burst through the membrane between the conscious world and the unconscious world of the dream and the spirit. (389)

One of Vincent's favorite authors, Jules Michelet wrote in a book appropriately titled *The Bird* that included the following description of crows:

> They interest themselves in everything, and observe everything. The ancients, who lived far more completely than ourselves in and with nature, found it no small profit to follow, in a hundred obscure things where human experience as yet affords no light, the directions of so prudent and sage a bird. (390)

Supposedly, the gun Vincent used to shoot himself he borrowed on the ploy that he needed it to scare away the crows that were bothering him as he painted. Just how closely he connected such actions to the symbolism is impossible to know, but the connections come pouring out—the intertwining strands suggesting he was unconsciously and probably consciously living out (or rather ending) his life in line with what had become for him a strongly symbolic existence. Heike E. Stucke, in pushing us to see the painting as foreshadowing his death and his views on it, highlights that the unfinished letter that was found on him included the line, "And truly, we can only make our paintings speak," the suggestion being that we need to "read" the painting, that a symbolic expression is consciously being created. (391)

Ingo F. Walther notes that the normal geometric perspective is inverted, the three separating roads come to their focus in front of the painting rather than in the depth of the painting. All he gains from this is that Vincent was expressing his loneliness and a sorrow, but what is more important is that it means Vincent is seeing a larger world beyond the one he inhabits physically, the world of the mind or spirit, which literally flows off the edges of the painting. (392)

Meyer Schapiro is more in line with this, describing the scene as "a field opening out from the foreground by way of three diverging paths. . . . [T]he lines, like rushing streams, converge toward the foreground from the horizon, as if space had suddenly lost its focus and all things turned aggressively upon the beholder. . . . [T]he great shining sun has broken up into a dark scattered mass. . . . [T]he crows unite] in one transverse movement the contrary directions of the human paths and the symbols of death. . . . It is as if [the artist] . . . saw an ominous fate approaching. The painter-spectator has become the object, terrified and divided, of the oncoming crows whose zigzag form, we have seen, recurs in the diverging lines of the three roads . . ." (393)

Humberto Nagera knows Meyer Schapiro's work, but disagrees, "Only his last work 'Crows in the Field' painted shortly before his suicide has signs of a loss of command and control; the composition and perspective are not right, according to Van Gogh's previous style, and there are two suns in the sky, but in spite of this the painting still makes a powerful impression. (394) As I have been saying throughout, Humberto Nagera is uneven, has some truly valuable insights, but sometimes is simply lost.

His close follower, Lubin, after quoting Schapiro's description, which he says sees it as a gloomy prediction of Vincent's suicide, writes:

> Professor Schapiro noted that the mood of despair in the painting was also present in the letter in which Vincent mentions it, but the despair is the voice of Christ. Vincent began the letter by telling his sister-in-law that a letter she had sent him was "like a gospel." A deliverance from the anguish which has been caused by the hours I had shared with you." [Letter 649] Christ as delivered from his anguish—the Agony in the Garden—by the Crucifixion. So we may guess that the painting depicts Vincent's own fantasized Crucifixion in the image of Christ. The two lateral paths represent the horizontal arms of the Cross and the central path the upper portion of the vertical axis of the Cross. The unseen head of the painter-Christ, lying at the convergence of the horizontal and vertical axis outside the painting, is looking toward heaven. Vincent had put himself in a similar position outside the canvas in earlier works, for instance in *The Cradle*,

where the beloved child-artist is in an unseen cradle, and in his modified version of Rembrandt's *The Raising of Lazarus*. The "broken-up" sun noted by Professor Schapiro is Vincent's "good god sun," the God who had momentarily abandoned him.

Seen in this light, *Crows over the Wheatfield* recalls that moment during the Crucifixion in which "there was darkness over all the land," and Christ cried out, "My God, my God, why hast thou forsaken me?" The painting remains a gloomy warning of impending doom. But as a Crucifixion, it served a more important, wish-fulfilling function for Vincent. It heralded the joyful rebirth and the welcomed ascension into heaven of God's favorite child. Like the gloomy aspects of the sermon themes, it was tolerable—indeed, it was welcomed—because of Vincent's conviction that only through suffering would he finally be beloved of God.

The painting seems not only to converge onto the spectator but to draw him into the sky as well. Students have debated whether the crows are flying toward the observer or toward the sky. Seen as a crucifixion, this apparent confusion is clarified. For the focus in the painful yet hopeful even t must remain divided between the death of the martyr on the cross and the everlasting joy that awaits him and all who believe in him. The crows descend to seal his fate, but they also accompany him to his eternity in heaven.

At last Vincent has made up for his mother's neglect. In dying as God's favorite child, he has finally outdone the first Vincent. (395)

Again, there is an uneven quality, and the concluding sentence really undermines the value of the painting, just as its basis in the Replacement Child Syndrome undermines his and Nagera's books in general. However, the whole connection with Christ works. This is the key, and it leads us directly back to the discussions in the Richard Dadd section about the artist as savior.

Graetz writes:

In those days he wrote his last letter to his mother and sister: "As for me, I am entirely absorbed by the immense plain with wheat fields against the hills, large as a sea, fine yellow, fine tender green, fine purple of a plowed and weeded piece of ground, regularly checkered by the green of flowering potato plants, everything under a sky with fine blue, white, pink, violet tones. I am entirely in a mood of an almost too great calmness, in a mood to paint this." (Letter 650)

The *almost too great calmness* was like a hint of a calm before a storm. It didn't take long in coming. We see the storm break in "Wheatfield with Crows." A billowing field of wheat in bright and chrome yellow rises mountain high to both sides of a path in the center, which runs from the foreground through the field until it is submerged under the spreading wheat. Two other grass-grown tracks turn to the sides, where they are cut off by the edges of the canvas. Large jet black birds are swarming over the field. Two light clouds stand out against the deep, almost black blue of a storm-laden sky. The line of the horizon is windswept and flooded by the mounting waves of the blazing wheat, *large as a sea.* The path in the center—as if to symbolize Vincent's own road—has come to a dead end, barred by the fire which has taken possession of the whole field.

The dark silhouettes of the crows were preceded by the loose branches above the house in the "Cottage at Auvers." In their movement against the sky they recalled the gesticulating branches of the tree which, since his earliest drawings, had symbolized Vincent's struggle. They are now completely detached and have become identical with the birds, traditional messengers, which were nearly always somber in his painting like the thoughts they carried. The few little birds we once saw in the "Sower", the counterpart to the wheat fields, have become the stormy swarm of the huge crows, sinister and foreboding in their intense blackness. In the center of that field was a large furrow. Its position under the enlivening morning sun resembles that of the path, again directly below the light cloud and in the center of this last of his wheat fields. In the earlier painting the furrow had been dug to prepare the earth for the sower's task and came towards us together with the striding man. Now the path goes away and disappears form our eyes.

Vincent painted the "Wheatfield with Crows" in the very fields where, it is reported, he put an end to his life. His leitmotiv of struggle, seen at the outset of his work in the little drawing "En Route," is brought to a climax here. He is at the end of his road and ceases his work in fire and storm. The glimmer of the miner's lamp, the gay light of the blossoming orchards in the spring of Provence, the glow of the summer sun in Arles and St. Remy have burst their bounds and become a huge wild fire. The overripe wheat is the prey of the raging blaze and the low-flying crows sharply accentuate the *image of death.*

His drive to the sun ends in his own fire, which consumes him as it overwhelms and burns his *image of humanity.* A letter which he had written to Theo years back, in 1882, comes to our mind: "*I feel the power in me which I must develop, a fire that I may not quench, but must keep ablaze, though I do not know to what end it will lead me, and shouldn't be surprised if it were a somber one. In times like these, what should one wish for? What is relatively the happiest fate? In some circumstances it is better to be the vanquished than to be the victor—for instance, better to be Prometheus than Jupiter. Well, it is an old saying, Let comet what may.*" It was like a premonition, as was the assumption he made in the following year that *his body would hold out a certain number of years, between six and ten for instance.* He then commented further on his working capacity: "*If one consumes oneself very much in those year one does not get over 40 . . . My intention is not to spare myself, nor to avoid any emotios or difficulties—for it is relatively immaterial to me whether I live longer or shorter . . .*"

> In the "Wheatfield with Crows" another presentiment would seem to have come true, one which Vincent had felt at the time when he painted a cemetery in Nuenen. He then wanted to express in that work "*. . . how very simple death and burial are, as natural as the fall of an autumn leaf,*" and further, *"The fields around where the grass of the cemetery ends, beyond the little wall—they make a last line against the horizon—like a horizon of a sea."* In Auvers he again compares the fields to *a sea, the fields beyond the little wall,* where he and Theo were to be put to rest.
>
> Yet in this final painting too, is the vision of resurrection. Vincent is no longer portrayed in his earthly existence—his soul rises to the sky, symbolized by the light cloud just above the road which has come to its end. The two clouds bring to mind the two clouds in the drawing of the "Starry Night," in which pillars of smoke rose from the earth to the sky to flow into the interlocked cloud. Earlier, in the "Self-Portrait with the Bandaged Ear" spirals of smoke and breath had risen above the horizon. As the detached tree branches have become swarming birds, so have the rising spirals—symbols of his spiritual force—left the earth and become the light clouds in the sky.
>
> The death of Theo a few months later is as if foretold by the second, more distant cloud; the crow within its contours enhances the ominous impact. The two clouds may also recall the "Sunflowers," where the two falling flowers had conveyed Vincent's awareness—if not presentiment—of death. By stressing in his last letters to Theo *their fragile existence, his life too being attacked at the very root, his step too being unsteady* he implied that he was afflicted as was Theo. (396)

This is the right idea, though I would not say it expresses his own road coming to an end, but rather his own road leading straight into the painting, and the world of the painting has become one with the literal world. The painting, indeed, IS the connecting door between the two worlds. He has realized this for many years, has spent his life trying to get other people to understand his realization, and now is connecting them dramatically in his suicide. He is going into that frightening world beyond conscious, literal existence, that world of the spirit.

Cliff Edwards in his determined effort to put a *spiritual* spin, and especially a Buddhist spin on all of Vincent's work, has the right idea. He stresses that Vincent saw the *Wheatfield* as both a literal place and a symbolic place where literal human existence comes together with the *eternal*:

> It is in the presence of "death," in the shared center of the impermanence of all things that wheat and humanity interpenetrate. Just as Vincent was fascinated by the potato eaters who planted, dug, and ate potatoes on the same earth where they were born and where they would one day be planted, so he was moved by the resonance between humans and wheat, in

> their seedtime and harvest, their deep patterns of cooperative sympathy in the field, the gathered sheaves, the loaf of bread. When searching for an example of truly symbolic, sacred art, Vincent had rejected Bernard's *Adoration of the Magi* in favor of Millet's painting of peasants carrying a new-born calf to their home, where mother and children stood waiting to care for it. Vincent, in *Wheatfield with Crows,* provided a similar image of deep religious import, the impermanence of wheat and humanity which is both our shared fragility and our shared Eternity. The transitoriness of things is intensified in Vincent's work through the quickly passing summer storm, the crows disappearing into the sky, and the wheat so soon to be stacked in sheaves and ground for bread or stored for a new seedtime and harvest. The painting itself enters the mode of being of all thing in their impermanence yet transformation, becoming a koan that poses the Zen Master's question: "If you call this wheat, you cling to it; it you do not call it wheat you depart from the facts. So what do you call it then?" The wheat of Vincent's painting is wheat "in such a way as to involve at the same time a deliverance of the self." The impermanence of humanity and wheat is freedom, creativity, the Eternal. But in the ambiguity of time, those who have not been transformed by the experience may well find only violence, foreboding, and chaos. (397)

It is a painting of a place of transformation, and of a coming together of two worlds, one the "impermanent world" of the body, and the other "eternity." Earlier in his book, claiming that Western theories are limiting, Cliff Edwards clarifies his views in terms of Eastern Buddhist beliefs, particularly those of Keiji Nishitani, an expert in the Kyoto School of philosophy that began in Japan with Kitaro Nishida in the 1800s. Opposing what he considers a subject-object dualist reduction of reality with humans at the center, as the object, whether it is seen through *reason* or *sensation* is nothing more than *how it is understood by human consciousness*. This is one of those claims that suggests an interesting perspective but is also in danger of misrepresenting itself. For example, Carl Jung, whom he lists as one of the Western thinkers, and in truth the entire field of psychology, stresses the opposite of this, stresses the reality beyond human consciousness. He does not include Carl Jung in his immediate list of thinkers here, Kierkegaard, Nietzsche, Sartre, and Heidegger, but rather dismisses Carl Jung because his "focus on psychic depth almost to the exclusion of the reality of nature finally limit[s] … [his] usefulness." (398) Interestingly, Jung strongly embraced this same world of Eastern Buddhism and stressed its concepts and symbols in his explanations. While all theoretical systems, all explanations are limited, psychology admits this; in fact, it stresses it and gives Western thought a doorway beyond it. This is true also of all of the theories that are often categorized under *romanticism*, which posits a truth beyond explanation, including those of the four mentioned above. Still, I like the theory he is about to put forth here, just wish he had seen the interpenetrations of it with those he contrasts to it. According to Cliff Edwards, Keiji Nishitani replaces the Western reduction of the world through reason and rationality to nihilism or emptiness with *sunyata*, absolute emptiness, which moves beyond relative emptiness into a world without limitations, one that includes both nothingness and being, one where illusion ("the impermanent nature of all things") is known, "yet where the center of all things is everywhere and all things interpenetrate."

Earlier I called this the *shamanic realization*, connecting it clearly to the experience of Black Elk. Cliff Edwards immediately connects his claims to the following statement by Keiji Nishitani:

> That is to say, on the field of sunyata, *the center is everywhere.* Each thing in its own selfness shows the mode of being of the center of all things. Each and every thing becomes the center of all things and, in that sense, becomes an absolute center. This is the absolute uniqueness of things, their reality. (399)

This cannot be more exactly what Black Elk gave us. Edwards elaborates, "Each thing in its uniqueness, or, in Buddhist terminology, in its *tathata,* ("suchness") interpenetrates all other things, and finds its center not in itself in isolation, but in absolute emptiness or sunyata where the center has no limits." (400) Furthermore, both Black Elk and Nishitani are saying in their own references that each center interpenetrates with all other centers, that in effect each center intertwines creating a huge web of existence with each other center.

Edwards the offers the following wonderful quote from Nishitani:

> Goethe says that things that will pass are metaphors of the Eternal. . . . yet so long as there is nothing like an eternal thing to serve as its archetype, the metaphor as such is the primal reality or fact. It is metaphor even as primal fact, and primal fact even as metaphor. A Zen master extends his staff and says: "If you call this a staff you cling to it; if you do not call it a staff you depart from the facts. So what should you call it then?" . . . The fact that the staff is this staff is a fact in such a way as to involve at the same time a deliverance of the self. In this the fact appears as a primal factuality. The point at which this fact can be comprehended in a primal manner is the point of deliverance where one becomes a Son of God, a Son of Buddha." (401)

Edwards then goes into an explanation of Nishitani's stress that what we must do is to not just observe existence but enter it. Using a haiku about a pine tree by Basho, Nishitani explains: "He does not simply mean that we should 'observe the pine tree careful.' Still less does he mean for us to 'study the pine tree scientifically.' He means for us to enter into the mode of the being where the pine tree is the pine tree itself. . . He calls on us to betake ourselves to the dimension where things become manifest in their suchness . . . the selfness of the bamboo." (402)

This is very similar to the views of the Romantic Movement out of England in the late 1700s, where the poet is said to be able to see beyond the physical object to its essence, and then to reveal that to others, to in a sense lead others to also partake of the essence beyond the literal. Thus, art all forms of expression are the windows to the higher truths, Picasso's lies that reveal the truth.

The Buddhist term for reaching such a consciousness is *samadhi*, and this is what Edwards claims Vincent experienced, and he quotes numerous passages from Vincent's letters revealing how Vincent tried to explain this, including "Come now, isn't it almost a true religion which these simple Japanese teach us, who live in nature as though they themselves were flowers? (Letter 542)" and "I have a terrible lucidity at moments, these days when nature is so beautiful, I am not conscious of myself any more, and the picture comes to me as in a dream. (Letter 543)"

Now Nishitani, and Edwards through him, move into the two realms of time, the one rectilinear (that continuous ticking of time on a line), the other circular or simultaneous (the eternity of now).

Nishitani explains:

> The idea of a stratified formation of simultaneous time systems necessitates the idea of an infinite openness at the bottom of time, like a great expanse of vast, sky like emptiness that cannot be confined to any systematic enclosure. Having such an openness at its bottom, each and every now, even as it belongs to each of the various layers accumulated through the total time system, is itself something new and admits of no repetition in any sense. The sequence of "nows" is really irreversible. Accordingly, in the true sense, each now passes away and comes into being at each fleeting instant. It is, in other words, something *impermanent* in the fullest sense of the word. As such a succession of nows with an infinite openness beneath it, time must be conceived of without beginning of end. Conversely, only when so conceived is it possible for every now to come about as a new now and as impermanent. Moreover, in time this newness and impermanency are tied to one another inseparable. (403)

Though Edwards doesn't go into it, this unity of the *linear time* (*impermanence)* and *eternity* is represented in the image of Shiva (Siva), a form of *Ishvara* (God) in later Vedic scriptures of Hinduism. In some views he is one of the *Trimurit* (Hindu trinity), where Brahma is the creator, Vishnu the preserver, and Shiva the destroyer (of Evil), and destruction is seen as positive, as it is followed by creation. One interpretation of this has Shiva creating Vishnu, who in turn creates Brahma, and thus creation begins. Beyond his role in the eternal cycle of the coming into being, being, and ending of being, Shiva takes on many other roles, or inflections of this role. Most important here is his representation as *Shiva Nataraga* (in Sanskrit *nata* means dance, and *raja* means king).

The most well known story of Shiva's beginnings as the *Nataraga* is set in a dense forest in South India where dwelt a number of heretical sagas. Shiva and Vishnu, disguised as a beautiful woman, traveled there to confront them. At first the sagas were led to a violent dispute among themselves, but then they turned their anger toward Shiva and attempted to destroy him with incantations. First they created a ferocious tiger in the sacrificial fires, and it rushed at him. But Shiva seized it and with the nail of his smallest finger stripped off its skin and wrapped it about him like a silken garment.

Then they created a monstrous serpent. However, Shiva seized it and wrapped it about his neck like a scarf.

Then he began to dance. A final monster in the form of an evil dwarf rushed toward him. But Shiva stepped on him and broke his back, leaving him writhing on the ground, and continued the dance.

Shiva, the arch-yogi of the gods, then, is also the god of the dance, and dancing is the art form where the artist, the dancer, and the art created are one and the same, resurrecting the oneness of God and Creation. Dance induces a trance, same as yoga, where the dancer experiences ecstasy and the divine. Thus, the dance, seemingly in opposition to it, goes hand-in-hand with the austerities of yoga meditation (fasting, introversion, and so on).

The symbolism of Shiva's dance is intricate and details out a complex mythology. Shiva is depicted with four arms, representing the four cardinal directions. Since hands are the messengers of the heart or soul, each hand has strong symbolic qualities. The upper right hand holds an hourglass drum, the symbol of creation, the ticking of time in the sensual universe (Kant's forms of sensibility). It is the pulse of the universe, the beat of Shiva's dance, and sound is considered the first and most pervasive element in the unfolding universe. In fact, when Shiva gave Panini (the

Sanskrit grammarian), wisdom, the entire system was included in this drumbeat. The grammar is called Shiva sutra.

The hourglass drum also symbolizes male and female principles (Jung's animus and anima). The two triangles penetrate each other to form a hexagon. When they part, the universe dissolves. Thus, only in the unity of male and female principles does sensate existence take place.

The opposite hand, the upper left hand, holds a tongue of flames. Fire is the element that destroys the universe (it is the symbol of violent transformation). For Hindus, the end of the world will take place through fire.

Thus, these two hands represent creation and destruction, sound opposed by flames, birth and death.

The second right hand with its open palm presents the *abhaya* pose (fearless), a gesture of protection, the idea of god as protector.

The left leg rises towards and across the right leg, and the left hand stretches across the body and points to the left leg in imitation of an elephant's trunk, symbolic of Ganesha, Shiva's son (the remover of obstacles), and symbolizing a release from birth and death.

Depicting the original story, Shiva dances on the body of the dwarf *Apasmarapurusha* (the man of forgetfulness), who represents indifference, ignorance and laziness, thus representing that creation and all creative energy is possible only when inertia is overcome and suppressed. Thus, each person must overcome the negative qualities of the dwarf to participant in the energy of the cosmos.

The ring of fire and light circles the dance and identifies it with the dance of the universe. The lotus pedestal serves as the floor of the dance and locates it in each person's individual consciousness.

Here we have a dance representing the paradox of time and eternity. The energetic arms and legs are contrasted to the immobile mask of the face, serine and silent. Shiva is both *Kala* (time) and *Maha Kala* (eternity).

His long, sensuous hair is loosened and waving in a violent frenzy to the energetic dance, embracing the world of time and space. Hair is of the world of growth and life, and such wild, untouched hair represents the power of the body (same as is expressed in the Christian story of Samson). This is also strongly connected to the woman as giver of life. When she lets down her hair she invites the act of creation. Shaving one's hair simulates sterility, and is symbolic of the life of asceticism, a life beyond the seasons of growth and change. (404)

Joseph Campbell discusses this:

> There is a very important Buddha figure [Shiva] who is shown holding a flaming sword over his head—and so what is that sword for? It is the

sword of discrimination, separating the merely temporal from the eternal. It is the sword distinguishing that which is enduring from that which is merely passing. The tick-tick-tick of time shuts out eternity. We live in this field of time. But what is reflected in this field is an eternal principle made manifest. . . .

Shiva's dance is the universe. In his hair is a skull and a new moon, death and rebirth at the same moment, the moment of becoming. In one hand he has a little drum that goes tick-tick-tick. That is the drum of time, the tick of time which shuts out the knowledge of eternity. We are enclosed in time. But in Shiva's opposite hand there is the flame which burns away the veil of time and opens our minds to eternity. . . .

In some of his manifestations he is a really horrendous got, representing the terrific aspects of the nature of being. His is the archetypal yogi, canceling the illusion of life, but he is also the creator of life, its generator, as well as illuminator. (405)

This commingling of the two aspects of time and existence gets a Christian spin in the *Round Dance*. Elaine Pagels explains:

In the Round Dance, which is found in the *Acts of John*, a second-century collection of stories and traditions inspired by John's gospel, John begins the story of Jesus' final night where the gospel account leaves off, and says that Jesus invited his disciples to dance and sing with him:

"Before he was arrested . . . he assembled us all, and said, 'Before I am delivered to them, let us sing a hymn to the Father, and so go to meet what lies before us." So he told us to form a circle, holding one another's hands, and he himself stood in the middle and sait, "Answer Amen to me."

Then, as the disciples circled him, dancing, Jesus began to chant a hymn in words that echo the Gospel of John:

"Glory to you, Father." And we, circling around him, answered him, "Amen."

"Glory to you, Logos; glory to you, Grace." "Amen."

"Glory to you, Spirit; glory to you, Holy One. . . ." "Amen."

"We praise you, Father; we thank you, Light, in whom dwells no darkness." "Amen."

"I am a light to you who see me." "Amen."

"I am a mirror to you who know me." "Amen."

"I am a door to you who knock upon me." "Amen."

"I am a way to you, the traveler." "Amen."

Although the phrase about the mirror could have come straight from the *Gospel of Thomas*, the primary source for the last two, as well as many of the others, is the *Gospel of John*.

Whoever composed this hymn, then, clearly found in John's gospel inspiration for the kind of teaching we more often associate with Thomas; for here Jesus invites his disciples to see themselves in him:

> "*[W]hich I am about to suffer is your own.* For you could by no means have understood what you suffer, unless I had been sent to you as word [logos] by the Father . . . if you knew how to suffer, you would be able not to suffer."
>
> Thus, in the Round Dance of the Cross, Jesus says that he suffers in order to reveal the nature of human suffering, and to teach the paradox that Buddha also taught: that those who become aware of suffering simultaneously find release from it. Yet he also tells them to join in the cosmic dance: "'Whoever dances belongs to the whole'. Amen. 'Whoever does not dance does not know what happens.' 'Amen.'"
>
> Those who love the Acts of John apparently celebrated the Eucharist by chanting these words, holding hands, and circling in this dance to celebrate together the mystery of Jesus' suffering, and their own—and some Christians celebrate it thus to this day. (406)

Joseph Campbell elaborates on this dance as a celebration, the celebration of the transformation from this world into the world beyond, and equates it with an Iroquois ritual:

> Moyers: Do you think it is true that he who loses his life gains his life?
>
> Campbell: That is what Jesus says.
>
> Moyers: Do you believe it is true?
>
> Campbell: I do—if you lost it in the name of something. There is a report by the seventeenth-century Jesuit missionaries in eastern Canada of a young Iroquois brave who has just been captured by an enemy tribe. He is being brought to be tortured to death. The Northeastern Indians had a custom of systematic torture of their male captives. The ordeal was to be suffered without flinching. That was the final test of real manhood. And so this young Iroquois is being brought in to endure this horrible ordeal; but, to the Jesuits' amazement, it is as though they were his welcoming hosts and he their honored guest. And he is playing the game along with them, knowing all the while to what end he is being conducted. The French priests describing the occasion are simply appalled by what they interpret as the heartless mockery of such a reception, characterizing the youth's captors as a company of savage brutes. But no! Those people were to be the young brave's sacrificial priests. This was to be a sacrifice of the altar and, by analogy, that boy was the like of Jesus. The French priests themselves, every day, were celebrating Mass, which is a replication of the brutal sacrifice of the cross.
>
> There is an equivalent scene described in the apocryphal Christian Acts of John, immediately before Jesus goes to be crucified. This is one of the most moving passages in Christian literature. In the Matthew, Mark, Luke and John gospels, it is simply mentioned that, at the conclusion of the celebration of the Last Supper, Jesus and his disciples sang a hymn before he went forth. But in the Acts of John, we have a word-for-word account of the whole singing of the hymn. Just before going out into the garden at

> the end of the Last Supper, Jesus says to the company, "Let us dance!" And they all hold hands in a circle, and as they circle around him, Jesus sings, "Glory be to thee, Father!"
>
> To which the circling company responds, "Amen."
>
> "Glory be to thee, Word!"
>
> And again, "Amen."
>
> "I would be born and I would bear!"
>
> "Amen."
>
> "I would eat and I would be eaten!"
>
> "Amen."
>
> "Thou that dancest, see what I do, for thine is this passion of the manhood, which I am about to suffer!"
>
> "Amen."
>
> "I would flee and I would stay!"
>
> "Amen."
>
> "I would be united and I would unite!"
>
> "Amen."
>
> "A door am I to thee that knocketh at me. . . . A way am I to thee, a wayfarer."
>
> And when the dance is ended, he walks out into the garden to be taken and crucified.
>
> When you go to your death that way, as a god, in the knowledge of the myth, you are going to your eternal life.
>
> So what is there in that to be sad about? Let us make it magnificent—as it is. Let us celebrate it."
>
> Moyers: The god of death is the lord of the dance." (407)

With the Buddhist unity of the eternal and the impermanent in place Edwards then goes into a discussion of how the views of other critics about *Wheatfield with Crows* are wrong to stress how it is a "doom-filled painting with threatening skies and the ill-omened crows," "sinister and foreboding," and "image of death" where "space had suddenly lost its focus and all things turned aggressively upon the beholder." He concludes his argument:

> For Vincent, a great unifying takes place at that point where shared impermance and the Eternal are one, and anguish and joy, life and death find their meaning together. It is interesting that the critics who cannot imagine peace and restoration in a scene with so much energy of color and stroke seek to find an escape from the impermanence of the wheat into some kind of resurrection or vague merging of things in the sky. For Vincent, motion and stillness, the impermanent and the Eternal, unify wheat and sky, and either of them is both reality and symbol of our gateway to salvation, the deep roots from which meaning wells up, the sun and stars from which the "ray from on high" visits us. (408)

What Edwards needed to dig into more is not that the scene is *not* a terrifying one, one filled with the frightening side of Shiva or dangers of the Shadow represented in the frenetic crows and the flaming wheat and dark energy of the sky, the swirling, energetic *dance of life and death*, or the coming together of the *eternal* and the *impermanent*, but that he was facing that Shadow, was incorporating it, was about to walk into that reality. He needed to do it. He wanted to do it. But it is not done without awareness of the dark side. There is no calmness here. It is a violent encounter.

This has been described as an experiencing of the *sublime*, where the overwhelming monstrous appears in the form of unimaginable energy or vast expanses of space, powers beyond human control, evil that explodes any sense of order or ethical existence. In Buddhist imagery, it is the experience of Vishnu at the end of the world, destroying with fire and flood until nothing is left but ashes. This is the experience of god as destroyer of all of existence. It is beyond ethics and aesthetics. And yet, it is at the same time the beginning of bliss, of rebirth.

Kathleen Powers Erickson gives a similar view of *Crows over the Wheatfield* (her title for the same painting), only stressing the Christian symbolism. First, she connects the *roads* in the painting to Vincent's sermon based on his reading of *Pilgrim's Progress*, and thus to the spiritual journey, in the painting to not just the end of this life but to the extended journey to the *Eternal City*. Then she brings in the *Wheatfield with Reaper* painting and Vincent's comments that he sees in this reaper "nothing sad in this death, it goes its way in broad daylight, with a sun flooding everything with a light of pure gold." Thus, she says, "Taking into consideration van Gogh's view of death, the images of crows and wheat and country roads begin to come together to show what van Gogh describes as the "healthful and restorative forces" of the French countryside." (p. 164) Her views tie into Edward Fell's in this Christian sense, and she sees in the "vast, turbulent skies of deep royal blue" an evocation of infinity, of the union of the soul with the "infinite God, Kempis' mystical embrace and Bunyan's union with Christ in the Celestial City." (409)

Again, this interpretation is a reduction, and the tremendous, horrific, overwhelming encounter with the *sublime* or the *numinous* gets lost by reducing it. The denial of the realization of the fearful, macabre, evil aspects of an encounter with this other reality undermines the power of the painting.

Nevertheless, Edwards and Erickson are on the right track and help open some wonderful doors into the depths of the work.

Before we can began to understand why or how Vincent stepped through the doors of perception and realized the invisible world of meaning and value beyond mere physical existence, we need to believe it exists ourselves, whether or not we can follow him. Some of us do. Some of us do not. Some of us pretend to and reduce it to either a religious or a scientific explanation, hiding behind the façade of logic, refusing to admit that the greatest logicians, mathematicians, and scientists of all time have been unable to discover or create a logical frame, a mathematical system, a science that explains away the underlying mystery of existence.

In the end, if everything can be explained, then there is nothing more, then life becomes meaningless, for it is all pre-determined by something outside of the human mind, and the human mind is not responsible for what it does.

Ken Kesey writes: "The answer is never the answer. The need for mystery is always greater than the need for answers." (410) John Keats writes: "Heard melodies are sweet, but those unheard / are sweeter; therefore ye soft pipes play on." (411)

The light of the sun reflecting off the moon dances on the surface of the water. Beneath the dance of shadow and and light the deep waters hold the mysteries of eternity, and above the dance, the night sky of the cosmos stretches into infinity.

Notes

1. Note, while the letters contain materials not directly related to the precise points being made, I am including most, and in many cases all of the writing in the ones I use because the sometimes seemingly irrelevant passages help reveal the mind, personality, and current emotional state of Vincent, which is ultimately the point. There are numerous translations of Van Gogh's letter, but the main one I am using is the one based on translations by Mrs. Johanna van Gogh-Bonger, edited by Robert Harrison, published in *The Complete Letters of Vincent van Gogh*,Bulfinch, 1991, number 129. URL: http://webexhibits.org/vangogh/letter/8/129.htm.This source states the translations can be reproduced for standard scholarship publications, and most of the translations are taken from it. However, many other translations of the letters have also been consulted, including *The Complete Letters of Vincent van Gogh*. Boston: New York Graphic Society, 1978. ISBN: 0-8212-0735-0; *The Letters of Vincent van Gogh,* translated by Arnold Pomerans, edited by Ronald de Leeuw, Penguin (this does not have the complete letters, but does contain a good deal of commentary with the ones it highlights); Auden, W.H. *Van Gogh: A Self-Portrait; Letters revealing his life as a painter, selected by W.H. Auden*. Greenwich, Connecticut: New York Graphic Society, 1961. L.C. Card Number: 61-8632; and the Chuck Avoub version available online at http://www.vggallery.com/letters/ main.htm.

2. Translated by Mrs. Johanna van Gogh-Bonger, edited by Robert Harrison, published in *The Complete Letters of Vincent van Gogh*, Publisher: Bulfinch, 1991, number 089. URL: http://webexhibits.org/vangogh/letter/5/089.htm.
3. Translated by Mrs. Johanna van Gogh-Bonger, edited by Robert Harrison, published in *The Complete Letters of Vincent van Gogh,* Publisher: Bulfinch, 1991, number 126. URL: http://webexhibits.org/vangogh/letter/7/126.htm.
4. Translated by Robert Harrison, edited by Robert Harrison, URL: http://webexhibits.org/vangogh/letter/8/etc-fam-1878b.htm.
5. Translated by Mrs. Johanna van Gogh-Bonger, edited by Robert Harrison, published in *The Complete Letters of Vincent van Gogh,* Publisher: Bulfinch, 1991, number 127. URL: http://webexhibits.org/vangogh/letter.
6. Translated by Robert Harrison, edited by Robert Harrison, URL: http://webexhibits.org/vangogh/letter/8/etc-fam-1879.htm.
7. Translated by Robert Harrison, edited by Robert Harrison, URL: http://webexhibits.org/vangogh/letter/8/etc-fam-1879.htm.
8. Translated by Robert Harrison, edited by Robert Harrison, URL: http://webexhibits.org/vangogh/letter/8/etc-fam-1879.htm.
9. Translated by Robert Harrison, edited by Robert Harrison, , URL: http://webexhibits.org/vangogh/letter/8/etc-fam-1879.htm.
10. Translated by Robert Harrison, edited by Robert Harrison, URL: http://webexhibits.org/vangogh/letter/8/etc-fam-1879.htm.
11. Auden, W.H. *Van Gogh: A Self-Portrait; Letters revealing his life as a painter, selected by W.H. Auden*. Greenwich, Connecticut: New York Graphic Society, 1961. L.C. Card Number: 61-8632, p. 66.
12. Translated by Robert Harrison, edited by Robert Harrison, URL: http://webexhibits.org/vangogh/letter/8/etc-fam-1879.htm.
13. Translated by Robert Harrison, edited by Robert Harrison, URL: http://webexhibits.org/vangogh/letter/8/etc-fam-1879.htm.
14. Sweetman, David. *Van Gogh: His Life and Art.* New York: Crown, 1990. ISBN: 0-517-57406-3, p. 114.
15. Translated by Robert Harrison, edited by Robert Harrison, URL: http://webexhibits.org/vangogh/letter/8/etc-fam-1879.htm.
16. Translated by Robert Harrison, edited by Robert Harrison, URL: http://webexhibits.org/vangogh/letter/8/etc-fam-1879.htm.
17. Translated by Mrs. Johanna van Gogh-Bonger, edited by Robert Harrison, published in *The Complete Letters of Vincent van Gogh, Publisher: Bulfinch, 1991*, number 133.URL: http://webexhibits.org/vangogh/letter/8/133.htm.
18. Lawrence & Elizabeth Hanson, *Passionate Pilgrim: The Life of Vincent Van Gogh*, NY, Random House 1955, p. 27.
19. Mendes da Costa. Letter to Het Algemeen Handelsblad. Written December 2 1910 in Amsterdam. Translated by Mrs. Johanna van Gogh-Bonger, edited by Robert Harrison, published in *The Complete Letters of Vincent van Gogh, Publisher: Bulfinch, 1991*, number htm. URL: http://webexhibits.org/vangogh/letter/6/etc-122a.htm.
20. Bruce Publ., Milwaukee, 1945;

http://www.ccel.org/ccel/kempis/imitation.ONE.2.html; Kempus, Thomas a, *The Imitation of Christ*, translated with an introduction by Leo Sherley-Price, New York: Penguin, 1952, p. 11 is where the claim for it being second most read book is stated. Other translations include, Kempis, Thomas à. *The Imitation of Christ*. Trans. and Introduction by Leo Sherley-Price. New York: Penguin, 1952. *The Imitation of Christ*. E-text. Trans. Aloysius Croft and Harold Bolton. Digitized by Harry Plantinga, 1994. 31 Dec. 2004.

21. http://www.ccel.org/ccel/kempis/imitation.html.

22. Lawrence & Elizabeth Hanson, *Passionate Pilgrim: The Life of Vincent Van Gogh*, NY, Random House 1955, p. 47.

23. Albert J. Lubin, *Stranger on the earth: a Psychological Biography of Vincent van Gogh*, Holt, Rinehart and Winston, 1972, pp. 27-28.

24. Translated by Mrs. Johanna van Gogh-Bonger, edited by Robert Harrison, published in *The Complete Letters of Vincent van Gogh, Publisher: Bulfinch, 1991*, number 133.URL: http://webexhibits.org/vangogh/letter/8/133.htm.

25. Ken Wilkie, *In Search of Van Gogh,* Prima Printing, 1991, in researching Vincent's life at this time, talked with Jean Richez, nephew of Jean-Baptiste Denis, the baker with whom Vincent lodged in the mining village of Petit Wasmes, and who remembered: "Aunt Esther said she used to hear Monsieur Van Gogh crying all night in his hut outside. He made a very deep impression on her. None of the miners who knew him ever forgot Vincent. They called him the Christ of the Coalmine."

26. Translated by Mrs. Johanna van Gogh-Bonger, edited by Robert Harrison, published in *The Complete Letters of Vincent van Gogh Publisher: Bulfinch, 1991*, number 347. URL: http://webexhibits.org/vangogh/letter/14/347.htm.

27. Written August 1872 in The Hague. Translated by Mrs. Johanna van Gogh-Bonger, edited by Robert Harrison, published in *The Complete Letters of Vincent van Gogh, Publisher: Bulfinch, 1991*, number 001. URL: http://webexhibits.org/vangogh/letter/1/001.htm.

28. Translated by Mrs. Johanna van Gogh-Bonger, edited by Robert Harrison, published in *The Complete Letters of Vincent van Gogh, Publisher: Bulfinch, 1991*, number 009. URL: http://webexhibits.org/vangogh/letter/2/009.htm.

29. *The Letters of Vincent van Gogh*, Penguin: London, 1996, p. 4.

30. Translated by Mrs. Johanna van Gogh-Bonger, edited by Robert Harrison, published in *The Complete Letters of Vincent van Gogh, Publisher: Bulfinch, 1991*. URL: http://webexhibits.org/vangogh/letter/2/009a.htm.

31. Written 20 July 1873 in London. Translated by Mrs. Johanna van Gogh-Bonger, edited by Robert Harrison, published in *The Complete Letters of Vincent van Gogh, Publisher: Bulfinch, 1991*, number 010. URL: http://webexhibits.org/vangogh/letter/2/010.htm.

32. Written 7 August 1873 in London. Translated by Mrs. Johanna van Gogh-Bonger, edited by Robert Harrison, published in *The Complete Letters of Vincent van Gogh, Publisher: Bulfinch, 1991*, number. URL: http://webexhibits.org/vangogh/letter/2/010a.htm.

33. *Jo van Gogh-Bonger's Memoir of Vincent van Gogh*; www.vggallery.com); also located in Johanna van Gogh-Bonger, *Memoirs of Vincent van Gogh*, URL:

http://webexhibits.org/.htm. Also emphasized in Ken Wilkie's biography *The Van Gogh Assignment*, Paddington Press, 1978; republished: *The Van Gogh File. A Journey of Discovery*, Souvenir Press, 1990, ISBN 0-285-62965-4.
34. *Van Gogh*, PRC Publ, London, 2004, p. 8.
35. Written 13 September 1873 in London. Translated by Mrs. Johanna van Gogh-Bonger, edited by Robert Harrison, published in *The Complete Letters of Vincent van Gogh, Publisher: Bulfinch, 1991*, number 011. URL: http://webexhibits.org/vangogh/letter/2/011.htm
36. *The Letters of Vincent van Gogh*, Penguin: London, 1996, p. 9.
37. Translated by Robert Harrison, edited by Robert Harrison. URL: http://webexhibits.org/vangogh/letter/2/etc-Anna-Theo.htm.
38. *Jo van Gogh-Bonger's Memoir of Vincent van Gogh*; www.vggallery.com; also located in Johanna van Gogh-Bonger, *Memoirs of Vincent van Gogh*, URL: http://webexhibits.org/.htm.
39. Written 31 July 1874 in London. Translated by Mrs. Johanna van Gogh-Bonger, edited by Robert Harrison, published in *The Complete Letters of Vincent van Gogh, Publisher: Bulfinch, 1991*, number 020. URL: http://webexhibits.org/vangogh/letter/2/020.htm.
40. Written 10 August 1874 in London. Translated by Mrs. Johanna van Gogh-Bonger, edited by Robert Harrison, published in *The Complete Letters of Vincent van Gogh,* Publisher: Bulfinch*, 1991*, number 021. URL: http://webexhibits.org/vangogh/letter/2/021.htm.
41. Translated by Mrs. Johanna van Gogh-Bonger, edited by Robert Harrison, published in *The Complete Letters of Vincent van Gogh, Publisher: Bulfinch, 1991*, number 026. URL: http://webexhibits.org/vangogh/letter/2/026.htm.
42. Translated by Robert Harrison, edited by Robert Harrison. URL: http://webexhibits.org/vangogh/letter/3/etc-1875.htm.
43. Translated by Robert Harrison, edited by Robert Harrison. URL: http://webexhibits.org/vangogh/letter/3/etc-1875.htm.
44. Translated by Robert Harrison, edited by Robert Harrison. URL: http://webexhibits.org/vangogh/letter/3/etc-1875.htm.
45. Written 13 August 1875 in Paris. Translated by Mrs. Johanna van Gogh-Bonger, edited by Robert Harrison, published in *The Complete Letters of Vincent van Gogh, Publisher: Bulfinch, 1991*, number 033. URL: http://webexhibits.org/vangogh/letter/3/033.htm.
46. Written August 1875 in Paris. Translated by Robert Harrison, edited by Robert Harrison, published in *Facsimile of original letter*, number 034. URL: http://webexhibits.org/vangogh/letter/3/034.htm.
47. Written 17 September 1875 in Paris. Translated by Mrs. Johanna van Gogh-Bonger, edited by Robert Harrison, published in *The Complete Letters of Vincent van Gogh, Publisher: Bulfinch, 1991*, number 038. URL: http://webexhibits.org/vangogh/letter/3/038.htm.
48. Written 25 September 1875 in Paris. Translated by Mrs. Johanna van Gogh-Bonger, edited by Robert Harrison, published in *The Complete Letters of Vincent van Gogh, Publisher: Bulfinch, 1991*, number 039. URL: http://webexhibits.org/vangogh/letter/3/039.htm.

49. Translated by Mrs. Johanna van Gogh-Bonger, edited by Robert Harrison, published in *The Complete Letters of Vincent van Gogh, Publisher: Bulfinch, 1991*, number 39b. URL: http://webexhibits.org/vangogh/letter/3/039b.htm.
50. Written 6 October 1875 in Paris. Translated by Mrs. Johanna van Gogh-Bonger, edited by Robert Harrison, published in *The Complete Letters of Vincent van Gogh, Publisher: Bulfinch, 1991*, number 041. URL: http://webexhibits.org/vangogh/letter/3/041.htm.
51. Benedikt: Taschen Verlag GmbH, 1990; rpt., Taschen: London, 2006. www.taschen.com.
52. *Van Gogh*, Wellfleet Press, 2004, "Book Jacket."
53. *Van Gogh*, Wellfleet Press, 2004, p. 27.
54. *Van Gogh*, Wellfleet Press, 2004, pp. 27-28.
55. *Van Gogh*, Wellfleet Press, 2004, p. 28.
56. *Van Gogh*, Wellfleet Press, 2004, p. 28.
57. *Van Gogh*, Praeger, 1975.
58. *Van Gogh: A Study of his Life and Work*, Praeger, 1958.
59. *Vincent Van Gogh: A Life*, Philip Callow, Ivan R. Dee; rpt., Elephant Paperback, 1996, pp. 34-35.
60. *Vincent Van Gogh: The Complete Paintings,* Taschen, 1990; rpt., 2006, pp. 34-5.
61. *Vincent Van Gogh: A Psychological Study,* George Allen and Unwin, London, 1967, p. 13-14.
62. *Vincent Van Gogh: A Psychological Study,* George Allen and Unwin, London, 1967, p. 14.
63. *Vincent Van Gogh: A Psychological Study,* George Allen and Unwin, London, 1967, p. 128.
64. *Vincent Van Gogh: A Psychological Study,* George Allen and Unwin, London, 1967, pp. 128-129.
65. *Vincent Van Gogh: A Psychological Study,* George Allen and Unwin, London, 1967, pp. 162-164.
66. *Vincent Van Gogh: A Psychological Study,* George Allen and Unwin, London, 1967, pp. 25-26.
67. *Vincent Van Gogh: A Psychological Study,* George Allen and Unwin, London, 1967, p. 29.
68. *Vincent Van Gogh: A Psychological Study,* George Allen and Unwin, London, 1967, pp. 29-30.
69. *Vincent Van Gogh: A Psychological Study,* George Allen and Unwin, London, 1967, p. 36.
70. *Vincent Van Gogh: A Psychological Study,* George Allen and Unwin, London, 1967, pp. 36-37.
71. *The Power of Myth: Joseph Campbell with Bill Moyers*, ed., by Sue Flowers, Anchor Books, 1988, p. 261.
72. *The Courage to Create*, 1975, Bantam, pp. 135-136.
73. *The Courage to Create,* 1975, Bantam, pp. 136-138.
74. *The Courage to Create*, 1975, Bantam, p. 131.
75. Doubleday, 1980, pp. 11-12.

76. Trans. By Holroyd-Reece, The Literary guild of America, 1933, p. xvi.
77. *Stranger on the earth: A Psychological Biography of Vincent Van Gogh,* Holt, Rinehart and Winston, 1972, p. 2.
78. *Stranger on the earth: A Psychological Biography of Vincent Van Gogh,* Holt, Rinehart and Winston, 1972, p. 38.
79. *Stranger on the earth: A Psychological Biography of Vincent Van Gogh,* Holt, Rinehart and Winston, 1972, pp. 104-105
80. *Vincent Van Gogh: A Psychological Study,* George Allen and Unwin, London, 1967, pp. 45-46.
81. *Vincent Van Gogh*, p. 44, Viking, New York, 1969, p. 44.
82. *Vincent Van Gogh*, Viking, New York, 1969, p. 46.
83. *Stranger on the earth: A Psychological Biography of Vincent Van Gogh, Holt, Rinehart and Winston,* 1972, p.36.
84. *Stranger on the earth: A Psychological Biography of Vincent Van Gogh, Holt, Rinehart and Winston,* 1972, p. 77.
85. *Stranger on the earth: A Psychological Biography of Vincent Van Gogh, Holt, Rinehart and Winston,* 1972, p. 84.
86. *Stranger on the earth: A Psychological Biography of Vincent Van Gogh, Holt, Rinehart and Winston,* 1972, p. 84.
87. *Stranger on the earth: A Psychological Biography of Vincent Van Gogh, Holt, Rinehart and Winston,* 1972, p. 85.
88. *Stranger on the earth: A Psychological Biography of Vincent Van Gogh, Holt, Rinehart and Winston,* 1972, p. 85.
89. *Stranger on the earth: A Psychological Biography of Vincent Van Gogh, Holt, Rinehart and Winston,* 1972, pp. 85-86.
90. *Stranger on the earth: A Psychological Biography of Vincent Van Gogh, Holt, Rinehart and Winston,* 1972, p. 86.
91. *Stranger on the earth: A Psychological Biography of Vincent Van Gogh, Holt, Rinehart and Winston,* 1972, p. 86.
92. *Stranger on the earth: A Psychological Biography of Vincent Van Gogh, Holt, Rinehart and Winston,* 1972, p. 86-87.
93. *Stranger on the earth: A Psychological Biography of Vincent Van Gogh, Holt, Rinehart and Winston,* 1972, p. 87.
94. *Stranger on the earth: A Psychological Biography of Vincent Van Gogh, Holt, Rinehart and Winston,* 1972, p. 88.
95. *Stranger on the earth: A Psychological Biography of Vincent Van Gogh, Holt, Rinehart and Winston,* 1972, p. 89.
96. *Stranger on the earth: A Psychological Biography of Vincent Van Gogh, Holt, Rinehart and Winston,* 1972, p. 95.
97. *Stranger on the earth: A Psychological Biography of Vincent Van Gogh, Holt, Rinehart and Winston,* 1972, p. 95.
98. *In Search of Van Gogh: A Journalist's Revealing New Findings about the Artist's Life and Death*, Prima Pub, 1992, pp. 21-22. This book, with slight variations, has also been published as *The Van Gogh Assignment* and most recently *The Van Gogh File*. While this final title is the most recent, it doesn't add anything to the two most intriguing mysteries raise in the one used, i.e., whether or not

Vincent had a child with Sien, and whether or not there is another Van Gogh painting buried in Antwerp.
99. *In Search of Van Gogh: A Journalist's Revealing New Findings about the Artist's Life and Death*, Prima Pub, 1992, p. 36.
100. *In Search of Van Gogh: A Journalist's Revealing New Findings about the Artist's Life and Death*, Prima Pub, 1992, p. 59.
101. www.psychematters.com.
102. *Van Gogh's Women: Vincent's Love Affairs and Journey into Madness*, Carroll & Graf, 2004, p. 10.
103. *Van Gogh's Women: Vincent's Love Affairs and Journey into Madness*, Carroll & Graf, 2004, pp. 10-11.
104. *Van Gogh's Women: Vincent's Love Affairs and Journey into Madness*, Carroll & Graf, 2004, p. 11.
105. *Van Gogh's Women: Vincent's Love Affairs and Journey into Madness*, Carroll & Graf, 2004, 11-12.
106. *Van Gogh's Women: Vincent's Love Affairs and Journey into Madness*, Carroll & Graf, 2004, 12.
107. *Van Gogh's Women: Vincent's Love Affairs and Journey into Madness*, Carroll & Graf, 2004, 12-13.
108. *Van Gogh's Women: Vincent's Love Affairs and Journey into Madness*, Carroll & Graf, 2004, 14.
109. *Van Gogh's Women: Vincent's Love Affairs and Journey into Madness*, Carroll & Graf, 2004, pp. 24-25.
110. *Vincent and Theo Van Gogh: A Dual Biography*, Jan Hulsker, James M. Miller, Editor, Fuller Publications: Ann Arbor, 1990.
111. "Introduction: Freud as Literature," *Freud: A Collection of Critical Essays*. ed. by Perry Meisel, Prentice-Hall: Englewood Cliffs, N. J., 1981, pp. 34-35.
112. It can be found in *Freud*, Standard Edition, volume 1, 1895, pp. 283-391; also in Peter Gay's The Freud Reader, Yale Univ., Norton: New York, 1989, pp. 86-89.
113. Quoted from a letter to Fliess, *The Origins of Psychoanalysis: Letters to Wilheilm Fliess, Drafts and Notes*, 1887-1902; trans. Eric Mosbacher and James Strachey, NewYork: Basic Books, 1954, pp. 215-218.

In 1925, Freud wrote:

> Before going further into the question of infantile sexuality I must mention an error into which I fell for a while and which might well have had fatal consequences for the whole of my work. Under the pressure of the technical procedure which I used at that time, the majority of my patients reproduced from their childhood scenes in which they were sexually seduced by some grown-up person. With female patients the part of seducer was almost always assigned to their father. I believed these stories, and consequently supposed that I had discovered the roots of the subsequent neurosis in these experiences of sexual seduction in childhood. My confidence was strengthened by a few cases in which relations of this kind with a father, uncle, or elder brother had continued up to an age at which memory was to be trusted. If the reader feels inclined to shake his head at my credulity, I cannot altogether blame him; though I may plead

that this was at a time when I was intentionally keeping my critical faculty in abeyance so as to preserve an unprejudiced and receptive attitude towards the many novelties which were coming to my notice every day. When, however, I was at last obliged to recognize that these scenes of seduction had never taken place, and that they were only *phantasies* which my patients had made up or which I myself had perhaps forced upon them, I was for some time completely at a loss. My confidence alike in my technique and in its results suffered a severe blow; it could not be disputed that I had arrived at these scenes by a technical method which I considered correct, and their subject-matter was unquestionably related to the symptoms from which my investigation had started. When I had pulled myself together, I was able to draw the right conclusions from my discovery: namely, that the neurotic symptoms were not related directly to actual events but to *phantasies* embodying wishes, and that as far as the neurosis was concerned psychical reality was of more importance than material reality. *Sigmund Freud: An Autobiographical Study*, 1925; trans., James Strachey, 1936; rpt., W.W. Norton: New York, 1963, pp.62-64.

114. See *The Interpretation of Dreams*, especially "VII, The Psychology of the Dream-Processes," and *The Unconscious, Standard Edition of the Complete Psychological Works of Sigmund Freud*, tr., under the general editorship of James Strachey in collaboration with Anna Freud, assisted by Alex Strachey and AlexTyson, 24 vols., 1953-1974 XIV, 161; rpt., *The Freud Reader*, ed. Peter Gay, W.W. Norton: New York, 1989, pp. 573-584.

115. *Standard Edition of the Complete Psychological Works of Sigmund Freud*, tr. under the general editorship of James Strachey in collaboration with Anna Freud, assisted by Alex Strachey and Alex Tyson, 24 vols., 1953-1974 ; rpt., *The Freud Reader*, ed. Peter Gay, W.W. Norton: New York, 1989.

This brief overview of Freud's mapping of the psyche over-simplifies the complexity he suggests. For his more in depth discussion see *The Interpretation of Dreams,* Standard Edition, vol. 5, 1900, 339-622; (reprinted, slightly revised in several editions through the first part of the twentieth century); *Five Lectures on Psychoanalysis,* Standard Ed., vol. 11, 1910, pp. 1-56; *Formations Regarding the Two Principles in Mental Functioning*, Standard Ed., vol. 11, 1911, pp. 409-417; *Repression*, Standard Ed., vol. 14, 1915, pp. 141-158; *The Unconscious*, St. Ed., vol. 14, 1915, pp. 159-202; *The Ego and the Id*, St. Ed., Bk., 19, 1923, pp. 1-59; *Civilization and it's Discontents*, 1930; rpt., trans. James Strachey, Norton: New York, 1962.

In *On Narcissism*, 1913, *The Freud Reader*, pp. 545-562] Freud proposed that a developmental stage of narcissism be inserted between an infant's primitive auto-eroticism and object love, and more dramatically, asserted that there is an ego-libido as well as an object-libido, undercutting the topographical view that separated ego and libidinal drives, suggesting that all drives are libido. This new structuring moved further toward his structural model with *Beyond the Pleasure Principle*, 1920, [*The Freud Reader*, pp. 594-626]. Here the struggles between the ego and the id are replaced with struggles between the drives to life and death (Eros and Thanatos). He also suggested that there is an "ego-ideal," and the stage was set for a complete restructuring of the psyche, which he presented in *The Ego and the Id*, [*The Freud Reader*, pp. 629-658]. In this view, the ego is no longer in opposition to the unconscious, but largely exists in the unconscious.

116. The quote is from Kafka's diary, Sept. 23, 1912; *Tagebucher*, ed. Max Brod (ed., 1967), p. 210. According to Ronald Hayman, *Kafka: A Biography*, 1981, p. 1, "Without Freud he [Kafka] might never have come to take so much interest in his own dreams."

117. Review of Freud's Psychopathology of Everyday Life, 1914, *The New English Weekly*; quoted in *Freud: A Collection of Critical Essays*, ed. Perry Meisel, Prentice-Hall, p. 6, 1981.

118. "Freud and Literature," *The Saturday Review of Literature*, Oct. 4, 1924.

119. "Postscript" to *An Autobiographical Study*, 1935, p. 139.

120. "Postscript" to *An Autobiographical Study*, 1935, 20:8.

121. "Freud and the Future," 1937; rpt., Knopf, 1965; Rpt, from *Essays of Three Decades*, Thomas Mann, trans., H. T. Lowe-Porter, New York: Knopf, 1947.

122. Ibid.

123. It should be noted that Jung and Freud had long ago separated, and Jung's theories, while partaking of the underlying premise that dreams have meaning, and that the meanings work symbolically, established a different theoretical basis for the mental processes. During the years of 1911-1913, both Alfred Adler and Carl Jung, once central followers of Freud, had reinterpreted several key elements of his theories. Jung had moved personalized analysis to a more impersonal, universal, non-historical realm, and relegated infantile sexuality and the Oedipus complex to lesser positions. Adler had eliminated sexuality altogether, and relegated all human character and neuroses solely to the desire for power. Both had gathered sizeable followings and, thus, established important separate schools of psychology. In 1914, Freud published "History of the Psycho-Analytic Movement," claiming they had both substituted watered down theories for the more scientific truths of psychoanalysis. However, by the time of Mann's comments, it is possible to see how Freud has moved more into the mythical extensions, as the mentioned publication, *Totem and Taboo*, clearly demonstrates.

124. From *The English Auden: Poems. Essays, and Dramatic Writings*, 1927-39, ed., Edward Mendelson (New York: Random House, 1977), pp. 332-342.

125. Kenneth Burke, "Freud—and the Analysis of Poetry," *The American Journal of Sociology*, 45 (1939), pp. 391-417; rpt., *The Philosophy of Literary Form: Studies in Symbolic Action*, 3rd ed., (Berkeley and Los Angeles: Univ. of Calif., 1973, pp. 258-292.

126. *Freud: Within and Beyond Culture*; originally published as *Freud and the Crisis of Our Culture*, Boston: Beacon, 1955).

127. "Freud and Literature," *The Kenyon Review*, spring, 1940, pp. 152-173; rpt., in *The Liberal Imagination,* 1950.

128. *Revolution in Poetic Language*, Margaret Waller, trans., Columbia Univ., 1984. This is but one of several books by this French theorist.

129. "L'Ecriture et la difference, Paris: Editions de Seuil," 1967; trans., Alan Bass, "Freud and the Scene of Writing," *Writing and Difference*, Chicago: Univ. of Chicago, 1978. Derrida is the most important figure in what has become

designated as 'deconstructionism,' a form of literary criticism that rejects the concept of a fixed meaning, but rather suggests that a text is always in a state of flux, always changing meanings, open ended, and continually morphing. Thus, literary criticism is really nothing more than a form of literature itself, participating in the ever-changing meanings of a work. Since language continually changes meanings, there is no fixed meaning, only an ever-changing map of meanings, which, at best, can only expand our thought, our consciousness of the inherent play of language as thought. If these views are put into the neurological views of two forms of human thinking, then they began to clarify themselves. There is the left brain form of literal meaning based on logic and explanation (which would be of the world of literary criticism), and there is the right brain world of symbolic and metaphoric meaning, meaning beyond explanation. As these two levels of meaning interact they constantly evolve, each infusing the other in an ever changing world of knowing that both includes explanation and that which is beyond explanation. This does not mean, as some deconstructionists seem to claim, and many anti-deconstructionists charge, that meaning does not exist. It simply means that meaning lives, grows, becomes more complex, i.e., is not stagnant (meaning takes shape in the human head where mind and brain come together). Furthermore, since the meanings expressed in a work of literature are capable of causing literal, physical responses (i.e., tears and laughter), meaning must exist beyond explanation, meaning that the human mind/body comprehends but cannot fully explain (as an explanation of the very writing that caused the emotional/physical reaction cannot cause the same reaction). In this way, language is capable of reaching beyond itself to spark kinds of knowing beyond language, truths of the body, the total organism.

For Derrida's major writings, see *Speech and Phenomena: And Other Essays on Husserl's Theory of Signs* (trans. David B. Allison, Evanston: Northwestern Univ., 1973; *Of Grammatology* (trans. Gayatri Chakravorty Spivak, Baltimore: Johns Hopkins Univ., 1974, 1976), and *Writing and Difference* (trans. Alan Bass, Chicago: Univ. of Chicago, 1978).

Other important deconstructionists include Harold Bloom, Eugenio Donato, Geoffrey H. Hartman, J. Hillis Miller, Joseph Riddel, Barbara Johnson, Shoshana Felman, Jeffrey Mehlman, and Gayatri Chakravorty Spivak.

130. *Die Medizin der Gegenwart in Selbstdarstellungen*, Leipzig: Felix Meiner, vol iv; *Sigmund Freud: An Autobiographical Study*, 1927, trans. James Strachey, New York: Brentano); rpt., 1936; 1963, W.W. Norton: New York, pp. 52-56.

131. see "Repression" and "The Unconscious" 1915; rpt., *The Freud Reader*, ed., Peter Gay, Yale Univ.: Norton, pp. 568-577.

132. "Freud and the Poetic Sublime: A Catastrophe Theory of Creativity," originally delivered as a speech to the William Alanson White Psychoanalytic Society, Sept. 23, 1977; first publ., *Antaeus*, Spring, 1978, pp. 355-377; rpt., *Freud: A Collection of Critical Essays*, ed. Perry Meisel, Englewood Cliffs, N. J.: Prentice-Hall, 1981.

133. *Freud: A Collection of Critical Essays*, Englewood Cliffs, N.J.: Prentice-Hall, 1981, p. 34.

134. *An Autobiographical Study*, 1935; rpt. W.W. Norton, 1952, pp. 119-124.

135. *Vincent Van Gogh,* Marc Edo Tralbaut, Lausanne: Edita Lausanne, 1969, p. 23.
136. *Starry Starry Night: Life and psychiatric history of Vincent van Gogh,* Benecke N. I., Amsterdam, 2003, p. 31.
137. *Starry Starry Night: Life and psychiatric history of Vincent van Gogh,* Benecke N. I., Amsterdam, 2003, p. 35.
138. *Starry Starry Night: Live and psychiatric history of Vincent van Gogh,* Benecke N. I., Amsterdam, 2003, p. 136.
139. *Starry Starry Night: Live and psychiatric history of Vincent van Gogh,* Benecke N. I., Amsterdam, 2003, p. 136.
140. *Starry Starry Night: Live and psychiatric history of Vincent van Gogh,* Benecke N. I., Amsterdam, 2003, p. 71.
141. *The Empty Fortress: Infantile Autism and the Birth of the Self,* Bruno Bettelheim, 1967; rpt., The Free Press, 1972.
142. *The Letters of Vincent van Gogh,* selected and edited by Ronald de Leeuw, translated by Arnold Pomerans, Penguin Books, 1996, p. 3.
143. *Vincent Van Gogh: The Complete Paintings,* Taschen, 1990; rpt., 2006, pp. 15-18.
144. *Van Gogh*, Jean Leymarie, translated by James Emmons, 1968; rpt., Rizzoli Int'l Publ., 1977, p. 11.
145. *Van Gogh*, Jean Leymarie, translated by James Emmons, 1968; rpt., Rizzoli Int'l Publ., 1977, p. 7.
146. *Vincent Van Gogh,* Marc Edo Tralbaut, Lausanne: Edita Lausanne, 1969, p. 23.
147. *Van Gogh and God: A Creative Spiritual Quest,* Cliff Edwards, Loyola Press, 1989, p. 19.
148. Translated by Mrs. Johanna van Gogh-Bonger, edited by Robert Harrison, published in *The Complete Letters of Vincent van Gogh*, Publisher: Bulfinch, 1991, number 079. URL: http://webexhibits.org/vangogh/letter/5/089.htm; *Van Gogh and God: A Creative Spiritual Quest,* Cliff Edwards, Loyola Press, 1989, p. 1.
149. *Van Gogh and God: A Creative Spiritual Quest,* Cliff Edwards, Loyola Press, 1989, p. 1-2.
150. Frank Elgar, *Van Gogh*, trans. J. Cleugh, Frederick A. Praeger Publishers, 1958, p. 18; *Van Gogh and God: A Creative Spiritual Quest,* Cliff Edwards, Loyola Press, 1989, p. 2-3.
151. Albert Lubin, *Stranger on the Earth,* Holt, Rinehart and Winston, Inc., 1972, p. 36; *Van Gogh and God: A Creative Spiritual Quest,* Cliff Edwards, Loyola Press, 1989, p. 3.
152. *Vincent Van Gogh,* Marc Edo Tralbaut, Lausanne: Edita Lausanne, 1969, p. 46; *Van Gogh and God: A Creative Spiritual Quest,* Cliff Edwards, Loyola Press, 1989, p. 3.
153. Van Gogh-Bonger, "Memoir of Vincent Van Gogh" in *Complete Letters,* 1: xxiv-xxvi; *Van Gogh and God: A Creative Spiritual Quest,* Cliff Edwards, Loyola Press, 1989, p. 4.

154. *Van Gogh*, Pierre Cabanne, trans. D. Woodward, Thames and Hudson, 1986, p. 14; *Van Gogh and God: A Creative Spiritual Quest,* Cliff Edwards, Loyola Press, 1989, p. 4.
155. Frank Elgar, *Van Gogh*, trans. J. Cleugh, Frederick A. Praeger Publishers, 1958, p. 18; *Van Gogh and God: A Creative Spiritual Quest,* Cliff Edwards, Loyola Press, 1989, p. 5.
156. *Vincent Van Gogh*, Marc Edo Tralbaut, Lausanne: Edita Lausanne, 1969, p. 44; *Van Gogh and God: A Creative Spiritual Quest,* Cliff Edwards, Loyola Press, 1989, p. 5.
157. *Van Gogh and God: A Creative Spiritual Quest,* Cliff Edwards, Loyola Press, 1989, p. 5.
158. *Van Gogh and God: A Creative Spiritual Quest,* Cliff Edwards, Loyola Press, 1989, p. 6.
159. *Van Gogh and God: A Creative Spiritual Quest,* Cliff Edwards, Loyola Press, 1989, p. 6.
160. *Van Gogh and God: A Creative Spiritual Quest,* Cliff Edwards, Loyola Press, 1989, p. 7.
161. *Van Gogh and God: A Creative Spiritual Quest,* Cliff Edwards, Loyola Press, 1989, p. 7.
162. *Van Gogh and God: A Creative Spiritual Quest,* Cliff Edwards, Loyola Press, 1989, p. 7.
163. *Van Gogh and God: A Creative Spiritual Quest,* Cliff Edwards, Loyola Press, 1989, p. 13.
164. *Van Gogh and God: A Creative Spiritual Quest,* Cliff Edwards, Loyola Press, 1989, p. 19.
165. Written c. 11 December 1882 in The Hague. Translated by Mrs. Johanna van Gogh-Bonger, edited by Robert Harrison, published in *The Complete Letters of Vincent van Gogh, Publisher: Bulfinch, 1991*, number 252. URL: http://webexhibits.org/vangogh/letter/11/252.htm.
166. September 24, 1880, sent from Cuesmes, translated by Mrs. Johanna van Gogh-Bonger, edited by Robert Harrison, published in *The Complete Letters of Vincent van Gogh*, Bulfinch, 1991, number 136. URL: http://webexhibits.org/vangogh/letter/8/136.htm.
167. *At Eternity's Gate: The Spiritual Vision of Vincent van Gogh,* Kathleen Powers Erickson, William B. Eerdmans, Grand Rapids, Michigan, 1998, p. 7.
168. *At Eternity's Gate: The Spiritual Vision of Vincent van Gogh,* Kathleen Powers Erickson, William B. Eerdmans, Grand Rapids, Michigan, 1998, p. 4.
169. Kathleen Powers Erickson's discussion of all of this takes place in Chapter 1 of *At Eternity's Gate: The Spiritual Vision of Vincent van Gogh,* Kathleen Powers Erickson, William B. Eerdmans, Grand Rapids, Michigan, 1998. Her sources and other, more detailed sources are: *The Articles of the Synod of Dort,* trans. Thomas Scott (Utica, N. Y.: Williams, 1831, pp. 93-120); *Crisis in the Reformed Churches, Essays in commemoration of the Synod of Dort, 1681-1619,* Grand Rapids: Reformed Fellowship, 1958, pp. 207-9; *The Five Points of Calvinism—Defined, Defended, Documented,* Steele and Thomas; *Five Points of Gospel Truth Asserted and Defended,* Moses Roney, 1836; and "Calvinism," *Catholic Encyclopedia.*

170. Walter Van Beselaere, *De Hollandsche Periode in het Werk van Vincent van Gogh* (Antwerp: De Sikkel, 1937), p. 4.
171. *At Eternity's Gate: The Spiritual Vision of Vincent van Gogh,* Kathleen Powers Erickson, William B. Eerdmans, Grand Rapids, Michigan, 1998, p. 17.
172. *Der Christliche Glaube*, 1820–21, 2nd ed. 1830–1), tr. H. R. MacKintosh, J. S. Stewart, editor. T. & T. Clark Publishers, Ltd. 1999 paperback: ISBN 0-567-08709-3.
173. "Address on Religion," 1799; quoted in Elie Kedourie. *Nationalism*, p. 26. Praeger. 1961, ISBN 0090534441.
174. Amsterdam: J. H. van Heteren, 1884; quoted in *At Eternity's Gate: The Spiritual Vision of Vincent van Gogh,* Kathleen Powers Erickson, William B. Eerdmans, Grand Rapids, Michigan, 1998, pp. 34-35.
175. *At Eternity's Gate: The Spiritual Vision of Vincent van Gogh,* Kathleen Powers Erickson, William B. Eerdmans, Grand Rapids, Michigan, 1998, p. 60.
176. Reprinted from *Het Algemeen Handelsblad* (leading Amsterdam newspaper) of December 2, 1910. Translated by Mrs. Johanna van Gogh-Bonger, edited by Robert Harrison, published in The Complete Letters of Vincent van Gogh, Publisher: Bulfinch, 1991, number htm.
URL: http://webexhibits.org/vangogh/letter/6/etc-122a.htm.
177. Bibliography: UNGER, *Die Flagellanten* (1902); COOPER (pseudonym), *Flagellation and the Flagellants, A History of the Rod, etc.* (new ed., London, 1896), an anti-Catholic and biased work; Barney, *Circumcision and Flagellation among the Filipinos* (Carlisle, Pa., 1903); Calmet's *Dict. of the Bible,* s. v. *Scourging;* KITTO, *Cyclop. of Biblical Lit.,* s. v. *Punishment.*
178. *The Catholic Encyclopedia*, Volume VI, New York: Robert Appleton Company. Nihil Obstat, September 1, 1909, Remy Lafort, Censor, Imprimatur, and John M. Farley, Archbishop of New York; John Allen, Jr., Opus Dei: an Objective Look Behind the Myths and Reality of the Most Controversial Force in the Catholic Church, Doubleday, 2005, ISBN: 0-385-51449-2; Peter Berglar, *Opus Dei. Life and Work of its Founder*. Scepter, 1994; Joan Estruch, *Saints and Schemers: Opus Dei and its paradoxes,* 1995; Noam Friedlander, *What Is Opus Dei? Tales of God, Blood, Money and Faith*, Collins and Brown, 2005; Scott Hahn, *Ordinary Work, Extraordinary Grace: My Spiritual Journey in Opus Dei, Random House,* ISBN: 978-0-385-51924-3; Massimo Introvigne, "Opus Dei and the Anti-cult Movement," *Cristianita,* 229, pp. 3-12; John Paul II, "Sacred Congregation for Bishops," August 23, 1982; James Martin, "Opus Dei in the United States," *America Magazine,* February 25, 1995; William O'Connor, *Opus Dei: An Open Book. A Reply to "The Secret World of Opus Dei" by Michael Walsh*, Mercier Press, Dublin, 1991; Michael Walsh, *Opus Dei: An Investigation into the Powerful Secretive Society within the Catholic Church.* Harper San Francisco. 2004.
179. July 1880, Cuesmes. Translated by Mrs. Johanna van Gogh-Bonger, edited by Robert Harrison, published in *The Complete Letters of Vincent van Gogh, Publisher: Bulfinch, 1991*, number 133. URL:
http://webexhibits.org/vangogh/letter/8/133.htm.

180. *Jo van Gogh-Bonger's Memoir of Vincent van Gogh*; www.vggallery.com; also located in Johanna van Gogh-Bonger, *Memoirs of Vincent van Gogh*, URL: http://webexhibits.org/.htm.
181. Translated by Mrs. Johanna van Gogh-Bonger, edited by Robert Harrison, published in The Complete Letters of Vincent van Gogh, Publisher: Bulfinch, 1991, number 136.
182._Written 12 November 1881 in Etten. Translated by Mrs. Johanna van Gogh-Bonger, edited by Robert Harrison, published in *The Complete Letters of Vincent van Gogh, Publisher: Bulfinch, 1991*, number 157. URL: http://webexhibits.org/vangogh/letter/10/157.htm.
183. *Vincent Van Gogh: A Psychological* Study, George Alen and Unwin, London, 1967, p. 63.
184. *In Search of Van Gogh: A Journalist's Revealing New Findings about the Artist's Life and Death*, Prima Pub, 1992, pp. 79-81. This book, with slight variations, has also been published as *The Van Gogh Assignment* and most recently *The Van Gogh File*. While this final title is the most recent, it doesn't add anything to the two most intriguing mysteries raise in the one used, i.e., whether or not Vincent had a child with Sien, and whether or not there is another Van Gogh painting buried in Antwerp.
185. *Vincent Van Gogh: A Psychological* Study, George Alen and Unwin, London, 1967, pp. 76-77.
186. Translated by Mrs. Johanna van Gogh-Bonger, edited by Robert Harrison, published in *The Complete Letters of Vincent van Gogh, Publisher: Bulfinch, 1991*, number 164.
187. Translated by Arnold Pomerans, edited by Ronald de Leeuw, published in *The Letters of Vincent Van Gogh,* Penguin Classics, Letter 192, May 3-12, 1882.
188. *Lotgenoten,* Jan Hulsker, Agathon/Unieboek bv, Weesp, 1885; *Vincent and Theo Van Gogh,* Translated and edited by James M. Miller, Fuller Publ.: Ann Arbor, 1990, p. 140.
189. Written 30 April 1883 in The Hague. Translated by Mrs. Johanna van Gogh-Bonger, edited by Robert Harrison, published in *The Complete Letters of Vincent van Gogh, Publisher: Bulfinch, 1991*, number 281. URL: http://webexhibits.org/vangogh/letter/12/281.htm.
190. Written 9 or 10 May 1883 in The Hague. Translated by Mrs. Johanna van Gogh-Bonger, edited by Robert Harrison, published in *The Complete Letters of Vincent van Gogh, Publisher: Bulfinch, 1991*, number 284. URL: http://webexhibits.org/vangogh/letter/12/284.htm.
191. Written 3 June 1883 in The Hague. Translated by Mrs. Johanna van Gogh-Bonger, edited by Robert Harrison, published in *The Complete Letters of Vincent van Gogh, Publisher: Bulfinch, 1991*, number 288. URL: http://webexhibits.org/vangogh/letter/12/288.htm.
192. *Vincent Van Gogh: A Psychological* Study, George Alen and Unwin, London, 1967, p. 59; the letter referred to is Letter 27.
193. *Stranger on the earth: A Psychological Biography of Vincent Van Gogh,* Holt, Rinehart and Winston, 1972, pp. 63-64.

194. *Stranger on the earth: A Psychological Biography of Vincent Van Gogh,* Holt, Rinehart and Winston, 1972, p. 64.
195. *Vincent Van Gogh: A Psychological* Study, George Alen and Unwin, London, 1967, pp. 83-84.
196. Translated by Mrs. Johanna van Gogh-Bonger, edited by Robert Harrison, published in *The Complete Letters of Vincent van Gogh, Publisher: Bulfinch, 1991*, number 423. URL: http://webexhibits.org/vangogh/letter/15/423.htm.
197. Written Summer/fall 1887 in Paris. Translated by Robert Harrison, edited by Inge Geerlings, published in *The Letters of Vincent van Gogh, 1886 - 1890, Publisher: Scholar Press, 1977*, number W01. URL: http://webexhibits.org/vangogh/letter/17/W01-nl.htm; also Translated by Arnold Pomerans, edited by Ronald de Leeuw, published in *The Letters of Vincent Van Gogh,* Penguin Classics, Letter W1 [D].
198. *In Search of Van Gogh: A Journalist's Revealing New Findings about the Artist's Life and Death*, Prima Pub, 1992, p. 122.
199. *In Search of Van Gogh: A Journalist's Revealing New Findings about the Artist's Life and Death*, Prima Pub, 1992, pp. 136-137.
200. *In Search of Van Gogh: A Journalist's Revealing New Findings about the Artist's Life and Death*, Prima Pub, 1992, p. 142.
201. *In Search of Van Gogh: A Journalist's Revealing New Findings about the Artist's Life and Death*, Prima Pub, 1992, p. 146.
202. *In Search of Van Gogh: A Journalist's Revealing New Findings about the Artist's Life and Death*, Prima Pub, 1992, p. 156.
203. *Lotgenoten,* Jan Hulsker, Agathon/Unieboek bv, Weesp, 1885; *Vincent and Theo Van Gogh,* Translated and edited by James M. Miller, Fuller Publ.: Ann Arbor, 1990, p. 101.
204. Translated by Arnold Pomerans, edited by Ronald de Leeuw, published in *The Letters of Vincent Van Gogh,* Penguin Classics, number 204, June 1 or 2, 1882.
205. Written c. 6-7 December 1883 in Nuenen. Translated by Robert Harrison, edited by Robert Harrison, published in *The Complete Letters of Vincent van Gogh, Publisher: Bulfinch, 1991*, number 345. URL: http://webexhibits.org/vangogh/letter/14/345.htm.
206. Written c. 17 December 1883 in Nuenen. Translated by Mrs. Johanna van Gogh-Bonger, edited by Robert Harrison, published in *The Complete Letters of Vincent van Gogh, Publisher: Bulfinch, 1991*, number 346. URL: http://webexhibits.org/vangogh/letter/14/346.htm.
207. Written c. 17 December 1883 in Nuenen. Translated by Mrs. Johanna van Gogh-Bonger, edited by Robert Harrison, published in *The Complete Letters of Vincent van Gogh, Publisher: Bulfinch, 1991*, number 346. URL: http://webexhibits.org/vangogh/letter/14/346.htm.
208. *Lotgenoten,* Jan Hulsker, Agathon/Unieboek bv, Weesp, 1885; *Vincent and Theo Van Gogh,* Translated and edited by James M. Miller, Fuller Publ.: Ann Arbor, 1990, p. 171.

209. Written c. 22-25 December 1883 in Nuenen. Translated by Mrs. Johanna van Gogh-Bonger, edited by Robert Harrison, published in *The Complete Letters of Vincent van Gogh, Publisher*: Bulfinch, 1991, number 349. URL: http://webexhibits.org/vangogh/letter/14/349.htm.
210. Written late December 1883 in Nuenen. Translated by Mrs. Johanna van Gogh-Bonger, edited by Robert Harrison, published in *The Complete Letters of Vincent van Gogh, Publisher: Bulfinch, 1991*, number. URL: http://webexhibits.org/vangogh/letter/14/193a.htm.
211. Written late December 1883 in Nuenen. Translated by Mrs. Johanna van Gogh-Bonger, edited by Robert Harrison, published in *The Complete Letters of Vincent van Gogh, Publisher: Bulfinch, 1991*, URL: http://webexhibits.org/vangogh/letter/14/193a.htm.
212. Found in *Lotgenoten,* Jan Hulsker, Agathon/Unieboek, Weesp, 1885; *Vincent and Theo Van Gogh,* Translated and edited by James M. Miller, Fuller Publ.: Ann Arbor, 1990, p. 173.
213. Written 13 February 1884 in Nuenen. Translated by Mrs. Johanna van Gogh-Bonger, edited by Robert Harrison, published in *The Complete Letters of Vincent van Gogh, Publisher: Bulfinch, 1991*, number 356. URL: http://webexhibits.org/vangogh/letter/14/356.htm.
214. Written c. 1 March 1884 in Nuenen. Translated by Mrs. Johanna van Gogh-Bonger, edited by Robert Harrison, published in *The Complete Letters of Vincent van Gogh, Publisher: Bulfinch, 1991*, number 358. URL: http://webexhibits.org/vangogh/letter/14/358.htm.
215. Written c. 11 March 1884 in Nuenen. Translated by Mrs. Johanna van Gogh-Bonger, edited by Robert Harrison, published in *The Complete Letters of Vincent van Gogh, Publisher: Bulfinch, 1991*, number 361. URL: http://webexhibits.org/vangogh/letter/14/361.htm.
216. These positive comments can be found in two subsequent letters, Written October 22, 1884 in Nuenen. Translated by Mrs. Johanna van Gogh-Bonger, edited by Robert Harrison, published in *The Complete Letters of Vincent van Gogh, Publisher: Bulfinch, 1991*, number 382.
URL: http://webexhibits.org/vangogh/letter/14/382.htm; Written late October 1884 in Nuenen. Translated by Mrs. Johanna van Gogh-Bonger, edited by Robert Harrison, published in *The Complete Letters of Vincent van Gogh, Publisher: Bulfinch, 1991*, number 383.
URL: http://webexhibits.org/vangogh/letter/14/383.htm.
217. In the newspaper *De Amsterdammer*, April 14 and 21, 1912.
218. Written c. 30 April 1885 in Nuenen. Translated by Mrs. Johanna van Gogh-Bonger, edited by Robert Harrison, published in *The Complete Letters of Vincent van Gogh, Publisher: Bulfinch, 1991*, number 404. URL: http://webexhibits.org/vangogh/letter/15/404.htm.
219. Written c. 30 April 1885 in Nuenen. Translated by Mrs. Johanna van Gogh-Bonger, edited by Robert Harrison, published in *The Complete Letters of Vincent van Gogh, Publisher: Bulfinch, 1991*, number 404. URL: http://webexhibits.org/vangogh/letter/15/404.htm.

220. Written c. 30 April 1885 in Nuenen. Translated by Mrs. Johanna van Gogh-Bonger, edited by Robert Harrison, published in *The Complete Letters of Vincent van Gogh, Publisher: Bulfinch, 1991*, number 404. URL: http://webexhibits.org/vangogh/letter/15/404.htm.
221. Written c. 30 April 1885 in Nuenen. Translated by Mrs. Johanna van Gogh-Bonger, edited by Robert Harrison, published in *The Complete Letters of Vincent van Gogh, Publisher: Bulfinch, 1991*, number 404. URL: http://webexhibits.org/vangogh/letter/15/404.htm.
222. Written 24 May 1885 in Nuenen. Translated by Mrs. Johanna van Gogh-Bonger, edited by Robert Harrison, published in *The Complete Letters of Vincent van Gogh, Publisher: Bulfinch, 1991*, number. URL: http://webexhibits.org/vangogh/letter/15/R51a.htm.
223. Written 2nd half August 1885 in Nuenen. Translated by Mrs. Johanna van Gogh-Bonger, edited by Robert Harrison, published in *The Complete Letters of Vincent van Gogh, Publisher: Bulfinch, 1991*, number R57. URL: http://webexhibits.org/vangogh/letter/15/R57.htm.
224. Written July 1885 in Nuenen. Translated by Mrs. Johanna van Gogh-Bonger, edited by Robert Harrison, published in *The Complete Letters of Vincent van Gogh, Publisher: Bulfinch, 1991*, number 418. URL: http://webexhibits.org/vangogh/letter/15/418.htm.
225. Written October 1885 in Nuenen. Translated by Mrs. Johanna van Gogh-Bonger, edited by Robert Harrison, published in *The Complete Letters of Vincent van Gogh, Publisher: Bulfinch, 1991*, number 427. URL: http://webexhibits.org/vangogh/letter/15/427.htm.
226. *Van Gogh*, Jean Leymarie, translated by James Emmons, 1968; rpt., Rizzoli Int'l Publ., 1977pp. 52-55
227. *Van Gogh's Women: Vincent's Love Affairs and Journey into Madness, Carroll & Graf, 2004*, pp. 65-66.
228. Albert Lubin, *Stranger on the Earth,* Holt, Rinehart and Winston, Inc., 1972, p. 36; *Van Gogh and God: A Creative Spiritual Quest,* Cliff Edwards, Loyola Press, 1989, p. 99.
229. Albert Lubin, *Stranger on the Earth,* Holt, Rinehart and Winston, Inc., 1972, p. 36; *Van Gogh and God: A Creative Spiritual Quest,* Cliff Edwards, Loyola Press, 1989, pp. 99-100.
230. Albert Lubin, *Stranger on the Earth,* Holt, Rinehart and Winston, Inc., 1972, p. 36; *Van Gogh and God: A Creative Spiritual Quest,* Cliff Edwards, Loyola Press, 1989, p. 100.
231. Albert Lubin, *Stranger on the Earth,* Holt, Rinehart and Winston, Inc., 1972, p. 36; *Van Gogh and God: A Creative Spiritual Quest,* Cliff Edwards, Loyola Press, 1989, pp. 100-101.
232. Albert Lubin, *Stranger on the Earth,* Holt, Rinehart and Winston, Inc., 1972, p. 36; *Van Gogh and God: A Creative Spiritual Quest,* Cliff Edwards, Loyola Press, 1989, pp. 101.
233. Albert Lubin, *Stranger on the Earth,* Holt, Rinehart and Winston, Inc., 1972, p. 36; *Van Gogh and God: A Creative Spiritual Quest,* Cliff Edwards, Loyola Press, 1989, 102.

234. *The Symbolic Language of Vincent Van Gogh,* McGraw-Hill, 1963, p. 43.
235. *The Symbolic Language of Vincent Van Gogh,* McGraw-Hill, 1963, p. 35.
236. *The Symbolic Language of Vincent Van Gogh,* McGraw-Hill, 1963, pp. 35-36.
237. *The Symbolic Language of Vincent Van Gogh,* McGraw-Hill, 1963, p. 39.
238. Written 28 November 1885 in Antwerp. Translated by Mrs. Johanna van Gogh-Bonger, edited by Robert Harrison, published in *The Complete Letters of Vincent van Gogh, Publisher: Bulfinch, 1991*, number 437. URL: http://webexhibits.org/vangogh/letter/16/437.htm.
239. Written c. 19 December 1885 in Antwerp. Translated by Mrs. Johanna van Gogh-Bonger, edited by Robert Harrison, published in *The Complete Letters of Vincent van Gogh, Publisher: Bulfinch, 1991*, number 441. URL: http://webexhibits.org/vangogh/letter/16/441.htm.
240._Written in Antwerp. Translated by Mrs. Johanna van Gogh-Bonger, edited by Robert Harrison, published in *The Complete Letters of Vincent van Gogh, Publisher: Bulfinch, 1991*, number 445. URL: http://webexhibits.org/vangogh/letter/16/445.htm.
241 Written c. 18 February 1886 in Antwerp. Translated by Mrs. Johanna van Gogh-Bonger, edited by Robert Harrison, published in *The Complete Letters of Vincent van Gogh, Publisher: Bulfinch, 1991*, number 456. URL: http://webexhibits.org/vangogh/letter/16/456.htm.
242. Written c. 15 December 1885 in Antwerp. Translated by Mrs. Johanna van Gogh-Bonger, edited by Robert Harrison, published in *The Complete Letters of Vincent van Gogh, Publisher: Bulfinch, 1991*, number 440. URL: http://webexhibits.org/vangogh/letter/16/440.htm.
243. Quoted in *Lotgenoten,* Jan Hulsker, Agathon/Unieboek, Weesp, 1885; *Vincent and Theo Van Gogh,* Translated and edited by James M. Miller, Fuller Publ.: Ann Arbor, 1990, p. 219.
244. Quoted in *Lotgenoten,* Jan Hulsker, Agathon/Unieboek, Weesp, 1885; *Vincent and Theo Van Gogh,* Translated and edited by James M. Miller, Fuller Publ.: Ann Arbor, 1990, p. 219.
245. Written 8-15 December 1885 in Antwerp. Translated by Mrs. Johanna van Gogh-Bonger, edited by Robert Harrison, published in *The Complete Letters of Vincent van Gogh, Publisher: Bulfinch, 1991*, number 439. URL: http://webexhibits.org/vangogh/letter/16/439.htm.
246. *Lotgenoten,* Jan Hulsker, Agathon/Unieboek, Weesp, 1885; *Vincent and Theo Van Gogh,* Translated and edited by James M. Miller, Fuller Publ.: Ann Arbor, 1990, p. 217.
247. *Lotgenoten,* Jan Hulsker, Agathon/Unieboek, Weesp, 1885; *Vincent and Theo Van Gogh,* Translated and edited by James M. Miller, Fuller Publ.: Ann Arbor, 1990, p. 218.
248. For more discussion see Malcolm Easton, *Artists and Writers in Paris. The Bohemian Idea, 1803–1867*, Arnold, 1964; César Graña, *Bohemian versus Bourgeois: French Society and the French Man of Letters in the Nineteenth Century*, Basic Books, 1964; Elizabeth Wilson, *Bohemians: The Glamorous Outcasts*, Rutgers Univ., 2001; and Christine Stansell, *American Moderns: Bohemian New York and the Creation of a New Century,* Henry Holt, 2000.

249. There many books on the birth of Impressionism and other artistic developments in France at this time, including such large anthologies as H. H. Arnason, *History of Modern Art: Painting, Sculpture, Architecture,* Abrams; Marilyn Stokstad, *Art History,* Vol. Two, Pearson, 2005; David Piper, *The Illustrated History of Art,* Bounty Books,1991; Sam Hunter and John Jacobus, *Modern Art: Painting, Sculpture, Architecture,* Abrams, 3rd ed., 1992; and more focused works such as Felix Baumann, Marianne Karabelnik, et al, *Degas Portraits*. London: Merrell Holberton, 1994; Bernard Denvir, *The Thames and Hudson Encyclopaedia of Impressionism*. London: Thames and Hudson, 1990; Robert Gordon and Andrew Forge, *Degas*. New York: Harry N. Abrams, 1988: Sylvie Patin and John Rewald, *Cezanne: The Early Years 1859-1872*. New York: Harry N. Abrams, 1988; Ira Moskowitz and Maurice Serullaz, *French Impressionists: A Selection of Drawings of the French 19th Century*. Boston and Toronto: Little, Brown and Company, 1962; John Rewald, *The History of Impressionism* (4th, Revised Ed.). New York: The Museum of Modern Art, 1973; John Richardson, *Manet* (3rd Ed.). Oxford: Phaidon Press Ltd., 1976, and Robert Rosenblum, *Paintings in the Musée d'Orsay*. New York: Stewart, Tabori & Chang, 1989.

250. H. R. Graetz, *The Symbolic Language of Vincent Van Gogh,* McGraw-Hill, 1963, p. 48.

251. Quoted in *Vincent Van Gogh,* Marc EdoTralbaut, Viking Press, 1969, p. 216.

252. Quoted in *Vincent Van Gogh,* Marc EdoTralbaut, Viking Press, 1969, p. 200.

253. *Vincent Van Gogh,* Marc EdoTralbaut, Viking Press, 1969, pp. 223- 224.

254. Written 24 September 1888 in Arles. Translated by Mrs. Johanna van Gogh-Bonger, edited by Robert Harrison, published in *The Complete Letters of Vincent van Gogh, Publisher: Bulfinch, 1991*, number 542. URL: http://webexhibits.org/vangogh/letter/18/542.htm.

255. *Vincent Van Gogh,* Marc EdoTralbaut, Viking Press, 1969, p. 214.

256. Translated by Robert Harrison, edited by Robert Harrison, published in *Letters of Vincent van Gogh 1886-1890. Scolar Press, 1977*, number 463. URL: http://webexhibits.org/vangogh/letter/18/463.htm.

257. September 29, 1888; Quoted in *Lotgenoten,* Jan Hulsker, Agathon/Unieboek, Weesp, 1885; *Vincent and Theo Van Gogh,* Translated and edited by James M. Miller, Fuller Publ.: Ann Arbor, 1990.

258. *Vincent Van Gogh,* Marc EdoTralbaut, Viking Press, 1969, p. 325.

259. Quoted in *Vincent Van Gogh,* Marc EdoTralbaut, Viking Press, 1969, p. 325.

260. *Vincent Van Gogh,* Marc EdoTralbaut, Viking Press, 1969, p. 240.

261. *Art History, Volume Two*, rev., Marilyn Stokstad and David Cateforis, Pearson, 2005, pp. 998-999.

262. *Vincent Van Gogh,* Marc EdoTralbaut, Viking Press, 1969, pp. 253-254.

263. Quoted in *Vincent Van Gogh,* Marc EdoTralbaut, Viking Press, 1969, p. 256.

264. Quoted in *Vincent Van Gogh,* Marc EdoTralbaut, Viking Press, 1969, p. 256.

265. *Van Gogh in Arles,* Alfred Nemeczek, Prestel, 1999, pp. 100-101.

266. Translated by Mrs. Johanna van Gogh-Bonger, edited by Robert Harrison, published in *The Complete Letters of Vincent van Gogh, Publisher: Bulfinch, 1991*, number 563. URL: http://webexhibits.org/vangogh/letter/18/563.htm.

267. Written 17 January 1889 in Arles. Translated by Mrs. Johanna van Gogh-Bonger, edited by Robert Harrison, published in *The Complete Letters of Vincent van Gogh, Publisher: Bulfinch, 1991*, number 571. URL: http://webexhibits.org/vangogh/letter/19/571.htm.
268. Written 10 or 11 February 1890 in Saint-Rémy. Translated by Mrs. Johanna van Gogh-Bonger, edited by Robert Harrison, published in *The Complete Letters of Vincent van Gogh, Publisher: Bulfinch, 1991*, number. URL: http://webexhibits.org/vangogh/letter/20/626a.htm.
269. H. R. Graetz, *The Symbolic Language of Vincent Van Gogh,* McGraw-Hill, 1963, p. 139.
270. *Vincent Van Gogh: A Psychological Study,* Humberto Nagera, George Allen and Unwin, London, 1967, pp. 138-9.
271. *Vincent Van Gogh: A Psychological Study,* Humberto Nagera, George Allen and Unwin, London, 1967, p. 140.
272. *Vincent Van Gogh: A Psychological Study,* Humberto Nagera, George Allen and Unwin, London, 1967, pp. 140-141.
273. Written 9 September 1888 in Arles. Translated by Mrs. Johanna van Gogh-Bonger, edited by Robert Harrison, published in *The Complete Letters of Vincent van Gogh, Publisher: Bulfinch, 1991*, number 534. URL: http://webexhibits.org/vangogh/letter/18/534.htm.
274. *Avant et Après,*, Paul Gauguin, 1903; *Paul Gauguin's Intimate Journals,* trans. 1923, Van Wyck Brooks, 1923; rpt., Dover, 1997, ISBN 0-486-29441-2; quoted in *Vincent Van Gogh,* Marc EdoTralbaut, Viking Press, 1969, p. 259.
275. A letter from Bernard to Albert Aurier, Jan. 1, 1889; rpt. in *Van Gogh in Arles*, Alfred Nemeczek, Prestel,1999, p. 96.
276. *Le Forum Republicaain,* Dec. 30, 1888; translated and included in numerous biographies of Vincent, including Albert Lubin, *Stranger on the Earth,* Holt, Rinehart and Winston, Inc., 1972.
277. "Memoir of Vincent van Gogh," *The Complete Letters of Vincent van Gogh,* Vol. I, American edition, Greenwich, Connecticut: New York Graphic Society, 1958, p. xlv; reprinted in Albert J. Lubin, p. 156.
278. A. J. Westerman Holstijn, "The Psychological Development of Vincent van Gogh," *American Imago,* VIII, 1950, pp. 239-273.
279. Daniel E. Schneider, *The Psychoanalyst and the Artist,* New York: International Universities Press, 1950, pp. 227-239.
280. Jacques Schnier, "The Blazing Sun: A Psychoanalytic Approach to Van Gogh," *American Imago,* VII, 1950, pp. 143-162.
281. Frank Elgar, *Van Gogh: A Study of his Life and Work*, New York, Praeger, 1958, pp. 202-203.
282. Translated and included in numerous biographies of Vincent, including Albert Lubin, *Stranger on the Earth,* Holt, Rinehart and Winston, Inc., 1972, p. 158.
283. H. R. Graetz, *The Symbolic Language of Vincent Van Gogh,* McGraw-Hill, 1963, p. 152.
284. H. R. Graetz, *The Symbolic Language of Vincent Van Gogh,* McGraw-Hill, 1963, p. 152.

285. H. R. Graetz, *The Symbolic Language of Vincent Van Gogh,* McGraw-Hill, 1963, p. 153.
286. Written 9 April 1888 in Arles. Translated by Mrs. Johanna van Gogh-Bonger, edited by Robert Harrison, published in *The Complete Letters of Vincent van Gogh, Publisher: Bulfinch, 1991*, number 474. URL: http://webexhibits.org/vangogh/letter/18/474.htm.
287. Written 9 April 1888 in Arles. Translated by Robert Harrison, edited by Robert Harrison, published in *The Complete Letters of Vincent van Gogh, Publisher: Bulfinch, 1991*, number B03. URL: http://webexhibits.org/vangogh/letter/18/B03.htm.
288. Albert Lubin, *Stranger on the Earth,* Holt, Rinehart and Winston, Inc., 1972, p. 159-160.
289. Albert Lubin, *Stranger on the Earth,* Holt, Rinehart and Winston, Inc., 1972, p. 179.
290. *Vincent Van Gogh,* Marc EdoTralbaut, Viking Press, 1969, p. 260.
291. Quoted in Erwin van Meekeren, *Starry starry night: Life and psychiatric history of Vincent van Gogh*, Benecke, N. I., Amsterdam; trans., Willem Hurkmans and Jackie Kelly, 2nd ed., 2003, p.118.
292. Erwin van Meekeren, *Starry starry night: Life and psychiatric history of Vincent van Gogh*, Benecke, N. I., Amsterdam; trans., Willem Hurkmans and Jackie Kelly, 2nd ed., 2003, p.118.
293. Letter 590; trans., Arnold Pomerans, *The Letters of Vincent Van Gogh,* ed., Ronald de Leeuw, Penguin Books, pp. 427-8.
294. Trans., Arnold Pomerans, *The Letters of Vincent Van Gogh,* ed., Ronald de Leeuw, Penguin Books Trans. Leeuw, pp. 428-430.
295. *Vincent Van Gogh,* Marc EdoTralbaut, Viking Press, 1969, pp. 254-265.
296. Letter 576F; trans., Arnold Pomerans, *The Letters of Vincent Van Gogh,* ed., Ronald de Leeuw, Penguin Books p. 432.
297. H. R. Graetz, *The Symbolic Language of Vincent Van Gogh,* McGraw-Hill, 1963, p. 146.
298. H. R. Graetz, *The Symbolic Language of Vincent Van Gogh,* McGraw-Hill, 1963, p. 147.
299. Erwin van Meekeren, *Starry starry night: Life and psychiatric history of Vincent van Gogh*, Benecke, N. I., Amsterdam; trans., Willem Hurkmans and Jackie Kelly, 2nd ed., 2003, p. 119.
300. Written 19 March 1889 in Arles. Translated by Mrs. Johanna van Gogh-Bonger, edited by Robert Harrison, published in *The Complete Letters of Vincent van Gogh, Publisher: Bulfinch, 1991*, number 579. URL: http://webexhibits.org/vangogh/letter/19/579.htm.
301. Albert Lubin, *Stranger on the Earth,* Holt, Rinehart and Winston, Inc., 1972, p. 184-185.
302. Written 19 March 1889 in Arles. Translated by Mrs. Johanna van Gogh-Bonger, edited by Robert Harrison, published in *The Complete Letters of Vincent van Gogh, Publisher: Bulfinch, 1991*, number 579. URL: http://webexhibits.org/vangogh/letter/19/579.htm.
303. Quoted in *Vincent Van Gogh,* Marc EdoTralbaut, Viking Press, 1969, p. 270.

304. Written c. 21 April 1889 in Arles. Translated by Mrs. Johanna van Gogh-Bonger, edited by Robert Harrison, published in *The Complete Letters of Vincent van Gogh, Publisher: Bulfinch, 1991*, number 585. URL: http://webexhibits.org/vangogh/letter/19/585.htm.
305. H. R. Graetz, *The Symbolic Language of Vincent Van Gogh,* McGraw-Hill, 1963, p. 146.
306. Originally published as part 5 of *Der gottliche Schelem,* by Paul Radin, with commentaries by C. G. Jung and Karl Kerenyi, Zurich, 1954; rpt., "On the Psychology of the Trickster Figure," *The Archetypes and the Collective Unconscious,* 2nd ed., trans., R. F.C. Hull, Princeton, 1968, p. 256.
307. *Joseph Campbell: The Power of Myth with Bill Moyers,* ed., Sue Flowers, Anchor Books, 1988, p. 109.
308. John G. Neihardt, *Black Elk Speaks: Being the Life Story of a Holy Man of the Oglala Sioux,* Pocket Books, 1959, pp. 15-16.
309. John G. Neihardt, *Black Elk Speaks: Being the Life Story of a Holy Man of the Oglala Sioux,* Pocket Books, 1959, p. 21-22.
310. John G. Neihardt, *Black Elk Speaks: Being the Life Story of a Holy Man of the Oglala Sioux,* Pocket Books, 1959, pp. 24-5.
311. John G. Neihardt, *Black Elk Speaks: Being the Life Story of a Holy Man of the Oglala Sioux,* Pocket Books, 1959, p. 29.
312. John G. Neihardt, *Black Elk Speaks: Being the Life Story of a Holy Man of the Oglala Sioux,* Pocket Books, 1959, p. 36.
313. This famous statement by Picasso comes from an interview with Marius de Zayas, translated with Picasso's approval and published as "Picasso Speaks," *The Arts* (New York), May 1923, pp. 315-326, included in *Theories of Modern Art: A Source Book by Artists and Critics*, edited by Herschel B. Chipp, Univ. of Calif. Press: Los Angeles, 1969.
314. H. R. Graetz, *The Symbolic Language of Vincent Van Gogh,* McGraw-Hill, 1963, pp. 294-295.
315. Letters 581a and 590a.
316. Written 24 March 1889 in Arles. Translated by Mrs. Johanna van Gogh-Bonger, edited by Robert Harrison, published in *The Complete Letters of Vincent van Gogh, Publisher: Bulfinch, 1991*, number 581. URL http://webexhibits.org/vangogh/letter/19/581.htm.
317. H. R. Graetz, *The Symbolic Language of Vincent Van Gogh,* McGraw-Hill, 1963, p. 151.
318. *Van Gogh in Arles*, Alfred Nemeczek, trans., Fiona Elliott, Prestel, Munich, 1999, pp. 110-111.
319. Translated by Mrs. Johanna van Gogh-Bonger, edited by Robert Harrison, published in *The Complete Letters of Vincent van Gogh, Publisher: Bulfinch, 1991*, number 582. URL: http://webexhibits.org/vangogh/letter/19/582.htm.
320. Translated by Mrs. Johanna van Gogh-Bonger, edited by Robert Harrison, published in *The Complete Letters of Vincent van Gogh, Publisher: Bulfinch, 1991*, number 583. URL: http://webexhibits.org/vangogh/letter/19/583.htm.
321. Albert Lubin, *Stranger on the Earth,* Holt, Rinehart and Winston, Inc., 1972, p. 189.

322. Albert Lubin, *Stranger on the Earth,* Holt, Rinehart and Winston, Inc., 1972, p. 189-190.
323. Albert Lubin, *Stranger on the Earth,* Holt, Rinehart and Winston, Inc., 1972, pp. 190-191.
324._Written 22 August 1889 in Saint-Rémy. Translated by Robert Harrison, edited by Robert Harrison, published in *The Complete Letters of Vincent van Gogh, Publisher: Bulfinch, 1991*, number 601. URL: http://webexhibits.org/vangogh/letter/20/601.htm.
325. Written 3 or 4 September 1889 in Saint-Rémy, Translated by Mrs. Johanna van Gogh-Bonger, edited by Robert Harrison, published in *The Complete Letters of Vincent van Gogh, Publisher: Bulfinch, 1991*, number 602, URL: http://webexhibits.org/vangogh/letter/20/602.htm.
326. Written 5 or 6 September 1889 in Saint-Rémy. Translated by Mrs. Johanna van Gogh-Bonger, edited by Robert Harrison, published in *The Complete Letters of Vincent van Gogh, Publisher: Bulfinch, 1991*, number 604. URL: http://webexhibits.org/vangogh/letter/20/604.htm.
327. Written 7 or 8 September 1889 in Saint-Rémy. Translated by Mrs. Johanna van Gogh-Bonger, edited by Robert Harrison, published in *The Complete Letters of Vincent van Gogh, Publisher: Bulfinch, 1991*, number 605. URL: http://webexhibits.org/vangogh/letter/20/605.htm.
328. Written 2 February 1890 in Saint-Rémy. Translated by Mrs Johanna van Gogh-Bonger, edited by Robert Harrison, published in *The Complete Letters of Vincent van Gogh, Publisher: Bulfinch, 1991*, number 625. URL: http://webexhibits.org/vangogh/letter/20/625.htm.
329. *Vincent Van Gogh: A Psychological Study,* Humberto Nagera, George Allen and Unwin, London, 1967, pp. 128-129.
330. Written 23 April 1890 in Saint-Rémy. Translated by Mrs. Johanna van Gogh-Bonger, edited by Robert Harrison, published in *The Complete Letters of Vincent van Gogh, Publisher: Bulfinch, 1991*, number T32. URL: http://webexhibits.org/vangogh/letter/20/T32.htm.
331. Written 30 April 1890 in Saint-Rémy. Translated by Mrs. Johanna van Gogh-Bonger, edited by Robert Harrison, published in *The Complete Letters of Vincent van Gogh, Publisher: Bulfinch, 1991*, number 629. URL: http://webexhibits.org/vangogh/letter/20/629.htm.
332. Letter W 11; written 30 April 1889 in Arles, trans. by Mrs. Johanna van Gogh-Bonger, edited by Robert Harrison, published in *The Complete Letters of Vincent van Gogh, Publisher: Bulfinch, 1991*, number W11, URL: http://webexhibits.org/vangogh/letter/19/W11.htm.
333. Rpt., *Vincent Van Gogh: A Psychological Study,* Humberto Nagera, George Allen and Unwin, London, 1967, p. 129.
334. Written 20 May 1890 in Auvers-sur-Oise. Translated by Robert Harrison, edited by Robert Harrison, published in *Letters of Vincent van Gogh, 1886 - 1890: Publisher: Scolar Press, 1977*, number 635.URL: http://webexhibits.org/vangogh/letter/21/635.htm.

335. Remo Fabbri, "Dr. Paul-Ferdinand Gachet: Vincent van Gogh's Last Physician," *Transactions of the College of Physicians* 4, series 32-33, 1964, pp. 202-208.
336. Written 30 April 1889 in Arles. Trans., Johanna van Gogh-Bonger, edited by Robert Harrison, published in *The Complete Letters of Vincent van Gogh, Publisher: Bulfinch, 1991*, number 588. URL: http://webexhibits.org/vangogh/letter/19/588.htm.
337. Translated by Mrs. Johanna van Gogh-Bonger, edited by Robert Harrison, published in *The Complete Letters of Vincent van Gogh, Publisher: Bulfinch, 1991*, number 604. URL: http://webexhibits.org/vangogh/letter/20/604.htm.
338. Written 7 or 8 September 1889 in Saint-Rémy. Trans., Johanna van Gogh-Bonger, edited by Robert Harrison, published in *The Complete Letters of Vincent van Gogh, Publisher: Bulfinch, 1991*, number 605.URL: http://webexhibits.org/vangogh/letter/20/605.htm.
339. Written 30 or 31 December 1889 in Saint-Rémy. Translated by Mrs. Johanna van Gogh-Bonger, edited by Robert Harrison, published in *The Complete Letters of Vincent van Gogh,* Publisher: Bulfinch*, 1991*, number .URL: http://webexhibits.org/vangogh/letter/20/622a.htm.
340. *Vincent Van Gogh,* Marc EdoTralbaut, Viking Press, 1969, p. 324.
341. A letter to Dr. A. Bredius, publ. in *Oud-Holland,* 1934, rpt. *Vincent Van Gogh,* Marc EdoTralbaut, Viking Press, 1969, p. 328.
342. B. G. P., *Diccionario universal de la mitologia,* Barcelona, 1835; J. E. Cirlot, *A Dictionary of Symbols,* 2nd ed., trans., Jack Sage, Routledge & Kegan Paul, 1971, p. 228.
343. J. E. Cirlot, *A Dictionary of Symbols,* 2nd ed., trans., Jack Sage, Routledge & Kegan Paul, 1971, p. 228
344. J. E. Cirlot, *A Dictionary of Symbols,* 2nd ed., trans., Jack Sage, Routledge & Kegan Paul, 1971, p. 76.
345. Jean Chevalier and Alain Gheerbrant, *A Discionary of Symbols,* trans., John Buchanan-Brown, Penguin, 1994, p. 701.
346. Sigmund Freud, *Complete Works,* p. 123; rpt., Jean Chevalier and Alain Gheerbrant, *A Discionary of Symbols,* trans., John Buchanan-Brown, Penguin, 1994, p. 311.
347. Carl Jung, *The Archetypes and the Collective Unconscious,* 2nd ed., trans., R. F. C. Hull, Bollinger Series XX, Princeton Univ., 1968, p. 21.
348. *Joseph Campbell: The Power of Myth with Bill Moyers,* ed., Sue Flowers, Anchor Books, 1988, p. 41.
349. Jean Chevalier and Alain Gheerbrant, *A Discionary of Symbols,* trans., John Buchanan-Brown, Penguin, 1994, pp. 311-312.
350. Written c. 18 June 1888 in Arles, translated by Mrs. Johanna van Gogh-Bonger, edited by Robert Harrison, published in *The Complete Letters of Vincent van Gogh, Publisher: Bulfinch, 1991*, number B07. URL: http://webexhibits.org/vangogh/letter/18/B07.htm.
351. *Vincent Van Gogh,* Marc EdoTralbaut, Viking Press, 1969, p. 279.
352. *Vincent Van Gogh,* Marc EdoTralbaut, Viking Press, 1969, pp. 280-281.

353. Kathleen Powers Erickson, *At Eternity's Gate: The Spiritual Vision of Vincent van Gogh,* William B. Eerdmans, 1998, pp. 171-172.
354. Kathleen Powers Erickson, *At Eternity's Gate: The Spiritual Vision of Vincent van Gogh,* William B. Eerdmans, 1998, p. 172.
355. Kathleen Powers Erickson, *At Eternity's Gate: The Spiritual Vision of Vincent van Gogh,* William B. Eerdmans, 1998, p. 172.
356. Kathleen Powers Erickson, *At Eternity's Gate: The Spiritual Vision of Vincent van Gogh,* William B. Eerdmans, 1998, p. 156; taken from Tuskasa Kodera, *Vincent van Gogh: Christianity versus Nature*, Amsterdam: John Benjamins, 1990, p. 35.
357. Kathleen Powers Erickson, *At Eternity's Gate: The Spiritual Vision of Vincent van Gogh,* William B. Eerdmans, 1998, p. 156.
358. Kathleen Powers Erickson, *At Eternity's Gate: The Spiritual Vision of Vincent van Gogh,* William B. Eerdmans, 1998, p. 170.
359. Kathleen Powers Erickson, *At Eternity's Gate: The Spiritual Vision of Vincent van Gogh,* William B. Eerdmans, 1998, p. 176.
360. *Vincent Van Gogh,* Marc EdoTralbaut, Viking Press, 1969, p. 283.
361. *Vincent Van Gogh,* Marc EdoTralbaut, Viking Press, 1969pp. 282-283.
362. H. R. Graetz, *The Symbolic Language of Vincent Van Gogh*, McGraw-Hill, 1963, p. 199.
363. *Memories, Dreams, Reflections*, ed., Aniela Jaffe, Pantheon, 1973, pp. 87-88.
364. Carl Jung, "Commentary on the Secret of the Golden Flower," *The Collected Works of Carl Jung: Alchemical Studies*, ed., Robert Read, Michael Fordham, Gerhard Adler and William MacGuire, trans., R. F. C. Hull, Bollington Series XX, Pantheon, 1967, pp. 38-39.
365. *The Secret of the Golden Flower, a Chinese Book of Life,* trans., Richard Wilhelm, commentary by Carl Jung, Routledge, 1931, p. 19.
366. H. R. Graetz, *The Symbolic Language of VincentVan Gogh*, McGraw-Hill, 1963, p. 213.
367. Marc Edo Tralbaut lists several names, *Vincent Van Gogh*, Viking Press, 1969, p. 287.
368. A study by J. K. Arenberg, L. F. Countryman, L. H. Bernstein and G. E. Shambaugh Jr., of the International Meniere's Disease Research Institute, Colorado Neurologic Institute, Englewood, *Journal of the American Medical Association*, Vol 264, No. 4, July 25, 1990.
369. Harry S. Abram, MD, *American Journal of Psychiatry*, 1986.
370. "Het ziektebeeld van Vincent van Gogh," *Soma en Psyche*, 16, Jan 1 and April 2; referred to by Erwin van Meekeren, *Starry starry night: Life and psychiatric history of Vincent van Gogh,*Benecke N. I., 2003, p. 131.
371. Frank Milner, *Van Gogh*, PRC Publ., 1990; rpt., 2004.
372. A. J. Westerman Holstijn, "The Psychological Development of Vincent van Gogh," *American Imago*, VIII, 1951, pp. 239-273.
373. Daniel E. Schneider, *The Psychoanalyst and the Artist,* International Universities Press, 1950, pp. 227-239.
374. Jacques Schnier, "The Blazing Sun: A Psychoanalytic Approach to Van Gogh," *American Imago,* VII, 1950, pp. 143-162.

375. Hans Prinzhorn, *Bildnerei der Geistesdranden,* 1922; *Artistry of the Mentally Ill*, trans., James L. Foy, Springer-Verlag, 1972.
376. Taken from Marc Edo Tralbaut, *Vincent Van Gogh*, Viking Press, 1969, p. 287.
377. Karl Jaspers, *Philosophie*, Berlin: Springer; rpt., *Philosophy*, trans. E. B. Ashton, Chicago: Chicago University Press, 1969–1971.
378. Karl Jaspers, *Strindberg and Van Gogh: An Attempt at a Pathographic Analysis with Reference to Parallel Cases of Swedenborg and Hölderlin*, Translated by Oskar Grunow and David Woloshin. Tucson: University of Arizona Press, 1977.
379. http:// www.nimh.nih.gov /health/publications/ schizophrenia/what-are-the-symptoms-of-schizophrenia.shtml.
380. Derek Fell, *Van Gogh's Women: Vincent's Love Affairs and Journey into Madness*, Carroll & Graf, 2004, p. 240.
381. Derek Fell, *Van Gogh's Women: Vincent's Love Affairs and Journey into Madness*, Carroll & Graf, 2004, p. 241.
382. http:// www.nimh.nih.gov/ health/ publications/ bipolar-disorder/ complete-publication.shtml. This publication, written by Melissa Spearing of NIMH, is a revision and update of an earlier version by Mary Lynn Hendrix. Scientific information and review were provided by NIMH Director Steven E. Hyman, M.D., and NIMH staff members Matthew V. Rudorfer, M.D., and Jane L. Pearson, Ph.D. Editorial assistance was provided by Clarissa K. Wittenberg, Margaret Strock, and Lisa D. Alberts of NIMH. NIH Publication No. 3679, Printed 2002.
383. Heike E. Stucke, http://heikestucke.com/vangogh.htm.
384. http://www.nimh.nih.gov/health/publications/borderline-personality disorder. shtml.
385. *Starry Starry Night: Live and psychiatric history of Vincent van Gogh,* Benecke N. I., Amsterdam, 2003, pp. 140-141.
386. *Journal of Integrative Neuroscience*, Vol 2, No. 2, 2003, pp.149-158.
387. "Brain Damage Can Make You Brilliant," J. R. Minkel, p. 18, Oct., 2006.
388. *Van Gogh: The Man and His Work*, Prentice-Hall, 1961.
389. Van Gogh, Thames & Hudson, 1966.
390. Jules Michelet, *The Bird*, Wildwood House, 1981, p. 16.
391. Letter 652; Heike E. Stucke http://heikestucke.com/vangogh.htm.
392. *Van Gogh: The Complete Paintings*, Benedikt Taschen Verlag, September, 1997.
393. Meyer Schapiro, "On a Painting of Van Gogh," *View,* I, 1952, pp. 9-14. also in *Exploring the Arts,* ed., D. Wolberg, S. Burton, and J. Tarburton, Visual Arts Press, 1969, pp. 154-166.
394. Humberto Nagera, *Vincent Van Gogh: A Psychological Study,* George Allen and Unwin, London, 1967, p. 124.
395. Albert Lubin, *Stranger on the Earth,* Holt, Rinehart and Winston, Inc., 1972, pp. 239-240.
396. H. R. Graetz, *The Symbolic Language of Vincent Van Gogh*, McGraw-Hill, 1963, pp. 277-278.

397. *Van Gogh and God: A Creative Spiritual Quest,* Cliff Edwards, Loyola Press, 1989, pp. 150-151.
398. *Van Gogh and God: A Creative Spiritual Quest,* Cliff Edwards, Loyola Press, 1989, p. 135.
399. *Religion and Nothingness,* Keiji Nishitani, trans., Jan Van Bragt, Berkeley, 1982, p. 146.
400. *Van Gogh and God: A Creative Spiritual Quest,* Cliff Edwards, Loyola Press, 1989, p. 136.
401. *Religion and Nothingness,* Keiji Nishitani, trans., Jan Van Bragt, Berkeley, 1982, p. 157.
402. *Religion and Nothingness,* Keiji Nishitani, trans., Jan Van Bragt, Berkeley, 1982, p. 128.
403. *Religion and Nothingness,* Keiji Nishitani, trans., Jan Van Bragt, Berkeley, 1982, p. 219.
404. See *The Dance of Shiva*, Ananda K. Coomaraswamy; Noonday Press, Rev. Ed. 1972; rpt. as *The Dance of Siva*, Dover, 1985, ISBN: 0486248178. *Shiva: The Wild God of Power and Ecstasy*,Wolf-Dieter Storl, Inner Traditions, 2004, ISBN: 159477014X. *Siva Sutras: The Supreme Awakening*, Swami Lakshmangoo, Authorhouse, 2002, ISBN: 0759654573.
405. *Joseph Campbell: The Power of Myth with Bill Moyers,* ed., Sue Flowers, Anchor Books, 1988, pp. 279-280.
406. *Beyond Belief: The Secret Gospel of Thomas*, Random House, 2003, pp. 123. *The Acts of John* (2nd century) is a collection of stories and traditions inspired by John's gospel. "The Round Dance of the Cross," is in Acts of John 94.1-4. For a recently edited Greek text with French translation and notes, see E. Junod and J. P. Kastli, *Acta Johannis: Praefatio-Texus, in Corpus Christinaorum* (Turnhout, 1983), English trans. Barbara E. Bowe, "Dancing into the Divine: The Hymn of the Dance in the Acts of John," *Journal of Early Christian Studies* 7:1 (1999), 83-104. According to *Irenaeus, Libros Quinque Adversus Haereses*, ed. W. W. Harvey, Cambridge, 1851, an anonymous follower of Valentinus wrote the Round Dance, suggesting it was a scene left off of John's gospel, in which Jesus danced with his disciples the night he was betrayed.
407. *The Power of Myth with Bill Moyers*, Doubleday, 1988, pp. 135-137.
408. *Van Gogh and God: A Creative Spiritual Quest,* Cliff Edwards, Loyola Press, 1989, p. 151-152.
409. *At Eternity's Gate: The Spiritual Vision of Vincent van Gogh,* Kathleen Powers Erickson, William B. Eerdmans, 1998, p. 165.
410. Ken Kesey, *Spit in the Ocean, #7*, ed., Ed McClanahan, Penguin, 2003.
411. John Keats, *Ode on a Grecian Urn,* 1819; rpt., *The Norton Anthology of Poetry,* ed., Arthur M. Eastman, et al, 1970, pp. 698-699.

Chapter Three

Broken Windmills

At that moment they caught sight of some thirty or forty windmills, which stand on that plain, and as soon as Don Quixote saw them he said to his squire: "Fortune is guiding our affairs better than we could have wished. Look over there, friend Sancho Panza, where more than thirty monstrous giants appear. I intend to do battle with them and take all their lives. With their spoils we will begin to get rich, for this is a fair war, and it is a great service to God to wipe such a wicked brood from the face of the earth."

"What giants?" asked Sancho Panza.

"Those you see there," replied his master, "with their long arms. Some giants have them about six miles long."

"Take care, your worship," said Sancho; "Those things over there are not giants but windmills, and what seem to be their arms are the sails, which are whirled round in the wind and make the millstone turn."

—Cervantes
Don Quixote, Part I, viii

Don Quixote, Pablo Picasso
© Succession Picasso/DACS 2008

Part I

Like a circle in a spiral
Like a wheel within a wheel
Never ending or beginning
On an ever spinning reel

As the images unwind
Like the circles
That you find
In the windmills of your mind

—Michel and Alan Bergman
The Windmills of Your Mind
EMU Music

For those embracing the bohemian life, the restaurant-concert hall at *Moulin de la Galette* and other such gathering places in Montmartre seethed with this rich alternative life style of the times. His correspondence reveals that Vincent must have enjoyed discussions there, arguments about art. The *Moulin de la Galette*, often used in mistaken contexts was actually a restaurant-concert hall built around three windmills The *Blute-Fin*, the *Radet*, and the *Debray* or *Poivre*. Several of Vincent's paintings and drawings from this time are of this group or various parts of it and include: *Moulin de la Galette, View of Montmartre with Windmills, Windmills at Montmartre, Windmill at Montmartre with Quarry,* and *View of Montmartre, Montmartre, the Windmills*. Not a surprising subject. Such gathering places were popular subjects for Vincent's current community of artists. August Renoir painted several, including *Dance Hall at the Moulin de la Galette*. Another one of these gathering places, *The Folies-Bergere,* featuring operettas, comic opera, popular song, gymnastics, and bawdy dancing, was the setting for Edouard Manet's appropriately named painting *A Bar at the Folies-Bergere* depicting a bar-girl, one of the *demimondaine*, standing before a mirror, completed in 1882. In 1889, Josep Oller, owner of the Paris Olympia, built the famous cabaret, *Moulin Rouge* (Red Windmill), so named because of a large, red windmill on its roof, in the "red-light" district of Pigalle, at the bottom of the hill that held the three windmills that made up *Moulin de la Galette,* and it became the subject of many paintings by Toulouse Lautrec. Vincent's paintings, however, are of the windmills from the outside and depict them more as traditional windmills, rough and without the excitement of the life that was going on inside them.

The mixture of real life taking place inside these windmills, filled with music, dance, laughter, wine, and an explosion of art, with the harsh conditions outside the windmills, and the history of them, going all the way back to the medieval world of Paris, where they were working windmills grinding out the flour to feed the city gives them a rich symbolism, one that applies to Vincent. It is important to note that he painted the harsh outsides of them, not indicating the intoxicating interiors he was frequenting, but rather connecting them to the world of the poor coal miners he once tried to comfort and for which he felt such great compassion.

By the time Vincent and his fellow artists frequented them, most of them had been destroyed, but at least four still stood, the *Moulin Rouge* transformed near the end of Vincent's stay into its famous cabaret status but perhaps never attended by Vincent, and the three at the top of the hill that Vincent hung out at and painted, the three that made up the *Moulin de la Galette* In the early 1800s, the Dubray family had come into possession of the windmills and surrounding farmland, and converted the *Blute-Fin* mill into a dance hall and bar, partly indoors and partly outdoors. The entire area became known as the *Moulin de la Galette*. The *Blute-Fin* was the largest of the windmills and the only one still having complete sails, or at the full ribs of sails, and the only one that still could be rotated about its axis. The other two mills just had the main spines of their sails, just skeletons against the sky. The *Radet* could no longer be rotated and permanently faced south, with its sails in a fixed X position. The *Poivre*, the smallest of the three, was to the east, with sails also frozen, this time looking like a plus sign.

So the old, broken down windmills, once the source of food for a busy city, now became the symbols of a new vision of life, born out of the poverty of a true red light district of prostitution and debauchery, a new vision that was about to transform the world of art and with it the entire European civilization. And among those artists was an artist that even in a world of misfits stood out, difficult to get along with, misunderstood and condemned, an outsider among outsiders, an artist painting visions that were not Realistic works, not Neo-Raphaelite works, not Impressionist works, not Symbolist works, not and not and not.

A broken windmill, sails flapping erratically, rotating sporadically, dilapidated, barely standing, yet filled with creative energy on the inside.

The winds of nature turn the towering windmills, and at least for the moment, humans gain access to archetypal powers. It is worth pausing to take note. Such structures no long seem so grandiose, not today, not when oil and electricity and the very atoms of life can be employed. There is, however, a sense of connection, of beauty in a landscape with but a few old windmills surrounded by rippling waves of grain, perhaps a few trees, a distant mountain, all resting peacefully beneath a blue sky holding a quiet, cumulous cloud that constantly mutates in response to the fantasies of the mind, a dragon, a pirate ship, a castle, whatever the imagination allows.

But sometimes even the most beautiful constructions of the human mind break, something goes wrong, and the connections between the wind and the mill stop or only function in a frustratingly erratic sputter, momentarily joining in wonderful harmony only to quickly separate, leaving the promise unfulfilled.

Part II

> Mild autism can give you a genius like Einstein. If you have severe autism, you could remain nonverbal. You don't want people to be on the severe end of the spectrum. But if you got rid of all the autism genetics, you wouldn't have science or art. All you would have is a bunch of social 'yak yaks.'
>
> —Temple Grandin

He was not normal, not handsome. He was sickly and considered to be an idiot. He limped, drooled, stuttered, and was often ill. His mother referred to him as "simple," and saw to it he was kept out of the public eye. (1)

R. K. Sherk (2) and E. M. Smallwood (3) describe the abnormal movements of his head and hands. D. Braund describes the hypertrophy of his neck muscles, and the unseemly laughter and anger. (4) Though it's hard to look back through the dark lenses of time, modern neurology suggests he had some form of mental illness. E. Rice suggests it was a secondary dystonia. (5) J. Mottershead (6) and John M. S. Pearce (7) think it was probably the athetoid variant of cerebral palsy.

However, for an idiot, Claudius' accomplishments are impressive. Born in Comta, what is now Lyon, France, August 1, 10 B.C., the son of Nero and Antonia, thus of the most royal blood, he is reported to have published many works, including forty-three books of Roman history, twenty-one books of Etruscan history, eight books of Carthaginian history, a book on philology, and a rhetorical defense of Cicero. (8)

He also demonstrated tremendous savvy for the internal politics of Rome, possibly arranging the murder of Gaius and his own ascendancy to the role of Caesar, a task for only the most skilled of politicians. Once in power he was ruthlessly practical, immediately executing those who had assassinated Gaius. It mattered not that they had been working for him; he knew that he needed to establish that he was not involved in any plots to murder an emperor and family member. Then he immediately moved his armies into England and began a lengthy attempt to annex it, more for his own political image than anything else, for he knew that he needed the army on his side, and the battles in England kept him in the position of being a strong military leader.

Once but a sickly, deformed, mentally ill, and clearly rejected child, Claudius had outwitted and out-powered the most cunning minds of his time.

In retrospect, he appears to have been capable of great cruelty and deception, while also able to be kind and generous. Whichever of these qualities gets stressed, one thing is certain, for a child categorized as mentally deficient, an idiot, he demonstrated an exceptional intelligence.

Donna Williams was "a severely withdrawn and bewildered autistic toddler," (9) whose mother and brother called her names, "a nut, a retard, a spastic," a child who "threw 'mentals' and couldn't act normal. 'Look at her, look at her,' they would say about a child who, to them, was either 'a retard' when I was in my own world or 'a nut' when I was in theirs. I couldn't win." (10).

She was considered either schizophrenic or autistic or both and more, certainly a problem child, and sent to several schools, where she failed miserably, then to special schools, where she also failed.

She had frightening dreams.

> I once woke up after being bitten by a beautiful blue-eyed kitten that had suddenly turned into a rat as I went to pat it. I had, during the nightmare, gone downstairs and played the scene in the living room, before waking up as I switched on the light. I stood there screaming as the blood dripped from my hand; then, like magic, it disappeared and everything in the room changed back to how it was in reality.
>
> Another night I woke up standing in the wardrobe doorway, rigid with horror, glaring at a doll that had suddenly returned to normal. Seconds before, I had seen it come to life, hands outstretched, its lips eerily mouthing words I could not hear, like some scene from a macabre horror film.
>
> I had become literally terrified of falling asleep. I would wait until everyone was asleep and then, frightening as my mother had become, I would go into her room and stand there watching her sleep, feeling secure in the knowledge that if anything was to get me it would have to get her, too, and she knew how to fight. From sheer tiredness, I began to hallucinate. Images on the walls would move. Unknown to my mother, I

lay stiff and silent under her bed, almost too afraid to breathe. Tears rolled down my face. I made no sound. (11)

She took on different personas to remove herself from herself, to hide behind.

> I took to sleeping under the bed and I became Willie. By this time I was three years old. Willie became the self I directed at the outside world, complete with hateful glaring eyes, a pinched-up mouth, a rigid corpselike stance, and clenched fists. Willie stamped his foot, Willie spat when he didn't like things, but the look of complete hatred was the worst weapon and Donna paid the price. His name probably derived from my own surname, and some of Willie's behavior was certainly modeled upon and in response to my violator: my mother.
>
> Willie learned to turn other people's phrases back at them in some sort of meaningful yet attacking way, thought silence still seemed a far deadlier weapon." (12)

She met the real Carol while playing at the park. Carol took Donna home to her mother, and she immediately was immediately attracted to them. "As I got older, I would compulsively bring home kitten after kitten, reenacting the way
Carol had taken me home, and I would wait and wonder when my mother would become Carol's mother. She never did. (13)

Then, in a wonderful coincidence (or is it?) with Lewis Carroll's *Through the Looking Glass*, the Carol within her "came in through the mirror" in her room, and that mirror became the entrance into Carol's world (just as it did for Alice to enter into Carroll's imaginative world; and the coincidence extends to the name Carol/Carroll; and even further to the connection with a kitten). Here is Alice's entrance from Carroll:

> But this is taking us away from Alice's speech to the kitten. 'Let's pretend that you're the Red Queen, Kitty! Do you know, I think if you sat up and folded your arms, you'd look exactly like her. Now do try, there's a dear!' And Alice got the Red Queen off the table, and set it up before the kitten as a model for it to imitate: however, the thing didn't succeed, principally, Alice said, because the kitten wouldn't fold its arms properly. So, to punish it, she held it up to the
> Looking-glass, that it might see how sulky it was, '—and if you're not good directly,' she added, 'I'll put you through in to Looking-glass House. How would you like that?
>
> 'Now, if you'll only attend, Kitty, and not talk so much, I'll tell you all may ideas about Looking-glass House. First, there's the room you can see through the glass—that's just the same as our drawing-room, only the

> things go the other way. I can see all of it when I get upon a chair—all but the bit just behind the fireplace. Oh! I do so wish I could see that bit! I want so much to know whether they've a fire in the winter: you never can tell, you know, unless our fire smokes, and then smoke comes up in that room too—but tha may be only pretence, just to make it look as if they had a fire. Well then, the books are something like our books, only the words go the wrong way: I know that, because I've held up one of our books to the glass, and then they hold up one in the other room.
>
> 'How would you like to live in Looking-glass House, Kitty? I wonder if they'd give you milk in there? Perhaps Looking-glass mild isn't good to drink— but, oh, Kitty, now we come to the passage. You can just see a little peep of the passage in Looking-glass House, if you leave the door of our drawing-room wide open and it's very like our passage as far as you can see, only you know it may be quite different on beyond. Oh, Kitty, how nice it would be if we could only get through into Looking-glass House! I'm sure it's got, oh! Such beautiful things in it! Let's pretend there's a way of getting through into it, somehow, Kitty. Let's pretend the glass has got all soft like gauze, so that we can get through. Why, it's turning into a sort of mist now, I declare! It'll be easy enough to get through—' She was up on the chimney-piece while she said the, though she hardly knew how she had got there. And certainly the glass was beginning to melt away, just like a bright silvery mist. (14)

When Donna saw her Carol in her mirror, she found that "Carol looked just like me, but the look in her eyes betrayed her identity. It was Carol all right. I began to talk to her, and she copied me. I was angry. I didn't expect her to do that. My expression asked her why, and hers asked me. I figured that the answer was a secret.

I decided that Carol understood that no one else was allowed to see me communicating with her and that this was her way of protecting me. I began to whisper to her, putting my face very close to hers and wondering why she didn't turn to hear.

When I was not in front of the mirror, Carol would disappear. I would feel deserted. When I walked in front of it, she would come back, and I'd try to look behind the mirror to find if she had gone through the door to my brother's room after all. It was Carol's house! The room I saw her in, in the mirror, was only a secret. If I could get through that room, I could leave with her, into her world. The only problem was how to get into the mirror." (15)

These and other personas allowed Donna's "real" self to hide, to "disappear." "I began to sit in a huddled ball inside the cupboard," she wrote. "I would close my eyes and try hard to lose all sensation of my own existence so that I could get into Carol's world in my mind. I became angry at any need to go to the toilet or eat, or any call to participate in the family. ... In short my humanness, my mere physical existence, was my failing.

In the darkness of that cupboard I found Carol within myself.

Carol was everything that people liked. Carol laughed a lot. Carol made friends. Carol brought things home. Carol had a mother ....

In the meantime . . . Donna had disappeared. I was by this time five years old." (16)

Donna was not born of royal bloodlines, not given the position that Claudius had, yet perhaps partially because she did not have such opportunities, her achievements are as great as or greater than his. This person, inhabiting "a place of chaos, cacophony, and dancing light—where physical contact is painful and sights and sounds have no meaning. . . labeled, at times, deaf, retarded, or disturbed . . .," by her own words "autistic," and "afflicted by a baffling condition of heightened sensory perception that imprisons the sufferer in a private, almost hallucinatory universe of patterns and colors" (back cover), not only wrote a stunning autobiography, *Nobody Nowhere*, and then followed it up with *Somebody Somewhere*, and two other autobiographical works, *Like Color to the Blind* and *Everyday Heaven*, but several other works, both non-fiction and fiction, film scripts, music, poetry, paintings, sculpture, and the list goes on. At the time of this writing, she has an online site filled with her impressive activities and accomplishments (www.donnawilliams.net).

While reading about her fantasy world, peopled by characters as real, in some ways more real than her own self, and especially while reading the passages about her interactions with Mary, an understanding psychiatrist who allowed her to keep her inner world, I was immediately reminded of the famous autobiography of Joanne Greenberg *I Never Promised You a Rose Garden* (originally published under the pseudonym Hannah Green and referring to herself as Deborah Blau), and the discussion of it by Rollo May. (17)

Joanne's private dream world did not have a clever mirror doorway. However, in another interesting coincidence, at times she entered it the same way as Alice did in Lewis Carroll's *Alice in Wonderland*. Here is Carroll's version:

> Alice had not a moment to think about stopping herself before she found herself falling down what seemed to be a very deep well.
>
> Either the well was very deep, or she fell very slowly, for she had plenty of time as she went down to look about her, and to wonder what was going to happen next. First, she tried to look down and make out what she was coming to, but it was too dark to see anything: then she looked at the sides of the well, and noticed that they were filled with cupboards and

bookshelves: here and there she saw maps and pictures hung upon pegs. She took down a jar from one of the shelves as she passed: it was labelled 'ORANGE MARMALADE' but to her great disappoint ment it was empty: she did not like to drop the jar, for fear of killing somebody underneath, so managed to put it into one of the cupboards as she fell past it.

'Well!' thought Alice to herself. 'After such a fall as this, I shall think nothing of tumbling down-stairs! How brave they'll all think me at home! Why, I wouldn't say anything about it, even if I fell off the top of the house!' (Which was very likely true.)

Down, down, down. Would the fall never come to an end? 'I wonder how many miles I've fallen by this time?' she said aloud. 'I must be getting somewhere near the centre of the earth. Let me see: that would be four thousand miles down, I thing—' (for, you, see, Alice had learnt several things of this sort in her lessons in the school-room, and though this was not a very good opportunity for showing off her knowledge, as there was o one to listen to her, still it was good practice to say it over) '—yes, that's about the right distance— but then I wonder what Latitude or Longitude I've got to?' (Alice had not the slightest idea what Latitude was, or Longitude either, but she thought they were nice grand words to say.)

Presently she began again. 'I wonder if I shall fall right through the earth! How funny it'll seem to come out among the people that walk with their heads downwards! The Antipathies, I think—' (she was rather glad there was no one listening, this time, as it didn't sound at all the right word) '—but I shall have to ask them what the name of the country is, you know. Please, Ma'am, is this New Zealand? Or Australia?' (and she tried to curtsey as she spoke—fancy, curtseying as you're falling through the air! Do you think you could manage it?) 'And what an ignorant little girl she'll think me for asking! No, it'll never do to ask: perhaps
I shall see it written up somewhere.'

Down, down, down. There was nothing else to do, so Alice soon began talking again. 'Dinah'll miss me very much to-night, I should think!" (Dinah was the cat.) 'I hope they'll remember her saucer of mild at tea-time. Dinah, my dear! I wish you were down here with me! There are no mice in the air, I'm afraid, but you might catch a bat, and that's very like a mouse, you know. But do cats eat bats, I wonder?' And here Alice began to get rather sleepy, and went on saying to herself, in a dreamy sort of way, 'Do cats eat bats? Do cats eat bats?' and sometimes 'Do bats eat cats?' for, you see, as she couldn't answer either question, it didn't much matter which way she put it. She felt that she was dozing off, and had just begun to dream that she was walking hand in hand with Dinah, and was saying to her, very earnestly, 'Now, Dinah, tell me the truth: did you ever eat a bat?' when suddenly, thump! Thump! Down she came upon a heap of sticks and dry leaves, and the fall was over. (18)

Hiding behind her Deborah character, Joanne Greenberg writes:

> She began to fall, going with Anterrabe through this fire-framed darkness into Yr. This time the fall was far. There was utter darkness for a long time and then a grayness, seen only in bands across the eye. The place was familiar; it was the Pit. In this place gods and Collect moaned and shouted, but even they were unintelligible. Human sounds came, too, but they came without meaning. The world intruded, but it was a shattered world and unrecognizable. (19)

It is interesting to see how both Joanne and Donna find ways of denying the self. Donna writes:

> Not surprisingly . . . I referred to myself as "you." This was because "you" logically captured my relationship to myself. One develops an "I" in interaction with "the world." Donna didn't interact; the characters did. Mary would try to get me to clarify whether I was implying that the events I was talking about applied to her, as she was the only "you" in the room. I tried to explain that this was how I described things. She continued. I reverted to using "I" in a conforming effort to avoid her pedantic emphasis on the pronouns I was using. Her efforts to get me to refer to these incidents in a personal manner overlooked the fact that my use of the word "you" captured the impersonal way in which I had experienced the incidents at the time they happened. She probably felt that she had to help me to overcome my depersonalization, as thought this were some recently developed defense reaction. I don't think she realized that I had actually experienced life this way since the creation of Willie and Carol, and my subsequent ability to communicate through them, thirteen years before. (20)

Yet, while she went to these extremes to hide, perhaps the greatest strength of her autobiography is that it is so incredibly personal, so revealing, opening up to the entire world at a level few others would do. Joanne, on the other hand, hid behind the literary devices of a pseudonym and a fictionalized persona, not revealing the autobiographical nature of the story until years later. It's also clear that she "bent" some truths for her fictionalized biography. How many of these "slight changes" are really important ultimately cannot be resolved. She has created a work of fiction to give her life a pattern. All biographies do that. And what's more important than whether her version of reality fits the literal reality better or worse than the rest of us is that she is expressing the deeper truths of the psyche, and has done that well.

Donna's first person narrative is so bluntly literal that similar questions can be raised from the opposite direction. Just how true is this seemingly straightforward account? It is an amazing accomplishment for a child as completely incapable of the very basics of language as she was, and the book itself suggests she was able to write it rather rapidly, sparked by her recent stumbling on some library books about autism. Given her penchant for assuming "roles" to hide behind, and her excellent skills at using various personas to fool people, it must be at least considered that she is not quite so innocently recounting her autism and childhood experiences as she claims. Nevertheless, the same entrance into the more important truths comes through.

These two adults, functioning on very high levels and demonstrating exceptional abilities, were once broken windmills, unable to connect with the normal winds of the mind, sputtering and stuttering and simply lost.

And they did it by entering the part of the mind that works through fiction, through the imagination. This is the place where meaning and value get mapped out, and it is not a reach at all to say that it is the natural place for people desperate for such truths to shine, because that is where the "self" is found.

At the time the Joanne Greenberg's book first came out (1964), and even at the time of Rollo May's reference to it in *The Cry for Myth* (1992), (21) Joanne Greenberg's condition was defined as schizophrenia, a term used to classify a wide variety of mental illnesses at that time, what is sometimes referred to as a "trashcan" diagnosis. (22) Current research supports redefining this initial classification of Joanne's illness from schizophrenia to perhaps obsessive-compulsive disorder or some form of autism. Whatever conclusions are reached about the classification, amazing similarities are to be found between Joanne Greenberg and Donna Williams.

Both retreated from the literal world into a fantasy world. Joanne Greenberg's world, instead of consisting of multiple personalities to disappear into, was made up of various "judges," godlike characters that generally took on super-ego qualities. Anterrabae (originally Antilobia), based on an illustration of Milton's Satan from *Paradise Lost*, falls eternally through the darkness, his hair and fingertips flickering in the wind. Latamaeon is the black, sarcastic god. Censor, originally a god meant to protect Deborah, became a demanding dictator intent on taking over complete control of her. The Collect of the Others consists of a group of critical voices that constantly condemns her. There are two main landscapes. The Kingdom of Yr (originally Iria) is the huge alternate world she escapes into. The Pit is a place of punishment in Yr, a place without meaning, will, or feeling, and it is especially frightening because the return from it begins with the *need for meaning* first, before the actual meanings return. (23)

Joanne Greenberg's story takes place at a time when psychoanalysis was a more accepted treatment, and in real life that is exactly what she got, and it apparently was successful. Her real life psychiatrist was Frieda Fromm-Reichmann, who, along with Harry Stack Sullivan, and Clara Thompson, operated out of the Chestnut Lodge Hospital, Rockville, Maryland, and was successful in treating several patients using nothing but psychoanalytical psychotherapy, a Freudian based system that consisted largely of simply sitting and listening to a patient talk out his or her cure.

This, however, is not so simple as it might at first appear, and the novel clearly demonstrates the difference between it being done correctly and incorrectly. Early in the sessions, Dr. Frieda (the fictionalized doctor) concludes a session: "Our time is over . . . You have done well to tell me about the secret world. I want you to go back and tell those gods and Collect and Censor that I will not be cowed by them and that neither of us is going to stop working because of their power." (24). This seemingly casual remark is a key to unfolding the treatment, as it is a strong assurance of acceptance of the patient's need for the fantasy world, not a rejection of exactly the form of thinking needed to work through the problems.

And this becomes extremely clear when Dr. Frieda leaves Deborah with a different doctor for a summer, one not understanding of this simple truth:

> From the silent self-conscious hall of B ward, she went to see the New One.
>
> She found Dr. Royson sitting stiffly in his chair in one of the offices on the main floor. "Come in," he said. "Sit down."
>
> She sat down.
>
> "Your doctor has told me a lot about you," he said. Deborah turned her mind for something to reply, thinking only: How stiffly he sits; I told her I would be fair . . . I told her I wold [typo for would] try as hard with this one. . . .
>
> "Yes," she said. He was not a friendly person. She understood and set out to try the first directions. "You're from England, aren't you?"
>
> "Yes."
>
> "I like the accent," she said.
>
> "I see."
>
> *This is one-by-one from the jawbone*! Anterrabae groaned a little scornfully.
>
> After a short silence the doctor said, "Tell me what you are thinking." It seemed to come like a demand.
>
> "About dentistry," Deborah said.
>
> "And what thoughts do you have about dentistry?" he said in his unchanging tone.

"That it can be more expensive than we think it will be," Deborah said. She caught herself and tried again. "I'm out of Novocain because Furii took it away with her."

"Who is that" Who took it away?" He jumped on it as if it were some prize.

"The doctor—Dr. Fried."

"You called her something else—what else did you call her?" The same demand, like a pickax.

"Just another name."

"Oh, the Secret Language," and he leaned back. Comfortably on safe ground, it looked to her. It was in the book on page ninety-seven. It was All Right. "Dr. Fried told me that you had a secret language."

Withdraw! Anterrabae said. He used the poetic Yri form and in her heartsickness it seemed newly beautiful—*Te quaru*: be as the sea and ebb and leave only a moment of the sandshine. *But I promised her*, Deborah insisted to the firelit falling god in the black place.

She is dead, Lactamaeon said on the other side of her.

"Tell me one of your words in that language," the outside voice insisted.

"*Quaru*," she said absently.

"What does it mean?"

"What?" She came to look at him suddenly and at the brutally hard lines of his disapproving face. He even sat austerely.

"What does it mean, that word you spoke? What was it?"

"*Quaru* . . ."she repeated. She was flustered with the confrontation, and she heard her own voice tell the gods, *But I promised* . . . "It means . . . well, it means wavelike, and it can imply something more of the sea, sometimes the coolness, or that soft, swishing sound, too. It means acting the way a wave acts."

"Why don't you merely say wavelike then?" he said.

"Well . . ." She was beginning the black sweat that was prelude to the Punishment. "You use it for anything that is wavelike, but it gives the sea-connotation with it and sometimes that can be very beautiful."

"I see," he said. She knew that he didn't.

"You can use it for the way the wind is blowing sometimes, or beautiful long dresses, or hair that is rippling, or . . . or leaving."

"It also means leaving?"

"No, . . ." Deborah said, ". . . there is another word that means leaving."

"What word?" He demanded.

". . . It depends on whether one has the intention of coming back . . ." she said miserably.

"Very interesting," he said.

"There is also a saying—" (She had made it up that minute to try to save herself and them.) "It is: don't cut bangs with a hatcher."

"Cut bangs?" he said.

> An Americanism, perhaps, so she tried again. “Don’t do brain surgery with a pickax.”
>
> “And what does that signify to you?” he said, perhaps forgetting that if she could speak truly to the world, she would not be a mental patient.
>
> “It suffered and died in translation,” she said.
>
> There followed a long silence between them, and though she tried at the next hour and the next and the next, his humorless and automatic responses brought down the muteness like a night. He worked hard to convince her that Yri was a language formulated by herself and not sent with the gods as a gift. He had taken the first words she gave him and shown her the roots of them from scraps of Latin, French, and German that a nine- or ten-year-old could pick up if she tried. He analyzed the structure of the sentences and demanded that she see that they were, with very few exceptions, patterned on the English structure by which she, herself, was bound. His work was clever and detailed and sometimes almost brilliant, and she had many times to agree with him, but the more profound he was the more profound was the silence which enveloped her. She could never get beyond the austerity of his manner or the icy logic of what he had proven, to tell him that his scalpels were intrusions into her body, and that furthermore, his proofs were utterly and singularly irrelevant. At the end she marshaled all of her strength, and with as a clarity as she could give him, she said, “Please, Doctor, my difference is not my sickness.” It was a last cry and it went unheard. (25)

Not surprisingly, Joanne Greenberg’s book suggests in standard Freudian analytical theory that a parent has been at least a part of the reason for the problems. In this case a brief family history is explored beginning with her mother’s father:

> Pop had come from Latvia. He had a clubfoot. Somehow these two things represented him more fully than his name or occupation. He had come to America a young man, poor and foreign and lame, and he had borne down on he new life as if it were an enemy. In anger her had educated himself; in anger he had gone into business, failed, succeeded, and made a fortune. With his fortune and his anger he had bought a great home in an old neighborhood of the inbred and anciently rich. His neighbors had every manner he admired, and in turn they despised his religion, his accent, and his style. They made the lives of his wife and children miserable, but he cursed them all, the neighbors and wife and children, in the crude, blunt words of his abhorrent past. (26)

And this man, Joanne’s grandfather had despised the man who would become her father, a poor man forced to take charity from the very man who hated him. And he, in turn, had been filled with anger, unable to “open his heart” to his daughter. So when he dropped Joanne off at the psychiatric ward he was too knotted up in his own emotions to reach out, to show his love. "He was a man of tempers and now he needed a rage that was cleansing, simple, and direct. But the anger here was so laced with pity, fear, and love that he did not know how he could free himself of it. It lay writhing and stinking inside him, and he began to feel the old, slow waking ache of his ulcer." (27)

It's not surprising that, when the parents come to visit her, Joanne refuses to see her father. She fears his own fears will be too much for her to deal with.

Donna Williams stresses the more current autistic explanations for her problems, but her discussions of her mother are actually far more filled with hate and a very, very troubled relationship, her mother's behavior simply criminal. One passage should suffice:

> My aunty had recalled many memories of my early childhood, but none had really struck a chord. Willie stood there remembering many of these, but did so without any feeling for the self who had experienced them. Then, as my aunty recalled an event when I was three, it triggered something, and my mind replayed it in all its vividness and horror.
>
> I was back there. I could see my aunty across the room. I could hear the pleading tone in her voice and sensed danger. I was watching everything around me happen as though in slow motion, though still moving far too fast for me to be able to respond in time.
>
> I look up at the figure of my mother. I shot silent glances in the direction of the pleading voice from across the room. I looked down at the opened tin of cold spaghetti in front of me and was aware of the fork in my hand.
>
> I had not heard the introduction: the threat of death against my spilling a single drop of food. I never connected the repeated slapping with the event. It was just something that came to me from out of the blue as a series of shocks.
>
> I felt the dishcloth being forced into my mouth. It made me gag. I was choking as I vomited up against it.
>
> The pleading voice was at war with the cutting snarl of my mother's voice. I glanced at the black and white striped cord as though it was a snake. It began to whip my face. I could not cry, or speak, or scream. I looked at my aunty and collapsed on the cold smooth surface of the table in front of me and vomited through my nose. I thought I had drowned. (28)

At the same time, both also suggest possible physical problems for their psychological struggles. Certainly, one contributing factor for Donna was the result of diet and allergies: "At the first clinic, using a series of blind tests, they found that I was allergic to all meat except beef, all dairy products, eggs, soy products, potatoes, tomatoes, and corn. I was also put through a six-hour blood-sugar test and found to be suffering from severe hypoglycemia. Another clinic found her to be "allergic to a group of chemicals called phenolics and salicylates, which are common to many fruits, vegetables, herbs, and spices, and almost all prepackaged foods." (29) And we cannot ignore the fact that Joanne had a brain tumor.

Whatever the diagnosis, yet again a broken mind has found a way to catch the power of the wind. Joanne Greenberg's accomplishments easily compare with Donna Williams. In addition to earning a degree in anthropology and literature from American University, she has published several novels and collections of short stories, receiving awards for her work, tutored Latin and Hebrew, taught cultural anthropology and creative writing, and appeared around the world at various writers' seminars and workshops. In 1977 *I Never Promised You a Rose Garden* was made into a film by Imorh Productions, unfortunately, not a great one, as it omits the whole anti-Semitism center and changes the psychological diagnosis to a retreat from reality as a result of a bout with urethral cancer.

Part III

Feral Children

SHE dropped the bar, she shot the bolt, she fed the fire anew,
For she heard a whimper under the sill and a great grey paw came through.
The fresh flame comforted the hut and shone on the roof beam,
And the Only Son lay down again and dreamed that he dreamed a dream.
The last ash fell from the withered log with the click of a falling spark,
And the Only Son woke up again, and called across the dark:—
"Now was I born of womankind and laid in a mother's breast?
For I have dreamed of a shaggy hide whereon I went to rest.
And was I born of womankind and laid on a father's arm?
For I have dreamed of clashing teeth that guarded me from harm.

And was I born an Only Son and did I play alone?
For I have dreamed of comrades twain that bit me to the bone.
And did I break the barley-cake and steep it in the tyre?
For I have dreamed of a youngling kid new-riven from the byre.
For I have dreamed of a midnight sky and a midnight call to blood
And red-mouthed shadows racing by, that thrust me from my food.
'Tis an hour yet and an hour yet to the rising of the moon,
But I can see the black roof-tree as plain as it were noon.
'Tis a league and a league to the Lena Falls where the trooping blackbuck go;

But I can hear the little fawn that bleats behind the doe.
'Tis a league and a league to the Lena Falls where the crop and the upland meet,
But I can smell the wet dawn-wind that wakes the sprouting wheat.
Unbar the door, I may not bide, but I must out and see
If those are wolves that wait outside or my own kin to me!"

She loosed the bar, she slid the bolt, she opened the door anon,
And a grey bitch-wolf came out of the dark and fawned on the Only Son!

This is the poem Rudyard Kipling used to introduce the world to his feral child, Mowgli, raised by wolves in the story "In the Rukh," (1892), and included in his book, *Many Inventions* (1893). He would expand on this in *The Jungle Book* (1894) and *The Second Jungle Book* (1895). It is generally believed that Mowgli is the inspiration for Edgar Rice Burroughs stories of Tarzan.

Mowgli by John Lockwood Kipling, Rudyard Kipling's father, for *The Second Jungle Book*, 1895. In the public domain.

Both were preceeded by a history of real feral children.

Peter the Wild Boy achieved a greater degree of fame than either Joanne Greenberg or Donna Williams, mainly because he never did anything as sophisticated or intellectual or artistic as either of them, and though he was taken in by King George of England to run about the royal court of Kensington Palace and written about by most of the learned scholars of his time, he was at the time and remains today an enigma.

In July of 1725, the burghers of Hamelin found him, a naked brownish black-haired boy of about thirteen hiding in the safety of a tree hollow, where he was apparently foraging for nuts, raw vegetables and birds, and fled on all fours to clamber up a tree when discovered. Another version of his discovery claims that Meyer Jurgen, a local farmer, found him sucking from the teat of a cow and coaxed him inside with an apple. Both stories have a folk tale quality to them, but that only begins the fantastic story that unfolds.

Unable to speak, walking on all fours, and prone to simply running away, he was placed in prison in the nearby town of Zell and given the name of Peter. It was soon discovered that once cleaned up he very much resembled a normal human boy (though there was a bit of webbing between two of his fingers). However, he continued his animal ways, refusing to eat cooked food, ripping off any clothes placed on him, and sleeping on the floor even when put in a bed. And though, later, when his legend had already been firmly established, it would be discovered that, instead of a feral child raised in the wild, he had actually been the child of an abusive father and a step-mother who simply threw him out of the house, and he had probably only wandered about the woods for about a year, the half animal/half human designation given him would linger, and even if the attractive legend was an embellished account, the real story was more pertinent in terms of current autistic theories.

At the time of his capture no one stepped forward to claim him, and tales of his curious habits began to spread. In 1714 the elector of Hanover had become George I, king of Britain. He was not happy with his life in England, and whenever possible he would return to his homeland. It was only natural that this strange animal/boy would be brought to the king's attention, and in November of 1725 that's exactly what happened. At the meal, Peter was oblivious to any of the accepted rules of civilization and proceeded to crawl about, eating whatever nuts and onions and asparagus he could find. While George's servants were horrified and attempted to remove the boy, George himself was amused, and ordered that he be treated well, in whatever way would best benefit society.

Princess Caroline was even more taken with Peter, and in March of 1726, he was brought to London, to St. James' Palace, where George spent his winter months, dressed in a blue outfit, and put on display in the Great Drawing-Room. According to *Wye's Letter*, a London newspaper, he was "uneasy" in his clothes, but was "much pleased" with the striking sound of a gold watch and with a chance to put on the Princess' white glove. Apparently, he also had a penchant for taking things out of the pockets of those attending. (30)

He was put in the care of Dr. Arbuthnot, the court physician and a man whose daily encounters included the intellectual elite of England and by extension all of Europe, such luminaries as Isaac Newton, John Radcliffe, Edmond Halley, Jonathan Swift, Alexander Pope, Joseph Addison, Richard Steele, and Georg Handel, who found this strange boy an object of constant speculation about the entire human condition. Upon seeing him, French nobleman Cesar de Saussure wrote, "His eyes were haggard, and did not rest on any object, and he looked so wild and extraordinary I cannot describe the impression he made. He frightened me." (31) Daniel Defoe wrote *Mere Nature Delineated: or a Body without a Sou*l, using Peter as an example of the negative qualities of humans in nature, "passive, weak and foolish," in need of civilization, of education, of the need of "mere Nature" of the "help of Art to bring it to perfection." (32) Jonathan Swift's encounters with Peter resulted in his anonymous satire *It Cannot Rain but it Pours; or, London Strew'd with Rarities*, where he takes the strange natural boy, the object of both ridicule and admiration, a curiosity of the court as in some ways representing himself, and as a good creature for ridiculing the civilized court. Peter becomes a likely prototype of the same satire found in the fourth portion of *Gulliver's Travels*. First satirizing John Locke's currently popular views that humans are born as a *tabula rasa*, pure in nature, Swift begins with Peter's first visit to the Court, where he licks people's hands, climbs upon people's heads, steals things from people's pockets, and is very guarded about his own possessions, all examples of an innate ambition. Then the irony switches, and Peter becomes an innocent victim, a catalyst for the corrupt civilized court, where Princess Caroline's ladies in waiting have sexual fantasies about the naked savage and are disappointed in their desires, where Peter's primitive language undermines the social pretense, calling one young woman a peacock, some older women " magpyes" and owls, young man in a toupee and dressed in bright frivolous clothes a monkey and a turd, and most pointedly in relation to the soon to be written *Gulliver's Travels* prefers the company of horses to men and expresses his joy by neighing. (33)

Others also saw Peter as a perfect way to satirize the corruption of "civilized man." An anonymous poem titled "The Savage" warns that though Peter's friends at court are "Adorn'd with each politer Grace / Above the rest of human Race," the language they speak—and the morals it implies—is deemed "lustful," "lawless," and blinder to reason than the "brutish Converse" of his former life (34) *The Manifesto of Lord Peter*, attributed to Peter's philosopher father claims that he abandoned the boy in order to "convince the World, how much a nobler Creature a Wild Man was than a Tame one." (35) *The Most Wonderful Wonder That Ever Appeared to the Wonder of the British Nation* (1726) has an account of a Dutchman tracking down the She Bear that suckled Peter, and a reunion of him with his adopted mother. (36) Attr. to John Arbuthnot and/or Jonathan Swift), in Aitken, *The Life and Works of John Arbuthnot*, p. 474. In *The Most Wonderful Wanderer*, Peter and an intelligent She-bear companion satirize humans from an animal's perspective, including the human penchant for killing and eating animals. While he had become an obvious curiosity, there was little done to attempt to understand him from a neurological perspective. And though it might seem obvious to us today that he suffered more from some sort of autism than that he was a missing link between animal and human, we have to remember that studies of the brain and mind were at the most primitive levels. Even the crude beginnings of the now discredited field of phrenology by Franz Joseph Gall would not appear for another eighty years, both psychology and neurology were two to three times that distant in the future, and Charles Darwin's theories of evolution were over a century away.

Interesting proto theories of these fields, however, can be seen. Lord James Burnett, who adopted the title Lord Monboddo (after the name of his father's estate), is the most famous to connect Peter (and other *Homines feri*, most notably Marie-Angelique) with evolution, stressing a comparison especially with orangutans, which he suggested represented "the first stage of human progression." (37) In another of his books, *Ancient Metaphysics*, he connects evolution with the stages of a human life. "In the womb, man is no better than a vegetable." (38) Then, from helpless infants (the stage exemplified by Peter), where we crawl on all fours and are unable to articulate language, we evolve into being biped and having some rudiments of the social and language abilities (the stage of the orangutan), move into the stage of Marie-Angelique, who could swim, use weapons, and language, (39) and finally cultivate the highest levels of a gentleman. Thus, he reasoned, the taming of wild children exhibits how the path to human perfection might also work with sub-human species. "I am persuaded, it is with wild men, as with wild fruits, which we know will not lose their savage nature at the first remove, but can only be tamed by continued culture for a succession of generations." (40) He remained stuck, however, as has the debate over Darwin's subsequent theories right down to today, with getting it to match his religious beliefs.

Others put a slightly different spin on Peter, seeing him as an example of primitive man representing man's innate ignorance and weakness saw in Peter the need for education, and found him to be an idiot mainly because of his isolation as a child. (41) Blumenbach took this even further, stating that Peter was "nothing more than a dumb imbecile idiot," and that "neither Peter nor any other 'Homo sapiens ferus' of Linnaeus can serve as a specimen of the original Man of Nature." (42)

Richard Lovell Edgeworth, and Maria Edgeworth, *Practical Education*, 1798, described their observations of Peter's intelligence, finding that he "had all his senses in remarkable perfection" but could only articulate "imperfectly a few words, in particular, "King George," which words he always accompanied with an imitation of the bells, which rang at the coronation of George the Second; he could in a rude manner imitate two or three common tunes, but without words." (vol. 1, p. 63) Determined to use Peter to demonstrate their theories that perfectibility is possible, given the right education, they developed several educational tests for Peter, none of which improved him any, and ultimately they were forced to simply say he was human, but of a lower level.

Theories over Peter the Wild Boy of Hamelin have resurfaced as theories of autism emerge, some claiming Peter was the first example we find in history of an autistic child; others dismissing him and other feral children as misrepresenting this condition. Paul Collins, *Not Even Wrong*, interweaves his research on Peter with his own struggles to deal with his son's autism, suggesting or more accurately simply accepting the autistic qualities in Peter, i.e., "Using gesture or pointing instead of words," "He simply did not want to talk." (p. 29); "Non responsive to verbal cues; acts as if deaf, although hearing tests in normal range," "He could hear, but he didn't listen. In fact, he could hear very well: he was passionately fond of music and would clap and stamp his feet until ready to drop from exhaustion; long afterward, he could be heard humming tunes over and over until he'd got them right." (p 38) "Repeating words or phrases in place of normal, responsive language," "He was then asked to name the family horse. 'Cuckow." This was always Peter's answer to this question, even though there was no horse on the farm by that name. It was his own word." (p. 44)

And just as autistic children today with their mixture of ability and disability confound experts, "The wild boy had been confounding philosophers from the day he first set foot in London; he was a perplexing combination of intelligence and obliviousness." (p. 40)

Jill Dawson (also raising an autistic son, one with Asperger syndrome) has written *Wild Boy*, a similar attempt to connect a historic feral child with autism. (43) In this case Victor the Wild Boy of Aveyron, thought to have been raised by wolves.

He was found wandering the woods near Saint Sernin sur Rance, France (near Toulouse) in 1797, captured, escaped, captured again, and kept in the care of a local woman for a week before escaping again, finally willingly coming out of the woods into civilization, January 8, 1800. As was the case with Peter, his age was guessed to be about twelve, he had not speaking abilities, and he became an example for the various prevailing theories about human existence.

Once again, the ability to speak a language as a distinguishing difference between human and animal became a key. He was put in the care of Jean-Marc Gaspard Itard, a student at the National Institute of the Deaf (though he wasn't deaf), and attempts were made to teach him to speak. They were, however, almost totally a failure. The only words he is reported to have learned wither lait (milk) and Oh Dieu (oh God).

Interestingly, the spark for Dawson's book came from an account by Uta Frith, a senior scientist in the Cognitive Development Unit of the Medical Research Council in London, whose book, *Autism: Explaining the Enigma*, has a chapter devoted to drawing parallels between autistic children and feral children, suggesting some, such as Victor probably were autistic. (44)

Others disagree with the comparison of feral and autistic children. Andrew Teo writes:

> I cannot bring myself to believe that autistic children and feral children are one and the same, neurologically. While they seem overtly similar, there seems to be subtle differences. Poor social interaction and lack of empathy seem to be the only thing the [that] both have in common, really, and that's the thing people notice the most. There do not seem to be any other traits of autism present in feral children such as the rote repetitive movements, sensory difficulties, inflexibility and uncreativity."
>
> Furthermore, cognitive retardation and the "irreversible inability to learn language" cease to exist at the higher end of the autistic spectrum, where intellect is on par with most people, but the core deficits of autism are still present. Also, people with high functioning autism and Asperger's syndrome are rather adept with the written word and verbal communication, but still suffer from social interaction deficits due to poor body language, i.e., nonverbal communication. . . .

> It is really quite difficult to draw any real conclusion from this, however, I still don't believe that there is a real connection between autism and feral children. In feral children I think the social interaction deficit is by nurture, rather than in autism, where it is (apparently) hardwired. (45)

Even Andrew Teo's concession of the similar inability of feral children and autistic children to show affection gets discounted by the Autism Society of America (46), which stresses: "One of the most devastating myths about autistic children is that they cannot show affection."

Here we begin immediately to meet the problems with autistic theories, which is simply that they haven't yet figured it out and end up not being able to come up with something that actually fits the many categories or types of children the autism umbrella is attempting to cover. The recently mentioned Uta Frith's main claim to fame in this debate is the idea that autism is a biological defect where the autistic individual cannot bring together pieces of information and construct a whole. Thus, thoughts remain fragmented. Furthermore, she refers to Lorna Wing's studies at the Medical Research Council's Social Psychiatry Unit in London to come up with three intertwining features of autism, "impairments in communication, imagination, and socialization." (47), which might apply to feral children, or at least many of them, but don't fit all of the artistically gifted children now often categorized as autistic (such as Donna Williams discussed above). In fact, there are a lot of correspondences between autistic children and creativity to the point where some, such as Michael Fitzgerald, *Autism and Creativity*, even claim that autistic children are or can be more creative than the non-autistic child or that at least there are some really interesting parallels to be drawn between the two. (48)

However, just as schizophrenia was the "trashcan" term for most anything strange in the psyche half a century ago, the new "in" term is autism. Actually, the term is not so new. It refers back to the work of the first American Child Psychiatrist Leo Kanner of John Hopkins Children's Psychiatric Clinic, and his paper from 1943, "Autistic Disturbances of Affective Contact," (49) and to similar work being done by Hans Asperger of the University Pediatric Clinic in Vienna. Both chose the term from adult psychiatry, where it was being used to refer to adult schizophrenics exhibiting a progressive withdrawl from the outside world.

Kanner's paper resulted from studying eleven children all ten years or younger he thought could be grouped under four common traits: the desire to be alone, the desire for sameness, the desire for elaborate routines, and the possession of some remarkable abilities in stark contrast with their disabilities.

In his report, he goes through each of the eleven in turn, simply titling them "case one," "case two," and so on. The first and most often referred to case is about Donald T., a boy Kanner first met at the age of five. Before this initial meeting the boy's father had sent a thirty-three page history that Kanner found filled with "obsessive detail." In it the father indicated that Donald's birth was normal, his breast feeding normal (though later he simply didn't show the appetite of other children), his teeth developed normally, and that he walked at thirteen months. By age one he hummed and sang tunes well, demonstrated an unusually good memory for faces and names and memorizing passages of the Bible. He had good enunciation, and memorized the alphabet and could count to 100. However, he did not ask the normal questions nor show any curiosity beyond his specific interests, and he demonstrated an extraordinary desire to simply be alone, even to the point of being indifferent to whether either father or mother was home. He was "self-satisfied." If he was interfered with, he had tantrums.

At age two, he was found to be "horrified" of self-propelled vehicles such as tricycles and of the playground slide, and he developed a strong attraction to spinning pans, blocks and other round objects.

At the age of four he was placed in a *tuburculosis preventorium*, where he continued to display his desire to simply be left alone "perfectly oblivious" to all that went on around him. He continued his obsession to spinning objects, and developed the habit of shaking his head from to side. Dr. Eugenia S. Cameron and George Frankl examined him over a two week period at age five at the Child Study Home of Maryland and found in addition to the above that there was a "marked limitation of spontaneous activity." He liked elaborate rituals or repetitions, and he made up his own ways of communicating. For example, "At meantime, repeating something that had obviously been said to him often, he said to his mother, 'Say eat it or I won't give you tomatoes, but if you don't eat it I will give you tomatoes', or say 'If you drink to there I'll laugh and I'll simle.'" And if his mother didn't go along with this ritual, he would throw a tantrum. All of the words he used had specific meanings to him, often far removed from their normal meanings.

By the following year, he was showing some movement to a more normal child, including more imagination, but was still dominated by the above symptoms.

The final report on him takes place when he is eight, concluding: "He was still extremely autistic. His relation to people had devoloped only in so far as he addressed them when he needed or wanted to know something. He never looked at the person while talking and did not use communicative gestures. Even this type of contact ceased the moment he was told or fiven what he had asked for." (p. 6)

And the report finishes with a follow up letter from his mother:

> Don is still indifferent to much that is around him. His interests change often but always he is absorbed in some kind of silly, unrelated subject. His literal-mindedness is still very marked, he wants to spell words as they sound and to pronounce letters consistently. Recently I have been able to have Don do a few chores around the place to earn picture show money. He really enjoys the movies now but not with any idea of a connected story. He remembers them in the order in which he sees them. Another of his recent hobbies is with old issues of Time magazine. He found a copu of the first issue of March 3, 1923, and has attempted to make a list of the dates of publication of each issue since that time. So far he has gotten to April, 1934. He has figured the number of issues in a volume and similar nonsense. (49)

At this time and for several decades after, autism was thought to have resulted from the same kinds of negative childhood experiences that caused other neurosis, and the once highly acclaimed, but now often discounted theories of Bruno Bettelheim put forth most completely in his book *The Empty Fortress: Infantile Autism and the Birth of Self*, which popularized the phrase "refrigerator moms" for mothers of such children, suggesting that it is indeed the result of poor parenting, and that what is needed is to get the children away from these bad environments. (50) It's easy to see how such views would bring a storm of protest from the parents of autistic children, especially because psychology can be and has been much oversimplified and misinterpreted from the beginnings of Freudian theory. In this case the central controversy has been oversimplified into one of blaming the parents, especially the mothers for the autism, which makes for good drama. And even though Bettelheim, in truth, does not blame the mothers, just the opposite, a more careful study of his views doesn't make for good drama, but rather makes for more work, too sophisticated, too complex, too much into the depths of psychology for the average person. Instead, let's just keep it simple. Bettelheim is saying autism is caused by refrigerator moms!

Furthermore, since Bettelheim became such a media sensation (appearing on the *Today Show*, getting extremely positive write ups in *The New Yorker* and *The New Republic*, and even having the rock opera *Tommy* by the Who based on his theories) he was a perfect target. Also, the growing field of neurology was looking to replace psycho-analysis with biologically based theories, and Bettelheim was the one they needed to bring down. The mud-slinging was as emotionally charged as any done during a Presidential Election. It really wasn't that hard to do. Bettelheim, it was pointed out, had not received any formal degrees in psychology (i.e., was a quack!). Furthermore, he committed suicide at the end of his life (obviously neurotic himself!). And let's go to the sources themselves; three of his former patients were found to have doubts about his work and even called him a tyrant (ran his own clinic like a concentration camp!). Now, how convenient, on the word of three disgruntled patients his years of interment at Dachau and Buchenwald concentration camps could be turned against him (and it would have been seen as politically incorrect to question their motives or their own psychological conditions). The most well known book ripping him to shreads is Richard Pollak's *The Creation of Dr. B: A Bibliography of Bruno Bettelheim*, and it's hard to come away from that book without hating Bettelheim.

Perhaps it's worth noting at this point that Bettelheim does continue to have his supporters among professionals in both pyschology and neurology. The point here is not so much whether or not Bettelheim's views are correct or incorrect or somewhere inbetween, but rather simply that the riddles remain.

Today autism is considered to be one of five or six categories of Pervasive Developmental Disorders (PDD), also referred to as Autistic Spectrum Disorders: Autistic Disorder, Asperger's Disorder, Childhood Disintegrative Disorder (CDD), Multiplex Disorder, Rett's Disorder, and PDD-not otherwise specified (PDD-NOS)—flexible categories that are constantly changing.

The Yale Development Disorders Clinic lists all six.

> 1. Autism (also referred to as Early Infantile Autism, Autistic Disorder, and Kanner's syndrome) . . . is characterized by marked problems in social interaction (autism), as well as by delayed and deviant communication development (speech is absent in about 50 percent of cases) and various other behaviors which are usually subsumed in the term 'insistence on sameness.' Such behaviors include stereotyped motor behaviors (hand flapping, by rocking), insistence on sameness and resistance to change. . . . Many individuals with autism exhibit mental retardation on the basis of their full-scale (or averaged) IQ score; however, unlike most people with primary mental retardation, those with autism often have marked scatter in their development, so that some aspects of the IQ, particularly nonverbal skills, may be within the normal range."
>
> 2. In Asperger's Disorder "affected individuals are characterized by social isolation and eccentric behavior in childhood. There are impairments in two-sided social interaction and non-verbal communication. Though grammatical, their speech is peculiar due to abnormalities of inflection and

a repetitive pattern. Clumsiness is prominent both in their articulation and gross motor behavior. They usually have a circumscribed area of interest which usually leaves no space for more age appropriate, common interests. Some examples are cars, trains, French Literature, door knobs, hinges, cappuccino, meteorology, astronomy or history. . . .In people with Asperger's Syndrome, "deficits in social interaction and unusual responses to the environment, similar to those in autism, are observed. Unlike in autism, however, cognitive and communicative development are within the normal or near-normal range in the first years of life, and verbal skills are usually an area of relative strength. Idiosyncratic interests are common and may take the form of an unusual and/or highly circumscribed interest (e.g., in train schedules, snakes, the weather, deep-fry cookers, or telegraph pole insulators)."

3. Childhood Disintegrative Disorder is a rare condition originally "described many years before autism (Heller, 1908) but has only recently been 'officially' recognized. With CDD children develop a condition which resembles autism but only after a relatively prolonged period (usually 2 to 4 years) of clearly normal development (Volkmar, 1994). This condition apparently differs from autism in the pattern of onset, course, and outcome (Volkmar, 1994)." It "develops in children who have previously seemed perfectly normal. Typically language, interest in the social environment, and often toileting and self-care abilities are lost, and there may be a general loss of interest in the environment. The child usually comes to look very 'autistic', i.e., the clinical presentation (but not the history) is then typical of a child with autism. . . .

4. Rett's Disorder is included as a Pervasive Developmental Disorder because there is some potential confusion with autism - particularly in the preschool years (Tsai, 1992). Otherwise the course and onset of this condition is very distinctive. In people with Rett's Disorder (first reported by Rett in 1966), very early development is normal. Head growth then decelerates, usually in the first months of life, and a loss of purposeful hand movements occurs. Motor involvement is quite striking and profound mental retardation is typical. Characteristic hand-washing stereotypes develop. While the DSM-IV does not list male sex in the exclusionary criteria, the existing literature on Rett's syndrome documents the condition primarily in girls. The DSM-IV field trial sample included only girls and a recent, very well executed epidemiological investigation documented a prevalence of 3.8 per 10,000 girls; boys were not included. Since the discovery of the MECP2 gene, responsible for Rett's, variants of the syndrome have been reported in males who have mutations of MECP2, with some overlap in the symptomatology observed in girls (Amir, Van de Veyver, Wan, Tran, Franke, & Zoghbi, 1999; Schwartzman, Zatz, Vasquez, Gomes, Koiffman, Fridman & Otto, 1999; Schanen, Kurczynski, Brunelle, Woodcock, Dure, & Percy 1998).

5. Ever since autism was first recognized, its continuity with schizophrenia has been a matter of debate. In fact, until the late 1970s, children with autism were often labeled as having "childhood schizophrenia." In the last thirty years, however, the term "childhood schizophrenia" has been displaced. Diagnostic criteria for autism have been established that rely solely on social, communicative and sensorimotor symptoms, without reference to the thought disorders typical of schizophrenia. Nevertheless, there are some children who display the severe, early-appearing social and communicative deficits characteristic of autism who ALSO display some of the emotional instability and disordered thought processes that resemble schizophrenic symptoms. Cohen, et al. (1986) coined the term MultiplexDevelopmental Disorder (MDD) to describe these children, although they are often given a diagnosis of PDD-NOS by clinicians who may be unfamiliar with this terminology. Unlike schizophrenia, MDD symptoms emerge in earliest childhood, often in the first years of life, and persist throughout development. Diagnostic criteria for MDD include:" Impaired social behavior and sensitivity (such as social disinterest and withdrawl), similar to autism, affective symptoms (such as recurrent panic and anxiety), though disorder symptoms (including confusions between reality and fantasy).

6. "Pervasive Developmental Disorder, Not Otherwise Specified (PDD-NOS) is a 'subthreshold' condition in which some - but not all - features of autism or another explicitly identified Pervasive Developmental Disorder are identified.

PDD-NOS is often incorrectly referred to as simply "PDD." The term PDD refers to the class of conditions to which autism belongs. PDD is NOT itself a diagnosis, while PDD-NOS IS a diagnosis. The term Pervasive Developmental Disorder - Not Otherwise Specified (PDD-NOS; also referred to as "atypical personality development," "atypical PDD," or "atypical autism") is included in DSM-IV to encompass cases where there is marked impairment of social interaction, communication, and/or stereotyped behavior patterns or interest, but when full features for autism or another explicitly defined PDD are not met.

It should be emphasized that this "subthreshold" category is thus defined implicitly, that is, no specific guidelines for diagnosis are provided. While deficits in peer relations and unusual sensitivities are typically noted, social skills are less impaired than in classical autism. The lack of definition(s) for this relatively heterogeneous group of children presents problems for research on this condition. The limited available evidence suggest that children with PDD-NOS probably come to professional attention rather later than is the case with autistic children, and that intellectual deficits are less common." be classified as having MDD. (51)

There are other categories that often get mixed into these, such as Fragile X syndrome, often called Martin-Bell Syndrome, a genetic disorder transmitted from the mother to her sons resulting in autistic behavior such as poor eye contact, hand-flapping and other random movements, poor sensory skills, and delayed speech abilities. There are also several unusual physical features, including strabismus (lazy eye), large ears, a long face, poor muscle tone, flat feet, and large testicles.

Part IV

Through the blur, I wondered if I was alone or if other parents felt the same way I did—that everything involving our children was painful in some way. The emotions, whether they were joy, sorrow, love or pride, were so deep and sharp that in the end they left you raw, exposed and yes, in pain. The human heart was not designed to beat outside the human body and yet, each child represented just that--a parent's heart bared, beating forever outside its chest.

—Debra Ginsberg

In reference to Einstein's definition of insanity . . . No Mr. Einstein, that is not insanity, that is autism.

—Eileen Miller, *The Girl Who Spoke with Pictures: Autism through Art*

What would happen if the autism gene was eliminated from the gene pool? You would have a bunch of people standing around in a cave, chatting and socializing and not getting anything done.

—Temple Grandin, *The Way I See It: A Personal Look at Autism & Asperger's*

On the other hand, I think cats have Asperger's. Like me, they're very smart. And like me, sometimes they simply need to be left alone.

—Jodi Picoult, *House Rules*

Autists are the ultimate square pegs, and the problem with pounding a square peg into a round hole is not that the hammering is hard work. It's that you're destroying the peg

—Paul Collins

I know of nobody who is purely autistic, or purely neurotypical. Even God has some autistic moments, which is why the planets spin.

—Jerry Newport, *Your Life is Not a Label*

You know, everybody's ignorant, just on different subjects.

—Will Rogers

If I could snap my fingers and be nonautistic, I would not. Autism is part of what I am.

—Temple Grandin

Think of it: a disability is usually defined in terms of what is missing. … But autism … is as much about what is abundant as what is missing, an over-expression of the very traits that make our species unique.

—Paul Collins, *Not Even Wrong: Adventures in Autism*

Autism, is part of my child, it's not everything he is. My child is so much more than a diagnosis.

—S.L. Coelho, *The World According to August, One Good Friend*

I like it that order exists somewhere even if it shatters near me.

—Elizabeth Moon, *The Speed of Dark*

But the Beast was a good person...the Prince looked on the outside the way the Beast was on the inside. Sometimes people couldn't see the inside of the person unless they like the outside of a person. Because they hadn't learned to hear the music yet.

—Karen Kingsbury, *Unlocked*

I believe there is a reason such as autism, severe manic-depression, and schizophrenia remain in our gene pool even though there is much suffering as a result.

—Temple Grandin, *Thinking in Pictures: My Life with Autism*

Art can permeate the very deepest part of us, where no words exist.

—Eileen Miller, *The Girl Who Spoke with Pictures: Autism through Art*

Why was it considered normal for a girl to live for fashion and makeup, but not car engines or bugs? And what about sports fanatics? My mom had a boyfriend who would flip out if he missed even a minute of a football game. Wouldn't that be what doctors considered autistic behavior?

—Tara Kelly, *Harmonic Feedback*

Autism: Where the "randomness of life" collides and clashes with an individual's need for the sameness.

—Eileen Miller, *The Girl Who Spoke with Pictures: Autism through Art*

I myself am opaque, for some reason. Their eyes cannot see me. Yes, that's it: The world is autistic with respect to me.

—Anne Nesbet, *The Cabinet of Earths*

Go to the edge of the cliff and jump off. Build your wings on the way down.

—Ray Bradbury, T.K. Thorne

Wearing a cloak is on Rose's list of the thousand things she hates most. The problem is that each of the thousand problems is ranked number one.

'But Dr. Rannigan says you must and anyway, it hardly weighs a thing, it's so full of holes.' I swung mine round my shoulders. Rose hates any bit of clothing that constricts, but I say Chin up and bear it. Life is just one great constriction.'Ventilated,' I said, 'that's the word. Our cloaks are terrifically ventilated.

—Franny Billingsley, *Chime*

Universities are renowned for their tolerance of unusual characters, especially if they show originality and dedication to their research. I have often made the comment that not only are universities a 'cathedral' for worship of knowledge, they are also 'sheltered workshops' for the socially challenged.

—Tony Attwood, *The Complete Guide to Asperger's Syndrome*

Do not fear people with Autism, embrace them, Do not spite people with Autism unite them, Do not deny people with Autism accept them for then their abilities will shine

—Paul Isaacs

Probably the most famous autistic today is Temple Grandin. Her resume is easily as impressive as those of Donna Williams and Joanna Greenberg, including two autobiographical works, *Emergence: Labelled Autistic* (co-written with Margaret Scariano) (52) and *Thinking in Pictures: and Other Reports from My Life with Autism.* (53) She earned a B.A. at Franklin Pierce College an M.S. in Animal Science at Arizona State University, a Ph.D. in Animal Science from the University of Illinois in 1989, and has become perhaps the most important animal scientist in her field, having designed 1/3 of the livestock-handling facilities in the United States. In North America, almost half of the cattle meat plants now use a center track restrainer system designed by her, cattle curved chute and race systems of her designed are used worldwide, and her writings on animal behavior have been central to animal welfare advocacy movements (her book *Animals in Transition* was a *New York Times* best seller).

She has appeared on such national television shows as *20/20, 48 Hours, Prime Time Live,* and the *Today Show*, been featured in such magazines as *U.S. News, World Report, Time, The New York Times, Forbes*, and *Discover*, and produced the videos *Visual Thinking, Careers, and Medications* on VHS and *Dr. Temple Grandin* on DVD which can be obtained from

Future Horizons. She is the focus of a semi-biographical HBO film, entitled *Temple Grandin*, starring Claire Danes as Grandin. The movie was released in 2010, was nominated for 15 Emmys, and received five awards, including Outstanding Made for Television Movie and Best Actress in a Drama. Grandin was on stage as the award was accepted, and spoke briefly to the audience. At the 2011 Golden Globes, Claire Danes won a Golden Globe for Best Performance by an Actress in a Mini-Series or Motion Picture Made for Television. Grandin was featured in *Beautiful Minds: A Voyage Into the Brain*, a documentary produced in 2006 by colourFIELD tell-a-vision, a German company. She appeared in a 2011 documentary on Sci Channel, "Ingenious Minds" She was named one of 2010's one hundred most influential people in the world by *Time* magazine. She also was interviewed by Michael Pollan in his best-selling book, *The Omnivore's Dilemma,* where she discussed the livestock industry.

Though she was well known in autistic circles, Oliver Sacks, prompted by Dr. Uta Firth to visit her, first brought Temple Grandin to public attention when he named one of his books, *An Anthropologist on Mars: Seven Paradoxical Tales,* on a phrase she used to describe herself, and concluded the book with a discussion of his visit. (54)

In his preface to Grandin's *Thinking in Pictures: and other Reports from my Life with Autism*, Oliver Sacks points out that she was, in fact, the spark that pushed neurologists to see autism in a new way. Her first autobiography, *Emergence,* presented an autistic life that was much more fully rounded, more capable of the deeper human experiences than the previous views of "a child mute, rocking, screaming, inaccessible, cut off from human contact," or of the autistic savant, similar to that portrayed in *Rain Man*, a person filled with uncanny powers of "calculation, memory, drawing, whatever," yet "cut off from normal life." Temple Grundin while obviously somehow different was at the same time not devoid of standard human needs and desires and abilities. Not only was she highly accomplished, and in the end capable of complex thinking and writing, but she did have a need for being "hugged."

And yet perhaps this need for being hugged can open the door to demonstrate how she was both the same and different. She writes:

> From as far back as I can remember, I always hated to be hugged. I wanted to experience the good feeling of being hugged, but it was just too overwhelming. It was like a great, all-engulfing tidal wave of stimulation, and I reacted like a wild animal. Being touched triggered flight; it flipped my circuit breaker. I was overloaded and would have to escape, often by jerking away suddenly. (p. 62)

Yet, as she points out, she was one of many autistic children who "crave pressure stimulation even though they cannot tolerate being touched." As a child of six, she would wrap herself in blankets and get under sofa cushions because the pressure relaxed her. She dreamed of building a machine that would supply this pressure. This dream began to take on real possibilites when she first saw a squeeze chute used to hold cattle in place when they were given vaccinations at her aunt's ranch.

> I asked Aunt Ann to press the squeeze sides against me and to close the head restraint bars around my neck. I hoped it would calm my anxiety. At first there were a few moments of sheer panic as I stiffened up and tried to pull away from the pressure, but I couldn't get away because my head was locked in. Five seconds later I felt a wave of relaxation, and about thirty minutes later I asked Aunt Ann to release me. For about an hour afterward I felt very calm and serene. This was the first time I ever felt really comfortable in my own skin. (p. 63)

After she discusses other autistics also having this need for touch stimulation, along with the dangers of the wrong forms of touching (the anxiety and violent reactions that can result from over-stimulation), and her own developments of a personal "squeeze machine," she writes:

> As I developed my squeeze machine, I designed it to enhance the feeling of being embraced. Now, if I suddenly resist, I cannot pull my head out of the softly padded neck opening. In order to open the latch, I have to relax and lean forward. I am never locked in the machine, but I am prevented from suddenly pulling away from the soothing pressure. At all times I am in control of the amount of pressure applied to my body. The new design has enabled me to give in completely to the gentle feeling of being held. (p. 80)

And concludes: "Help autistic children fulfill that most basic human need, the comfort of touch, is like taming an animal. At first they pull away, but then they learn that touching feels good." (p. 81)

As she admits, the squeeze machine does seem a strange desire, and those who knew her kept trying to get her to stop using it. The next chapter of her book takes this curious hate/love of being embraced to a deeper level. She begins it:

> To have feeling of gentleness, one must experience gentle bodily comfort. As my nervous system learned to tolerate the soothing pressure from my squeeze machine, I discovered that the comforting feeling made me a kinder and gentler person. It was difficult for me to understand the idea of kindness until I had been soothed myself. It wasn't until after I had

> used the modified squeeze machine that I learned how to pet our cat gently. He used to run away from me because I held him too tightly. Many autistic children hold pets too tightly, and they have a disproportionate sense of how to approach other people or be approached. After I experienced the soothing feeling of being held, I was able to transfer that good feeling to the cat. As I became gentler, the cat began to stay with me, and this helped me understand the ideas of reciprocity and gentleness.
>
> From the time I started using my squeeze machine, I understood that the feeling it gave me was one that I needed to cultivate toward other people. It was clear that the pleasurable feelings were those associated with love for other people. (p. 82)

As an aside here, this leads into a brief discussion of her now famous interactions with animals, how she connects up with them through touching, establishing an almost mystical unity. She also throws in a brief comment on Bettelheim's theories, following it with the comment, "We now know that autism is caused by neurological abnormalities that shut the child off from normal touching and hugging. It is the baby's abnormal nervous system that rejects the mother and causes it to pull away when touched." (p. 85)

This chapter, tellingly titled "Learning Empathy" is an interesting attempt to explain how autistics can logically or intellectually understand the deeper emotions, yet not experience them as emotions. She concludes:

> Yet, it has only been during the last two or three years that I have discovered that I do not experience the full range of emotions. My first inkling that my emotions were different came in high school, when my roommate swooned over the science teacher. Whatever it was she was feeling, I knew I didn't feel that way toward anyone. But it was years before I realized that other people are guided by their emotions during most social interactions. Frl me, the proper behavior during all social interactions had to be learned by intellect. I became more skilled at social interactions as I became more experienced. Throughout my life I have been helped by understanding teachers and mentors. People with autism desperately need guides to instruct and educate them so they will survive in the social jungle. (p. 95)

However, just as we begin to see her as missing the deeper, more complex emotions of adults, we come across this report from Oliver Sacks:

> Temple is an intensely moral creature. She has a passionate sense of right and wrong, for example, in regard to the treatment of animals; and law, for her, is clearly not just the law of the land but, in some far deeper

> sense, a divine or cosmic law, whose violation can have disastrous effects—seeming breakdowns in the course of nature itself. "You've read about action at a distance, or quantum theory," she said. "I've always had the feeling that when I go to a meat plant I must be very careful, because God's watching. Quantum theory will get me."
>
> Temple started to become excited. "I want to get this out before you get to the airport," she said, with a sort of urgency.
>
> She had been brought up an Episcopalian, she told me, but had rather early "given up orthodox belief"—belief in any personal diety or intention—in favor of a more "scientific" notion of God. "I believe there is some ultimate ordering force for good in the universe—not a personal thing, not Buddha or Jesus, maybe something like order out of disorder. I like to hope that even if there's no personal afterlife, some energy impression is left in the universe. . . . Most people can pass on genes—I can pass on thoughts or what I write.
>
> "This is what I get very upset at. . . ." Temple, who was driving, suddenly faltered and wept. "I've read that libraries are where immortality lies. . . . I don't want my thought to die with me. . . . I want to have done something. . . . I'm not interested in power, or piles of money. I want to leave something behind. I want to make a positive contribution—know that my life has meaning. Right now, I'm talking about things at the very core of my existence." (p. 296, *An Anthropologist from Mars: Seven Paradoxical Tales*)

In his introduction to her book *Thinking in Pictures: And Other Reports from my Life with Autism*, Oliver Sacks repeats this passage and concludes: "Thus, in my brief (but very full) few days with Temple, I had a revelation of how, while in many ways so flat and constricted, her life was in other ways full of health, of depth, of deep human striving." (p. 15)

Theories abound. Brilliant minds struggle to explain minds that somehow just don't fit the general category of "normal." It's always dangerous to enter into such debates, but at the same time, whether or not we can explain such categories, we certainly can sense them. No matter how much we encounter, read about and study such children as Donna Williams, Joanna Greenberg, and Temple Grandin we cannot deny that they are somehow not normal children. They have qualities that just don't fit the category of normal. This is not a negative statement, simply a statement of fact.

Psychology and neurology and related fields have worked hard to sort out just what makes such children unique. Perhaps, today, certain things have been discovered, and at least some hesitating categories, causes, and treatments can be put forth. However, the field is still young, and contradictions rule. Even just in the small arena of creativity and imagination in such children, the studies and theories abound, and offer most every possible conclusion on the spectrum from poor to exceptional. In 1999, Jamie Craig and Simon Baron-Cohen published findings demonstrating "impoverished creativity" in autistic children. (55) In the same journal six years later Neil Gordon published "Unexpected discoveries of Artistic Talent," where he found "The development of exceptional and unexpected artistic skills . . . can occur among young children with severe learning difficulties, especially if they are autistic." (56) A. Chetterjee suggested that creativity was neither more nor less in autistic children (or children with other mental malfunctions), but rather was simply brought about through different pathways. (57) C. Hou, B. L. Miller, J. L. Commings, M. Goldberg, P. Mychack, V. Bottino, and D. F. Benson studied five autistic savants, concluding that "Savants exhibit extraordinary visual talents along with profound linguistic and social impairment." (58) Autisim and genius is one of the hot topics. Frank Klein suggests that Albert Einstein was exactly this. (59) In "Illness: the Pathway to Creative Genius," Rodger Dobson claims that "Disease, rather than being a barrier to greatness, may be its wellspring," and suggests that "Einstein, Warhol, Newton, Cézanne, Goya, Michelangelo, Turner and Berlioz are among many whose achievements are now thought to have been influenced by disease." In fact, he claims that "Conditions such as depression, autism, myopia, anxiety, chronic pain, gout, stroke and dementia heavily influenced their paths to creativity." (60) Clara Claiborne Park has two books on her autistic, yet exceptionally artistic daughter. (61) Lorna Selfe has an excellent book on *Nadia: a case of extraordinary drawing ability in an autistic child.* (62) Tito Rajarshi Mukhopadhyay has written *The Mind Tree: A Miraculous Child Breaks the Silence of Autism*, a book similar to Donna Williams in that it is a well written autobiography by an autistic person. (63) These, however, are the exceptions, stunning because of their sharp contrast with the norm.

But what is the norm? Just what does it mean to be autistic? Some, such as Bruno Bettelheim, think it is at least partially determined by the environment, especially the mother-child relationship. Others, such as Uta Firth, think it is biological. Some think it should be divided up into five or six categories, the final category simply a catch-all one for children "sort of but not quite" fitting the others. Some think it promotes creativity and "genius," others that it is characterized by a lack of imagination. Many think much of what was once diagnosed as schizophrenia really fits autism better. Feral children, once thought to be animal-like because of growing up outside of civilization (for better or worse), now are more likely to be thought of as autistic. And while reasoned and friendly debates are possible and do take place, the field is often characterized by emotional diatribes. The Wikipedia site online has a section titled "Controversies in autism," and finds the topic so explosive that it includes a highlighted hand on a stop sign with the disclaimer "The neutrality of this article is disputed." (64)

There are several emotionally charged groups working in the field today. As long ago as 1962, a group of parents in the United Kingdom who were frustrated by the lack of support and understanding of their own children with autism, created The Society for Autistic Children, which later was renamed The National Autistic Society (NAS). While their focus is help for the families dealing with autism, another organization, The National Alliance for Autism Research (NAAR) focuses on biomedical research into autism, and provides grants for research. It has, in turn, formed an alliance with Autism Speaks, an organization founded by Suzanne and Bob Wright as a result of their personal experiences with an autistic grandson. The Autism Society of America (ASA) was founded in 1965 by Bernard Rimland, Ph.D., and claims to be the oldest and largest grassroots organization within the autism community, currently having more than 120,000 members and supporters connected through a working network of nearly 200 chapters nationwide. As with the others, it is dedicated to increasing public awareness about autism and the day-to-day issues faced by individuals with autism, their families and the professionals with whom they interact.

So many people trying to understand, trying to help. Yet the silence, the separation remains. And while there are the wondrous exceptions to the general lack of ability, even they are informed by a quiet sorrow, a shadowy world inside the mind unable to find the normal passages, unable to take the same paths others do, forced to create their own maps of meaning, perhaps, at times, brilliant new maps, but still maps across lonely landscapes that cannot help but express an isolation. Bruno Bettelheim describes Laurie:

> When Laurie was about two and a half, the young nursemaid left suddenly. She was replaced with an older woman who took care of Laurie until she was four years old. This woman, as well as other caretakers who followed her, never seemed of much importance to Laurie who, with the leaving of her original nursemaid, began to give up what she had learned.
>
> Within a few days the mother noticed a great change in Laurie. She stopped saying the few words she had known and all talking was replaced by peculiar clucking sounds. One day, after repeatedly making what sounded to the mother like loud animal-like noises, the mother became very angry, spanked her, and told her to be still. Laurie then stopped talking and has not spoken since.
>
> Soon Laurie gave up bowel control. A while later began an ever more severe withdrawal from the world, which by the age of six had reached

> such proportions that for long time periods she seemed blind, deaf, and unable to move on her won. Most of the time she spent her days motionless, staying wherever she was put, sitting in a chair, on the floor, or on the toilet, until bodily moved by someone to some other place. The small remainder of her days she spent in an empty turning of magazine pages without looking at them, or in tearing them into tiniest pieces.
>
> The only activities Laurie engaged in spontaneously were destructive. She ripped buttons off her clothes, tore her sheets, ripped wallpaper off the walls. If she got hold of a piece of ribbon, she shredded it to a mass of fine fuzzy matter until it looked like a ball of absorbent cotton. Similarly she tore shag rugs to shreds, or her blankets. This is mentioned in contrast to an elaborate and complex tearing of paper that characterized her later months with us.
>
> When spoken to, Laurie turned her face away. When annoyed or scolded, she simply stared at her hand or into empty space. As time passed, she withdrew more and more. (65)

I watch a steady wind rippling gently through the yellow-green prairie grass beneath a warm, light blue sky. A beautiful cumulous cloud floats in a soft, contented way on the fresh country air, its gentle shadow drifting peacefully over the Edenic meadow, and I think about how it all comes together, a natural world were everything feels right. In the distance, a few broken-down windmills haphazardly catch then miss then catch the wind. One has warped blades that spin in an uneven rhythm. Another has two blades missing, causing it to stumble hesitantly, as if uncertain about each new movement. Yet another has a blade only partially attached, and it flaps unnaturally against the general flow as the rest of the blades take up the slack and pull it forward in its appointed rounds.

Part V

The Scientific Odyssey

> Biology gives you a brain. Life turns it into a mind.
> —Jeffrey Eugenides, *Middlesex*

> "Rabbit's clever," said Pooh thoughtfully.
> "Yes," said Piglet, "Rabbit's clever."
> "And he has Brain."
> "Yes," said Piglet, "Rabbit has Brain."
> There was a long silence.
> "I suppose," said Pooh, "that that's why he never understands anything."
>
> —A.A. Milne, *Winnie-the-Pooh*

> I consider that a man's brain originally is like a little empty attic, and you have to stock it with such furniture as you choose. A fool takes in all the lumber of every sort that he comes across, so that the knowledge which might be useful to him gets crowded out, or at best is jumbled up with a lot of other things, so that he has a difficulty in laying his hands upon it. Now the skillful workman is very careful indeed as to what he takes into his brain-attic. He will have nothing but the tools which may help him in doing his work, but of these he has a large assortment, and all in the most perfect order. It is a mistake to think that that little room has elastic walls and can distend to any extent. Depend upon it there comes a time when for every addition of knowledge you forget something that you knew before. It is of the highest importance, therefore, not to have useless facts elbowing out the useful ones."
>
> —Arthur Conan Doyle, *A Study in Scarlet*

Ministers, preachers, evangelists, the entire hierarchy of a church--these are people who protect, support, spread the word of a religion. They are the administrators of a system of belief, a *religion*. On rare occasions they also enter the doors of perception and become the containers of *faith*, the sacred people upon whom a religion is based, the shamans, in the Catholic religion the *saints* who have experienced that which is beyond explanation.

Two forms of understanding life are at work here, those that are based on logic and reason, and those that are based on the kinds of knowing *beyond logic and reason*.

The beginnings of modern neurology are neither distant nor dignified, sputtering to an uneven ignition in the eighteen-hundreds and revealing more about the human penchant for prejudice and superstition than for clarity.

Franz Joseph Gall, was born in Tiefenbronn (Baden), March 9, 1758, the son of an Italian merchant. In 1777, he began medical studies with Jean Hermann in Strasbourg, and there displayed an affinity for research, particularly in comparative anatomy, and after earning his diploma in Vienna, began his practice as a physician there in 1785. But what most interested him was a scientific pursuit of a belief that a person's intellectual abilities could be inferred from the physical appearance of the skull, resulting in a field of study called phrenology (originally called cranioscopy). Gall was certain that a powerful memory could be deduced from prominent eyes, and as he applied his theories he began to map out other mental characteristics indicative of such things as a talent for painting or for musical ability or mechanical adeptness. He claimed that his primary goal was to develop a functional anatomy and physiology of the brain and a revised psychology of personality, what he called "organology." He ultimately identified 27 discrete brain "centers" of behavior, 25 of which have never been confirmed to exist. However he did connect on two, language and word memory, and that is why he is important.

By 1796 he was giving a private series of lectures on phrenology, and gaining an audience, resulting in censorship in the form of Emperor Francis I, who demanded Gall stop his research because it was contradicting moral and religious beliefs, resulting in "materialism" and "fatalism." The problems stemmed mainly from the current controversy over just what the *spirit* or *soul* was, some suggesting that instead of the *soul* being the driving force of life, some physical part of the human brain might be at the center of behavior and cognition. The Church leaders feared this might undermine standard Catholic faith.

Refusing to be intimidated, in March of 1805, Franz Joseph Gall and his associate Johann Gasper (Christoph) Spurzheim left Vienna on a tour of Germany and Europe in general, where Franz Joseph Gall's lectures excited the public and created controversy among fellow scientists. Some of the intellectuals of the time became staunch supporters, among them Goethe, credited by some with the first detailed description of aphasia in his *Wilhelm Meister's Apprenticeship,* published in 1842, and Stendhal, for whom Franz Joseph Gall served as a personal physician. Others, such a Napoleon, whom Gall thought less of an intellectual because of his small head, despised him. In 1807, at the peak of his fame, he established himself in Paris as physician, writer, and lecturer. Several books followed laying out his theories: *Introducion au cours de physiologie du cerveau*, 1808, *Recherches sur le système nerveux en général, et sur celui du cerveau en particulier* 1809, co-authored with Spurzheim and originally appearing before the Institute of France in March 1808, and the first installment of the *Anatomie et physiologie du système nerveux en général, et du cerveau en particulier, avec des observations sur la possibilité de reconnaître plusieurs dispositions intellectuelles et morales de l'homme et des animaux par la configuration de leurs têtes*, 1810, the first two volumes co-authored with Johann Gasper Spurzheim. This final work was completed in 1819, and appeared in a second edition of six volumes in 1822-25. In 1811 Franz Joseph Gall also responded to a charge of Spinozism or atheism, which had been laid against him in a treatise entitled *Des dispositions innées de l'âme et de l'esprit*, which he afterwards incorporated with his greater work. In 1819 he became a naturalized French subject, but his efforts two years afterwards to obtain admission to the Academy of Sciences, although supported by E. Geoffrey Saint-Hilaire, were unsuccessful.

Johann Gasper Spurzheim split with Franz Joseph Gall in 1813, and presented his own version of phrenology two years later. He never gained the notoriety of Gall, but claimed throughout his life to have been equally responsible, actually even more responsible for the views than Gall. (66) Spurzheim quickly gained the following of George Combe, who was one of the founders of the Edinburgh Phrenological Society (EPS) in February 1820, and was a strong advocate of phrenology until his death in 1858. Others promoting phrenology included H.C. Watson, Charles Caldwell, and John Elliotson.

Gall's theories were based on three principles: one, the brain is the organ of the mind; two, the brain is a composite of parts, each of which serves a distinct mental faculty; and three, the size of the different parts of the brain corresponds to the relative strength of the different faculties served. As unscientific as Gall's views of phrenology might seem today, they were, in fact, a major step forward, because they represent the first theory of cerebral localization, and even though Gall was only right on two of the areas of localization he put forth, he was right on two important ones, those involving aphasia, the loss of the ability to speak, and alexia, the loss of the ability to read. (67)

Incidents of aphasia are recorded in the *Smith Surgical Papyrus* as early as 3500 B.C. In 400 B.C. the *Hippocratic Corpus* discusses several different forms of aphasia, most resulting from some form of brain injury. Hippocrates, generally considered to be the father of medicine, believed that the brain was the organ of intellect, and that it controlled the senses and movement, and noted that lesions produced a contra-lateral effect. Around 350 B.C. Aristotle suggested that mind and body were merely two aspects of the same entity, the mind one of the body's functions, and he localized the *soul* in the heart. Around 300 B.C. Herophilus, an Alexandrian physician who was an early performer of public dissections on human cadavers during the brief period in Greek history when the ban on human dissection was lifted, and often called the father of anatomy, suggested that the brain, not the heart, was the organ of intellect, and that the third ventricle was responsible for cognition, the fourth ventricle the seat of the soul.

As unscientific and crude as the beginnings were, nevertheless, they were a movement toward an explanation for how the brain works. In the first and second centuries A.D., Greek and Roman physicians discussed speech impairments resulting from paralysis, possibly connected to memory loss. Valerious Maximum described a man hit in the head by a stone and subsequently losing his memory for letters, but not his speech and other languages skills, possibly the first reference to a case of alexia.

In 150 A.D. Galen of Pergamum, a Greek physician who founded experimental physiology (based on his experiments on animals), concluded it was not the ventricles but the brain itself which was important, and said the *soul* resided in the frontal lobes.

As with many things, there is little indication or discussion of aphasia from medieval times, but the Renaissance brought a number of theories of brain localization. At first anatomists built on Galen's anatomy, offering large general localization theories, usually with perception in the anterior, reason in the middle, and memory in the posterior of the brain. Theories also returned to the positioning of functions in the ventricles. Antonio Guainerio, discussed two patients under this model, one who could only say three words, the other unable to name things, and attributed their problems to an "excessive accumulation of phlegm in the posterior ventricle."

In 1543, Joannes Oporinus published Andreas Vesalius' book *De humani corporis fabrica*, a classic work of human anatomy based on the study of human cadavers that condemned and corrected the errors of Galen's anatomy, which was highly flawed as a result of guessing at human anatomy through animal dissection. In 1584, Johannes Schenck von Grafenberg, a pioneer of neurolinguistics, published *Observationes medicae de capite humano*, and later *Observationum medicarum rariorum, libri VII*, a seven-volume compendium that described pathological conditions concerning all the parts of the human body. Information in these books was derived from his own medical experiences, those of his contemporaries, and medical observations taken from sources dating back to antiquity. He attributed a possible lingual motor component to loss of speech, and proposed a separate kind of "central, brain-related" type of speechlessness, recording observations of patients who could not speak, but who could move their tongues, and attributing their problems to loss of memory. The 16th century also produced a number of documented surgical cases of patients who were "cured" of their aphasias after removal of bone fragments from their skulls.

In the first half of the 1600s, Rene Descartes, considered the father of modern philosophy, localized the interaction of the mind and brain in the pineal gland. He disagreed with the Platonic tripartite soul, and considered the soul to be a *res cogitans*, unitary mind, distinct from the body, but interacting with it. As wrong as this was in terms of anatomy, it was an important reopening of philosophic approaches to understanding the connections of the brain and the mind.

Nevertheless, scattered scientific, neurological evidence for localization theories was slowly gathering. In 1664, Thomas Willis wrote *Anatomy of the Brain, with a Description of the Nerves and Their Function*, the most complete account of the nervous system to date.

In 1673, Johann Schmidt described an acquired alexia with writing preserved and distinguished a paraphasia as distinct from a simple motor aphasia. In 1683, Peter Rommel provided a detailed description of a patient who could produce serial speech while at the same time suffering from severe auditory comprehension problems. (68)

In 1750, Luigi Galvani studied electricity in animals and discovered that electrical stimulation causes muscles to contract. In 1770, Johann A. P. Gesner published several discussions of aphasia, and attributed aphasia to a specific impairment of verbal memory. In 1789, Van Goens wrote about a lack of awareness of the deficit in cases of paraphasia and jargon aphasia. In 1798, Alexander Crichton extended this by suggesting aphasia is the result of a "defect of that principle, by which ideas, and their proper expressions, are associated."

By the late 1800s the groundwork was in place, centuries of slowly accumulating the data and fumbling searches for an explanation was ready for a new theory, one with a more scientific explanation. The next step was to connect up aphasia to a specific neuropathology.

In a letter to Retzer in 1798, Franz Joseph Gall outlined four theories that were to serve as the foundation for his studies: 1. Moral and intellectual qualities are innate; 2. Their functioning depends on organic supports; 3. The brain is the organ of all faculties, all tendencies, and all feelings (it is the organ of the soul); and 4. The brain is composed of as many organs as there are faculties, tendencies and feelings. (69) The key here is the break away from the current beliefs espousing a psychic or environmental basis for speech (words are not dependent on ideas) to an anatomic and functional basis. For Gall, innate qualities came from God, not experience or intention.

Gall's theories for the localization of word memory (word retrieval) and for speech and language abilities were based more on real anatomical study than most of his other localization claims. In fact, he is credited with the first complete description of expressive aphasia due to a wound of the brain. The evidence for his theories concerning speech, language and word memory comes from observed points of wound entry, from dissections of brains which had suffered gunshots, stabbings, and perhaps stroke, and from individuals diagnosed with retardation and dementia with likely frontal lobe damage. In the study of Rampan, a man suffering from a foil sword into his posterior left frontal lobe, Gall provided a detailed description including the many signs and symptoms that are currently associated with the effects of stroke in that region, including right hemiparesis and anomia. He reported that Rampan had trouble remembering proper names, but no trouble recalling object labels (perhaps the first recorded case of a "dissociation" language disorder). Such studies resulted in his claim that "there exists a form of partial insanity limited to the faculty of speech . . . (a phenomenon) impossible if the faculty of spoken language was not the function of a particular part of the brain." (70)

Most of the world of science, however, disavowed Gall's theories, mainly because Gall's research was suspect. However, another physiologist, Marie-Jean-Pierre Flourens had already been put onto the task by the Academy of Sciences of Paris on the order of Napoleon Bonaparte, and in 1825 Flourens carried out localized lesions of the brain in living pigeons and rabbits, observing the effects on motricity, sensibility, and behavior. These experiments demonstrated convincingly for the first time that the main divisions of the brain were indeed responsible for largely different functions. When the cerebral hemispheres were removed, for example, all perceptions, motricity, and judgment ended. When the cerebellum was removed, the animal's equilibrium and motor coordination suffered. When the brainstem was removed, the result was death. Thus, Marie-Jean-Pierre Flourens concluded that the cerebral hemispheres are responsible for higher cognitive functions, the cerebellum for regulating and integrating movements, and the medulla for vital functions, such as circulation, respiration and general bodily stability. On the other hand, he was unable to find specific regions for memory and cognition, resulting in a view that such functions do not have specific locals, but are diffused throughout the brain. In other words, though Marie-Jean-Pierre Flourens used a more scientific approach to brain localization (at least a more sanctioned approach), his conclusions about memory and cognition were incorrect, whereas Franz Joseph Gall, caught up in phrenology, since proven a misguided theory, and ostracized by the sanctioned scientists (certainly at least partially because of his political views) was correct in assigning such mental tasks as language and memory specific areas of the brain.

At the same time another French neurologist Jean-Baptiste Bouillaud published a collection of over 100 cases of frontal lobe damage resulting in loss of speech, not enough to convince the anti-localization followers of Jean-Pierre-Marie Flourens (and his powerful support from the government) at the time, but important beginnings that would serve to support future research. (71)

This, then, was the state of neurology in 1840, the year Richard Dadd was diagnosed as insane and committed to Bethlem, the year Soren Kierkegaard wrote *Fear and Trembling*, thirty-nine years prior to the failed attempts of Vincent van Gogh to save the *souls* of the miners of Wasmes through evangelism. Phrenology was at the cutting edge.

Other early scientific evidence for the location of various brain functions in specific areas of the brain is credited to Marc Dax, a country doctor, who read one paper at a medical society meeting in Mintpellier, France, 1836, where he pointed out that all of his patients suffering from loss of speech, from aphasia, had experienced damage to the left side of the brain, and, thus, that speech is controlled by the left hemisphere of the brain. His findings were not well received. After all, at that time, it was the wonderful symmetry of the brain that impressed scientists.

Dax died a year later. (72)

In 1844, A. L. Wigan argued that "the mind is essentially dual, like the organs by which it is exercised," (73) and that he could prove "1. That each cerebrum is a distinct and perfect whole as an organ of thought. 2. That a separate and distinct process of thinking or ratiocination may be carried on in each cerebrum simultaneously." (74) He defended his views in several ways, including the simple realization that humans can think two thoughts, even opposing thoughts, at exactly the same time, which he said made any other hypothesis "utterly inexplicable."

His major basis for his claims came from some autopsies he observed:

> One hemisphere was entirely gone—that was evident to my senses; the patient, a man about 50 years of age, had conversed rationally and even written verses within a few days of his death. (75)
>
> Dr. Connolly mentions the case of a gentleman who had so serious a disease that it spread through the orbit into the cerebrum, and by very slow degrees destroyed his life. He was a man of family and independence . . . on examining the skull, one brain was entirely destroyed—gone, annihilated—and in its place (in the narrator's emphatic language) "a yawning chasm." All of his mental faculties were apparently quite perfect and his mind was clear and undisturbed to within a few hours of his death. (76)
>
> Dr. James Johnson mentions to me another example of a gentleman under his care, who retained the entire possession of his faculties until the last day of his existence, yet on opening the skull, one cerebrum was reduced by absorption to a thin membrane—the whole solid contents of the one half of the cranium, above the tentorium, absolutely gone. (77)

Thus, A. L. Wigan concluded:

> If for example, as I have so often stated, and now again repeat, one brain be a perfect instrument of thought—if it be capable of all the emotion, sentiments, and faculties, which we call in the aggregate, mind—

> then it necessarily follows that Man must have two minds with two brains; and however intimate and perfect their unison in their natural state, they must occasionally be discrepant when influenced by disease, either direct, sympathetic or reflex. (78)

In 1848, while these primitive beginnings of neurology were trying to find a niche in the world of medicine and science, perhaps the most famous case of brain damage ever recorded was taking place just outside of Cavendish, Vermont. On September 13 of that year, Phineas Gage, a well liked and respected railroad foreman, survived an explosion that drove a three-foot-long, one-and-a-fourth inch wide, iron rod completely through his skull! Amazing! Even more amazing, he retained his consciousness, and was reported to have said to Edward Williams, the first doctor on the scene, in a somewhat humorous, laughing manner, “Here’s work enough for you, doctor.” The doctor cleaned out his wound, actually touching left and right fingers clear through Phineas’ head.

However, soon after, Phineas’ temperament changed dramatically. The once even-tempered, responsible man now became wild, and amoral. His erratic behavior grew increasingly irritating, unpredictable, and unsocial. At times, he would lie in bed intermittently screaming and laughing. And this continued into the world outside the bedroom. His public behavior would quickly turn from being overly friendly and loving to threats and profanity (even in front of women, not acceptable at that time).

George Page, narrator of *The Brain*, an acclaimed video series, suggests that what happened to Phineas Gage was that the limbic system, the place where emotions arise, got cut off from the frontal cortex, the place where such emotions are censored. (79)

Hanna Damasio and Thomas Grabowski, employing our contemporary knowledge and technology, have been able to recreate a three–dimensional image of Phineas Gage’s damaged brain, and speculate that the areas most likely to have been damaged by the accident are the left anterior prefrontal cortices and the ventromedial prefrontal cortex, generally thought to be areas crucial to decision-making. (80)

Obviously, such sophisticated techniques were not possible at the time, though it is possible to find some of their beginnings.

Soon, Marc Dax’s localization theories were to resurface, this time with more evidence, by Paul Broca, who, in 1861, spoke twice before the Paris Society of Anthropology, pointing out how a patient who had suffered from loss of speech and paralysis on one side of his body had been shown to have a region of damaged tissue in his brain in a postmortem examination, suggesting that this region controlled speech, that, in fact, specific areas of the brain controlled specific mental functions.

In 1863, Gustave Dax, the son of Marc Dax, published the 1836 memoir of Marc Dax, his deceased father, and additional clinical observations of his own on 140 patients. His contribution received a negative appraisal by the Academy. (81) In 1865, Broca published *General Instructions on Anthropological Research*, and a controversy was sparked. (82)

Paul Broca employed a clinical approach to several aphasic patients who could not talk or could talk very, very little. The most famous could only utter the word "tan." When he died, Paul Broca studied his brain and found a small portion of the left hemisphere had been destroyed by neurosyphilus and deduced that this is the area of the brain responsible for speech.

After further study, Paul Broca was able to state more definitively:

> I have been struck with the fact that in my first aphemics the lesion always lay not only in the same part of the brain but always the same side—the left side. Since them, from many postmortems, the lesion was always left sided. One has also seen many aphemics alive, most of them hemiplegics, and always hemiplegic on the right side. Furthermore, one has seen at autopsy lesions on the right side in patients who had shown no aphemia. It seems from all this that the faculty of articulate language is localized in the left hemisphere, or at least that it depends chiefly upon that hemisphere. (83)

The times were ripe for an acceptance of this scientific view of the human brain. An unpopular, conservative Catholic Church had entered into an alliance with an unpopular, conservative French monarchy, expressing a belief in the transcendent unity and immateriality of the human spirit (i.e., it is beyond logic and reason, but is of the world of faith). Science and naturalist philosophies were a possible way to challenge these powers and perhaps undermine the views of these traditional authorities. If, just if, everything in the universe, including the human soul, is "natural," i.e., can be explain scientifically, then those views of the church must give way to a new rational order (84)

Paul Broca, it is worth noting, was the founder of the Paris Society of Anthropology, a focus for anticlerical activity, and pushed to ground philosophy firmly in a material base. (85) And he was well aware that, if he could localize the *soul* in different portions of the brain, it would be a powerful tool for these political views. (86) To further Broca's new discoveries, students in the Third Republic seized on the new ideas because they wanted a scientific theory for their rebellion against the prevailing generation. (87)

The location of language in the frontal lobes also carried with it a strong prejudice in favor of the prevailing racist views. Anne Harrington summarizes, "Gall's original claim for a link between intelligence and frontal lobe functioning (and, conversely, between instinct, passion, and posterior lobe functioning) had been buttressed since the early nineteenth century by a wide range of studies that compared the brains of different human racial groups with their 'known' intellectual capacities. Virtually all of these had concluded that the white European races, which everyone 'knew' stood biologically and culturally at the peak of human evolution, also possessed a considerably more developed frontal area than the 'primitive' nonwhite human races. Pierre Gratiolet had even gone so far as to classify the Caucasian, Mongoloid, and Negroid races in terms of their alleged dominant brain regions; viz., "frontal race," "parietal race," and occipital race," respectively. It is worth noting in passing that these were not the conclusions Franz Joseph Gall came to. Rather, Franz Joseph Gall, who had very liberal views in general for his time, thought his phrenology supported the opposite view, stating that "All men have the same brains, therefore the same faculties and tendencies . . . a Negro and a European stand on the same level of the scale of the animal kingdom."

However, just as Franz Joesph Gall's own views were discounted early-on, more for his negative comments about Napoleon's phrenology then for scientific reasons, and the embracement of Flourens' views was done more in that same political spirit, this new wave of scientific insights was driven more by social, religious, sexist, racist, and political beliefs than science. Anne Harrington states:

> The first venture was largely shaped by French and, to a lesser extent, by German researchers during a time when the brain sciences were fascinated by problems in comparative neuroanatomy, (racialist) anthropology, and the unusual consciousness associated with what would today be called dissociative personality disorder, but was then called hysteria or dual personality. Research questions at the time took their starting point and justification from a tendency to turn to classic brain localization theory as a basic approach to most questions about mind-brain relations, from differences in intelligence, to outright madness, to subtle personality changes. (88)

In spite of the less-than-scientific political and social forces behind the new views, Paul Broca "opened the way to a cerebral neurology, which made it possible, over the decades, to 'map' the human brain, ascribing specific powers—linguistic, intellectual, perceptual, etc.—to equally specific 'centers' in the brain." (89) The particular area of the brain that served as the focus of Broca's research has ever since been called Broca's Area by neurologists. And it is now universally accepted that it is the area that contributes the correct words to express oneself, and damage to it results in expressive dysphasia, i.e., the inability to find the correct words to express one self.

While the beginnings of neurology and psychology and anthropology were still mired in phrenology, racism, and political concerns that had little to do with science, Adolf Bastian, an etnoloogist, something of a polymath, was to put in place the foundations for uniting all of them, comparative anthropology (including comparative mythology), Carl Jung's archetypal theories of a collective unconscious, and neurology's localization theories.

Bastain earned his degree in medicine from Charles University, Prague in 1850 and became a ship's doctor, where he spent the next eight years traveling around the world to Australia, Peru, the West Indies, Mexico, China, the Malay Archipelago, India and Africa. When he returned to Germany in 1859, he published *Der Mensch in der Geschichte (Man in History)*, 1860. From 1861-1865, he continued his travels through Asiatic countries, resulting in his book, *Die Volkerdes Ostlichen Asien (The People of East Asia)*. After his return to Germany, he began establishing ethnological institutions in Berlin, contributing many of the artifacts he had collected to Berlin's Royal Geographical Society of Germany. In the 1870s he again traveled, this time through Africa and the Americas, dying in Trinidad and Tobago in 1905.

His encounters and study of various groups of people, his ethnology, led him to the important underlying theory that all humans share a basic mental framework, a psychic unity. He thought that the world could be divided into different geographical regions, each evolving its own culture traits, which did not diffuse into one-another. He also rejected philosophic speculation in favor careful observation, thus denouncing Darwin's theories because they had not been empirically observed. Howefver, the fact that he was more interested in gathering artifacts and preserving traditional cultures before they vanished rather than in strict scientific study, brought criticism of his random, rather than coherent approach.

The key, however, is in the theory of what he called *elementargedanken* (elementary ideas), which states that the mental acts of all humans result from physiological mechanisms common to the species. In other words, every human inherits species specific elementary ideas, and thus the minds of all people, regardless of race of culture, work the same way. The environment puts its unique spin on these universal elementary ideas, what he called *volkergedanken* (folk ideas). Each culture moves from more basic, simple socio-cultural systems to more complex organizations.

What he put forth was the basis for a psycho-biologically grounded, cross-cultural social psychology that would collect as much ethnographic data as possible (untainted by contact with Europe), and study the psychological laws of mental development that are revealed in the diverse geographical settings, i.e., comparative mythology.

While Bastian was laying the foundations for universal mental structures, and initial localizations by broca and others were being discovered in the left hemisphere, John Hughlings Jackson, of considered the true founder of neurology, was suggesting a larger perspective. He wrote in 1864: "If, then, it should be proved by wider evidence that the faculty of expression resides in one hemisphere, there is no absurdity in raising the question as to whether perception—its corresponding opposite—may not be seated in the other." (90)

His theory was supported by what he saw in a patient with a right hemisphere tumor:

> She did not know objects, persons and places . . . there was what I would call "imperceptions," a defect as special as aphasia. I think as Bastian does, that the posterior lobes are the seat of the most intellectual processes. This is in effect saying that they are the seat of visual ideas. I think too that the right posterior lobe is the "leading" side, the left the more automatic.
>
> This is analogous to the difference I make as regards use of words, the right is the automatic side for words, and the left side for that use of words which is speech. (91)

By 1870 "Karl Wernicke, a German neurologist, was able to demonstrate that "damage to the back part of the temporal lobe of the left hemisphere could produce receptive dysphasia, difficulties in understanding speech." (92) These patients were able to hear sounds just fine, but could not recognize words. As with Broca, Wernicke's area has since been established as the name for a particular function of the brain, an area located slightly behind Broca's area. At about this same time, Gustav Fitsch and Eduard Hitzig stimulate with electricity the brain surfaces of awake dogs, proving that stimulation of different areas produced contractions in different parts of the body, beginning a more detailed mapping out of brain localization. Their work was extended by a serious of experiments done on dogs and monkeys by David Ferrier, who managed to locate fifteen different areas related to the precise control of movement by stimulating the cortical gyri with electricity. Furthermore, he was able to accurately predict how these areas related to human neurological diagnosis and the location of brain tumors. It is worth highlighting that Ferrier's views were in direct opposition to Flouren's and to others currently denying cortical localization, most famously to Fredrich Goltz, who was unable to duplicate Ferrier's findings.

In other words, in 1870, the year Vincent was struggling to figure out how to have both *faith* and *religion*, the neurological understanding of brain localization was much the same as it had been thirty years previously, at least on the surface. But the birthing process was in motion.

In 1880, localization theory in neurology is still in its infancy, barely beginning the long journey out of the darkness of prejudice and superstition. But, you'll recall, some scientists are opening interesting doors. Franz Joseph Gall's initial attempts at phrenology have led through A. L. Wagin, Marc Dax and others to Paul Broca's proof that a specific area of the left hemisphere of the brain controls speech, at least some aspects of it; in 1877 C. E. Brown-Sequard's studies lead him to support and even quote A. L. Wigin, "I have come to the conclusion that we have two brains, perfectly distinct the one from the other" (93); and in 1880, the same year Vincent goes through his dramatic rejection of religion, Karl Wernicke demonstrates that there is a specific area of the left hemisphere of the brain connected to understanding speech.

In 1884, Victor Horsley begins using intra-operative electrical stimulation of the cortex for the localization of epileptic foci in humans and states: "We are not single animals: We are really two individuals joined together in the middle line." (94)

In 1885, German physician Ludwig Lichtheim helps to support the theories of Broca and Wernicke, and demonstrates "that, although [language] deficits are not due to a general inability to understand speech, they are associated with injury to the left hemisphere," further concluding that the "left hemisphere controls 'purposeful' movements as well as language but that the specific areas of the left hemisphere involved are different in the two cases" (95) He also provides a diagram, affectionately known as Lichtheim's Lighthouse, to attempt to explain just how it all works.

Six years later, Sigmund Freud, at the time mainly experimenting with the use of hypnosis to treat patients with nervous disorders, writes: "An invitation which I received in the same year [1891] to contribute to an encyclopedia of medicine led me to investigate the theory of aphasia. This was at the time dominated by the views of Wernicke and Lichtheim, which laid stress exclusively upon localization. The fruit of this enquiry was a small critical and speculative book, *Zur Auffassung der Aphasien*." (96)

This study suggested a physiological basis for aphasia and other disorders of recognition and perception (i.e., they could be explained scientifically by neurology). (97) This, of course, needs to taken in the context of Freud's ultimate embracement of, creation of psychoanalysis theory, which is a very different brain/mind model.

Theories at the time are even more complicated. Just prior to Sigmund Freud's supporting claim for brain localization, in 1886 D. Ferrier writes: "The brain as an organ of motion and sensation, or presentative consciousness, is a single organ composed of two halves; the brain as an organ of ideation, or representative consciousness is a dual organ, each hemisphere complete in itself." (98)

Furthermore hemispherectomies done on animals suggest, not that the left hemisphere is dominant, but that "either" side can function upon the removal of the other. In 1888 F. Goltz writes:

> I will begin by relating an experiment which I hope will be acclaimed by all true friends of science. I succeeded in observing for 15 months an animal in which I had taken away the whole left hemisphere. (99)
>
> We have seen that a dog without a left hemisphere can still move voluntarily all parts of his body and that from all parts of his body, action can be induced which can only be the consequence of conscious sensation. This is incompatible with that construction of centers which assumes that each side of the body can serve only those conscious movements and sensations which concern the opposite half of the body. (100).
>
> Finally, as far as Man is concerned, the fact that a dog after an extirpation of a whole hemisphere shows essentially the same personality with only slightly weakened intelligence might make it possible to take out even very large tumors if they are confined to one half of the brain. (101)

There is a danger here. If either side can function equally upon the removal of the other, that works against brain localization. At the same time, if there are specific locations, they seem to all be in the left hemisphere, and the conclusions being reached are that the right hemisphere is either simply a duplicate of the left hemisphere or is in fact worthless. (102)

Another major problem with neurological research in the later part of the nineteenth century and first half of the twentieth century is that the field has moved so far into the occult that it has lost what little creditability it had. One need only look at Sigmund Freud's own overview of the general "scientific" beliefs at this time in relation to his own emerging theories of dreams in the first chapter of his seminal book *The Interpretation of Dreams* to see how little there is of science in them, and these are the more creditable ones. (103)

As early as 1866, Dr. Edward Sequin published *Idiocy: Its Treatment by the Physiological Method* (104), a holistic approach to what he called "idiotic geniuses," where his description of the early lives of two of the more famous cases serve as the basis for future biographies, that of James Henry Pullen (discussed later), and that of Blind Tom, such a remarkable case study that it bares some elaboration.

His story begins at a slave auction in Georgia, 1850, where his mother was sold as a slave by Perry Oliver to Colonel Bethune, and her fourteenth child was include for free because he was blind and thought to be of no value. His new master named him Thomas Greene Bethune, and allowed him to wander about the mansion. It was noticed that he seemed attracted to sounds, and would gyrate to rain or the grating of corn, and most importantly to music. By the age of four, though he showed no other signs of intelligence, he could sit at the piano and beautifully play any songs he had heard. One night, Colonel Bethune heard music coming from the piano, and thinking it was his daughters, came down to discover Blind Tom (at age four) playing perfectly a Mozart sonata he had heard earlier. By age six, Blind Tom could improvise at will as well as repeat, and by age seven he was performing sold out concerts across the country. It was reported that he could hear any piece of music and play it note for note without error. At age eleven he played at the White House for President Buchanan. The following day, several musicians, feeling that there must be some trick to the performance, tested him, supplying two completely new compositions, one thirteen pages long, the other twenty pages long. In both cases, he played to perfection upon hearing them but once. At age fourteen he was presented with a fourteen page original composition (not to read, he couldn't read, but to hear) and asked to play secondo with the composer while he played the treble part (meaning he had to improvise the entire secondo part in step with the musician's performance of the first part). He did it perfectly, then pushed aside the composer and played both parts at the same time. This obviously confirmed that Blind Tom could create, and create on a high level. While on tour in Europe, his perfect pitch was tested. As he listened, two pianos were hammered on in a noisy and haphazard manner, while a third did a run of twenty notes. He was then asked to play the twenty note run, and did so flawlessly. Sequin gave the following description of Blind Tom in concert:

He is led by the hand or sleeve before an audience, and begins by presenting himself in the third person, and in a few words thrown away, rather than spoken, saying, "Blind Tom will play this or that piece for you," etc. after which he begins the piano. His execution is sometimes sweet, oftener of an unknown force, which manifestly proceeds from powers higher up than his wrist. When he sends certain clangorous agonies his shoulder-blades bear as it were directly on the keys, his whole frame vibrates with the instrument. If some person of the company is invited to play a new tune that the sable artist will have to repeat, he being used to it, understands what is the matter, and shows his satisfaction by his countenance, a laughing, stooping, with various rubbings of the hand, alternating with an increase of the sideway swinging of his body, and some uncouth smiles. As soon as the new tune begins, Tom takes some ludicrous posture, expressive of listening, but soon lowering his body and rising on one leg, so that both are perfectly horizontal, and supported upon the other leg, representing the letter T, he moves upon that improvised axis like the pirouette dancer, but indefinitely. These long gyrations are interrupted by other spells of motionless listening, with or without change of posture, or persevered in and ornamented with spasmodic movements of the hands; this is his studying posture. When the stranger is through, Tom stops, seats himself at the piano, and reproduces the musical idea perfectly, if the piece was entirely new to him; but reproduces tune for tune, note for note, if he only heard it previously two of three times.

Tom's concert performances continued until he was 53, when Colonel Bethune died, and apparently left such a void in Tom's life that he fell into a "sullen belligerent state of mind." He died in 1908, lonely and alone.

While Sequin's case studies and compassionate views of such people went largely unnoticed, another would take up the cause.

In 1887, J. Langdon Down, the superintendent of Earlswood Asylum in England, presented the highly prestigious Lettsomian Lectures before the Medical Society of London, and used these lectures to as a "convenient place to treat an interesting class of cases for which the term 'idiot savants' has been given, and of which a considerable number have come under my observation." In addition to giving this name to the field, he offered a definition, "Children who, while feebleminded, exhibit special faculties which are capable of being cultivated to a very great extent." Interestingly, even while giving this two term designation to such children, he condemned it; "I have no liking for the term 'idiot.' It is so frequently a name of reproach. No one likes the name, and no mother will admit that her child deserves the title." (As an aside, it might be mentioned that his suggestion of 'feebleminded' as a replacement is not much of an improvement; but his intentions were in the right place.) In addition to a careful and detailed discussion of such children, he also gave a meticulous description of the category that has since been named for him, Downs Syndrome. (105)

Alfred Binet and Jean Martin Charcot worked with lightning and exceptional calendar calculators. Interestingly in terms of subsequent developments in psychology, by 1894, Benet was suggesting that "the unconscious which is within us, and which psychology has in recent years often succeeded in illuminating, is perhaps capable of foreseeing the solution to a problem or long arithmetic operation without carrying the details of the calculations." (106)

During this same decade, William W. Ireland published *The Mental Affections of Children: Idiocy, Imbecility, and Insanity*, a close study of several patients at his institution in Ireland, pushing to understand these people who where so deficient mentally nevertheless had some extraordinary abilities, suggesting perhaps that some of those with "special talents" have resulted in a "very careful cultivation" of those talents, "and while they have been cultivated the other faculties have been frequently neglected." (107)

Joseph Jules Dejerine, one of the pioneers of localization study, demonstrated in 1891 and again along with Nehemie Vialet in 1893 that word blindness might occur as a result of lesions of the supramarginal and angular gyri. In 1906, he would work with his student Andre-Thomas to show that thalamic syndrome is the result of a pathologic lesion.

Starting in 1900 Hugo Karl Liepmann, Carl Wernicke's assistant in the 1890s, did an extensive study of apraxia (the inability to act or move different parts f the body in a purposeful manner, even though there is no physical reason), and came to the conclusion that it is the result of damage to the parietal lobe. As a result of his studies, he divided apraxia into three forms: ideational (object blindness), where the patient is incapable of making appropriate use of familiar objects upon command; ideomotor, where the patient is not able to follow verbal commands of mimic an action; and kinetic, where the patient's physical actions are clumsy and there is no physical diagnosis, no paralysis, muscle weakness, or sensory loss.

In 1899, Theodore Simon, an admirer of Binet, became an intern at Perray-Vaucluse, where he worked with abnormal children. His work drew the attention of Binet, and the two of them began working on an evaluation method for intelligence, resulting in the Binet-Simon Intelligence Scale, which Binet first published in his journal *L'anneé psychologique* in 1905, and they continued to revise through 1911. It was development mainly out of Binet's studies of children ages 3-15 where he compared "normal" children to "mentally challenged" children through one-on-one interviews to reach a determination of the intellectual level they had achieved.

This opened an important door for education to help understand and monitor differences in mental development and teaching methods. However, there are also dangers in such a test, as Simon realized and worked to prevent after Binet's death, urging psychiatrists not to over use it or replace other methods of understanding childhood development.

The most important revision of this test came from Lewis Terman in 1916, and was titled the "Stanford Revision of the Binet-Simon Scale" or the "Stanford-Binet" I.Q. test. Soon Robert Yerkes, president of the American Psychological Association, finding it so valuable, began using it to develop the *Army Alpha* and *Army Beta* tests to be used to classify recruits.

As of 2000, the test had gone through five updates, and the one currently used is appropriately titled the Standard-Binet 5. It justifies its accuracy by retesting individuals several times, finding little variation in their scores, and has made great strides in its attempts to reach across gender, race, and various other environment factors. Five categories are tested, fluid reasoning, knowledge, quantitative reasoning, visual-spatial processing, and working memory. Each of these is tested in two separate domains, verbal and nonverbal, in order to accurately assess individuals with deafness, limited English, or communication disorders.

Dangers, however, are evident. Intelligence is not such a narrowly defined situation and people who have high levels of intelligence but do not function well in an organized test situation will not receive proper credit. The reverse is also true, people capable of scoring high in a test situation or in a "book learning" format, might well have extremely low intelligence in other areas. Brilliant right cerebrum thinkers are especially likely to poorly on such tests, and powerful left cerebrum thinkers are likely to score very high, even though they are deficient in forms of right brain thinking.

A. F. Tredgold, with a list of credits in the field of amentia, including his roles as consulting physician to the National Association for the Feeble-Minded and the Littleton Home for Defective Children, lecturer for the Medical Graduates' College of London, medical expert to the Royal Commission on the Feeble-Minded, and Research Scholar in Insanity and Neuropathology of the London County Council and Assistant in the Claybury Pathological Laboratory, worked with all forms of mental disease, trying to find scientific explanations, studying brain cells to find abnormalities, recording both the medical data and the life histories, and in his role on the Royal Commission, working to improve the laws, helping to pass the Mental Deficiency Act of 1913 in Parliament to provide at better supervision for the mentally deficient. His book *Mental Deficiency*

(Amentia) is a detailed discussion of the entire scientific field at the time, and includes in depth discussions of many cases, including some of those that fall under the same category as the idiot savants discussed above (in fact, is an important source for the discussions of Gottfried Mind, and the blind woman of Salpetriere mentioned earlier).

He elaborated on another interesting case, one mention earlier as first detailed out by Edward Sequin, that of J. H. Pullen, "The Genius of Earlswood Asylum," a man born deaf in 1835 (his parents, interestingly, were first cousins), who did not began talking at all until the age of seven, never attended school, and was admitted to the Earlswood Asylum at age fifteen. At the time, he was in good health, in possession of all of his senses (except for hearing, and very poor speech), had initiative, imagination, resourcefulness, an attention span far superior to the other inmates, and demonstrated an amazing ability at drawing and mechanical invention.

Tredgold discusses him:

> One of the most wonderful of his works, and the one of which he was the most proud, is the model of a steamship, which he has named the Great Eastern. This, I think, he rightly regards as his magnum opus, and it attracted universal admiration at the Fisheries Exhibition, where it was shown in the year 1883. It took him three years and three months to complete, and every detail, including brass anchors, screw, pulley-blocks, and copper paddles, were actually made by the patient from careful drawings, which he prepared beforehand. The planks of this leviathan are fixed to the ribs by wooden pins to the number of nearly a million and a quarter. All of these were made by Pullen in a special instrument, which in turn he also planned and made. He also devised and executed a strong carriage on four wheels for the conveyance of the ship. The model is 10 feet long, 18 5/8 inches wide, and 13 5/8 inches in depth. It contains 5,585 copper rivets, and there are thirteen lifeboats hoisted on complete davits, each of which is a perfectly finished model. It is fitted with paddles, screw, and engines, and it contains state cabins, which are decorated and furnished with chairs, tables, beds, and bunks. In fact, the whole thing is complete to the most minute detail, and will bear the closest inspection. He has invented and attached an arrangement of pulleys by which the whole upper deck many be raised so as to show the parts below. I believe that when first put into water the huge model capsized, but that has since been remedied. It is perhaps hardly to be expected that a person with no knowledge of practical boat-building should succeed in making a vessel that would be really navigable, but as a highly finished model it is unmatched in its completeness.

This one creation is just that, one of hundreds.

> Pullen, however, also had the uneven abilities of such people. He was prone to violence, even attempting to kill a steward of the asylum with a carefully planned and created guillotine like instrument (only barely failing in the attempt). He once decided his wanted to marry a woman he had chanced to see outside of the museum and insisted he was going to marry her. Only after a number of other ploys had failed was he lured away from this plan by being offer a pretend commission as Admiral in the Navy, which carried with it a resplendent naval uniform. This did the trick, and he never again referred to any interest in marriage, but could often be seen wearing the uniform, especially on ceremonious occasions. (108)

Demonstrating his even-handed approach to such cases, Tredgold considered the insanity aspects were probably more the result of Pullen's deafness than any kind of inherent occult or mystical insanity, or even a physically damaged brain, and suggested that the deprivation that resulted from his insanity and his isolation resulted in his curious personality.

By the 1920s, neurology is slowly moving into ever more logical, scientific directions. The editor of *Brain*, a famous neurological journal, from 1910 to 1925, Sir Henry Head, included diagrams and schematics with his less scientific discussions of "feeling-tone." (109) It is, in fact, interesting to note, that, in building on and attempting to rephrase Jackson's views, he objected to the use of "prepositional," (110) and suggested it be replaced by "symbolic formulation and expression." (111) Jackson's point had been that the two hemispheres each used words, but used them in different ways, the left in what he referred to as a prepositional manner, which he stated "is not a mere sequence . . . [but] consists of words referring to one another in a particular manner [so that each] modifies the meaning of the other." (112)

Looking back, it is possible now to see how these ideas relate to two important aspects of language, its sequential, literal qualities, and its holistic, symbolic qualities. Indeed, neurology, though it is still putting forth the dominance of the left hemisphere, is beginning to build a foundation for a clearer understanding of the right hemisphere's contributions and value. Indeed, the right hemisphere seems to be where humans understand the higher truths, those beyond the literal truths of physical existence, the symbolic truths of the arts and the truths of meaning and value.

However, at the time, the right hemisphere was still being given little attention. Even such strong localizers as S. E. Henschen considered the right hemisphere nothing more than a "regressing" or "reserve" organ, and gave it only a compensatory role after left hemisphere lesions, where in each case it "shows a manifest inferiority as compared with the left, and plays an automatic role only." (113)

Lord Brain wrote that "the posterior half of the left cerebral hemisphere is thus the site of those neurotic linkages which underlie the elaboration of meanings in response to auditory and verbal stimuli." (114) R. R. Grinker, P. C. Bucy, and A. L. Sahs stated that "the hemisphere which controls handedness, expression and comprehension is known as the dominant hemisphere." (115) W. Ritchie Russell said, "The processing of past and present information arriving in the dominant hemisphere seems to provide a scaffolding on which thought activity depends." (116)

Henry Head himself, disagreeing with the localization theories of Broca and Wernicke, largely ignored the right hemisphere and the whole concept of the duality of the brain. In his important series of articles on neurology, Joseph E. Bogen included the following:

> In spite of the fact that Head was one of Jackson's most fervent admirers, he gave the right hemisphere short shrift. In the nearly one thousand pages of his 1926 classic, Head largely ignored the duality of the brain which so intrigued Jackson. Head discussed "imperception" for thirteen pages, including four pages devoted to von Stauffenberg's 1913 case. This was a woman with cerebral arteriosclerosis who had left hemianopia for more than six years; and whose recurrent paralyses were always of the left limbs; and who could write but could hardly copy even simple drawings. When she finally died, she was found to have injury in both hemispheres. So Head concluded that to produce imperception, "the destruction of tissue must be severe, extensive, and in most of the best reported cases, was bilateral. (117)

Nevertheless, the field was coiling for an explosive insight. Important scientific tools were getting put in place. Most notably in the 1920s was Hans Berger's invention of the electro-encephalograph, which allowed the study of electronic activity in specific parts of the brain.

More detailed case studies were also being published. In 1925, Geza Revesz published the first careful study of a child musical genius. (118) In it he applied the current neurological and psychological theories to the first eleven years of Erwin Nyiregyhazi, who, by the age of seven had already gained an international reputation for his extra-ordinary abilities to perform, improvise, and compose at the piano. In this particular case, his talents were not in the context of noticeable deficiencies in other mental abilities. In fact, through a series of tests, Revesz was able to conclude that Erwin was at least slightly above average in his general mental abilities across the board. But his musical abilities were in a range that can be defended as extraordinary. He had perfect pitch, apparently from birth, is said to have actually begun attempting to imitate melodies before reaching the age of one, and by two to correctly reproduce melodies (well before he was able to speak), by three to reproduce pitches on the piano and even write his own music, and by four to play whatever he heard. In a concentrated study, Revesz not only tested Erwin over a period of about six years, but researched his genealogy, and even put him in the context of other musical geniuses (from Mozart, who is reported by his father to have been able to write out the entire score of Allegri's *Miserere* after only one hearing (119) to Bach to Handel to Mendelssohn and more). In fact, the brief discussions of these famous musicians has an appeal all of its own. (120)

By the late 1920s, Kurt Goldstein (ironic, since Goldstein would oppose his views, a student and friend of C. Wernicke) had become actively involved in efforts to do away with Cartesian dualism and promote a psychosomatic approach to neurology. His holistic theories developed out of his own clinical work at The Institute for Research into the Consequences of Brain Injuries beginning during WWI, where he worked with Gestalt psychologist Adhemar Gelb for the practical purpose of getting brain damaged soldiers back to a condition of independent or at least semi-independent existence. This experience would give Goldstein hands-on knowledge of brain damage from which to draw his theories.

The most important patient was a twenty-four-year-old soldier named Schneider. In 1915, Schneider had suffered two wounds to the back of his head, penetrating to the occipital lobes. Once he had healed sufficiently, Goldstein and Gelb were able to administer a number of tests, the standard ones demonstrating no permanent damage. However, they went further, and observed that, indeed, Schneider was not back to normal. What they discovered, instead, was that he had found ways to compensate for his mental losses. For example, he had found a new way of reading by using a "series of minute head and hand movements" to write "with his hand what his eyes saw." If he was prevented from doing this, "all he saw were individual lines and tracks without any overall pattern of meaning." Interestingly, he was unaware he had made this compensation. (121)

According to A. R. Luria (122), the true beginning of neuropsychology was Goldstein's paper in 1925 that provided a *philosophical* basis for an analysis of psychic problems beyond the assumption that they resulted from damage to a localized function. (123) Goldstein's most important book *Der Aufbau des Organismus* (*The Organism*) presents an "organismic approach" to aphasia, where symptoms are regarded as part of the total human, an approach that challenges reductive approaches and approaches that deal with "localized" symptoms. He believed the organism could not be divided into "mind" and "body" because nothing is independent within the organism. It is the whole that reacts to the environment, and goes on to explain the holistic approaches of healing not coming through "repair" but adaptation. An organism cannot return to the state it was before the event changed it; instead it must adapt to the new conditions caused by the new state. He also emphasized the ability of the organism to react and adjust to catastrophic situations of mental or physical functions. These concepts of biological knowledge, self-actualization and abstract thinking allowed him to challenge approaches to "localized" symptoms, such as certain parts of the brain being affected and where reflexes stem from.

In 1935, T. Weisenberg and K. E. McBride published *Aphasia: A Clinical and Psychological Study* (New York: Commonwealth Fund), a study of over 200 patients who had suffered brain damage involving over nineteen hours of testing for each patient, and revealing that, not only did the expected damage to the left hemisphere result in poor performance on tests involving verbal abilities, but that damage to the right hemisphere resulted in poor performance on non-verbal tests involving such things as manipulation of geometrical figures, puzzle assembly, completion of missing parts of patterns and figures, and other tasks involving form, distance and space relationships.

Furthermore, these tests revealed or at least suggested that language involves a complex interaction of several levels of thinking, some of which are in the right hemisphere. (124)

In September of 1948, Theophile A. J. Alajouanine published "Aphasia and Artistic Realization," a lecture he had recently delivered for the Harveian Society. (125) Focusing on three major artists, a writer, a musician, and a painter, whose careers he followed before and after their aphasia, he discussed what happens to artistic abilities as a result of it. The writer was a well-established poet, novelist, and critic before his sudden hemiplegia (paralysis) of the right side, including the loss of language abilities generally designated as aphasia. At first, he uttered "Bonsoir les choses d'ici bas" involuntarily whenever he wanted to respond to anything, supporting the previous views of John Hughlings Jackson that such a condition often results in the use of a singular expression, a leitmotiv. As time went on, he was able to abandon this automatic response with more normal responses, though his syntactical construction remained a problem. His writing followed a similar partial recovery. His reading rapidly resumed its normal abilities, though now done more slowly. Almost needless to say, his artistic production was no longer possible. However, his memory, his literary judgment, and his aesthetic tastes were not impaired. Alajouanine's conclusion was that this is an instance of damage to Broca's portion of the brain, which, in turn, results in loss of artistic ability.

The musician he studied is Maurice Ravel (1875-1937). (126) By 1928, Maurice Ravel was considered the foremost composer in France. He had already written numerous highly acclaimed pieces: from his early *Pavane pour une Infante defunte* and *Jeux d' eau,* to his Rapsodie espagnole, *L'Heure espagnole, Ma mere l'oye, Gaspard de la nuit, Balses nobles et sentimentales,* to his ballet *Daphnis et Chloe* performed at the Theatre Chatelet, in 1912, and his "choreographic poem" *La Balise*, and opera *L'Enfant et les sortileges*.

In 1928, he went on a highly acclaimed tour of America, thirty-one performances in all. In the same year, he was commissioned by the dancer Ida Rubinstein to write a ballet for her. The resulting piece, *Bolero*, became his most popular. When it received its first American performance on November 20, 1929, by the New York Philharmonic-Symphony, Toscanini conducting, it became a huge hit, spawning six immediate recordings, and becoming a standard on the radio, adapted for jazz bands, used in a Broadway revue, and in a cabaret. He was paid a large sum of money for the film rights, and even though he was used to the public eye and had a high opinion of himself, he was actually a bit overwhelmed by it all.

Then he suffered a stroke, presumed by Alajouanine to have been in the left hemisphere, and the right hemisphere abilities remained, including the vitality of many musical skills, i.e., the ability to recognize melodies, notice even minor errors in performances, and hear pitch, but the left hemisphere ability to read music was lost. Alajouanine considered it a form of "Wernicke aphasia of moderate intensity, without any trace of paralysis, without hemianopia, but with an ideomotor apractic component," and thought, "The cause, though indefinite," belonged "to the group of cerebral atrophies, there being a bilateral ventricular enlargement."

He found that oral and written language were only slightly impaired, memory, judgment, affectivity, and aesthetic taste remained strong, and the understanding of language was better than oral and written abilities.

Alajouanine believed that Ravel's brain damage was aphasia of Wernicke's area. He found that Ravel retained his ability to recognize even subtle changes in his own music, both rhythm and pitch. However, he could no longer recognize musical dictation, or any kind of analytic discussion of notes. He also lost the ability to play the piano, as he could no longer control his hands well enough. They no longer recognized the correct keys to play. As with the writer, aesthetic appreciation remained strong, but his own creativity was now impossible.

The study reveals much more to neurologists today than it did to Alajouanine in 1948, and the analysis would involve a much stronger consideration of the aspects of music retained or lost in relation to left or right hemisphere abilities. Nevertheless, Alajouanine's discussion is couched in an article attempting to comprehend how damage to certain areas of the brain affect artistic abilities, and this localization of differing functions of the brain in relation to mental abilities is important and forward looking.

The painter also suffered a form of aphasia of the Wernicke type. He had trouble with spoken language, and a less severe loss of abilities with the written language. However, in contrast to the first two, both his aesthetic judgment and his own creativity remained strong.

Alajouanine's conclusion was that writing and music require a form of language, which the aphasia has damaged, but that artistic expression does not require this language patterning. He wrote, "This is not the use of a language made up of symbols, it is the use of a plastic reality, more or less transposed by particular technique into a climate which, in the painter, is the exact equivalent of the language." It is, in other words, a direct encounter, "without any symbolic medium."

Alajouanine's case studies represent the status of neurology at this time. His attempts to clarify his findings have obvious flaws, and the assumed localizations demonstrate how little is actually known. In truth, what we find is that no real mapping has been done since Broca and Wernicke. And their discoveries are being stretched to try to explain more than they can account for. Nevertheless, the still young field of study is starting to find its way toward an explosion of activity and accomplishment.

Aphasia, sometimes referred to as aphemia, is the inability to produce or comprehend language the result of some injury to the specific regions of the brain. Depending on the area and extent of the damage, the victim will exhibit different forms of language loss, such as being able to sing but not to speak. Dysarthria and apraxia (problems with pronunciation) are related to this and might occur at the same time. While, as Paul Broca and others have demonstrated, these aphasias are normally the result of damage to areas of the left hemisphere of the brain, there are a few instances where it gets reversed and takes place in the right hemisphere. The damage can occur either through a dramatic event, such as a stroke or some kind of traumatic brain injury, or it can occur slowly, the result of a brain tumor or of a progressive neurological disease. Epilepsy, migraines, and various benzodiazepines can also cause it. Symptoms include the inability to comprehend language, to pronounce words, to speak spontaneously, to name object, to repeat phrases, to use correct grammar, to give words the correct stress and rhythm, to complete sentences, to read, and to write.

Locationist models attempt to classify the various forms of aphasia by their major characteristics and then link these characteristics to specific locations in the brain. Since Paul Broca and Carl Wernicke provided the initial locations in the late 1800s, other researchers have built on their models, most notably Harold Goodglass and Edith Kaplin, who came up with what is referred to as the Boston-Neoclassical Model. Harold Goodglass was the one most involved with the development of a standardized aphasia test known as the Boston Diagnostic Aphasia Examination, which has been translated into many languages. (127) Nina Dronkers, using imaging and lesion analysis, has demonstrated that patients with Broca's Aphasia have lesions to the medial insular cortex, something Paul Broca missed.

While neurologists have been slowly mapping out the specifics and specific kinds of aphasia, there is still much to be learned, and the various classifications in place must be considered as fluid and only generally correct. Furthermore, though the locationist model is the original model and retains merit, modern anatomical techniques and analyses have shown that *precise* connections between brain regions and symptom classification don't exist. Rather, the neural organization of language is complicated, language is a comprehensive and complex behavior, and it makes sense that it isn't the product of some small, circumscribed region of the brain. Thus, no classification of patients in subtypes and groups of subtypes can be seen as absolute or providing the full picture. For example, only about 60% of patients fit in a classification scheme such as fluent/nonfluent/pure aphasia. Furthermore, there is a huge variation among patients with the same diagnosis, and aphasias can be highly selective. For instance, patients with naming deficits (anomic aphasia) might show an inability only for naming automobiles, or people, or colors. Having said this, here are the general categories in place at the moment:

Individuals with Wernicke's aphasia may speak in long sentences that have no meaning, add unnecessary words, and even create new words. Their auditory and reading comprehension will be poor, and they will be fluent, but nonsensical in oral and written expression. They normally will have difficulty understanding the speech of both themselves and others and will therefore often be unaware of their mistakes. They also are also sometimes unaware of their surroundings, and may present a risk to themselves and others around them.

Transcortical Sensory Aphasia (TSA) is a form of Wernicke's Aphasia where those having it have trouble with comprehension, but can speak fluently, communicate and repeat well. The main problem for them is that they have Semantic paraphasia, which means that they do not use the correct word, but rather substitute one with similar content, such as mistaking banana for apple.

Conduction Aphasia is caused by damage to the arcuate fasciculus, the structure that transmits information between Broca's area and Wernicke's area, and results in poor repetition ability, though auditory comprehension is near normal and oral expression is good with an occasional paraphasic error.

Anomic Aphasia involves a difficulty with naming. The patient may have difficulties naming certain words, linked by their grammatical type (e.g. difficulty naming verbs and not nouns) or by their semantic category (for example. difficulty naming words relating to sports but nothing else) or a more general naming difficulty. Patients tend to produce grammatical, yet empty speech. Auditory comprehension is generally still good.

Broca's Aphasia is characterized as a nonfluent aphasia, where patients speak with difficulty in short yet meaningful phrases. Affected people often omit small words such as "is," "and," and "the." For example, a person with Broca's aphasia may say, "Walk store" meaning, "I will take a walk to the store." However, it could also mean "You walk to the store," or "I walk at the store." Since individuals with Broca's aphasia are able to understand the speech of others to varying degrees, they are often aware of their difficulties and become frustrated. Interestingly, it is associated with right hemiparesis, meaning that there will be paralysis of the patient's right arm, leg, and face.

Transcortical Motor Aphasia is similar to Broca's aphasia, except repetition ability remains intact. Auditory comprehension is generally fine for simple conversations, but declines rapidly for more complex conversations. Again, it is associated with right hemiparesis, meaning that there will be paralysis of the patient's right arm, leg, and face.

Global Aphasia results in severe communication difficulties where patients are extremely limited in their ability to speak or comprehend language. They may be totally nonverbal or only use facial expressions and gestures to communicate. Again, it is associated with right hemiparesis, meaning that there will be paralysis of the patient's right arm, leg, and face.

Transcortical Mixed Aphasia is similar to global aphasia, except that repetition ability remains intact.

Finally, Subcortical Aphasia is a less clear category where the individual's characteristics and symptoms depend upon the site and size of subcortical lesion. Possible sites of lesions include the thalamus, internal capsule, and basal ganglia.

Today, Localization Theory of language processing sometimes refers to itself as the Cognitive Neuropsychological Model, where language processing is broken down into modules, each with a specific function. Thus, there is a module that deals with phonemes as they are spoken, and a different module that stores formulated phonemes before they are spoken. (128)

Famous cases of aphasia besides Maurice Ravel include Antony Flew, Sir John Hale, Joesph Chaikin, Robert E. Lee, Ralph Waldo Emerson, Sven Nykvist, and Jan Berry (of the music duo Jan and Dean).

During World War II, a number of Russian scientists (e.g., Leontev, Anokhin, Bernstein, and most importantly A. R. Luria) began a more formal scientific study of the brain, what they called neuro-psychology. It was all set out systematically in A. R. Luria's *The Higher Cortical Functions in Man*, first translated into English in 1966, and put forth in the form of a biography/autobiography/journal or pathography in *The Man with the Shattered Brain*, first translated into English in 1972, both dealing with the resulting losses of left hemisphere damage to the brain. (129)

Zasetsky's life in Russia began much as any other. He wrote in his journal, "In 1941, right before the war began, I finished my third year of courses at a polytechnic institute and hoped soon to get some practical experience in a specialized plant. . . . For some reason, even as a child I was fascinated by science, by knowledge in general, and greedily devoured any information I could pick up . . . Before I was two years old my father died suddenly in a coal mine where he worked as an engineer. After his death my mother had a rough time of it with four young children, since she was illiterate and didn't know how to go about getting a pension for her children. . . . Soon, I thought, I'll be graduating from the institute. Two years left to go? That's nothing! What can possibly stand in my way now? And as soon as I get my degree, I'll start giving my mother a hand, it's about time she had a chance to rest!"

Then came WWII. Zasetsky's plans dissolved as the war demanded soldiers. He reported with good detail his fighting on the western front, being wounded in the temple, but not seriously, and returning to the Russian offensive in 1943, the battle of Smolensk, the flame-throwers, the wait for the order to attack. Then the moment, the order to move forward, the clank of armor, the momentary stillness, the sudden burst of machine guns. Waking somewhere in a tent, head empty, no thoughts, no memory, just a buzz, a dizzy feeling. A man with a broad, thick-set face and angry eyes studying through glasses. People in bright white coats and gauze masks rushing about. Being held down to the operating table, screaming, gasping for breath, warm blood running down ears and neck, a salty taste in the mouth.

Later, waking, he felt his head was a blank emptiness, no memory. At first, he could not even recognize himself when a doctor spoke to him, "Comrade Zasetsky?" As the days passed, he began to have what he called memory fragments. "Right after I was wounded, I seemed to be some newborn creature that just looked, listened, observed, repeated, but still had no mind of its own." These scattered, disconnected memories came and went. At times he could remember having a mother and two sisters, and a brother.

Since that initial confusion, up to the time of his journal entries he found that he was "in a kind of fog all the time, like a heavy half-sleep," his memory "a blank, unable to "think of a single word," just flashes of "images, hazy visions that suddenly appear and just a suddenly disappear, giving way to fresh images," but unable to "understand or remember what these mean."

He certainly still had self-awareness, felt that others realized his worthlessness, felt that he was fit for nothing, that in truth he had died, a man killed in the war. He wrote, "I miraculously remained alive. Still, even though I seem to be alive, the burden of this head wound gives me no peace. I always feel as if I'm living out a dream—a hideous, fiendish nightmare—that I'm not a man but a shadow, some creature that's fit for nothing."

He wondered if it was all a dream, not real, but a nightmare he just could not wake up from. Yet it could not be. He was aware enough, at least for moments, to understand the therapist, who told him he had been wounded. Finally, after being moved from one hospital to another, he ended up at the rehabilitation hospital in the Urals. He found it a "lovely, quiet place, a refuge with a magnificent view of "an enormous lake surrounded by evergreens." He could remember the trip there with clarity, yet he wrote in his journal that he still had "to read syllable by syllable like a child," that he was still "plagued by amnesia," could not remember "words or meanings," was overcome by "aphasia," and could not recover his memory.

He had continual thoughts of the meaninglessness of his life, of how he should just die, but at the same time fought them with a desire to live and have a meaningful life.

When A. R. Luria first met Zasetsky, he found him young looking, "scarcely more than a boy . . . with a puzzled smile" and a need to tilt his head, since the vision on his left side was gone. The initial interview revealed a person unable to remember relatives, where he lived, how he had been wounded, read, add, find the words to answer even the simplest questions. He could list the months of the year, but not respond to which month precedes another, not indicate which hand was right or left. He enjoyed nature greatly, listened intently to the sounds of birds, noted the smoothness of the lake.

The official record states that the bullet penetrated the left parieto-occipital region of the cranium, that it was followed by a prolonged coma, that inflammation complicated matters by causing adhesions of the brain to meninges and changes in adjacent tissues, that scar tissue pulled the left lateral ventricle up and produced an incipient atrophy of the medulla.

Luria's medical knowledge of the structure of the brain leads him to postulate what has been lost do to the wound, and what retained, and his conclusions lead to a chilling realization, for he knows that the frontal cortex has been spared, and with it the ability to understand the damage and want to overcome it, the awareness of what it means to be human, the need to find meaning and value. In other words, Zasetsky was not only severely brain damaged, but he was still able to step outside of himself and realize his damaged state. Zasetsky writes, "It was depressing, unbearable to realize how miserable and pathetic my situation was. You see, I'd become illiterate, sick, had no memory."

As the journal entries continue, they trace out some of the strange physical problems outlining his mental wound. He writes: "Ever since I was wounded I haven't been able to see a single object as a whole—not one thing. Even now I have to fill in a lot about objects, phenomena, or any living thing from imagination . . . Even now I still don't see entire objects, things, or people as I did before—only part of them. When I look at a spoon, at the left tip, I'm amazed. I can't figure out why I only see the tip and not the whole spoon. . . . Through and beyond the objects I see there are endless numbers—a myriad really—of tiny, shifting swarms of midges that made it hard for me to look at the objects themselves. . . . Once when I left my room and was walking in the corridor, I'd no sooner taken a few steps than I suddenly banged my right should and the right side of my forehead against the wall and got a huge bump on my forehead. I was furious; I simply couldn't understand why I'd suddenly bumped into the wall. I should have seen it. Just then I happened to look down—at the floor and at my feet—and I shuddered. I couldn't see the right side of my body. My hands and feet had disappeared."

His body often seemed to have disappeared. "I move the fingers of my left hand, feel them, but can't see the fingers of my right hand and somehow I'm not even aware they're there. . . . I can't get used to that idea, so often I'm terrified when part of my body disappears. . . . Sometimes when I'm sitting down I suddenly feel as though my head is the size of a table—every bit as big—while my hands, feet, and torso become very small. . . . Sometimes, when I'm sitting on a chair, I suddenly become very tall, but my torso becomes terribly short and my head very, very tiny."

Space also became an ever-shifting mystery. He wrote, "Ever since I was wounded I've had trouble sometimes sitting down in a chair or on a couch. I first look to see where the chair is, but when I try to sit down I suddenly make a grab for the chair since I'm afraid I'll land on the floor. Sometimes that happens because the chair turns out to be further to one side than I thought"

The tragedy, as tragedies often do, also reveals a tremendous resilience, that force of life that drives humans--that powerful need for mapping out meaning and value, shines forth even more clearly in such a darkness. As the decades passed, and Luria continued to work with Zasetsky, he witnessed some amazing recoveries. Slowly, with years of effort, Zasetsky learned to live with his mental loss of vision on the right side, with his visual hallucinations, with his inability to even sense parts of his body, much less control them, with his spatial disorientation, and even amazingly with his loss of reading and writing abilities. From not even being able to remember the alphabet, from not being able to recall even the most basic words, he actually gained the ability to write out sentences and express deep human concerns (as this journal demonstrates).

The major breakthrough here came from kinetic-motor memory. He wrote: "Each time I wanted to think of a particular letter I'd have to run through the alphabet until I found it. But one day a doctor I'd come to know well, since he was always very informal with me and the other patients, asked me to try to write automatically—without lifting my hand from the paper. I was bewildered and questioned him a few times before I could even begin. But I finally picked up the pencil and after repeating the word 'krov' [blood] a few times, I quickly wrote it. I hardly knew what I'd written since I still had trouble reading—even my own writing . . . after the doctor showed me how to write quickly, automatically, and not letter by letter, I was able to get one word down after the other without having to think about it . . . even though I couldn't read my own writing."

And the fact that he forced himself to write out a journal, putting a pattern on his life, tells us even more about the human need for such a mapping out of meaning.

Beneath the factual, literal, scientific truths to be found in this case study lie those intuitive human truths, the very truths of meaning and value Zasetsky so desperately sought, the truths that draw us into the study, the truths that make us emphasize with him, see him as more than just a case study, see him as a human. These truths are captured in a strikingly similar fictional scenario, Dalton Trumbo's *Johnny Got His Gun*, where Joe Bonham also gets seriously wounded in war, this time not a wound to the brain, but a wound that separates his brain, and thus his mind from contact with the outside world. In this case an exploding bomb causes Joe Bonham to lose his arms, legs, and face. He is kept alive only for medical study, and assumed to have no higher thought functioning. We then experience the horror of a man who, in fact, does have such higher powers of thought, but who is unable to connect them with the physical world about him. We understand, not just logically, but intuitively, feelingly, how dispirit such a situation is for humans. There is no denying the terrible tragedy of the physical deformations, but that becomes secondary to the deeper tragedy of the loss of a means to give meaning and value to life.

He is closeted off from the world, literally put in a storage room without light, without anything other than the basic physical care needed to survive.

When his hospital bed is finally put in front of a window by a humane nurse, and he can feel the sunlight on his skin, we know the joy of his mind connecting with a means to tell the passage of days, to figure out how to map out even this simple bit of a pattern for his life. Such seemingly simple desires, such enormously important human needs. But communication is still denied. The isolation, the loneliness, the need to communicate is overpowering, so powerful, he cannot give up trying to satisfy it. And he searches his mind for a means. Empathizing, we search with him, trying to figure out some way for him to call out from his prison. Then comes that Zasetsky-like discovery, that alternate way around the obsticales:

> When he had thought for an instant that his mother and his sisters and Kareen might be standing beside the bed he had wanted to hide. But now that he had generals and big guys he felt a sudden fierce surging desire for them to see him. Just as before he had started to reach for the medal without an arm to reach with so now he began to blow the mask off his face without having mouth and lips to blow with. He wanted them to get just one look at that hole in his head. He wanted them to get their fill of a face that began and ended with a forehead. He lay there blowing and then he realized that the air from his lungs was all escaping through his tube. He began to roll again from shoulder to shoulder hoping to dislodge the mask.
>
> While he lay there rolling and puffing he felt a vibration way down in his throat a vibration that might be a voice. It was a short deep vibration and he knew that it was making a sound to their ears. Not a very big sound

not a very intelligent sound but it must seem to them at least as interesting as the grunting of a pig. And if he could grunt like a pig why then he was accomplishing a great thing because before he had been completely silent. So he lay thrashing and puffing and grunting like a pig hoping that they would see damned well how much he appreciated their medal. While he was in the middle of this there was an indefinite churning of footsteps and then the departing vibrations of his guests. A moment later he was all alone in the blackness in the silence. He was all alone with his medal.

Suddenly he quieted. He was thinking about the vibrations of those footsteps. He had always carefully felt for vibrations. He had measured the size of his nurses and the dimensions of his room by them. But suddenly to feel the vibrations of four or five people tramping across the room made him think. It made him realize that vibrations were very important. He had thought of them up to this time only as vibrations coming to him. Now he began to consider that also there could be vibrations going from him. The vibrations which he received told him everything—height weight distance time. Why shouldn't he be able to tell something to the outside world by vibrations also?

In the back of his mind something began to glimmer. If he could in some way make use of vibrations he could communicate with these people. Then the glimmer became a great dazzling white light. It opened up such breathless prospects that he thought he might suffocate from sheer excitement. Vibrations were a very important part of communication. The fall of a foot on the floor is one kind of vibration. The tap of a telegraph key is simply another kind.

When he was a kid way back maybe four year ago or five he had a wireless set. He and Bill Harper used to telegraph each other. Dot dash dot dash dot. Particularly on rainy nights when their folks wouldn't let them go out and there was nothing to do and they just lounged around the house and got in everybody's way. On such nights he and Bill Harper used to dot and dash at each other and they had a hell of a good time. He still remembered the Morse code. All he had to do in order to break through to people in the outside world was to lie in bed and dot dash to the nurse. Then he could talk. Then he would have smashed through his silence and blackness and helplessness. Then the stump of a man without lips would talk. He had captured time and he had tried to figure geography and now he would do the greatest thing of them all he would talk. He would give messages and receive messages and he would have made another step forward in his struggle to get back to people in his terrible lonely eagerness for the feel of people near him for the things that were in their minds for the thoughts they might give him his own thoughts were so puny so unfinished so incomplete. He would talk." (130)

He begins tapping his head, tapping and tapping and tapping, dispirit for the nurse, for anyone to catch on to what he had discovered. Time passes, days, weeks, months, perhaps years, he does not know. But he keeps on tapping. Then, a new nurse, a nurse with young hands, a nurse not so certain he is nothing more than a piece of meat without a mind, takes over his care, and discovers how to communicate with him by writing letters on his chest, fittingly writing Merry Christmas, and he is ecstatic, and we rejoice at his realization that now at least he can place his map of time in real time, can measure the days by the alternating warmth of the sun and cool of the night from Christmas and know at least what time of year his is in. But still, he cannot respond, cannot talk back to her.

He keeps tapping, tapping, tapping.

Then comes the break-through, the military representative, the Morse code expert, and the contact, the chance at interaction. We feel his joy as the question is tapped out: "What do you want?"

> It is a good question. But he has been so focused on the simple desire to communicate he has not thought about it, not considered just what he now wanted, given the limited options.
>
> They should know what he wanted the silly bastards and they should know they couldn't give it to him. He wanted the things they took for granted the things nobody could ever give him. He wanted eyes to see with. Two eyes to see sunlight and moonlight and blue mountains and tall trees and little ants and houses that people live in and flowers opening the morning and snow on the ground and streams running and trains coming and going and people walking and a puppy dog playing with an old shoe worrying it and growling at it and backing away from it and frowning and wiggling its bottom and taking the shoe very seriously. He wanted a nose so that he could smell rain and burning wood and cooking food and the faint perfume that stays in the air after a girl has passed by. He wanted a mouth so he could eat and talk and laugh and taste and kiss. He wanted arms and legs so that he could work and walk and be like a man like a living thing.
>
> What did he want what was there for him to want what was there left that anybody could give him?
>
> It came over him rushing and howling like a torrent of water from behind a dam that has broken. He wanted to get out. (131)

Let me out, let me out, let me out. Put me on display. Let me have some purpose, some value. Give my life meaning. We hear his unspoken words. We feel his loneliness, his pain, his absolute entrapment. We anxiously with him for the reply, wanting some level of human existence to touch him, wanting some reason for him to continue his life.

But the cold, logical military does not, apparently cannot get beyond its rational, left hemisphere rules. The Morse code reply deadens us: "What you ask is against regulations!" The movie version Trumbo directed pushes it even further. In it Joe simply requests to be put to death, as pure physical existence is meaningless, and a meaningless existence is worse than no existence at all. The cold reply is the same. He will be forced to continue isolated, in as pure a Hell as can be imagined. (132)

The impact of this novel comes not from the scientific explanations that might accompany Joe Bonham's condition, but from the intuitive knowing of the horror of it. The same is true with Zasetsky. While we read his "story," we are feeling his suffering, understanding the agony, the tremendous loss, the sublime struggle to regain the human part of his life.

Luria interrupts Zasetsky's journal writings in the book to make comments, and goes off on a few more lengthy discussions of the neurological reasons for his condition, pointing the discipline of neurology forward to a more complex understanding of the brain/mind interactions, but still unable to go much beyond the two established areas of the left side of the brain (Broca and Wiernicke areas). Notable here is that there is no emphasis put on the two separate halves of the brain having important differences in terms of thinking abilities.

Nonetheless, while Luria was studying Zasetsky, other scientific evidence for the location of separate and complementary modes of human thinking in the cerebral cortex was accumulating from a study of other brain lesions, accidents or strokes in the left hemisphere of the neocortex that resulted in diminished ability to read, write, speak, or perform simple mathematical tasks, and accidents or strokes in the right neocortex that resulted in a loss of pattern recognition, musical ability, and holistic reasoning abilities, even, in some cases, the ability to recognize one's own face in a mirror.

Such detailed, personalized studies were revealing compelling evidence. Even though, as with Zasetsky, the studies focused on and were mainly successful in dealing with what today is established as left brain thinking, and right brain thinking remained still much more in the dark, it was about to become obvious that, in the complex, intertwining operations of the human mind and brain, two separate forms of higher level thinking were taking place.

The key to it all is the corpus callosum, a mass of white transverse fibers, a bundle of nerves that carries information from one hemisphere of the brain to the other. In the 1940s, neurosurgeons developed a technique, technically titled corpus colostomy, less technically referred to as split-brain operation, for severing this information highway, as a means of helping patients who suffered epileptic seizures, and though it was done in a very crude, experimental fashion, it worked, or at least helped.

As is often the case, the practical solution resulted in some unexpected side effects, which, in turn, would open unexpected doors to higher understandings. Patients, while no longer burdened with seizures, began instead to experience other strange forms of thinking. One complained that the left half of his body often did unusual things that he simply could not understand. Another reported that, while his right hand was reaching into a drawer to retrieve a sock, his left hand slammed the drawer shut. Another complained that his left hand kept unbuttoning his shirt as he was attempting to button it with his right. Another told of his right hand placing objects in a shopping cart, only to have his left hand remove them and place them back on the shelf:

> According to Norman Geschwind, one split-brain patient complained that his left hand on several occasions suddenly struck his wife—much to the embarrassment of his left (speaking) hemisphere. In another case, a patient's left hand attempted to choke his own throat and had to be wrestled away. The patient expressed complete shock and surprise regarding these incidents and claimed that his left hand acted of its own accord.
>
> Another split-brain patient, described by S. J. Dimond, remarked that, on several occasions when she had overslept, she was suddenly awakened by her left hand, which had slapped her until she woke up.
>
> J. E. Bogen reported that almost all of his complete commissurotomy patients manifested some degree of intermanual conflict in the early postoperative period. One patient, Rocky, for years complained of difficulty getting his left leg to go in the direction he (or rather his left hemisphere) desired. Another patient often referred to the left half of her body as "my little sister" when complaining of its peculiar and independent actions.
>
> Indeed, the French neurosurgeons Brion and Jedynak reported that this type of independent left-sided (right hemisphere) activity was common in their split-brain patients and termed it the "alien hand." (133)

The dual mind concept becomes bizarrely clear in extreme cases of conflict between left and right hemispheres. A. J. Akelaitis wrote about a patient who "in tasks requiring bimanual activity" found that "the left hand would frequently perform oppositely to what she desired to do with the right hand. For example, she would be putting on clothes with her right and pulling them off with her left, opening a door or drawer with her right hand and simultaneously pushing it shut with the left." Another patient, who was recently divorced, when walking about town "found himself forced by the left half of his body to go some distance in another direction," which was later discovered to be where his wife lived, suggesting his right brain wanted to go where his left brain did not. (134)

Clearly this corpus colostomy procedure was entering neurology into as of yet misunderstood territory, and also clearly it could be used to study the brain from a new perspective.

The now classic experiment was worked out by Roger W. Sperry (who later would win the 1981 Nobel Prize in Physiology or Medicine for his work) and Ronald Myers in 1953, and it would not be overly bold to give this as the date contemporary neurology began. For now, for the first time, definite proof was initiated that the two halves of the brain work in two separate arenas. The initial research was done on a cat, and it began with an unassuming report:

> Through a transbuccal approach, the optic chiasma was sectioned in the midsagittal plane thereby restricting the central projection of retinal patterns to the homolateral sides of the brain. The cats were then taught a simple form discrimination (circle vs. square) with a mask covering the left eye. After overtraining, the mask was shifted to the right eye for critical trials. "Interocular transfer" of the habit was found to be effected with no significant deteration in level of performance (two cases). In one of these cases the transfer to the untrained eye survived subsequent section of the corpus callosum in the midsagittal plane; in the other it did not. Transfer failed in another case in which the corpus callosum was sectioned just before the critical tests after the completion of training. "Interocular transfer" was demonstrated in two cases in which the corpus callosum as well as the chiasma had been sectioned prior to training.
>
> However, three later cases with the corpus callosum cut failed to show transfer. This discrepancy in the data may be due to incomplete section of the chiasma in the earlier cats. Anatomical check of the lesions has been deferred until further tests on the cats have been completed. The effects of eliminating other commissural systems and brain centers in various combinations are currently being investigated with particular regard to the location and nature of the mnemonic traces. (135)

Now, for the first time, definite proof was appearing that the two halves of the brain work in two separate arenas. This initial research done on a cat suggested that, with its corpus callosum and anterior commissar severed, the cat was unable to pass information learned from one eye to the other eye. In fact, each eye needed to be trained completely from the beginning. In effect, this can be said to be when the concept of two totally separate brains within one, or double-brain theory, came to the forefront of neurology. Again, the connecting corpus callosum is the key. With it cut, the human brain becomes two separate brains with two separate consciousnesses. And soon it would be discovered that these two separate consciousnesses think in two very distinct ways

Sperry and his associates continued their experiments. The first experiments on cats were then duplicated on monkeys with similar results. (136) Already by 1955, H. L. Teuber was able to discuss "double-dissociation," a method where comparison of abilities not affected by injury to the left-hemisphere are subsequently found to be affected by right-hemisphere injury is used to confirm theories of hemisphere specific functions. (137) J. E. Bogen, M. S. Gazzaniga, J. Levy, and other neurologists joined in. The research quickly gained momentum, and articles detailing and attempting to explain the data from this new asymmetrical approach filled the neurological publications. (138) By 1962, C. B. Trevarthen was able to demonstrate that the two hemispheres cannot only work independently but also simultaneously. (139) By 1966, M. S. Gazzaniga and R. Sperry demonstrated that humans can solve two independent problems by the two hands independently simultaneously. (140)

By the late 1960s and into the early 1970s, Roger W. Sperry, and others at the California Institute of Technology were able to help victims of grand mal epilepsy by cutting the corpus callosum in a much more sophisticated and refined manner than what had been done in the 1940s, and demonstrated that, not only did this procedure help tremendously in solving the grand mal epilepsy problems, but the split-brain patients were able to function normally, suggesting that the two sides of the brain can work independently. (141)

Already by 1969, Bogen was able to offer a well supported claim for the specific, complementary interactions between the two hemispheres of the brain, the left supplying the needed logical, "prepositional," structures, the right the needed symbolic, "appositional" content, concluding:

> One of the most obvious and fundamental features of the cerebrum is that it is double. Various kinds of evidence, especially from hemispherectomy, have made it clear that one hemisphere is sufficient to sustain a personality or mind. We may then conclude that the individual with two intact hemispheres has the capacity for two distinct minds. This conclusion finds its experimental proof in the split-brain animal whose two hemispheres can be trained to perceive, consider, and act independently. In the human, where prepositional thought is typically lateralized to one hemisphere, the other hemisphere evidentially specializes in a different mode of thought, which may be called appositional. . .
>
> The belief that man is possessed of two ways of thought, occasionally conflicting, is common in everyday speech where it often takes the form of supposing a struggle between "reason" and "emotion," or between "the mind" and "the heart." (142)

The early 1970s saw a rapid development in use of brain scanning techniques. Robert Ornstein and David Galin, at the Langley Porter Neuropsychiatric Institute, San Francisco, found that the EEG (electroencephalogram), which had originally be developed nearly a half century earlier, through offering poor spatial resolution, perhaps, at best, a few inches, does offer excellent temporal resolution, and indicates activity in normal people moves from one hemisphere to the other in relation to what form of thinking is occurring. For example, alpha activity (8-12 Hz) is suppressed in the left hemisphere more than the right during several verbal tasks. (143)

By 1978, Sperry was able to write: "Everything we have seen so far indicates that the surgery has left these people with two separate minds, that is, two separate spheres of consciousness. . . . Each brain half, in other words, seems to have its own largely separate cognitive domain with its own private perceptual, learning, and memory experiences, all of which are seemingly oblivious to corresponding events in the other hemisphere." (144)

George Ojemann, building on the electrical stimulation of portions of the brain experiments of Wilder Penfield, at the Montreal Neurological Institute, conducted elaborate brain mapping. (145)

John Wada used anesthetics (the Wada amytal test), injecting sodium amobarbital into one side of the brain or the other (the left or right carotid artery, which supplies blood to the ipsilateral hemisphere), to put one side or the other to sleep, and then record what functions are affected. (146)

Others demonstrated that isotopic scanning, the injection of radioactive isotopes (i.e., technetium 99 or iodine 131) into the blood vessels to the cranial region, can locate and provide information about the size, etc., of tumors and blood vessel lesions. This is now generally done with CAT (computerized axial tomography) scans, which allow x-rays from multiple angles to provide a 3D view, and which have been used to offer spatial resolution in the 1-2mm range, but have the disadvantage of slower temporal resolution (30 seconds plus). Furthermore, they cannot provide a dynamic picture of how activation changes during task performance.

Others used PET (positron emission tomography) scans, introducing a radioactive tracer-labeled compound and tracking its behavior. It was discovered that NMR or NURI (nuclear magnetic resonance imaging) scans can provide maps of slices of the brain, and other rCBF (regional cerebral blood flow) measures can offer dynamic pictures of brain activity. (147)

And these pictures, in turn, suggested specific areas of left or right hemisphere location for specific tasks. For example, during verbal tasks rCBF is stronger in the left hemisphere than the right hemisphere. (148)

Other tests, in addition to reaffirming that the left hemisphere is involved with verbal ability, suggested that the right hemisphere involves the ability to manipulate geometric figures, assemble puzzles, complete patterns, and, in general work with forms, distances, and space relationships. (149)

Tests were then done to attempt to determine how the two separate hemispheres comprehended the world around them, and, though initially the assumption still remained that the two halves thought in similar ways, only the right hemisphere less adept at it, a mapping of what was to emerge as two distinct forms of knowing began in earnest.

In one test, a word such as "hatband" was placed on a screen, "hat" visible only to the left eye, "band" visible only to the right eye. When asked to identify the word, the split-brain patient was only able to verbally communicate what the left hemisphere received (a note, the hemispheres process information from the opposite eyes, left hemisphere from the right eye, right hemisphere from the left), i.e., only the word "band." When asked what kind of "band," the patient could only guess. In contrast, when asked to write what he saw, with his hand hidden from sight, the patient wrote only "hat." Thus, he could write, but not verbalize, what he saw with his left eye. Furthermore, when asked what he wrote, he would give a totally incorrect response. (150)

As the brain mapping continued, elaborate distinctions were made between the functions of the left and right hemispheres. For instance, most physical activities, highlighted in athletics, have been found to be mainly right hemisphere functions, and involving left hemisphere thinking can actually get in the way. In other words, if an athlete thinks too much about making a free throw in basketball or hitting a golf ball it can cause him to mess up. This is why opposing teams call time out before a place kicker in football attempts his kick.

Musical ability also has mainly right hemisphere origins, and patients with damage to the right temporal lobe or right hemispherectomies demonstrate severe loss of musical ability, though, curiously, not a loss in verbal ability.

M. P. Bryden, R. G. Sey, and J. H. Sugarman demonstrated that the right hemisphere is better than the left hemisphere at comprehending the emotional aspects of music, and, in fact relates to music as emotional more so than the left hemisphere does. (151)

Such findings make sense out of earlier mysteries such as a report by O. Dalin, in 1745:

> He had an attack of a violent illness, which resulted in a paralysis of the entire right side of the body and complete loss of speech. He can sing certain hymns, which he had learned before he became ill, as clearly and distinctly as any healthy person. . . . Yet this man is dumb, cannot say a single word except "yes" and has to communicate by making signs with his hand. (152)

Not long ago a confusing, perhaps demonic condition, now, an easily explained loss of the left hemisphere's control of speech and of the right half of the body while maintaining the right hemisphere control of musical abilities.

Such brain-damaged individuals offer dramatic evidence for the separation of the two hemispheres, the two forms of human thinking. But, the theories needed to be confirmed in non-damaged brains. This has been accomplished by M. S. Gazzaniga, and others, who have confirmed, among other things, that most language processing takes place in the left hemisphere. (153)

As the mapping continued, it became clear that the left hemisphere processes information in an analytic manner, the right brain in a holistic manner. And, in conjunction with this, the left hemisphere processes information sequentially, the right hemisphere synthetically. (154)

Another interesting difference involves emotional responses, which also appear to be largely a right hemisphere function. Joan C. Borod suggests that the reason that the right hemisphere is better at emotional processing is because of the "strategies and functions" involved, i.e., "strategies termed nonverbal, synthetic, integrative, holistic, and Gestalt, and functions such as pattern perception, bisuospatial organization, and visual imaging." (155)

Dr. Rhawn elaborates:

> Infants and children are more emotional in their psychological orientation. Verbal thinking and the maturation of the conscious mind (which is heavily dependent on language and linguistic processing) develop much later. Emotion and non-conscious mental functioning are present from the very beginning.
>
> It is within this emotional, non-conscious psychic realm that our early experiences are stored. It is this same realm that later gives birth to emotional conflicts, intuitive leaps of the imagination, creativity, slips of the tongue, "thoughtless" behavior, daydreams, and even the melodies to which we dance and sing. As we shall see, this non-conscious domain is

maintained and mediated by the right half and limbic regions of the brain. (156)

However, these findings need to be placed in more precise terms. Recent research finds that particular emotions are attached to the left hemisphere. Richard J. Davidson states:

> Activation in the left anterior region is associated with approach-related emotions; deficient activation in this region is associated with emotion-related phenomena that might best be described as reflecting approach-related deficits such as sadness and depression; and activation in the right anterior region is associated with withdrawal-related emotions such as fear and disgust and withdrawal-related psychopathologic states such as anxiety. (157)

He further goes on to argue a difference in the perception of emotion (which he admits is a right hemisphere function) and the experience or expression of emotion.

Some suggestive research has been done on the relation of dreams to the two hemispheres indicating that the imagery, fantasy and imaginative aspects of dreams are related to the right hemisphere. (158)

Carl Sagan suggests that "perhaps we are today able to sense directly the operations of the right hemisphere mainly when the left hemisphere has 'set'—that is, in dreams," a time when studies indicate several left-hemisphere functions, such as reading, writing, arithmetic, and verbal recall, suffer. "In addition to the symbolic content of dreams," he continues, "other aspects of dream imagery point to a neo-cortical presence in the dream process." And after giving an example, he continues, "While the content of many dreams seems haphazard, others are remarkably well structured; these dreams have a remarkable resemblance to drama. We now recognize the very attractive possibility that the left hemisphere of the neocortex is suppressed in the dream state, while the right hemisphere—which has an extensive familiarity with signs but only a halting verbal literacy—is functioning well. It may be that the left hemisphere is not entirely turned off at night but instead is performing tasks that make it inaccessible to consciousness: it is busily engaged in data dumping from the short-term memory buffer, determining what should survive into long-term storage." (159)

Eric Fromm claims that these dream insights come from what he calls "the forgotten language," the origin of not only dream imagery, but also that of fairy tales and myths. (160) This is similar to Carl Jung's theories of the universal unconscious, and K. D. Hoppe, L. Miller, and R. Ornstein have posited a relationship of left and right hemispheric functioning to Sigmund Freud and Carl Jung's theories. (161) R. Joseph has directly related the unconscious mind to the right hemisphere and limbic regions of the brain and the conscious mind to the left hemisphere. (162) D. Galin has related the processes of the two hemispheres to Freud's theories of conscious and unconscious thought: "Certain aspects of right hemisphere functioning are congruent with the mode of cognition psychoanalysts have termed primary process, the form of thought that Freud originally assigned to the system Ucs (unconscious)." (163)

As by now is obvious, this mapping has resulted in some general distinctions. Sally P. Springer and Georg Deutsch (164) list what they call the five main groups of characteristics usually used to distinguish the two hemispheres:

Left Hemisphere	Right Hemisphere
Verbal	Nonverbal, visuospatial
Sequential, temporal, digital	Simultaneous, spatial, analogical
Logical, analytical	Gestalt, synthetic
Rational	Intuitive
Western thought	Eastern thought

And they provide a quote from Aurobindo:

> The intellect is an organ composed of several groups of functions, divisible into two important classes, the functions and faculties of the right hand, the functions and faculties of the left. The faculties of the right hand are comprehensive, creative, and synthetic; the faculties of the left hand critical and analytic . . . . The left limits itself to ascertained truth, the right grasps that which is still elusive or unascertained. Both are essential to the completeness of the human reason. These important functions of the machine have all to be raised to their highest and fastest working-power, if the education of the child is not to be imperfect and one sided. (165)

Joseph Rhawn writes:

> In addition to controlling the right half of the body, the left hemisphere controls talking, reading, writing, spelling, speech comprehension, linguistic and verbal thinking, verbal intelligence, verbal memories, dreaming in words, thought dreams, rhythm, temporal and sequential information, processing, keeping score of a football game, math, marching, grammar, typing, logical and analytical reasoning, confabulation, and perception of details; in addition to controlling the left half of the body, the

> right hemisphere controls unconscious awareness, perception of faces, comprehension of body language, creative thinking, emotional and melodic speech, comprehension of music and emotion, singing and swearing, insight and intuitive reasoning, visual-spatial processing, throwing and catching a football, riding a bicycle, dancing, visual closure, Gestalt formation, perception of environmental sounds, and social-emotional nuances, emotional and childhood memories, intuitive reasoning (i.e., seeing the whole picture, reading between the lines), initial self-concept, dreaming in visual-emotional images, and geometric thinking. (166)

He elaborates:

> The human brain is organized so that two potentially independent mental systems coexist, literally side by side: one psychic system within the right, the other within the left half of the brain. In fact, over the course of evolution, and particularly over the last fifty thousand years, each half of the brain has developed its own strategy for perceiving, processing, and expressing information, as well as specialized neuroanatomical interconnections that assist in mediating these functions.
>
> For instance, it is well known that, in the majority of the population, the left half of the brain controls the ability to speak, write, and comprehend spoken and written language and to perform arithmetical operations. It is responsible for naming, spelling, writing, and counting. The left brain also controls the right hand, as well as the ability to perform skillful sequential movements such as those involved in manual tool making, typing, and the ability to communicate via sign language.
>
> The right half of the brain controls the left hand and is good at manipulating and constructing complex shapes, puzzles, and block designs (as in the skills of carpentry). It also controls the perception of visual, spatial, and geometric relationships, such as depth, distance, location, movement, and motion. It is more concerned with the processing and mimicking of emotional, musical, and environmental sounds (i.e., wind, rain, thunder, and animal cries) than the left brain as well. The right brain, moreover, is responsible for nonlinguistic forms of communicating, such as facial expression, body language, and the expression of sounds that convey emotional nuances. In fact, the right brain is associated with the capacity to sing, swear, and even pray. (167)

Joseph Hellige organizes the research and findings into five thematic centers: 1., that there are indeed several behavioral asymmetries that can be attributed to cerebral asymmetries (though, the finding suggest both hemispheres can, in different ways, and less efficiently, perform tasks attributed to the other hemisphere); 2., that the two hemispheres must be understood in the context of the total brain, and the differing process each hemisphere uses nonetheless contributes to a complementary duality; 3., that the asymmetries are not unique to humans, and evidence supports a theory of evolution to present day humans; 4., that the variations in individual humans, though they may be minor, need to be taken into account; and 5., that asymmetric differences unfold over time, both in an individual life, and for the species, and that the seeds of this unfolding are sown before birth. (168)

An interesting interaction takes place in how the right hemisphere and left hemisphere handle aspects of mathematics, geometric space, and music, mathematics (with geometry as one of its categories) generally considered the ultimate left hemisphere activity, music a right hemisphere activity, even though there are curious correspondences, and seeming cross-overs between mathematics and music.

Pythagoras (c. 580-500 B.C.), and his followers, defended the numerical basis of sound, and even translated music into numbers and geometric proportions, finding pleasing musical intervals matching mathematical ratios (1:2 an octave, 2:3 a fifth, 3:4 a fourth, 4:5 a major third, and 5:6 a minor third), these ratios becoming the basis of western music harmony relationships; and they further found that these same ratios could be reproduced geometrically, and could reproduce themselves within themselves, thus resulting in what the Classical Greeks called the golden ratio or golden rectangle, the same relationship found reproduced throughout nature. Thus, through the evidence to be found in nature, and the logic of explanation, Pythagoras was able to extend the mathematics of sound relationships into all of his mathematical, geometrical explanations. As a result, the brilliant mathematician given credit for founding arithmetic, geometry, and trigonometry, believed music was geometric. (169)

But, the key to unraveling the mysterious relationship of that right hemisphere world of music with the left hemisphere world of mathematics involves careful attention to aspects of the two worlds, more precisely, the kinds of mathematics involved. Geometry is concerned with space, shape, points, lines, surfaces, and configurations—all a form of right hemisphere thinking, which also includes the perception of movement, speed, distance, depth, geometric gestalts, angles, and visual relationships in general. (170)

In contrast to the right hemisphere mathematical concerns, the mathematics the left hemisphere employs is temporal-sequential.

The intertwining of both hemispheres for music appreciation (and ultimately all forms of knowing) gets even more complex, because each hemisphere provides but a part of the final knowing. In music, for example, it has been found that this temporal-sequential aspect relates to sequential and rhythmical aspects of music, and the more these are emphasized the more the left hemisphere becomes activated. (171)

The same confusion occurs with language, generally considered a left hemisphere activity. However, it is in the right hemisphere that the "feel" of language, the way something is expressed, the context of the expression, the rate, amplitude, pitch, inflection, timbre, melody, stress, and context of the statement is understood, not the denotative meanings but the connotative values, not the grammar but the emotional tone. People incapable of "reading between the lines," of "getting the drift," of "catching the humor," of knowing when others are kidding, being deceitful, being sarcastic are simply not using their right hemisphere correctly. (172)

This leads to an interesting, and by now obvious, particularity in relation to right/left hemisphere functions involving the recognition of metaphors and puns (i.e., the entire world of symbolic meaning!). E. Winner and H. Gardner discovered that right-brain damaged patients tended to be overly literal in their interpretation of metaphorical statements (i.e., "sour grapes") and clichés (i.e., "A penny saved is a penny earned"). (173) Furthermore, working with N. S. Foldi and M. Cicone, H. Gardner found that right-brain damaged patients often chose completely inappropriate endings to cartoon strips, as if the humor is in the surprise ending. (174)

This confabulation can become more elaborate and bizarre, very similar to a dream or nightmare. (175) Rhawn Joseph discusses a 24-year-old patient, a grocery store clerk, who had been shot in the right hemisphere, resulting in the destruction of the right inferior convexity and orbital portions. He attributed his hospitalization to a ploy by the government to steal his inventions and ideas. When it was explained to him that the surgery was done to remove the bullet and bone fragments, his response was, "That's how they're stealing my ideas!" Another patient with a degenerative disturbance in the right frontal lobe randomly claimed to be a police officer, a doctor, or even married to various members of the staff. After several explanations of his true situation, he responded, "I'm a doctor. I'm here to protect people." (176)

The darkened theater: a soft, comfortable seat; the controlled temperature and humidity; and thousands of shadowy images moving across the screen—these strongly suggest the dream world each of us occupies while sleeping. It is possible to think of cinema as the deliberate manufacture of dreams. Psychological experiments reveal that individuals deprived of dreams—allowed to sleep but awakened when they begin to dream—move into a psychotic state. Clearly, dreaming is a biological and cultural necessity. In some way, our human psychobiological equipment is purged and regenerated while we sleep. Sanity is maintained, it seems, by the curious, illogical flow of images we call dreams.

The opening camera shot moves from the artificial, impersonal, exterior world of the hot, blinding sun and cold brick buildings, into a window, and we are aware that the camera and our own curiosity to know what is on the other side of that window take us from the conscious world into a world behind the physical shell, an invisible world of the mind, a world where symbolic meanings are to be found, a world that has not invited us in, one that is unaware of us, but one that will seduce us, harm us if we are not ready for the experience, one that will almost certainly change us. The unnatural feel of the camera, the quiet, the slow, steady pacing all tell us we are doing something wrong, are intruding, are entering some uncertain, twilight reality.

Then we get through the window, the threshold, and see the half naked woman and man. The heat, the colorless, documentary feel of the picture, the halfway experience, both sensual and incomplete, both driven by desire and yet matter-of-fact, both stripping and yet not stripping create an uneasy mixture of sin and sadness, of union and yet even more of separateness.

This is a dangerous world. We're not sure just what the rules are, just where the ethos resides. The discussion has a cold, even calculating quality. The man's divorce, the money needed for their own projected marriage, the underlying monotony of a very unromantic and seemingly mundane adultery, the sadness and lower-class quality of this meeting in a cheap motel room during a lunch break rob the scene of any heroic qualities. These are two lost souls, not to be emulated, not called to some higher adventure, but, rather, seeming to have stumbled into a dead wasteland, a lifeless existence where even the challenging of its barrenness is, inevitably, just a fruitless act.

Marion Crane, while somewhat endearing in her simplicity, is nevertheless certainly not admirable. There is an annoyingly prim, tough, shallow need-to-be-respectable quality about her, not at all one with the romantic, heroic, or roguish quality of the best of us. She does not seem to be in this motel room, this world, because she has realized a higher truth, because she has connected with her own superior qualities demanding she break the rules and enter a higher spirituality at any cost. Truth is, she is tempted to simply break off the affair because of the lack of money. She is not here because she has entered into a form of unconditional, higher love that demands she take risks. In fact, curiously, the meeting doesn't really seem all that risky, perhaps because we sense that the two lives are so unremarkable that discovery would do little to hurt them.

Sam Loomis is even less attractive. He is petulant, weak, and inconsiderate. It's hard to believe he is capable of true love. We're not sure if we want the affair to end or not. After all, even if they do find the energy to get through their problems, it's hard to see an eventual marriage lasting for more than a short time. Certainly, there is no "happy ever after" suggestion.

Marion Crane is in a state of transformation, always represented literally, immediately by her half dressed initial appearance, to be followed by her leaving her known physical situation behind, crossing a threshold, i.e., stealing the money and becoming a fugitive (by definition a person on the run). She has lost her real life innocence, is no longer pure, but has taken self-responsibility in going against the rules (in spite of her inability to comprehend the consequences). She is a criminal in the eyes of her culture/her childhood training/and her spirituality. She has broken the rules on all levels. She is Eve disobeying God, biting the apple and falling from her innocence. Noticeably, the white undergarments she wore in the opening scene get discarded for black.

We don't blame her. After all, that innocent world of appearances is also a shallow world, just as the Garden of Eden is a meaningless world until the apple from the tree of the knowledge of good and evil is consumed against the rules of God. This same initiation of human existence runs through all of us, and we hunger for it. We cannot be saved, cannot be redeemed, resurrected, rebirthed until we "fall" into ourselves, until we exist in the human sense, in the ethical, spiritual sense. We cannot become adults until we leave childhood behind.

And this world of innocence is represented to us in the most disgusting manner. Marion's "job" at the "office" in the "business" "building" in the "city" (notice all of the symbols of the conscious world) denies any intuitive, or unconscious life-giving qualities. Immediately she has to deal with a fat client who makes vulgar attempts to flirt with her (not a sensitive, loving, caring relationship, not a personal relationship, not a relationship of the heart, just the opposite, an impersonal, instinctual, id driven crude attempt at satisfying crude desires, not through empathy, but through a cold purchase, through the insensitive use of money).

To emphasize the universal insensitive acceptance of such a low level of existence, the office clerk who works with Marion fells slighted that Marion has been the "object" of the unethical attempt at a "transaction." To comfort herself, she says, "I expect he saw my wedding ring," a simple sentence that carries several levels of meaning, reminding both Marion and us of the recent similar relationship we just saw between her and Sam Loomis, involving money, sex, marriage, adultery, love (or, rather, the lack of it), and the overriding atmosphere of an *ignorance* of any higher values. Here we have an entire city of Adam and Eve innocence, though not so pure, because these people are living a post fall existence, and there just is no salvation likely for any of them. Adam and Eve, when the opportunity is offered, take advantage of it. These people have not and will not. They may eat apples, but the apples will not be from any tree of the knowledge of good and evil. Nor is there going to be any Christ in their future to save them. We needn't know any more about them to justify that claim. The shallow, two-dimensional qualities are obvious.

Already, within the first few moments of the film, Alfred Hitchcock has immersed us in an uneasy world of shadowy, unrealized people, vaguely familiar people, perhaps even more than vaguely familiar people, and we are uneasy, anxious to know what Marion will do, wanting her, in spite of her less admirable qualities, and partially because of these qualities (they do, after all, make her uncomfortably normal), to somehow find a way to rise above herself, rise above this common, unethical, unspiritual, uninspiring existence, hoping she will, in fact, do something against the meaningless rules that keep such an existence in place.

Alfred Hitchcock has taken us into the world of the dream, an existence beyond logic, beyond reason, a place of realizations that can only partially be explained. We know it exists, know that there are truths beyond what math and science can explain. (177)

William Faulkner would have said it expressed "the old verities and truths of the heart." (178)

Sigmund Freud would have told us it reached beyond our conscious, touched chords in the unconscious, where the depths of meaning reside. Perhaps he would have gone about the task of interpreting the specific chords it had touched. If so, he would have seen it all symbolically, the manifestation of latent needs and desires. (179)

As we have seen, science in the form of neurology is coming to some startlingly similar conclusions. The leading neurologists, especially since the mid-twentieth century, have found increasing evidence to support the dual brain theory, the idea that the two sides of the cerebrum of the brain are not symmetrical, but think in very different ways, the left side using logic, reason, sequential processes, and seeing the literal truths, the right side using intuition, feelings, and holistic processes to see symbolic truths, the truths *beyond explanation*, the truths of the world of the dream.

Todd E. Feinberg, neurologist at Beth Israel Medical Center in New York City, building on the works of such important neurologists as Richard Sperry, Michael S. Gazzaniga, and Joseph Bogen (previously discussed), and working with his own neurologically damaged patients, has pushed these divisions of the brain to an attempt to discover that elusive "I" within:

> Early in my career as a medical student I was required to perform a dissection of the brain using an old atlas as a step-by-step guide. The brain was bathed in a jar of preservative and wrapped in white cheesecloth. I took it out carefully and placed it on an orange cafeteria tray. The first procedure called for the removal of the gray matter in order to see the underlying brain structures. Beginning in a section that the atlas identified as the temporal lobe, I slowly and gently scraped away its cortical surface using a beveled wooded probe. Having barely exposed a centimeter of brain matter, I was struck by the realization of what I was actually doing. Had this person's memory of his childhood just been scraped away? Perhaps his recall of his whole family now was gone, or simply his recollection of a family outing on a warm summer day in July. Lying inert on this cafeteria tray was the substrate of this person's mind. I picked up the brain again, held it in my hands, and looked at it anew. This was a person's essence, humanity, and entire life experience now frozen in space and time. Who might he or she have been? It became apparent that studying the physical brain would allow me to explore some fascinating questions such as: *What is the self? Where is it located in the brain? How does the brain produce a unified self? What is the relationship between the brain and the mind?* The answers to these questions reflect the knowledge I have gained over the years as a neurologist and psychiatrist from patients with disorders of the self, and are the subject of this book. (180)

Though Feinberg puts forth a theory of mind/brain based on emergence (i.e., a hierarchically of ever more complexity resulting in a level where the totality is greater than the sum of its parts resulting from purpose), he also leans toward the right brain as dominant in the sense of *self*. (181)

Julian Keenan goes further, claiming the right hemisphere is essential to self. "Just a tenth of an inch beneath the furrowed ridges of gray matter that cover the right front side of the brain is a layer of tangled cell tissue that makes us uniquely human." (182)

While conducting postdoctoral research in behavioral neurology at Harvard Medical School, Keenan did the following face-recognition experiment to test self-awareness. He had five people about to undergo brain surgery for severe epilepsy at Boston's Beth Israel Deaconess Medical Center have their two hemispheres anesthetized, one at a time, while they were awake and alert. He then showed each a morphed image of him or her self blended with the face of a famous person. After the surgery, each patient was asked which face he or she remembered, his or her own fact, or the face of the famous person. With the right hemisphere anesthetized, four of the five remembered only seeing the famous person; with the left hemisphere anesthetized, all five remembered only themselves. The conclusion: it is in the right hemisphere that self-recognition takes place.

Of course, Feinberg and Keenan are not the first to attempt to figure out the mind/brain mystery, what is often referred to as *the ghost in the machine*. (183) Certainly, the very definition of human has to include some form of *self-realization*, some sense that there is a self beyond the physical. Anthropologists, in fact, first claim human appearance in evolution, not when crude tools or weapons are found, but when cave wall paintings are discovered (thus exhibiting a belief in the invisible world of the mind, the world beyond the physical world, coupled with a belief that this invisible world can influence the physical world).

France is the landscape. The first to emerge is Altamira. In the 1860s, a poacher on the estate of Senor Don Marcelino Santiago Tomas Sanz de Sautuola followed his dog, who had gotten lost in pursuit of a wounded fox, into a small hole. Within it stretched the huge, painted caves of the beginnings of human consciousness. There was no immediate rush to explore and unravel this hidden world. However, a few years later Don Marcelino was made aware of the caves and decided to play at archeology and geology. After having the entrance enlarged, he was able to recover some bone fragments of long extinct animals, which he showed to his friend, Vilanova y Piera, Professor of Geology at the University of Madrid, who confirmed that the bones were from the European Stone Age, and had been split by prehistoric man to extract their marrow.

Again, there was no rush to further explore the caves, but in 1879, Don Marcelino realized that the prehistoric objects on display at the Universal Exhibition of Paris from the Dordogne were of the same materials his own caves held. The next spring, he further explored his caves, discovering among the bones flints that he recognized as belonging to Lartet's Magdelanian Culture, and further back, the bones of a giant bear, some marks on the walls, and what appeared to be the pigment that made them in a shell.

While he was exploring, his daughter Maria with her dog Robot wandered off to entertain herself and discovered a huge cave vault with paintings on the ceiling of a herd of bison. Realizing that these bison were extinct, thus confirming a prehistoric origin of the paintings, Don

Marcelino hired an artist to make copies and in 1880 published a scientific monograph on his amazing cave.

However, the initial reaction from the scientific community was one of disbelief, and a suggestion that it was all a hoax. Don Marcelino had little interest in taking on the academic community, and his friend Vilanova y Piera was unable to persuade it to reconsider its initial reaction.

It was not until 1895 that French historian Emile Riviere was able to respark interest. He noticed that the markings in Don Marcelino's caves matched those found in a Dardogne cave, La Mouthe, that had been unopened until its excavation, thus, giving it a stronger proof against tampering, and a new rush of archeology began to discover and open up other similar caves. The only ones of these new ones to emerge on as grand a scale as Altamera were the caves at Lascaux (coincidentally, once again, discovered as the result of following a dog).

Now the evidence was overwhelming, and the scientific community had to accept the artistic imagination of these early humans. (184)

Jacob Bronowski, in attempting to comprehend these cave wall paintings, to make sense out of why these early humans would bother to go to such extremes (the paintings are, after all, done deep in dark caves on ceilings that had to be nearly impossible to even touch, much less paint), writes, "The cave paintings act as a kind of telescope tube of the imagination: they direct the mind from what is seen to what can be inferred or conjectured. Indeed, this is so in the very action of painting; for all its superb observation, the flat picture only means something to the eye because the mind fills it out with roundness and movement, a reality by inference, which is not actually seen but is imagined." (185)

Campbell calls them temple caves, for "a temple is a landscape of the soul." He explains:

> When you walk into a cathedral, you move into a world of spiritual images. It is the mother womb of your spiritual life—mother church. All the forms around are significant of spiritual value. Now, in a cathedral, the imagery is in anthropomorphic form. God and Jesus and the saints and all are in human form. And in the caves the images are in animal form. But it's the same thing, believe me. . . . The message of the caves is of a relationship of time to eternal powers that is somehow to be experienced in that place. (186)

Furthermore, he elaborates:

> If a *differentiating* feature is to be named, separating human from animal psychology, it is surely this of the subordination in the human

> sphere of even economics to mythology. And if one should ask why or how any such unsubstantial impulsion ever should have become dominant in the ordering of physical life, the answer is that in this wonderful human brain of ours there has dawned a realization unknown to the other primates. It is that of the individual, conscious of himself as such, and aware that he, and all that he cares for, will one day die. (187)

In spite of the prior millennium of human existence defined by the natural, undisciplined, casual, or even spiritual recognition of the visible and invisible worlds somehow interacting in the brain/mind, the serious, organized approach to the duality is generally credited to Descartes (1596-1650), sometimes considered the first psychologist. (188) He theorized that mind is connected to the soul, and thus partook of its oneness, and of its intangibility:

> Because it is of a nature which has no relation to extension, nor dimensions, nor other properties of the matter of which the body is composed, but only to the whole conglomerate of its organs, as appears from the fact that we could not in any way conceive of the half or the third of a soul, nor of the space it occupies and because it does not become smaller owing to the cutting off of some portion of the body, but separates itself from it entirely when the union of its assembled organs is dissolved. (189)

The brain, on the other hand, is physical, matter, and, thus, divisible. This then led him to see the mind and the brain as separate realities, the mind indivisible and nonmaterial (res cogitans), and the brain material and divisible (*res extensa*). In attempting to find a place the two separate worlds connected he came up with the pineal gland:

> The reason that persuades me that the soul cannot have any other place in the whole body than this gland, where it immediately exercises its functions, is that I consider that the other parts of our brain are all double so that we have two eyes, two hands, two ears, and, finally all the organs of our external senses are double; and, that inasmuch as we have only one solitary and simple thought of one single thing during the same moment, it must necessarily be that there is some place where the two images which come from the two eyes, or the two other impressions which come from a single object by way of the double organs of the other senses, may unite before they reach the soul, so that they do not present to it two objects instead of one. It can easily be conceived how these images or other impressions could unite in this gland through the mediation of the spirits that fill the cavities of the brain. There is no other place in the body where they could be thus united unless it be in the gland. (190)

Of course, modern neurology easily discounts the pineal gland as the center of human conscious. It is simply a neuroendocraine organ that has no real role in thinking at all. Nevertheless, the mind/brain duality, or, perhaps better stated, unity remains.

Sir Charles Sherrington, considered the father of modern neurophysiology, came at it through a consideration of how the two eyes resolve two separate images into one in the brain. On the one hand, except in unusual circumstances, he found:

> Our binocular visual field is shown by analysis, to presuppose outlook from the body by a single eye centred at a point in the midvertical of the forehead at the level of the root of his nose. It, unconsciously, takes for granted that it's seeing is done by a cyclopean eye having a center of rotation at the point of intersection just mentioned. (191)

However, he could not find such a central entity or place for this, writing: "Where it is a question of "mind" the nervous system does not integrate itself by centralization upon one pontifical cell. Rather it elaborates a million fold democracy whose each unit is a cell," (192) and ended up with a paradox.

Again, as with Descartes, he concluded that the integration of the mind and brain is mental, not physical.

In 1947, British philosopher Gilbert Ryle rejected this whole idea. There is, he proclaimed, no intangible self, no 'homunculus," or miniature man, directing a person's thoughts and actions. Rather the entire world is processed completely by the gelatinous gray-and-white matter within our heads. There is no *ghost in the machine*. (193)

Part VI

Insanity and Genius

> Men have called me mad; but the question is not yet settled, whether madness is or is not the loftiest intelligence–whether much that is glorious–whether all that is profound– does not spring from disease of thought–from moods of mind exalted at the expense of the general intellect.
>
> —Edgar Allan Poe, *Complete Tales and Poems*

> Insanity is relative. It depends on who has who locked in what cage.
>
> —Ray Bradbury

> There's a fine line between genius and insanity. I have erased this line.
>
> —Oscar Levant

Genius is the recovery of childhood at will.

—Arthur Rimbaud

No great mind has ever existed without a touch of madness.

—Aristotle

Genius is nothing more nor less than childhood recaptured at will.

—Charles Baudelaire, *The Painter of Modern Life and Other Essays*

The public is wonderfully tolerant. It forgives everything except genius.

—Oscar Wilde, *The Artist as Critic: Critical Writings of Oscar Wilde*

The true genius shudders at incompleteness—imperfection—and usually prefers silence to saying the something which is not everything that should be said.

—Edgar Allan Poe, *Marginalia*

Mediocrity knows nothing higher than itself; but talent instantly recognizes genius.

—Arthur Conan Doyle, *The Valley of Fear*

The truly creative mind in any field is no more than this: A human creature born abnormally, inhumanly sensitive. To him... a touch is a blow, a sound is a noise, a misfortune is a tragedy, a joy is an ecstasy, a friend is a lover, a lover is a god, and failure is death. Add to this cruelly delicate organism the overpowering necessity to create, create, create--so that without the creating of music or poetry or books or buildings or something of meaning, his very breath is cut off from him. He must create, must pour out creation. By some strange, unknown, inward urgency he is not really alive unless he is creating.

—Pearl S. Buck

When a great genius appears in the world you may know him by this sign; that the dunces are all in confederacy against him.

—Jonathan Swift, *Abolishing Christianity and Other Essays*

Neither a lofty degree of intelligence nor imagination nor both together go to the making of genius. Love, love, love, that is the soul of genius.

—Wolfgang Amadeus Mozart

It takes a lot of time to be a genius. You have to sit around so much, doing nothing, really doing nothing.

—Gertrude Stein

Madness is the acme of intelligence.

—نجيب محفوظ

These are the times in which a genius would wish to live. It is not in the still calm of life, or the repose of a pacific station, that great characters are formed. The habits of a vigorous mind are formed in contending with difficulties. Great necessities call out great virtues. When a mind is raised, and animated by scenes that engage the heart, then those qualities which would otherwise lay dormant, wake into life and form the character of the hero and the statesman.

—Abigail Adams

The most tragic thing in the world is a man of genius who is not a man of honor.

—George Bernard Shaw

The reason I talk to myself is because I'm the only one whose answers I accept.

—George Carlin

The statistics on sanity are that one out of every four people is suffering from a mental illness. Look at your 3 best friends. If they're ok, then it's you.

—Rita Mae Brown

I became insane, with long intervals of horrible sanity.

—Edgar Allan Poe

It is sometimes an appropriate response to reality to go insane.

—Philip K. Dick, *VALIS*

One person's craziness is another person's reality.

—Tim Burton

In individuals, insanity is rare; but in groups, parties, nations and epochs, it is the rule.

—Friedrich Nietzsche

I'll take crazy over stupid any day.

—Joss Whedon

Mad Hatter: "Why is a raven like a writing-desk?"

"Have you guessed the riddle yet?" the Hatter said, turning to Alice again.

"No, I give it up," Alice replied: "What's the answer?"

"I haven't the slightest idea," said the Hatter.

—Lewis Carroll, *Alice in Wonderland*

Dispute not with her: she is lunatic.

—William Shakespeare, Act 1, Scene iii, 258, *Richard III*

THE EDGE, there is no honest way to explain it because the only people who really know where it is are the ones who have gone over.

—Hunter S. Thompson

A casual stroll through the lunatic asylum shows that faith does not prove anything.

—Friedrich Nietzsche

We do not have to visit a madhouse to find disordered minds; our planet is the mental institution of the universe.

—Johann Wolfgang von Goethe

If you're going to be crazy, you have to get paid for it or else you're going to be locked up.

—Hunter S. Thompson

One ought to hold on to one's heart; for if one lets it go, one soon loses control of the head too.

—Friedrich Nietzsche

Sweet, crazy conversations full of half sentences, daydreams and misunderstandings more thrilling than understanding could ever be.

—Toni Morrison, *Beloved*

First sign of madness, talking to your own head.

—J.K. Rowling, *Harry Potter and the Order of the Phoenix*

I mean, maybe I am crazy. I mean, maybe. But if this is all there is, then I don't want to be sane.

—Neil Gaiman, *Neverwhere*

Do you think it's possible for an entire nation to be insane?

—Terry Pratchett, *Monstrous Regiment: The Play*

I do not suffer from insanity, I enjoy every minute of it.

—Edgar Allan Poe

Richard Dadd planned and executed a cold blooded murder of his father, a man who had always been kind to him and one he had no reason to harm. He thought he was doing it on the command of the Egyptian god

Osirus and it is evident he was planning on murdering others at this god's command.

These are not the actions of a *sane* man.

Vincent van Gogh exhibited traits throughout his life, certainly during the later portions of his life that also strongly suggest he was not *sane*. *Sane* men do not cut off their ears or commit suicide.

If strict demands for being open-minded and non-judgmental are employed, it is possible to say that in certain circumstances such unusual actions might be justified, but in both cases it is nevertheless clear that something about these men was outside of the *normal* range of human thinking and behavior.

Both were designated as *insane* and institutionalized for it. In Dadd's case, the confinement was easy to defend, as he was certainly a danger to others. In Van Gogh's case, it was more that he needed help. There was little reason to think he was likely to hurt others, though obviously he was likely to harm himself.

What does it mean to be *insane*? Keeping in mind Juliet Capulet's famous line that "a rose by any other name would still smell as sweet," that a person is the same regardless what he is named, that a word is a name for something but is not the thing itself, it is nonetheless both necessary and worthwhile to explore the word. While a dictionary is not the place to end a discussion on the meanings of words, it works well as a place to begin, if nothing else, a way to ground the exploration in the popular, general frame for the current useage. The online *Merrium-Webster Dictionary* offers the following:

In·san·i·ty: *noun,* \in-ˈsa-nə-tē\; *plural* in·san·i·ties

1. severe mental illness : the condition of being insane
2. something that is very foolish or unreasonable

Full Definition:

1. a deranged state of the mind usually occurring as a specific disorder (as schizophrenia)

2. such unsoundness of mind or lack of understanding as prevents one from having the mental capacity required by law to enter into a particular relationship, status, or transaction or as removes one from criminal or civil responsibility
3. *a.* extreme folly or unreasonableness; *b.* something utterly foolish or unreasonable

Examples of *insanity:*

- She was found not guilty by reason of *insanity*.
- His friends thought his decision to quit his job was pure *insanity*.
- Please, no more violence. It's time to stop this *insanity*.
- the *insanities* of modern life

First known use of *insanity:* 1590

Synonyms:

aberration (archaic), dementia, derangement, lunacy, madness, mania, rage.

Antonyms:

mind, saneness, sanity

Related Words:

neurosis, psychosis; instability, irrationality, unreasonableness; delirium, frenzy, hysteria; hallucinosis, hypomania, paranoia, schizophrenia; senile dementia; delusion, hallucination; monomania, obsession, phobia; abnormality, dementedness, unsoundness

Near Antonyms:

lucidity, rationality, rationalness, reasonability, reasonableness; normality, soundness

Other Psychology Terms:

fetish, hypochondria, intelligence, mania, narcissism, neurosis, pathological, psychosis, schadenfreude, subliminal

Medical Definitions:

1: a severely disordered state of the mind usually occurring as a specific disorder (as paranoid schizophrenia)

2: unsoundness of mind or lack of the ability to understand that prevents one from having the mental capacity required by law to enter into a particular relationship, status, or transaction or that releases one from criminal or civil responsibility
Legal Definition:

In criminal law, a disease, defect, or condition of the mind that renders one unable to understand the nature of a criminal act or the fact that it is wrong. Tests of insanity are not intended as medical diagnoses but rather only as determinations of whether a person may be held criminally responsible for his or her actions. The most enduring definition of insanity in Anglo-American law was that proposed by Alexander Cockburn (1843). Many U.S. states and several courts have adopted a standard under which the accused must lack "substantial capacity either to appreciate the criminality of his conduct or to conform his conduct to the requirements of the law." Some states have abolished the insanity plea, and others allow a finding of "guilty but mentally ill." *See also* diminished responsibility.

The etymology of a word is also a place to ground it. And *insanity* is a word that happens to be the antonym of another, *sanity*. So what does *sanity* mean?

The same online dictionary offers the following:

San·i·ty: *noun* \ˈsa-nə-tē\:

1. the condition of having a healthy mind: the condition of being sane
2. the condition of being based on reason or good judgment

Full Definition of *sanity:*

1. the quality or state of being sane; *especially* : soundness or health of mind

Examples of *sanity:*

1. People have begun to doubt his *sanity*.
2. She is the mother of six children but somehow keeps her *sanity*.
3. The *sanity* of the decision was never in question.

Origin of *sanity:*

Middle English *sanite,* from Anglo-French *sanité,* from Latin *sanitat-, sanitas* health, sanity, from *sanus* healthy, sane

First known use: 15th century

Synonyms:

daylights, head, marbles, reason, saneness, mind, wit(s)

Antonyms:

dementia, derangement, insanity, lunacy, madness, mania, unreason

Related Words:

rationality, reasonableness, sense; health, healthfulness, healthiness, wholesomeness; clearheadedness, lucidity, lucidness, normalcy, normality, soundness; wisdom

Near Antonyms:

delusion, hallucination; delirium, frenzy, hysteria

Other Psychology Terms:

fetish, hypochondria, intelligence, mania, narcissism, neurosis, pathological, psychosis, schadenfreude, subliminal

Medical Definition of *sanity:* the quality or state of being sane; *especially* soundness or health of mind.

As expected, these dictionary definitions don't help much beyond presenting the general, common understanding of the terms, which is that to be *sane* falls into the idea of being mentally normal, mentally healthy, and its opposite, *insane*, means to be not mentally normal or healthy.

There is a consenses that being normal means being able to think in a rational, reasonable manner, including self-awareness and being aware of how one fits into the environment. *Insanity*, then, is the opposite of this. It means that a person is not able to think in a rational, reasonable manner, not mentally healthy but suffering some form of mental illness, a *dis*ease, a sickness. It is also important to point out that intelligence is not the determining factor (i.e., it is possible to be both highly intelligent and insane).

It is worth emphasizing that *sanity* and *insanity* are understood in the contexts of *what the culture believes to be normal, acceptable, rational, and reasonable behavior*. In other words, as with everything, these terms *mean* in context. And the often ignored aspect of language is that it is composed of arbitrary sounds that *humans have decided mean something* beyond mere noise. Certainly it is possible to trace a number of words to their onomatopoetic origins (bark, meow, buzz), and certainly the sound of a word suggests basic meanings. Harsh words suggest harsh meanings; soft, soothing words suggest soft, soothing meanings; and the pitch, tone, timbre and other aspects of the sound of a word influence its meaning.

However, while admitting that certain types of sound are more likely to mean certain things, the emphasis should be on the opposite, on the truth that words mean in context, and that discussions such as this one take place in the context of a cultural worldview, an assumption that there is a common ground, a general agreement on what it means to be normal and mentally healthy.

For example, this discussion you are reading is taking place in the context of a European civilization that has led the views of human culture for over 2,000 years, and its assumed concepts of normal behavior, of healthy thoughts, might be valueable, might be enlightened, might be right in some context, but they should not be seen as absolute. However, to be understood as *sane* is to fit into these views of normalcy, and to be *insane* is to fall outside of them in what is considered a negative way.

Richard Dadd and Vincent van Gogh both did this, both were and still are seen as having some form of *insanity*.

But they also fell outside of normalcy in a positive way. The paintings they created exhibit mental abilities beyond those of a normal human. The word for this is *genius*.

How does our Merrium-Webster online dictionary define this word?

Ge·nius: *noun* \ˈjēn-yəs, ˈjē-nē-əs\; *plural* ge·nius·es *or* ge·nii:

1. a very smart or talented person: a person who has a level of talent or intelligence that is very rare or remarkable
2. a person who is very good at doing something
3. great natural ability: remarkable talent or intelligence

Full Definition of *genius:*

1. a. *plural genii*: an attendant spirit of a person or place; b. *plural usually genii*: a person who influences another for good or bad
2. a strong leaning or inclination: penchant
3. *a*. a peculiar, distinctive, or identifying character or spirit; b. the associations and traditions of a place; c. a personification or embodiment especially of a quality or condition
4. *plural usually genii*: spirit, jinni
5. *plural usually geniuses*: *a*. a single strongly marked capacity or aptitude; had a *genius* for getting along with boys — Mary Ross; *b*. extraordinary intellectual power especially as manifested in creative activity; c. a person endowed with transcendent mental superiority; *especially* a person with a very high IQ.

Examples of *genius:*

1. Albert Einstein and Isaac Newton were great scientific *geniuses*.
2. You don't have to be a *genius* to see that this plan will never work.
3. He was a *genius* at handling the press.
4. She's now widely recognized as an artist of *genius*.
5. He's admired for his comic *genius*.
6. My plan is simple—that's the *genius* of it.
7. The *genius* of these new computers is their portability.

Origin of *genius:* Latin, tutelary spirit, natural inclinations, from *gignere* to beget

First known use: 1513

Synonyms:

brain, brainiac, intellect, thinker, whiz, wiz, wizard

Antonyms:

blockhead, dodo, dolt, dope, dumbbell, dummy, dunce, fathead, goon, half-wit, hammerhead, idiot, imbecile, knucklehead, moron, nitwit, numskull (*or* numbskull), pinhead

Related Words:

polyhistor, polymath, Renaissance man; blue, bluestocking, highbrow, intellectual; sage, savant; egghead, geek, longhair, nerd; master, virtuoso; ace, crackerjack (*also* crackajack), natural

Near Antonyms:

ignoramus, illiterate, know-nothing, lowbrow; anti-intellectual, philistine; ass, donkey, fool, jackass; beast, boor, cad, churl, clown, creep, cretin, cur, heel, jerk, louse, lout, skunk, snake, stinker

Medical Definition of *genius*:

1. extraordinary intellectual power, especially as manifested in creative activity.
2. a person endowed with transcendent mental superiority; *specifically* a person with a very high IQ.

There are a couple of interesting nuiances here. The Latin origin connects it to *beget* (to create, to bring about, to give birth to). Thus, a *genius* is a person who brings something new into existence. Again, this is going beyond the *normal*, the current views of what it means to use reason and rationality (though, in the disciplines that employ these qualities, noteably math and science, it's not so much rejecting reason and rational thinking as it is redefining them).

Also, there are connections to spiritual beings, to that *other* world. So there is at least a sense in which *genius* partakes of existence beyond the *normal* in the manner of the invisible, numinous world. In other words, though most commonly used today to mean a person who is exceptionally good at something, *genius* involves more. It is better used to mean a person who breaks the continuum of intelligence, who is something other than a very intelligent person. Einstein is a perfect example. He was not exceptional at math. As biographers have delighted in pointing out, he was poor at math as a child, he had trouble following the quick mathematical discussions of young, intelligent mathematicians who sought him out, and even the brilliant theories he came up with are filled with mathematical mistakes. Yet, when people give examples of a *genius*, he is almost certainly going to be on the list.

Both *genius* and *insanity* share this quality. Though they get used to represent people who are on opposite ends of the intellectual or ethical continuums, that is exactly wrong. Instead, they represent people who have some kind of mental condition beyond it or other than it. It is possible, and not uncommon, for both a *genius* and an *insane* person to demonstrate poor mental abilities in *normal* mental aspects of life, so much so that they suffer a common web of stereotypes—the absent-minded, eccentric, odd-ball, childlike, socially inept painter, mathematician, scientist, musician (whatever discipline).

They are to be grouped with other categories of people who fall outside the norm, those discussed in the section on Richard Dadd—the shamans and the saviors, people who have experienced the numinous plane of existence.

Richard Dadd and Vincent van Gogh are interesting because they exhibit the dual aspects of such people, but the studies of them are hampered by time. When they were alive, the disciplines for understanding their mixture of insanity and genius barely existed, were extremely primitive by today's standards, and attempts to sort through what records remain are clouded in guesswork.

Part VII

Frankenstein

> Did I request thee, Maker, from my clay
> To mould Me man? Did I solicit thee
> From darkness to promote me?
>
> John Milton, *Paradise Lost* (X.743–5)

In her introduction to the 1831 edition of Frankenstein, Mary Shelley wrote:

> "Perhaps a corpse would be re-animated . . . galvanism had given token of such things." It was after midnight before they retired, and unable to sleep, she became possessed by her imagination as she beheld the grim terrors of her "waking dream":
>
> I saw the pale student of unhallowed arts kneeling beside the thing he had put together. I saw the hideous phantasm of a man stretched out, and then, on the working of some powerful engine, show signs of life, and stir with an uneasy, half vital motion. Frightful must it be; for supremely frightful would be the effect of any human endeavour to mock the stupendous mechanism of the Creator of the world.

CBS seemed to have figured out the formula, premiering its January 2014 offerings with *Intelligence*, clocking nearly seventeen million viewers nestled between *NCIS* and *Person of Interest*, to become the most watched series premiere of the season.

It was a tried and true formula, a James Bond like high-level action espionage series, except with the more Americanized, thus less sophisticated leading man. And though the Bond movies are able to pair him up with a new leading lady each movie, a television series works better by having a more permanent relationship with some beautiful woman, who it also just so happens in the currently politically correct world is also able to outfight anyone in the fun action scenes (à la Kate Becket in the series *Castle*).

Latest masculine heartthrob Gabriel Vaughn (Josh Holloway) is a high-tech intelligence operative enhanced with a super-computer microchip in his brain. With this implant, he is the first human ever to be connected directly into the globalized information grid. He can get into any of its data centers and access key intelligence files in the fight to protect the United States from its enemies. Lillian Strand (Marg Helgenberger), the director of the United States Cyber Command who supports Gabriel and oversees the unit's missions, assigns Riley Neal (Meghan Ory), a Secret Service agent, to protect Gabriel from outside threats, as well as from his appetite for reckless, unpredictable behaviors and disregard for protocols. Meanwhile, Gabriel takes advantage of his chip to search for his wife who disappeared years ago after being sent by the C.I.A. to infiltrate and prevent the Lashkar-e-Taiba from carrying out a terrorist attack in Mumbai, India. No need to worry about the possible problems of good hearted Gabriel flirting with his new partner while still yearning to find his missing wife, as she blows herself up in the second episode (though, of course, knowing how television works, no one would be overly surprised to find her somehow having survived or for whatever reason being found alive in some future episode).

If these remarks seem glib, it's because the show was poorly done, and even those wanting such a show turned against it immediately. The clichés already meantioned are matched by many more, and the dialogue and acting is clumbsy to the point of embarrassment.

So why bother discussing it?

For the same reason it got such a huge initial audience.

Putting aside the overused tried-and-true formulas and insistence on being politically correct to the point of insulting any thinking person, it grasped one thing that is at the core of the new world--cybernetics, robotics and artificial intelligence. By whatever name, and in whatever particular discipline, human existence is changing at an incredible rate, and the changes go to the centers of what it means to be human.

What this show did to pull in an initial audience was to offer a new twist on the old formulas, a human with a "super-computer microchip implanted in his brain," one that allowed him to connect directly to the "global information grid."

Of course, this is not the first time science fiction has explored this, generally emphasizing the negative aspects, the use of the chip to control the human, or in more interesting ways how the chip might affect and change the basic qualities of being human (in negative ways).

Examples immediately come to mind, such cyborgs and androids as RoboCop, Terminators, Transformers, Evangelion, The Six Million Dollar Man, Replicants from Blade Runner, Daleks and Cybermen from Doctor Who, the Borg from Star Trek, Darth Vader and General Grievous from Star Wars, Inspector Gadget, and Cylons from Battlestar Galactica. From manga and anime are characters such as 8 Man (the inspiration for RoboCop), Kamen Rider, Ghost in the Shell's Motoko Kusanagi, as well as characters from western comic books like Ironman Tony Stark (after his Extremis and Bleeding Edge armor) and Victor "Cyborg" Stone, even such characters as Robbie the Robot and the Tin Man from the Wizard of Oz.

Why books, television shows and movies exploring this latest twist on the Frankenstein monster are increasing is that science fiction is quickly becoming reality. The human mind is escaping the restrictions that once defined human existence.

At the time of writing this, two recent publications of *The New Yorker* magazine offered glympses of the emerging realities. "The Body Electric: A scientist takes computing power under the skin," by Kin Tingley, goes directly to the real world of creating computer chips with organic materials, biological electric circuits. The language for this rapidly developing field is also rapidly developing, and so the terminology can be difficult to follow, but the idea is to connect hard science (in this case, the inorganic computer chip or micro-robot to human cells, to organic materials). This article focused on John Rogers, a Professor of Engineering at the University of Illinois. He has been on the cutting edge of research resulting in a series of successful experiments involving electronic and photonic systems in living tissue, pushing the interphase to where biological materials (human cells) can be programmed in the same way as a typical non-biological computer. (194)

In 2011 he worked with others to produce a silicon circuit with the properties of skin (an epidermal electronics system). He has also worked on an "artificial pericardium," a filmy device that can be slipped over the heart like a sock, run off the heart's normal motions and be able to monitor and even repair cardiovascular irregularities (similar to the typical pacemaker, except, instead of two electrodes, it would have hundreds and not be run off a battery but be self-generating and be a computer inside or around the heart to keep it monitored and running smoothly). It is expected to be available within twenty years. This same process can be used for other organs as well, and can become more sophisticated, as it is almost certain that a way will be found to use a delivery system that will dissolve in the future, basically leaving the electronic circuit spread out over the natural organic systems, completely organic electronic webs that can be controlled by automatically programmed and consciously controlled human decisions. And the electronics in the brain, the neurons firing off and on in patterns can be affected through "optogenetics," the use of photons—to cure serious mental illnesses, to increase mental ability, and theoretically to program the brain/mind.

While such possibilities have been surmised and projected in numerious fictional futures, they are no longer just interesting speculations, but real world possibilities.

The other article, "Letter from Shenzhen—The Gene Factory: A Chinese firm's bid to crack hunger, illness, evolution—and the genetics of human intelligence" by Michael Specter, comes at it from the opposite direction, the human genetic structure—DNA. As the title says, this article focuses on a leading company, the Beijing Genomics Institute (B.G.I.). (195)

It's all been projected before, the latest way to solve hunger, combat and eliminate disease, and even increase intelligence. But now it's gone from science fiction to science reality.

On the east edge of the Chinese factory town of Shenzhen sits an eight-story building that once was a shoe factory, but now houses the largest genetic research factory in the world. It holds seventy-eight machines that are busy sequencing the precise order of the billions of molecules of chemicals in each molecule of DNA, a least a quarter of the the entire world's genomic data, more than Harvard University, the National Institutes of Health, or any scientific institute.

Science has long known the nucleotides that make up the code of life—adenine, thymine, cytosine, and guanine. These, in turn, make up the three billion pairs of molecules in the human genome that are arranged to make up the genes that are necessary to existence. Its four thousand employees wear parkas because the machines need to be kept in frigid rooms to preserve the chemicals necessary to process the DNA and sit in identical, powder-blue cubicles, each bathed in an eerie glow from their computer monitors.

In 1953, James Watson and Francis Crick found the final structure for human DNA, the double helix, and since then they and others have been scrambling to map it out. In 1995, Craig Venter was the first to sequence a bacterium, and now B.G.I. has already mapped out, not only many strains of rice, cucumber, chidkpea, the giant panda, the Arabian camel, the yak, and forty types of silkworm, but it has mapped out over sixty thousand people.

Jian Wany, the company's president, and co-founder Huanming Yang, the chairman, began their company affiliated to the Chinese Academy of Sciences, but they were soon kicked out and went off on their own. They wanted the Chinese government's support, but it wouldn't fund such a radical project, and in the end they had to go it alone, and at first they literally were working out of cardboard boxes. However, in 2000, Bill Clinton highlighted their work and that gave them the push they needed to expand, and they quickly became central the world's network of scientists involved with this research, including James Watson.

While it took over a decade and three billion dollars to for an international team of scientists to map out the first human genome, by now it has already reached the point where it can been done in about four days for less than $4,000 (and no doubt by the time this book is printed, that will have decreased noticeably). The hope is to become a reference library freely accessible for scientists and anyone wanting access to the rapidly expanding maps of human, animal, and plant genomes.

These maps, in turn, will reveal the origins and evolution of all living life, increase life span and global food production, decode genetic diseases, provide information about autism, and drasticly reduce birth defects.

More controversial doors will also be opened. One of its projects, the Cognative Genomics project, is looking for the genetic basis for human intelligence.This is one of the central hot topics, a form of eugenics, the idea of creating a super human race, and mixed with the idea of humans playing the role of God. Of course, any kind of human intervention into the "natural" happenings in nature fall into this category, and humans have been debating them ever since the biting of that first forbidden fruit from the tree of the knowledge of good and evil. But when the focus is on human intelligence, the use of science to affect it becomes a more obvious intrusion of human intelligence into what some consider God's rightful domain. Nevertheless, the research continues, whether or not countries such as the United States of America will allow it, and the world web of scientists are all sharing in the exciting science unfolding.

Also controversial, but likely necessary in the world's rapidly growing population and depletion of resourses, the mapping out of the building blocks of the plants and animals necessary to feed the world is rapidly unfolding new varieties of food, not necessarily genetically engineered, though that is possible and likely, but even just learning what existing species work best. Instead of suggesting that someday future generations will find the planet has been used up by previous generations, such research is offering strong possibilities that science will solve this problem for good.

If all of the variables of human DNA can be mapped and identified, then they can all be manipulated. Is this not the blueprint for making humans from scratch, as one would build a toy tower from an erector set or construct a building out of the same clay God used to create the first humans?

As always, such discussions find their way back to Mary Shelley's *Frankenstein*, the first well known work to include the concept of a human creating a human through science, however crude and minimal the science.

And she does suggest that the created monster not only lives on basic animal levels, but it has feelings, human feelings and desires. Is this the key distinction? If humans can create humans through artificial means, are these artificial humans really human? And as such should they have rights similar to human rights?

This might seem the stuff of science fiction, as, of course, it is. Three of the numerious recent movies exploring it are *I Robot*, *Blade Runner* and *A.I.: Artificial Intelligence*, each of them with cyborgs that have the ability to feel human emotions. But such science fiction, once understood as theoretical, perhaps possible in a distant future or "other" reality, is rapidly edging into the real present. As laughable as it might seem on first thought, "robot rights" are currently being seriously considered by, for example, California's Institute for the Future (IFTF), a Palo Alto, California based think tank established in 1968, as a spin-off from the RAND Corporation, to help organizations plan for the long-term future, a subject known as futures studies, and they are discussed in depth in the 2010 documentary film *Plug & Pray*. (196)

Human Intelligence and Artificial Intelligence—what is the difference? Artificial Intelligence by the traditional definitions of the two words means a form of intelligence that is not real, more precisely a form of human intelligence that is not real, or a human like intelligence.

In the disciplines involved with Artificial intelligence (AI), it is defined as the intelligence exhibited by machines or software, and the branch of computer science that develops machines and software with intelligence. Major AI researchers and textbooks define the field as "the study and design of intelligent agents," (197) where an intelligent agent is a system that perceives its environment and takes actions that maximize its chances of success. (198) John McCarthy, who coined the term in 1955. defines it as "the science and engineering of making intelligent machines." (199)

The main goals of AI research include reasoning, knowledge, planning, learning, communication, perception and the ability to move and manipulate objects. (200) General intelligence (referred to as "strong AI") is still among the field's long term goals. (201) Currently popular approaches include statistical methods, computational intelligence and traditional symbolic AI. There are an enormous number of tools used in AI, including versions of search and mathematical optimization, logic, methods based on probability and economics, and many others.

More to the point, what artificial intelligence suggests is that a central property of humans, intelligence—the sapience of *Homo sapiens*—can be created by humans so precisely that it can be done by a machine. (202)

Yet again, we face the central questions, the conflict between religious beliefs and science. Can humans create humans? Or is this on the province of a God?

Political scientist Charles T. Rubin believes that AI can be neither designed nor guaranteed to be friendly, (203) arguing that "any sufficiently advanced benevolence may be indistinguishable from malevolence." Humans should not assume machines or robots would treat us favorably, because there is no *a priori* reason to believe that they would be sympathetic to our system of morality, which has evolved along with our particular biology (which AIs would not share).

Joseph Weizenbaum writes that Artificial Intelligence applications cannot by definition successfully simulate genuine human empathy and that viewing the human mind as nothing more than a computer program (computationalism) is wrong, pointing out that AI research devalues human life. (204)

On the other hand futurists such as Ray Kurzweil have used Moore's law (which describes the relentless exponential improvement in digital technology) to calculate that desktop computers will have the same processing power as human brains by the year 2029, and also predict that by 2045 artificial intelligence will reach a point where it is able to improve *itself* at a rate that far exceeds anything conceivable in the past, a scenario that science fiction writer Vernor Vinge named the "singularity." (205)

Robot designer Hans Moravec, cyberneticist Kevin Warwick and inventor Ray Kurzweil have predicted that humans and machines will merge in the future into cyborgs that are more capable and powerful than either.[185] This idea, called *transhumanism*, which has roots in Aldous Huxley and Robert Ettinger, has been illustrated in fiction as well, for example in the manga *Ghost in the Shell* and the science-fiction series *Dune*. In the 1980s artist Hajime Sorayama's Sexy Robots series were painted and published in Japan depicting the actual organic human form with life-like muscular metallic skins; and later "the Gynoids" book was used by or influenced movie makers including George Lucas and other creatives. Sorayama never considered these organic robots to be real part of nature but always unnatural product of the human mind, a fantasy existing in the mind even when realized in actual form.

Almost 20 years later, the first AI robotic pet, AIBO, came available as a companion to people. AIBO grew out of Sony's Computer Science Laboratory (CSL). Famed engineer Toshitada Doi is credited as AIBO's original progenitor: in 1994 he had started work on robots with artificial intelligence expert Masahiro Fujita, at CSL. Doi's, friend, the artist Hajime Sorayama, was enlisted to create the initial designs for the AIBO's body. Those designs are now part of the permanent collections of Museum of Modern Art and the Smithsonian Institution, with later versions of AIBO being used in studies in Carnegie Mellon University. In 2006, AIBO was added into Carnegie Mellon University's "Robot Hall of Fame."

Edward Fredkin argues that "artificial intelligence is the next stage in evolution," an idea first proposed by Samuel Butler's *Darwin among the Machines* (1863), and expanded upon by George Dyson in his book of the same name in 1998. (206)

Notes

1. Suetonius, E. Rice, Apr. 2000, J. R. S. M. 93, 198-202; rpt., *Fragments of Neurological History,* John M. S. Pearce, Imperial College Press: London, 2003, p. 590.
2. *The Roman Empire: Augustus to Hadrian*, Cambridge, 1988.
3. *Documents Illustrating the Principates of Gaius, Claudius, and Nero*, Cambridge, 1967.
4. *Augustus to Nero: A Sourcebook on Roman History*, London, 1985.
5. Suetonius, Apr. 2000, J. R. S. M. 93, 198-202; rpt., *Fragments of Neurological History*, John M. S. Pearce, Imperial College Press: London, 2003, p. 590. .
6. Suetonius, Claudius, Bristol Classical Press, 1986, pp. 145-8.
7. Fra*gments of Neurological History*, John M. S. Pearce, Imperial College Press: London, 2003, p. 590.
8. Claudius' full name was Tiberius Claudius Drussus Nero Germanicu, son of Nero, Drussus and Antonia (daughter of Antony and Octavia).
9. Bernard Rimland, "Foreword," *Nobody Nowhere*, 1992; rpt, HarperCollins, 2002.
10. Donna Williams, *Nobody Nowhere*, 1992; rpt, HarperCollins, 2002, p. 11.
11. Donna Williams, *Nobody Nowhere*, 1992; rpt, HarperCollins, 2002, p. 47.
12. Donna Williams, *Nobody Nowhere*, 1992; rpt, HarperCollins, 2002, p. 12.
13. Donna Williams, *Nobody Nowhere*, 1992; rpt, HarperCollins, 2002, p. 17.
14. "Through the Looking-Glass," *The Works of Lewis Carroll*, ed. Roger Lancelyn Green, Spring Books/Hamlyn Publ., 1968, p. 120.
15. Donna Williams, *Nobody Nowhere*, 1992; rpt, Harper Collins, 2002, p. 18.
16. Donna Williams, *Nobody Nowhere*, 1992; rpt, Harper Collins, 2002, pp. 19-20.
17. *I Never Promised You a Rose Garden*, Hannah Green, Signet, 1964; rpt., Joanna Greenberg, 1989; *The Cry for Myth*, Rollo May, Delta, 1991, pp. 17-21.
18. "Alice's Adventures in Wonderland," *The Works of Lewis Carroll*, ed. by Roger Lancelyn Green, Hamlyn Publ., 1965, pp. 24-25.
19. *I Never Promised You a Rose Garden*, Hannah Green, Signet, 1964; rpt., Joanna Greenberg, 1989, p. 31.
20. Donna Williams, *Nobody Nowhere*, 1992; rpt, HarperCollins, 2002, p. 10
21. Rollo May used *I Never Promised You a Rose Garden* to demonstrate the relationship of individualized myths with archetypal myths and how they both

supply the necessary landscapes for psychic health; put quote from May, p. 1920 to demonstrate connection with myths, concluding his discussion: "Myth making thus is central in psychotherapy. It is of the essence that the therapist permit the client to take his or her myths seriously, whether the myths come up in dreams or in free assiciation or in fantasy. Every individual who needs to bring order and coherence into the streams of her or his sensations, emotions, and ideas entering consciousness from within and without is forced to do deliberately for himself what in previous ages had been done for him by family, custom, church, and state. In therapy myths may be a reaching out, a way of trying out new structures of life, or a desperate venture at rebuilding his or her broken way of life. Myths, as Hannah Green put it, are "sharers of our loneliness." (p. 21)

22. When a medical or psychiatric diagnosis can't come up with a specific condition it is sometimes referred to as a "trashcan" diagnosis. In other words, something is obviously wrong, but just what it is cannot be determined. This has been and remains a common situation for psychology and neurology. For a number of decades in the mid-1900s, schizophrenia was the trashcan term, taking on a somewhat glamorous quality and being applied to many forms of mental illness now classified outside of it. The current "trashcan" term (though, of course, since it is the "in" thing, would not be so easily dismissed as such) is autism.

23. It would be an interesting exercise to compare the various characters in Joanna Greenberg's fantasy world with those in Lewis Carroll's. In both cases, these entities are far from friendly, and seem to be operating out of some illogical sense of how the world works, have some system of ethics, but it either simply doesn't match that of the real world or is expressed in ways that make it hard to sort out.

24. *I Never Promised You a Rose Garden*, Hannah Green, Signet, 1964; rpt., Joanna Greenberg, 1989, p. 54.

25. *I Never Promised You a Rose Garden*, Joanne Greenberg, 1989, pp. 152-156.

26. *I Never Promised You a Rose Garden*, Joanne Greenberg, 1989, pp. 34-35.

27. *I Never Promised You a Rose Garden*, Joanne Greenberg, 1989, pp. 13-14.

28. *Nobody Nowhere*, 1992; rpt, HarperCollins, 2002, p. 130.

29. *Nobody Nowhere*, 1992; rpt, HarperCollins, 2002, p. 160.

30. Reprinted in *Not Even Wrong: Adventures in Autism*, Paul Collins, Bloomsbury, 2004, p. 21.

31. *A Foreign View of England in 1725-29: The Letters of Monsieur Cesar de Saussure to His Family*. London: Caliban Books, 1995.

32. London: T. Warner, 1726; discussed in the context of the entire European view of feral children in "Homo ferus: Between Monster and Model," *Eighteenth Century Life* 21-2, p. 179, 1997.

33. (originally attributed to John Arbuthnot but now generally thought to have come from Swift), in George A. Aitken, *The Life and Works of John Arbuthnot* (Oxford: Clarendon, 1892), p. 471.

34. Can be found in *Miscellaneous Poems by Several Hands*, ed. David Lewis, London: J. Watts, 1726, p. 305.

35. London: J. Roberts, 1726, p. 3.

36. Attributed to John Arbuthnot and/or Jonathan Swift in Aitken, *The Life and Works of John Arbuthnot*, p. 474.

37. *The Origin and Progress of Language*, J. Balfour, 1774, 1:242-69.
38. Edenburgh, 1779-99, vol. 4, p 32
39. vol. 4, pp. 25-42.
40. *Of the Origin*, vol. 1, p. 300.
41. see Montesquieu, *De l'esprit des lois*, 1734. As the century progressed, others, such as Manon Roland, *Letters de Madame Roland*. Nouvelle series, 17671780, ed. Claude Perroud. Paris: Imprimerie Nationale, 1913, 1:215, trans.
by J. Douthwaite, *Homo ferus: Between Monster and Model. Eighteenth-Century Life* 21:2, 1997.
42. *Beytrage*, pp. 334, 339, trans. By Douthwaite, "Homo ferus: Between Monster and Model,". *Eighteenth-Century Life* 21:2, 1997.
43. *Scepter*, 2003. Victor's life was dramatized in Francois Truffaut's film, *l,Enfant Sauvage*, marketed in English as *The Wild Child.*
44. *Autism: Explaining the Enigma*. Blackwell, 1992.
45. www.Feralchildren.com, August 6, 2006.
46. www.autism-society.org, August 6, 2006.
47. www.ourworld,compuserve.com/homepahes/mdenoncourt/page1.htm, August 6, 2006.
48. Brunner-Routledge, 2004.
49. *Nervous Child*, 2, pp.217-250.
50. The Free Press/Simon & Schuster, 1972. New York: Simon and Schuster, 1997.
51. Yale Development Disorders Clinic,
www.med.yale.edu/chldstdy/autism/pddinfl/html, August 9, 2006.
52. Coauthored with Margaret M. Scariano, Warner Books, 1986.
53. Vintage Books, 1995.
54. Vintage Books, 1995.
55. "Creativity and Imagination in Autism and Asperger Syndrome," *Journal of Autism and Developmental Disorders*, August 29, 1999, pp. 319-326.
56. *Journal of Autism and Developmental Disorders*, December 8, 2005, pp. 753-755.
57. "The Neuropsychology of Visual Artistic Production," *Neuropsychologia*, 2004 (42:11), pp. 1568-1583.
58. *Neuropsychiatry, Neuropsychology, and Behavorial Neurology*, January, 2000, pp. 29-38.
59. home.att.net/~ascaris1/genius.html.
60. "Body and Soul," *The Times*, December 1, 2005.
61. *The Siege: A Family's Journey into the World of an Autistic Child*, Little, Brown and Company, 1967, and *Exiting Nirvana: A Daughter's Life with Autism*, Little, Brown and Company, 2001.
62. Harcourt Brace Jovanovich, 1977.
63. Arcade Publishing, 2003.
64. www.answers.com/topic/controversies-in-autism.
65. *The Free Press.*

66. See *Phrenology in connexion with the study of physiognomy.* 2 volumes, London and Edinburgh, 1826; *Phrenology: or the Doctrine of the Mental Phenomena in Two Volumes*, 4th edition, Boston, 1833.
67. For Franz Joseph Gall's seminal work see "*The Anatomy and Physiology of the Nervous System in General, and of the Brain in Particular*"; For discussions of Gall's work see: E. H. Ackerknecht and H. V. Vallois, *Franz Joseph Gall inventory of phrenology and his collection,* Madison: Univ. of Wisconsin Medical School; Department of neurology, Univ. of Illinois, Dr. Franz Gall, http://www.uic.edu/depts/mcne/founders/page0038.html; C. Heeschen, "Franz Josef Gall," *Reader in the history of aphasia,* pp. 1-28, Amsterdam: John Benjamin's Publ. Co., 1994, J. W. Brown and K. L. Chobor, "Phrenological studies of aphasia before Broca: Broca's aphasia or Gall's aphasia?" *Brain and Language, 43, pp. 475-486; Franz Josef Gall;* for comprehensive overviews of the early history of aphasia, see A. L. Benton, "Contributions to aphasia before Broca," *Cortex*, 1, 1964, pp. 314-327; A. L. Benton and R. J. Joynt, "Early descriptions of aphasia," *Archives of Neurology*, 3, 1960, pp. 205-221; R. H. Wozniak, *Mind & Body: Rene Descartes to William James, Ch. 5, http://serendip.brynmawr.edu.*
68. Stanley Finger, *Origins of neuroscience: A History of Explorations into Brain Function*, p. 372.
69. Ackerknecht, E. H., & Vallois, H. V. (1956). *Franz Joseph Gall, inventory of phrenology and his collection*, Madison: Univ. of Madison Medical School.
70. F. J. Gall (1825), cited in "Phrenological studies of aphasia before Broca: Broca's aphasia or Gall's aphasia?" *Brain and Language, 43, pp. 475-486,* J. W. Brown & K. L. Chobor, 1992.
71. "*Traite clinique et physiologique de l'encephalite ou inflammation du cerveau, et de des suites*," Paris, 1825.
72. Dax, M. *Lésions de la moitié gauche de l'encéphale coïncident avec l'oubli des signes de la pensée (lu à Montpellier en* 1836*). Bulletin hebdomadaire de médecine et de chirurgie, 2me série,* 1865, 2, 259-62; Joynt, R.J.; Benton, A.L. *The memoir of Marc Dax on aphasia. Neurology.* 1964 Sep;14:851-4; Critchley, M. *La controverse de Dax et Broca. Revue neurologique*, 1964, 110, 553-57 (English translation in Critchley, *The Divine Banquet of the Brain and Other Essays*, New York: Raven Press, 1979, pp. 72-82); Benton A. "Hemispheric dominance before Broca," *Neuropsychologia*. 1984;22(6):807-11.
73. *The Duality of the Mind.* Longman: London, 1844, p. 4. Some references to Wigan's theories can be found in the second half of the 19th century. C. E. Brown-Sequard even quoted him in *Dual Character of the Brain,*Toner Lecture #2, Smithsonian Misc. Collections #291; Washington, 1877; and in 1886 D. Ferrier wrote "The brain as an organ of motion and sensation, or presentative consciousness, is a single organ composed of two halves; the brain as an organ of ideation, or representative consciousness is a dual organ, each hemisphere complete in itself." (*The Functions of the Brain*, 2nd ed., p. 426, G. P. Putnam's Sons: New York) However, these dual brain theories faded as Broca and others discovered some of the central language functions of the left hemisphere, and

theories of cerebral dominance emerged, reducing the right hemisphere to an almost meaningless position.
74. *The Duality of the Mind.* Longman: London, 1844, p. 26.
75. *The Duality of the Mind.* Longman: London, 1844, p. 40.
76. *The Duality of the Mind.* Longman: London, 1844, p. 41.
77. *The Duality of the Mind.* Longman: London, 1844, p. 42.
78. *The Duality of the Mind.* Longman: London, 1844, p. 271.
79. The Annenberg/CPB Collection, P.O. Box 2345, S. Berlington, VT 05407-2345, 1984. This video presentation has a re-enactment of the case. See Antonio R. Damasio, *Descartes' Error: Emotion, Reason, and the Human Mind.* Putnam, 1994, for a further discussion of the case of Phineas P. Cage.
80. Ratey, John. *A User's Guide to the Brain: Perception, Attention, and the Four Theatres of the Brain*, Pantheon Books: New York, pp. 231-2.
81. *M. Dax soumet au jugement de l'Académie un Mémoire intitulé: `Observations tendant à prouver la coïncidence constante des dérangements de la parole avec une lésion de l'hémisphère gauche du cerveau'." Compt.rend.hebdom.séan.l'Acad Science*, 1863, 56, 536. Roe D., Finger S., "Gustave Dax and his fight for recognition: an overlooked chapter in the early history of cerebral dominance." *J History of Neuroscience*, 1996 Dec;5 (3):228-40.; Finger S.; Roe, D., "Does Gustave Dax Deserve to Be Forgotten? The Temporal Lobe Theory and Other Contributions of an Overlooked Figure in the History of Language and Cerebral Dominance," *Brain and Language*, 1999; 69(1), 16-30.; Finger S, Roe D. "Gustave Dax and the early history of cerebral dominance," *Arch Neurol.* 1996 Aug; 53 (8): 806-13; Cubelli, R, Montagna, C. G., "A reappraisal of the controversy of Dax and Broca," *J Hist Neurosci,* Oct, 1994, pp. 215-26.

According to authors Cubelli and Montagna, the Broca's theory should be renamed: "Probably, Broca was aware of the paper prior to 1865, but he never acknowledged Dax's original theoretical contribution. On the contrary, he always claimed to be the first to espouse the theory of left hemisphere dominance for language and never quoted Marc Dax (Broca, 1877 p 536), *I do not like dealing with the questions of priority concerning myself. That is the reason why I did not mention the name of Dax in my paper.* In our opinion, the weight of evidence reported here suggests that the theory of the left hemisphere dominance for speech must be attributed equally to Dax and Broca, and henceforth should be called *the theory of Dax-Broca*."
82. For political and religious implications, see L. S. Jacyna, "The physiology of mind, the unity of nature, and the moral order in Victorian thought," *British Journal of the History of Science*, 14, pp. 109-131; for a discussion of philosophic considerations, see M. Hammond, "Anthropology as a weapon of social combat in late nineteenth century France," *Journal of the History of the Behavioral Sciences*, 16, pp. 118-132, 1980.
83. cited in M. Critchley, *Aphasiology and Other Aspects of Language*, London: Edward Arnold, 1970; see also P. Broca, "*Remarques sur le siege da la faculte du language articule, suive d'une observation d'aphemie*," 1861; trans. By J. Kann, *Journal of Speech and Hearing Disorders*, 15, pp. 16-20, 1950.

84. L. S. Jacyna, 1981, "The physiology of mind, the unity of nature, and the moral order in Victorian thought," *British Journal of the History of Science*, 14, 109-132.
85. M. Hammond, 1980, "Anthropology as a weapon of social combat in late nineteenth century France, *Journal of the History of the Behavioral Sciences*, 16, pp. 118-132).
86. Broca, 1861, Ch 2, "*Remarques sur le siege de la faculte du langage articule, suivies d'uine observation d' aphemie (perte de la parole)." Bulletins de la Societe Anatomique* 36, 330-357.
87. Pierre, Marie, "*Revision de la question de l'aphasie: l'aphasie de 1861 a 1866; essai de critique historiaue sur la genese de la doctrine de Broca.* Semaine Medicale, pp. 565-571.
88. Harrington, 1987, p. 42, *Medicine, Mind, and the Double Brain: A Study in Nineteenth Century Thought.* Princeton, NJ: Princeton Univ. Press, 1987. This book contains a lengthy discussion of the surrounding social, religious, and political factors. For a less detailed discussion see "Unfinished Business: Models of Laterality in the Nineteenth Century," *Brain Asymmetry*, ed. Richard J. Davidson and Kenneth Hugdahl, MIT Press, 1995. Actually, the overlap of neurology and the occult is still with us, and libraries include in their neurology sections books that might better be placed under less scientific categories, including books that mix neurology and telepathy, extra-sensory-perception, channeling, and various other spiritual or religion phenomena. I recently serendipitously picked one up in my library's neurology section titled *A Child of Eternity: An Extraordinary Young Girl's Message from the World Beyond* (Andriana Rocha and Kristi Jorde, Ballantine Books: New York, 1995). It is a well written account of a mother attempting to come to terms with the fact that her daughter is autistic, and has a convincing discussion of the slow discovery of the daughter's condition, and the search through the world of neurology and other fields for answers. It ends up leading the readers into the world of Facilitated Communication, where an autistic child rests his or her hand on the arm of a facilitator, who then, supposedly, can feel the arm being directed to spell out words on a computer, and the blocked communication from speech is opened through this. There is some evidence that the arm and hand can connect up to parts of the brain the mouth cannot. However, in this case, this highly doubtful form of communication ends up with the autistic child being able to travel through time, to communicate with famous people from the past, and so on, to in fact, have a higher consciousness than other humans. As I said, it is well written, and, of course, we'd like for such mystical communications to be possible, but it is beyond science, and, in truth, is something other than neurology. All that it really tells us is that the mysteries of the mind/brain connection are still mysteries.
89. Sacks, Oliver. *The Man who Mistook his Wife for a Hat and other Clinical Tales*. Simon & Schuster, 1998, p. 3.
90. Jackson, J. H., *Selected Writings of John Hughlings Jackson,* London; Staples Press, *Vol I: On Epilepsy and Epileptiform Convulsions, and Vol II: Evolution and Dissolution of the Nervous System Speech; Various Papers, Addresses and Lectures,* 1931, ed. J. Taylor; rpr., New York: Basic Books, 1958, p. 220. Jackson can be considered the true founder of neurology as a science. When he first came

to London in 1859, at the age of 25, he was already leaning towards a study of diseases of the nervous system, especially convulsions of a limited range, and his early work in the hospitals exposed him to many such cases. What is important is that he was able to take the clinical studies and suggest larger philosophical implications, including suggestions that the brain worked asymmetrically.

91. Ibid, p. 148.

92. See C. Wernicke, *The Symptoms of Complex Aphasia*, 1874; trans. A. Church *Disorders of the Nervous System*, Appleton-Century-Crofts.

93. *Dual Character of the Brain*. Tonber Lecture #2, Smithsonian Misc. Coll. #291, Washington, 1877.

94. *Sir Victor Horsley*, biography by S. Paget, Constable: London, 1919; Victor Horsley, *Functions of the Marginal Convolutions* (1884) and, as a co-author, *Experiments upon the Functions of the Cerebral Cortex* (1888).

95. Sally P. Springer and Georg Deutsch, *Left Brain, Right Brain*, 4th ed., W. H. Freeman and Company, 1993, pp. 12-13.

96. *Die Medizin der Gegenwart in Selbstdarstellungen*, 1925; trans., *The Freud Reader*, Ed., Peter Gay, New York: Norton, 1995, p. 10.

97. Sacks, Oliver. *The Man who mistook his Wife for a Hat and other Clinical Tales*, New York, 1998, p. 3.

98. *The Functions of the Brain*. 2nd ed. p. 426, Putnam: New York, 1886.

99. F. Goltz, "On the Functions of the Hemispheres"; trans. G. von Bonin, *Some Papers on the Cerebral Cortex,* Springfield, Ill; C. C. Thomas, 1960, p. 118.

100. F. Goltz, "On the Functions of the Hemispheres"; trans. G. von Bonin, *Some Papers on the Cerebral Cortex,* Springfield, Ill; C. C. Thomas, 1960, p. 130.

101. F. Goltz, "On the Functions of the Hemispheres"; trans. G. von Bonin, *Some Papers on the Cerebral Cortex,* Springfield, Ill; C. C. Thomas, 1960, p. 158.

102. For early discussions of left hemisphere localization see H. Monk, "On the Functions of the Cortex," trans, G. von Bonin, *Some Papers on the Cerebral Cortex*, C. C. Thomas: Springfield, 1960; D. Denny-Brown, "The Psychological Basis of Perception and Speech," in *Problems of Dynamic Neurology*, Hebrew Univ., Hadassah Medical School: Jerusalem, 1963. Joseph E. Bogen offers a good summary of the dominance of this view from the late 1800s through the mid 1900s in "The Other Side of the Brain II: An Appositional Mind," *Bulletin of the Los Angeles Neurological Societies,* Vol. 34, July, 1969, Number 3, pp. 135-137.

103. Freud, Sigmund. *The interpretation of Dreams*, originally published as *Die Traumdeutung*, 1899, and subsequently revised by Freud and reprinted several times through 1932; trans., James Strachey, Avon Books: New York,1965.

104. Rpt., New York, Kelley, 1971.

105. The lectures were reproduced in *On Some Mental Affections of Childhood and Youth* by J. L. Down, London: Churchill, 1887.

106. A. Binet, *Psychologie des Grands Calculateurs et Joueurs d'echecs*, Hachette, Paris, 1894.

107. W. W. Ireland, Philadelphia: P. Blakiston's Son & Co., 1891, rpt. 1900.

108. *Mental Deficiency*, New York: William Wood, 1914. E. Sequin, *Idiocy and its Treatment by the Physiological Method*, 1866; rpt. New York: Kelley, 1971, and F. Sano, "James Henry Pullen, the Genius of Earlswood," *Journal of Mental*

Science 64:251-67, 1918, (who actually did a postmortem mapping of Pullen's brain) also wrote about him, and concluded, contrary to Tredgold, that the mental condition was biologically based.
109. Henry Head, *Studies in Neurology*, 2 vols. Oxford, and *Aphasia and Kindred Disorders of Speech*. 2 vols. Cambridge, 1926; Henry Head had previously discussed John Hughlings Jackson's theories of speech in *Brain*, vol. xxxviii, 1915, p.1, and vol. xvi, 1923, p. 355.
110. Henry Head, *Studies in Neurology*, 2 vols. Oxford, and *Aphasia and Kindred Disorders of Speech*, 2 volumes, Cambridge, 1926, p. 130.
111. Henry Head, *Studies in Neurology*, 2 vols. Oxford, and *Aphasia and Kindred Disorders of Speech*, 2 volumes, Cambridge, 1926, p. 42.
112. Henry Head, *Studies in Neurology*, 2 vols. Oxford, and *Aphasia and Kindred Disorders of Speech*, 2 volumes, Cambridge, 1926, p. 130.
113. S. E. Henschen, "On the function of the right hemisphere of the brain in relation to the left in speech, music and calculation," *Brain*, 49:110-123. 1926.
114. *Diseases of the Nervous System*, 6th ed., p. 81 Oxford Univ.: London, 1962.
115. *Neurology*, 5th ed., p. 621. Thomas: Springfield, Ill., 1959.
116. *Some Anatomical Aspects of Aphasia*. Lancet, June 1, 1963, p. 1173-1177.
117. *Bulletin of the Los Angeles Neurological Society*, there is an article titled "The Other Side of The Brain II: An Appositional Mind," Vol. 34, No 3, July, 1969, pp. 1135-1161. Von Stauffenberg's 1913 case comes from W. von Stauffenbery, "Uber Seelenblindheit. Arbeit. A. D. Hirnanatom Institute, Zurich. 8:1-212, 1913; cited by Henry Head, *Aphasia and Kindred Disorders of Speech*, Vol. 1. Hafner: New York, 1963, p. 106; originally publ., 1926; the Head quote from (footnote 59) p. 116.
118. *The Psychology of a Musical Prodigy*; rpt., Books for Libraries Press: Freeport, N. Y., 1970.
119. E. Anderson, *The Letters of Mozart and his Family*, London: Macmillan, 1966.
120. John A. Sloboda refers to this study as the most complete and accurate account given prior to his own study of a musical idiot savant (NP) in 1985; "An Exceptional Musical Memory," J.A. Sloboda, B. Hermelin, and N. O'Connor, *Music Perception*, 3:155-170, 1985.
121. Adhemar Gelb and Kurt Goldstein*, 1918, Zur Psychologie des optischen Wahrnehmungs-und Erkennungsvorganges (Psychologische Analyse hirnpathologischer Falle auf Grund von Untersuchungen Hirnverletzter) I, Zeitschrift f. d. ges. Neurol. U. Physiol.*, 41: 1-143; rpt. in *Sourcebook of Gestalt Psychology*, trans., by W. D. Ellis, 315-25, London: Routledge & Kegan Paul Ltd., 1938. See also: Kurt Goldstein, *Der Aufbaudes OrGanismus,* 1934; trans., *The Organism: A Holistic Approach to Biology Derived from Pathological Date in Man*, New York, 1939.
122. A. R. Luria, "Kurt Goldstein and Neuropsychology," *Neuropsychologia*, 4: pp. 311-313, 1966.
123. K. Goldstein, "*Das Symptom, seine Entstehung und Bedeutung fur unsere Auffassung vom Bau und von der Funktion des Nervensystems,*" *Arch. Psych.,* Nervenkrank, 76:84-108, 1926.

124. T. Weisenbury and K. E. McBride, "Aphasia: A Clinical and Psychological Study," Commonwealth Fund; New York, 1935.
125. T. Alajouanine, "Aphasia and Artistic Realization," *Brain* 71 (Part 3, Sept, 1948, pp. 229-241). For other studies on the right hemisphere and music, see A. Gates and J. Bradshaw, "The Role of the Cerebral Hemispheres in Music," *Brain and Language* 4 (1977), pp. 403-431; S. Blumstein and W. E. Cooper, "Hemispheric Processing of Intonational Contours," *Cortex* 10 (1974), pp.146-158; F. Boller, M. Cole, P. B. Vrtunski, M. Patterson, and V. Kim,"Paralinguistic Aspects of Auditory Comprehension in aphasia," *Brain and Language* 9 (1979), pp. 164-174; D. Breitling, W. Guenther, and P. Randot, "Auditory Perception of Music Measured by Brain Electrical Activity Mapping," *Neuropsychologia* 25 (1987), pp. 765-774; M. P. Bryden, R. G. Ley, and J. H. Sugarman, "A Left-Ear Advantage for Identifying the Emotional Quality of Tonal Sequences," *Neuropsychologia* 20 (1982), pp. 83-87; A. Carmon and I. Nachshon, "Ear Asymmetry in Perception of Emotional Non-Verbal Stimuli," *Acta Psychologica* 37 (1973), pp. 351-357; H. W. Gardan, "Hemispheric Asymmetries in the Perception of Musical Chords," *Cortex* 6 (1970), pp. 387-398; H. R. McFarland and D. Fortin, "Amusia due to Right Temporal-Parietal Infarct," *Archives of Neurology* 39 (1982), pp. 725-727; K. M. Heilman, D. Bowers, L. Speedie, and H. B. Coslett, "Comprehension of Affective and Noneffective Prosody," *Neurology* 34 (19834), pp. 917-921; K. Heilman, R. Scholes, and R. T. Watson, "Auditory Affective Agnosia," *Journal of Neurology, Neurosurgery and Psychiatry.* 38 (1975), pp. 69-72; F. L. King and D. Kimura, "Left Ear Superiority in Dichotic Perception of Nonverbal Sounds in Boys and Girls," *Neuropsychologia* 8 (1970), pp. 227-237.
126. Since Ravel was such an important composer, there are many discussions of his life and music. Brief overviews can be found in Milton Cross' *Encyclopedia of the Great Composers and Their Music*, Milton Cross and David Ewen, New Revised Ed., Vol II, Doubleday: New York, 1953, pp. 609-621; and *The Lives of the Great Composers*, Harold C. Schonberg, Norton: New York, 1997, who wrote: "In 1932, perhaps as a result of an automobile accident, perhaps from a disease more deeply functional, Ravel had a nervous breakdown. His injuries from the crash appeared to be superficial, and Ravel made light of them: 'a few cracks, and arched nose to persuade the Americans of my Hebrew origin, but particularly some bruises on my chest that force me to cough in a crooked way.' The following year, 1933, he began to lose control of his arms and legs. This was followed by memory loss and an inability to coordinate. Though his mind remained normal, he could not compose or play the piano. In 1937 he underwent brain surgery, from which he never recovered. His malady has been kept secret."

Oliver Sacks suggests Ravel might have had Pick's disease, "a degenerative brain disorder more commonly affecting the frontal and temporal lobes." *The New Yorker*, Oct 7, 2002, p. 68.
127. Worthwhile readings on aphasia include: R. Chapey, ed., 2001. *Language Intervention Strategies in Aphasia and Related Neurogenic Communication Disorders* (Fourth Edition), Philadelphia: Lippincott, Williams & Wilkins; Goodglass, H. & Kaplan, E., 1972. *Assessment of Aphasia and Related Disorders*,

Philadelphia: Lea and Febinger; Kay, J., Lesser, R., & Coltheart, M., 1992, *Psycholinguistic Assessments of Language Processing in Aphasia (PALPA)*. Hove, Erlbaum; Spreen, O. & Risser, A.H. (2003). *Assessment of Aphasia*, New York: Oxford University Press; Delis, D. C., Kramer, J. H., & Kaplan, E., 2001. *The Delis-Kaplan Executive Function System,* San Antonio, TX: The Psychological Corporation; Armengol, C., Kaplan, E., & Moes, E. (Eds., 2001. *The consumer oriented neuropsychological report.* Odessa, FL: Psychological Assessment Resources; Kaplan, E. (2002). *Serendipity in science: A personal account.* In T. Stringer, E. Cooley, & A.L. Christensen, Eds. *Pathways to prominence in neuropsychology: Reflections of twentieth century pioneers*. New York: Psychology Press.

128. Tim Shallice, *From Neuropsychology to Mental Structure*, Cambridge Univ. Press, 1988, details out the history and theory underlying this approach up to the date of his publication.

129. Lynn Solotaroff, New York, 1972. It is a detailed journal of a man's attempt to overcome his damaged brain and give meaning and value to his life, with numerous interspersed comments by Luria attempting to explain the neurological implications. Oliver Sacks considers Luria's works (especially this case study) to contain the greatest treasures of all of neurological thought; Luria's other works offer a more scientific discussion of the field of neurology. Some of the more important are: *The Mind of a Mnemonist*, New York, 1968; *Higher Cortical Functions in Man*, 2nd ed., New York, 1980, and the condensed version, *The Working Brain*, Harmondsworth, 1973.

130. *Johnny Got His Gun,* Bantam: New York, 1939. Dalton Trumbo directed a film of the novel in 1971 for which he received International Film Critics and Cannes Film Festival awards. Later, Metallica wrote and performed *One,* a song based on the film, and a video that incorporated clips from the film, Elektra Records, New York, 1989, pp. 160-162.

131. *Johnny Got His Gun,* Bantam: New York, 1939, pp. 221-222.

132. *Johnny Got His Gun,* Bantam: New York, 1939.

133. Dr. R. Joseph, *The Right Brain and the Unconscious: Discovering the Stranger Within*, New York: Plenum Press, 1992, p.104. The only one of these Joseph includes a source for is J. E. Bogen, "The Callosal Syndrome," *Clinical Neuropsychology*, K. M. Heilman and E. Valenstein, Eds. (pp. 308-358), New York: Oxford Univ. Press, 1979.

134. A. J. Akelaitis, "Studies on the Corpus Callosum: 4, Diagnostic Dyspraxia in Epileptics Following Partial and Complete Section of the Corpus Callosum,"*American Journal of Psychiatry* 101 (1945), 594-599.

135. R. E. Myers and Roger Sperry, "Interocular transfer of a visual form discrimination habit in cats after section of the optic chiasma and corpus callosum," *Anatomical Record*, 115: 351-352, 1953.

136. Here are a few of the actual clinical publications from the fifties (listed by date of publication): R. E. Myers, "Interocular transfer of pattern discrimination in cats following section of crossed optic fibers," *Journal of Comparative Physol. Psychology*, 48, pp. 470-473, 1955; R. E. Myers, "Neural basis of bilateral perceptual integration," *Science*, 122: pp. 877, 1955; R. E. Myers, "The corpus

callosum and hemispheric interaction," Ph. D. Thesis, Univ. of Chicago, 1955; Wilder Penfield, "Mechanisms of voluntary movement," *Brain*, 77, pp 1-17, 1955; R. E. Myers, "Function of corpus callosum in interocular transfer," *Brain*, 79, pp. 358-363, 1956; R. W. Sperry, "Experiments on perceptual integration in animals," *Psychait. Res. Rep.*, 6, pp. 151-159, 1956; R. W. Sperry, J. S. Stamm, and Nancy Miner, "Relearning tests for interocular transfer following division of optic chiasma and corpus callosum," *Journal of Comp. Physiol. Psychol.*, 49, pp. 529-533, 1956; R. W. Sperry and J. S. Stamm, "Function of corpus callosum in contralateral transfer of somesthetic discrimination in cats," *Journal of Comp. Physiol. Psychol.*, 50, pp. 138-143, 1957; R. W. Sperry, "High-order integrative functions in surgically isolated somatic cortex in cat," *Anat. Rec.*, 127, p. 371, 1957; R.E. Myers and R. W. Sperry, "Interhemispheric Communication Through the Corpus Callosum, Mnemonic Carry-Over Between the Hemispheres," *Archives of Neurology and Psychiatry* 80, 1958, pp. 298-303; R. W. Sperry, J. S. Stamm, and Nancy Miner, "Relearning tests for interocular transfer following division of optic chiasma and corpus callosum," *Journal of Comparative Physiol. Psychology*, 49, 1959, pp. 529-533; R. J. White, L. H. Schreiner, R. A. Hughes, C. S. MacCarty, and J. H. Grindlay, "Physiologic consequences of total hemispherectomy in the monkey," *Neurology*, 9, pp. 149-159, 1959.

137. "Physiological psychology," *Annual Review of Psychology*, 6, pp. 267- 296, 1955.

138. Here are some of the lab reports coming out of the sixties, listed by date: J. E. Bogen and B. Campbell, "Total hemispherestomy in the cat," *Surg. Forum*, 11: 381-383, 1960; R. Sperry, "Cerebral organization," *Science*, 1331749-1757, 1961; R. Sperry, "Some developments in brain lesion studies of learning," *Fed. Proc.*, 20:609-616, 1961; M. S. Gazzaniga, J. E. Bogen, and R. W. Sperry, "Some functional effects of sectioning the cerebral commissures in man," Proc. *Nat. Acad. Sci.*, 48: 1765-9, 1962; J. E. Bogen and P. J. Vogel, "Treatment of generalized seizures by cerebral commissurotomy," *Surg. Forum*, 14: 431-433, 1963; M. S. Gazzaniga, J. E. Bogen, and R. W. Sperry, "Laterality effects in somesthesis following cerebral commissurotomy in man," *Neuropsychologia* , 1:209-215, 1963; R. W. Sperry, "The great cerebral commissure," *Sci Amer.*, 210:42-52, 1964; R. W. Sperry, "Problems outstanding in the evolution of brain function," James Arthur Lecture, American Museum of Natural History; New York, 1964; J. E. Bogen, E. D. Fisher, and P. J. Vogel, "Cerebral commissurotomy: A second case report," *Journal of the American Medical Association.*, 194: 1328-1329, 1965; R. E. Myers, "The Neocortical Commissures and Interhemispheric Transmission of Information," *Functions of the Corpus Callosum*. Ciba Fndn. Study Group No. 20. Ettlinger, E. G., Reuck, A. V. S., and Porter, R., eds., Little, Brown and Co.: Boston, 1965; M. S. Gazzaniga, J. E. Bogen, and R. W. Sperry, "Observations on visual perception after disconnection of the cerebral hemispheres in man," *Brain*, 88:221-236, 1965; J. E. Bogen, M. S. Gazzaniga, "Cerebral commissurotomy in man: Minor hemisphere dominance for certain visuospatial functions," *Journal of Neurosurgery.*, 23: 394-399, 1965; R. W. Sperry, "Brain bisection and mechanisms of consciousness," pp. 442-468, *Semaine d'Etude Sur Cerveau et Experience Consciente.*, J. Eccles, ed., Pontificae Acad. Scient. Scripta varia, Rome, 1965; R.

W. Sperry and M. S. Gazzaniga, "Language Following Surgical Disconnection of the Hemispheres," *Brain Mechanisms Underlying Speech and Language*, ed., F. L. Darley, Grune and Stratton: New York, 1966; M. S. Gazzaniga and R. W. Sperry, "Simultaneous double discrimination response following brain bisection," *Psychon. Sci.*, 4:261-262, 1966; M. S. Gazzaniga, and E. D. Young, "Effects of commissurotomy on the processing of increasing visual information." *Experimental Brain Research*, 3: 368-371, 1967; M. S. Gazzaniga and R. W. Sperry, "Language after section of the cerebral commissures," *Brain,* 90:131-148, 1967; M. Gazzaniga, "The split-brain in man," *Scientific American*, 217:24-29, 1967; M. S. Gazzaniga, J. E. Bogen, and R. W. Sperry, "Dyspraxia following division of the cerebral commissures," *Arch. Neurol.*, 16:606-612, 1967; M. S. Gazzaniga, "Short-term memory and brain-bisected man," *Psychon. Sci.*, 12:161-162, 1968; S. R. Butler and U. Norrsell, "Bocalization possibly indicated by the minor hemisphere," *Nature*, 220:793-794, 1968; R. W. Sperry, "Hemisphere deconnection and unity in conscious awareness," *American Psychologist,* 23:723-733, 1968; R. W. Sperry, "Mental unity following surgical disconnection of the cerebral hemispheres,"Harvey Lectures, Series 62, N. Y.; *Academic Press*, 1968; B. Milner, L. Taylor, and R. W. Sperry, "Lateralized suppression of dichotically presented digits after commissural section in man," *Science*, 161:184-6, 1968; J. Levy-Agresti, "Ipsilateral projection systems and minor hemisphere function in man after neocommissurotomy," *Anatomy Record*, 160:3844, 1968; J. E. Bogen, R. W. Sperry, and P. J. Vogel, "Commissural Section and the Propagation of Seizures," *Basic Mechanisms of the Epilepsies*, H. H. Jaspher, A. A. Ward, and A. Pope, eds., Little, Brown: Boston, 1969; J. E. Bogen, "The corpus callosum, the other side of the brain, and pharmacologic opportunity," *Drugs and Cerebral Function*, W. L. Smith, ed., Thomas: Springfield, Ill, 1969; Here are some of the reports from the 1970s: Michael S. Gazzaniga, *The Bisected Brain*, Appleton-Century-Crofts, New York, 1970 (an overview of the findings to this date by one of those involved in the research); M. Kinsbourne, "The cerebral basis of lateral asymmetries in attention," *Acta Psychologica*, vol. 33, pp. 193-201, 1970; J. Levy, C. Trevarthen, and R. W. Sperry, "Perception of Bilateral Chimeric Figures Following Hemispheric Deconnection," *Brain* 95, pp. 61-78, 1972; R. Sperry, "Lateral Specialization in the Surgically Separated Hemispheres," *The Nurosciences: Third Study Program*, F. O. Schmitt and F. G. Worldern, Eds., Cambridge: MIT Press, pp. 1-12, 1974; M. Kinsbourne, "The mechanism of hemispheric control of the lateral gradient of attention," *Attention and Performance*, vol. 5, ed. P.M.A. Rabbitt and S. Dornic, New York: Academic Press, 1975; J. Levy and C. Trevarthen, "Metacontrol of Hemispheric Function in Human Split-Brain Patients," *Journal of Experimental Psychology: Human Perception and Performance 2*, pp. 299-312, 1976; H. A. Whitaker and G. Ojemann, "Lateralization of higher cortical functions: A critique," *Annals of the New York Academy of Science*, vol. 299, pp. 459-473, 1977; J.E. Bogen, "The Callosal Syndrome," *Clinical Neuropsychology*, K. M. Heilman and E. Valenstein, Eds., New York: Oxford Press, pp. 308-358, 1979; J. Levy, "Language, Cognition, and the Right Hemisphere," *American Psychologist* 38, pp. 538-541, 1983.

139. C. B. Trevarthen, "Double visual learning in split-brain monkeys," *Science*, 136:258-259, 1962.
140. M. S. Gazzaniga and R. W. Sperry, "Simultaneous double discrimination response following brain bisection," *Psychon. Sci.*, 4: 261-262, 1966.
141. See "Interhemispheric relationships: the neocortical commissures, syndromes of hemispheric disconnection," *Handbook of Clinical Neurology, Vol. 4, Disorders of Speech, Perception and Symbolic Behavior*, ed. P. J. Vinken and G. W. Bruyn. Amsterdam: Elsevier/North Holland Biomedical Press, pp. 145-153, 1969; see "Brain Bisection and Consciousness," *Brain and Conscious Experience*, ed. J. Eccles, New York: Springer-Verlag, 1966; see "Perception in the Absence of the Neocortical Commissures, Perception and its Disorders," *Research Publication of the Association for Research in Nervous and Mental Diseases*, Vol. 48, 1970.
142. J. E. Bogen, "The Other Side of the Brain," p. 157-8, *Bulletin of the Los Angeles Neurological Societies* 34, pp. 135-162, 1969; Bogen also includes a list of famous people's dual terms for the two forms of thought, which I list here in paragraph form: C. S. Smith, Atomistic/Gross; Price, Analytic or reductionist/Synthetic or concrete; Wilder, Numerical/Geometric; Head, Symbolic or systematic/Perceptual or non-verbal; Goldstein, Abstract/Concrete; Rluesch, Digital or discursive/Analogic or eidetic; Bateson & Jackson, Digital/Analogic; J. Z. Young, Abstract/Map-like; Pribram, Digital/Analogic; W. James, Differential/Existential; Spearman, Eduction of relations/Eduction of correlates; Hobbes, Directed/Primary process; Pavlov, Second signaling/First signaling; Freud, Secondary process/Primary process; Sechenov (Luria), Successive/ Simultaneous; Levi-Strauss, Positive/Mythic; Bruner, Rational/Metaphoric; Akhilinanda, Buddhi/Manis, Radhakrishnan, Rational/Integral.
143. See "Lateral Specialization of Cognitive Mode: An EEG Study," *Psychophysiology*, Vol. 9, pp. 412-418, 1972; see *The Psychology of Consciousness*, 2nd ed., New York: Harcourt Brace Jovanovich, 1977; see "The Split and Whole Brain," *Human Nature 1*, 1978
144. Roger Sperry, "Brain Bisection and the Neurology of Consciousness," *Brain and Conscious Experience*, J. C. Eccles, Ed. (pp. 298-313) (New York: Springer Verlag, 1978.
145. See W. Penfield and L. Roberts, *Speech and Brain Mechanisms*, Princeton, N.J.: Princeton Univ. Press, 1959; see G. A. Ojemann, "Brain Organization for Language from the Perspective of Electrical Stimulation Mapping," *The Behavioral and Brain Sciences 6*, 1983, pp. 235-238; see A. S. Gevins, "Brain potential (BP) evidence for lateralization of higher cognitive functions," *Cerebral Hemisphere Asymmetry: Method, Theory and Application*, J. B. Hellige, New York: Praeger, 1983.
146. See J. Risberg, J. H. Halsey, E. L. Wills, and B. M. Wilson, "Hemispheric specialization in normal man studied by bilateral measurements of the regional cerebral blood flow," *Brain*, 98, pp. 511-524, 1975; M. I. Posner, S. E. Peterson, P. T. Fox, and M. E. Raichle, "Localization of cognitive operations in the human brain," *Science*, 240, pp. 1627-1631, 1988.
147. For further discussion, see F. Wood, "Laterality of cerebral function: Its investigation by measurement of localized brain activity," *Cerebral Hemisphere*

Asymmetry: Method, Theory, and Application, ed. J. B. Hellige, pp. 383-410, New York: Praeger, 1983; J. L. Bradshaw, *Hemispheric Specialization and Psychological Function*, Chichester, England: Wiley, 1989; F. B. Wood, D. L. Flowers, and E. E. Naylor, "Cerebral laterality in functional neuroimaging," *Cerebral Laterality: Theory and Research*, ed. F. L. Kitterle, pp. 103-116, Hillsdale, N. J.: Erlbaum, 1991.

148. See J. Risberg, J. H. Halsey, E. L. Wills, and B. M. Wilson, "Hemispheric specialization in normal man studied by bilateral measurements of the regional cerebral blood flow," *Brain*, 98, pp. 511-524, 1975; M. I. Posner, s. e. Peterson, P. T. Fox, and M. E. Raichle, "Localization of cognitive operations in the human brain," *Science*, 240, pp. 1627-1631, 1988.

149. See H. Hecaen and M. Albert, *Human Neuropsychology*, New York: Wiley, 1978.

150. For more discussion of the studies from 1980-1990, see: E. Zaidel and A. M. Peters, "Phonological encoding and ideographic reading by the disconnected right hemisphere; Two case studies," *Brain and Language*, vol 14, pp. 205-234, 1981; J. G. Beaumont, *Divided Visual Field Studies of Cerebral Organization*, New York: Academic Press, 1982; M. P. Bryden, *Laterality: Functional Asymmetry in the Intact Brain*, New York: Academic Press, 1982; J. J. Sidtis and M. S. Gazzaniga, "Competence versus performance after callosal section: Looks can be deceiving," *Cerebral Hemisphere Asymmetry: Method, Theory and Application*, Hellige, pp. 152-176, 1983; E. Zaidel, "Disconnection syndrome as a model for laterality effects in the normal brain," *Cerebral Hemisphere Asymmetry: Method, Theory, and Application*, ed. J. B. Hellige, New York: Praeger, pp. 95-151, 1983; M. Kinsbourne and M. Hiscock, "Asymmetries of dual-task performance," *Cerebral Hemisphere Asymmetry: Method, Theory and Application*, Hellige, New York: Praeger, pp. 255-334, 1983; M. P. Bryden, "Handedness and cerebral organization: Data from clinical and normal populations," *Duality and Unity of the Brain: Unified Functioning and Specialization of the Hemispheres*, ed. D. Ottoson, vol. 55-70, London: Macmillan, 1987; *Duality and Unity of the Brain*, ed. D. Ottoson, London: Macmillan, pp. 454-465, 1987; M. W. O'Boyle, F. van Wyhe-Lawler, and D. A. Miller, "Recognition of letters traced in the right and left palms: Evidence for a process-oriented tactile asymmetry," *Brain and Cognition*, vol. 6, pp. 474-494, 1987; K. Hugdahl, *Handbook of Dichotic Listening: Theory, Methods and Research*, New York: Wiley, 1988; J. B. Hellige and D. W. Kee, "Asymmetric manual interference as an indicator of lateralized brain function," *Cerebral Control of Speech and Limb Movements*, ed. G. R. Hammond, Amsterdam: North-Holland, pp. 635-660, 1990; R. Efron, *The Decline and Fall of hemispheric Specialization*, Hillsdale, N. J.:Erlbaum, 1990; A. Bouma, *Lateral Asymmetries and Hemispheric Specialization: Theoretical Models and Research*, Amsterdam: Swets and Zeitlinger, 1990.

151. "A left-ear advantage for identifying the emotional quality of tonal sequences," *Neuropsychologica,* 20, 1982, pp. 83-87.

152. Cited in A. I. Benton and R. J. Joynt, "Early Descriptions of Aphasia," *Archives of Neurology* 3, 1960, pp. 205-222.

For other studies on the right hemisphere and music, see: M. L. Albert, R. Sparks, & N. Helm, "Melodic intonation therapy for aphasia," *Archives of Neurology*, 29, 1973, pp. 334-339; S. Blumstein and W. E. Cooper, "Hemispheric Processing of Intonational Contours," *Cortex* 10 (1974), pp.146-158; F. Boller, M. Cole, P. B. Vrtunski, M. Patterson, and V. Kim, "Paralinguistic Aspects of Auditory Comprehension in aphasia," *Brain and Language 9* (1979), pp. 164-174; D. Breitling, W. Guenther, and P. Randot, "Auditory Perception of Music Measured by Brain Electrical Activity Mapping," *Neuropsychologia 25* (1987), pp. 765-774; M. P. Bryden, R. G. Ley, and J. H. Sugarman, "A Left-Ear Advantage for Identifying the Emotional Quality of Tonal Sequences," *Neuropsychologia 20* (1982), pp. 83-87; A. Carmon and I. Nachshon, "Ear Asymmetry in Perception of Emotional Non-Verbal Stimuli," *Acta Psychologica 37* (1973), pp. 351-357; F. K. Curry, "A comparison of left-handed subjects on verbal and non-verbal dichotic listening tasks," *Cortex*, 3, 1967, pp. 343-352; R. Day, J. E. Cutting, & P. Copeland, "Perception of linguistic and non-linguistic dimensions of dichotic stimuli," *Status Report of Haskins Laboratories*, 27, 1971, pp. 1-6; W. Freeman & J. M. Williams, "Hallucinations in Braille," *Archives of Neurology and Psychiatry*," 70, 1953, pp. 630-634; H. Gardner, *The Shattered Mind*, New York: Vintage, 1975; A. Gates and J. Bradshaw, "The Role of the Cerebral Hemispheres in Music," *Brain and Languag"e 4* (1977), pp. 403-431; K. Goldstein, "After effects of brain injuries in war," Orlando, Fl: Grune & Stratton,1942; H. W. Gordan, "Hemispheric Asymmetries in the Perception of Musical Chords," *Cortex 6* (1970), pp. 387-398; H. W. Gordan, & J. E. Bogen, "Hemispheric lateralization of singing after intracarotid sodium amylobarbitone," *Journal of Neurology, Neurosurgery and Psychiatry 37*, 1987, pp. 727-737; K. M. Heilman, D. Bowers, L. Speedie, and H. B. Coslett, "Comprehension of Affective and Noneffective Prosody," *Neurology 34* (19834), pp. 917-921; K. Heilman, R. Scholes, and R. T. Watson, "Auditory Affective Agnosia," *Journal of Neurology, Neurosurgery and Psychiatry 38* (1975), pp. 69-72; N. Helm-Estabrooks, "Exploring the right hemisphere for language rehabilitation: Melodic intonation therapy," *Cognitive Processing in the Right Hemisphere*, ed., E. Perecman, Orlando, Fl: Academic Press, 1983, pp. 165-190; R. Joseph, "The neuropsychology of development: Hemispheric laterality, limbic language and the origin of thought," *Journal of Clinical Psychology*, 38, 1982, pp. 4-33; D. Kimura, "Left-right differences in the perception of melodies," *Quarterly Journal of Psychology 16*, 1964, pp. 355-358; C. Knox & D. Kimura, "Cerebral processing of nonverbal sounds in boys and girls," *Neuropsychologia, 8*, 1970, pp. 227-237; F. L. King and D. Kimura, "Left Ear Superiority in Dichotic Perception of Nonverbal Sounds in Boys and Girls," *Neuropsychologia 8* (1970), pp. 227-237. H. R. McFarland and D. Fortin, "Amusia due to Right Temporal-Parietal Infarct," *Archives of Neurology 39* (1982), pp. 725-727; B. Milner, "Laterality effect in audition," *Interhemispheric Relations and Cerebral Dominance*, ed. V. Mountcastle, Baltimore: John Hopkins Press, 1962, pp. 173-201; D. Molfese, R. B. Freeman, & D. S. Palermo, "The ontogeny of brain lateralization for speech and nonspeech stimuli," *Brain and Language*, 2, 1975, pp. 356-368; D. M. Piazza, "The influence of sex and handedness in the hemispheric specialization of verbal and nonverbal tasks," *Neuropsychologia*, 18,

1980, pp. 163-176; H. H. Reese, "The relation of music to diseases of the brain," *Occupational Therapy and Rehabilitation*, 27, 1948, pp 12-18; S. J. Segalowitz and P. Plantery, "Music draws attention to the left and speech draws attention to the right," *Brain and Cognition,* 4, 1985, pp. 1-6; A. Smith, "Speech and other functions after left (dominant) hemispherectomy," *Journal of Neurology, Neurosurgery, and Psychiatry* 29, 1966, pp. 467-471; A. Smith & C. W. Burklund, "Dominant hemispherectomy," *Science,* 153, 1966, pp. 1280-1282; F. Spellacy, "Lateral preference in the identification of patterned stimuli," *Journal of the Acoustical Society of America,* 47, 1970, pp. 574-578; L. P. Swisher, J. G. Dudley, & D. G. Doehring, "Influence of contralateral noise on auditory intensity discrimination," *Journal of the Acoustical Society of America*, 45, 1969, pp. 1532-1536; T. Tsunoda, "Functional differences between right-and-left-cerebral hemispheres detected by the key-tapping method," *Brain and Language* 2, 1975, pp. 152-170; A. Yamadori, U. Osumi, S. Mashuara, & M. Okuto, "Preservation of singing in Broca's aphasia," *Journal of Neurology, Neurosurgery and Psychiatry*, 40, 1977, pp. 221-224; E. B. Zurif, "Auditory lateralization: Prosodic and syntactic factors," *Brain and Language 1*, 1974, pp. 391-404.

There have also been a number of cases where exceptional musical ability occurs along with mental deficiencies in other areas. Some, such as Blind Tom, have been discussed in the manuscript (see E. Podolsky, *Encyclopedia of Aberrations*, New York, Citadel Press, 1953; and R. M. Goldensen, *Mysteries of the Mind*, New York, Harper & Row, 1976, for additional discussion of him). Others include "Blind Joe" and the case of a "blind, imbecile girl" (see D. C. Rife, and L. H. Snyder, "Studies in human inheritance VI: a genetic refutation of the principles of 'behavioristic' psychology," *Human Biology*, 3:547-59, 1931); Leslie Lemke (blind, retarded, and suffering cerebral palsy) and Ellen Boudreaux (born prematurely and completely blind), Darold A. Treffert, *Extraordinary People: Understanding 'Idiot Savants'*, Harper & Row, New York, 1988. W. A. Owens and W. Grimm discuss a girl from Faribault, Minnesota, with an I.Q. of 23, who could play any tune hummed to her immediately on the piano ("A note regarding exceptional musical ability in low grade imbecile)," *Educational Psychology* 32:636-37, 1941). Kurt Goldstein, Martin Scheerer, and Eva Rothmann discuss a five year study of a boy who, in additional to his musical ability, was a lightning and calendar calculator ("A case of 'idiot savant': an experimental study of personality organization,") *Psychology Monograph* 8:1-63, 1945. B. Rimland lists 63 autistic people with exceptional musical abilities "Savant capabilities of autistic children and their cognitive implications," *Cognitive Defects in the Development of Mental Illness*, New York: Brunner/Mazel, 1978). B. Hermelin, N. O'Connor, and S. Lee did a study of five such individuals ("Musical Inventiveness of five idiot savants," *Psychological Medicine*, 17:685-94, 1987).

At times the loss of abilities from some disease has been a factor in the emergence of exceptional abilities. See B. M. Minogue, "A Case of secondary mental deficiency with musical talent," *J. Applied Psychology* 7:349-57, 1923; H. J. Rothstein discusses a 42 year old male who was normal until he contracted spinal meningitis a age 3, and then rapidly developed exceptional musical skills, ("A study of aments with special abilities," masters thesis, Columbia Univ., 1942);

and A. Anastasi and R. Levee, "Intellectual defect and musical talent: a case report," *Am. J. Mental Deficiency* 64:695-703, 1960.

153. See "The split Brain in Man," Scientific American, Vol. 217, pp. 24- 29,1967; and "Consistency and Diversity in Brain Organization," Proceedings of the Conference on Evolution and lateralization of the Brain, Annuals of the new York Academy of Sciences, 1977); see J. E. Bogen and M. S. Gazzaniga, "Cerebral commissurotomy in man; Minor hemisphere dominance for certain visuo-spatio functions," Journal of Neurosurgery, 23, pp. 394-399, 1965; see M. S. Gazzaniga, *The Bisected Brain*. New York: Appleton-Century-Crofts, 1970; "Right hemisphere language following brain bisection: A 20-year perspective," *American Psychologist*, 38, pp. 525-537, 1983; *The Social Brain*, New York: Basic Books, 1985; see Gazzaniga and J. E. LeDoux, *The Integrated Mind*, New York: Plenum Press, 1978. For studies that most language processing takes place in the left hemisphere, see R. S. Berndt, A. Caramazza, and E. Zurif, "Language functions: Syntax and semantics," *Language Function and Brain Organization*, ed. S. J. Segalowitz, New York: Academic Press, 1983, pp. 5-28; J. L. Bradshaw, "The evolution of human lateral asymmetries: New evidence and second thoughts," *Journal of Human Evolution*, vol. 17, pp. 615-637, 1988; J. L. Bradshaw, *Hemispheric Specialization and Psychological Function*, Chechester, England: Wiley, 1989; J. L. Bradshaw and N. C. Nettleton, "The nature of hemispheric specialization in man," *Behavioral and Brain Sciences*, vol. 4, pp. 51-91, 1981; J. L. Bjradshaw and N. C. Nettleton, *Human Cerebral Asymmetry*, Englewood Cliffs, N. J.: Prentice-Hall, 1983; A. W. Ellis, A. W. Young, and C. Anderson, "Modes of Word recognition in the left and right cerebral hemispheres," Brain and Language, vol. 35, pp. 254-273, 1988; and S. J. Segalowitz, and M. P. Bryden, "Individual differences in hemispheric representation of language," *Language Functions and Brain Organization*, ed. By S. J. Segalowitz, New York: Academic Press, pp. 341-372, 1983.

154. See J. Levy, "Psychobiological Implications of Bilateral Asymmetry," *Hemispheric Function in the Human Brain*, Dimond and Beaumont; J. Levy, "The Regulation and Generation of Perception in the Asymmetric Brain," Brain Circuits and Functions of the Mind—Essays in Honor of Roger Sperry, ed. C. Trevarthen, Cambridge: Cambridge Univ., 1990.

155. See D. M. Bear, "Hemispheric specialization and the neurology of Emotion," *Archives of Neurology*, 40 (1983), pp. 195-202; K. M. Heilman, R. Scholes, and R. T. Watson, "Auditory Affective Agnosia: Disturbed Comprehension of Affective Speech," *Journal of Neurology, Neurosurgery and Psychiatry 38*, 1975, pp. 69-72; D. M. Tucker, R. T. Watson, and K. M. Heilman, "Affective Discrimination and Evocation in Patients with Right Parietal Disease," *Neurology 27*, 1977, pp. 947-950; J. C. Borod, F. Andelman, L. K. Obler, J. R. Stweedy, and J. Welkowitz, "Right Hemisphere Specialization for the Appreciation of Emotional Words and Sentences: Evidence from Stroke Patients," *Neuropsychologia 30*, 1992, pp. 827-844; D. Van Lancker and J. J. Sidtis, "Identification of Affective-Prosodic Stimuli by Left and Right Hemisphere Damaged Subjects: All Errors are not Created Equal," *Journal of Speech and Hearing Research 35*, 1992, pp. 963-970; H. A. Sackheim, R. C. Gur, and M. Saucy, "Emotions Are Expressed More Intensely on

the Left Side of the Face," *Science 202*, 1978, pp. 434-436; J. C. Borod and H. S. Caron, "Facedness and Emotion Related to Lateral Dominance, Sex, and Expression Type," *Neuropsychologia 18*, 1980, pp. 237-242; "Interhemispheric and Intrahemispheric control of Emotion: A Focus on Unilateral Brain Damage," *Journal of Consultation and Clinical Psychology* 60, 1992, pp. 339-348.
156. *The Right Brain and the Unconscious: Discovering the Stranger Within,* Plenum Press: New York, 1992, p.7.
157. "Cerebral Asymmetry, Emotion, and Affective Style," *Brain Asymmetry*, Cambridge, Mass.: MIT Press, 1995, pp. 363-364.
158. See P. Greenwood, D. H. Wilson, and M. S. Gazzaniga, "Dream Report Following Commissurotomy," *Cortex* 13, 1977, pp. 311-316; William C. Dement, *Some Must Watch while Some Must Sleep*, W. H. Freeman: San Francisco, 1974; K. D. Hoppe, "Split Brains and Psychoanalysis," *The Psychoanalytic Quarterly* 46, 1977, pp. 220-224; H. M. Ehrlichman, J. S. Antrobus, and M. Wiener, "EEG Asymmetry and Sleep Mentation during REM and NREM," *Brain and Cognition* 4 (1985), pp. 477-485; Ernest L. Hartmann, *The Functions of Sleep*, Yale Univ. Press, New Haven, Conn., 1973; D. Hodoba, "Paradoxic Sleep Facilitation by Interictal Epileptic Activity of Right Temporal Origin," *Biological Psychiatry 21* (1986), pp. 1267-1278; N. H. Kerr and D. Foulkes, "Right Hemisphere Mediation of Dream Visualization: A Case Study," CORTEX 17 (1981), pp. 603-611; L. Goldstein, N. W. Stoltzfus, and J. F. Gardocki, "Changes in Interhemispheric Amplitude Relationships in the EEG during Sleep," *Physiology and Behavior* 8 (1972), pp. 811-815); R. Broughton, "Human Consciousness and Sleep/Waking Rhythms: A Review and Some Neuropsychological Considerations," *Journal of Clinical Neuropsychology* 4 (1982), pp. 193-218; J. P. Banquet, "Inter-and-Intrahemispheric Relationships of the EEG Activity during Sleep in Man," *Electroencephalography and Clinical Neurophysiology* 55 (1983), pp. 51-59.
159. *Dragons of Eden*, Random House: New York, 1977, pp. 177-178; for a more detailed discussion, see chapter "Tales of Dim Eden," pp. 135-160.
160. *The Forgotten Language: An Introduction to the Understanding of Dreams, Fairy Tales and Myths*, Grove Press: New York, 11951.
161. "Split-Brains and Psychoanalysis," *Psychoanalytic Quarterly 46,* 1977, pp. 220-244; L. Miller, Inner Natures, New York: St. Martin's Press, 1990; and *Freud's Brain*, New York: Guilford Press, 1991; and R. Ornstein, *The Psychology of Consciousness*, San Francisco: Freeman, 1972.
162. "The Neuropsychology of Development," *Journal of Clinical Psychology*. 38 (1982), pp. 4-33, and "The Right Cerebral Hemisphere: Emotion, Music, Visual-spatial Skills. Body Image, Dreams, and Awareness," *Journal of Clinical Psychology* 44 (1988), pp. 630-673, and "The Limbic System: Emotion, Id and Unconscious Mind," *The Psychoanalytic Review*, 79, 1992.
163. "Implication for Psychiatry of Left and Right Cerebral Specialization," *Archives of General Psychiatry*, 31, 1974, pp. 572-583.
164. *Left Brain, Right Brain*, W. H. Freeman and Comp.: New York, 1993, p. 273.
165. Taken from J. E. Bogen, "The Other side of the Brain, VII: Some Educational Aspects of Hemispheric specialization," *UCLA Educator* 17, 1975, pp. 24-32.

166. *The Right Brain and The Unconscious: Discovering the Stranger Within*, p. 116, an elaboration on theories he previously put forth in *Neuropsychiatry, Neuropsychology, and Behavioral Neurology*, New York: Plenum Press, 1990.
167. *The Right Brain and the Unconscious: Discovering the Stranger Within*, pp. 221-224. For additional research and theory by others, see E. De Renzi, *Disorder of Space Exploration and Cognition* (New York: Wiley, 1982); H. Hecaen & M. L. Albert, *Human Neuropsychology*, Bollet, "Visuoperceptual and Visuomotor Abilities and Locus of Lesion," *Neuropsychologia* 2 (1984), pp. 177-185); D. Breitling, W. Guenther, and P. Rondot, "Auditory Perception of Music Measured by Brain Electrical Activity Mapping," *Neuropsychologia* 25 (1987), pp. 765-774; M. P. Bryden, R. G. Ley, and J. H. Sugarman, "A Left-Ear Advantage for Identifying the Emotional Quality of Tonal Sequence," *Neuropsychologia* 20 (1982), pp. 83-87; E. Knox and D. Kimura, "Cerebral Processing of Nonverbal Sounds in Boys and Girls," *Neuropsychologia* 8 (1970), pp. 227-237; S. J. Segalowitz and P. Plantery, "Music Draws Attention to the Left and Speech Draws Attention to the Right," *Brain and Cognition 4* (1985), pp. 1-6; A. Smith, "Speech and Other Functions after Left Hemispherectomy," *Journal of Neurology, Neurosurgery, and Psychiatry* 29 (1966), pp. 467-471; A. Yamadori, et al., "Preservation of Singing in Broca's Aphasia," *Journal of Neurology, Neurosurgery, and Psychiatry* 40 (1977).
168. See *Hemispheric Asymmetry: What's Right and What's Left*, Harvard Univ.: Cambridge, Mass., 1993, for an overview, including a discussion of four categories where asymmetries appear: 1., handedness and the control of motor activities; 2., language; 3., visuospatial processing (including the processing of human faces); and 4., emotion.
169. W. Durant, *The Life of Greece*, New York: Simon & Schuster, 1939; E. G. McClain, *The Pythagorean Plato*, Denver, Co: Great Eastern Books, 1978; K. H. Worner, *History of Music*, New York: Free Press, 1973.
170. See D. Benson and M. Barton, "Disturbances in Constructional Ability," *Cortex* 6 (1970), pp. 19-46; A. Benton, "Visuoperceptive, visuospatial and visuoconstructive disorders," *Clinical Neuropsychology*, K. M. Heilman and E. Valenstein, eds., Oxford: Oxford Univ. Press, 1979, pp. 186-232; F. W. Black and B. A. Bernard, "Constructional Apraxia as a Function of Lesion Locus and Size in Patients with Focal Brain Damage," *Cortex 20* (1984), pp. 111-120; R. Calvanio, P. N. Petrone and D. N. Levine, "Left Visual Spatial Neglect is both Environment-Centered and Body-Centered," *Neurology* 37 (1987), pp. 1179-1183; E. De Renzi, *Disorder of Space Exploration and Cognition* (New York: Wiley, 1982; Y. Kim, L. Morrow, D. Passafiume, and E. Boller, "Visuoperceptual and Visuomotor Abilities and Locus of Lesion," *Neuropsychologia* 2 (1984), pp. 177-185.)
171. D. Breitling, W. Guenther, & P. Rondot, "Auditory perception of music measured by brain electrical activity mapping," *Neuropsychologia* 25, 1987, pp. 765-774; Y. Halperin, I. Nachshon, & A. Carmon, "Shift of ear superiority in dichotic listening to temporally patterned nonverbal stimuli," *Journal of Experimental Psychology*, 101, 1973, pp. 46-54; see A. Gates & F. L. Bradshaw, "The role of the cerebral hemispheres in music," *Brain and Language*, 3, 1977, pp. 451-460, for and overview of what the two hemispheres contribute. For studies

related specifically to the right brain's ability to distinguish vocal inflectional nuances, intensity, stress, pitch, timbre, cadence, emotional tone, frequency, amplitude, melody, duration, and intonation see: S. Blumstein & W. E. Cooper, "Hemispheric processing of intonational contours." *Cortex*, 10, 1974, pp. 146-158; D. Bowers, B. Coslett, R. Bauer, L. J. Speedie, & K. Heilman, "Comprehension of emotional prosody following unilateral hemisphere lesions: Processing defect versus distraction defect." *Neuropsychologia*, 25, 1987, pp. 317-328; H. H. Brownell, H. H. Potter, & A. M. Bihrle, "Inference deficits in right brain-damaged patients," *Brain and Language*, 27, 1986, pp. 310-321; A. Carmazza, J. Gordon, E. B. Zurif, & D. DeLuca, "Right hemispheric damage and verbal problem solving behavior," *Brain and Language*, 3, 1976, pp. 41-46; D. C. Delis, L. C. Robertson, & R. Efron, "Hemispheric specialization of memory for visual hierarchical stimuli," *Neuropsychologia*, 24, 1986, pp. 410-433; V. DeUrso, G. Denes, S. Testa & C. Semenze, "The role of the right hemisphere in processing negative sentences in context" *Neuropsychologia*," 24, 1986, pp. 289-292; J. W. Dwyer & W. E. Rinn, "The role of the right hemisphere in contextual inference," *Neuropsychologia*," 19, 1981, pp. 479-482; N. S. Foldi, M. Cicone & H. Gardner, "Pragmatic aspects of communication in brain-damaged patients," pp. 230-250, *Language Functions and Brain Organization*, ed. by S. S. Segalowitz, Orlando, Fl: Academic Press, 1983 H. Gardner, H. H. Brownell, W. Wapner, & D. Michelow, "Missing the point: The role of the right hemisphere in the processing of complex linguistic materials," pp. 201-244, *Cognitive Processing in the Right Hemisphere*, Ed., E. Perceman, Orlando, Fl: Academic Press, 1983; K. M. Heilman, D. Bowers, L. Speedie, & H. B. Coslett, "Comprehension of affective and nonaffective prosody," *Neurology* 34 (New York), pp. 917-921; K. M. Heilman, R. G. Scholes, and R. T. Watson, "Auditory affective agnosia," *Journal of neurology, Neurosurgery and Psychiatry*, 38, 1975, pp. 69-72; R. J. Ley & M. P. Bryden, "Hemispheric asymmetry in the perception of emotional sounds," *Brain and Language*, 7, 1979, pp. 127-138; A. M. Mahoney & R. S. Sainsbury, "Hemispheric asymmetry in the perception of emotional sounds," *Brain and Cognition*, 6, pp. 216-233; E. Ross, "The aprosodias: Functional-anatomic organization of the affective components of language in the right hemisphere." *Archives of Neurology*, 38, 1981, pp. 561-589; M. Safer & H. Leventhal, "Ear differences in evaluating emotional tones and verbal content," *Journal of Experimental Psychology, Human Perception and Performance*, 3, 1977, pp. 75-82; B. F. Shapiro & M. Danly, "The role of the right hemisphere in the control of speech prosody in prepositional and affective contexts," *Brain and Language*, 1, 1985, pp. 111-139; D. Tucker, R. T. Watson, & K. M. Heilman, "Affective discrimination and evocation of affectively toned speech in patients with right parietal disease," *Neurology*, New York, 27, 1977, pp. 947-950; W. Wapner, S. Hamby & H. Gardner, "The role of the right hemisphere in the apprehension of complex linguistic materials," *Brain and Language*, 14, 1981, pp. 15-33; S. Weintrabu, M. M. Mesulam, & L. Kramer, "Disturbances in prosody: A right hemisphere contribution to language," *Archives of Neurology*, 38, 1981, pp. 742-744.

172. See H. Gardner, H. H. Brownell, W. Wapner, and D. Michelow, "Missing the point: The role of the right hemisphere in the processing of complex linguistic

materials," *Cognitive Process in the Right Hemisphere*, ed. E. Perecman, New York: Academic Press, pp. 169-192, 1983; H. H. Brownell, T. L. Simpson, A. M. Bihrle, H. H. Potter, and H. Gardner, "Appreciation of metaphoric alternative word meanings by left and right brain-damaged patients," *Neuropsychologia* 28, pp. 375-384, 1990; M. S. Hough, "Narrative comprehension in adults with right and left hemisphere brain-damage: Theme organization," *Brain and Language*, 38, pp. 253-277, 1990; J. A. Kaplan, H. H. Brownell, J. R. Jacobs, and H. Gardner, "The effects of right hemisphere damage on the pragmatic interpretation of conversational remarks," *Brain and Language*, 38, pp. 315-333, 1990; R. G. Ley and M. P. Bryden, "A dissociation of right and left hemispheric effects for recognizing emotional tone and verbal content," *Brain and Cognition*, 1, pp. 3-9, 1982; M. P. Bryden and L. MacRae, "Dichotic laterality effects obtained with emotional words," *Neuropsychiatry, Neuropsychology, and Neurology*, 1, pp. 171-176, 1989.

For studies related specifically to the right brain's ability to distinguish the emotional aspects of language see: M. Cicone, W. Wapner, & H. Gardner, "Sensitivity to emotional expressions and situations in organic patients," *Cortex*, 16, 1980, pp. 145-158; J. Day, "Right hemisphere language processing in normal right-handers," *Journal of Experimental Psychology: Human Perception and Performance,* 3, 1977, pp. 518-528; G. Deloche, X. Serion, G. Scius, & J. Segui, "Right hemisphere language processing: Lateral difference with imageable and nonimageable ambiguous words," *Brain and Language*, 30, 1987, pp. 197-207; H. D. Ellis & J. W. Shepherd, "Recognition of upright and inverted faces presented in the left and right visual fields," *Cortex,* 11, 1975, pp. 3-7; D. Hines, "Recognition of verbs, abstract nouns and concrete nouns from the left and right visual half-fields," *Neuropsychologia*, 14, 1976, pp. 211-216; P. B. Goerlick & E. D. Ross, *Journal of Neurology, Neurosurgery and Psychiatry*, 37, 1987, pp. 727-737; K. M. Heilman, D. Bowers, L. Speedie, & H. B. Coslett, "Comprehension of affective and nonaffective prosody," *Neurology* (New York), 34, 1984, pp. 917-921; K. Heilman, R. Scholes, & R. T. Watson, "Auditory affective agnosia," *Journal of neurology, Neurosurgery and Psychiatry*, 38, 1975, pp. 69-72; R. Joseph, "The right cerebral hemisphere, *Journal of Clinical Psychology*, 44, 1988, pp. 630-673; R. Joseph, "Dual mental functioning in a split-brain patient," *Journal of Clinical Psychology*, 44, 1988, pp. 770-779; T. Landis, R. Graves, & H. Goodglass, "Aphasic reading and writing: possible evidence for right hemisphere participation *Cortex*, 18, 1982, pp. 105-112; H. R. Manhaupt, "Processing of abstract and concrete nouns in lateralized memory-search tasks," *Psychological Research*, 45, 1983, pp. 91-105; P. E. Roland, E. Skinhoj, & N. A. Lassen, "Focal activation of human cerebral cortex during auditory discrimination," *Journal of Neurophysiology*, 45, 1981, pp. 1139-1150; E. Ross, "The Aprosodias: Functional-anatomic organization of the affective components of language in the right hemisphere *Archives of Neurology*, 38, 1981, pp. 561-589; B. E. Shapiro & M. Danly, "The role of the right hemisphere in the control of speech prosody in prepositional and affective contexts," *Brain and Language*, 1, pp. 111-139; D. Tucker, R. T. Watson, & K. M. Heilman, "Affective discrimination and evocation of affectively toned speech in patients with right parietal disease," *Neurology* (New York, 27, 1977, pp. 947-950;

173. "The Comprehension of Metaphor in Brain-Damaged Patients," *Brain* 100, 1977, pp. 717-729.

174. For studies specifically related to speech and linguistic forms of thinking in the left hemisphere see: Articles on the left hemisphere's importance in providing a neural foundation for verbal perception, verbal comprehension, verbal differentiation, verbal identification, the linguistic identification of visual, auditory, and somesthetic information, and in the perception and processing of rhymes, numbers, word lists, single phonemes, Morse code, and the transitional elements of speech: S. Blumstein & W. E. Cooper, "Hemispheric processing of intonational contours," *Cortex*, 10, 1974, pp. 146-158; J. E. Cutting, "Two left hemisphere mechanisms in speech perception," *Perception and Psychophysics*,16, 1974, pp. 601-612; D. Kimura, "Cerebral dominance and the perception of verbal stimuli," *Canadian Journal of Psychology*, 15, 1961, pp.156-171; D. Kimura & S. Folb, "Neural processing of backward speech sounds," *Science*, 161, 1968; J. Levy, "Psychological implications of bilateral asymmetry," *Hemispheric Function in the Human Brain*, London: Paul Elek, Ltd., Ed., S. Diomond & J. G. Beaumont, 1974, pp. 121-132; L. Mills & G. B. Rollman, "Hemispheric asymmetry for auditory perception of temporal order," *Neuropsychologia,* 18, 1980, pp. 41-47; G. Papcun, S. Drashen, D. Terbeek, "Is the left hemisphere specialized for speech, language and-or something else?," *Journal of the Acoustical Society of America*, 55, 1974, pp. 319-327; D. Shankweiler & M. Studdert-Kennedy, "Lateral differences in perception of dichotically presented synthetic consonant-vowel syllables and steady-state vowels," *Journal of the Acoustical Society of America*, 39, 1966, p. 1256A; D. Shankweiler & M. Studdert-Kennedy, "Identification of consonants and vowels presented to left and right ears," *Quarterly Journal of Experimental Psychology*, 19, 1966, pp. 59-63; M. Studdert-Kennedy & D. Shankweiler, "Hemispheric specialization for speech perception, *Journal of the Acoustical Society of America*, 48, 1970, pp. 579- 594. For studies relating to the left hemisphere's importance in reading, writing, speaking, spelling, naming, grammar, syntax, rhythm, analytical reasoning, verbal memory and verbal concepts see: M. L. Albert, R. Sparks, T. von Strockert & D. Sax, "A case of auditory agnosia linguistic and nonlinguistic processing," *Cortex,* 8, 1972, pp. 427-443; A. Carmazza & E. B. Zurif, "Dissociation of algorithmic and huristic process in language comprehension: Evidence from aphasia," *Brain and Language*, 3, 1976, pp. 572-582; E. DeRenzi, A. Zambolini, & G. Crisi, "The pattern of neuropsychological impairment associated with left posterior cerebral artery infarcts," *Brain*, 110, 1987, pp. 1099-1116; R. Efron, "The effect of handedness on the perception of simultaneity and temporal order," *Brain*, 86, 1963, pp. 261-284; H. Goodglass & E. Daplan, *Boston Diagonistic Aphasia Examination*, Philadelphia: Lea & Febiger, 1972; H. Hecaien & M. L. Albert, *Human Neuropsychology*, New York: John Wiley, 1978; K. Heilman & R. J. Scholes, "The nature of comprehension errors in Broca's conduction, and Wernicke's aphasia," *Cortex*, 12, 1976, pp. 258-265; A. Kertesz, "Localization of lesions in Wernicke's aphasia," pp. 150-170, and "Right-hemisphere lesion constructional

apraxia and visuospatial deficit," pp. 301-318, and D. N. Levine & E. Sweet, "Localization of lesions in Broca's motor aphasia," pp. 185-207, and L. Vignolo, "Modality-specific disorders of written language," *Localization in Neuropsychology*, New York: Academic Press, 1983; A. Luria, *Higher Cortical Functions in Man*, New York, Basic Books, 1980; B. Milner, "Memory and the medial temporal regions of the brain," *Biology of Memory*, ed. K. Pribram & D. E. Broadbent, New York: Academic Press, 1970; E. B. Zurif & G. Carson, "Dyslexia in relation to cerebral dominance and temporal analysis," *Neuropsychologia*, 8, 1970, pp. 239-244. For the left brain's dominance in the sequential aspects of language see: R. Joseph, "The neuropsychology of development: Hemispheric laterality, limbic language and the origin of thought," *Journal of Clinical Psychology*, 38, 1982, pp. 4-33; R. Joseph, "Confabulation and delusional denial: Frontal lobe and lateralized influences," *Journal of Clinical Psychology*, 42, 9186, pp. 507-518; D. Kimura, "Acquisition of a motor skill after left-hemisphere damage," *Brain*, 100, 1977, pp., 527-542; E. Lenneberg, *Biological Foundations of Language*, New York: Wiley & Sons, 1967; A. Luria, Higher Cortical Functions in Man, New York: Basic Books, 1980; "Pragmatic Aspects of Communication in Brain Damaged Patients," *Language Functions and Brain Organization*, ed. S. J. Segalowitz, New York: Academic Press, 1983.

175. For more neurological discussion of dreams and their relationship to the hemispheres of the brain see: J. P. Banquet, "Inter-and intra-hemispheric relationships of the EEG activity during sleep in man," *Electroencephalography and clinical Neurophysiology*, 55, 1983, pp. 51-59; M. I. Botez, M. Oliver, J. L. Vezina, T. Botez, & B. Kaufman, "Defective revisualization: Dissociation between cognitive and imagistic thought case report and short review of the literature *Cortex*, 21, 1985, pp. 375-389; R. D. Cartwright, L. W. Tipton, & J. Wicklund, "Focusing on dreams," *Archives of General Psychology*, 37, 1980, pp. 275-288; W. D. Foulkes, "Dream reports from different stages of sleep," *Journal of Abnormal and Social Psychology*, 65, 1962, pp. 14-25; L. Goldstein, N W. Stoltzfus, & J. F. Gardocki, "Changes in interhemispheric amplitude relationships in the EEG during sleep," *Physiology and Behavior*, 8, 1972, pp. 811-815; D. R. Goodenough, A. Shapiro, M. Holden, & R. Steinschriber, "Comparison of "dreamers" and "Non-dreamers," *Journal of Nervous and Mental Disease*, 59, 1959, pp. 295-302; J. A. Hobson, R. Lydic, & H. A. Baghdoyan, "Evolving concepts of sleep cycle generation: From brain centers to neuronal populations," *Behavioral and Brain Sciences*, 9, 1986, pp. 371-448; D. Hudcba, "Paradoxic sleep facilitation by interictal epileptic activity of right temporal origin," *Biological Psychiatry*, 21, 1986, pp. 1267-1278; M. E. Humphrey & C. L. Zangwell, "Cessation of dreaming after brain injury," *Journal of Neurology, Neurosurgery and Psychiatry*, 14, 1951, pp. 322-325; R. Joseph, *Neuropsychology, Neuropsychiatry, and Behavioral Neurology*, Plenum Press: New York, 1989, pp. 31-34; J. Kamiyz, "Behavioral, subjective and physiological aspects of drowsiness and sleep," *Function of Varied Experience,* ed., D. W. Fiske & S. R. Maddi, pp. 145-174, Homewood, IL: Dorsey Press, 1961; N. H. Kerr & D. Foulkes, "Reported absence of visual dream imagery in a normally sighted subject with Turner's syndrome," *Journal of Mental Imagery*, 2, 1978, pp. 246-264; N. H. Kerr & D.

Foulkes, "Right hemisphere mediation of dream visualization: A case study," *Cortex*, 17, 1981, pp. 603-611; M. Kramer, R. M. Whitman, B. J. Baldridge, & L. M. Lansky, "Patterns of dreaming: The interrelationship of the dreams of a night," *Journal of Nervous and Mental Disease*, 139, 1964, pp. 426-439; J. S. Meyer, Y. Ishikawa, T. Hata, & I. Karacan, "Cerebral blood flow in normal and abnormal sleep and dreaming," *Brain and Cognition*, 6, 1987, pp. 266-294; B. Monroe, A. Rechtschaffen, D. Foulkes, & J. Jensen, "Discriminablility of REM and NREM reports," *Personality and Social Psychology* 2, 1965, pp. 456-460; L. Murri, R. Arena, G. Siciliano, R. Mazzotta, & A. Muratorio, "Dream recall in patients with focal cerebral lesions," *Archives of Neurology*, 41, 1984, pp. 183-185.

176. *Neuropsychology, Neuropsychiatry, and Behavioral Neurology*, Plenum Press: New York, 1989, p. 7; for additional studies on confabulation, see: R. Joseph, "Confabulation and delusional denial: Frontal lobe and lateralized influences," *Journal of Clinical Psychology*, 42, 1986, pp. 507-518; B. E. Shapiro, M. P. Alexander, H. Gardner, & B. Mercer, "Mechanisms of confabulation," *Neurology* (New York), 31, 1981, pp. 1070-1076; D.T. Stuss, M. P. Alexander, A. Liebermann, & H. Levine, "An extraordinary form of confabulation," *Neurology* (New York), 28, 1978, pp. 1166-1172.

177. *Psycho*, Paramount Pictures, 1960.

178. Speech of Acceptance upon the Award of the Nobel Prize for Literature, Stockholm, Dec. 10, 1950.

179. See *The Interpretation of Dreams*, especially "III "A Dream is the Fulfillment of a Wish," "Distortion in Dreams," (where latent and manifest content are discussed), and VII, "The Psychology of the Dream- Processes," and "The Unconscious," *Standard Edition of the Complete Psychological Works of Sigmund Freud*, tr., under the general editorship of James Strachey in collaboration with Anna Freud, assisted by Alex Strachey and Alex Tyson, 24 vols., 1953-1974 XIV, 161; rpt., *The Freud Reader*, ed. Peter Gay, W.W. Norton: New York, 1989, pp. 573-584.

180. *Altered Egos: How the Brain Creates the Self,* Univ. Press, 2001, p. 1.

181. "Explorers seek the 'I' in 'I think, therefore I am'," by Amy Ellis Nutt, *The Ann Arbor News*, Dec. 31, 2002, Sec. E.

182. Amy Ellis Nutt, "Explorers seek the 'I' in 'I think, therefore I am'," *The Ann Arbor News*, Dec. 31, 2002, Sec. E. More in depth discussion is promised in a book by him not yet published at the time of this writing.

183. See Arthur Koestler, *The Ghost in the Machine*, Harmondsworth: Hutchinson Publ. Corp.,1967, and an overview of all his theories in *Janus: A Summing Up*, New York: Random House, 1978.

184. John Romer, *The History of Archeology: Great Excavations of the World*, Checkmark Books: New York, 2001,

185. *The Ascent of Man*, Little, Brown: Toronto, p. 56, 1973.

186. *Power of Myth*, Doubleday: New York, p. 101, 1988.

187. *Myths to Live By*, p. 20; Bantam Books, New York, 1972.

188. Edwin G. Boring, *A History of Experimental Psychology,* Appelton-Century, 1929; rpt., 1959.

189. Descartes, 1649; 1: XXX; *Les Passions de l'ame*, Pt. 1, article 30; trans., Beakley and Ludlow, 1992, p. 111.
190. Descartes, *Les Passions de l'ame*, Pt. 1, article 32, 1649, trans. by Clarke and O'Malley, 1996, p. 471.
191. C. Sherrington, *The Integrative Action of the Nervous System*. New Haven: Yale Univ., 1947, p. xiv.
192. *Man on His Nature*, New York: Macmillan, p.277, 1941.
193. *The Concept of Mind*, Hutchinson, 1949.
194. Nov., 25, 2013, pp. 78—83; see also http://rogers.matse.illinois.edu/; and http://web.mit.edu/newsoffice/2011/rogers-lemelson-500k.html.
195. January 6, 2014, pp. 34-43.
196. Martin R. Ford, *The Lights in the Tunnel: Automation, Accelerating Technology and the Economy of the Future*, Acculant Publishing, 2009.
197. David Poole, Alan Mackworth and Randy Goebel, *Computational Intelligence: A Logical Approach*, New York: Oxford University Press, p.1.
198. Stuart J. Russell and Peter Norvig, *Artificial Intelligence: A Modern Approach* (2nd ed.), Upper Saddle River, New Jersey: Prentice Hall, 2003
199. Stuart J. Russell and Peter Norvig, *Artificial Intelligence: A Modern Approach* (2nd ed.), Upper Saddle River, New Jersey: Prentice Hall, 2003, p. 55.
200. Nils Nilsson, *Artificial Intelligence: A New Synthesis*. Morgan Kaufmann, 1998.
201. Russell & Norvig, pp. 27, 32–58, 968–972.
202. Poole, Mackworth & Goebel 1998, pp. 7–21.
203. Luger & Stubblefield 2004, pp. 235–240.
204. McCarthy, 2007.
205. Edward Fredkin is quoted in McCorduck (2004, p. 401).
206. Samuel Butler, "Darwin among the Machines," *The Press,*" Christchurch, New Zealand, Letter to the Editor, June 13, 1863; George Dyson, *Darwin among the Machines*, Allan Lane, *Science*, 1998.

CHAPTER FOUR

DON QUIXOTE

Honoré Daumier, Don Quixote and Sancho Panza (ca. 1850) Black crayon and wash 16 x 22 cm, The Metropolitan Museum of Art. New York. This copyright free version uploaded from: http://commons.wikimedia.org/wiki/Image:Honore-Daumier-Don-Quixote.jpg

A crisp, clean November afternoon gave the campus a still, silent quality. Except for the shoveled sidewalks surrounding and crisscrossing the open ground between the buildings, a thin sheet of crusted snow covered the ground, as if, at least for the time, the world had paused to reflect upon itself. Even the occasional student walking rapidly from building to building could not disturb the feeling of a scene momentarily frozen in time.

The seminar room had a full wall of windows jutting out from the side of the building, offering a panoramic view of nature's meditation, and the silence surrounding us gave our conversation a forbidden quality. It was just the two of us, Dr. Norton Kinghorn, who was the chair of the English department, and me. The conversation turned to the one great work of literature to come out of Spain, the work often said to have fathered the modern novel.

Norton smiled. He had a wonderful smile, and a certain Don Quixote childlike gleam that often filled his eyes. "I once had a professor who said Don Quixote was cracked in the head," he began and stopped in mid thought. The pregnant pause, the endearing smile, the mischievous eyes—Norton should have been an entertainer, for he knew how to capture the moment.

Again the silence filled the room, and the high ceiling gave it a certain spiritual quality. I waited for the punch line.

"But the crack let out a beautiful light." (1)

***

In 1857, Charles Pierre Baudeliare, perhaps the greatest French poet of the nineteenth century, shocked the literary world with the tortured, morbid, even perverse content of his meticulously crafted verse. His initial collection of poems, *Flowers of Evil*, was prosecuted for blasphemy and obscenity, and his publisher was forced to suppress six of the poems, two of which were explicit depictions of lesbianism. Those published offered a nightmarish landscape of "lovers of Dementia," of "woman, a vile slave, proud in her stupidity, / Self-worshipped, without the least disgust," of "Man, greedy, lustful, ruthless in cupidity, / Slave to a slave, and sewer to woman's lust," of "Sanctity who treasures, / . . . a fragile drone, / in horsehair, nails, and whips, his highest pleasures," and of "Prating Humanity" crying forth "'O God, my Lord and likeness, be thou cursed!" (2)

Though Baudeliare's form and carefully crafted vocabulary harkened back to the Classicism of Racine, embracing a long established traditional format, and opposing the new less structured Romanticism of his contemporaries, his preoccupation with sex and death, dark views about the corruption and evil inherent in human endeavor, and personal feelings of despair set him up to be seen as a macabre agent of darkness. Thus, in 1866, at the peak of his career, when he was stricken with what today is designated as an aphasia and hemiplegia, but then was more simply seen as insanity, leaving the once powerful poet now unable to participate in an ordinary conversation, it was proof for the censors that indeed God had had enough of him.

Perhaps, in a way, they were right, for this real life Don Quixote, so brilliant in his creation of a sublime commingling of horror and beauty, had been forced to enter his own malevolent nightmare, a world without poetic beauty and only the frustrating fragments of meaning, scattered and always disturbing the mind, but no longer able to be brought to coherence, no longer able to be formed into a world of significance and value.

Rebecca was nineteen when she was first brought to the clinic where Oliver Sacks worked. But, as her grandmother said when she dropped her off, she was "'just like a child in some ways.'" Indeed, she had trouble putting on her clothes, often putting them on back-to-front or inside out, got lost trying to navigate herself around a simple city block, and never could figure out how to open a door with a key. She was clumsy, had poor coordination, and fit the general category of a klutz (though importantly, when she danced, all of this general klutziness disappeared). (3)

She had a partial cleft palate, resulting in a whistling noise when she spoke, short, stumpy fingers, blunt, deformed nails, and a degenerative myopia requiring thick glasses—all the features working together to make her a perfect target for teasing, and she felt very much that she had always been the object of ridicule, resulting in a painful shyness.

When he first met her, Oliver Sacks found her to be "a casualty, a broken creature, . . . a multitude of apraxias and agnosias, a mass of sensorimotor impairments and breakdowns," with "limitations of intellectual schemata and concepts similar (by Piaget's criteria) to those of a child of eight. . . . A poor thing . . . with perhaps a 'splinter skill', a freak gift, of speech; a mere mosaic of higher cortical functions, Piagetian schemata—most impaired."

Rebecca could not count change, do any kind of simple math, read, write, or score higher than a 60 on an IQ test. She was a poor, lost soul, wandering through a maze of handicaps and incapacities, living out a nightmare existence of frustrations and failure.

But wait, I thought. Not so fast. Not so simple. And indeed, the unseen depths of her mind would emerge for those open to the light. Within that nightmare, that world of crude imperfections, filled with the most complete darkness, was a wonderful light, a deep, richly mapped world of meaning and value. The next time Oliver Sacks came across her, he began to see through the flaws. This time he came upon her, not in a laboratory or classroom setting, but, rather, sitting quietly on a bench, gazing at the April world of nature that surrounded her. He wrote:

> As I approached, she heard my footsteps and turned, gave me a broad smile, and wordlessly gestured. "Look at the world," she seemed to say. "How beautiful it is." And then there came out, in Jacksonian spurts, odd, sudden, poetic ejaculations: "Spring," "birth," "growing," "stirring," "coming to life," "seasons," "everything in its time." I found myself thinking of Ecclesiastes: "To everything there is a season, and a time to every purpose under heaven. A time to be born, and a time to die: a time to plant, and a time . . ." This was what Rebecca, in her disjointed fashion, was ejaculating—a vision of seasons, of times, like that of the Preacher. "She is an idiot Ecclesiastes," I said to myself. And in this phrase, my two visions of her—as idiot and as symbolist—met, collided and fused. She had done appallingly in the testing—which, in a sense, was designed, like all neurological and psychological testing, not merely to uncover, to bring out deficits, but to decompose her into functions and deficits. She had come apart, horribly, in formal testing, but now she was mysteriously 'together' and composed. (pp. 180-181)

Oliver Sacks now saw something beyond what the clinical tests and other obvious limitations in her had revealed. Now he could remember how she had exhibited such a "warm, deep, even passionate" attachment to her grandmother (who had raised her since she was three, when both of her parents died), how fond she had always been of nature, of the city parks, of the botanic gardens, how she had loved, had a "hunger," for stories and poetry, how she had "the world re-presented to her in verbal images, in language, and seemed to have little difficulty following the metaphors and symbols of even quite deep poems," how "the language of feeling, of the concrete, of image and symbol, formed a world she loved and, to a remarkable extent, could enter," how "metaphors, figures of speech, rather striking similitudes, would come naturally to her . . . as sudden poetic ejaculations or allusion," how she shared her grandmother's quiet religious devotion, "loved the lighting of the Sabbath candles, the benisons and orisons which thread the Jewish day, . . . loved going to the synagogue, . . . fully understood the liturgy, the chants, the prayers, rites and symbols of which the Orthodox service consists, . . . how she had a "strangely moving, poetic power . . . and could "organize herself in a narrative way." (pp. 178-182)

Then her grandmother died. When Oliver Sacks visited her, he found that, though the joy he had seen in her in April had now turned to deepest grief, she nevertheless conducted herself with a "great dignity," a great "ethical depth." "'Why did she have to go?' she cried; and added, 'I'm crying for me, not for her.' She's gone to her Long Home. . . . I'm so cold. . . . It's not outside, it's winter inside. Cold as death. . . . She was a part of me. Part of me died with her." (p. 182)

Oliver Sacks found in her grief that her deeper, moral, ethical world was rich and complete, not mentally defective. Later, in the same conversation, she rose above her sorrow to say: "It is winter. I feel dead. But I know the spring will come again." (p. 182)

Helped by a "sympathetic and supportive great aunt . . . the synagogue, and religious community," especially by some of the "rites" involved with mourning, and interestingly, by "dreams," Rebecca was able to work through her grief. (4)

She never was able to function in the various workshops and classes set up to push the "clients" into, finally using her power of metaphor to tell her misguided caretakers: "'I want no more classes, no more workshops,'" . . . 'They do nothing for me. They do nothing to bring me together. . . . I'm like a sort of living carpet. I need a pattern, a design, like you have on that carpet. I come apart, I unravel, unless there's a design. . . . I must have meaning. . . . The classes, the odd jobs have no meaning . . . What I really love . . . is the theatre.'"

She was removed from the workshops and allowed to perform in a theatre group, and, according to Oliver Sacks, if one were to see "Rebecca on stage, for theatre and the theatre group soon became her life, one would never even guess that she was mentally defective." (pp. 184-185)

Oliver Sacks wrote that he saw in Rebecca that humans have two "wholly different, wholly separate, forms of thought and mind, 'paradigmatic' and 'narrative' (in Bruner's terminology)." (p. 183) (5)

And later he added: "It was perhaps fortunate that I chanced to see Rebecca in her so-different modes—so damaged and incorrigible in the one, so full of promise and potential in the other—and that she was one of the first patients I saw in our clinic. For what I saw in her, what she showed me, I now saw in them all." (p.182)

When I think of Rebecca, I am reminded of Don Quixote's cracked skull, his insane embracement of a now lost world of chivalric romance, and how he was depicted as but a caricature, a parody. I remember his failed adventures, the cloud of dust he took for a huge battle, which he fearlessly charged into, never realizing it was but a herd of sheep, and ending up getting knocked to the ground with most of his teeth missing by the annoyed herder's stones, the barber he deprived of a brass basin which he insisted was the golden helmet of Mambrino, the gang of chained convicts he freed, only to be mauled by the very men he had freed, the penance in the mountains, consisting of near naked gambols and insane, self-inflicted tortures. Again and again, he managed to misrepresent the most common happenings, turning them into wild adventures, which inevitably resulted in his own humiliation.

Yet so completely in tune was he with his illusions that I knew he had somehow risen above his own faulty reasoning, or, perhaps, simply showed me how the reasoning that would condemn him itself is faulty. I remembered Norton Kinghorn's clever insight, the damaged brain that let in the beautiful light.

I considered. In the end, who is being deceived here? Is it Don Quixote? Certainly Don Quixote is neither a rich nor a physically endowed knight, certainly not a knight of some mythic fraternity whose very appearance inspires awe, even a spiritual worship, in those who gaze upon him. Rather, he is an impoverished country gentleman who has read so many stories about such men that he's lost touch with reality. His armor doesn't gleam with a blinding light. It is but an old suit given to him by his great-grandfather and left to rust in a corner for years. His helmet is but a visor-less headpiece, a morion, to which he has attached a kind of half-helmet made from cardboard. His horse is best called a nag, for it is more a patchwork of skin and bones than a fine steed. His lady-in-waiting, for each knight must serve some great lady, is but a farm girl living nearby, known more for her ability to salt pork than her refined nobility. All-in-all, he is but a comic parody, someone to be laughed at. And he is, time and again.

Yet, as he charges upon those windmills, convinced they are giants, in spite of the attempts by his more reasonably inclined squire Sancho Panza to dissuade him, runs his lance into the mast of the first of them, is hurled to the ground, and even caught in the most obvious proof of his mistake, insists they were just transformed into windmills by some necromancer, refusing to give up the illusion, even though there can be no denying the literal truth, I begin to sense something behind my laughter, a suggestion that, if it is but an insane dream, it would be still a good dream, one worth a bit of humiliation.

And then, as the surreal adventures continue, and his insanity seems to insist he is but a pitiful clown lost in some deranged nightmare, I began to see a coherence, a method to the madness, for it is always driven by a feeling of moral greatness, of a universe endowed with some empowering vision of meaning and value, some invisible stage hidden behind the crude appearances of the literal world, and a strange reversal begins. The literal, physical world of logic may win in fact, but perhaps, just perhaps, Don Quixote has the higher victory, perhaps, from a moral perspective, from an aesthetic vision, he has the last laugh. Even if his actions endlessly make him out the fool, his vision is, after all, a noble one. Is not the redressing of wrongs a good thing? Is it not admirable to come to the aid of the afflicted, the suffering? Should we not stand up to those who would bring torture and pain? Is not integrity to be respected?

I notice that the more sensible Sancho Panza also cannot help but be converted to this insanity, this crazy illusion, in the end begging his master not to die but to continue the fantasy, the colorful, thrilling world of his visions.

I still remember Norton Kinghorn's endless stories, filled with humor, filled with life. He always insisted that he was not to be trusted, not to be believed. Yet the lies held truths that still crystallize when the time is right for me to grasp them. And I can still see the sparkle in his eyes, the same light that must have emanated from the eyes of Don Quixote.

And I know, without a doubt, that these stories, merely fiction, merely lies, non-existent outside of the human mind, create a reality far more real and important than mere brute bodily existence, more filled with truth and meaning than that literal truth of reason and logic that the neo-classicists so proclaimed. These stories, by definition lies, are the truth that Norton so enjoyed, so lived, the beautiful light that emitted from that cracked head of the old man who would be Don Quixote de la Mancha and go on quests in the name of the wonderful Dulcinea del Toboso. Yoda, the physically and literally comic entity of the *Star Wars Trilogy*, and yet obviously something much more than that, the psychic/mythic/spiritual life force of the great archetypal swamp world of Dagobah, was right, "Luminous beings are we." Not just "crude matter." (6) Rebecca also knew this world, lived it. When she danced, she entered it. When she performed on stage, she entered it. She was Don Quixote, a comic, sad figure in the literal, physical world, but able to transcend its limitations into a higher level of existence, a luminous level.

Since he never mentioned it in reference to her, it's interesting to note that, in his "Postscript" to his discussion of Rebecca, Oliver Sacks plunged immediately into a discussion of music, bridging the discussion with the understated claim that "the power of music, narrative and drama is of the greatest practical and theoretical importance." And he led it up to the final paragraph, where he claimed that, "What we see, fundamentally, is the power of music to organize—and to do this efficaciously (as well as joyfully!), when abstract or schematic forms of organization fail." (pp. 185-186)

Perhaps this postscript was meant as a bridge to the next patient he discussed, but another Don Quixote, another inept, annoying, even disgusting clown in the literal world, yet who, when given the chance to entire the symbolic world of the mind, the world where connotation replaces denotation, was able to reach beyond his prosaic deficiencies and demonstrate clearly that music is also of this form of higher truth.

Martin A. was slow, dreamy, and an incompetent worker, spastic, impulsive, and prone to seizures, all presumably the result of a nearly fatal meningitis as a child.

However, his obvious retardation was only half the story. The other half was equally amazing in the other direction. "I know more than 2,000 operas," he said truthfully, not the result of an extensive study of scores of sheet music, but simply because he had the capability to retain an entire opera after a single hearing. This ability also extended to a memory of every detail of every performance he had ever seen, and he was considered something of an idiot savant for this wild disparity in his mental abilities. He did gain a bit of fame for his prodigious memory, and enjoyed it in a child-like way, but what centered his life was performing music, singing in the church choirs and cathedrals. He even managed to sing in the choirs at the Met. However, unfortunately, perhaps because of his childhood disease, he did not have a very good voice, and, thus, only could enjoy this life-giving activity when his voice was, in effect, drowned out by the rest of the choir. Nevertheless, as Oliver Sacks says, when he "soared up into the music Martin forgot that he was 'retarded', forgot all the sadness and badness of his life, sensed a great spaciousness enfold him, felt himself both a true man and a true child of God." (p. 188)

Most of his huge memory banks seemed to mean little more to him than they might to a computer. There was no real attachment, no feeling, no human center or relationship to them. However, there was one curious exception. He had, indeed, memorized the entire huge, nine-volume edition of Grove's *Dictionary of Music and Musicians*, published in 1954, all six thousand pages. His father, once a famous vocalist at the Met, now growing too old to continue his career, spent most of his time at home, listening to records, and reading aloud to his thirty-two-year old son the entire dictionary, and Martin, illiterate himself, had imprinted it in his brain along with the his memory of his father, of his father's voice sharing it with him. This work, then, did have emotional, human connections.

Outside of his world of music, Martin was retarded, prone to childish tantrums, spiteful, dirty, and inclined to show-off. In clinical language, he was filled with "adjustment difficulties." But then, when Oliver Sacks visited him the following January, the solution was found. "'I've got to sing,' he said hoarsely, 'I can't live without it. And it's not just the music—I can't pray without it.' And then, suddenly, with a flash of his old memory: '"Music, to Bach, was the apparatus of worship". Grove article on Bach, page 304 . . . I've never spent a Sunday,' he continued, more gently, reflectively, 'without going to church, without singing in the choir. I first went there, with my father, when I was old enough to walk, and I continued going after his death in 1955. I've got to go,' he said fiercely. 'It'll kill me if I don't."

Seems so simple. He was allowed to return to his church, where he was welcomed back, welcomed back into the same position of the brains and adviser of the choir his father had been before him. Now he could sing.

And now, with his brilliant, life-giving form of thought in place, he could rise above his 'adjustment difficulties.' "'You see,'" he told Oliver Sacks on his next visit, in a simple, modest manner, "'they know I know all of Bach's liturgical and choral music. I know all the church cantatas—all 202 that Grove lists—and which Sundays and Holy Days they should be sung on. We are the only church in the diocese with a real orchestra and choir, the only one where all of Bach's vocal works are regularly sung. We do a cantata every Sunday—and we are going to do the *Matthew Passion* this Easter!'" (p. 191)

Certainly, the respect Martin received from his congregation was important to him, but what Oliver Sacks was most impressed with "was to see Martin when he was actually singing, or in communion with music—listening with an intentness which verged on rapture—'a man in his wholeness wholly attending'. . . in a word, transformed. All that was defective or pathological fell away, and one saw only absorption and animation, wholeness and health." (p. 190)

Oliver Sacks even suggested that "Martin's musical intelligence was fully up to appreciating much of the technical complexity of Bach; but, more than this—that it wasn't a question of intelligence at all. Bach lived for him, and he lived in Bach."

So, yet again, this time in the aesthetic world of music, a real life Don Quixote was able, through an honest, sincere, perhaps even desperate embracement of a world of higher meaning and value, to achieve dignity. And I wonder, who is the more mentally defective, the man who can calculate, and manipulate, and achieve wealth and power, but cannot relate to music, and dance, and drama, and story, the man who would laugh at Don Quixote, think the *Stars Wars Trilogy* nothing more than a simple morality play, or the man who is inept when it comes to adding and subtracting and manipulating, cannot handle practical existence at all, but who can enter the realms of meaning expressed in a fugue by Johann Sebastian Bach.

And when it comes to a definition of human, which man comes closest? Which offers the most worthy image?

William Wordsworth wrote endlessly of the distinction between the unpoetic soul, the Peter Bell for whom "A primrose by a river's brim / a yellow primrose was to him, / And it was nothing more," and the sensitive, artistic soul, who, Wassily Kandinsky points out "is similar in many ways to the child throughout his life, [and] can often arrive at the inner resonance of things more easily than someone else." (7)

Martin was unusual, but not singular. David S. Viscott leaves us an account of another such quixotic soul. (8)

Harriet G. was the sixth of seven children, the daughter of a domineering mother, much more prone to threats than praise, and a father who was content to let the mother have her way. In the ninth month of pregnancy the mother experienced a tremendous scare viewing the near accident of her oldest daughter, was told by her doctor that her organs had "retracted from the fright," and assumed that Harriet would be born somehow deformed. Though there was no noticeable damage to Harriet upon birth, and in fact the birth went well, the mother continued to believe a defect would show itself. For this reason, she kept Harriet separated from the other children. The mother had trained at a music conservatory in Italy, and now earned her living as a music teacher and singing coach. Thus, Harriet's crib was placed against the grand piano in the studio where the music lessons were given, and she spent the first years of her life mainly left alone in her crib, unattended, except for her breast feeding, and with no stimuli except for the non-stop music lessons going on next to her. When the lessons were done for the day, Harriet was left alone in her crib until the next day's lessons began. In other words, there was no sensory stimulation nor any companionship or caring given her, except for the music that surrounded her. Soon Harriet did, indeed, began to exhibit the unnatural behavior the mother expected, banging her head and rocking back and forth, and the like, convincing the mother she had been right in her suspicions, and causing her to be even more restrictive.

At the age of seven months, the father was startled to hear Harriet, lying on her back in her crib, actually humming in perfect pitch, tempo, and phrasing, the "*Caro Nome*" from *Rigoletto*. Harriet also began singing the various vocal performances she had been hearing her mother's students performing, all with perfect pitch, proper accents, and in all keys, major and minor. The mother immediately assumed that Harriet was a genius, had inherited from her (after all, she was an accomplished musician herself). This, again, led to even more isolation, even more of a life of simply lying in her crib, lasting until she was two years old.

At age two, Harriet had a younger sister born, this one deformed physically, her right hand. The mother could not accept a deformed child and never gave it a real name. Now, Harriet was allowed out of the studio to interact with the other children, but immediately and consistently exhibited violent behavior, breaking everything she could get her hands on, squeezing the pet cats until they cried, and then laughing about it. She refused to be toilet trained, and spent a good deal of time walking about the house with a stick, pounding on the floor, which later was thought to be in an attempt to make music.

By three she began on her own to learn the piano, and by four could play all of the arias her mother's students sang, using both hands, proper fingering, and even creating harmony. She refused to speak, only once under severe penalty of starvation demanding milk. In fact, she did not become toilet trained until she was nine.

At age five, her parents decided they could no longer handle her, and they brought her to a home for disturbed children, the kind of place they had often threatened her about, and she assumed this was a terrible punishment. She could not comprehend time (which caused her to fear the length of her imagined, or, perhaps real, imprisonment), hated taking the baths that were forced on her, and when finally returned home reported having been anesthetized and operated upon, giving precise details. A scar on her abdomen supported her claims, and it was thought her Fallopian tubes had been tied off, though, curiously, apparently no one bothered to press the issue. In general, the reasons for her return home from the institution were never clarified.

By the time she was six, she was just beginning to speak, and it was discovered she had an amazing memory, especially for dates At seven she was taken to a "philosopher in psychology," who thought her a genius and wanted to take her and give her special training. Her father, however, believing she was better off at home, refused. Then, with her mother's moving of the studio outside of the home, Harriet regressed into wild, animal behavior, destroying household articles, even killing two of the cats. However, by the age of nine, she had progressed enough to actually be sent to school, though the first attempt did not work, and she did not return until age eleven, this time continuing until age sixteen. At age eighteen, she was given a kitchen job at a hospital, where she developed a strict routine and continued for many years. At age thirty-five, during a

Christmas dinner, she announced her first dream, one in which her father died, and amazingly, he did die that same day of a heart-attack.

When David S. Viscott first met her at the Boston State Hospital, he found her prone to talk continuously, and, though never looking directly at him, generally friendly, and hungry for affection. In his testing of her, he found her overall I.Q. to be 73, her attention span diminished, her ability for abstract reasoning noticeably impaired, her arithmetic abilities poor, her vocabulary poor, her memory amazing, and her knowledge and abilities in music "breathtaking."

Then she was asked to perform on the piano for a seminar at the Boston State Hospital. David S. Viscott writes, "She sat down demurely, stared quietly at the keyboard until we all grew silent, then brought her hands slowly to the keyboard and let them rest a moment. Then she nodded her head and began to play with all the feeling and movement of a concert performer. For that moment in time she was another person."

Indeed, her musical skills were extraordinary. She had perfect pitch, could name all of the notes played in a four note chord for less than half a second, name notes struck at random, even name notes struck at both ends of a piano keyboard with fists at the same time. She could change the key of a piece she was playing to any other key instantaneously, even shift into the subtle stylistic differences between composers with ease, and revise songs to fit various composers' styles. In contrast to her inability to hold deep emotions in general, she could express great feelings in music, and could comprehend and discuss the emotions and complex plots of operas.

Nonetheless, David S. Viscott always felt that even her great musical depth of emotion was lacking something, found it somehow shallow and childlike. He attributed her singular centering on music to maternal failure, suggesting that this "maternal failure did not allow for an orderly progression and development of a reality sense," that music served as a "transitional object" representing the mother, and that in creative persons artistic creation is actually a "re-creation" of the mother associated with pleasure. Furthermore, he theorized, since the culture applauds creative talents, the child gets the best of both worlds, the missing love of the parent and reinforcement from the society.

It is worth mentioning in passing that such conclusions fall in line with Sigmund Freud's views that creative activity is in actuality a substitutive activity for desires denied in the real world, and it is obvious that David S. Viscott has approached his study through the lenses of Freudian psychology.

Others have suggested similar connections between creativity, the workings of the unconscious, and early childhood experiences with the mother. Heinz Kohut wrote that superego formation is affected not only by what the parent says, but by the quality and tone of the parent's voice, concluding that the deeper layers of the superego "are related to a preverbal acoustic sphere," the "tone or the voice of conscience," and music can serve as this function. (9)

Lawrence H. Rockland, concluding his study of a patient who constantly experienced and used fragments of popular songs to attempt to communicate, to work his way through his psychological anxiety, suggested that what the patient "was experiencing was related to wishes in the transference for the early symbiotic maternal object, and the "melodies—as opposed to the lyrics—can be viewed as derivatives of the early cry of the helpless infant," (10) and supported these psychological interpretations of what he considered "regression," by referring to Theodor Reik, who suggested that a tune, a "haunting melody," an "obsessional thought" satisfies a "purifying, cathartic function, calling up dark emotions which cannot be verbalized," (11) and to Kurt Goldstein, one of the most important early neurologists, who stated, "According to the more primitive character of singing and the close relationship of singing to expressive movements and emotional language, it can be assumed that in a brain damage, singing will be preserved longer than language." (12)

Though these beginning correspondences between psychology and neurology are of the hit-and-miss variety, they are, nonetheless, demonstrating a relationship between the study of the mind and that of the brain. And other conclusions by David S. Viscott suggest connections with Carl Jung's theories, for, among his conclusions, David S. Viscott suggests that "the creative process is a passageway to the unconscious which has been kept open . . . a distinct mental process in its own right. . . . a sense of repetition of the life cycle in the birth of every new work and a sense of loss and mourning appearing as each is completed, and given up. This depression stems in part from the artist giving form to his deepest feelings." And then, once the work is finished, once, in effect, the artist has given birth, this artistic child must be exposed to the world to fend for itself. And so the creative moment passes, and the artist must again start it over, always, as it is a means for him or her to experience for a fleeting moment the beautiful yet terrible combining of the infinite with the finite. Here, though David S. Viscott is still couching it all in his surrounding comments about the relationship with the mother, he is also suggesting the larger mythological landscape of the archetypal psyche, the coming together of the invisible world of meaning and value of the mind with the visible, physical world in the human brain.

Frances Hannett does the same, suggesting that song lyrics that come to mind in a patient during psychoanalysis are "a voice of the preconscious, and must be understood in the same way as a dream fragment, a fantasy, or a repetitive act," because "popular lyrics seem to recast in existential terms the ancient fantasy of the Great Mother who controls the fate of man." (13)

And this huge hall of theory echoes with the words of Vincent Van Gogh to clarify *The Night Café*:

> In my picture of the "Night Café" I have tried to express the idea that the café is a place where one can ruin oneself, go mad or commit a crime. So I have tried to express, as it were, the powers of darkness in a low public house, by soft Louis XV green and malachite, contrasting with yellow-green and harsh blue-greens, and all this in an atmosphere like a devil's furnace, of pale sulphur. And all with an appearance of Japanese gaiety, and the good nature of Tartarin . . .
>
> I have tried to express the terrible passions of humanity by means of red and green.
>
> The room is blood red and dark yellow with a green billiard table in the middle; there are four citron-yellow lamps with a glow of orange and green. Everywhere there is a clash and contrast of the most disparate reds and greens in the figures of little sleeping hooligans, in the empty, dreary room, in violet and blue. The blood-red and the yellow-green of the billiard table, for instance, contrast with the soft tender Louis XV green of the counter, on which there is a pink nosegay. The white coat of the landlord, awake in a corner of that furnace, turns citron-yellow, or pale luminous green. (14)

Such words and explanations are always inadequate to the task, yet fumbling towards spiritual heights, words to try to help Theo understand, feeble substitutes for the actual experience, yet perhaps supporting and even helping to empower the experience, "I have tried to express the terrible passions of humanity by means of red and green." So simple, so filled with meaning.

In homage to Van Gogh's demanding spiritual and artistic struggle, Don McLean used his own form of expression, and wrote *Vincent*, particularizing the struggle to Vincent Van Gogh's most famous painting, *The Starry Night*, and concluding his own struggle to understand the humanity of the colors, and the hues, and the shadows, the "flaming flowers that brightly blaze," the "swirling clouds in violet haze," the "colors changing hue," the "morning fields of amber grain," he sang, "And now I think I know what you tried to say to me / How you suffered for your sanity / How you tried to set them free. / They would not listen, they're not listening still / Perhaps they never will." (15)

Katherine Paterson, highly acclaimed author of such award winning works as *Bridge to Teribethia* and *Jacob Have I Loved*, began her keynote speech with several kind references to my own previously presented views on literature and literary theory. That, of course, pleased me, but I was far more interested in other aspects of her discussion. (16)

One of her stories was about Eugene, a boy she had spent four years with at the Calvin H. Wiley School, the other 'weird' student in her class.

Since she had just returned with her missionary parents from Hong Kong, because WWII had broken out, she was looked upon by the other children as somehow different and perhaps even one whose sympathies were with the enemy. Eugene was seen as different because of his own staunchly stated choice. He was most certainly of the Don Quixote mode, for he had chosen, even though he was, as Katherine Paterson put it, "a perfectly round little boy," to become a ballet dancer! When she told us this in her speech, the auditorium filled with laughter. How could one not laugh at such a comic desire, such an obviously illogical dream?

The two of them became and remained friends.

Then Katherine's family moved. As the years passed she wondered about her "chubby little friend," wondered how he had faired with such an impossible dream, worried about him. Generally, as she later was to point out, such friendships, such real-life stories remain without closure. However, in this case, the unusual happened, for, as an adult, Katherine was accidentally, and gratefully, able to satisfy her curiosity. When her own son was in high school, and seeking a career on stage, he found that it would be necessary for him to learn how to dance to get the roles he sought. In trying to help him find a dance school without letting his friends on the soccer team know about it, Katherine was led back to Eugene, who had by now established himself as one of the finest dance teachers around.

The telephone reunion was a fulfilling one, the satisfying completion of a circle. But there was more. As it turned out, Eugene, that fat, little boy, who now was a huge man, had, amazingly, once been a totally physically fit ballet dancer, a highly respected dancer, performing on the New York and London stages. He had, indeed, fulfilled his impossible dream.

How wonderful! How heart-warming! Here, I thought to myself, the world of the mind, of the dream, had literally conquered the crude literal world of the brain. Here, the human spirit, soul, mind had proven itself capable of succeeding against the logical restrictions of the physical world. The underdog had won. It was possible. Perhaps rare, but still possible, given a sincere commitment to the dream. And, I thought, perhaps, just perhaps, in truth, here is proof, or if not proof, certainly support for the view that life is more than what John Watson, B. F. Skinner, and their behaviorist philosophies would have us believe. (17) Here, Don Quixote had won!

In another section of her speech, Katherine discussed respected British musician and director, Clive Wearing, who was struck with a strange inflammation of the brain, virtually destroying his memory, and leaving him in a state of continual reawakening into life. "It's as if he's wearing blinders," his wife said. Every moment of life is separated from the next. He cannot make connections in time. He writes continually in his diary, desperate to preserve each fleeting moment, yet the next moment he wonders how the words in his own handwriting got on the page.

His intellect was otherwise untouched, and this made his condition even more tragic, as he could still stand outside himself and judge the state of his life. Thus, as one would imagine, he was often deeply depressed.

Two aspects of his once rich life remained. He could still feel the deep love of his wife, and every time he saw her he would throw his arms around her as if it had been a long time since he last saw her, even if it had really only been five minutes. He wrote constantly in his diary, "I will love Debbie forever and ever," even though, for him, there was no ever.

And he still retained his musical ability. In the PBS program, *The Search for Mind*, he is shown being taken into a room by Debbie where there is a piano and a small group of singers. He is confused, says he doesn't know how to play. But Debbie urges him to try.

Katherine Paterson continued, "We watch him sit down, hesitant, and then strike a few keys. The choir begins to sing, and immediately Clive Wearing's face lights up, and he sings and directs Mozart's *Ave Verum*," an appropriately deeply spiritual piece.

His memory has been destroyed, yet somehow love and music have survived. "But perhaps this miraculous survival in his mind and heart of love and music deepen the tragedy of his life, which is that he has no story, but at the same time it makes us marvel at the basic properties of that which we call mind. In Clive Waring's damaged brain a mind which knows love and creates beauty survives."

Deborah Wearing begins her book *Forever Today: A Memoir of Love and Amnesia*:

> Clive had no idea that Tuesday 26 March 1985 would be his last day of conscious thought. We weren't ready. Did he feel his brain disappearing that night? Why didn't he wake me? By morning he could not answer a simple question or remember my name. The doctor said it was flu and lack

of sleep causing the confusion. He tucked him up with a temperature of 104 degrees and a bottle of sleeping pills.

"No need to stay home," he said to me. "These'll knock him out for eight hours. Go to work."

I went to work. When I came home that night the bed was empty. His empty pyjamas were collapsed in the middle of the bare sheet. I screamed his name. Running the length of the flat, I already knew something bad had happened. I knew it in my screaming soul before I knew it in my head.

"I'm never ill," Clive used to say. And he never was. Then all of a sudden he was. But instead of a normal illness the doctors stand some chance of recognizing, this one is rare, sneaky. Nobody knew what was wrong with him. At the hospital, they thought he was a goner. Only they didn't put it like that. They told me what they thought he might have, and said it had a high mortality rate. I didn't know if that meant probably live or probably die and I didn't like to ask. Eventually I understood that they didn't expect him to last more than a few days because of signs they'd seen on a brain scan they did not show me that evening. Then they said, go home. We were standing in the doorway to the ward and I could see the shape of my husband under a white blanket behind them. He and I had been inseparable since first meeting and now, this night when he might be going to die, I am told, go home and don't come back until three o'clock tomorrow—visiting hours. What were they thinking of?

A sudden virus had caused holes in Clive's brain; memories fell out. But nothing could touch what was in his heart. He could not remember a single thing that had ever happened to him, but he remembered me and knew that he loved me. (18)

Clive's sudden, inexplicable illness and resulting loss of memory are eerily foreshadowed in a subsequent chapter of the book where Deborah recalls the following:

The first weekend of New Year 1985 we took a day off and put our feet up. Clive had just produced an enormous series, eight programmes broadcast over ten days through Christmas, and we were exhausted. Sunday 6 January 1985 was everything a day off ought to be—late breakfast, croissants and jam, then retiring to the drawing room with newspapers and more coffee. "*Re-lax, honey*!" Clive said in faux American. He sat on the sofa with the *Observer* scattered at his feet. I sat across from him with other papers. We read and rustled.

"Darling," he said suddenly. "You've got to read this!" There was something about his voice. He lifted his arm for me to come under and closed me to his side, and automatic gesture.

The *Observer Review* lead story was *The Lost Mariner*, an extract from a book of case studies by Dr Oliver Sacks. The cartoon at the head of the page showed a young man looking in a mirror, the reflection of an old

> man looking back at him. Clive waited until I got to the foot of the page, turned over and we read on together.
>
> The story concerned one Jimmie G, a man with amnesia. He could remember nothing beyond 1945. No new information stuck. The doctor had visited many times, but remained a stranger. Jimmie still thought of himself as a nineteen-year-old GI. One day, out of curiosity perhaps, Sacks handed him a mirror. The cartoon captured the horror of that moment as Jimmie caught sight of himself. The 'young' man saw his own reflection as an old man with a bush of grey hair. He turned ashen, gripped the sides of his chair. "What's going on?" he asked. "What's happened to me? Am I going crazy?" There was no satisfactory explanation. He was all at sea. The mercy of amnesia is that the episode is soon forgotten and when Sacks came back into the room minutes later it was as if it had never happened. Jimmie did not remember seeing him before. Sacks wrote, "It was heartbreaking, it was absurd, it was deeply perplexing, to think of his life lost in limbo, dissolving . . . He is, as it were . . . isolated in a single moment of being . . . a man without a past (or future), stuck in a constantly changing, meaningless moment." He wondered how to establish some continuity for this man. He wondered, did Jimmie G have a soul?
>
> Clive and I could not get this story out of our heads and talked about it for days. The saying goes: "that which you fear most comes upon you." Maybe fear creates a chink in the armour, a vulnerability, predisposition. And there was already underlying weakness from Clive's overwork—compulsion overriding capacity. In remembering the picture of Jimmie G, we were staring into a mirror of our own future. (pp. 70-71)

But why should we focus on such brain-damaged individuals? What, Katherine Paterson asked, does it all have to do with us. Later in her speech, she offered an answer in her conclusion to a discussion of *Lyddie*, her most recent publication at the time. "I was very fearful for this story," she said. "I loved the main character so fiercely and yet I've given her so little to make others love her. She is not fair of face or manner. She is not only inarticulate; she is, certainly in the beginning, barely literate. And even as she struggles to educate herself, she is still so narrowly focused in her need to make money that she ignores injustices and fails to examine her own prejudices. So why, as one reviewer asked, should you want to read about such a character, why, which is more to the point, would I want to write about such a person. I can't really explain it, except to say that somehow in her story everything extraneous seemed to have been striped away and I saw as the poet said, beauty bare."

And then she concluded her speech, "When God created the world the Bible says, he saw all that he had made and beheld it was good, and I think we need to add that that means it was very beautiful."

And, I tell myself, if we are open to it, if we can truly understand the meanings behind the words, those poetic levels of meaning, we cannot help but be reminded of John Keats' famous couplet from *Ode on a Grecian Urn*: "'Beauty is truth, truth beauty,—that is all / Ye know on earth, and all ye need to know.'" (19)

As has already become clear, Clive Wearing was unusual, but not alone in his demonstration of this naked beauty.

Theatre, dance, music, then, are all of this world of meaning and value beyond the cold, physical world of literal meanings. Oliver Sacks also found such strange dichotomies of intelligence in "retardates," autistic children, who, though they could not manage the simplest of daily tasks, nevertheless understood visual art on a high level.

At age 21, Jose was prone to violent seizures and considered hopelessly retarded. Sacks gave him a pocket watch and asked him to draw it. Immediately, the distracted, restlessness ceased. Jose took the watch, set it in front of him, and began studying it intently.

The attendant quickly cut in, "'He's an idiot. Don't even ask him. He don't know what it is—he can't tell time. He can't even talk. They says he's 'autistic', he's just an idiot.'" (pp. 214-215)

But Oliver Sacks encouraged Jose, and Jose focused, lost his previous confliction, and drew swiftly and confidently, in a clear, bold line, without erasures. And the result was a remarkably detailed reproduction, with "its 'feel' . . . strikingly brought out," especially if, as the attendant had said, Jose had no knowledge of time at all.

Oliver Sacks was confused. And idiot, autism, no, he thought, something else was involved here. He arranged a second visit.

This time Jose's initial dull, indifferent look flashed a smile when he was again asked to draw a picture. Perhaps, Oliver Sacks thought, Jose couldn't understand the words, but he could grasp the tone, and he took the pen. This time he was asked to reproduce a picture from a copy of Arizona Highways, a scene of two men canoeing against a background of mountains. When Jose began with a careful detailing of the foreground, Oliver Sacks told him to leave that out and go directly to the men in the canoe. With focus and speed, Jose completed a dramatization of the men, one that Oliver Sacks found filled with the feeling, the involvement, the purpose necessary to a work of art. This was no mere copying, but held perception and meaning. This was a person with imagination and creativity.

Oliver Sacks turned the pages of the magazine to a picture of a trout. Jose smiled and began again. This time the result was richly expressive, animated fish, a humorous fish, a nursery character, one Oliver Sacks found of the same world as the frog-footman in *Alice in Wonderland*.

Oliver Sacks was excited, for here was a brain damaged person, an autistic person, who could do more than simply reproduce, more than simply draw mechanical copies. He began to research Jose's life, and found a mass of data since Jose's original illness at the age of eight, which began with a high fever, and the onset of continuing seizures of many differing types. Dr. Rapin, the doctor in charge, hypothesized that perhaps an encephalitis had taken place, and that the temporal-lobe disorders had also caused damaged auditory capacities, especially those associated with perception and production of speech, perhaps even a verbal auditory agnosia, an inability to recognize speech.

At the time, his schooling was discontinued, and, in essence, he became a prisoner of his own house. His mother feared taking him anywhere, as he was suffering from twenty to thirty convulsions a day. This lasted for fifteen years, and there are few records from this time. Various drugs were used unsuccessfully, and there is some evidence that he did, during those moments between seizures, do a bit of drawing. Then there was the sudden, violent rage, involving the smashing of objects, and resulting in him being brought to the hospital, and Oliver Sacks encounter with him.

Now the powerful new drugs were able to be administered and keep his seizures under control, giving him a bit more freedom, and hope. He began to emerge.

The next time Oliver Sacks visited, Jose jumped up, and "eagerly, hungrily, followed" to the art room. Oliver Sacks asked Jose to draw the fish again, this time from memory, and there emerged an even more personalized vision, this time a fish with human lips, and a scene, with another fish, and a large wave. And it was followed by a cry, perhaps an attempt at speech. Oliver Sacks suggests that just perhaps the entire drawing was a form of symbolic communication, that, just perhaps, the little fish and the big fish were Jose and Oliver Sacks interacting. It seems to me almost certain that communication was meant, that the scene was more than a reproduction, that the language was forming, perhaps already formed, for a patterning out of meaning and value, that Jose, autistic, mentally retarded in many ways, was desperately struggling to give his world, his mental world life.

Oliver Sacks next asked Jose to draw a picture based on a nearby Christmas card, a picture of a robin on a tree, snow, and stark twigs for the scene. Again, the drawing was eagerly completed, and the most exciting aspect was the change from winter to spring, the winter twig now a powerful florid open bloom. Oliver Sacks does not postulate on the symbolism, but it rushes out to me. Here was Jose communicating his own rebirth, his own new spring, his own bursting into life.

The final visit Oliver Sacks discusses occurred some time later. Jose had been removed to a special ward in the hospital. This time, when Oliver Sacks approached him, Jose pointed to the locked door and indicated he wanted to go outside. Once outside in the sunlit garden, he immediately began drawing--clovers, grass, dandelions—each with great detail. And Oliver Sacks makes an important observation. Jose was attracted to, understood, "the concrete, the particular, the singular. . . . It is a mode of mind at the opposite extreme from the generalizing, the scientific, but still 'real', equally real, in a quite different way."

Then, without making a direct statement connecting ethical and spiritual worlds with this, Oliver Sacks, perhaps without consciously realizing the connections, makes an amazing suggestion. Could Jose "with his peculiarities, his idiosyncrasy, do drawings for fairy tales, nursery tales, *Bible* tales, myths?"

And Jose, though having unusual artistic abilities for an autistic child, is not alone. Lorna Selfe published a highly illustrated book about Nadia, an extraordinarily artistically gifted autistic child. (20)

Nadia was born to Ukrainian émigré parents in Nottingham, England, Oct. 24, 1967, the second of three children. The other two children had a normal, bilingual language development. Nadia, however, was nearly mute. When Lorna Selfe first began working with Nadia, when Nadia was 6 ½, she found her lethargic, impassive, clumsy, poorly coordinated, and slow in her movements. She had almost no vocabulary at all.

However, at the age of 3 ½ Nadia, contrary to everything else about her, had begun drawing at an extraordinary level, demonstrating not only a manual dexterity far beyond her years, but an artistry and expressive ability beyond most adults.

In observing her, Lorna Selfe found that Nadia drew with a small ball point pin held in a practiced, controlled manner, contrasting the clumsy manner other children her age held such writing implements, placed her eyes very close to the paper (though she had no sight problems), drew swiftly and confidently, able to stop lines assuredly at the exact intersections despite her speed, and quickly executed her work and then stopped abruptly. She had no interest in color, drew from memory, and often changed the perspective and size of the original picture, greatly improving the dynamism through an amazing use of perspective, foreshadowing, and the illusion of movement. Because of this, it is generally hard to even know what the original picture or source of the drawing was. Her most common subject matter was horses and men riding horses. However, other animals, human figures, and even trains also appeared.

The drawings demonstrate some complex aspects that many adults are unable to achieve: geometric perspective, even a sense of lines creating a horizon, and distant clouds, a difficult angle or view, a sense of motion, as if the horse and rider are coming right off the page, attention to details, even the buttons on the shirt, and, for better or worse, a strange squirrel and grotesque head on the side of the horse.

Nadia had a normal pregnancy and birth experience, and there was nothing during her pre and postnatal periods to suggest any potential problems or special abilities. However, her development was slow, and her mother said that she was different than other babies. She was unresponsive, lethargic, and had poor muscle tone.

There was no trouble with breast or bottle-feeding and she was weaned at six months without any problems. Her first words appeared at nine months, the standard mama, dada, grandma, goodnight, and so on. She stood with support at one year, but did not walk alone until two years. Her speech did not improve, and her single words came less and less frequently. She had a bad attack of measles at age two, and began to withdraw from the family. She also became increasingly difficult to control and heedless of danger. At age three, she was left in the care of her grandmother for three months while her mother was in the hospital dealing with breast cancer, and the grandmother did not give her much attention. She was overjoyed when her mother returned. At this time, the mother was able to spend a great deal of time with her, and her drawings began to appear, impressive from the beginning.

When she was 4 ½ the local General Practitioner advised special education, and she was put into a Special School for the Severely Sub Normal. The headmistress there found her slow and passive, though at times destructive and having uncontrollable attacks of screaming. These occasional attacks, however, were not enough to warrant special care, and she was able to enter one of their normal classes, and was not given special training. While she was there, both the headmistress and the teachers noticed her exceptional drawing abilities.

Her language skills remained undeveloped, a vocabulary of less than ten words. However, she did show interest in and some abilities at various perceptual toys and puzzles. Her general passive behavior continued, as did her occasional temper tantrums.

When she was 5 years and 3 months a Senior Clinical Psychologist visited with her and found her something of an enigma, concluding his report by saying "Her language skills are severely retarded and in addition she shows a number of unusual features that indicate full psychiatric assessment." A month later this thorough examination was done at the Hospital for Sick Children, Great Ormond St., London, and concluded with a similar expression of confusion: "Our speech therapist thought that her features were somewhat autistic and although she could not put any specific label to Nadia's behaviour she was impressed by her exceptional drawing skills and felt that she required extensive language stimulation."

Tests done by the Child Development Research Unit, Nottingham University when Nadia was 6 ½ showed only slight improvements. They found her still generally inattentive and non-responsive, slow at dressing and prone to putting on her clothes inside out and backward. Her language skills showed very little development, though she had gone from being so unresponsive that she was thought perhaps deaf to now demonstrating a slight interest in music and being able to sing parts of melodies. Still, the one thing that jumped out was her extraordinary drawing ability.

Curiously, by the time Nadia had reached the age of seven, she had begun to lose her extraordinary artistic abilities, and reverted more and more to an average child's style and ability. She slowly improved her language and social skills, though they remained far behind the normal child of her age.

Elizabeth Newson described Nadia at age nine:

> Nadia is coping well with simple number work, and her number concepts 1-10 are well-established. She can manage simple addition and subtraction, and has begun working with money. She is making progress with reading and writing: she can now orally construct simple sentences to describe pictures that she draws, and will then copy-write them, and read them back with understanding.
>
> Sadly, Nadia seldom draws spontaneously now, although from time to time one of her horses appears on a steamed-up window! If asked, however, she will draw: particularly, portraits of the 30 or so adults and children in the school. These portraits may or may not be posed (though she gives little attention to the sitter), and they are recognizable likenesses; in her most productive period, between six and seven, she drew only two portraits from life, and those barely recognizable. In style, Nadia's portraits are much more economical than her earlier drawing, with much less detail; often they have a Thurberesque quality. Occasionally, Nadia has produced at home a drawing that shows traces of her original interests. . . . The fact that Nadia at eight and nine can produce recognizable drawings of the people around her still makes her talent a remarkable one for her age: but one would not longer say that it is unbelievable.

Is this a tragedy? For us, who love to be astonished, maybe. For Nadia, perhaps it is enough to have been a marvelous child. If the partial loss of her gift is the price that must be paid for language—even just enough

> language to bring her into some kind of community of discourse with her small protected world—we must, I think, be prepared to pay that price on Nadia's behalf. (pp. 130-131)

This study of Nadia is especially valuable and intriguing because Lorna Self gives a careful and well-researched consideration of the numerous possible explanations by experts in the corresponding fields, including neurology, and considers the possibilities of bilateral brain theory. In the end, she concludes, with the data available on Nadia, what this extraordinarily gifted child must remain is an enigma.

Clara Claiborne Park found similar abilities in her daughter Jessy (called Ella in the initial book), and has written a beautiful two-book account of her experiences discovering the autism and raising this child, who has become a functioning adult with some excellent artistic abilities. (21) It is perhaps the single best account of such an experience. Interestingly, Clara found, in addition to art, music was a useful way of supplementing or even replacing standard language communication. It is also worth mentioning as an aside that initial experiences with psychiatrists was disappointing, but once Clara took Jessy to Anna Freud's Hempstead Clinic in Rome, she discovered what an enlightened psychological approach could accomplish.

Again, there were no pregnancy or post birth problems. However, again, similar lack of development took place. At twenty-two months Jessy did not walk, talk, or respond to speech, so much so that doctors took her in for observation for three days in the hospital. They reported that she remained remote, and suggested she seemed to be a child raised alone. This was not the case, as she had three older siblings and received a great deal of attention. There was not enough evidence, they decided, to assume she was abnormal, and the decision was made to wait six months and reassess. Reassessment brought the same inconclusive decision, wait and see. Another six months brought the same.

By the time Jessy was three, her parents decided it was time to get a better diagnosis. A close friend suggested Dr. Blank, a nationally known pediatrician. He offered a possibility they had not been aware of, autism, a category just then beginning to be understood.

Later, Clara Claiborne Park would read in Dr. Blank's report:

> In many ways she seems to be an autistic child, relating poorly to the other children. Her apparent early normal development reaching a plateau, her interest in small objects and their orderly arrangement, her failure in speech—all would fit into such a picture. On the other hand, she enjoys

being played with physically, she likes to have her parents enter into the 'games' she uses, and these findings do not fit well.

Dr. Blank had seen similar children under the care of Leo Kanner at Johns Hopkins, in Baltimore, children who did not fit the established categories of psychotic, neurotic, brain-damaged or retarded children. Leo Kanner was a leader in this new field, and had been studying such children for twenty years. Though others in the field had labeled this condition "childhood schizophrenia," he found it different, a distinct condition, and named it Early Infantile Autism.

Though her mother noticed she was attracted to drawing, Jessy did not exhibit her extraordinary artistic abilities as young as did Nadia. Looking back, her mother tried to piece together her artistic development, from her very childish pictures of humans, to two high school girls who helped her learn some of the techniques of drawing. But Jessy's artistic interests were random, and her geometric, abstract, surrealism at the time little more than simple geometric exercises.

The breakthrough came accidentally. Her mother writes:

> We couldn't have guessed how time, and luck, would bring everything together—luck and the principle of numerical reinforcement. We took Jessy to an autism meeting where I was making a speech. She was already twenty-one, too old for the children's activities provided, and I suggested she sketch to keep her busy. She made an accurate, ugly sketch of the ugly building we were meeting in, a man who'd heard my speech offered five dollars for it, and that's what started her career. . . .
>
> Money didn't mean [anything to her]. But numbers did, and she liked to see them rise in her checkbook. The staff at the Society for Autistic Children were very kind; they gave her a little exhibition, and sketches and school paintings were sold for small sums. The glorious colors began to come back, and then to proliferate. Perhaps she remembered a school exercise form years before, when she had been told to paint a snow scene, first in its natural white and evergreen, then in whatever colors fantasy might suggest. Who knows? At any rate, Jessy was drawing again, not because she was told to but because she wanted to. Once more she was finding her own subject matter. She drew, then painted, not snow landscapes, still less portraits or even buildings; she drew radio dials, speedometers and mileage gauges, clocks, heaters, and electric blanket controls. People with autism like such things. Jessy's fascination gave these new paints an intensity that her academic drawings had lacked. Not that these weren't realistic, but a dial is more than a dial when it is realized in apricot and turquoise. Jessy's dials and gauges dazzled; her heaters throbbed with color as in a dream, transfiguring the simple grid perceived by her geometrizing eye. Sometimes they achieved an instant surrealism;

> what more natural than to honor three enthusiasms together? So against an electric blue she combined a rock group logo, an album title, and a heater, to yield the bizarrerie of *Boekamp Heater with Women and Children First.*
>
> Jessy had reverted to the abstract patterns of her childhood. But now they were abstractions in the true sense—patterns perceived in, drawn from, abstracted from, the visible world. There was a window in a house near us; through it, by some architectural quirk, a chimney could be seen, right up against the glass. Fascinated by the pattern of the bricks, Jessy painted it four times: first just the chimney; then the chimney and the window; then the chimney, window, and roof; finally chimney, window, roof, and the night sky with stars. (pp. 126-127)

The literature is full of similar artistic abilities in autistic children, not that they are all artistic geniuses, but that they often do have highly developed abilities, far beyond their other below average abilities, and above the general population.

Lewis Hill found 52 studies of gifted autistic individuals, or idiot savants, many of them with remarkable artistic talents. (22) He briefly discusses a few of them, including Gottfried Mind, "a cretin imbecile who obtained European fame for his paintings of children and animals." He had such an amazing ability to draw cats that he was deemed "The Cat's Raphael." Borne in Berne in 1768, he was unable to read or write, had no idea of the value of money, and fit the appearance of a mentally deformed person enough that he was jeered by crowds of children whenever he walked the streets. His hands were unusually large, and had a rough appearance, yet he demonstrated artistic talent at a young age, and thus was given some training in drawing by his father's employer. His drawings of not only cats, but other animals and even children were so lifelike he gained a European fame, and one of his pictures was even purchased by King George IV.

Lewis Hill also mentioned briefly one who "played the piano at rehearsals for a leading chamber music orchestra." The reports on her came down through several scholars from Dr. Trelat (La Folie Lucide," Paris, 1861), who wrote that, "they had in the Salpetriere an imbecile born blind, affected with rickets, and crippled, who had great musical talents. Her voice was very correct, and whenever she had sung or heard some piece she knew perfectly well the words and the music. As long as she lived they came to her to correct the mistakes in singing of her companions; they asked her to repeat a passage which had gone wrong, which she always did admirably. One day, Geraldy, Liszt and Meyerbeer came to the humble singing-class of our asylum to bring her their encouraging consolations.

And the distinctions between idiot savants and "normal" genius become ever more unclear when we look at Dr. P.

According to Oliver Sacks, "Dr. P. was a musician of distinction, well-known for many years as a singer, then, at the local School of Music, as a teacher. . . . He was a man of great cultivation and charm who talked well and fluently, with imagination and humour." Sacks saw no trace of dementia, no reason why Dr. P. had been referred to the clinic at all.

However, he did sense something wasn't quite the norm, as if Dr. P. "faced" him "more with his 'ears'" than "with his eyes." (p. 9).

When Oliver Sacks questioned him about why he was there, about what he thought might be wrong with him, Dr. P. replied, "'Nothing that I know of, . . . but people seem to think there's something wrong with my eyes.'" And though Dr. P. didn't notice anything wrong, he did admit that he "occasionally" made "mistakes."

Oliver Sacks found him to be a delightful man.

However, in the process of his initial, routine examination, he asked Dr. P. to remove a shoe so his foot reflex could be tested. Afterwards, Oliver Sacks told him to put the shoe back on. However, Dr. P. did not understand, could not grasp the whole concept of replacing his shoe, and finally was found unable to distinguish between his shoe and his foot, was, in fact, totally disorientated by the whole event.

Oliver Sacks was baffled. Dr. P. was unconcerned, and actually just didn't grasp that there was any problem. Oliver Sacks could see that Dr. P. really did not have a problem with his eyes. His physical vision was fine, and yet there was something wrong with his mental vision. A copy of *National Geographic Magazine* was produced, and Dr. P. was asked to describe it. Oliver Sacks noticed immediately that the description was not what a normal person would give. Rather, it was something of a description of odds-and-ends, of details, but not of the entire picture, not of a landscape or scene, but of unrelated features of the scene. It was even more obvious when he was asked to describe another picture, this one an unbroken expanse of Sahara dunes. Here, because the picture was so bland, so without particulars, the man simply made up the particulars. Dr. P. was, apparently, unaware of his strange inventiveness.

Sensing that the examination was over, Dr. P. began looking around for his hat, reached out, and took hold of his wife's head, and tried to pull it off! He thought it was his hat!

Oliver Sacks was lost as to what to make of it. Dr. P. was a highly valued teacher at the music school. The next meeting took place in Dr. P.'s study at home. Oliver Sacks, knowing that Dr. P enjoyed it, brought a score of Schumann's *Dichterliebe*, and since it turned out that Dr. P. could no longer read music, played it for him on the piano, while Dr. P. sang it, exhibiting musical abilities that Oliver Sacks found extraordinary. It was easy to see why the school kept him on.

Oliver Sacks began to test Dr. P's visual abilities, first with various geometric shapes, then a deck of cards, then cartoons. In each case, Dr. P. did fine. However, Oliver Sacks noted, all of these are stylized, formal, schematic shapes.

Next he turned on the television, left the sound off, and asked Dr. P. about an old Betty Davis love scene that happened to be on. Dr. P. was lost. Now what? Oliver Sacks next tried a stack of family photographs, and the humor became tragic. He could not identify any of them. A few objects were produced. A rose Oliver Sacks had purchased on his way there. Dr. P. felt it and guessed at what it was but could make nothing of it, until he smelled it, and then knew immediately. A glove proved equally mystifying. He felt at it and guessed, perhaps a change purse.

Next, perhaps, a test of his memory and imagination. He was asked to imagine entering a nearby square, to describe it. What resulted was a description of all of the building on the right side, but nothing on the left. The memory of Tolstoy's *Anna Karenina* resulted a similar contrast. He could remember the plot, character's words, but not the visual narrative, not the scenes, not the visual characteristics. How about chess, a highly schematic game. He was very good at this.

His wife called them to lunch, ad Dr. P. began eating hungrily, humming as he did. Everything was normal and seemed fine. But then there was a knock at the door, and Dr. P. became disoriented, lost, until his wife poured him some coffee, and the smell brought him back to reality.

Dressing, Oliver Sacks was told, was much the same. His wife would leave his clothes out for him, and he would dress while singing to himself. But if the he got interrupted, he was lost.

Oliver Sacks noticed pictures on the walls and found out that Dr. P. was also a highly respected painter. Even more interesting was the progression of the style. The earlier ones were highly detailed, naturalistic and realistic, then, as time passed, ever more abstract, and finally simply chaotic lines (though his wife saw them as representative of his progress artistically).

Through it all Dr. P was seemingly, pleasantly unaware of his literal failings.

Certainly, we have here a real life Don Quixote, a brilliant man, seemingly totally unaware of his failings on the literal level, and if not better off for having them, clearly better off than the man who would be fully competent or even superior at precisely what Dr. P. cannot comprehend but unable to enter the higher levels of music and art. And Dr. P. is so exceptional in his higher reality that those Sancho Panzas who surround him and know of his failings in the more literal world nonetheless desire to be near him, willingly giving up practical, logical existence, or at least putting it in a lower position and embracing Dr. P for the higher levels he exists in and opens up for all who are capable of experiencing. (23)

Perhaps, if pushed to defend himself, Dr. P. might have uttered Don Quixote's great speech after his comic encounter with the lions, where he insisted a lion cage be open that he might prove his worth against such terrifying beasts. Though their owner had built them up to be the most ferocious of all lions, and, in truth, because of this very claim, and though Sancho Panza, as always, had urged against this needless, reckless adventure, Don Quixote was determined to prove his mettle by having the lions loosed that he might face them.

In the end, Don Quixote won the argument, and the huge male lion's cage was opened. However, the prefigured ferocious lion showed no interest in engaging in battle, and after gazing about for a time the great lion lazily turned his back on Don Quixote and calmly lay down. It was clear no battle was going to take place.

But it mattered not to Don Quixote, for he had, in all truth, faced the fear and danger, and thus, as he pointed out was common among knights-errant, he assumed the title of The Knight of the Lions.

Don Diego de Miranda saw it all, and thought to himself that perhaps Don Quixote was sane, for his words were consistent, elegant, and well put, yet perhaps he was mad, for his actions were wild and foolish. Then came the great speech:

> "No doubt, Don Diego de Miranda, your worship considers me both foolish and mad. And it would be no marvel if you did, for my deeds testify no less. But, for all that, I wish your worship to take note that I am not so mad or so lacking as I must have seemed to you. It is a brave sight to see a gallant gentleman, beneath the eyes of his King, deliver a well aimed lance-thrust against a brave bull in the midst of a great square. It is a brave sight to see a knight, armed in shining armour, pace the lists in merry jousts before the ladies. And it is a brave sight to see all those knights, in military exercises or the like, entertain, cheer and, if one may say so, grace the courts of their princes. But braver than all these it is to see a knight errant traveling through desert and waste, by crossroads, forests and mountains, to seek perilous adventures, in order to bring them to a fortunate and happy conclusion only for the sake of glory and lasting renown. It is a braver sight, I say, to see a knight errant succouring a widow in some lonely spot than a courtier knight wooing a maiden in the cities. All knights have their particular offices. Let the courtier serve ladies,

> lend splendour to the King's court with his liveries, support poor knights at his splendidly appointed table, arrange jousts, maintain at tourneys, and show himself grand, liberal, magnificent and, above all, a good Christian; for in this way he will comply with his precise obligations. But let the knight errant search the corners of the world, penetrate the most intricate labyrinths, at every step encounter the impossible, a midsummer brave the burning rays of the sun on high and desert wastes, and in winter the harsh inclemency of winds and frosts. Let no lions alarm him, nor hobgoblins daunt him, nor dragons affright him; for to seek them, attack them and conquer them all are his chief and proper exercises. I, since it has been my lot to be of the number of knights errant, cannot, then, fail to attempt everything which seems to me to fall within the bounds of my duty. And so to attack these lions whom I just now attacked was my rightful concern, although I knew it to be an excessive temerity. For I well know what valour is—a virtue placed between the two extremes of vice, cowardice and foolhardiness. But it is better for the brave man to rise to the height of rashness than to sink into the depths of cowardice. For just as it is easier for the spendthrift to be liberal than for the miser, so it is easier for the rash than for the cowardly to climb to true valour. And in this matter of encountering adventures, believe me, Don Diego, it is better to lose the game by a card too many than by a card too few, for such a knight is rash and foolhardy sounds better in the hearer's ears than such a knight is timid and cowardly." (24)

The deep importance of this crazed view of life becomes more evident when compared with another literary misfit of the eighteenth century, Candide, who, bolstered by the philosophy of his mentor, Dr. Panglass, wandered through a life of random sufferings, but who continually put forth the view that "this is the best of all possible worlds," and that individual suffering is fine, since, one way or another, it is really for the best of all. As we read about Candide's often comic, usually horrifying, adventures, we can easily find similarities to those of Don Quixote, but they take on opposite attitudes towards the physical level of existence, for Don Quixote stood staunchly against such a random, cruel, meaningless world, proclaiming a world of meaning and value, regardless of the physical suffering that was necessary to defend it. As indicated above, the highly developed world of knightly values, of honor and chivalry, while it may have had flaws, nevertheless was an ethical world, a world where various values were simply more important than the instinctual world of physical comfort and survival, a world where humans do take on the responsibility of giving physical actions meaning and value, do transcend the physical limitations and exist in an invisible world far more powerful than the visible world. Candide had no such world to enter, for from his perspective the physical world is all that there is. There is no room for making judgments and attempting to improve upon basic instincts, as it is ipso facto the best of all possible worlds no matter what happens.

In his introduction to the Eaton Press edition, Paul Morand comments: "Are they happy in the end? No, they are bored; indeed one day the old woman makes bold to speak as follows: 'I'd like to know which is worse, to be raped a hundred times by Negro pirates, to have a buttock cut off, to run the gauntlet among the Bulgars, to be whipped and hanged in an auto-da-fe, to be dissected, to row in a galley—in short to endure all the miseries through which we have passed—or to remain here doing nothing?' 'Tis a great question,' was Candide's reply." (25)

Later, the final seeming moral of the book leaves us with a more ironic statement of the same question: "'All events are beautifully linked up in this best of all possible worlds,' said Pangloss. 'For, if you had not gone through all these terrible misfortunes, would you now be here, eating candied citrons and pistachios?' 'Tis well said,' Candide answered, 'but we must cultivate our gardens.'" (26)

Voltaire, of course, did not embrace such a world, was not of the same mind as his fictional scholar Dr. Panglass. The narrative was meant to be a satire, a put-down of such a world, a world similar to the world of Don Quixote. However, Voltaire's central character Candide was not supposed to show us the beauty of a cracked skull, but rather to show us the folly of blind acceptance, of not trusting one's own experiences and thoughts, of not becoming an adult human and experiencing the grand call to the adventure of being alive in the human sense of accepting self-responsibility and the human task of mapping out meaning and value. We do not look at Candide as someone to emulate but with the insight of the horror of such emulation.

In his PBS video interviews with Bill Moyers, Joseph Campbell says "The influence of a vital person vitalizes, there's no doubt about it. The world is a spiritual wasteland. People have the notion of saving the world by shifting things around, changing the rules, and who's on top, and so forth. No, no! Any world is a valid world if it's alive. The thing to do is to bring life to it, and the only way to do that is to find in your own case where the life is and become alive yourself." (27)

Don Quixote was alive, and struggled to bring life to the wasteland that surrounded him. That life was what so attracted Sancho Panza, the logical onlooker who sensed and desperately needed that life force, totally illogical as it was. Candide, more like Sancho Panza than Don Quixote, perhaps more intelligent than Don Quixote in certain logical levels, could not bring his world to life, but simply let the winds of his wasteland blow him about, a victim of the uncaring environment rather than a life force within it. Even the concluding escape into simply tending his garden is, as any farmer will tell you, not a final release into paradise or a return to the Garden of Eden, but, rather, a submission to a harsh garden world ruled by the uncontrollable weather and other events of nature and fate (which Stephan Crane shows us dramatically in his short story *The Open Boat* have no concern for the human condition). An embracement of this is an embracement of meaninglessness.

In our encounters with fictional and real world Don Quixotes, certain forms of thinking, of knowing, of mapping out meaning emerge. Rebecca was unable to put on her clothes, navigate a standard city block, or figure out how to unlock a door with a key. She was unable to read, write, or even talk in normal sentences. She was, as Oliver Sacks said, a broken creature, unable to score higher than 60 on a standard IQ test. Yet she thirsted for stories, understood deep emotional depths, and overcame her poor coordination when she danced or performed in theatre. Martin was slow, dreamy, incompetent, a poor worker, prone to sudden fits. Yet his musical IQ was close to Bach's. Harriet G. was an awkward, ungainly, overgrown child, but when put in front of a piano she could play with the feeling and movement of a concert pianist. Jose was unable to speak, prone to seizures, lacking the discipline to concentrate. Yet, when given the chance he could draw with great accuracy, and, more importantly, infuse his work with meaning, even humor. Clive Wearing couldn't remember one minute to the next, had no past, no future. Yet he was able to perform difficult pieces of music, and experienced great depths of both sorrow and love. Dr. P. was a highly successful and respected teacher of music and an acclaimed artist, yet he thought his wife's head was a hat. Helen Keller, deaf and blind from nineteen-months, left with only smell and touch to try to make sense of the physical world around her, nevertheless graduated cum laude from Radcliffe College, and knew rich spiritual feelings. "As time went on," she wrote, "my thoughtless optimism was transmuted into that deeper faith that weighs the ugly facts of the world, yet hopes for better things and keeps on working for them even in the face of defeat."(28)

Again and again these people, deficient in some ways, demonstrate the most wealthy possible artistic, emotional, and ethical depths, even more dramatic and touching because they are so psychologically naked.

Notes

1. This conversation took place on the University of North Dakota campus in 1981. Norton Kinghorn was a special man, filled with such insights, a man who not only was extremely intelligent, but who was also wise, and to top it off had a great sense of humor.
2. Enid Starkie, *Baudelaire*, 1962 (a biography); Martin Turnell, *Baudelaire*, 1954, (literary criticism). The quoted phrases come from "The Voyage," the final poem in *Flowers of Evil*, translated by me out of the French. It concludes with the Captain, Death, taking over and leading us into a vague "Through the unknown, we'll find the new" view of afterlife.
3. Sacks, Oliver. *The Man who mistook his Wife for a Hat and other Clinical Tales*, Simon & Schuster: New York, 1998. The immediately following quotes and references to him all come from this source.
4. Peters, L. R. "The role of dreams in the life of a mentally retarded individual," *Ethos* (1983): 49-65.
5. Brunner, J. "Two Modes of Thought," *Actual Minds: Possible Worlds*. Boston: 1986, pp. 11-43.
6. *The Empire Strikes Back*. Lucas Film Ltd, 1980. This quote can also be found in *Star Wars: The Annotated Screenplays*, Laurent Bouzereau, Ballantine Books, 1997, p.187.
7. The main character in "Peter Bell, A Tale in Verse" written 1798; first published in 1819; rpt. *The Longer Poems of Wordsworth*, New York: Putnam, no date; p. 49-77, the lines are on page 55; the quote from Kandinsky comes from "Uber die Formfrage," *Der Blaue Reiter*, Munich: R. Piper, 1912; trans, "On the Problem of Form," by Kenneth Lindsey; included in *Theories of Modern Art: A source Book by Artists and Critics*, ed. by Herschel B. Chipp, Univ. of Calif. Press, 1969.
8. Viscott, David. "A Musical idiot savant: a psychodynamic study, and some speculation on the creative process." *Psychiatry,* 1970, 33 (4), pp. 494-515. The following discussion of this patient is taken from this article.
9. "Observations on the Psychological Functions of Music," *Journal of American Psycholanalysis Association,* 1957, p. 396.
10. "What Kind of Fool Am I?: A Study of Popular Songs in the analysis of a Male Hysteric," *Psychiatry*, 1970, 33, 4, pp.524-525.
11. "The Haunting Melody," *Psychoanalic Experiences in life and Music*, Farrar Straus and Young, 1953, p. 216.
12. *Language and Language Disturbances*. Grune & Stratton, 1948, pp. 146-7.
13. "The Haunting Lyric: The Personal and Social Significance of American Popular Songs," *Psychoanalysis Quarterly*, 1964, 33: 226-269.
14. *The Complete Letters of Vincent Van Gogh*. 3 vols. Greenwich, Conn.: The New York Graphic Society, 1958; trans., Herschel B. Chipp, *Theories of Modern Art: A Source Book by Artists and Critics*, Univ. of Calif. Press: Berkeley, 1969, pp. 36-37. The first half of the quote, up to the ellipsis comes from a letter to Theo, circa Sept., 1888; the second half of the quote comes from a letter to Theo, Arles, Sept. 8, 1888.

15. Recorded on United Artists Records, Copyright, 1971, Mayday Music, Inc. and Yahweh Tunes, Inc., all rights administrated by Unart Music Corporation, New York.
16. Katherine Paterson's speech took place at Eastern Michigan University, March 13, 1992. Her numerous novels have been published by Thomas Y. Crowell, New York. All of her comments on Clive Wearing are taken from this lecture. I have been unable to locate the PBS series source she got her information from, though it might be that the program she mentioned was simply titled differently and is actually the same program as the film *Portrait of an Autistic Young Man.* I have also found a review for *The Man with the 7 Second Memory* (a CBC documentary, Nov. 2005), and a reference to "The man who keeps falling in love with his wife," *The Daily Telegraph*, Dec., 2005.
17. J. B. Watson, *Behavioralism*, New York: Norton, 1930; B. F. Skinner, *Science and Human Behavior*, New York: Macmillan, 1953.
18. Doubleday, 2005, ISBN: 0385606265.
19. Written May, 1819; rpt., *The Norton Anthology of Poetry*, ed. Arthur M. Eastman, 1970, p. 698.
20. Selfe, L. *Nadia: A Case of Extraordinary Drawing Ability in an Autistic Child*, London, 1977.
21. Park, C. C. *The Siege: A Family's Journey into the World of an Autistic Child.* New York, 1967; paperback: Boston and Harmondsworth, 1972; and the sequel *Exiting Nirvana: A Daughter's Life with Autism.* Little Brown: New York, 2001. In one of his series of videos, *The Mind Traveler*, Oliver Sacks has a program on her as well.
22. Lewis Hill, "Idiot Savants: A Categorization of Abilities," Mental Retardation. Dec. 1974, pp. 12-13. The reference to the idiot savant who played the piano refers to A. Anastasi and R. F. Levee's article, "Intellectual defect and musical talent: A case report," *American Journal of Mental Deficiency*, 1960, 64 (6), 695-703; a brief discussion of her can also be found in A. F. Tredgold, *Mental Deficiency (Amentia),* London: Bailliere, Tindall and Cox, 1914, p. 304. The reference to Godfried Mind also comes from Tredgold, who quotes from W. W. Ireland (p. 306-7), who refers to E. G. Fairholme's account published in *The Animal World*, January, 1909.
23. Oliver Sacks discusses a similar case "The Case of Anna H." in "A Neurologist's Notebook," *The New Yorker*, Oct. 7, 2002, pp. 63-73, a brilliant pianist who went through a slow losing of her abilities to recognize objects or even read music, yet was still able to perform from memory and even continue to teach music, and, who, strangely could often recognize particular details in the visual world randomly while being unable to grasp the difference between a violin and a banjo, or a glove and a statue.
24. *The Adventures of Don Quxoite*, Miguel de Cervantes Saavedra, trans. by J. M. Cohen. Penguin Books, Maryland, 1950, rpt. 1967, pp. 587-589.
25. Voltaire, *Candide*, Caron de Beaumarchais, 1759; trans. by Richard Aldington, Easton Press: Norwalk, Conn., 1977. This and the following quotes come from Morand's discussion in the "Preface," p. xv. The quotes actually occur on p. 127 and p. 131 in the novel.

26. This final sentence has raised a lot of discussion, as a "garden" has strong multifaceted, life-giving symbolism. In fact, Pangloss has just point that out, claiming it is reminiscent of the Garden of Eden, where Adam and Eve enjoyed perfect bliss, and only upon disobedience of God were punished by being kicked out of the Garden. Thus, the ending becomes unclear. Is this Garden positive? Has Candide finally realized some higher truth? Considering the obvious condemnation of Plangloss' philosophy throughout the novel, one thing that seems clear is that there is a rejection of anything he says, so it's hard to argue his views are being promoted. In fact, it's fairly clear that by putting this suggestion in his words, the novel is condemning it, and by extension condemning the whole world of philosophic discourse. This fits the rest of the novel, and working one's garden becomes simply accepting the world as meaningless, only practical, physical work having any value. One might say that this physical work will result in real rewards, food, and even a sense of community. However, it is clear these are not enough. First, in truth, tending a garden is not easy work and does not necessarily result in a fair reward. Nature easily wrecks havoc on the best of human intentions, and a well kept garden can easily be destroyed in one storm. Second, the characters make it clear they are not happy. In fact, ironically, it seems they would rather return to the senseless, often horrific trials they've just been through than the peace of the garde.

27. *Joseph Campbell: The Power of Myth with Bill Moyers*, ed. Sue Flowers. Doubleday: New York, 1988, pp. 183-184. This book statement is a rough translation of the video statement (which, for example, does not state "without spirit," but, perhaps infers it).

28. Midstream, p. 158, 1929; quoted in the introduction of *The Story of My Life*, Doubleday: New York, 1954, p. 16.

INDEX